Nonequilibrium Statistical Physics

Second Edition

Statistical mechanics is hugely successful when applied to physical systems at thermodynamic equilibrium; however, most natural phenomena occur in nonequilibrium conditions, and more sophisticated techniques are required to address this increased complexity. This second edition presents a comprehensive overview of nonequilibrium statistical physics, covering essential topics such as Langevin equations, Lévy processes, fluctuation relations, transport theory, directed percolation, kinetic roughening, and pattern formation. The first part of the book introduces the underlying theory of nonequilibrium physics, the second part develops key aspects of nonequilibrium phase transitions, and the final part covers modern applications. A pedagogical approach has been adopted for the benefit of graduate students and instructors, with clear language and detailed figures used to explain the relevant models and experimental results. With the inclusion of original material and organizational changes throughout the book, this updated edition will be an essential guide for graduate students and researchers in nonequilibrium thermodynamics.

Roberto Livi is an honorary professor of theoretical physics at the University of Florence and an associate member of the National Institute of Nuclear Physics and of the Institute for Complex Systems of the National Research Council. His research is focused on nonequilibrium statistical physics, and he has extensive experience teaching courses on statistical physics. He is the current president of the Italian Society of Statistical Physics.

Paolo Politi is a research director at the Institute for Complex Systems of the National Research Council and a fellow of the Marie Curie Association, the Alexander von Humboldt Foundation, and the Japan Society for the Promotion of Science. He currently teaches a course on stochastic processes and nonequilibrium statistical physics at the University of Florence. He was awarded the Outreach Prize of the Italian Physical Society.

CAMBRIDGE
UNIVERSITY PRESS

Shaftesbury Road, Cambridge CB2 8EA, United Kingdom

One Liberty Plaza, 20th Floor, New York, NY 10006, USA

477 Williamstown Road, Port Melbourne, VIC 3207, Australia

314–321, 3rd Floor, Plot 3, Splendor Forum, Jasola District Centre, New Delhi – 110025, India

103 Penang Road, #05–06/07, Visioncrest Commercial, Singapore 238467

Cambridge University Press is part of Cambridge University Press & Assessment,
a department of the University of Cambridge.

We share the University's mission to contribute to society through the pursuit of
education, learning and research at the highest international levels of excellence.

www.cambridge.org
Information on this title: www.cambridge.org/9781316512302

DOI: 10.1017/9781009058230

© Cambridge University Press & Assessment 2025

First published 2025

A catalogue record for this publication is available from the British Library

A Cataloging-in-Publication data record for this book is available from the Library of Congress

ISBN 978-1-316-51230-2 Hardback

Dedicated to the memory of
Angelo Baracca (1939–2023)
and
Jacques Villain (1934–2022)

Contents

The first edition of this book was published in 2017, and it was motivated by the desire to provide a textbook rather than a monograph. For this reason, its writing was led by the ambition to be as didactic as possible. On the other hand, the subject covered by its title, *Nonequilibrium Statistical Physics*, is potentially boundless: We made the choice to balance standard topics (Brownian motion, Langevin and Fokker–Planck equations, and linear response theory), more modern but inescapable subjects (transport processes and nonequilibrium phase transitions), and a personal selection of other topics (kinetic roughening, phase-ordering kinetics, and pattern formation).

Years after its publication, having discussed with many colleagues and having had the opportunity to use the book for our second-level master course, we thought that an enlarged and improved second edition might make sense from a scientific/pedagogical point of view. Fortunately, the first edition was successful enough to also make sense from an editorial point of view. We have therefore proposed a second edition whose main differences with the first edition are the addition of two new chapters and the heavy rewriting of four old chapters.

The new material is now contained in Chapters 1 and 3. The first chapter, "Kinetic Theory and the Boltzmann Equation" is completely new, while the third one, "Fluctuations and Their Probability," contains pedagogical introductions to large deviations and to stochastic thermodynamics. It also contains a discussion on generalized random walks, which is now framed in a more homogeneous context. Chapters 2 and 4, the core of standard topics of the first edition, are reissued with few changes.

The two old chapters on nonequilibrium phase transitions have instead been reorganized so as to distinguish between driven lattice models, Chapter 5, and absorbing phase transitions, Chapter 6. The next two chapters, Chapter 7 on kinetic roughening and Chapter 8 on phase ordering, have been rewritten to make them clearer and more fluid. Finally, Chapter 9 on pattern formation is almost unchanged.

Each chapter is accompanied by an opening, followed by an introductory section, and is closed by a final section with bibliographic notes. The reading of opening and introduction is recommended for those who aim at framing the great deal of methods and concepts contained in the book and specifically in each chapter. The final section lists some bibliographic advices for the specific chapter. A few suggested, general readings are: P. M. Chaikin and T. C. Lubensky, *Principles of Condensed Matter Physics* (Cambridge University Press, 2000); D. Chandler, *Introduction to Modern Statistical Mechanics* (Oxford University Press, 1987); P. L. Krapivsky, S. Redner, and E. Ben-Naim, *A Kinetic View of Statistical Physics* (Cambridge University Press, 2010); L. Peliti, *Statistical Mechanics in a Nutshell* (Princeton University Press, 2011); and J. P. Sethna, *Statistical Mechanics: Entropy, Order Parameters, and Complexity* (Oxford University Press, 2006).

The book is complemented by a dedicated website: https://sites.google.com/site/nespbook/. Readers are encouraged to report any errata or share their comments via the email address provided on the site.

Acknowledgments

Some colleagues have read and commented on parts of the book. So it is a great pleasure to thank Federico Corberi, Joachim Krug, and Alessandro Sarracino. Discussions with many colleagues helped us to clarify several questions. We are especially grateful to Franco Bagnoli, Filippo Colomo, Peter Grassberger, and Ruggero Vaia. Filippo Cherubini is acknowledged for having produced many figures of the book. PP thanks the Alexander von Humboldt Foundation for financial support, allowing him to spend a month at the Institute for Theoretical Physics of the University of Cologne to work on the present edition of this book.

Throughout this book, we assume the Boltzmann constant ($K_B = 1.380658 \times 10^{-23}\,\mathrm{J\,K^{-1}}$) to be equal to 1, which corresponds to measuring the temperature in joules or energy in kelvin.

We often use the symbol $N_A = 6.022141 \times 10^{23}$ to indicate the Avogadro number.

The space dimension is indicated by d.

As for the Fourier transform, we use the same symbol in the real and dual spaces, using the following conventions:

$$h(\mathbf{k}) = \int d\mathbf{x}\exp(-i\mathbf{k}\cdot\mathbf{x})h(\mathbf{x}),$$

$$h(\mathbf{x}) = \frac{1}{(2\pi)^d}\int d\mathbf{k}\exp(i\mathbf{k}\cdot\mathbf{x})h(\mathbf{k}).$$

Similarly, for the Laplace transform,

$$\omega(s) = \int_0^\infty dt\,e^{-st}\omega(t),$$

$$\omega(t) = \frac{1}{2\pi i}\mathrm{PV}\int_{a-i\infty}^{a+i\infty} ds\,e^{st}\omega(s).$$

Here, PV is the principal value of the integral and $a > a_c$, with a_c being the abscissa of convergence.

The friction coefficient of a particle of mass m, that is, the ratio between the force F_0 acting on it and its terminal drift velocity v_∞, is indicated by the symbol $\tilde{\gamma} = F_0/v_\infty$, but we frequently use the reduced friction coefficient, $\gamma = \tilde{\gamma}/m$, which has the dimension of the inverse of time. Similar reduced quantities are used in the context of Brownian motion.

The Helmholtz free energy is indicated by the symbol $F = U - TS$, and the free-energy density is indicated by f. We often use a free-energy functional, also called pseudo-free-energy functional or Lyapunov functional, and it is indicated by $\mathcal{F}$. It is the space integral of a function f, $\mathcal{F} = \int d\mathbf{x}f$. The susceptibility, indicated by χ, may be an extensive as well as an intensive quantity, depending on the context.

We use O and o to indicate the big and small O notations. For example, $\sin x = x + o(x)$ or $\sin x = x + O(x^3)$ for vanishing x.

When in the main text we cite a scientist, at the first occasion, we add their given name and nationality. If the family name of a scientist is only used to define an effect, an equation, or a model, we generally do not add further information.

ASEP	asymmetric simple exclusion process
BD	ballistic deposition
BTW	Bak–Tang–Wiesenfeld
CA	cellular automaton
CDP	compact directed percolation
CH	Cahn–Hilliard
CIMA	chlorite-iodide-malonic-acid
CTRW	continuous time random walk
DK	Domany–Kinzel
dKPZ	deterministic Kardar–Parisi–Zhang
DLA	diffusion limited aggregation
DLG	driven lattice gas
DP	directed percolation
DyP	dynamical percolation
DW	domain wall
emf	electromotive force
EW	Edwards–Wilkinson
GL	Ginzburg–Landau
GOE	Gaussian orthogonal ensemble
GUE	Gaussian unitary ensemble
HD	high density
KLS	Katz–Lebowitz–Spohn
KMC	kinetic Monte Carlo
KPZ	Kardar–Parisi–Zhang
LD	low density
LG	lattice gas
LW	Lévy walk
MC	maximal current
MF	mean field
NESS	nonequilibrium steady state
PC	parity conserving
pdf	probability distribution function
PV	principal value
QFT	quantum field theory
RD	random deposition

RDR	random deposition with relaxation
RG	renormalization group
RSOS	restricted solid on solid
SH	Swift–Hohenberg
SOC	self-organized criticality
SS	single step
TASEP	totally asymmetric simple exclusion process
TDGL	time-dependent Ginzburg Landau
TW	Tracy–Widom
WV	Wolf–Villain

Kinetic Theory and the Boltzmann Equation

In Section 1.2, we describe the approach to thermodynamics of elementary kinetic theory, which is based on the model of the ideal gas as an ensemble of classical point-like particles of equal mass. We also illustrate how elementary kinetic theory provides a phenomenological approach to transport phenomena, which exemplifies out-of-equilibrium stationary processes and allows us to introduce a definition of transport coefficients.

Boltzmann's approach represents a deep refinement of this elementary kinetic theory. The basic idea is that the ideal gas, as a collection of a gigantic number of mechanical particles, should be more properly described by a distribution function, rather than by the single trajectories of the particles (Section 1.3). In principle, these trajectories could be computed by the laws of mechanics, but in practice no human being can have at disposal the computational facilities to successfully accomplish this task. Boltzmann's proposal amounts to overtake this obstacle by replacing the mechanical equations of the evolution in time of all the gas particles with only one equation describing the evolution in time of their distribution function. The mathematical procedure adopted for obtaining this equation makes use of many physically plausible hypotheses, which are carefully illustrated in Section 1.4 . Among them, the hypothesis of molecular chaos (*Stosszahlansatz*) plays a crucial role: It is based on the assumption that colliding particles in a gas have no memory of their previous history, because of the gigantic number of collisions each particles goes through, before colliding again with a particle already met in the past. This assumption amounts to admit that in a gas, the states of two colliding particles are "statistically independent" of each other. This hypothesis introduces into the Boltzmann equation an effective representation of binary collisions in terms of "products of individual distribution functions," which is at the origin of the irreversible nature of this equation, despite the hypothesis that collisions between particles have to be conservative, that is, reversible and mechanical processes.

Boltzmann's equation is quite a complicated nonlinear integrodifferential equation and a general theorem, stating the existence and uniqueness of its solution, is not available. This notwithstanding, we can prove the so-called H-theorem, which allows us to conclude that the asymptotic evolution in time of the distribution function converges to the thermodynamic equilibrium, while identifying a quantity, that essentially amounts to the thermodynamic state function Entropy. In particular, in Section 1.5, we provide the proof of the H-theorem in the uniform case (i.e., in the simple situation, where the distribution function is independent of space coordinates), and we also show in Section 1.6 how this proof can be extended to the nonuniform case. In both instances, we obtain the explicit expression of the equilibrium distribution function, characterizing the thermodynamics of the ideal gas.

Boltzmann's equation allows us to work out also a fundamental theory of transport phenomena (see Section 1.7). In fact, by taking explicitly into account the quantities conserved in each binary collision process, namely, mass, momentum, and kinetic energy, we can derive the hydrodynamic equations, that account for the evolution in time of the densities of these locally conserved quantities. In the zero-order approximation, that is, when the average quantities present in the hydrodynamic equations are estimated at thermodynamic equilibrium, they describe the peculiar situation of an inviscid fluid. We conclude this chapter by showing that a perturbative approach allows us to improve the hydrodynamic equations to the first order of approximation, where they reproduce the Navier–Stokes equation and the heat equation.

1.1 Historical Perspective

The idea that thermodynamics could be related to a mechanical theory of matter dealing with a large number of particles, that is, atoms and molecules, was speculated on from the very beginning of kinetic theory in the middle of the nineteenth century. In a historical perspective, we could say that such an idea was a natural consequence of the formulation of the first principle of thermodynamics by the German natural philosopher Julius Robert von Mayer, establishing the equivalence between mechanical work and heat. This was checked in the famous experiment by the British James Prescott Joule and many contemporary physicists, among which the German Rudolf Clausius and August Karl Krönig, the British William Thomson (Lord Kelvin) and James Clerk Maxwell, and the Austrian Ludwig Eduard Boltzmann, devoted a good deal of their efforts to develop the foundations of kinetic theory.[1]

The mechanistic approach to thermodynamics was pushed to its extreme consequences in the work by Boltzmann in the last decades of the nineteenth century. His celebrated transport equation represents a breakthrough in modern science, and still today we cannot avoid expressing our astonishment about the originality and deep physical intuition of the Austrian physicist. Despite being inspired by a specific model, namely, the ideal gas, the main novelty of Boltzmann equation was that it represents the evolution of a distribution function, rather than the trajectories of individual particles in the gas. Boltzmann realized quite soon that the only way to describe the behavior of a large number of particles (a mole of a gas contains an Avogadro number of particles, $N_A \simeq 6.022 \times 10^{23}$) was to rely on a statistical approach, where the laws of probability had to be merged into the description of physical laws. We want to point out that the success of Boltzmann equation is not limited to establishing the foundations of equilibrium statistical mechanics. In fact, it also provides a description of the evolution toward equilibrium by the derivation of hydrodynamic equations associated with the conservation of mechanical quantities, that is, number, momentum, and energy of particles. These equations provide a mathematical basis for the theory of transport phenomena and a physical definition of transport coefficients in

---

[1] The reader should consider that all of these scientists were assuming the validity of the atomic hypothesis, despite no direct experimental evidence of the existence of atoms and molecules available at that time.

terms of basic quantities of kinetic theory, such as the mean free path, the average speed of particles, the heat capacity, and so on.

This notwithstanding, a good deal of contemporary scientists strongly criticized his equation as cumbersome or even paradoxical. In order to be properly interpreted, the reason for such a tough opposition to Boltzmann theory has to be framed into the cultural and philosophical debate of the time, when materialism was tightly conjugated to determinism and both of these philosophical categories had come to a deep crisis, in a rapidly changing world. Such cultural trends strongly influenced also the scientific community, where many distinguished physicists and chemists openly stated that they did not believe in the atomistic hypothesis, because it was considered a premise to a materialistic view of nature. In that boiling cultural environment, where different views and options were fighting daily inside the society, the debate among scientists such as Boltzmann, Ostwald, Loschmidt, Zermelo, and Poincaré[2] was certainly based on scientific arguments, but the aggressiveness of the contenders, emerging in some overheated disputes, frequently overtook the standards of a purely academic confrontation.

On a technical ground, it should be said that the main difficulties in understanding Boltzmann equation are not met in its mathematical derivation. They rather stem from finding a logical plausibility to the hypotheses that originate it. What appeared odd to his opposers was how one could obtain a time irreversible equation (consistent with the second law of thermodynamics) starting from time reversible dynamical rules at atomic level. Even the reader, which approaches for the first time this topic, might be puzzled by this seemingly contradictory scenario.

In order to dissipate doubts and cast the problem in the proper perspective, it is worth pointing out that the time irreversibility of Boltzmann equation has not a mechanical origin, but it is a consequence of what Boltzmann called the *Stosszahlansatz*. In modern words, we could translate it in "hypothesis of molecular chaos." It contains far from trivial concepts and one cannot fail to be astonished by the deep intuition of Boltzmann, who did not possess the refined mathematical tools that nowadays (after more than a century of mathematical progresses in ergodic theory) allow us to provide his *Stosszahlansatz* the support of a rigorous mathematical theory. We can safely state that if the main Boltzmann legacy to physical sciences was his equation and its applications, a not less relevant legacy to the development of mathematical sciences was the hint for building up a rigorous basis for his ergodic hypothesis.

We want to conclude by pointing out the truly paradoxical aspect of Boltzmann's life as a scientist. Many of his contemporaries considered him as the last priest of mechanicism, although, in a modern perspective, it can be easily realized that his work is the starting point of the end of mechanicism, intended as a purely deterministic approach to natural phenomena. In fact, his distribution function can be interpreted as a probability and his *Stosszahlansatz* is formulated on the basis of probabilistic assumptions. Unknowingly, he was paving the pathway that led to quantum mechanics. In fact, at the end of the nineteenth century his younger colleague and opposer, the German Max Planck, in order to provide

---

[2] Wilhelm Ostwald and Ernst Zermelo were German, Johann Josef Loschmidt was Austrian, and Henri Poincaré was French.

a theoretical explanation of the black-body spectrum, adapted Boltzmann statistical theory of the ideal gas to a gas of radiation at thermodynamic equilibrium: The main conceptual consequences were the existence of energy *quanta* and the identification of the scale of inaccuracy of a dynamical state in configuration space, given by the Planck's constant.

1.2 Kinetic Theory

1.2.1 The Ideal Gas

The basic model for understanding the mechanical foundations of thermodynamics is the ideal gas. It is a collection of N identical particles of mass m that can be represented geometrically as tiny homogeneous spheres of radius r. One basic assumption of the ideal gas model is that we are dealing with a diluted system; that is, the average distance δ between particles is much larger than their radius,

$$\delta = \left(\frac{1}{n}\right)^{\frac{1}{3}} \gg r, \tag{1.1}$$

where $n = N/V$ is the density of particles in the volume V occupied by the gas.[3] In the absence of external forces, particles move with constant velocity[4] until they collide pairwise, keeping their total momentum and energy constant (elastic collisions[5]). It can be easily realized that in such a diluted system, multiple collisions are such rare events that they can be neglected for practical purposes.

Now we want to answer the following question: What is the rate of these collisions and the average distance run by a particle between subsequent collisions? We can estimate these quantities by considering that a particle moving with velocity $\mathbf{v}$ in a time interval Δt can collide with the particles that are contained in a cylinder of basis $\sigma = 4\pi r^2$ (called cross section) and height $|\mathbf{v}|\Delta t$; see Fig. 1.1. For the sake of simplicity, we can assume that all the particles inside the cylinder are at rest with respect to the moving particle, so that we can estimate the number of collisions as

$$\mathcal{N}_{coll} = n\sigma|\mathbf{v}|\Delta t. \tag{1.2}$$

Accordingly, the number of collisions per unit time is given by the expression

$$\frac{\mathcal{N}_{coll}}{\Delta t} = n\sigma|\mathbf{v}| \tag{1.3}$$

---

[3] For a real gas of hydrogen molecules at room temperature (300 K) and atmospheric pressure (1 atm), $\delta \sim 10^{-6}$ m and $r \sim 10^{-10}$ m.

[4] One could argue that at least gravity should be taken into account, but its effects are generally negligible in standard conditions. An example where gravity has relevant, measurable effects will be studied in Section 2.2: It is the Brownian motion of colloidal particles (see Fig. 2.2).

[5] This hypothesis amounts to assuming that the particles of the gas are rigid spheres, so that they do not suffer any deformation in the collision process. In fact, in a real gas the energy transferred to the internal degrees of freedom of the molecules can be practically neglected in standard conditions.

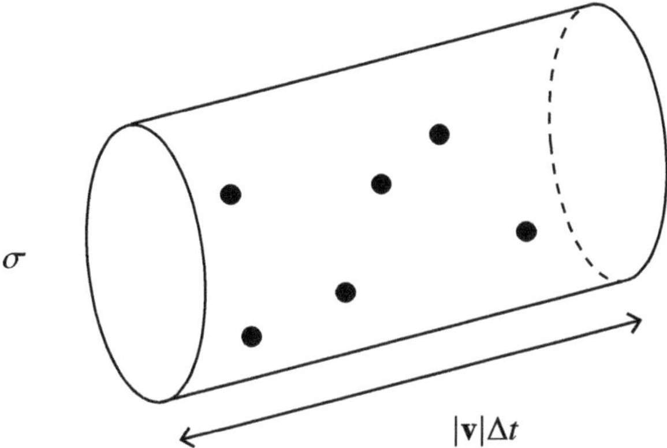

Fig. 1.1 Illustration of the concept of cross section. The black dots in the cylinder spanned by the cross section σ represent the centers of molecules hit in the time interval Δt by a molecule moving at speed v.

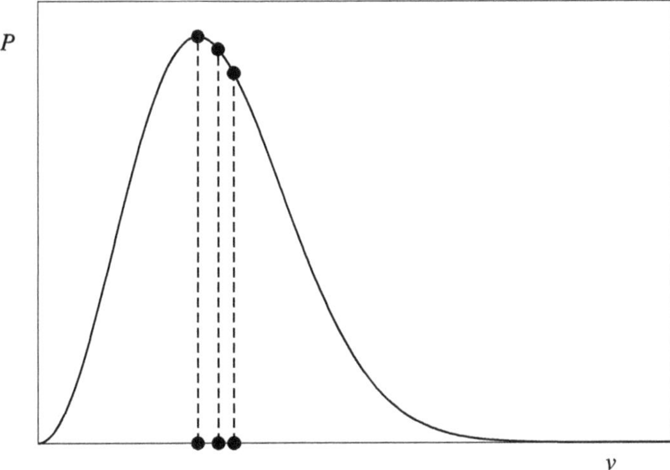

Fig. 1.2 The Maxwell distribution, Eq. (1.5). We indicate, from left to right, the most likely velocity v_{max}, the average velocity $\langle v \rangle$, and the square root of the average square velocity, $\langle v^2 \rangle^{1/2}$, whose expressions are given in Eq. (1.6).

and the average time between collisions reads

$$\tau \equiv \frac{\Delta t}{\mathcal{N}_{coll}} = \frac{1}{n\sigma|\mathbf{v}|}. \tag{1.4}$$

A quantitative estimate of τ can be obtained by attributing to $|\mathbf{v}|$ the value $\langle v \rangle$ of the equilibrium average of the modulus of the velocity of particles, v, in the ideal gas, according to Maxwell distribution (see Fig. 1.2),

$$P(v) = \frac{4}{\sqrt{\pi}} \left(\frac{m}{2T} \right)^{3/2} v^2 \exp\left(-\frac{mv^2}{2T} \right), \tag{1.5}$$

where T is the temperature of the ideal gas at equilibrium. Using such distribution, we obtain the expressions

$$v_{\max} = \sqrt{\frac{2T}{m}}, \qquad \langle v \rangle = \sqrt{\frac{8T}{\pi m}} = \frac{2}{\sqrt{\pi}} v_{\max}, \qquad \text{and} \qquad \langle v^2 \rangle^{1/2} = \sqrt{\frac{3T}{m}} = \sqrt{\frac{3}{2}} v_{\max}, \quad (1.6)$$

for the most likely velocity, the average velocity, and the square root of the average square velocity, respectively.

We can now rewrite (1.4) as

$$\tau = \frac{1}{n\sigma\langle v \rangle} \tag{1.7}$$

and determine the average distance run by a particle between two collisions, that is, its mean free path, by the expression

$$\lambda = \langle v \rangle \tau = \frac{1}{n\sigma}. \tag{1.8}$$

This formula corresponds to the case of a single moving particle colliding with target particles that are supposed to be immobile. But this is not the case, because in reality the target particles also move and a better estimate of τ and λ can be obtained using the formula

$$\tau = \frac{1}{n\sigma\langle v_r \rangle}, \tag{1.9}$$

where v_r is the modulus of the relative velocity $\mathbf{v}_r$, which follows the distribution

$$P_r(v_r) = \sqrt{\frac{2}{\pi}} \left(\frac{m}{2T} \right)^{3/2} v_r^2 \exp\left(-\frac{m v_r^2}{4T} \right). \tag{1.10}$$

This formula is a consequence of the general observation that the sum (or the difference) of two Gaussian variables is a Gaussian variable whose variance is the sum of their variances. In this case, $\mathbf{v}_r = \mathbf{v}_1 - \mathbf{v}_2$, with $\mathbf{v}_{1,2}$ satisfying the Maxwell distribution (1.5) and the doubling of the variance explains why the exponent $(mv^2/2T)$ in Eq. (1.5) now becomes $(mv_r^2/4T)$. Then, the prefactor changes accordingly, in order to keep $P_r(v_r)$ normalized.

With Eq. (1.10) at hand, we can evaluate

$$\langle v_r \rangle = \sqrt{\frac{16T}{\pi m}} = \sqrt{2}\langle v \rangle \tag{1.11}$$

and obtain

$$\tau = \frac{1}{\sqrt{2} n\sigma\langle v \rangle}, \tag{1.12}$$

from which we can evaluate the mean free path,

$$\lambda = \langle v \rangle \tau = \frac{1}{\sqrt{2} n\sigma}. \tag{1.13}$$

It is worth noting that the ratio between λ and τ gives $\langle v \rangle$, not $\langle v_r \rangle$, because one particle travels an average distance λ in time τ.

We can finally use the formula (1.13) to evaluate the mean free path for a gas at room temperature and pressure. In this case, λ is typically $O(10^{-7}\mathrm{m})$, which is three orders of magnitude larger than the typical size r of a particle, $O(10^{-10}\text{m})$, and one order of magnitude smaller than the distance between particles, $\delta = (1/n)^{1/3} = O(10^{-6}\text{m})$.

1.2.2 Transport Phenomena

Transport processes concern a wide range of phenomena in hydrodynamics, thermodynamics, physical chemistry, electric conduction, magnetohydrodynamics, and so on. They typically occur in physical systems (gases, liquids, or solids) made of many particles in the presence of inhomogeneities. Such a situation can result from nonequilibrium conditions (e.g., the presence of a macroscopic gradient of density, velocity, or temperature), or simply from fluctuations around an equilibrium state.

The kinetic theory of transport phenomena provides a unified phenomenological description of these apparently unlike situations. It is based on the assumption that even in nonequilibrium conditions, gradients are small enough to guarantee that local equilibrium conditions still hold. In particular, the kinetic approach describes the natural tendency of the particles to transmit their properties from one region to another of the fluid by colliding with the other particles and eventually establishing global or local equilibrium conditions.

The main success of the kinetic theory is the identification of the basic mechanism underlying all the above-mentioned processes: the transport of a microscopic quantity (e.g., the mass, momentum, or energy of a particle) over a distance equal to the mean free path λ of the particles, that is, the average free displacement of a particle between two subsequent collisions (see Eq. (1.13)). By this definition, we are implicitly assuming that the system is a fluid, where each particle is supposed to interact with the others only through mutual collisions.

Here we assume that we are dealing with a homogeneous isotropic system, where λ, the mean free path, is the same at any point and in any direction in space. Without prejudice of generality, we consider a system where a uniform gradient of the quantity $A(\mathbf{x})$ is established along the z-axis, and $A(x, y, z) = A(x', y', z) = A(z)$ for any x, x', y, and y'. In particular, we assume that $A(z)$ is a microscopic quantity, which slowly varies at constant rate along the coordinate z of an arbitrary Cartesian reference frame. We consider also a unit surface S_1 located at height z and perpendicular to the z-axis; see Fig. 1.3(a). Any particle crossing the surface S_1 last collided at an average distance $\pm\lambda$ along the z-axis, depending on the direction it is moving. The net transport of the quantity $A(z)$ through S_1 amounts to the number of crossings of S_1 from each side in the unit time. Consistently with the assumption of local equilibrium we attribute the same average velocity $\langle v \rangle$ to all particles crossing S_1. Isotropy and homogeneity of the system imply also that one-third of the particles move on average along the z-axis, half of them upward and half downward. Accordingly, S_1 is crossed along z in the unit time interval by $\frac{1}{6}n\langle v \rangle$ particles in each direction.

The net flux of $A(z)$ through S_1 is given by

$$\Phi(A) = \frac{1}{6}\langle v \rangle \left[n(z - \lambda)A(z - \lambda) - n(z + \lambda)A(z + \lambda) \right]. \tag{1.14}$$

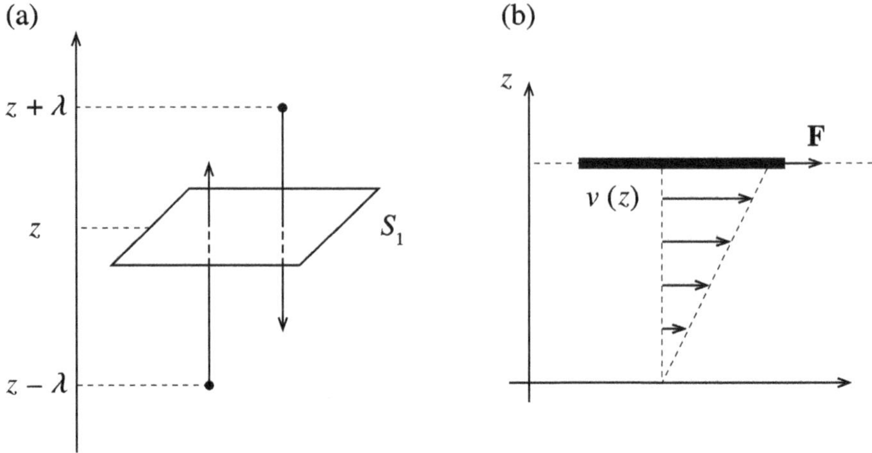

Fig. 1.3 (a) The surface S_1, normal to the $\hat{z}$ axis, is crossed by particles from both sides. Assuming particles move along the $\hat{z}$ axis, their most recent collision occurred at height $z - \lambda$ $(z + \lambda)$ if their speed is positive (negative). (b) A gradient of velocity along the z-axis in a liquid, produced by a plate at the liquid surface that is constrained to move at a finite speed by the application of a force F.

Since n and A vary weakly on the scale λ, one can use a first-order Taylor expansion and rewrite Eq. (1.14) as

$$\Phi(A) = -\frac{1}{3}\langle v\rangle\lambda\frac{\partial(nA)}{\partial z}. \tag{1.15}$$

This calculation can be performed more carefully by introducing explicitly the Maxwell distribution function of particle velocities at equilibrium. Nonetheless, one recovers the same result.

It is worth noting that the density n can be assumed to be constant in some circumstances. If so, since λ is inversely proportional to n the flux $\Phi(A)$ and the resulting kinetic coefficient are independent of the density n. Let's start with the simplest case, the transport of mass, where such assumption is basically wrong because the current is induced by a density gradient along the z-axis. In this case, $A(z)$ is a constant and

$$\Phi(n) = -\frac{1}{3}\langle v\rangle\lambda\frac{\partial n}{\partial z} = -D\frac{\partial n}{\partial z}, \tag{1.16}$$

where the quantity $D = \frac{1}{3}\langle v\rangle\lambda$ defines the diffusion coefficient of particles inside the fluid. This expression is equal to the definition of D through Eq. (2.19), $D = \lambda\langle v\rangle/d$, because in the calculation here above $d = 3$. In a real physical situation, D depends both on the diffusing substance and the medium of diffusion. At room temperature, a gas in air typically has $D \simeq 0.3$ cm^{2}s^{-1}; the diffusion coefficient of a liquid in water is typically of the order $D \simeq 10^{-5}$ cm^{2}s^{-1}; a gas in a solid has a much smaller diffusivity of the order $D \simeq 10^{-9}$ cm^{2}s^{-1}.

Other cases of physical interest correspond to the situations where a gradient of velocity or temperature is present and the density n is assumed to be constant, so $\Phi(A) = -\frac{1}{3}n\langle v\rangle\lambda\frac{\partial A}{\partial z}$. If there is a gradient of velocity, we assume that the fluid flows with constant macroscopic velocity $v(z)$ parallel to the (x, y)-plane. In such a situation, there is a net transport of kinetic momentum $mv(z)$ (m is the mass of a particle), yielding a shear stress $\Phi(mv(z))$ between the fluid layers laying on the (x, y) plane (see Fig. 1.3(b)):

$$\Phi(mv(z)) = -\frac{1}{3}nm\langle v\rangle\lambda\frac{\partial v(z)}{\partial z} = -\eta\frac{\partial v(z)}{\partial z}, \tag{1.17}$$

where the quantity $\eta = \frac{1}{3}nm\langle v\rangle\lambda = \frac{m\langle v\rangle}{3\sqrt{2}\sigma}$ defines the viscosity of the fluid. As anticipated, the viscosity of an ideal fluid turns out to be independent of the density n, therefore of the pressure. This counterintuitive conclusion was first derived by Maxwell, and its experimental verification sensibly contributed to establish in the scientific community a strong consensus on the atomistic approach of kinetic theory. It is worth stressing that such a conclusion does not hold when dealing with very dense fluids. At room temperature, diluted gases typically have η of order 10 μPa·s, while in water and blood, η is of the order of few millipascal-second and honey at room temperature has $\eta \approx 1$ Pa·s (called Poiseuille, Pl).

It remains to consider the case when $A(z)$ is the average kinetic energy of particles $\bar{\epsilon}(z)$. At equilibrium, the energy equipartition condition yields the relation $n\bar{\epsilon}(z) = \rho C_V T(z)$, where $\rho = mn$ is the mass density of particles, C_V is the specific heat at constant volume, and $T(z)$ is the temperature at height z. The net flux of kinetic energy $\Phi(\bar{\epsilon})$ can be read as the heat transported through the fluid along the z-axis,

$$\Phi(\bar{\epsilon}) = -\frac{1}{3}n\langle v\rangle\lambda\frac{\partial\bar{\epsilon}}{\partial z} = -\frac{1}{3}\rho C_V\langle v\rangle\lambda\frac{\partial T(z)}{\partial z} = -\kappa\frac{\partial T(z)}{\partial z}, \tag{1.18}$$

where the quantity $\kappa = \frac{1}{3}\rho C_V\langle v\rangle\lambda = \frac{mC_V\langle v\rangle}{3\sqrt{2}\sigma}$ defines the heat conductivity. Also κ is found to be independent of n. The variability of κ in real systems is less pronounced than for other kinetic coefficients: In fact, a very good conductor such as silver has $\kappa \simeq 400$ Wm^{-1}K^{-1}, while for cork, an effective heating insulator, it drops down to 4×10^{-2} in the same units.

One can conclude that the transport coefficients, that is, the diffusion constant D, the viscosity η, and the heat conductivity κ, are closely related to each other and depend on a few basic properties of the particles, such as their mass m, their average velocity $\langle v\rangle$, and their mean free path λ. For example, by comparing the definitions of κ and η, one finds the remarkable relation

$$\frac{\kappa}{\eta} = \alpha C_V, \tag{1.19}$$

with $\alpha = 1$. In real systems, the constant α takes different values, which depend on the presence of internal degrees of freedom (e.g., $\alpha = \frac{5}{2}$ for realistic models of monoatomic gases).

The conceptual relevance of the relation (1.19) is that it concerns quantities that originate from quite different conditions of matter. In fact, on the left-hand side, we have the ratio of two transport coefficients associated with macroscopic nonequilibrium conditions, while on the right-hand side, we have a typically equilibrium quantity, the specific heat at constant volume. After what has been discussed in this section, this observation is far from mysterious: By assuming that even in the presence of a macroscopic gradient of physical quantities equilibrium conditions set in locally, the kinetic theory provides a unified theoretical approach for transport and equilibrium observables.

1.3 Distribution Function in Molecular Space

A real gas is a collection of a large number N of molecules (typically 10^{19} particles/cm^3) interacting by short-range forces. In order to simplify the problem, we assume that all molecules are identical (in particular, they have the same mass m), that the laws of their mutual interactions are ruled by suitable short-range conservative forces, and, moreover, that quantum and relativistic effects can be neglected. One further simplification amounts to assume that each molecule can be represented as a point particle (i.e., we ignore its internal degrees of freedom), obeying the laws of Newtonian dynamics in the three-dimensional space

$$\dot{\mathbf{r}}_i(t) = \mathbf{v}_i(t) \tag{1.20a}$$

$$m\dot{\mathbf{v}}_i(t) = \mathbf{F}_i(t), \tag{1.20b}$$

with $i = 1, \cdots, N$ (we have used the shorthand notation $\dot{a}$ for the derivative of a with respect to time). This is a system of $6N$ first-order differential equations, where $\mathbf{r}_i$ and $\mathbf{v}_i$ are the position and the velocity vectors of the i-th particle; $\mathbf{F}_i$ is the total conservative force acting of the i-th particle, which results from some conservative pairwise interaction potential between particles, gravity, inertial forces, and those exerted by the walls of the container, which confine the portion of space available to the gas. In practice, we are assuming that the total conservative force $\mathbf{F}_i$ acting on each particle contains a complete information about the physical properties of the system.

In principle, the time evolution of the gas could be determined by solving the $6N$ equations of motion (1.20) for any given set of initial conditions $\mathbf{r}_i(0)$ and $\mathbf{v}_i(0)$. In practice, it is evident that solving this problem for a macroscopic gas is far beyond any realistic possibility.[6]

But do we really need to know the dynamical state of all the molecules in the gas at any given time in order to obtain physical inferences about its thermodynamic properties? In order to extract relevant physical information, that is, those accessible to experimental observations, we are rather interested in measuring some observables, which involve only macroscopic average properties of the gas. For instance, what we operatively define as a measure of the pressure of a gas is a space and time average of the instantaneous force that each single molecule exerts on the walls of the container. It is important to observe that in a microscopic perspective, such a quantity may greatly vary both in space and time; nonetheless for the sake of physical significance, we have to assume that all instantaneous microscopic processes combine, yielding well-defined average macroscopic quantities (irrespectively of equilibrium or nonequilibrium macroscopic states of the gas). Upon these remarks, we are naturally led to consider a less detailed description of the dynamical state of a gas than the one contained in Eq. (1.20).

---

[6] Nowadays, the computational discipline named *molecular dynamics* tackles explicitly the problem of integrating very large sets of dynamical equations relying upon the power of modern computers. Anyway, even if large, the number of dynamical equations which can be integrated by this method remains many orders of magnitude smaller than N_A.

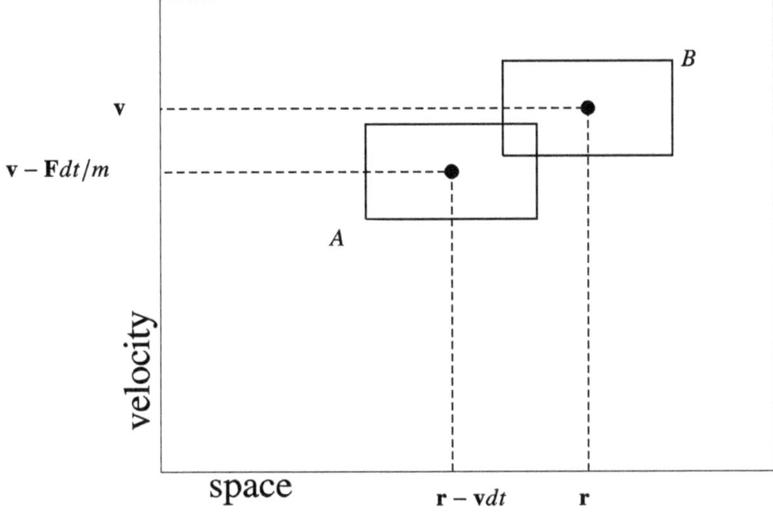

Fig. 1.4 The volume elements in μ-space at time $t - \delta t$ (A) and at time t (B). In this "planar" representation of the μ-space, each axis corresponds to a three-dimensional space of position coordinates (horizontal axis) and velocity coordinates (vertical axis) of the particles of a gas. The number of particles in the small volume of dimension $d\mathbf{r}d\mathbf{v}$ at time t is given by expression (1.21).

For this purpose, let us introduce a distribution function $f(\mathbf{r}, \mathbf{v}; t)$ summarizing the microscopic knowledge of the state of the gas in such a way that the expression

$$f(\mathbf{r}, \mathbf{v}; t)d\mathbf{r}d\mathbf{v} \tag{1.21}$$

is the number of particles which at time t are in the small volume $d\mathbf{r}d\mathbf{v}$ of the six-dimensional space engendered by the position and velocity coordinates, the so-called molecular space or, in a shorthand notation, μ-space, where each point represents the dynamical state of a particle (see Fig. 1.4). In the μ-space, the microscopic state of a gas composed of N particles is represented by a collection of N points.

We can interpret the following quantities:

$$n(\mathbf{r}; t) = \int d\mathbf{v} f(\mathbf{r}, \mathbf{v}; t) \tag{1.22}$$

and

$$N = \int d\mathbf{r} \int d\mathbf{v} f(\mathbf{r}, \mathbf{v}; t) \tag{1.23}$$

as the number of molecules per unit volume at the point $\mathbf{r}$ and the total number of molecules,[7] respectively. The integration over the space coordinates $\mathbf{r}$ extends to the volume V containing

---

[7] Eq. (1.23) has to be properly read as a conservation constraint: In fact, the total number of particles in the gas is assumed to be the same at any time t.

the gas.[8] In order to simplify the notation in what follows, we avoid to indicate explicitly the support of the integral over the space coordinates, namely, $\int d\mathbf{r} \equiv \int_V d\mathbf{r}$.

The integral over the velocity coordinate $\mathbf{v}$ in Eq. (1.22) should be defined on a finite support, limited by the finiteness of the total energy of the gas. On the other hand, for any practical purpose, we can extend the support of this integral over $\mathbb{R}^3$, provided we assume that the distribution function $f(\mathbf{r}, \mathbf{v}; t)$ vanishes sufficiently fast for large values of $\mathbf{v}$. We shall comment more precisely on this delicate point at the end of Section 1.4. In what follows, we simplify the notation by avoiding to indicate explicitly the support of the integral over the velocity coordinates, namely, $\int d\mathbf{v} \equiv \int_{-\infty}^{+\infty} d\mathbf{v}$.

We can conclude these preliminary considerations about the definition of the distribution function in μ-space by defining the local and global average values of any dynamical observable $A(\mathbf{r}, \mathbf{v}; t)$ as follows:

$$\langle A(\mathbf{r}; t)\rangle = \frac{\int d\mathbf{v} A(\mathbf{r}, \mathbf{v}; t) f(\mathbf{r}, \mathbf{v}; t)}{\int d\mathbf{v} f(\mathbf{r}, \mathbf{v}; t)} = \frac{1}{n}\int d\mathbf{v} A(\mathbf{r}, \mathbf{v}; t) f(\mathbf{r}, \mathbf{v}; t) \tag{1.24}$$

$$\bar{A}(t) = \frac{1}{N}\int d\mathbf{r}\int d\mathbf{v} A(\mathbf{r}, \mathbf{v}; t) f(\mathbf{r}, \mathbf{v}; t). \tag{1.25}$$

These definitions imply that for any given dynamical observable $A(\mathbf{r}, \mathbf{v}; t)$, the distribution function $f(\mathbf{r}, \mathbf{v}; t)$ must be such that the integration over space and velocity coordinates is convergent at any time t.

1.4 Time Evolution of the Distribution Function

In this section, we want to determine the time evolution of $f(\mathbf{r}, \mathbf{v}; t)$,

$$\frac{\partial f(\mathbf{r}, \mathbf{v}; t)}{\partial t}.$$

Consider the small volume $d\mathbf{r}d\mathbf{v}$ in μ-space centered at the point of coordinates $(\mathbf{r}, \mathbf{v})$. The number of gas particles inside this volume varies in time for the following reasons:
(i) each particle has a velocity $\mathbf{v}$ and consequently its position changes in time. Furthermore, in the presence of a (total) external conservative force $\mathbf{F}(\mathbf{r})$, the velocity coordinates evolve as well;
(ii) at some positions in space, particles may collide between themselves and with the walls of the volume containing the gas, thus changing instantaneously their velocities.

---

[8] In a real gas, any molecule occupies a finite, even if extremely small, portion of the physical space. In principle, this would demand to consider an excluded volume term in the integrals over space coordinates and, in any case, would prevent the definition of the density $n(\mathbf{r}; t)$ as a point function. On the other hand, for our present purposes we can assume the simplifying hypothesis that our gas is made of mass points. This is very reasonable if the gas is diluted, that is, the average distance between particles is much larger than their linear extension (see Eq. (1.1)).

According to these remarks, we can separate the time derivative of $f(\mathbf{r}, \mathbf{v}; t)$ into two contributions: a stream one due to (i) and a collision one due to (ii). In formulae:

$$\frac{\partial f(\mathbf{r}, \mathbf{v}; t)}{\partial t} = \left(\frac{\partial f(\mathbf{r}, \mathbf{v}; t)}{\partial t}\right)_s + \left(\frac{\partial f(\mathbf{r}, \mathbf{v}; t)}{\partial t}\right)_c . \tag{1.26}$$

Let us compute the first term on the r.h.s. of this equation. A particle with coordinates $(\mathbf{r}, \mathbf{v})$ at time t will have coordinates $(\mathbf{r} + \mathbf{v}dt, \mathbf{v} + \frac{\mathbf{F}}{m}dt)$ at time $t + dt$. As a consequence, during the small time interval dt, all particles that at time t had coordinates $(\mathbf{r} - \mathbf{v}dt, \mathbf{v} - \frac{\mathbf{F}}{m}dt)$ will enter the small volume of μ-space centered at the point $(\mathbf{r}, \mathbf{v})$:

$$f(\mathbf{r}, \mathbf{v}; t + dt) = f(\mathbf{r} - \mathbf{v}dt, \mathbf{v} - \frac{\mathbf{F}}{m}dt; t), \tag{1.27}$$

so that one can write

$$f(\mathbf{r}, \mathbf{v}; t + dt) - f(\mathbf{r}, \mathbf{v}; t) = -\nabla_\mathbf{r} f \cdot \mathbf{v}dt - \nabla_\mathbf{v} f \cdot \frac{\mathbf{F}(\mathbf{r})}{m}dt, \tag{1.28}$$

where

$$\nabla_\mathbf{r} = \left(\frac{\partial}{\partial x}, \frac{\partial}{\partial y}, \frac{\partial}{\partial z}\right), \qquad \nabla_\mathbf{v} = \left(\frac{\partial}{\partial v_x}, \frac{\partial}{\partial v_y}, \frac{\partial}{\partial v_z}\right). \tag{1.29}$$

Summarizing, we obtain the explicit form of the stream term

$$\left(\frac{\partial f(\mathbf{r}, \mathbf{v}; t)}{\partial t}\right)_s = -\mathbf{v} \cdot \nabla_\mathbf{r} f(\mathbf{r}, \mathbf{v}; t) - \frac{\mathbf{F}(\mathbf{r})}{m} \cdot \nabla_\mathbf{v} f(\mathbf{r}, \mathbf{v}; t). \tag{1.30}$$

This equation has an intuitive interpretation; if we disregard collisions, the gas should behave as an incompressible fluid for which

$$\frac{df(\mathbf{r}, \mathbf{v}; t)}{dt} = \frac{\partial f(\mathbf{r}, \mathbf{v}; t)}{\partial t} + \nabla_\mathbf{r} f(\mathbf{r}, \mathbf{v}; t) \cdot \frac{\partial \mathbf{r}}{\partial t} + \nabla_\mathbf{v} f(\mathbf{r}, \mathbf{v}; t) \cdot \frac{\partial \mathbf{v}}{\partial t} = 0, \tag{1.31}$$

since the total hydrodynamic time derivative is defined as the measure of the variation in time of any observable with respect to a point in μ-space moving with the same speed of the fluid.

It remains to specify the collision term. For this purpose, let us consider at time $(t - dt)$ a small volume A in μ-space centered at $(\mathbf{r} - \mathbf{v}dt, \mathbf{v} - \frac{\mathbf{F}}{m}dt)$ and its image B at time t centered at $(\mathbf{r}, \mathbf{v})$, see Fig. 1.4. In the absence of collisions, all particles that are in A at time $(t - dt)$ are found in B at time t. Conversely, if a particle in A at time $t - dt$ collides with any other particle, we assume that it cannot be found in B after the time interval dt. On the other hand, there are particles that, due to collisions, can reach B in the time interval dt, coming from small volumes other than A. We can formalize these considerations by introducing the expression

$$\left(\frac{\partial f}{\partial t}\right)_c d\mathbf{r}d\mathbf{v}dt = (\bar{R} - R)d\mathbf{r}d\mathbf{v}dt. \tag{1.32}$$

The quantities $Rd\mathbf{r}d\mathbf{v}dt$ and $\bar{R}d\mathbf{r}d\mathbf{v}dt$ represent the number of collisions in the time interval dt where one of the colliding particles is in A at time $(t - dt)$ and the number of collisions in the time interval dt for which one of the colliding particles is in B at t, starting at time

$(t - dt)$ from any volume, respectively: These two expressions can be interpreted as a *loss* term and a *gain* one.

Note that these definitions are not strictly rigorous: For instance, the definition of R assigns a double contribution to collisions between particles, which are both in A at time $t - dt$. On the other hand, the contribution coming from this kind of collision can be neglected, provided $d\mathbf{v}$ is sufficiently small, because particles with the same velocity cannot collide.

Still, Eq. (1.32) in its present form is not very useful: It is hopeless to find an explicit expression for R and $\bar{R}$ by taking into account all the details of the collision dynamics at a microscopic level (which would amount again to solve Eqs. (1.20)). So, we are obliged to introduce some simplifying hypotheses.

The first one is that the gas is diluted, that is, multiple collisions among particles can be considered extremely rare events with respect to binary collisions. Accordingly, we shall perform explicit calculations by considering only the contribution coming from binary collisions. We can represent formally a binary collision with the symbol

$$(\mathbf{v}, \mathbf{u}) \rightarrow (\mathbf{v}', \mathbf{u}'), \tag{1.33}$$

where $\mathbf{v}$ and $\mathbf{u}$ are the velocities of the particles just before the collision and $\mathbf{v}'$ and $\mathbf{u}'$ just after the collision (see Appendix A).

Consider first that colliding particles must be in the same volume element centered at $\mathbf{r}$ and dimension $d\mathbf{r}$ in the subspace of position coordinates of the μ-space. We can elaborate the expression of $Rd\mathbf{r}d\mathbf{v}dt$ introduced in Eq. (1.32) by schematizing the binary collisions in such a way that particles with velocity $\mathbf{u}$ are target particles, over which the particles with velocity $\mathbf{v}$ collide.

Let us denote with Φ the flux of particles with velocity $\mathbf{v}$ that collide per unit time in the volume centered in $\mathbf{r}$ with the target particles with velocity $\mathbf{u}$, in formulae

$$\Phi = f(\mathbf{r}, \mathbf{v}; t)|\mathbf{v} - \mathbf{u}|d\mathbf{v}. \tag{1.34}$$

By recalling that $f(\mathbf{r}, \mathbf{u}; t)$ is a density in the μ-space, it is immediate to check that Φ has actually the physical dimensions $[\ell^{-2}][t^{-1}]$. Accordingly, the number of particles scattered per unit time into the solid angle element $d\Omega$ around the direction Ω is given by the expression

$$\Phi\sigma(\mathbf{v}, \mathbf{u}; \Omega)d\Omega, \tag{1.35}$$

where $\sigma(\mathbf{v}, \mathbf{u}; \Omega)$ denotes the differential cross-section (i.e., the cross-section per unit solid angle) associated with the collision process. This quantity has the physical dimension $[\ell^2]$ and, in practice, it can be determined experimentally, or theoretically, if the nature of the interaction forces entering the collisions is known. For our present purposes it is not necessary to discuss this problem in detail. We would rather introduce a further simplifying hypothesis typical in classical mechanics that collisions between the gas particles amount to interactions between rigid bodies, ruled by central forces. In this case it can be argued that the differential cross-section σ depends only on $|\mathbf{v} - \mathbf{u}|$ and not on $\mathbf{v}$ and $\mathbf{u}$ separately (see Appendix A). Accordingly, the number of binary collisions in the time interval dt can be obtained by multiplying the quantity in Eq. (1.35) by dt and by the number of target particles in the small volume of μ-space, $d\mathbf{r}d\mathbf{u}$, centered around $(\mathbf{r}, \mathbf{u})$, namely

$$f(\mathbf{r}, \mathbf{v}; t)|\mathbf{v} - \mathbf{u}|dv\sigma(|\mathbf{v} - \mathbf{u}|; \Omega)d\Omega dt f(\mathbf{r}, \mathbf{u}; t)d\mathbf{r}d\mathbf{u}. \tag{1.36}$$

By summing this quantity over all possible scattering directions, Ω, and over all possible values of the velocities of the target particles, $\mathbf{u}$, we finally obtain an explicit expression for $Rd\mathbf{r}dvdt$, thus yielding

$$R = f(\mathbf{r}, \mathbf{v}; t) \int d\Omega \int d\mathbf{u}\sigma(|\mathbf{v} - \mathbf{u}|; \Omega)f(\mathbf{r}, \mathbf{u}; t)|\mathbf{v} - \mathbf{u}|. \tag{1.37}$$

Note that the physical dimension of the double integral on the r.h.s. of this equation is $[t^{-1}]$, because $f(\mathbf{r}, \mathbf{u}; t)$ is a density in the μ-space.

In order to obtain an explicit expression also for $\bar{R}d\mathbf{r}dvdt$, we have to consider the time-reversed collision process represented in (1.33), namely

$$(\mathbf{v}', \mathbf{u}') \rightarrow (\mathbf{v}, \mathbf{u}), \tag{1.38}$$

where particles with initial velocities $\mathbf{v}'$ and $\mathbf{u}'$ acquire the final velocities $\mathbf{v}$ and $\mathbf{u}$, respectively. The number of these "reversed" binary collisions in the time interval dt can be obtained, as in (1.36), by the following expression:

$$f(\mathbf{r}, \mathbf{v}'; t)|\mathbf{v}' - \mathbf{u}'|dv'\sigma(|\mathbf{v}' - \mathbf{u}'|; \Omega)d\Omega dt f(\mathbf{r}, \mathbf{u}'; t)d\mathbf{r}d\mathbf{u}'. \tag{1.39}$$

At this point we can use again the hypothesis that the collision process of the gas particles can be represented in the form of conservative binary collisions between hard spheres, ruled by central forces. Due to the time-reversibility of the laws of classical mechanics, the frequency of the collision $(\mathbf{v}, \mathbf{u}) \rightarrow (\mathbf{v}', \mathbf{u}')$ must be the same for the inverse collision $(\mathbf{v}', \mathbf{u}') \rightarrow (\mathbf{v}, \mathbf{u})$. Moreover, the velocities of the colliding particles $(\mathbf{v}', \mathbf{u}')$ are univocally determined functions of $(\mathbf{v}, \mathbf{u})$ and vice versa (see Eqs. (A.18) to (A.20) in Appendix A). The conservative nature of collisions between hard spheres yields also the important relation (see Eq. (A.4))

$$|\mathbf{v}' - \mathbf{u}'| = |\mathbf{v} - \mathbf{u}|, \tag{1.40}$$

which implies that the differential cross-section σ must be the same for both collision processes. Due to Liouville's theorem (see Eq. (A.12) in Appendix A), volumes in the subspace of velocities are also conserved, that is,

$$d\mathbf{v}'d\mathbf{u}' = d\mathbf{v}d\mathbf{u}. \tag{1.41}$$

In conclusion, (1.39) can be rewritten in the form

$$f(\mathbf{r}, \mathbf{v}'; t)|\mathbf{v} - \mathbf{u}|dv\sigma(|\mathbf{v} - \mathbf{u}|; \Omega)d\Omega dt f(\mathbf{r}, \mathbf{u}'; t)d\mathbf{r}d\mathbf{u}. \tag{1.42}$$

The last step for obtaining $\bar{R}d\mathbf{r}dvdt$ amounts to integrate Eq. (1.42) over all possible scattering directions, Ω, and over all possible values of the velocities of the target particles, $\mathbf{u}$, thus yielding

$$\bar{R} = \int d\Omega \int d\mathbf{u}\sigma(|\mathbf{v} - \mathbf{u}|; \Omega)f(\mathbf{r}, \mathbf{v}'; t)f(\mathbf{r}, \mathbf{u}'; t)|\mathbf{v} - \mathbf{u}|, \tag{1.43}$$

where both $\mathbf{v}'$ and $\mathbf{u}'$ are functions of $\mathbf{v}$ and $\mathbf{u}$ (see Eqs. (A.19) and (A.20)).

It is very important to observe that Eqs. (1.37) and (1.43) contain an implicit strong hypothesis, that is, the number of binary collisions is proportional to the product of two distribution functions. This statement amounts to the so-called hypothesis of molecular chaos, the *Stosszahlansatz* by Boltzmann, which cannot be assumed to be valid simply on the basis of dynamical arguments. In fact, this hypothesis implies that the velocities of any pair of colliding particles are statistically independent quantities, so that their distribution functions can be factorized as is usually made for independent events in probability theory. On the other hand, if we would rely upon a purely dynamical description of an ideal gas at thermodynamic equilibrium as an ensemble of particles interacting by conservative collisions, we should observe that a given pair of particles colliding at a given instant of time has already collided before and will collide later for an arbitrarily large number of times. In principle, this could be enough for arguing that the *Stosszahlansatz* is not a rigorous hypothesis.[9]

The argument by Boltzmann for justifying this hypothesis goes as follows. In a gas containing an extremely large number of particles, a pair of particles will come back to their own successive collision only after having collided *on average* with a huge number of other particles of the gas, thus effectively losing memory of their previous collision. Accordingly, it is the so-called *law of large numbers* which justifies the existence of individual distribution functions of the particle velocities and their effective statistical independence. The intuition of Boltzmann is that, being our computational ability unavoidably limited when describing the evolution of an astronomical number of dynamical degrees of freedom, we are obliged to turn to a statistical description, consistently with the very foundations of the Kinetic Theory of Gases. This heuristic procedure is at the origin of the successes and drawbacks of Boltzmann theory; while apparently contradicting the original assumption of a purely dynamical foundation of thermodynamics, it yields a probabilistic justification of the irreversibility.

This scenario is presumably quite disappointing for those who assign dominant relevance to mathematical rigor in a physical theory. In fact, as we mention in Section 1.1, already at the end of the nineteenth century, many scientists, among which Ostwald, Loschmidt, Zermelo and Poincaré, had pointed out the formal inconsistencies of the Boltzmann theory.

Here we want to remark only that there is no a priori justification for all of the assumptions that we have introduced up to here. Notice that, in principle, even the hypothesis that a gas can be described by the laws of classical mechanics as colliding hard spheres contains a degree of arbitrariness. What is even more important is that all the hypotheses made on the microscopic properties of the gas cannot be checked directly. In fact, the only statement that we can certainly affirm is that at any time we have an unavoidably partial or inaccurate knowledge of the microscopic dynamical state of a system composed of a macroscopic number of particles. This is a physical fact that cannot be contradicted, because it goes back to the operative definition of the measurement of any physical observable. Anyway, as we have anticipated in Section 1.1, the comparison between experimental measurements and

---

[9] For instance, if one studies numerically the dynamical model of a few units of particles colliding elastically in a box with reflecting walls, one will eventually realize that the system evolves to an almost recurrent state, because exact recurrence is prevented by numerical accuracy.

the predictions of the theory about macroscopic average properties of physical observables is the only a posteriori criterion for establishing the limits of validity of this theory.

Coming back to math, by substituting Eqs. (1.37) and (1.43) into Eq. (1.32), we obtain

$$\left(\frac{\partial f_\mathbf{v}}{\partial t}\right)_c = \int d\Omega \int \sigma(|\mathbf{v} - \mathbf{u}|; \Omega) \, (f_{\mathbf{v}'} f_{\mathbf{u}'} - f_\mathbf{v} f_\mathbf{u}) \, |\mathbf{v} - \mathbf{u}| d\mathbf{u}, \qquad (1.44)$$

where we have introduced the shorthand notation $f_\mathbf{v} \equiv f(\mathbf{r}, \mathbf{v}; t)$.

Finally, Eq. (1.26) can be written in the explicit form

$$\frac{\partial f_\mathbf{v}}{\partial t} = -\left(\mathbf{v} \cdot \nabla_\mathbf{r} + \frac{\mathbf{F}(\mathbf{r})}{m} \cdot \nabla_\mathbf{v}\right) f_\mathbf{v} + \int d\Omega \int \sigma(|\mathbf{v} - \mathbf{u}|; \Omega) \, (f_{\mathbf{v}'} f_{\mathbf{u}'} - f_\mathbf{v} f_\mathbf{u}) \, |\mathbf{v} - \mathbf{u}| d\mathbf{u}. \quad (1.45)$$

This is the Boltzmann transport equation, which was established by Boltzmann in 1872. Before concluding this section, we want to add some remarks about the physical and mathematical contents of Eq. (1.45).

(i) The hypothesis of purely binary collisions is quite reasonable if one considers that in a real gas at room temperature and pressure, 1 cm^3 contains 10^{19} atoms, whose total volume is 10^{-4} cm^3. In such conditions, multiple collisions can be actually considered as extremely rare events.

(ii) All the results that we are going to derive from Eq. (1.45) crucially depend on the hypothesis that the values taken by the velocities of colliding particles are statistically independent of each other (molecular chaos hypothesis).

(iii) External forces (although they are explicitly included in the stream term) are supposed to play no role in the collision process, which is simply governed by short-range intermolecular forces acting between particles. This also implies that particles collide (as we have supposed in deriving Eq. (1.45)) when they are found in the same small, but not infinitesimal, volume $d\mathbf{r}$, whose linear dimensions should be comparable to the range of intermolecular forces. We can assume, as well, that the interaction lasts over a very short interval of time. Since our aim is to obtain predictions about average macroscopic properties of the gas, it is evident that this is possible only by measuring those observables depending on length and time scales significantly larger than those involved in the collision process. In this case, $d\mathbf{r}$ can be considered as an infinitesimal quantity and the collision as an instantaneous event. All those phenomena that involve length and time scales smaller than or comparable with the abovementioned ones cannot be correctly predicted by the present approach.

(iv) From a mathematical point of view, Eq. (1.45) is still quite a complicated nonlinear integral-differential equation, also due to the nontrivial functional dependence of $\mathbf{v}'$ and $\mathbf{u}'$ on $\mathbf{v}$, $\mathbf{u}$ and on the geometric parameters involved in the collision process. A general solution of the initial value problem of this equation is not available. For our purposes, we can consider a class of initial conditions $f(\mathbf{r}, \mathbf{v}; 0) = f_{in}(\mathbf{r}, \mathbf{v})$ and external forces $\mathbf{F}(\mathbf{r})$ both satisfying continuity, derivability, and asymptotic properties, for which it seems plausible to assume the existence and unicity of the solution.

(v) The support of the integral over $d\mathbf{u}$ in Eq. (1.45) is not specified, although it is assumed to take values from $-\infty$ to $+\infty$ for any component of the velocity vector $\mathbf{u}$. This is justified by the fact that the distribution function $f_\mathbf{u}$ is assumed to vanish sufficiently fast when $|\mathbf{u}| \to \pm\infty$, not only to ensure the convergence of the integral but also to make negligible the probability that $|\mathbf{u}| \gtrsim c$, the light speed. In fact, this limit is extremely far from the typical values of the velocity of a gas particle in standard thermodynamic conditions.

1.5 The H-theorem in the Spatially Uniform Case

In this section, we aim at studying the solutions of Eq. (1.45) for a distribution function independent of the position coordinates $\mathbf{r}$, so that the shorthand notation introduced at the end of Section 1.4 further reduces to $f_{\mathbf{v}} \equiv f(\mathbf{v}; t)$. This is the so-called uniform case, which can be accomplished by assuming that any conservative force in (1.45) is absent, namely

$$\mathbf{F}(\mathbf{r}) = 0. \tag{1.46}$$

This condition implies that also the effect of the walls containing the gas is immaterial. In this sense, we can say that Eq. (1.46) amounts to the infinite–volume approximation or to periodic boundary conditions. In both cases, physical boundaries are absent, the stream term vanishes, and the Boltzmann equation (1.45) simplifies to

$$\frac{\partial f_{\mathbf{v}}}{\partial t} = \int d\Omega \int d\mathbf{u}\,\sigma(|\mathbf{v} - \mathbf{u}|, \Omega)|\mathbf{v} - \mathbf{u}|\,(f_{\mathbf{v}'} f_{\mathbf{u}'} - f_{\mathbf{v}} f_{\mathbf{u}})\,. \tag{1.47}$$

The quantities

$$n = \int d\mathbf{v}\, f(\mathbf{v}, t), \tag{1.48}$$

$$e = \int d\mathbf{v}\, \frac{1}{2} m v^2 f(\mathbf{v}, t), \tag{1.49}$$

$$\mathbf{p} = \int d\mathbf{v}\, m\mathbf{v} f(\mathbf{v}, t), \tag{1.50}$$

representing the number of particles, the energy, and the momentum per unit volume, respectively, are all independent of $\mathbf{r}$. As a consequence of the conservation laws of the number of particles, energy, and momentum for any collision process (see Appendix A), all of these quantities, resulting from an integration over all possible velocities, have to be constants of the motion; that is, they are independent of time (notice that in the infinite–volume limit, the total number of particles, the total energy, and the total momentum are by definition infinite quantities). In what follows we report the proof of the Boltzmann H-theorem, in the uniform case.

We first define the quantity

$$H(t) = \int d\mathbf{v}\, f(\mathbf{v}, t) \log\left(f(\mathbf{v}, t)\right). \tag{1.51}$$

The H-theorem states that $H(t)$ is a nonincreasing function of time. The corollaries of the H-theorem yield the following conclusions:

(i) $H(t)$ is bounded from below by the value H_0, which is attained in the limit $t \to \infty$;

(ii) H_0 corresponds to a time-independent distribution function $\bar{f}_\mathbf{v}$, which is just the Maxwell velocity distribution function, see Section 1.2.

In a physical perspective, we can extract from (i) and (ii) two main consequences. First, the existence of an equilibrium distribution function $\bar{f}_\mathbf{v}$ of the gas, which represents the cornerstone for the construction of equilibrium thermodynamics. Second, H is a quantity related to the entropy S of the gas through the simple relation $S = -H$, thus allowing us to reinterpret H-theorem as the derivation from kinetic theory of the second principle of thermodynamics, formulated as an empirical law half a century before Boltzmann by the French Sadi Carnot.

The proof of the Boltzmann H-theorem starts by computing the total time derivative of the quantity defined in Eq. (1.51) using Eq. (1.47):

$$\frac{dH}{dt} = \int d\mathbf{v} \frac{\partial}{\partial f_\mathbf{v}} (f_\mathbf{v} \log f_\mathbf{v}) \frac{\partial f_\mathbf{v}}{\partial t} = \int d\mathbf{v} \, (1 + \log f_\mathbf{v}) \frac{\partial f_\mathbf{v}}{\partial t} =$$
$$= \int d\Omega \int d\mathbf{v} \int d\mathbf{u} \sigma(|\mathbf{v} - \mathbf{u}|, \Omega)|\mathbf{v} - \mathbf{u}| \, (1 + \log f_\mathbf{v}) \, [f_{\mathbf{v}'} f_{\mathbf{u}'} - f_\mathbf{v} f_\mathbf{u}] \,.$$

$$(1.52)$$

An equivalent expression can be written by exchanging $\mathbf{v}$ with $\mathbf{u}$, since the last term is integrated over all the possible values of both velocity variables, and the cross-section is the same for the collision processes $\{\mathbf{v}, \mathbf{u}\}$ and $\{\mathbf{u}, \mathbf{v}\}$:

$$\frac{dH}{dt} = \int d\Omega \int d\mathbf{v} \int d\mathbf{u} \sigma(|\mathbf{v} - \mathbf{u}|, \Omega)|\mathbf{v} - \mathbf{u}| \, (1 + \log f_\mathbf{u}) \, [f_{\mathbf{v}'} f_{\mathbf{u}'} - f_\mathbf{v} f_\mathbf{u}] \,. \qquad (1.53)$$

Similarly, the r.h.s. of Eq. (1.52) has to remain invariant for the interchange between the velocities $\mathbf{v}, \mathbf{u}$ with $\mathbf{v}', \mathbf{u}'$, respectively, because for any collision process $\{\mathbf{v}, \mathbf{u}\} \rightarrow \{\mathbf{v}', \mathbf{u}'\}$, the inverse one $\{\mathbf{v}', \mathbf{u}'\} \rightarrow \{\mathbf{v}, \mathbf{u}\}$ has the same cross-section:

$$\frac{dH}{dt} = \int d\Omega \int d\mathbf{v}' \int d\mathbf{u}' \sigma(|\mathbf{v}' - \mathbf{u}'|, \Omega)|\mathbf{v}' - \mathbf{u}'| \, (1 + \log f_{\mathbf{v}'}) \, [f_\mathbf{v} f_\mathbf{u} - f_{\mathbf{v}'} f_{\mathbf{u}'}] =$$
$$= \int d\Omega \int d\mathbf{v} \int d\mathbf{u} \sigma(|\mathbf{v} - \mathbf{u}|, \Omega)|\mathbf{v} - \mathbf{u}| \, (1 + \log f_{\mathbf{v}'}) \, [f_\mathbf{v} f_\mathbf{u} - f_{\mathbf{v}'} f_{\mathbf{u}'}] \,,$$

$$(1.54)$$

where in the last equality we have used the Eqs. (1.40) and (1.41).

Following the same way of reasoning, we can conclude that the last double integral is equivalent by exchanging $\mathbf{v}'$ with $\mathbf{u}'$, thus yielding

$$\frac{dH}{dt} = \int d\Omega \int d\mathbf{v} \int d\mathbf{u} \sigma(|\mathbf{v} - \mathbf{u}|, \Omega)|\mathbf{v} - \mathbf{u}| \, (1 + \log f_{\mathbf{u}'}) \, [f_\mathbf{v} f_\mathbf{u} - f_{\mathbf{v}'} f_{\mathbf{u}'}] \,. \qquad (1.55)$$

By summing Eqs. (1.52), (1.53), (1.54), and (1.55), and dividing by the factor 4, we finally obtain

$$\frac{dH}{dt} = \frac{1}{4} \int d\Omega \int d\mathbf{v} \int d\mathbf{u} \sigma(|\mathbf{v} - \mathbf{u}|, \Omega)|\mathbf{v} - \mathbf{u}| \log \left(\frac{f_\mathbf{v} f_\mathbf{u}}{f_{\mathbf{v}'} f_{\mathbf{u}'}} \right) [f_{\mathbf{v}'} f_{\mathbf{u}'} - f_\mathbf{v} f_\mathbf{u}] \,. \qquad (1.56)$$

Notice that if x and y are positive quantities, the expression

$$\log \left(\frac{x}{y} \right) (y - x) \qquad (1.57)$$

is always negative, or null when $x = y$. Therefore, since all f's appearing in this expression are positive definite quantities (representing the number of particles per unit volume in μ-space), one has

$$\frac{dH}{dt} \leq 0, \tag{1.58}$$

with the equality implying $f_{v'} f_{u'} = f_v f_u$ for any binary collision. This concludes the proof of the H-theorem in the uniform case.

It is worth stressing again that this result is a direct consequence of the peculiar form of the collisional term appearing in the Boltzmann equation, which has been obtained by making use of the following basic assumption: Binary elastic collisions, which represent the only relevant process in a diluted gas, emerge from the contribution of statistically independent states of the pair of involved particles (molecular chaos).

Now we want to prove that the functional $H[f_v]$ has an absolute minimum and this will be done by finding the explicit expression of the function $\bar{f}_v$, minimizing H. The first step is to use variational calculus to minimize $H[f_v]$, taking into account the constraint that the quantities defined in Eqs. (1.48) to (1.50) are constants of motion. The Lagrange multiplier method yields the variational equation

$$\delta H + \alpha \delta n + \beta \delta e + \boldsymbol{\gamma} \cdot \delta \mathbf{p} = 0, \tag{1.59}$$

where δ is the functional variation w.r.t. $\bar{f}_v$, while α, β, and the three components of the vector $\boldsymbol{\gamma}$ are five scalar quantities to be determined in such a way that Eqs. (1.48) to (1.50) hold. Making use of Eq. (1.51), one can write Eq. (1.59) in the explicit form

$$\int d\mathbf{v}(1 + \log \bar{f}_v)\delta \bar{f}_v + \alpha \int d\mathbf{v}\delta \bar{f}_v + \beta \int d\mathbf{v}\frac{1}{2}mv^2\delta \bar{f}_v + \int d\mathbf{v}m(\boldsymbol{\gamma} \cdot \mathbf{v})\delta \bar{f}_v = 0. \tag{1.60}$$

By summing all contributions, one can write

$$\int d\mathbf{v}\left(1 + \log \bar{f}_v + \alpha + \frac{\beta}{2}mv^2 + m\boldsymbol{\gamma} \cdot \mathbf{v}\right)\delta \bar{f}_v = 0. \tag{1.61}$$

Due to the arbitrariness of $\delta \bar{f}_v$, Eq. (1.61) implies

$$1 + \log \bar{f}_v + \alpha + \frac{\beta}{2}mv^2 + m\boldsymbol{\gamma} \cdot \mathbf{v} = 0, \tag{1.62}$$

thus yielding an explicit expression for the distribution function minimizing H,

$$\bar{f}_v = A \exp\left[-\frac{1}{2}m\beta(\mathbf{v} - \mathbf{v}_0)^2\right], \tag{1.63}$$

where we have defined

$$\mathbf{v}_0 = -\frac{\boldsymbol{\gamma}}{\beta}, \tag{1.64}$$

$$A = e^{-(1+\alpha)}e^{\frac{1}{2}m\beta|\mathbf{v}_0|^2}. \tag{1.65}$$

Equation (1.63) is the so-called Maxwell-Boltzmann equilibrium distribution and the parameters $(A, \beta, \mathbf{v}_0)$ have to be determined in such a way that the conservation laws

(1.48) to (1.50) are satisfied. In Appendix B, we determine the constants $A = n\left(\frac{m\beta}{2\pi}\right)^{\frac{3}{2}}$ and $\beta = 1/T$, so that we can finally write

$$\bar{f}_{\mathbf{v}} = n\left(\frac{m\beta}{2\pi}\right)^{\frac{3}{2}} \exp\left[-\frac{1}{2}m\beta(\mathbf{v} - \mathbf{v}_0)^2\right].\tag{1.66}$$

In order to conclude that this expression corresponds to the distribution function of an ideal gas in thermodynamic equilibrium at temperature $T = 1/\beta$, it remains to prove that $H[\bar{f}_{\mathbf{v}}]$ is an *absolute* minimum of $H[f_{\mathbf{v}}]$, that is, that for any $f_{\mathbf{v}}$ satisfying Eq. (1.47), the following inequality holds

$$H[f_{\mathbf{v}}] \geq H[\bar{f}_{\mathbf{v}}].\tag{1.67}$$

By combining Eqs. (1.51) and (1.62), one obtains

$$\begin{aligned}
H[\bar{f}_{\mathbf{v}}] &= \int d\mathbf{v}\,\bar{f}_{\mathbf{v}} \log \bar{f}_{\mathbf{v}}\\
&= -\int d\mathbf{v}\,\bar{f}_{\mathbf{v}}\left(1 + \alpha + \frac{1}{2}m\beta v^2 + m\mathbf{v}\cdot\boldsymbol{\gamma}\right)\\
&= -\left[(1 + \alpha)n + \beta e + \boldsymbol{\gamma}\cdot\mathbf{p}\right].
\end{aligned}\tag{1.68}$$

Since n, e, and $\mathbf{p}$ are constants of the motion independent of the choice of $f_{\mathbf{v}}$, the solution of the Boltzmann equation, in the second expression of (1.68), we can substitute $\bar{f}_{\mathbf{v}}$ with a generic solution $f_{\mathbf{v}}$

$$\begin{aligned}
H[\bar{f}_{\mathbf{v}}] &= -\int d\mathbf{v}\,f_{\mathbf{v}}\left(1 + \alpha + \frac{1}{2}m\beta v^2 + m\mathbf{v}\cdot\boldsymbol{\gamma}\right) =\\
&= \int d\mathbf{v}\,f_{\mathbf{v}} \log \bar{f}_{\mathbf{v}}.
\end{aligned}\tag{1.69}$$

Finally, one obtains

$$\begin{aligned}
H[f_{\mathbf{v}}] - H[\bar{f}_{\mathbf{v}}] &= \int d\mathbf{v}(f_{\mathbf{v}} \log f_{\mathbf{v}} - f_{\mathbf{v}} \log \bar{f}_{\mathbf{v}})\\
&= \int d\mathbf{v}(f_{\mathbf{v}} \log f_{\mathbf{v}} - f_{\mathbf{v}} \log \bar{f}_{\mathbf{v}} - f_{\mathbf{v}} + \bar{f}_{\mathbf{v}}) \geq 0,
\end{aligned}\tag{1.70}$$

where we have used the relation

$$\int d\mathbf{v}\,f_{\mathbf{v}} = \int d\mathbf{v}\,\bar{f}_{\mathbf{v}} = n\tag{1.71}$$

and the inequality

$$x \log x - x \log y - x + y \geq 0,\tag{1.72}$$

which holds for any real positive quantities x and y.[10]

---

[10] Let us consider the function $g(x) = x \log x - x \log y - x + y$, which is defined only for positive values of the variable x and of the parameter y. The function $g(x)$ and its first derivative $\frac{dg(x)}{dx} = \log\left(\frac{x}{y}\right)$ vanish for $x = y$, while its second derivative, $\frac{d^2 g(x)}{dx^2} = \frac{y}{x} > 0$. Accordingly, $g(x)$ has an absolute minimum in $x = y$ and $g(x) > 0$ for any positive $x \neq y$.

This concludes the proof that $H(t)$ has an absolute minimum $H_0 \equiv H[\bar{f}_\mathbf{v}]$. Since we had already proved that $H(t)$ is a nonincreasing function of time, this implies also

$$\lim_{t \to +\infty} H(t) = H_0. \tag{1.73}$$

We want to point out that, while the H-theorem is a direct consequence of the hypothesis made on the collision process between particles, the last result follows from the conservation laws of the quantities defined in Eqs. (1.48) to (1.50).

1.6 The H-theorem in the Nonuniform Case

We come back to the general case where the distribution function of the ideal gas depends also on space coordinates, that is, $f_\mathbf{v} \equiv f(\mathbf{r}, \mathbf{v}, t)$. This is a necessary assumption as soon as the presence of external conservative forces has to be included in the theory. In particular, a special kind of external force is that exerted by the walls of a container, confining the gas inside a finite volume. These forces are short-range, and their molecular nature would make a detailed physical description quite complicated. We simply assume that collisions of particles with the walls can be thought of as elastic reflections. This assumption allows one to express in mathematical form the presence of forces exerted by the walls under the form of boundary conditions:

$$f(\mathbf{r} = \mathbf{r}_w, v_n, \mathbf{v}_p, t) = f(\mathbf{r} = \mathbf{r}_w, -v_n, \mathbf{v}_p, t), \tag{1.74}$$

where $\mathbf{r}_w$ represents the coordinates of any point on the wall, $v_n(\mathbf{r}_w) \equiv \mathbf{v} \cdot \mathbf{n}(\mathbf{r}_w)$ is the component of the particle velocity along the outgoing normal to the wall $\mathbf{n}(\mathbf{r}_w)$, and $\mathbf{v}_p(\mathbf{r}_w)$ represents the velocity component parallel to the wall. In fact, the reversibility of the particle/wall collisions amounts to impose that the distribution function at any point on the wall is an even function of v_n, since the result of a collision with the wall is $v_n \to -v_n$.

We define the generalized functional

$$\mathcal{H}(t) = \int d\mathbf{r} H(\mathbf{r}, t) = \int d\mathbf{r} \int d\mathbf{v} f_\mathbf{v} \log f_\mathbf{v}, \tag{1.75}$$

where the integral over the space coordinates extends to the volume V containing the ideal gas. Following the same procedure of Section 1.5 and using the Boltzmann Equation (1.45), one obtains

$$\begin{aligned}
\frac{d\mathcal{H}}{dt} &= \int d\mathbf{r} \int d\mathbf{v} \frac{\partial f_\mathbf{v}}{\partial t}(1 + \log f_\mathbf{v}) \\
&= \int d\mathbf{r} \int d\mathbf{v} \left[\left(\frac{\partial f_\mathbf{v}}{\partial t}\right)_c - \mathbf{v} \cdot \nabla_\mathbf{r} f_\mathbf{v} - \frac{\mathbf{F}}{m} \cdot \nabla_\mathbf{v} f_\mathbf{v} \right] (1 + \log f_\mathbf{v}) \\
&= \int d\mathbf{r} \int d\mathbf{v} \left(\frac{\partial f_\mathbf{v}}{\partial t}\right)_c (1 + \log f_\mathbf{v}) \\
&\quad - \int d\mathbf{r} \int d\mathbf{v} \left[\mathbf{v} \cdot \nabla_\mathbf{r} + \frac{\mathbf{F}}{m} \cdot \nabla_\mathbf{v} \right] (f_\mathbf{v} \log f_\mathbf{v}),
\end{aligned} \tag{1.76}$$

where the last expression is obtained by the identity $(\mathbf{b}\cdot\nabla_{\mathbf{a}}f_{\mathbf{v}})(1+\log f_{\mathbf{v}}) \equiv \mathbf{b}\cdot\nabla_{\mathbf{a}}(f_{\mathbf{v}}\log f_{\mathbf{v}})$, with $\mathbf{a}$ and $\mathbf{b}$ general vectors.

Using the divergence theorem, we obtain

$$\int d\mathbf{r}\int d\mathbf{v}\,\mathbf{v}\cdot\nabla_{\mathbf{r}}(f_{\mathbf{v}}\log f_{\mathbf{v}}) = \int dS\int d\mathbf{v}\,\mathbf{v}\cdot\mathbf{n}(f_{\mathbf{v}}\log f_{\mathbf{v}}) = 0, \qquad (1.77)$$

where the first integral on the r.h.s. of this equation extends over the surface S (whose infinitesimal element is denoted by dS), containing the volume V, $\mathbf{n}(\mathbf{r_w})$ is the external normal to S at $\mathbf{r_w}$ and the last equality is a consequence of the boundary conditions (1.74). In fact, on the surface S we can represent the measure of the second integral as $d\mathbf{v} = dv_n d\mathbf{v}_p$,[11] while the function to be integrated is an odd function of v_n, due to the presence of the factor $\mathbf{v}\cdot\mathbf{n}\equiv v_n$ ($f_{\mathbf{v}}$ being an even function of v_n on S, due to Eq. (1.74)). The last equality in Eq. (1.77) can be read as a no-flux condition through the wall, because every colliding particle is elastically reflected by a wall.

Also, the integral containing $\nabla_{\mathbf{v}}$ in the last term of Eq. (1.76) vanishes. In fact, the distribution function $f_{\mathbf{v}}$ must be such that the particle density, the components of the momentum density, and the energy density must be finite, in formulae:

$$n(\mathbf{r},t) = \int d\mathbf{v}f(\mathbf{r},\mathbf{v},t) < \infty, \qquad (1.78)$$

$$\mathbf{p}(\mathbf{r},t) = \int d\mathbf{v}m\mathbf{v}f(\mathbf{r},\mathbf{v},t) < \infty, \qquad (1.79)$$

$$e(\mathbf{r},t) = \int d\mathbf{v}\frac{1}{2}mv^2 f(\mathbf{r},\mathbf{v},t) < \infty. \qquad (1.80)$$

All of these conditions hold if in the limit $|\mathbf{v}|\to\infty$ the distribution function $f_{\mathbf{v}}$ vanishes faster than $|\mathbf{v}|^{-5}$, as one can conclude by simple power counting.[12] Then, we can write

$$\int d\mathbf{v}\frac{\mathbf{F}}{m}\cdot\nabla_{\mathbf{v}}(f_{\mathbf{v}}\log f_{\mathbf{v}})$$

$$= \int dv_x\int dv_y\int dv_z\left(\frac{F_x}{m}\frac{\partial}{\partial v_x}+\frac{F_y}{m}\frac{\partial}{\partial v_y}+\frac{F_z}{m}\frac{\partial}{\partial v_z}\right)(f_{\mathbf{v}}\log f_{\mathbf{v}})$$

$$= \int dv_y\int dv_z\frac{F_x}{m}[f_{\mathbf{v}}\log f_{\mathbf{v}}]\Big|_{v_x=-\infty}^{v_x=+\infty} + (x,y,z)\to(y,z,x) + (x,y,z)\to(z,x,y)$$

$$= 0, \qquad (1.81)$$

because

$$\lim_{|\mathbf{v}|\to\infty} f_{\mathbf{v}}\log f_{\mathbf{v}} = 0. \qquad (1.82)$$

As a consequence, the time evolution of $\mathcal{H}$ does not depend on the stream term and Eq. (1.76) simplifies to

$$\frac{d\mathcal{H}}{dt} = \int d\Omega\int d\mathbf{r}\int d\mathbf{v}\int d\mathbf{u}\sigma(|\mathbf{v}-\mathbf{u}|,\Omega)(1+\log f_{\mathbf{v}})[f_{\mathbf{v}'}f_{\mathbf{u}'}-f_{\mathbf{v}}f_{\mathbf{u}}]. \qquad (1.83)$$

---

[11] At each point on the surface S of volume V, v_n is a scalar, while $\mathbf{v}_p$ is a two-dimensionl vector.

[12] In the presence of long-range interactions, either between particles or induced by the boundaries, this condition may not be satisfied.

Now we can go through the same procedure reported in Section 1.5 and we finally obtain the H-theorem for the nonuniform case,

$$\frac{d\mathcal{H}}{dt} \le 0. \tag{1.84}$$

We can conclude, again, that irreversibility is due to binary collisions, and the condition $\frac{d\mathcal{H}}{dt} = 0$ still implies

$$f_{\mathbf{v}} f_{\mathbf{u}} = f_{\mathbf{v}'} f_{\mathbf{u}'}, \tag{1.85}$$

but it does not imply $\frac{\partial f_{\mathbf{v}}}{\partial t} = 0$, as in the uniform case, see Eq. (1.45), because of the stream term. Conversely, the condition $\frac{\partial f_{\mathbf{v}}}{\partial t} = 0$ implies $\frac{d\mathcal{H}}{dt} = 0$, as one can easily derive from the first line of Eq. (1.76).

The minimum f^* of the functional $\mathcal{H}[f_{\mathbf{v}}]$ can be obtained by following the same procedure outlined for the uniform case by the method of Lagrange multipliers, thus yielding

$$f^*(\mathbf{r}, \mathbf{v}, t) = n(\mathbf{r}, t) \left(\frac{m\beta(\mathbf{r}, t)}{2\pi} \right)^{\frac{3}{2}} \exp\left\{ -\frac{1}{2} m\beta(\mathbf{r}, t) \, [\mathbf{v} - \mathbf{v}_0(\mathbf{r}, t)]^2 \right\}. \tag{1.86}$$

Note that we have obtained a solution minimizing $\mathcal{H}$, which is formally equivalent to Eq. (1.66): The main difference is that in the nonuniform case, the particle density n, the inverse temperature β, and the center of mass velocity $\mathbf{v}_0$ are functions of $\mathbf{r}$ and t. This is why Eq. (1.86) is called the *local* Maxwell–Boltzmann distribution function, corresponding to the minimum value of $\mathcal{H}$, for $f_{\mathbf{v}} = f_{\mathbf{v}}^*$. It is worth observing that Eq. (1.86) yields the condition

$$\left(\frac{\partial f_{\mathbf{v}}^*}{\partial t} \right)_c = 0. \tag{1.87}$$

This is a necessary but not sufficient condition for the stationarity of $f_{\mathbf{v}}$ (see Eqs. (1.45) and (1.76)), which can be imposed by assuming the additional condition

$$\frac{\partial f_{\mathbf{v}}^*}{\partial t} = -\left(\mathbf{v} \cdot \nabla_{\mathbf{r}} f_{\mathbf{v}}^* + \frac{\mathbf{F}}{m} \cdot \nabla_{\mathbf{v}} f_{\mathbf{v}}^* \right) = 0. \tag{1.88}$$

In order to be consistent with the condition $\frac{d\mathcal{H}}{dt} = 0$, we have also to assume that the quantities n, β, and $\mathbf{v}_0$ still depend on $\mathbf{r}$, but they are independent of time t. Nonetheless, Eq. (1.88) imposes specific constraints on these parameters, implying that a stationary solution of the Boltzmann equation in the nonuniform case exists only if the external forces $\mathbf{F}$ are conservative, as we are now going to argue. This condition can be rewritten, dividing both of its sides by $f_{\mathbf{v}}^*$, as follows:

$$\mathbf{v} \cdot \nabla_{\mathbf{r}} \log f_{\mathbf{v}}^* + \frac{\mathbf{F}}{m} \cdot \nabla_{\mathbf{v}} \log f_{\mathbf{v}}^* = 0. \tag{1.89}$$

Making use of Eq. (1.86), we can write

$$\log f_{\mathbf{v}}^* = \log \left(n\beta^{\frac{3}{2}} \right) - \frac{1}{2}\beta m \, (\mathbf{v} - \mathbf{v}_0)^2 + \text{const}, \tag{1.90}$$

thus yielding

$$\frac{\partial \log f_{\mathbf{v}}^*}{\partial r_i} = \frac{\partial \log(n\beta^{\frac{3}{2}})}{\partial r_i} + m\beta \sum_{j=x,y,z}\left[(v_j - v_{0j})\frac{\partial v_{0j}}{\partial r_i}\right] - \frac{\partial \beta}{\partial r_i}\frac{1}{2}m\,(\mathbf{v}-\mathbf{v}_0)^2$$

$$\frac{\partial \log f_{\mathbf{v}}^*}{\partial v_i} = -m\beta(v_i - v_{0i}), \tag{1.91}$$

where the indices $i, j = x, y, z$ label the three components of the space and velocity vectors.

Accordingly, ordering by increasing powers of the velocity components v_i, Eq. (1.89) can be rewritten as follows:

$$\beta\mathbf{F}\cdot\mathbf{v}_0 + \sum_{i=x,y,z} v_i\left(\frac{\partial \log(n\beta^{\frac{3}{2}})}{\partial r_i} - m\beta \sum_{j=x,y,z} v_{0j}\frac{\partial v_{0j}}{\partial r_i} - \frac{1}{2}mv_0^2\frac{\partial\beta}{\partial r_i} - \beta F_i\right) +$$
$$+ \sum_{i,j=x,y,z} v_i v_j\left(m\beta\frac{\partial v_{0j}}{\partial r_i} + mv_{0j}\frac{\partial\beta}{\partial r_i}\right) - mv^2\sum_{i=x,y,z} v_i\frac{1}{2}\frac{\partial\beta}{\partial r_i} = 0. \tag{1.92}$$

This implies that the coefficients of all powers of the components v_i must vanish. For what concerns the highest power, one obtains the condition

$$\frac{\partial\beta}{\partial r_i} = 0, \quad i = x, y, z, \tag{1.93}$$

that is, $\beta = 1/T$ must be independent of the space coordinate $\mathbf{r}$. This indicates that the stationary condition for $\mathcal{H}$, as expected, does correspond to thermodynamic equilibrium conditions. By exploiting this result, one finds that the coefficients of the second order terms $v_i v_j$ vanish if

$$\frac{\partial v_{0j}}{\partial r_i} + \frac{\partial v_{0i}}{\partial r_j} = 0, \quad i, j = x, y, z, \tag{1.94}$$

where we have used the exchange symmetry between the indices i and j. This system of linear differential equations has the solution

$$\mathbf{v}_0(\mathbf{r}) = \mathbf{v}'_0 + \omega \times \mathbf{r}, \tag{1.95}$$

where $\mathbf{v}'_0$ and ω are constant vectors.[13] This is the expression of the velocity of a rigid body, whose center of mass translates with constant velocity $\mathbf{v}'_0$ and which rotates with angular velocity ω. For the sake of simplicity, one can take $\mathbf{v}'_0$ parallel to ω and both vectors directed along a specific axis, for example, the z-axis, in such a way that $\mathbf{v}'_0 = (0, 0, v'_z)$ and $\omega = (0, 0, \omega_z = \omega)$, so that the components of the vector $\mathbf{v}_0(\mathbf{r})$ simplify to

$$v_{0x} = -\omega y \quad , \quad v_{0y} = \omega x \quad , \quad v_{0z} = v'_z. \tag{1.96}$$

---

[13] In fact, explicit calculations yield $\frac{\partial v_{0x}}{\partial y} = \frac{\partial}{\partial y}(\omega_y z - \omega_z y) = -\omega_z$ and $\frac{\partial v_{0y}}{\partial x} = \frac{\partial}{\partial x}(\omega_z x - \omega_x z) = \omega_z$; by iterating these relations for a cyclic permutation of the space indices, one verifies Eq. (1.94).

Making use of these relations, one finds that the coefficients of the linear term in the velocity v_i of (1.92) vanish if

$$\frac{\partial \log n}{\partial x} - m\beta\omega^2 x - \beta F_x = 0 \tag{1.97}$$

$$\frac{\partial \log n}{\partial y} - m\beta\omega^2 y - \beta F_y = 0 \tag{1.98}$$

$$\frac{\partial \log n}{\partial z} - \beta F_z = 0. \tag{1.99}$$

These conditions can be expressed in a more compact form by introducing the centrifugal potential $\psi(\mathbf{r}) = \frac{1}{2}m\omega^2(x^2 + y^2)$,

$$\nabla_{\mathbf{r}} \left(\frac{1}{\beta} \log n(\mathbf{r}) + \psi(\mathbf{r}) \right) = \mathbf{F}. \tag{1.100}$$

Finally, it remains to impose that also the constant term in (1.92) vanishes, namely

$$\mathbf{F} \cdot \mathbf{v}_0 = 0. \tag{1.101}$$

Eq. (1.100) is quite important, because it tells us that a necessary condition for the existence of a stationary solution of the Boltzmann equation in the nonuniform case is that the external forces $\mathbf{F}$ must depend on a conservative potential $\mathcal{V}(\mathbf{r})$, that is,

$$\mathcal{V}(\mathbf{r}) = -\left(\frac{1}{\beta} \log n(\mathbf{r}) + \psi(\mathbf{r}) \right), \tag{1.102}$$

which yields the relation

$$n(\mathbf{r}) = \exp\left[-\beta \left(\mathcal{V}(\mathbf{r}) + \psi(\mathbf{r}) \right) \right]. \tag{1.103}$$

If $\omega = 0$, that is, the collective motion of the gas amounts to a pure translation with constant velocity (null if the center of mass is at rest), then also the centrifugal potential vanishes, that is, $\psi(\mathbf{r}) = 0$, and Eq. (1.103) simplifies to

$$n(\mathbf{r}) = \exp\left(-\beta\mathcal{V}(\mathbf{r}) \right). \tag{1.104}$$

It remains to discuss the consequences of constraint (1.101), which is trivially satisfied in the absence of external forces, while, in the general case where $\mathbf{F} \neq 0$, it implies that $\mathbf{v}_0$ must be orthogonal to $\mathbf{F}$, that is, tangential to the equipotential surface $\mathcal{V}(\mathbf{r}) = \text{const.}$

Further constraints on the form of $\mathbf{v}_0(\mathbf{r})$ originate from Eq. (1.74). By imposing that this equation is satisfied by the local Maxwell–Boltzmann distribution (1.86), the following condition must hold

$$\left(v_n(\mathbf{r}_w) - v_{0n}(\mathbf{r}_w) \right)^2 = \left(-v_n(\mathbf{r}_w) - v_{0n}(\mathbf{r}_w) \right)^2, \tag{1.105}$$

where $v_n(\mathbf{r}_w)$ and $v_{0n}(\mathbf{r}_w)$ are the components of $\mathbf{v}$ and $\mathbf{v}_0$ at position $\mathbf{r}_w$ on the wall, respectively, thus yielding

$$v_{0n}(\mathbf{r}_w) = 0. \tag{1.106}$$

We can conclude that in the presence of conservative external forces, the stationary solution of the Boltzmann equation in the nonuniform case reads

$$f^*(\mathbf{r}, \mathbf{v}) = n_0 \left(\frac{m\beta}{2\pi}\right)^{\frac{3}{2}} \exp\left\{-\beta\left[\frac{1}{2}m\,(\mathbf{v} - \mathbf{v}_0(\mathbf{r}))^2 + \mathcal{V}(\mathbf{r}) + \psi(\mathbf{r})\right]\right\}, \qquad (1.107)$$

where n_0 is a suitable normalization constant. This expression is similar to the one of the distribution function describing local equilibrium conditions reported in Eq. (1.86): The main differences here are that $f^*(\mathbf{r}, \mathbf{v})$ is independent of time, n_0 and β are constants, $\mathbf{v}_0(\mathbf{r})$ is given by Eq. (1.95), while satisfying the constraints (1.101) and (1.106). We want to point out also that the expression in square brackets in (1.107) is the total energy of a particle having coordinates $(\mathbf{r}, \mathbf{v})$ in μ-space, measured in a reference frame, where the center of mass of the gas is at rest.

In conclusion, we have proved that, in the limit $t \to +\infty$, $\mathcal{H}$ has a limit corresponding to the condition $\frac{d\mathcal{H}}{dt} = 0$, where the distribution function is given in Eq. (1.86). If we make the physically plausible hypothesis that for $t \to +\infty$ also $f_\mathbf{v} \equiv f(\mathbf{v}, \mathbf{r}, t)$ has a limit corresponding to the condition $\frac{df_\mathbf{v}}{dt} = 0$, then this limit is given by Eq. (1.107). The equilibrium thermodynamics of the ideal gas in the nonuniform case is summarized in Appendix C.1. We can apply the results found there to hydrogen in the interstellar space, where there is a density of 1 molecule per cubic centimeter, thus yielding $\lambda = 10^{15}$cm. For a gas at a pressure of 100 atm, λ is of the order of the radius of intermolecular forces (thus implying that Boltzmann's equation is no more valid). From this kind of estimate, one can reasonably expect that in hydrogen at room pressure and temperature, inhomogeneities occurring over distances of the order of 10^{-5}cm disappear after 10^{-11}s, while variations of density and temperature over macroscopic length scales are expected to persist for much longer times.

1.7 Evolution toward Equilibrium and Transport Phenomena

In Section 1.6, we have seen that one can associate to an ideal gas in the non-uniform case two different equilibrium distribution functions, namely (1.86) and (1.107). From a physical point of view, they correspond to quite different situations. Actually, it is clear that macroscopic equilibrium conditions hold only if the ideal gas is described by the global or stationary equilibrium distribution function (1.107). If this would be replaced by the local equilibrium distribution function (1.86), we should be faced with a gas where particle velocities obey a local Maxwell distribution, which may greatly vary in coordinate space over a length scale that is neither infinitesimal nor comparable to any macroscopic dimension. More precisely, this length scale is comparable with the mean free path λ (see Eq. (C.20)) run by the gas particles between collisions. For a diluted gas at room temperature and pressure $\lambda \approx 10^{-5}$cm, that is, three orders of magnitude larger than the particle linear dimension. In fact, only by experiencing a sufficient number of collisions the gas particles are expected to typically establish local equilibrium conditions. Once such conditions are achieved, the macroscopic quantities will necessarily continue to evolve in time, until a global equilibrium state is eventually attained. In this sense, Eq. (1.86) can be thought of as corresponding to the simplest nonequilibrium state of the gas.

Relying upon these observations, we assume that, almost independently of the initial nonequilibrium state, the gas will evolve spontaneously toward macroscopic equilibrium.[14] More precisely, we assume that the evolution toward macroscopic equilibrium goes through two main steps. After a relatively short time scale, comparable with the average collision time τ between the gas particles (for a gas at room pressure and temperature $\tau \approx 10^{-10}$s, see Appendix C.2 and Eq. (C.15)), local equilibrium conditions described by Eq. (1.86) are established in the space of velocities. This first step is dominated by the collision term, which is associated with the "fast" time scale τ and we can write the estimate

$$\left(\frac{\partial f(\mathbf{r}, \mathbf{v}; t)}{\partial t} \right)_c \tau \sim f(\mathbf{r}, \mathbf{v}; t), \tag{1.108}$$

meaning that τ corresponds to the time scale over which the variation of the distribution function $f(\mathbf{r}, \mathbf{v}; t)$, due to the collisional derivative (defined in Eq. (1.44)), is comparable with $f(\mathbf{r}, \mathbf{v}; t)$ itself. Accordingly, over the time scale τ, $f(\mathbf{r}, \mathbf{v}; t)$ converges to $f^*(\mathbf{r}, \mathbf{v}; t)$, see Eq. (1.86). The spatial inhomogeneities of local equilibrium conditions are explicitly accounted for by the dependence on space coordinates and time of the quantities $n(\mathbf{r}, t)$, $T(\mathbf{r}, t)$, and $e(\mathbf{r}, t)$ defined in Eqs. (1.78) to (1.80), together with $\mathbf{v}_0(\mathbf{r}, t)$.

For times $t \gg \tau$, we expect that these inhomogeneities will be progressively eliminated, eventually yielding equilibrium conditions also in configuration space, described by Eq. (1.107). In particular, after local equilibrium conditions have been attained on the time scale τ, the second addendum of the stream derivative in Eq. (1.30) can be neglected w.r.t. to the first one, because $f(\mathbf{r}, \mathbf{v}; t) \sim f^*(\mathbf{r}, \mathbf{v}; t)$ has become a weakly dependent function of $\mathbf{v}$, whose typical value is close to $\mathbf{v}_0$. We can obtain an estimate of the typical length scale L_0 of spatial inhomogeneities by first summing over all possible values of the velocities, the dominant contribution of the stream derivative (1.30) after local equilibration, in formulae

$$\int d\mathbf{v}\, \mathbf{v} \cdot \nabla_{\mathbf{r}} f(\mathbf{r}, \mathbf{v}; t) = \nabla_{\mathbf{r}} \cdot \int d\mathbf{v}\, \mathbf{v} f(\mathbf{r}, \mathbf{v}; t)$$
$$= \nabla_{\mathbf{r}} \cdot (n \langle \mathbf{v} \rangle)$$
$$= n \sum_{j=x,y,z} \left(\frac{1}{n} \frac{\partial n}{\partial r_j} + \frac{1}{\langle v_j \rangle} \frac{\partial \langle v_j \rangle}{\partial r_j} \right) \langle v_j \rangle, \tag{1.109}$$

where we have made use of the normalization condition (1.22) and of the Eq. (1.25), specialized to the case $A(\mathbf{r}, \mathbf{v}; t) = \mathbf{v}$. In fact, all the quantities appearing inside the sum identify the same length-scale L_0 as follows:

$$\frac{1}{n} |\nabla_{\mathbf{r}} n| \sim \frac{1}{\langle v_j \rangle} |\nabla_{\mathbf{r}} \langle v_j \rangle| \sim L_0^{-1}. \tag{1.110}$$

---

[14] We cannot exclude that peculiar out-of-equilibrium initial conditions could evolve for an exceedingly long time far from equilibrium; anyway, despite no rigorous statement is available, we assume that typical initial conditions guarantee that the system relaxes to thermodynamic equilibrium over a typical time scale.

In summary, this estimate procedure yields

$$\int d\mathbf{v}\,\mathbf{v}\cdot\nabla_{\mathbf{r}}f(\mathbf{r},\mathbf{v};t) \sim n\frac{1}{L_0}\langle v_j\rangle = \int d\mathbf{v}f(\mathbf{r},\mathbf{v};t)\frac{\langle v_j\rangle}{L_0}, \tag{1.111}$$

or, equivalently

$$\left(\frac{\partial f(\mathbf{r},\mathbf{v};t)}{\partial t}\right)_s \sim \mathbf{v}\cdot\nabla_{\mathbf{r}}f(\mathbf{r},\mathbf{v};t) \sim \frac{\langle v_j\rangle}{L_0}f(\mathbf{r},\mathbf{v};t). \tag{1.112}$$

We can conclude that the long time scale associated with the relaxation process to global equilibrium determined by the stream term amounts to

$$t_0 = \frac{L_0}{\langle v_j\rangle}. \tag{1.113}$$

L_0 is the length scale of spatial inhomogeneities and it must be much larger than the mean-free path of colliding particle, $L_0 \gg \lambda = \tau\langle v_j\rangle$, for the approximations above to be applicable. This means that $t_0 \gg \tau$.

1.7.1 Additive Invariants

In order to obtain from the Boltzmann Equation (1.45), the so-called hydrodynamic equations ruling the nonequilibrium evolution of the ideal gas, one can first consider the theorem of additive invariants. An additive invariant is a quantity $\phi(\mathbf{r},\mathbf{v})$, which is conserved during a binary collision, that is,

$$\phi(\mathbf{r},\mathbf{v}) + \phi(\mathbf{r},\mathbf{u}) = \phi(\mathbf{r},\mathbf{v}') + \phi(\mathbf{r},\mathbf{u}'), \tag{1.114}$$

where the velocities are associated with the conservative (i.e., reversible) collision process at position $\mathbf{r}$ between gas particles, $\{\mathbf{v},\mathbf{u}\} \iff \{\mathbf{v}',\mathbf{u}'\}$, detailed in Appendix A. In fact, Eq. (1.114) epitomizes the conservation rules of the collision process that are valid at any time t.

For any additive invariant $\phi(\mathbf{r},\mathbf{v})$, the following relation holds

$$\int d\mathbf{v}\phi(\mathbf{r},\mathbf{v})\left(\frac{\partial f_{\mathbf{v}}}{\partial t}\right)_c = 0, \tag{1.115}$$

where $\left(\frac{\partial f_{\mathbf{v}}}{\partial t}\right)_c$ is defined in Eq. (1.44). The proof of this theorem is an almost straightforward consequence of the symmetries ruling the conservative binary collision process. Due to hypothesis (1.114), one can write the obvious relation

$$0 = \int d\mathbf{v}\left[\phi(\mathbf{r},\mathbf{v}) + \phi(\mathbf{r},\mathbf{u}) - \phi(\mathbf{r},\mathbf{v}') - \phi(\mathbf{r},\mathbf{u}')\right]\left(\frac{\partial f_{\mathbf{v}}}{\partial t}\right)_c, \tag{1.116}$$

where the first addendum in the square parentheses reads explicitly

$$\int d\mathbf{v}\int d\mathbf{u}\phi(\mathbf{r},\mathbf{v})\int d\Omega\int \sigma(|\mathbf{v}-\mathbf{u}|;\Omega)\,(f_{\mathbf{v}'}f_{\mathbf{u}'} - f_{\mathbf{v}}f_{\mathbf{u}})\,|\mathbf{v}-\mathbf{u}| = \mathcal{I}. \tag{1.117}$$

If in this equation we exchange $\mathbf{v}$ with $\mathbf{u}$, we obtain that the second addendum in the square parentheses of (1.115) is equal to the first one, namely

$$\int d\mathbf{u} \int d\mathbf{v}\phi(\mathbf{r},\mathbf{u}) \int d\Omega \int \sigma(|\mathbf{u}-\mathbf{v}|;\Omega)\,(f_{\mathbf{v}'}f_{\mathbf{u}'}-f_{\mathbf{u}}f_{\mathbf{v}})\,|\mathbf{u}-\mathbf{v}| = I. \qquad (1.118)$$

Now we can consider the time-reversed collision process (i.e., we exchange $\mathbf{v}$ with $\mathbf{v}'$ and $\mathbf{u}$ with $\mathbf{u}'$ in Eq. (1.117)) and we can conclude that the third addendum in the square parentheses of (1.115) is again equal to the first one, namely

$$-\int d\mathbf{v} \int d\mathbf{u}\phi(\mathbf{r},\mathbf{v}') \int d\Omega \int \sigma(|\mathbf{v}-\mathbf{u}|;\Omega)\,(f_{\mathbf{v}'}f_{\mathbf{u}'}-f_{\mathbf{u}}f_{\mathbf{v}})\,|\mathbf{v}-\mathbf{u}|$$
$$=\int d\mathbf{v}' \int d\mathbf{u}'\,\phi(\mathbf{r},\mathbf{v}') \int d\Omega \int \sigma(|\mathbf{v}'-\mathbf{u}'|;\Omega)\,(f_{\mathbf{v}}f_{\mathbf{u}}-f_{\mathbf{v}'}f_{\mathbf{u}'})\,|\mathbf{v}'-\mathbf{u}'|$$
$$= I, \qquad (1.119)$$

which holds true, because the conservative nature of binary collisions, described in Appendix A, leaves invariant the measure in velocity space, due to Liouville's theorem (i.e., $d\mathbf{v}d\mathbf{u} = d\mathbf{v}'d\mathbf{u}'$), and the modulus of the relative velocity of colliding particles (i.e., $|\mathbf{v}-\mathbf{u}| = |\mathbf{v}'-\mathbf{u}'|$).

The same conclusion holds for the fourth addendum in the square parentheses of (1.115), which is obtained from (1.119) by exchanging $\mathbf{v}'$ with $\mathbf{u}'$.

By summing the four contributions, we finally obtain

$$4I = 4 \int d\mathbf{v}\phi(\mathbf{r},\mathbf{v}) \left(\frac{\partial f_{\mathbf{v}}}{\partial t}\right)_c = 0, \qquad (1.120)$$

which concludes the proof of the theorem.

The result of this theorem can be applied to obtain the general conservation laws associated with the Boltzmann equation: This aim is straightforwardly accomplished by multiplying Eq. (1.45) by the additive invariant $\phi(\mathbf{r},\mathbf{v})$ and integrating this expression over $d\mathbf{v}$, thus yielding

$$\int d\mathbf{v}\phi(\mathbf{r},\mathbf{v}) \left(\frac{\partial}{\partial t} + \mathbf{v}\cdot\nabla_{\mathbf{r}} + \frac{\mathbf{F}(\mathbf{r})}{m}\cdot\nabla_{\mathbf{v}}\right) f(\mathbf{r},\mathbf{v},t) \equiv I_1 + I_2 + I_3 = 0. \qquad (1.121)$$

It is useful to rewrite the three addenda I_i starting from

$$I_1 = \frac{\partial}{\partial t}\int d\mathbf{v}\phi(\mathbf{r},\mathbf{v})f(\mathbf{r},\mathbf{v},t) = \frac{\partial}{\partial t}\langle n(\mathbf{r},t)\phi(\mathbf{r},\mathbf{v})\rangle. \qquad (1.122)$$

The equality stems from Eq. (1.24), which defines the local average $\langle\bullet\rangle$ over the velocity coordinates, while the time dependence on the r.h.s. of this equation is contained in $n(\mathbf{r},t)$ only (see Eq. (1.22)), because any additive invariant $\phi(\mathbf{r},\mathbf{v})$ is by definition independent of time t.

The second addendum can be rewritten as follows:

$$I_2 = \nabla_{\mathbf{r}}\cdot\left(\int d\mathbf{v}\,\mathbf{v}\phi(\mathbf{r},\mathbf{v})f(\mathbf{r},\mathbf{v},t)\right) - \int d\mathbf{v}f(\mathbf{r},\mathbf{v},t)\mathbf{v}\cdot\nabla_{\mathbf{r}}\phi(\mathbf{r},\mathbf{v})$$
$$= \nabla_{\mathbf{r}}\cdot\langle\mathbf{v}n(\mathbf{r},t)\phi(\mathbf{r},\mathbf{v})\rangle - n(\mathbf{r},t)\langle\mathbf{v}\cdot\nabla_{\mathbf{r}}\phi(\mathbf{r},\mathbf{v})\rangle. \qquad (1.123)$$

Finally, the third addendum can be rewritten as follows:

$$I_3 = \int d\mathbf{v} \nabla_{\mathbf{v}} \cdot \left(\frac{\mathbf{F(r)}}{m} \phi(\mathbf{r},\mathbf{v}) f(\mathbf{r},\mathbf{v},t) \right) - \int d\mathbf{v}(\nabla_{\mathbf{v}}\phi(\mathbf{r},\mathbf{v})) \cdot \frac{\mathbf{F(r)}}{m} f(\mathbf{r},\mathbf{v},t)$$

$$= -\frac{n(\mathbf{r},t)}{m} \langle \nabla_{\mathbf{v}}\phi(\mathbf{r},\mathbf{v}) \cdot \mathbf{F(r)} \rangle, \tag{1.124}$$

where the first integral on the r.h.s. of this equation vanishes, because $\lim_{|\mathbf{v}| \to \infty} f(\mathbf{r},\mathbf{v},t) \to 0$.

In summary, Eq. (1.121) can be expressed in the more useful form

$$\frac{\partial}{\partial t}\langle n(\mathbf{r},t)\phi(\mathbf{r},\mathbf{v})\rangle + \nabla_{\mathbf{r}} \cdot \langle \mathbf{v}n(\mathbf{r},t)\phi(\mathbf{r},\mathbf{v})\rangle - n(\mathbf{r},t)\langle \mathbf{v} \cdot \nabla_{\mathbf{r}}\phi(\mathbf{r},\mathbf{v})\rangle - \frac{n(\mathbf{r},t)}{m}\langle \nabla_{\mathbf{v}}\phi(\mathbf{r},\mathbf{v}) \cdot \mathbf{F(r)}\rangle = 0. \tag{1.125}$$

For the ideal gas, the physical quantities conserved in each collision process, obeying Eq. (1.114), are five: the mass of the particles, $\phi_1 = m$, the momentum components, $\phi_{2,i} = mv_i$ $(i = x, y, z)$, and the kinetic energy $\phi_3 = \frac{1}{2}m|\mathbf{v}|^2$. In all cases ϕ does not depend on $\mathbf{r}$, therefore the third addendum in Eq. (1.125) vanishes and it simplifies to

$$\frac{\partial}{\partial t}\langle n(\mathbf{r},t)\phi(\mathbf{v})\rangle + \nabla_{\mathbf{r}} \cdot \langle \mathbf{v}n(\mathbf{r},t)\phi(\mathbf{v})\rangle - \frac{n(\mathbf{r},t)}{m}\langle \nabla_{\mathbf{v}}\phi(\mathbf{v}) \cdot \mathbf{F(r)}\rangle = 0. \tag{1.126}$$

Let us now specialize Eq. (1.126) to the invariants just introduced. The conservation of mass, $\phi_1 = m$, is associated with the so-called *continuity equation*

$$\frac{\partial \rho}{\partial t} + \nabla_{\mathbf{r}} \cdot (\rho\langle \mathbf{v}\rangle) = 0 \quad \text{(mass conservation)}, \tag{1.127}$$

where $\rho(\mathbf{r},t) = mn(\mathbf{r},t)$ is the mass density of the ideal gas.

The conservation of the momentum components, $\phi_{2,i} \equiv mv_i$, specializes Eq. (1.126) to the equations

$$\frac{\partial \rho\langle v_i\rangle}{\partial t} + \nabla_{\mathbf{r}} \cdot \langle \mathbf{v}\rho v_i\rangle = \rho\frac{F_i}{m}, \quad i = x, y, z. \tag{1.128}$$

Taking into account the equality $\langle v_i v_j\rangle = \langle (v_i - \langle v_i\rangle)(v_j - \langle v_j\rangle)\rangle + \langle v_i\rangle\langle v_j\rangle$, we can write the previous equation as follows:

$$\frac{\partial \rho\langle v_i\rangle}{\partial t} + \nabla_{\mathbf{r}} \cdot (\langle \mathbf{v}\rangle\rho\langle v_i\rangle) = \rho\frac{F_i}{m} - (\nabla_{\mathbf{r}} \cdot \hat{P})_i \quad , \quad i = x, y, z, \tag{1.129}$$

where $\hat{P}$ is the pressure tensor, whose components read (see also Eq. (C.9))

$$P_{ij} = \rho\langle (v_i - \langle v_i\rangle)(v_j - \langle v_j\rangle)\rangle, \quad i, j = x, y, z \tag{1.130}$$

so that $(\nabla_{\mathbf{r}} \cdot \hat{P})_i \equiv \sum_{j=x,y,z} \frac{\partial}{\partial r_j}\rho\langle (v_i - \langle v_i\rangle)(v_j - \langle v_j\rangle)\rangle$. We can use (1.127) for rewriting Eq. (1.129) as follows:

$$\left(\frac{\partial}{\partial t} + \langle \mathbf{v}\rangle \cdot \nabla_{\mathbf{r}} \right)\langle v_i\rangle = \frac{F_i}{m} - \frac{1}{\rho}(\nabla_{\mathbf{r}} \cdot \hat{P})_i, \quad i = x, y, z. \quad \text{(momentum conservation)} \tag{1.131}$$

It remains to specialize Eq. (1.126) to the conservation of kinetic energy, namely

$$\frac{\partial}{\partial t}\langle \rho\frac{1}{2}|\mathbf{v}|^2\rangle + \nabla_{\mathbf{r}} \cdot \langle \rho\mathbf{v}\frac{1}{2}|\mathbf{v}|^2\rangle - \frac{\rho}{m}\langle \mathbf{F} \cdot \mathbf{v}\rangle = 0. \tag{1.132}$$

This equation can be rewritten in a more physical form by introducing the local quantities

$$T(\mathbf{r}, t) = \frac{1}{3} m \langle |\mathbf{v} - \langle \mathbf{v} \rangle|^2 \rangle \tag{1.133}$$

$$\mathbf{q}(\mathbf{r}, t) = \frac{1}{2} \rho \langle (\mathbf{v} - \langle \mathbf{v} \rangle) |\mathbf{v} - \langle \mathbf{v} \rangle|^2 \rangle \tag{1.134}$$

which define the local temperature and the local heat flux, respectively. Note that both of these quantities have to be defined as functions of the relative velocity of particles $\mathbf{v} - \langle \mathbf{v} \rangle$, because the average displacement of particles determined by their average velocity $\langle \mathbf{v} \rangle$ does not contribute neither to the temperature nor to the heat current inside the gas.[15] It is also useful to introduce the symmetric tensor

$$\Lambda_{ij} = \frac{1}{2} m \left(\frac{\partial \langle v_i \rangle}{\partial r_j} + \frac{\partial \langle v_j \rangle}{\partial r_i} \right), \quad i, j = x, y, z. \tag{1.135}$$

Making use of the previous definitions and of Eqs. (1.127) and (1.131) and considering that the pressure tensor P_{ij} defined in (1.130) is symmetric, one can finally express (1.132) in the following form:

$$\frac{3}{2} \rho \left(\frac{\partial}{\partial t} + \langle \mathbf{v} \rangle \cdot \nabla_\mathbf{r} \right) T + \nabla_\mathbf{r} \cdot \mathbf{q} = - \sum_{i,j=x,y,z} (P_{ij} \Lambda_{ij}). \quad \text{(energy conservation)} \tag{1.136}$$

In summary, Eqs. (1.127), (1.131), and (1.136) express the three conservation equations associated with the additive invariants of the microscopic collision process between particles of an ideal gas. In order to obtain useful physical information from these equations, one should evaluate explicitly all the averages contained in these equations and solve them. Unfortunately, this task cannot be accomplished, because the general solution, $f(\mathbf{r}, \mathbf{v}, t)$, of the Boltzmann Equation (1.45) is unknown. This notwithstanding, in what follows we can work out an effective method for finding an explicit form of these equations at increasing levels of approximation. For pedagogical reasons, we first illustrate the so-called zero-order approximation, then the first-order approximation. A short summary of their generalization, known as the Chapman–Enskog method, is reported in Appendix D.

1.7.2 Hydrodynamics of the Inviscid Flow

The zero-order approximation to the conservation Equations (1.127), (1.131), and (1.136) amounts to assume that the ideal gas has approached local equilibrium conditions over a typical time scale τ, which is of the order of the average collision time between particles (see Eq. (C.18) in Appendix C.2). In practice, we approximate $f(\mathbf{r}, \mathbf{v}, t)$ with the Maxwell–Boltzmann local equilibrium distribution $f^*(\mathbf{r}, \mathbf{v}, t)$ (see Eq. (1.86)). As we are going to discuss, the corresponding hydrodynamic equations describe the evolution of an inviscid flow, that is, an ideal gas or, more generally, a diluted fluid with no viscosity. Despite such a description may seem quite far from any physical interest, it is pedagogically interesting,

---

[15] Eq. (1.133) is a compact way to express the equipartition theorem of kinetic energy in a gas, namely $\frac{3}{2} nT = \frac{1}{2} mn \langle |\mathbf{v} - \langle \mathbf{v} \rangle|^2 \rangle = \frac{1}{2} mn \langle |\mathbf{v}|^2 \rangle - \frac{1}{2} mn |\langle \mathbf{v} \rangle|^2$. The expression of the local heat flux stems from its very definition $\mathbf{q} = \frac{1}{2} m \int d\mathbf{v} f(\mathbf{r}, \mathbf{v}, t)(\mathbf{v} - \langle \mathbf{v} \rangle)|\mathbf{v} - \langle \mathbf{v} \rangle|^2$.

because it allows one to obtain a preliminary insight about the structure of the hydrodynamic equations.

Before entering calculations, let us point out that Eq. (1.86) is not a solution of the Boltzmann equation, because

$$\left(\frac{\partial}{\partial t} + \mathbf{v} \cdot \nabla_{\mathbf{r}} + \frac{\mathbf{F}(\mathbf{r})}{m} \cdot \nabla_{\mathbf{v}}\right) f^*(\mathbf{r}, \mathbf{v}, t) \neq 0. \tag{1.137}$$

On the other hand, when local equilibrium conditions have been established, that is, after a time $t > \tau$ and over a length scale larger than λ, where $(\frac{\partial f(\mathbf{r}, \mathbf{v}, t)}{\partial t})_c \approx 0$, we can write

$$\left(\frac{\partial}{\partial t} + \mathbf{v} \cdot \nabla_{\mathbf{r}} + \frac{\mathbf{F}(\mathbf{r})}{m} \cdot \nabla_{\mathbf{v}}\right) f^*(\mathbf{r}, \mathbf{v}, t) \approx 0. \tag{1.138}$$

This implies also that the quantities defined in Eqs. (1.78) to (1.80) are approximately conserved if $f(\mathbf{r}, \mathbf{v}, t)$ is replaced by $f^*(\mathbf{r}, \mathbf{v}, t)$. We can compute the zero-order approximations, $P_{ij}^{(0)}$ and $\mathbf{q}^{(0)}(\mathbf{r}, t)$, of the pressure tensor, defined in Eq. (1.130), and of the heat flux, defined in Eq. (1.134), respectively: In practice, the averages in both formulae are computed over the distribution function $f^*(\mathbf{r}, \mathbf{v}, t)$ defined in Eq. (1.86). Since this is symmetric with respect to each component of the vector $(\mathbf{v} - \langle \mathbf{v} \rangle)$, one can immediately conclude that

$$\mathbf{q}^{(0)}(\mathbf{r}, t) = 0, \tag{1.139}$$

that is, no heat flows through the gas, while

$$\begin{aligned}
P_{ij}^{(0)} &= \rho \left(\frac{m}{2\pi T}\right)^{\frac{3}{2}} \int d\mathbf{v}(v_i - \langle v_i \rangle)(v_j - \langle v_j \rangle) \exp\left[-\frac{m}{2T}(\mathbf{v} - \langle \mathbf{v} \rangle)^2\right] \\
&= \rho \left(\frac{m}{2\pi T}\right)^{\frac{3}{2}} \int d\mathbf{w} w_i w_j \exp\left(-\frac{m}{2T} \mathbf{w}^2\right) \\
&= \delta_{ij} P
\end{aligned} \tag{1.140}$$

where

$$P = \frac{1}{3} \rho \left(\frac{m}{2\pi T}\right)^{\frac{3}{2}} \int d\mathbf{w} \mathbf{w}^2 \exp\left(-\frac{m}{2T} \mathbf{w}^2\right) = nT. \tag{1.141}$$

P is the so-called local hydrostatic pressure and the last equation amounts to a local equation of state of an ideal gas, which has reached local equilibrium. Note that a vanishing heat flow implies no further evolution inside the gas: In the zero-order approximation, once the gas has reached local equilibrium conditions, it would remain frozen in this configuration, which corresponds to quite an unexpected scenario with respect to our everyday experience. In this approximation the continuity equation (1.127) still holds unchanged, while Eq. (1.131) simplifies to

$$\left(\frac{\partial}{\partial t} + \langle \mathbf{v} \rangle \cdot \nabla_{\mathbf{r}}\right) \langle v_i \rangle + \frac{1}{\rho} \frac{\partial P}{\partial r_i} = \frac{F_i}{m}, \quad i = x, y, z \tag{1.142}$$

which is also known as Euler's equation.

Eq. (1.136) becomes

$$\rho \left(\frac{\partial}{\partial t} + \langle \mathbf{v} \rangle \cdot \nabla_{\mathbf{r}}\right) T = -\frac{2}{3} P \sum_i \Lambda_{i,i} = -\frac{2}{3} Pm \nabla_{\mathbf{r}} \cdot \langle \mathbf{v} \rangle \tag{1.143}$$

where we have used (1.135). Taking into account also (1.141), we can rewrite the latter equation as follows:

$$\left(\frac{\partial}{\partial t} + \langle \mathbf{v} \rangle \cdot \nabla_{\mathbf{r}}\right) T + \frac{1}{mC_V} T \nabla_{\mathbf{r}} \cdot \langle \mathbf{v} \rangle = 0, \tag{1.144}$$

where $C_V = 3/(2m)$ represents the specific heat at constant volume of the ideal gas.

In order to simplify the notation, we introduce the so-called *material-derivative operator*, acting over the generic function $A(\mathbf{r}, t)$

$$\mathbb{D}A \equiv \left(\frac{\partial}{\partial t} + \langle \mathbf{v} \rangle \cdot \nabla_{\mathbf{r}}\right) A(\mathbf{r}, t) \tag{1.145}$$

and representing the time variation rate of $A(\mathbf{r}, t)$ experienced by an observer moving along a stream line, that is, with local velocity $\langle \mathbf{v} \rangle$. We can rewrite Eq. (1.127) as follows:

$$\mathbb{D}\rho = -\rho \nabla_{\mathbf{r}} \cdot \langle \mathbf{v} \rangle \tag{1.146}$$

and Eq. (1.144) as follows:

$$\mathbb{D}T + \frac{2T}{3} \nabla_{\mathbf{r}} \cdot \langle \mathbf{v} \rangle = 0. \tag{1.147}$$

Finally, by combining the last two equations, one obtains

$$\mathbb{D}(\rho T^{-\frac{3}{2}}) = 0. \tag{1.148}$$

This equation implies that the quantity $\rho T^{-\frac{3}{2}}$ is constant along a streamline. Taking into account (1.141), this condition can be rewritten in the following form:

$$P\rho^{-(C_P/C_V)} = \text{const.}, \tag{1.149}$$

which is the condition defining an adiabatic transformation in an ideal gas, where $C_P/C_V = \frac{5}{3}$ is the ratio between the specific heats at constant pressure, C_P, and constant volume, C_V. We can conclude that in the approximation of an inviscid fluid, the evolution along a stream line corresponds to an adiabatic transformation, that is, no heat transfer mechanism can be at work.

1.7.3 Hydrodynamics: First-order Approximation

The first-order approximation to the hydrodynamics of the ideal gas amounts to assume that the distribution function $f(\mathbf{r}, \mathbf{v}, t)$ is a small perturbation of the Maxwell–Boltzmann local equilibrium distribution $f^*(\mathbf{r}, \mathbf{v}, t)$, namely

$$f(\mathbf{r}, \mathbf{v}, t) = f^*(\mathbf{r}, \mathbf{v}, t) + g(\mathbf{r}, \mathbf{v}, t) \tag{1.150}$$

with $|g(\mathbf{r}, \mathbf{v}, t)| \ll f^*(\mathbf{r}, \mathbf{v}, t)$. As a first step, we can estimate the order of magnitude of $g(\mathbf{r}, \mathbf{v}, t)$ by substituting Eq. (1.150) into the collision term of the Boltzmann equation (1.44), while neglecting terms of order g^2:

$$\left(\frac{\partial f_{\mathbf{v}}}{\partial t}\right)_c = \int d\Omega \int \sigma(|\mathbf{v} - \mathbf{u}|; \Omega) \left[g_{\mathbf{v}'} f^*_{\mathbf{u}'} + g_{\mathbf{u}'} f^*_{\mathbf{v}'} - g_{\mathbf{v}} f^*_{\mathbf{u}} - g_{\mathbf{u}} f^*_{\mathbf{v}}\right] |\mathbf{v} - \mathbf{u}| d\mathbf{u}, \tag{1.151}$$

where we have reintroduced the shorthand notations $g_\mathbf{v} \equiv g(\mathbf{r}, \mathbf{v}, t)$, $f_\mathbf{v} \equiv f(\mathbf{r}, \mathbf{v}, t)$ and we have used the condition $\left(\frac{\partial f_\mathbf{v}^*}{\partial t}\right)_c = 0$.

All terms inside the square brackets in the integral of Eq. (1.151) are of the same order of magnitude and any one of them can be used to obtain an estimate of the perturbation, for example,

$$\int d\Omega \int \sigma(|\mathbf{v} - \mathbf{u}|; \Omega) g_\mathbf{v} f_\mathbf{u}^* |\mathbf{v} - \mathbf{u}| d\mathbf{u} \approx g_\mathbf{v} \int d\Omega \int \sigma(|\mathbf{v} - \mathbf{u}|; \Omega) f_\mathbf{u}^* |\mathbf{v} - \mathbf{u}| d\mathbf{u} \sim \frac{g_\mathbf{v}}{\tau},$$
(1.152)

where the last expression stems from Eq. (1.108), considering that, at leading order, $f_\mathbf{u} \sim f_\mathbf{u}^*$.

Accordingly, in the first-order approximation, we can assume the simplifying hypothesis

$$\left(\frac{\partial f_\mathbf{v}}{\partial t}\right)_c \approx \frac{g_\mathbf{v}}{\tau} = \frac{f_\mathbf{v} - f_\mathbf{v}^*}{\tau},$$
(1.153)

in such a way that we can reduce the transport Boltzmann equation (1.45) in the form

$$\left(\frac{\partial}{\partial t} + \mathbf{v} \cdot \nabla_\mathbf{r} + \frac{\mathbf{F}(\mathbf{r})}{m} \cdot \nabla_\mathbf{v}\right) f_\mathbf{v}^* \approx \frac{g_\mathbf{v}}{\tau},$$
(1.154)

where on the l.h.s. of this equation we have approximated $f_\mathbf{v}$ with $f_\mathbf{v}^*$. Taking into account the estimate Eq. (1.112) of the stream term $\mathbf{v} \cdot \nabla_\mathbf{r} f_\mathbf{v}^*$, the l.h.s. of the latter equation is of the order $\frac{f_\mathbf{v}^* \langle v_i \rangle}{L_0}$ and we can write

$$f_\mathbf{v}^* \frac{\langle v_i \rangle}{L_0} \sim \frac{g_\mathbf{v}}{\tau} \quad \rightarrow \quad \frac{g_\mathbf{v}}{f_\mathbf{v}^*} \sim \frac{\lambda}{L_0},$$
(1.155)

where we have used Eq. (1.13). The last relation implies also that $f_\mathbf{v}^*$ is a good approximation of $f_\mathbf{v}$ provided the mean-free path of particles λ is much smaller than the linear dimension L_0 of the volume containing the gas.[16] In a physical view, this statement is equivalent to the assumption that the characteristic length scale over which mass, momentum, and energy densities of the particles vary is much smaller than L_0.

Note that in the uniform case and in the absence of external forces, Eq. (1.153) simplifies to[17]

$$\frac{\partial f_\mathbf{v}}{\partial t} \sim -\frac{f_\mathbf{v} - f_\mathbf{v}^*}{\tau} \quad \rightarrow \quad f_\mathbf{v} - f_\mathbf{v}^* \sim \exp\left(-\frac{t}{\tau}\right),$$
(1.156)

where the last relation tells us that $f_\mathbf{v}$ converges exponentially fast to $f_\mathbf{v}^*$ over the time scale determined by the average collision time τ, consistently with the hypothesis that $f_\mathbf{v}^*$ is a suitable approximation of $f_\mathbf{v}$. Rewriting Eq. (1.154) in the form

$$g_\mathbf{v} \simeq -\tau \left(\frac{\partial}{\partial t} + \mathbf{v} \cdot \nabla_\mathbf{r} + \frac{\mathbf{F}(\mathbf{r})}{m} \cdot \nabla_\mathbf{v}\right) f_\mathbf{v}^*$$
(1.157)

provides us a formula for estimating explicitly $g_\mathbf{v}$ through the derivatives of $f_\mathbf{v}^*$, defined in Eq. (1.86).[18] After lengthy calculations, whose details are reported in Appendix E, one

---

[16] In the presence of an external force, L_0 can be determined by its spatial scale, which must be much larger than λ.

[17] A minus sign is necessary in order to obtain stability of the solution f^*.

[18] Since we have approximated $f_\mathbf{v}$ with $f_\mathbf{v}^*$, for consistency reasons we have to assume that in the former the dependence on $\mathbf{r}$ and t goes only through the quantities $\rho(\mathbf{r}, t)$, $T(\mathbf{r}, t)$, and $\langle \mathbf{v} \rangle(\mathbf{r}, t)$, as is for the latter. On a rigorous ground, such an assumption cannot be justified, but it is at least physically plausible and consistent with the perturbative approach inherent in the first-order approximation.

obtains

$$g_{\mathbf{w}} \simeq -\tau f_{\mathbf{w}}^{*} \left[\frac{1}{T}\left(\frac{m}{2T}|\mathbf{w}|^{2} - \frac{5}{2}\right)\mathbf{w}\cdot\nabla_{\mathbf{r}}T + \frac{m}{T}\sum_{i,j}\Lambda_{ij}\left(w_i w_j - \frac{1}{3}|\mathbf{w}|^{2}\delta_{ij}\right)\right] \tag{1.158}$$

where $\mathbf{w} = \mathbf{v} - \langle\mathbf{v}\rangle$ (see also Eq. (1.140)) and Λ_{ij} is defined in Eq. (1.135), with both i and j running over the space coordinates.

The next step for obtaining the hydrodynamic equations in the first-order approximation amounts to compute the pressure tensor defined in Eq. (1.130) and the heat flux defined in Eq. (1.134), where the averages are performed over the approximate distribution function $f_{\mathbf{v}} = f_{\mathbf{v}}^{*} + g_{\mathbf{v}}$. Let us first compute the heat flux, while recalling that $\mathbf{q}^{(0)}(\mathbf{r},t) = 0$:

$$\mathbf{q}^{(1)}(\mathbf{r},t) = \frac{m}{2}\int d\mathbf{w}\,\mathbf{w}|\mathbf{w}|^{2}g_{\mathbf{w}}. \tag{1.159}$$

Note that the second addendum of $g_{\mathbf{w}}$ in Eq. (1.158) gives no contribution to the integral on the r.h.s. of this equation, because Λ_{ij} (defined in Eq. (1.135)) and δ_{ij} are symmetric tensors, and, accordingly, this component of $g_{\mathbf{w}}$ yields integrals of odd powers of the integration variable $\mathbf{w}$, while $f_{\mathbf{w}}^{*}$, the local Maxwell–Boltzmann distribution, is an even function of $\mathbf{w}$. Thus, we obtain

$$\mathbf{q}^{(1)}(\mathbf{r},t) = -\frac{\tau m}{2}\int d\mathbf{w}\,\mathbf{w}|\mathbf{w}|^{2}\frac{1}{T}\left(\frac{m}{2T}|\mathbf{w}|^{2} - \frac{5}{2}\right)\mathbf{w}\cdot(\nabla_{\mathbf{r}}T)f_{\mathbf{w}}^{*}.$$

The proportionality between the heat flux and the thermal gradient appears when we remark that in the evaluation of $q_i^{(1)}$, only the i–th term of the product $\mathbf{w}\cdot(\nabla_{\mathbf{r}}T)$ gives a nonvanishing contribution to the integral. In fact,

$$q_i^{(1)} = -\frac{\tau m}{2}\int d\mathbf{w}\,w_i|\mathbf{w}|^{2}\frac{1}{T}\left(\frac{m}{2T}|\mathbf{w}|^{2} - \frac{5}{2}\right)\left(w_x\frac{\partial T}{\partial x} + w_y\frac{\partial T}{\partial y} + w_z\frac{\partial T}{\partial z}\right)f_{\mathbf{w}}^{*}$$

$$= -\frac{\tau m}{2}\int d\mathbf{w}\,w_i^{2}|\mathbf{w}|^{2}\frac{1}{T}\left(\frac{m}{2T}|\mathbf{w}|^{2} - \frac{5}{2}\right)f_{\mathbf{w}}^{*}\frac{\partial T}{\partial r_i}. \tag{1.160}$$

Since the integral multiplying $\frac{\partial T}{\partial r_i}$ does not depend on i, we can finally write

$$\mathbf{q}^{(1)}(\mathbf{r},t) = -\kappa(\nabla_{\mathbf{r}}T) \tag{1.161}$$

where

$$\kappa = -\frac{\tau m}{6T}\int d\mathbf{w}\,|\mathbf{w}|^{4}\left(\frac{m}{2T}|\mathbf{w}|^{2} - \frac{5}{2}\right)f_{\mathbf{w}}^{*} = \frac{5}{2}\frac{\tau n T}{m} \tag{1.162}$$

is the transport coefficient named *thermal conductivity*.

As anticipated, we now evaluate the components of the pressure tensor (see Eq. (1.130)) in the first-order approximation. They read

$$P_{ij}^{(1)} = m\int d\mathbf{w}\,w_i w_j(f_{\mathbf{w}}^{*} + g_{\mathbf{w}}) = \delta_{ij}P + \Pi_{ij} \tag{1.163}$$

where $P = nT$ (see Eq. (1.141)) and

$$\Pi_{ij} = m\int d\mathbf{w}\,w_i w_j g_{\mathbf{w}} = -\frac{\tau m^{2}}{T}\sum_{k,l}\Lambda_{kl}\int d\mathbf{w}\,w_i w_j\left(w_k w_l - \frac{1}{3}\delta_{kl}\mathbf{w}^{2}\right)f_{\mathbf{w}}^{*}. \tag{1.164}$$

The last equality is obtained by considering that the first addendum inside the square parentheses of Eq. (1.158), defining $g_\mathbf{w}$, yields vanishing contributions from integrals of odd powers of the components of $\mathbf{w}$.

An important property of the components Π_{ij} is that they define a symmetric tensor with vanishing diagonal elements. In fact,

$$\Pi_{ii} = \left(-\frac{\tau m^2}{T}\right) \sum_{k,l} \Lambda_{kl} \int d\mathbf{w}\, w_i^2 \left(w_k w_l - \frac{1}{3}\delta_{kl}|\mathbf{w}|^2\right) f_\mathbf{w}^* \tag{1.165}$$

$$= \left(-\frac{\tau m^2}{T}\right) \sum_{k} \Lambda_{kk} \int d\mathbf{w}\, w_i^2 \left(w_k^2 - \frac{1}{3}|\mathbf{w}|^2\right) f_\mathbf{w}^* \tag{1.166}$$

$$= 0, \tag{1.167}$$

where the passage from the first to the second line is due to the vanishing of the integral for parity reasons, when $k \neq l$. Instead, the passage from the second to the third line is due to isotropy.

As for the off-diagonal components of the pressure tensor, from Eq. (1.164) we easily conclude it must be $k = i, l = j$ or the other way round. Using that Λ_{kl} is a symmetric tensor (see Eq. (1.135)), we find that for $i \neq j$

$$\Pi_{ij} = -\frac{2\tau m^2}{T}\Lambda_{ij} \int d\mathbf{w}\, w_i^2 w_j^2 f_\mathbf{w}^* \equiv -2\eta \Lambda_{ij}, \tag{1.168}$$

where

$$\eta = \frac{\tau m^2}{T} \int d\mathbf{w}\, w_i^2 w_j^2 f_\mathbf{w}^* = \tau n T \tag{1.169}$$

represents the transport coefficient named *viscosity*.[19] Because of isotropy, the diagonal elements Λ_{ii} are equal and we can write

$$\Pi_{ij} = -2\eta \left[\Lambda_{ij} - \frac{1}{3}\delta_{ij}\left(\sum_k \Lambda_{kk}\right)\right] = -2\eta\left(\Lambda_{ij} - \frac{1}{3}\delta_{ij} m \nabla_\mathbf{r} \cdot \langle\mathbf{v}\rangle\right), \tag{1.170}$$

obtaining the final expressions

$$P_{ij}^{(1)} = \delta_{ij} P + \Pi_{ij} = \delta_{ij} n T - 2\eta\left(\Lambda_{ij} - \frac{m}{3}\delta_{ij} \nabla_\mathbf{r} \cdot \langle\mathbf{v}\rangle\right). \tag{1.171}$$

At this point, it is standard to point out that the following important relation holds for an ideal gas

$$\frac{\kappa}{\eta} = \frac{5}{2m} = \frac{5}{3}C_V, \tag{1.172}$$

which tells us that the ratio of the transport coefficients is related to an equilibrium quantity, namely the specific heat at constant volume C_V. Physically this is not surprising, if one considers that, in order to actually measure C_V for a gas or fluid, one has to gently perturb (by heating or cooling) the equilibrium state of the system and measure the amount of heat exchanged with the environment, after the system has come back to thermodynamic

---

[19] Isotropy makes the integral independent of $i, j \neq i$.

equilibrium, through its hydrodynamic evolution. All this process occurs because, in the first-order approximation, heat can flow, being absorbed or released, in a viscous fluid. For the peculiar ideal fluid described by the zero-order approximation, C_V is undefined and unmeasurable.

The final step for obtaining the explicit form of the hydrodynamic equations in the first-order approximation can be performed by substituting the Eqs. (1.161) and (1.171) into Eqs. (1.127), (1.131), and (1.136). Lengthy calculations, detailed in Appendix E, yield

$$\frac{\partial \rho}{\partial t} + \nabla_{\mathbf{r}} \cdot (\rho \langle \mathbf{v} \rangle) = 0 \tag{1.173}$$

$$\left(\frac{\partial}{\partial t} + \langle \mathbf{v} \rangle \cdot \nabla_{\mathbf{r}} \right) \langle \mathbf{v} \rangle = \frac{\mathbf{F}}{m} - \frac{1}{\rho} \nabla_{\mathbf{r}} \left(P - \frac{\eta}{3} \nabla_{\mathbf{r}} \cdot \langle \mathbf{v} \rangle \right) + \frac{\eta}{\rho} \nabla_{\mathbf{r}}^2 \langle \mathbf{v} \rangle \tag{1.174}$$

$$\left(\frac{\partial}{\partial t} + \langle \mathbf{v} \rangle \cdot \nabla_{\mathbf{r}} \right) T = -\frac{1}{C_V} (\nabla_{\mathbf{r}} \cdot \langle \mathbf{v} \rangle) T + \frac{\kappa}{\rho C_V} \nabla_{\mathbf{r}}^2 T. \tag{1.175}$$

While the continuity equation holds at any order of approximation, in the first-order approximation, the conservation equation for momentum yields the well-known Navier–Stokes equation for viscous fluids (1.174), obtained on the basis of heuristic arguments some decades before Boltzmann theory, while the conservation equation for energy turns out as the so-called heat equation (1.175), already established by the French Joseph Fourier more than half a century before Boltzmann theory. This is to stress that one of the main successes of Boltzmann was the recovery not only of equilibrium properties of the ideal gas but also of these transport equations. In this respect, it is important to point out that the perturbative approach to hydrodynamics, outlined here up to first order, can be further improved in a systematic way, while the heuristic equations derived before Boltzmann theory could not be amenable to such an important task. In particular, Boltzmann theory allows us to proceed to higher order of approximations by following a well-defined strategy, known as the Chapman–Enskog method, that we summarize in Appendix D.

1.8 Bibliographic Notes

A historical perspective about kinetic theory and statistical mechanics is contained in the book by S. G. Brush, *The Kind of Motion We Call Heat* (North Holland, 1976).

A pedagogical introduction to the Boltzmann transport equation and related topics can be found in the book by K. Huang, *Statistical Mechanics*, 2nd ed. (Wiley, 1987). A more detailed account of the many aspects related to this fundamental equation is contained in the book by C. Cercignani, *The Boltzmann Equation and Its Applications* (Springer, 1988).

2

Brownian Motion, Langevin, and Fokker–Planck Equations

In this chapter, first we present different theoretical approaches to the problem of Brownian motion (Section 2.2). Einstein's theory amounts to an effective phenomenological approach, combining basic thermodynamic and dynamical concepts. As a preliminary insight on the role of randomness (noise) in diffusion processes, we illustrate the simple model of the random walker, as a representation of the evolution of a gas particle, going through random collisions with the other gas particles. More refined dynamical models for the evolution of a Brownian particle are then introduced. In the mechanistic approach by Langevin, noise appears in the form of a random force: It represents the effect of incoherent fluctuations of the Brownian particle due to its collisions with the solvent particles at thermal equilibrium. The Fokker–Planck probabilistic approach provides an alternative description of the dynamics of a Brownian particle by representing the evolution of its space–time probability distribution.

In Section 2.3, we introduce the mathematical theory of Markov chains, which provides a general approach to stochastic processes in discrete time and state-space. In particular, we illustrate how Markov chains can be formulated in terms of a discrete-time master equation, which allows us to point out the importance of detailed balance conditions. These conditions amount to imposing a sort of time-reversibility, yielding a stochastic evolution toward an equilibrium state. Other basic concepts, such as the definition of ergodic Markov chains, are preliminary for various applications, including the Monte Carlo procedure for the statistical sampling of equilibrium configurations.

The useful extension of stochastic processes in continuous time is contained in Section 2.4. First, while maintaining a discretized state-space, we derive the continuous-time version of the master equation. In particular, we discuss its stationarity condition and how we can discriminate between a continuous-time stochastic evolution to a nonequilibrium steady state (NESS) or to an equilibrium state when also detailed balance conditions hold.

Taking inspiration from the Langevin equation of the Brownian particle, we introduce the basic elements of the general theory of stochastic differential equations, formulated in terms of Wiener processes. This class of equations describes stochastic processes evolving in continuous space and time. We also discuss the physical implications associated with the Itô and Stratonovich formulation. As an application, we show how detailed balance conditions can be recovered for the Langevin equation in the presence of a conservative force, provided Einstein's fluctuation-dissipation relation holds. Then, we discuss how the approach based on stochastic differential equations allows us to derive a general formulation of the Fokker–Planck equation, describing the evolution of the corresponding probability distribution in continuous space and time. Various applications are later discussed, showing also that the Fokker–Planck equation easily reproduces several results obtained by Markov chains.

We conclude this chapter (see Section 2.5) by introducing a more general formulation of the Fokker–Planck equation, based on the so-called forward- and backward-Kolmogorov equations. The asymmetry between these two equations points out that in this framework time-reversal yields different stochastic equations. The practical importance of this formulation is illustrated by showing that the solution of the classical first-passage problem, usually known as Arrhenius' formula, can be found making use of the backward-Kolmogorov equation.

2.1 The Origins of Stochastic Processes

Since 1827 the experiment performed by English botanist Robert Brown describing the phenomenon known as Brownian motion challenged the scientific community. A pollen particle suspended in water (or any similar solvent) was found to exhibit an erratic motion that, apparently, could not be reconciled with any standard mechanical description. Even assuming the atomistic hypothesis and modeling the motion of the pollen particle as a result of collisions with the atoms of the solvent could lead to an incorrect conclusion. In fact, at the microscopic level, one may argue that elastic collisions with the atoms of the solvent can transmit a ballistic motion to the pollen particle. However, one could naively conclude that the combined effect of all of these collisions, occurring for symmetry reasons in any direction, vanishes to zero. On the contrary, the experimental observation of the erratic motion of the pollen particle indicated that the average distance of the particle from its original position grew over sufficiently long time intervals as the square root of time, thus showing the diffusive nature of its motion. Repeating many times the same experiment, where the pollen particle, the solvent, and the temperature of the solvent are the same, the particle in each realization follows different paths, but one can perform a statistical average over these realizations that enforces the conclusion that the particle exhibits a diffusive motion. The universal character of this phenomenon is confirmed by the experimental observation that a diffusive behavior is found also when the type of Brownian particle, the solvent, and the temperature are changed, yielding different values of the proportionality constant between time and the average squared distance of the particle from its initial position.

A convincing explanation of Brownian motion had to wait for the fundamental contribution by the German Albert Einstein, which appeared in 1905, the same year as his contributions on the theories of special relativity and the photoelectric effect. Einstein's phenomenological theory of Brownian motion, relying on basic physical principles, inspired the French Paul Langevin, who proposed a mechanistic approach. The basic idea was to write a Newton-like ordinary differential equation where, for the first time, a force was attributed a stochastic nature, representing uncorrelated fluctuations in space and time. Some years later, the Dutch Adriaan Daniël Fokker and Max Planck proposed an alternative formulation of Brownian motion, based on a partial differential equation for the evolution of the probability distribution of finding a Brownian particle in position $\mathbf{x}$ at time t, in the same spirit as the Boltzmann's equation for an ideal gas. In fact, the Fokker–Planck equation was derived as a master equation, where the rate of change in time of the distribution function depends on favorable and unfavorable processes, described in terms of transition rates between different space–time configurations of the Brownian particle. Making use of some simplifying assumptions, this equation was cast into a form yielding a particularly simple solution, where the diffusive nature of Brownian motion emerges naturally.

On the side of mathematics, at the end of the nineteenth century, the Russian Andrej Andreevič Markov developed a new mathematical theory concerning stochastic processes, nowadays known as Markov chains. The original theory takes advantage of some simplifying assumptions, like the discreteness of time as well as the numerability of the possible states visited by the stochastic process. For the first time, a dynamical theory was assumed to depend on random uncorrelated events, typically obeying the laws of probability.

All these pioneering achievements paved the pathway to a more general formulation of stochastic processes, to which many other scientists, such as Smoluchowski, Bachelier, Wiener, Uhlenbeck, and van Kampen,[1] later contributed.

In fact, as it happens for a Brownian particle, the evolution in time of a large number of systems in nature is affected by random fluctuations, usually referred to as "noise." These fluctuations can be originated by many different physical processes, such as the interaction with thermal reservoirs, absorption and emission of radiation, chemical reactions at the molecular level, different behaviors of agents in a stock market, or of individuals in a population, etc. When we introduce the concept of "noise" in these phenomena, we are assuming that part of their evolution is ruled by "unknown" variables, whose behavior is out of our control, that is, they are unpredictable, in a deterministic sense. On the other hand, we can assume that such fluctuations obey general statistical rules, which still allow us to obtain the possibility of predicting the evolution of these systems in a probabilistic sense.

2.2 Brownian Motion

The basic example of transport properties emerging from fluctuations close to equilibrium is Brownian motion. This phenomenon had been observed by Robert Brown in 1827: small pollen particles (typical size, 10^{-3} cm) in solution with a liquid exhibited an erratic motion, whose irregularity increased with decreasing particle size. The long scientific debate about the origin of Brownian motion lasted over almost a century and raised strong objections to the atomistic approach of the kinetic theory (see Section 1.2). In particular, the motion of the pollen particles apparently could not be consistent with an explanation based on collisions with the molecules of the liquid, subject to thermal fluctuations. In fact, pollen particles have exceedingly larger mass and size with respect to molecules and the amount of velocity acquired by a pollen particle in a collision with a molecule is typically 10^{-6} m/s. In Fig. 2.1 we reproduce some original data by the French Jean Baptiste Perrin.

---

[1] Marian Smoluchowski was Polish, Louis Bachelier was French, Norbert Wiener was American, and George Uhlenbeck, and Nico van Kampen were Dutch.

The kinetic approach also implies that the number of collisions per second of the Brownian particle with liquid molecules is gigantic and random. Thermal equilibrium conditions and the isotropy of the liquid make equally probable the sign of the small velocity variations of the pollen particle. On observable time scales, the effect of the very many collisions should average to zero and the pollen particle should not acquire any net displacement from its initial position in the fluid. This way of reasoning is actually wrong, as Albert Einstein argued in his theory of Brownian motion.

Einstein's methodological approach, based on the combination of simple models with basic physical principles, was very effective: He first assumed that the Brownian macroscopic particles should be considered as "big mass molecules," so that the system composed by the Brownian particles and the solvent, in which they are suspended, can be considered a mixture of two fluids at thermal equilibrium. This implies the validity of van 't Hoff's law,

$$P(\mathbf{x}) = T\,n(\mathbf{x}),\tag{2.1}$$

where P is the osmotic pressure between the fluids, T is the equilibrium temperature of the solution, and n is the volumetric density of Brownian particles (the "solute" of the binary mixture). However, according to classical kinetic theory, one should not be allowed to write an equation of state like (2.1) for a collection of mesoscopic Brownian particles; they should be better described by the laws of dynamics, rather than by those of thermodynamics. Einstein's opposite point of view can be justified by considering that, despite their dimension, the Brownian particles are not subject to standard macroscopic forces: They experience microscopic forces originated by the thermal fluctuations of the fluid molecules and in this sense they are equivalent to particles in solution.

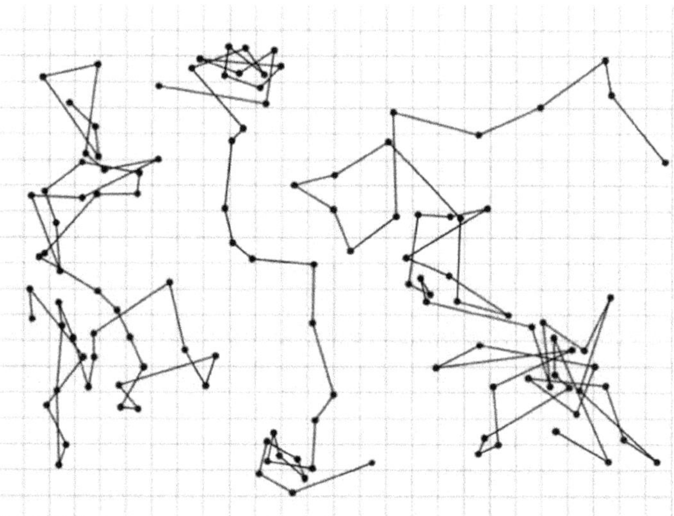

Fig. 2.1   Three tracings of the motion of colloidal particles of radius 0.53 μm, as seen under the microscope, are displayed. Successive positions every 30 s are joined by straight line segments (the mesh size is 3.1 μm). Reproduced from J. B. Perrin, *Les Atomes* (Librairie Félix Alcan, 1913). Wikimedia Commons.

Einstein guessed that Brownian motion looks random on a much larger time scale than the one needed for dissipating the energy acquired through collisions with molecules. In practice, this amounts to assuming that the dissipation mechanism should be described by the macroscopic Stokes's law

$$\mathbf{F} = m\,\frac{d\mathbf{v}}{dt} = -6\pi\eta R\mathbf{v}, \tag{2.2}$$

where $\mathbf{F}$ is the friction force proportional to the velocity $\mathbf{v}$ of the Brownian particle of radius R and η is the viscosity of the solvent. Therefore, the energy dissipation process of Brownian particles has to occur on the time scale

$$t_d = \frac{m}{6\pi\eta R}, \tag{2.3}$$

which must be much larger than the time τ between two collisions; see Eq. (1.7). If we consider a Brownian particle whose mass density is close to that of the solvent (e.g., water) and with radius $R \simeq 10^{-4}$ cm, at room temperature and pressure one has $t_d = O(10^{-7}$ s), while $\tau = O(10^{-11}$ s).

On a time scale much larger than t_d, the Brownian particles are expected to exhibit a diffusive motion, as a consequence of the many random collisions with the solvent molecules. On the other hand, the kinetic theory associates diffusion with the transport of matter in the presence of a density gradient (see Section 1.2.2). We should therefore induce such a gradient by a force. The Einstein approach will be exemplified by considering an ensemble of Brownian particles within a solvent, in a closed container and subject to gravity. If particles were macroscopic (and heavier than the solvent), they would sit immobile at the bottom of the container, the collisions with the atoms of the solvent having no effect. In the case of Brownian particles, we expect some diffusion, hindered by gravity. More precisely, we expect some profile of the density $n(z)$ of Brownian particles, with $\partial_z n < 0$ and $n(z) \to 0$ for $z \to +\infty$. This profile is the outcome of a stationary state that we are now going to describe in terms of equilibrium between currents and equilibrium between forces.

If $F_0 = -mg$ is the force of gravity acting on a Brownian particle of mass m, its velocity satisfies the equation[2]

$$m\frac{dv}{dt} = F_0 - 6\pi\eta Rv, \tag{2.4}$$

which leads to the asymptotic, sedimentation speed $v_0 = F_0/6\pi\eta R$. In the steady state, the resulting current $J_0 = n(z)v_0$ must be counterbalanced by the diffusion current, giving a vanishing total current,

$$-D\partial_z n(z) + n(z)v_0 = 0. \tag{2.5}$$

In terms of equilibrium between forces, the force related to the diffusive motion of Brownian particles is due to the osmotic pressure (2.1). Therefore, equilibrium implies

$$-\partial_z P + n(z)F_0 = 0. \tag{2.6}$$

---

[2] We can deal with relations among scalar quantities, because all of them are directed along the z-axis.

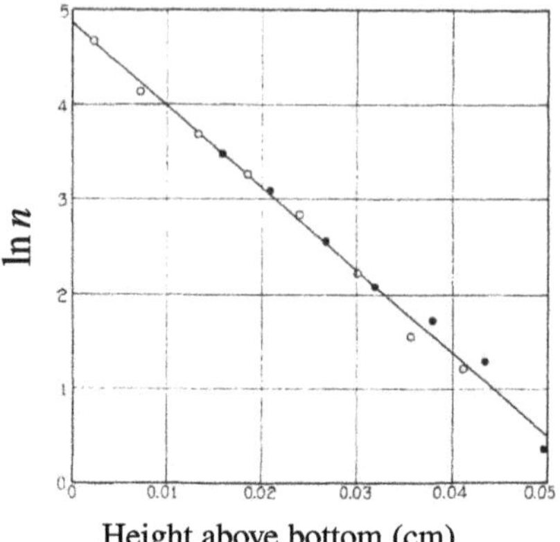

Height above bottom (cm)

Fig. 2.2 Height distribution of the density of gold sols, showing an exponentially decreasing function. From N. Johnston and L. G. Howell, Sedimentation Equilibria of Colloidal Particles, *Physical Review*, **35** (1930) 274–282.

Using the van 't Hoff law and the explicit expression of the sedimentation speed, from Eqs. (2.5) and (2.6), we find the same equation for the density profile,

$$-\partial_z n(z) + \frac{F_0}{6\pi\eta RD}n(z) = 0 \tag{2.7}$$

$$-\partial_z n(z) + \frac{F_0}{T}n(z) = 0, \tag{2.8}$$

which implies the remarkable relation, called Einstein's formula,

$$D = \frac{T}{6\pi\eta R} \equiv \frac{T}{\tilde{\gamma}}, \tag{2.9}$$

where we have defined the friction coefficient (see Eq. (2.4)), $\tilde{\gamma} = 6\pi\eta R$. Perrin showed that this formula provides very good quantitative agreement with experimental data.

The solution of the equation for $n(z)$ gives an exponential profile for the density of Brownian particles,

$$n(z) = n(0)\exp\{(-mg/T)z\}, \tag{2.10}$$

as attested by the experimental results reported in Fig. 2.2.

In summary, Einstein's theory of Brownian motion is based on the description of the Brownian particle in both microscopic and macroscopic terms: The microscopic description is employed when we use the van 't Hoff law to describe Brownian particles as a diluted gas in a solvent, while the macroscopic description is introduced via the Stokes law.

2.2.1 Random Walk: A Basic Model of Diffusion

We consider an ideal gas at thermal equilibrium with a heat bath at temperature T. If we fix our attention on one particle, we observe that collisions with the other particles produce a stepwise irregular trajectory, that is, a sort of random walk. Beyond this qualitative observation, we would like to obtain a quantitative description of this random walk. In

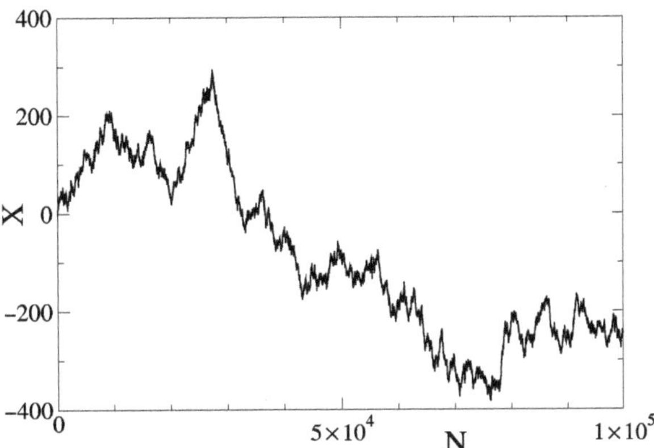

Fig. 2.3   Plot of a one-dimensional random walk, corresponding to $\ell = 1$.

principle the problem could be tackled by applying the laws of classical mechanics. In practice such a program is unrealistic, because one should know not only the initial velocity of the particle under examination but also the velocities of all the particles that it will collide with. Such a computation is practically unfeasible, if we have to deal with a very large number of particles, like those contained in a mole of a gas.

In order to overcome such a difficulty, we can introduce a suitable model, based on simplifying hypotheses. We assume that in between two collisions the observed particle keeps constant the modulus of its velocity, v. Moreover, the distance run by the particle between two collisions is also assumed to be constant and equal to ℓ. Finally, the direction along which the particle moves after a collision is completely uncorrelated with the one it was moving along before the collision. The latter hypothesis amounts to assuming that collisions can actually be considered as random events, thus contradicting the fully mechanical, that is, deterministic, origin of the problem.[3] Without prejudice to generality, we assume that the selected particle at time $t = 0$ is at the origin of a Cartesian reference frame and we call $\mathbf{X}(t)$ the vector that identifies its position at time t. After having gone through N collisions, we can write

$$\mathbf{X} = \sum_{i=1}^{N} \mathbf{x}_i, \tag{2.11}$$

where $\mathbf{x}_i$ is the ith segment run by the particle after the ith collision ($|\mathbf{x}_i| = \ell, \forall i$), whose direction is random, that is, uniformly distributed in the solid angle 4π. It is intuitive to conclude that as $N \to \infty$ the average value $\mathbf{X}/N \to 0$. For $d = 1$, in Fig. 2.3 we plot a typical realization of the space–time trajectory, corresponding to $\ell = 1$.

---

[3] We want to point out that in this way we introduce a statistical concept into the description of a purely mechanical process. This conceptual step has been at the origin of a long-standing debate in the scientific community over more than a century. Nowadays, it has been commonly accepted and it is a cornerstone of modern science. Anyway, this basic assumption still today relies more on its effectiveness in predicting observed phenomena, rather than on its logical foundations. On the other hand, the need for a stochastic approach could also be justified by invoking the contribution of dynamical details, such as the finite size of the particles or their internal rotational or vibrational degrees of freedom, that are usually neglected.

As for the square displacement, we can write

$$X^2 \equiv \mathbf{X} \cdot \mathbf{X} = \sum_{i=1}^{N} \sum_{j=1}^{N} \mathbf{x}_i \cdot \mathbf{x}_j = \ell^2 \sum_{i=1}^{N} \left(\sum_{j=1}^{N} \cos(\theta_{ij}) \right), \tag{2.12}$$

where θ_{ij} is the angle in between the directions of segments $\mathbf{x}_i$ and $\mathbf{x}_j$. If $j = i$, then $\theta_{ij} = 0$, that is, $\cos(\theta_{ij}) = 1$, and the previous equation can be written

$$X^2 = \ell^2 \sum_{i=1}^{N} \left(1 + \sum_{j \neq i} \cos(\theta_{ij}) \right). \tag{2.13}$$

The values taken by $\cos(\theta_{ij})$ can be thought of as random numbers, distributed in the interval $[-1, +1]$. If we compute the average of X^2 over a very large number of different realizations (replicas) of this random walk, the sum $\sum_{j \neq i} \cos(\theta_{ij})$ is negligible and one can finally write

$$\langle X^2 \rangle = \ell^2 N, \tag{2.14}$$

where the symbol $\langle \rangle$ denotes the average over replicas. Notice that the larger the number of replicas, the better the statistical estimate of $\langle X^2 \rangle$ for any N.

This result can be generalized by assuming the less strict hypothesis that the length of runs in between subsequent collisions is distributed according to some normalized distribution $g(\ell)$. A "real" example in two dimensions has been discussed in the context of Brownian motion, see Fig. 2.1. If $\mathbf{x}_i = \ell_i \hat{\mathbf{x}}_i$, where $\hat{\mathbf{x}}_i$ is the unit vector in the direction of $\mathbf{x}_i$, we can write

$$\langle \mathbf{x}_i \cdot \mathbf{x}_j \rangle = \langle \ell_i \ell_j \rangle \langle \hat{\mathbf{x}}_i \cdot \hat{\mathbf{x}}_j \rangle = \langle \ell_i^2 \rangle \delta_{ij} \tag{2.15}$$

and

$$\langle X^2 \rangle = \langle \ell^2 \rangle N. \tag{2.16}$$

If $g(\ell)$ is an exponential distribution, which corresponds to independent random events,

$$g(\ell) = \frac{1}{\lambda} \exp\left(-\frac{\ell}{\lambda} \right), \tag{2.17}$$

where λ is the mean free path defined in Eq. (1.13), we have to substitute ℓ^2 with $\langle \ell^2 \rangle = 2\lambda^2$ in Eq. (2.14), thus obtaining

$$\langle X^2 \rangle = 2\lambda^2 N. \tag{2.18}$$

Notice that $\lambda N = L = \langle v \rangle t$ is the total length run by the particle after N collisions, so we can write

$$\langle X^2 \rangle = 2\lambda \langle v \rangle t. \tag{2.19}$$

This relation indicates that in the random walk, the particle that was at the origin at time $t = 0$ is found at time t at an average distance from the origin, $\sqrt{\langle X^2 \rangle}$, that grows proportionally to $\sqrt{t}$. The proportionality constant between $\langle X^2 \rangle$ and t is usually written as $2\lambda \langle v \rangle = 2dD$, where D is the diffusion coefficient of the random walk in d space dimensions. Diffusion in real situations is quite a slow process. For instance, if we consider air molecules at $T = 20°C$, we have $\langle v \rangle \sim 450$ m/s, while $\lambda \sim 0.06$ μm. Accordingly, a diffusing air molecule in these conditions runs a distance of 1 m in approximately 5 h and a distance of 10 m in approximately 20 days (a quasi-static situation, if convective or turbulent motions do not occur).

2.2.2 The Langevin Equation for the Brownian Particle

A mechanical approach that is valid at both $t < t_d$ (ballistic regime) and $t > t_d$ (diffusive regime) was later proposed by Langevin. For the first time, deterministic and stochastic forces were introduced into the same equation. The deterministic component is the friction term $-\tilde{\gamma}\mathbf{v}$, acting on the Brownian particle: $\tilde{\gamma}$ is the friction coefficient, which amounts to $\tilde{\gamma} = 6\pi\eta R$ in the Stokes regime, and $\mathbf{v}$ is the three-dimensional velocity vector of the Brownian particle, whose components are denoted by v_i, $i = 1, 2, 3$. The stochastic component associated with the effect of collisions with the solvent molecules is a random, time-dependent force $\tilde{\boldsymbol{\eta}}(t)$, whose components are analogously denoted by $\tilde{\eta}_i$, $i = 1, 2, 3$. Symmetry arguments indicate that each one of these components can be assumed to be independent, isotropic, and uncorrelated in time (at least for $t \gg \tau$) and Gaussian, since it results from the combination of a very large number of collisions, which can be approximately considered as independent events. Accordingly, the average of the stochastic force is null and its time-correlation function can be approximated by a Dirac delta for $t \gg \tau$:

$$\langle \tilde{\eta}_i(t) \rangle = 0 \tag{2.20a}$$

$$\langle \tilde{\eta}_i(t)\tilde{\eta}_j(t') \rangle = \tilde{\Gamma}\delta_{ij}\delta(t - t'), \tag{2.20b}$$

where the brackets $\langle \ \rangle$ indicate the statistical average, $\tilde{\Gamma}$ is a suitable dimensional constant, and δ_{ij} is a Kronecker delta. The Langevin equation is a Newtonian equation containing the stochastic force $\tilde{\boldsymbol{\eta}}$. For each component of the three-dimensional velocity vector, it reads

$$m\frac{dv_i(t)}{dt} = -\tilde{\gamma}\,v_i(t) + \tilde{\eta}_i(t), \tag{2.21}$$

which is rewritten, for the sake of simplicity, as

$$\frac{dv_i(t)}{dt} = -\gamma\,v_i(t) + \eta_i(t), \tag{2.22}$$

where we have introduced the symbol $\gamma = \tilde{\gamma}/m = 1/t_d$ and the stochastic force per unit mass, $\eta = \tilde{\eta}/m$. The latter satisfies the same equations as Eq. (2.20), with $\tilde{\Gamma}$ replaced by $\Gamma = \tilde{\Gamma}/m^2$. Note that, as a consequence of Eq. (2.20b) where the physical dimension of the Dirac-delta distribution is $[t]^{-1}$, the physical dimensions of Γ are $[l]^2[t]^{-3}$.

Equation (2.22) can be formally integrated, yielding the solution

$$v_i(t) = \exp(-\gamma t)\left[v_i(0) + \int_0^t d\tau \exp(\gamma\tau)\,\eta_i(\tau)\right]. \tag{2.23}$$

Accordingly, $v_i(t)$ is a stochastic function with average value

$$\langle v_i(t) \rangle = v_i(0)\exp(-\gamma t) \tag{2.24}$$

and average squared value

$$
\begin{aligned}
\langle v_i^2(t) \rangle &= \exp\left(-2\gamma t\right) \left[v_i^2(0) + \int_0^t d\tau \int_0^t d\tau' \exp\left(\gamma(\tau + \tau')\right) \langle \eta_i(\tau)\eta_i(\tau') \rangle \right] \\
&= \exp\left(-2\gamma t\right) \left[v_i^2(0) + \int_0^t d\tau\, \Gamma \exp\left(2\gamma\tau\right) \right] \\
&= v_i^2(0) \exp\left(-2\gamma t\right) + \frac{\Gamma}{2\gamma} \left[1 - \exp\left(-2\gamma t\right) \right].
\end{aligned}
\tag{2.25}
$$

Earlier we first used Eq. (2.20a) to cancel out the terms that are linear in the noise, then we explicitly used the noise correlation function (2.20b).

As time t grows, the average velocity vanishes exponentially with rate γ, while the average squared velocity approaches the value

$$
\lim_{t\to\infty} \langle v_i^2(t) \rangle = \frac{\Gamma}{2\gamma}.
\tag{2.26}
$$

In order to bridge this purely mechanical approach with Einstein's theory, it remains to establish a relation with thermodynamics. This is naturally obtained by attributing a thermal origin to the fluctuations of the stochastic force. In particular, if we assume that the Brownian particles and the solvent are in thermal equilibrium at temperature T, the energy equipartition theorem establishes that, in the limit of very large times, the average kinetic energy per degree of freedom is proportional to the temperature,

$$
\lim_{t\to\infty} \left\langle \frac{1}{2} m\, v_i^2(t) \right\rangle = \frac{1}{2} T.
\tag{2.27}
$$

By comparison with Eq. (2.26), one obtains the formula,

$$
\Gamma = \frac{2\gamma T}{m},
\tag{2.28}
$$

or equivalently

$$
\tilde{\Gamma} = 2\tilde{\gamma} T,
\tag{2.29}
$$

which is a basic example of a fluctuation–dissipation relation.

From an experimental point of view, the fluctuations of the velocity of the particle cannot be measured because of the erratic trajectories, while it is easy to measure the mean square displacement of the Brownian particle from its initial position:

$$
\begin{aligned}
\langle (x_i(t) - x_i(0))^2 \rangle &= \left\langle \left[\int_0^t v_i(\tau) d\tau \right]^2 \right\rangle \\
&= \int_0^t d\tau \int_0^t d\tau' \langle v_i(\tau) v_i(\tau') \rangle.
\end{aligned}
\tag{2.30}
$$

The expression between brackets in the last integral is the velocity correlation function of the Brownian particle, which can be computed from Eq. (2.23), similarly to the derivation of Eq. (2.25). We obtain

$$
\langle v_i(\tau) v_i(\tau') \rangle = v_i^2(0) e^{-\gamma(\tau+\tau')} + \Gamma e^{-\gamma(\tau+\tau')} \int_0^\tau dt_1 \int_0^{\tau'} dt_2\, e^{\gamma(t_1+t_2)} \delta(t_1 - t_2).
\tag{2.31}
$$

The double integral depends on which variable between τ and τ' is the smallest; if $\tau < \tau'$, we must first integrate over t_2, obtaining

$$\Gamma e^{-\gamma(\tau+\tau')} \int_0^\tau dt_1 \int_0^{\tau'} dt_2 e^{\gamma(t_1+t_2)} \delta(t_1 - t_2) = \Gamma e^{-\gamma(\tau+\tau')} \int_0^{\min(\tau,\tau')} dt_1 e^{2\gamma t_1}$$

$$= \frac{\Gamma}{2\gamma} \left(e^{-\gamma|\tau-\tau'|} - e^{-\gamma(\tau+\tau')} \right),$$

a result that is valid in any case. So, Eq. (2.31) becomes

$$\langle v_i(\tau)v_i(\tau')\rangle = v_i^2(0)e^{-\gamma(\tau+\tau')} + \frac{\Gamma}{2\gamma}\left(e^{-\gamma|\tau-\tau'|} - e^{-\gamma(\tau+\tau')}\right). \tag{2.32}$$

Notice that for $\tau' = \tau$, the previous correlator simplifies to Eq. (2.25). Finally, substituting into Eq. (2.30), we obtain

$$\langle(x_i(t) - x_i(0))^2\rangle = \left(v_i^2(0) - \frac{\Gamma}{2\gamma}\right)\frac{(1 - e^{-\gamma t})^2}{\gamma^2} + \frac{\Gamma}{\gamma^2}t - \frac{\Gamma}{\gamma^3}\left(1 - e^{-\gamma t}\right), \tag{2.33}$$

where we have used the following relations:

$$\int_0^t d\tau \int_0^t d\tau' e^{-\gamma(\tau+\tau')} = \left(\frac{1 - e^{-\gamma t}}{\gamma}\right)^2$$

$$\int_0^t d\tau \int_0^t d\tau' e^{-\gamma|\tau-\tau'|} = 2\int_0^t d\tau \int_0^\tau d\tau' e^{-\gamma(\tau-\tau')}$$

$$= \frac{2}{\gamma}t - \frac{2}{\gamma^2}\left(1 - e^{-\gamma t}\right). \tag{2.34}$$

We can now evaluate Eq. (2.33) in the limits $t \ll t_d$ and $t \gg t_d$. In the former case, $\gamma t \ll 1$ and $1 - e^{-\gamma t} \simeq \gamma t - \gamma^2 t^2/2$; in the latter case, $\gamma t \gg 1$ and $1 - e^{-\gamma t} \simeq 1$. Therefore, in the two limits, the mean square displacement of the Brownian particle has the following expressions:

$$\langle(x_i(t) - x_i(0))^2\rangle = \begin{cases} v_i^2(0)t^2 & t \ll t_d \quad \text{ballistic regime} \\[2mm] \dfrac{\Gamma}{\gamma^2}t & t \gg t_d \quad \text{diffusive regime.} \end{cases} \tag{2.35}$$

Using Eq. (2.28) and the Einstein relation (2.9), we obtain $\Gamma/\gamma^2 = 2T/(m\gamma) = 2D$, so that $\langle(x_i(t) - x_i(0))^2\rangle = 2Dt$ in the diffusive regime. More generally, in d spatial dimensions, we have

$$\lim_{t\to\infty} \frac{\langle(\mathbf{x}(t) - \mathbf{x}(0))^2\rangle}{t} = 2dD. \tag{2.36}$$

As a final remark, we want to point out that the Langevin equation for the Brownian particle (2.21) can be generalized to the case in which it is subject to a force $F(x) = -\frac{U(x)}{dx} = -U'(x)$, generated by the conservative potential $U(x)$:

$$m\ddot{x}(t) = -U'(x) - \tilde{\gamma}\dot{x}(t) + \tilde{\eta}(t). \tag{2.37}$$

It is important to remark that, if Einstein relation (2.9) holds, this equation satisfies detailed balance conditions, that is, the Brownian particle evolves to a thermodynamic equilibrium state. This statement will be proved in Section 2.4.4. Moreover, the special case of the Brownian particle subject to an elastic force, known as the Ornstein–Uhlenbeck process, will be explicitly solved in Section 2.4.6. In order to obtain both of these results, we make use of the general formalism of stochastic differential equations, introduced in Section 2.4.3.

2.2.3 The Fokker–Planck Equation for the Brownian Particle

The statistical content of the Langevin approach emerges explicitly from the properties attributed to the random force in Eq. (2.20b). The statistical average $\langle \ \rangle$ can be interpreted as the result of the average over many different trajectories of the Brownian particle obtained by different realizations of the stochastic force components η_i in Eq. (2.22). By taking inspiration from Boltzmann kinetic theory, one can get rid of mechanical trajectories and describe from the very beginning the evolution of Brownian particles by a distribution function $p(\mathbf{x}, t)$, such that $p(\mathbf{x}, t)d\mathbf{x}$ represents the probability of finding the Brownian particle at time t in a small volume $d\mathbf{x}$ at $\mathbf{x}$. The evolution is ruled by the transition rate $W(\mathbf{x}', \mathbf{x})$, which represents the probability per unit time and unit volume that the Brownian particle "jumps" from $\mathbf{x}$ to $\mathbf{x}'$.

In this approach, any reference to instantaneous collisions with the solvent molecules is lost and one has to assume that relevant time scales are much larger than t_d. As proposed by Fokker and Planck, one can write the evolution equation for $p(\mathbf{x}, t)$ as a master equation, that is, a balance equation where the variation in time of $p(\mathbf{x}, t)$ emerges as the result of two competing terms: a gain factor, due to jumps of particles from any position $\mathbf{x}'$ to $\mathbf{x}$, and a loss factor, due to jumps from $\mathbf{x}$ to any $\mathbf{x}'$. Thus

$$\frac{\partial p(\mathbf{x}, t)}{\partial t} = \int_{-\infty}^{+\infty} d^3x' \, [\, p(\mathbf{x}', t)W(\mathbf{x}, \mathbf{x}') - p(\mathbf{x}, t)W(\mathbf{x}', \mathbf{x})]\,, \tag{2.38}$$

where $d^3x' W(\mathbf{x}, \mathbf{x}') \, (d^3x' W(\mathbf{x}', \mathbf{x}))$ represents the probability per unit time that the particle jumps from a neighborhood of $\mathbf{x}'$ to $\mathbf{x}$ (or vice versa). If we define $\chi = \mathbf{x}' - \mathbf{x}$ and we use the notation $W(\mathbf{x}; \chi) = W(\mathbf{x}', \mathbf{x})$, we obtain

$$\frac{\partial p(\mathbf{x}, t)}{\partial t} = \int d^3x' \, [\, p(\mathbf{x}', t)W(\mathbf{x}'; -\chi) - p(\mathbf{x}, t)W(\mathbf{x}; \chi)] \tag{2.39}$$

$$= \int d^3\chi \, [\, p(\mathbf{x} - \chi, t)W(\mathbf{x} - \chi; \chi) - p(\mathbf{x}, t)W(\mathbf{x}; \chi)]\,, \tag{2.40}$$

where in the second equality we passed from the variable $\mathbf{x}'$ to χ and in the first integral of (2.40) we have substituted χ with $-\chi$.

Since large displacements of the Brownian particles are very infrequent, one can reasonably assume that the rate functions $W(\mathbf{x} - \chi; \chi)$ and $W(\mathbf{x}; \chi)$ are significantly different from zero only for very small χ. This allows one to introduce a formal Taylor series expansion in Eq. (2.40) around $\chi = 0$. By considering only terms up to the second order, one obtains

$$\frac{\partial p(\mathbf{x}, t)}{\partial t} = \int d^3\chi \left[-\nabla\left(p(\mathbf{x}, t)W(\mathbf{x}; \chi) \right) \cdot \chi + \frac{1}{2} \sum_{i,j} \frac{\partial^2}{\partial x_i \, \partial x_j} \left(p(\mathbf{x}, t)W(\mathbf{x}; \chi) \right) \chi_i \chi_j \right].$$

$$\tag{2.41}$$

Equation (2.41) has been found with the Brownian particle in mind, but the variable $\mathbf{x}$ can be any vectorial quantity defining the state of a physical system, not only the spatial position of a particle. For this reason it has a range of applications going well beyond the motion of pollen particles in a solvent and it is worth making a further step before specializing in the Brownian motion. We can formally define the quantities

$$\alpha_i(\mathbf{x}) = \int d^3\chi\, W(\mathbf{x};\chi)\, \chi_i \tag{2.42}$$

$$\beta_{ij}(\mathbf{x}) = \int d^3\chi\, W(\mathbf{x};\chi)\chi_i\chi_j \tag{2.43}$$

so that Eq. (2.41) can be written as

$$\frac{\partial p(\mathbf{x},t)}{\partial t} = -\sum_i \frac{\partial}{\partial x_i}\left(\alpha_i(\mathbf{x})p(\mathbf{x},t)\right) + \frac{1}{2}\sum_{i,j}\frac{\partial^2}{\partial x_i \partial x_j}\left(\beta_{ij}(\mathbf{x})p(\mathbf{x},t)\right). \tag{2.44}$$

This is the celebrated Fokker–Planck equation and it will be used in different contexts in this book.

Let us now come back to the Brownian motion, in which case $\mathbf{x}$ is the spatial position of the particle and the quantities α and β_{ij} have a simple interpretation as average quantities. The quantity α is the average displacement of a Brownian particle per unit time[4]; that is, its average velocity is

$$\alpha = \frac{\langle \Delta\mathbf{x}\rangle}{\Delta t}. \tag{2.45}$$

In the absence of external forces, it vanishes, but in the presence of a constant force $\mathbf{F}_0$ (e.g., gravity), the average velocity is equal to the sedimentation speed $\mathbf{v}_0 = \mathbf{F}_0/\tilde{\gamma}$ (see below Eq. (2.4)), resulting from the balance of the external force acting on the Brownian particle with the viscous friction force produced by the solvent.

The quantity β_{ij} amounts to the average squared displacements per unit time,

$$\beta_{ij} = \frac{\langle \Delta x_i \Delta x_j\rangle}{\Delta t}. \tag{2.46}$$

In a homogeneous, isotropic, medium, it is diagonal and proportional to the diffusion constant D,

$$\beta_{ij} = 2\delta_{ij}D. \tag{2.47}$$

For the Brownian particle, we finally obtain

$$\frac{\partial p(\mathbf{x},t)}{\partial t} = -\mathbf{v}_0 \cdot \nabla p(\mathbf{x},t) + D\nabla^2 p(\mathbf{x},t). \tag{2.48}$$

In the absence of external forces, the viscous term can be neglected and Eq. (2.48) reduces to the diffusion equation,

$$\frac{\partial p(\mathbf{x},t)}{\partial t} = D\nabla^2 p(\mathbf{x},t), \tag{2.49}$$

---

[4] Note that W is a transition rate, that is, a probability per unit time and also per unit volume.

whose solution is given, for $d = 1$, by Eq. (M.11),

$$p(x, t) = \frac{1}{\sqrt{4\pi Dt}} \exp\left(-\left[\frac{(x - x_0)^2}{4Dt}\right]\right). \tag{2.50}$$

The average squared displacement is given by the variance of the Gaussian distribution, that is,

$$\langle x^2 \rangle - \langle x \rangle^2 = 2Dt. \tag{2.51}$$

Comparing with Eq. (2.35), we can conclude that the Fokker–Planck approach predicts the same asymptotic diffusive behavior as the Langevin approach.

2.3 Discrete Time Stochastic Processes

In this section, we want to introduce the reader to a general mathematical formulation of stochastic processes by short memory probabilistic evolution rules, known as Markov chains, where both time and state-space are discretized. Due to its simplicity, this approach is particularly useful for illustrating the physical interest in stochastic processes, beyond the problem of Brownian motion. Moreover, Markov chains allow us to establish a series of basic concepts, which also apply to the more refined approaches contained in Section 2.4, where we describe stochastic processes in continuous space and time.

2.3.1 Markov Chains

Markov chains are stochastic processes discrete in time and in the state-space, where the value assumed by each stochastic variable depends on the value taken by the same variable at the previous instant of time. Let us translate these concepts into a mathematical language. We assume that the stochastic variable $x(t)$ takes values at each instant of time t over a set of N states, $S = \{s_1, s_2, \ldots, s_{N-1}, s_N\}$. For the sake of simplicity, we also assume that $x(t)$ is measured at equal finite time intervals, so that time also becomes a discrete variable, equivalent, in some arbitrary time unit, to the ordered sequence of natural numbers, $t = 1, 2, \ldots, n, \ldots$. The basic quantity we want to deal with is the probability, $p(x(t) = s_i)$, that $x(t)$ is in state s_i at time t. If $p(x(t) = s_i)$ does not depend on the previous history of the stochastic process, we are dealing with the simple case of a sequence of independent events, like tossing a coin. In general, one could expect that, if some time correlation (i.e., memory) is present in the evolution of $x(t)$, then $p(x(t) = s_i)$ could depend also on the previous history of the stochastic process. In this case it is useful to introduce the conditional probability

$$\Omega(x(t) = s_i \mid x(t-1) = s_{i_1}, x(t-2) = s_{i_2}, \ldots, x(t-n) = s_{i_n}) \tag{2.52}$$

that $x(t)$ is in the state s_i at time t, given the evolution of the stochastic process backward in time up to $t - n$. In this case we say that the stochastic process has memory n: The case $n = 1$ defines a Markov process, where

$$\Omega(x(t) = s_j \mid x(t-1) = s_i) \tag{2.53}$$

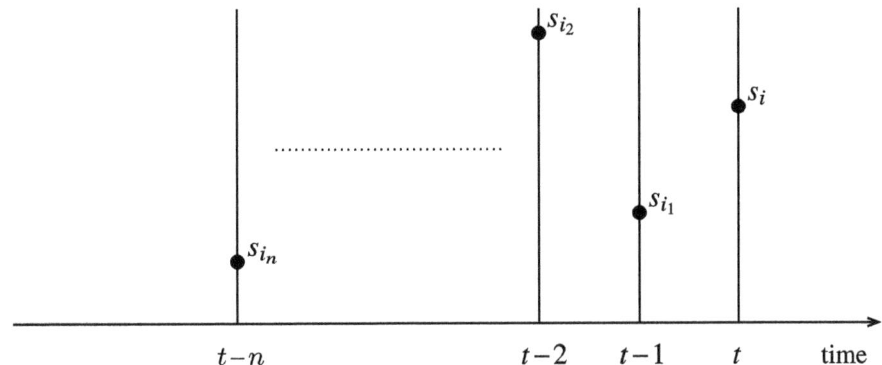

Fig. 2.4 The state-space at time t is represented by a vertical axis crossing the time axis in t. In the most general case, the probability to be in s_i at time t depends on the previous history at times $t-1, t-2, \ldots, t-n$.

is the transition probability in a unit time step from s_i to s_j; see Fig. 2.4.[5]

In general, $\Omega(\,x(t) = s_j \,|\, x(t-1) = s_i\,)$ is a function of time, but here we limit ourselves to consider stationary Markov processes, where this transition rate is independent of time. In this case we can deal with many interesting problems that can be discussed and solved with basic mathematical ingredients. The time-dependent case is certainly more general and of major interest, but even a short introduction would be hardly accessible for the audience to which this textbook is addressed.

Making use of the shorthand notations $p_i(t) \equiv p(\,x(t) = s_i\,)$ and $W_{ji} \equiv \Omega(\,x(t+1) = s_j \,|\, x(t) = s_i\,)$, these quantities must obey the following conditions:

$$p_i(t) \geq 0, \qquad \forall i, t \tag{2.54}$$

$$\sum_i p_i(t) = 1, \qquad \forall t \tag{2.55}$$

$$W_{ij} \geq 0, \qquad \forall i, j \tag{2.56}$$

$$\sum_i W_{ij} = 1, \qquad \forall j . \tag{2.57}$$

We can define the stochastic dynamical rule of the Markov chain as

$$p_j(t+1) = \sum_i W_{ji} p_i(t), \tag{2.58}$$

---

[5] An example of a non-Markovian process is a self-avoiding random walk. We can think of a random walker moving at each time step between nearby sites of a regular square lattice, choosing with equal probability any available direction. On the other hand, the walker cannot move to any site already visited in the past. The time evolution of the random process modifies the probability of future events, or said differently, the walker experiences a persistent memory effect, that typically yields long-time correlations. Such a situation is peculiar to many interesting phenomena concerning other domains of science, such as sociology or economics. For instance, the price X of a stock at time t is usually more properly represented as an outcome of a non-Markovian process, where this quantity depends on the overall economic activity $a(t)$ up to time t, that is, $X(a(t))$ rather than simply $X(t)$.

so one can easily realize that the W_{ij} can be viewed as the entries of an $N \times N$ matrix W, called stochastic matrix. Accordingly, Eq. (2.58) can be rewritten in vector form as

$$\mathbf{p}(t + 1) = W\mathbf{p}(t), \tag{2.59}$$

where $\mathbf{p}(t) = (p_1(t), p_2(t), \ldots, p_j(t), \ldots, p_N(t))$ is the column vector of the probability. This matrix relation can be generalized to obtain

$$\mathbf{p}(t + n) = W^n \mathbf{p}(t), \tag{2.60}$$

where W^n, the nth power of W, is also a stochastic matrix, since it satisfies the same properties (2.56) and (2.57) of W. This is easily proved by induction, assuming W^n is a stochastic matrix and showing that W^{n+1} is also stochastic. In fact,

$$(W^{n+1})_{ij} = \sum_k (W^n)_{ik} W_{kj} \geq 0, \qquad \forall i, j \tag{2.61}$$

because each term of the sum is the product of non-negative quantities, and

$$\sum_i (W^{n+1})_{ij} = \sum_{i,k} (W^n)_{ik} W_{kj} \tag{2.62}$$

$$= \sum_k \left(\sum_i (W^n)_{ik} \right) W_{kj} \tag{2.63}$$

$$= \sum_k W_{kj} \tag{2.64}$$

$$= 1. \tag{2.65}$$

The matrix relation (2.60) also leads to another important relation concerning the stochastic matrix, the Chapman–Kolmogorov equation,

$$p(t + n) = W^n p(t) = W^n W^t p(0) = W^{t+n} p(0). \tag{2.66}$$

It extends to stochastic processes the law valid for deterministic dynamical systems, where the evolution operator from time 0 to time $(t + n)$, $\mathcal{L}^{t+n}$, can be written as the composition of the evolution operator from time 0 to time t with the evolution operator from time t to time $t + n$, namely $\mathcal{L}^{t+n} = \mathcal{L}^n \circ \mathcal{L}^t$.

As usual for any $N \times N$ matrix, it is useful to solve the eigenvalue problem

$$\det (W - \lambda I) = 0,$$

where I is the identity matrix and λ is a scalar quantity, whose values solving the eigenvalue equation are called the spectrum of W.

Since W is not a symmetric matrix, its eigenvalues are not necessarily real numbers.[6] Moreover, one should distinguish between right and left eigenvectors of W. We denote the right ones as $\mathbf{w}^{(\lambda)} = (w_1^{(\lambda)}, w_2^{(\lambda)}, \ldots, w_j^{(\lambda)}, \ldots, w_N^{(\lambda)})$, so that we can write

$$W \, \mathbf{w}^{(\lambda)} = \lambda \, \mathbf{w}^{(\lambda)}. \tag{2.67}$$

The spectrum of W has the following properties, whose proof is given in Appendix F:

---

[6] Notice that the asymmetry of W implies that, in general, $\sum_j W_{ij} \neq 1$, at variance with Eq. (2.57).

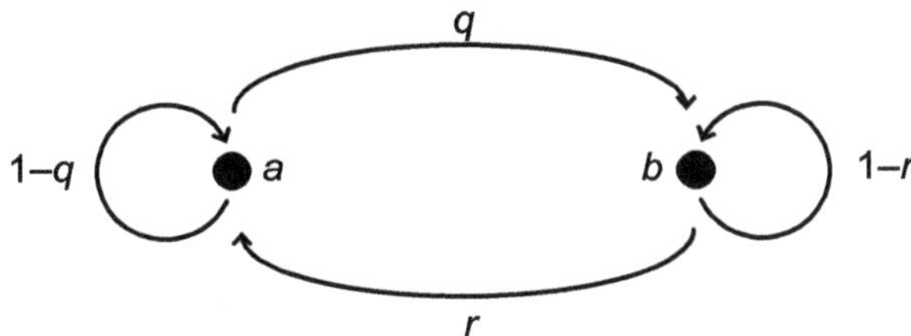

Fig. 2.5 Allowed transitions for a two-state Markov chain.

(a) $|\lambda| \le 1$;
(b) there is at least one eigenvalue $\lambda = 1$;
(c) $\mathbf{w}^{(\lambda)}$ is either an eigenvector with eigenvalue 1, or it fulfills the condition $\sum_j w_j^{(\lambda)} = 0$.

We conclude this section by introducing some definitions:

Definition 2.1 A state s_j is accessible from a state s_i if there is a finite value of time t such that $(W^t)_{ji} > 0$.

Definition 2.2 A state s_j is persistent if the probability of returning to s_j after some finite time t is 1, while it is transient if there is a finite probability of never returning to s_j for any finite time t. As a consequence of these definitions, a persistent state will be visited infinitely many times, while a transient state will be discarded by the evolution after a sufficiently long time.

Definition 2.3 A Markov chain is irreducible when all the states are accessible from any other state.

Definition 2.4 A Markov chain is periodic when the return times T_j on a state s_j are all (integer) multiples of a period T, that is, $(W^T)_{jj} > 0$.

2.3.2 Basic Examples of Markov Chains

The Two-State Case

We consider a Markov chain made of two states, $S = \{a, b\}$. The stochastic matrix has the form

$$W = \begin{pmatrix} 1 - q & r \\ q & 1 - r \end{pmatrix},$$
(2.68)

because condition (2.57) implies that the sum of elements of each column is equal to one. Accordingly, q is the probability rate per unit time of passing from a to b and $(1 - q)$ of remaining in a, while r is the probability rate per unit time of passing from b to a and $(1 - r)$ of remaining in b (see Fig. 2.5).

Let us denote with $p_a(t)$ the probability of observing the system in state a at time t: Relation (2.55) yields $p_b(t) = 1 - p_a(t)$ at any t. By applying the stochastic evolution rule

(2.58), one obtains the equation

$$p_a(t+1) = (1-q)p_a(t) + r(1-p_a(t)) = r + (1-r-q)p_a(t). \tag{2.69}$$

This equation also implies the similar equation for $p_b(t)$, which can be obtained from Eq. (2.69) by exchanging $p_a(t)$ with $p_b(t)$ and r with q. By simple algebra,[7] one can check that the explicit solution of Eq. (2.69) is

$$p_a(t) = \alpha + (1-r-q)^t \left(p_a(0) - \alpha\right), \qquad \alpha = \frac{r}{r+q}, \tag{2.70}$$

where $p_a(0)$ is the initial condition, that is, the probability of observing the state a at time 0.

There are two limiting cases: (i) $r = q = 0$, no dynamics occurs; (ii) $r = q = 1$, dynamics oscillates forever between state a and state b. In all other cases, $|1-r-q| < 1$ and in the limit $t \to \infty$, $p_a \to \alpha$ and $p_b \to (1-\alpha)$. More precisely, $p_a(t)$ converges exponentially fast to α with rate $\tau = -1/\ln|r+q-1|$, that is, the dynamics approaches exponentially fast a stationary state.[8] This simple Markov chain is irreducible and its states are accessible and persistent.

Notice that also the following relation holds

$$W_{ba}p_a(\infty) = W_{ab}p_b(\infty). \tag{2.71}$$

As we are going to discuss in more generality in Section 2.3.5, Eq. (2.71) is the detailed balance condition that establishes a sort of time reversibility of the stochastic process. Actually, Eq. (2.71) tells us that, in the stationary state, the probability of being in state a and passing in a unit time step to state b is equal to the probability of being in state b and passing in a unit time step to state a.

Random Walk on a Ring

The state-space is the collection of the nodes of a one-dimensional lattice, that is, $S = \{1, 2, \ldots, i, \ldots, N\}$, with periodic boundary conditions: A random walker moves along the ring by jumping in a unit time step from site i to site $i+1$ with rate r or to site $i-1$ with rate $1-r$, for any i (see Fig. 2.6). This Markov chain is described by the $N \times N$ tridiagonal stochastic matrix

$$W = \begin{pmatrix} 0 & 1-r & 0 & 0 & 0 & \cdots & 0 & 0 & r \\ r & 0 & 1-r & 0 & 0 & \cdots & 0 & 0 & 0 \\ 0 & r & 0 & 1-r & 0 & \cdots & 0 & 0 & 0 \\ 0 & 0 & r & 0 & 1-r & \cdots & 0 & 0 & 0 \\ \vdots & \vdots & \vdots & \vdots & \vdots & \ddots & \vdots & \vdots & \vdots \\ 1-r & 0 & 0 & 0 & 0 & \cdots & 0 & r & 0 \end{pmatrix} \tag{2.72}$$

---

[7] If we start iterating Eq. (2.69), we find $p_a(1) = r + \beta p_a(0)$, $p_a(2) = r(1+\beta) + \beta^2 p_a(0)$, $p_a(3) = r(1+\beta+\beta^2) + \beta^3 p_a(0), \ldots$, with $\beta = 1-r-q$. We can make the ansatz $p_a(t) = r\sum_{\tau=0}^{t-1}\beta^\tau + \beta^t p_a(0)$. Evaluating the summation explicitly, which is equal to $(1-\beta^t)/(1-\beta)$, we find the expression given in Eq. (2.70).

[8] We are allowed to speak about a stationary state, rather than an asymptotic evolution, because, after a time $t \gg \tau$, we make an exponentially small error in approximating $p_a(t)$ with α.

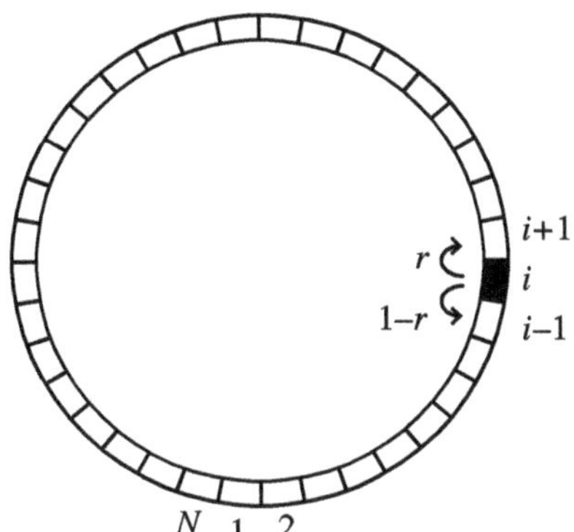

Fig. 2.6 Anisotropic diffusion on a ring, that is, a linear segment of N sites with periodic boundary conditions.

and Eq. (2.58) reads

$$p_i(t+1) = (1-r)p_{i+1}(t) + rp_{i-1}(t). \tag{2.73}$$

If a stationary state can be eventually attained, all probabilities should be independent of time and Eq. (2.73) simplifies to

$$p_i = (1-r)p_{i+1} + rp_{i-1}, \tag{2.74}$$

where we write p_i rather than $p_i(\infty)$. A straightforward solution of this equation is $p_i = $ const, independently of i. Due to Eq. (2.55), we finally obtain $p_i = 1/N$. This result could be conjectured also on the basis of intuitive arguments: In the long run the random walker will have lost any memory of its initial state and, since due to the lattice symmetry all sites are equivalent, the stationary state will correspond to an equal probability of visiting any site. As in the previous example, the Markov chain is irreducible and all its states are accessible and persistent.

The spectrum of W provides us with more detailed mathematical information about this problem. Eq. (2.67) reads

$$(1-r)w_{k+1} + rw_{k-1} = \lambda w_k, \tag{2.75}$$

which has the solution $w_k = \omega^k$, where ω should be determined by imposing a periodic boundary condition, which requires $w_{N+1} = w_1$, that is, $\omega^N = 1$. We therefore have N independent solutions,

$$\omega_j = \exp\left(\frac{2\pi i}{N}j\right), \qquad j = 0, 1, \ldots, N-1. \tag{2.76}$$

The corresponding eigenvalues λ_j are determined by Eq. (2.75), which rewrites

$$\lambda_j = (1-r)\omega_j + r\omega_j^{-1} = \cos\left(\frac{2\pi}{N}j\right) + i(1-2r)\sin\left(\frac{2\pi}{N}j\right), \quad j = 0, 1, \ldots, N-1. \tag{2.77}$$

Notice that for $r = 1/2$, the eigenvalues λ_j are all real, at variance with the case $r \neq 1/2$, where, apart $\lambda_0 = 1$, they are complex.

The corresponding eigenvectors are

$$\mathbf{w}^{(\lambda_j)} = \frac{1}{\sqrt{N}}(1, \omega_j, \omega_j^2, \ldots, \omega_j^{N-1})^T, \tag{2.78}$$

where we have introduced the normalization factor, $1/\sqrt{N}$, such that $\mathbf{w}^T\mathbf{w} = 1$. For $j = 0$, we have $\lambda_0 = 1$, independently of r, thus showing that the stationary solution is the same for the asymmetric and symmetric random walk on a ring. On the other hand, for $r \neq 1/2$, there is a bias for the walker to move forward ($r > 1/2$) or backward ($r < 1/2$). In fact, the walker moves with average velocity $v = r - (1 - r) = 2r - 1$, and the time it takes to visit all lattice sites is $O(N/v)$.[9]

Making reference to the basic model discussed in Section 2.2.1, we can guess that for $r = 1/2$ the walker performs a diffusive motion, and the time it takes to visit all lattice sites is $O(N^2)$. In this symmetric case, some additional considerations are in order. In fact, if N is an even number and at $t = 0$ the random walker is located at some even site $2j$, that is, $p_{2j}(0) = 1$, at any odd time step, t_o, $p_{2k}(t_o) = 0$, while at any even time step, t_e, $p_{2k+1}(t_e) = 0$. A similar statement holds if the random walker at $t = 0$ is on an odd site, by exchanging odd with even times and vice versa. This does not occur if N is odd and in this case, $\lim_{t \to +\infty} p_i(t) = 1/N$. Conversely, if N is even, this limit does not exist, but the following limit exists:

$$\lim_{t \to +\infty} \frac{1}{t} \sum_{\tau=1}^{t} p_i(\tau) = \frac{1}{N}. \tag{2.79}$$

This is the ergodic average of $p_i(t)$, which, by this definition, recovers the expected value of the stationary probability. As a final remark, we want to observe that for even N and $r = 1/2$, we have $\lambda_{N/2} = -1$, whose eigenvector is made by an alternation of 1 and -1 (see Eq. (2.78)). The existence of this peculiar eigenvalue and eigenvector can be related to the nonexistence of the limit $\lim_{t \to +\infty} p_i(t) = 1/N$.

2.3.3 Random Walk with Absorbing Barriers

The problem of the random walk with absorbing barriers has been widely investigated, because of its relevance for game theory and in several diffusion problems occurring in material physics. The random walker has an absorbing barrier if there is a state i such that in the corresponding Markov chain $W_{ii} = 1$ and, accordingly, $W_{ji} = 0$ for any $j \neq i$. We can sketch the problem as follows. A random walker moves on a lattice of N sites with fixed ends at states 1 and N, which are absorbing barriers (see Fig. 2.7). The stochastic dynamics on the lattice, apart at sites 1 and N, is the same as that of the random walker on a ring discussed in the previous example. The corresponding stochastic matrix has the same form of Eq. (2.72), apart from the first and the last columns, which change into $(1, 0, \ldots, 0, 0)$

---

[9] This is true for v fixed and diverging N, because for N fixed, the ballistic time N/v should be compared with the diffusive time, of order N^2.

and $(0, 0, \ldots, 0, 1)$, respectively, thus yielding

$$
W = \begin{pmatrix}
1 & 0 & 0 & 0 & 0 & \cdots & 0 & 0 & 0 \\
0 & 0 & 1-r & 0 & 0 & \cdots & 0 & 0 & 0 \\
0 & r & 0 & 1-r & 0 & \cdots & 0 & 0 & 0 \\
0 & 0 & r & 0 & 1-r & \cdots & 0 & 0 & 0 \\
\vdots & \vdots & \vdots & \vdots & \vdots & \ddots & \vdots & \vdots & \vdots \\
0 & 0 & 0 & 0 & 0 & \cdots & 0 & 0 & 1
\end{pmatrix}. \tag{2.80}
$$

Let us assume that at time 0 the walker is at site j $(1 < j < N)$: Due to the structure of the stochastic matrix, we can state that the walker will be eventually absorbed (with probability 1) either at 1 or N. We can conclude that the Markov chain is not irreducible; apart from the persistent states 1 and N, all the other states are transient.

We can also ask the more interesting question: What is the probability that a random walker starting at j will reach 1 without being first absorbed in N? In order to answer this question, we can introduce the time-independent conditional probability of being trapped at 1 starting at time 0 from site j, p_j. It obeys the equation ($j = 2, \ldots, N - 1$)

$$
\begin{aligned}
\mathrm{p}_j &= W_{j+1,\, j}\ \mathrm{p}_{j+1} + W_{j-1,\, j}\ \mathrm{p}_{j-1} \\
&= r\ \mathrm{p}_{j+1} + (1 - r)\ \mathrm{p}_{j-1},
\end{aligned} \tag{2.81}
$$

which exhibits the same formal structure of Eq. (2.74), but with boundary conditions $\mathrm{p}_1 = 1$ and $\mathrm{p}_N = 0$. The equality in (2.81) stems from the observation that a walker starting in j before being absorbed in 1 should visit either the site $(j - 1)$ or the site $(j + 1)$ and after one of these two events (occurring with probabilities $W_{j\pm1,\, j}$), the walker must be absorbed in 1 rather than in N (which occurs with probability $\mathrm{p}_{j\pm1}$).

In order to solve Eq. (2.81), we cannot use the same solution of Eq. (2.74), because of the presence of the absorbing barriers that impose the boundary conditions $\mathrm{p}_1 = 1$ and $\mathrm{p}_N = 0$. However, we use the same ansatz $\mathrm{p}_j = \psi^j$. By substituting into Eq. (2.81), one obtains

$$
\psi^j = r\psi^{j+1} + (1 - r)\psi^{j-1}, \tag{2.82}
$$

or $1 = r\psi + (1 - r)\psi^{-1}$, whose solutions are $\psi_1 = 1$, that is, $\mathrm{p}_j^{(1)} = 1$ and $\psi_2 = (1 - r)/r \equiv s$, that is, $\mathrm{p}_j^{(2)} = s^j$. These equations coincide for $r = 1/2$, that is, $s = 1$, in which case (2.81) reads

$$
\mathrm{p}_j = \frac{1}{2}(\mathrm{p}_{j+1} + \mathrm{p}_{j-1}). \tag{2.83}
$$

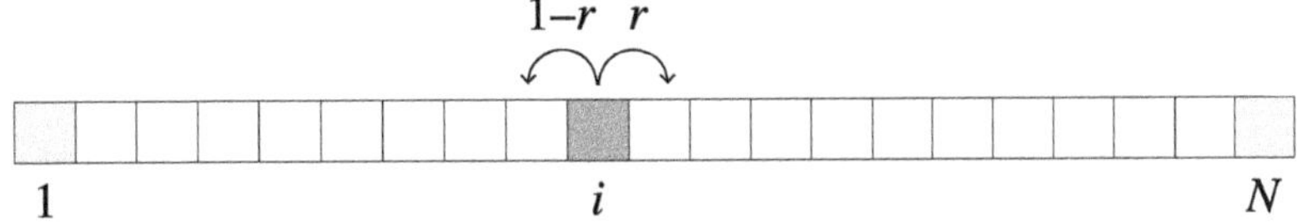

Fig. 2.7 Anisotropic diffusion on a segment with absorbing boundary conditions in $i = 1$ and $i = N$. In the main text, we also use the quantity $s = (1 - r)/r$ as a parameter to characterize the asymmetry.

The solution $p_j^{(1)} = 1$ still holds, but the second solution now reads $p_j^{(2)} = j$, as one can immediately check.

Since Eq. (2.81) is a finite-difference linear equation in the variable p_j, its general solution has to be a linear combination of $p_j^{(1)}$ and $p_j^{(2)}$. For $r \neq 1/2$, we have

$$p_j = As^j + B \tag{2.84}$$

with A and B constants to be determined by the boundary conditions,

$$p_1 = As + B = 1 \tag{2.85}$$

$$p_N = As^N + B = 0, \tag{2.86}$$

whose solution is

$$A = \frac{1}{s(1 - s^{N-1})}, \qquad B = -\frac{s^{N-1}}{(1 - s^{N-1})}. \tag{2.87}$$

Replacing in Eq. (2.84), we obtain

$$p_j = \frac{s^{j-1} - s^{N-1}}{1 - s^{N-1}}. \tag{2.88}$$

For $r = 1/2$ $(s = 1)$, we have instead

$$p_j = Aj + B \tag{2.89}$$

and boundary conditions are $p_1 = A + B = 1$ and $p_N = AN + B = 0$, that is,

$$A = -\frac{1}{N - 1}, \qquad B = \frac{N}{N - 1}. \tag{2.90}$$

Replacing in Eq. (2.89), we obtain

$$p_j = \frac{N - j}{N - 1}. \tag{2.91}$$

Let us now consider the interesting limit $N \to \infty$. For $r > 1/2$ $(s < 1)$, one has $A = 1/s$, $B = 0$ and

$$p_j = s^{j-1}, \tag{2.92}$$

that is, the probability of reaching the absorbing barrier at 1 is exponentially small with j, because the random walker has a bias toward the other absorbing barrier at infinity. For $r < 1/2$, $A = 0$, $B = 1$, and $p_j = 1$, that is, the absorbing barrier at 1 will be eventually reached with probability 1, because the random walker has a bias to move toward 1. The same conclusion can be drawn for $r = 1/2$, because, even if this case is unbiased, any site j is at a finite distance from the absorbing barrier at 1. Therefore, we can conclude

$$p_j = 1, \qquad r \leq \frac{1}{2} \quad \text{and} \quad N \to \infty. \tag{2.93}$$

This stochastic process describes a basic problem of game theory. Imagine that you are a player with an initial capital j playing a unit of your capital on "red" or "black" for each run at a roulette table. You would like to know what the probability is that you will eventually reach the situation of being left with one single unit of capital (the absorbing barrier at 1) before reaching a capital $N > j$, which may convince you to stop playing (perhaps because your profit is adequate to your expectation). What we have discussed before allows you to conclude that if the game is biased against the player, that is, you are in the case $r < 1/2$, you have a probability quite close to 1 of being eventually ruined. For instance, we can compute the probability $\mathfrak{p}_j$, with $j = 500$ units of capital, of eventually reaching a capital of a single unit, before having doubled your capital. The unfavorable bias seems to be weak, because the probability of losing on a single bet (19/37) is just greater than the probability of winning (18/37).[10] Despite this, starting with 500 tokens, the probability $\mathfrak{p}_{500}$ to be left with only one token before arriving at double the initial capital is $\mathfrak{p}_{500} \approx 1 - O(10^{-12})$. You can realize that this strategy of playing roulette is certainly time-consuming, but practically will lead you to ruin.

2.3.4 Ergodic Markov Chains

An important class of Markov chains is the ergodic ones. Let us consider a Markov chain with a finite state-space, that is, $S = \{s_1, s_2, \ldots, s_N\}$: It is ergodic if it is irreducible, nonperiodic, and all states are persistent (see Definitions 2.1–2.4 at the end of Section 2.3.1). The main property of ergodic Markov chains is that they determine a unique invariant, that is, stationary, probability distribution. This is given by the eigenvector $\mathbf{w}^{(1)}$, which is the solution of the eigenvalue problem with $\lambda = 1$:

$$W\mathbf{w}^{(1)} = \mathbf{w}^{(1)}. \tag{2.94}$$

The spectral property (b), proved in Appendix F, guarantees that such an eigenvector exists. For an ergodic Markov chain, it is also unique, because each state is accessible (i.e., the matrix is irreducible) and persistent, that is, it will be revisited with probability 1 after a finite lapse of time. If these properties do not hold, in general one is faced with several peculiar scenarios, for example, the reduction of the stochastic matrix into blocks, whose number equals the degeneracy of the eigenvalue $\lambda = 1$.

The stationary probability, $\mathbf{w}^{(1)}$, of an ergodic matrix will be eventually attained exponentially fast on a time scale, τ, independent of the initial conditions, namely

$$\mathbf{p}(t) \approx \mathbf{w}^{(1)} + \mathbf{A}e^{-t/\tau}, \tag{2.95}$$

where $\mathbf{A}$ is a suitable vector with constant components that sum up to zero to fulfill condition (2.55). In the limit $t \to \infty$, $\mathbf{w}^{(1)}$ is a true dynamical state of the stochastic process, and, accordingly, it has to obey conditions (2.54) and (2.55), that is, all its components are non-negative and it is normalized.

Since all states of an ergodic Markov chain are persistent, we would like to know the typical return time $\langle T_j \rangle$ to j (i.e., the average value of the return time T_j in the limit $t \to \infty$). The answer to this question is given by the Kac lemma, according to which

$$\langle T_j \rangle = \frac{1}{w_j^{(1)}}. \tag{2.96}$$

---

[10] The zero of the roulette determines the bias favorable to the casino.

A simple argument to explain this result is the following. Let us denote with $T_j^{(n)}$, with $n = 1, 2, \ldots$, the nth return time to j. The total time $\mathcal{T}_M$ needed for the ergodic Markov chain to come back M times to j is given by

$$\mathcal{T}_M = \sum_{n=1}^{M} T_j^{(n)}. \tag{2.97}$$

On the other hand, this means that the fraction of time $\phi_j(\mathcal{T}_M)$ spent by the stochastic process at j in the time $\mathcal{T}_M$ is given by

$$\phi_j(\mathcal{T}_M) = \frac{M}{\mathcal{T}_M}. \tag{2.98}$$

According to the interpretation of $\mathbf{w}^{(1)}$, $\phi_j(\mathcal{T}_M)$ has to converge to $w_j^{(1)}$ in the limit $M \to \infty$ (which is equivalent to the limit $t \to \infty$), and one can write

$$\langle T_j \rangle = \lim_{M \to \infty} \frac{1}{M} \, \mathcal{T}_M = \lim_{M \to \infty} \frac{1}{\phi_j(\mathcal{T}_M)} = \frac{1}{w_j^{(1)}}. \tag{2.99}$$

This relation points out that ergodicity amounts to the equivalence between ensemble and time averages.

Another important result is that the spectral properties of an ergodic Markov chain determine the time scale of convergence to the stationary probability. In fact, according to the results of Appendix F, the eigenvalues $\lambda^{(j)}$ of the stochastic matrix representing an ergodic Markov chain can be ordered as

$$\lambda^{(1)} = 1 > |\lambda^{(2)}| \geq |\lambda^{(3)}| \geq \cdots \geq |\lambda^{(N)}|. \tag{2.100}$$

Let us come back to Eq. (2.95) to explain its origin. According to the projection theorem of linear algebra, any probability on the state-space at time t, $\mathbf{p}(t)$, can be written as a suitable linear combination of the eigenvectors $\mathbf{w}^{(k)}$ of the stochastic matrix, which form a complete basis:

$$\mathbf{p}(t) = \sum_{k=1}^{N} a_k(t)\mathbf{w}^{(k)}, \qquad a_j(t) \in \mathbb{R}. \tag{2.101}$$

Let us consider the evolution of the ergodic Markov chain from time 0 to time t,

$$\mathbf{p}(t) = W^t \mathbf{p}(0) = W^t \sum_{k=1}^{N} a_k(0)\mathbf{w}^{(k)} = \sum_{k=1}^{N} a_k(0)(\lambda^{(k)})^t \mathbf{w}^{(k)} \equiv \sum_{k=1}^{N} a_k(t)\mathbf{w}^{(k)}. \tag{2.102}$$

Apart from $a_1(t)$, which does not change in time, because $\lambda^{(1)} = 1$, all the other coefficients evolve in time as

$$a_k(t) = a_k(0)(\lambda^{(k)})^t = (\pm)^t a_k(0)e^{-t/\tau_k}, \tag{2.103}$$

where $(\pm)$ is the sign of the eigenvalue $\lambda^{(k)}$ and

$$\tau_k = -\frac{1}{\ln |\lambda^{(k)}|}. \tag{2.104}$$

Therefore, for $k > 1$, $a_k(t)$ eventually vanish exponentially with rate τ_k.

Making use of property (2.100), we can conclude that the overall process of relaxation to equilibrium from a generic initial condition is dominated by the longest time scale, that is, the one corresponding to the eigenvalue $\lambda^{(2)}$, so that in Eq. (2.95) we have

$$\tau = \tau_2 = -\frac{1}{\ln|\lambda^{(2)}|}. \tag{2.105}$$

2.3.5 Master Equation and Detailed Balance

The dynamical rule of the Markov chain, Eq. (2.58), can be rewritten as

$$p_i(t+1) = p_i(t) - p_i(t) + \sum_j p_j(t)W_{ij} = p_i(t) - p_i(t)\sum_j W_{ji} + \sum_j p_j(t)W_{ij}, \tag{2.106}$$

where we have used condition (2.57). The previous formula can be recast in the form of a master equation,

$$p_i(t+1) - p_i(t) = \sum_j \left(W_{ij}p_j(t) - W_{ji}p_i(t) \right). \tag{2.107}$$

This equation tells us that the variation of the probability of being in state s_i in a unit time step can be obtained from the positive contribution of all transition processes from any state s_j to state s_i and from the negative contribution of all transition processes from state s_i to any other state s_j.

This form is particularly useful to define the conditions under which one can obtain a stationary probability, that is, all p_i are independent of time t: In this case the left-hand side of (2.107) vanishes and the stationarity condition reads

$$\sum_j \left(W_{ij}p_j - W_{ji}p_i \right) = 0, \quad \forall i, \tag{2.108}$$

where the p_js are the components of the stationary probability.

Notice that Eq. (2.108) is verified if the following stronger condition holds

$$W_{ij}p_j - W_{ji}p_i = 0, \quad \forall i, j, \tag{2.109}$$

which is called the detailed balance condition. A Markov chain whose stochastic matrix elements obey Eq. (2.109) is said to be reversible and it can be shown that it is also ergodic (see Appendix G), with $\mathbf{p} \doteq \mathbf{w}^{(1)}$ representing the so-called equilibrium probability.[11]

We have already discussed in Section 2.3.2 the simple example of a two-state Markov chain that trivially obeys detailed balance. As one can immediately realize, detailed balance does not hold in the asymmetric ($r \neq 1/2$) random walk on a ring, because the stationary probability is $w_j^{(1)} = 1/N$ for all j, while $r = W_{j+1,j} \neq W_{j,j+1} = 1 - r$. In this example we are faced with an ergodic Markov chain that is not reversible. Only if the symmetry of the

---

[11] In the case of reversible Markov chains, the stationary probability is more properly called equilibrium probability, because it is equivalent to the concept of thermodynamic equilibrium in statistical mechanics, where the equilibrium probability is determined by the Hamiltonian functional, which engenders a time-reversible dynamics through the Hamilton equations (see Eqs. (4.27a) and (4.27b)).

process is restored (i.e., $r = 1/2$) does the detailed balance condition hold and the invariant probability is an equilibrium one, although it is the same for the asymmetric case.

It is important to point out that the detailed balance condition can be used as a benchmark for the absence of an equilibrium probability in an ergodic irreversible Markov chain. Actually, in this case it is enough to find at least a pair of states for which Eq. (2.109) does not hold to conclude that the stationary probability $\mathbf{w}^{(1)}$ is not the equilibrium one. Many of the examples discussed in Section 2.3.2 belong to the class of stochastic processes that evolve to a stationary nonequilibrium probability. The difference between equilibrium and NESSs will be reconsidered in Section 2.4.2.

2.3.6 Monte Carlo Method

The Monte Carlo method is one of the most useful and widely employed applications of stochastic processes. The method aims at solving the problem of the effective statistical sampling of suitable observables by a reversible Markov chain.

More precisely, the problem we are discussing here is the following: how to estimate the equilibrium average of an observable O making use of a suitable, reversible Markov process. As in the rest of Section 2.3, we assume that the state-space is made by N microscopic states, $S = \{s_1, s_2, \ldots, s_{N-1}, s_N\}$, whose equilibrium probabilities are $(p_1, p_2, \ldots, p_N)$. We can write

$$\langle O \rangle_{eq} = \sum_{j=1}^{N} O_j \, p_j \tag{2.110}$$

where $O_j \doteq O(s_j)$ is the value taken by O in state s_j. The problem of computing such an equilibrium average emerges when the number of states N is exceedingly large and we want to avoid sampling O over equilibrium states, whose probabilities may be very small. This is the typical situation of many models of equilibrium statistical mechanics. For instance, the Ising model (see Section 5.2.2), originally introduced to describe the ferromagnetic phase transition, is made by locally interacting binary spin variables, $\sigma = \pm 1$, located at the nodes of a lattice. If the lattice contains L nodes, the total number of states is $N = 2^L$: Already for $L = 100$ one has $N \approx 10^{30}$, quite close to an astronomical number, which makes the computation of (2.110) practically unfeasible.[12]

In order to work out the Monte Carlo strategy, we have to assume from the very beginning that we know the equilibrium probabilities $(p_1, p_2, \ldots, p_N)$. In equilibrium statistical mechanics they are given by the Gibbs weight

$$p_i = \frac{e^{-\beta E_i}}{Z}, \tag{2.111}$$

where $\beta = T^{-1}$ is the so-called inverse reduced temperature, E_i is the energy of state s_i, and $Z = \sum_{i=1}^{N} e^{-\beta E_i}$ is the partition function. One could argue that there is something contradictory in the previous statement: In order to know p_i, one has to also compute Z,

---

[12] The value $L = 100$ is a very small number: in $d = 3$, a cube of only five nodes per side has more than 100 nodes!

which again is a sum over an astronomical number of states.[13] As we are going to show, the Monte Carlo procedure needs only local information, that is, the ratios of the equilibrium probabilities, $p_i/p_j = e^{-\beta(E_i - E_j)}$, which do not depend on Z.

The following step is how to define a reversible Markov chain that has the p_is as equilibrium probabilities. As we pointed out, a reversible Markov chain is also ergodic, that is, $\langle O \rangle_{eq}$ can be approximated by a sufficiently long trajectory $(s_1, s_2, \ldots, s_n)$ in the state-space of the Markov chain:

$$\sum_{j=1}^{N} O_j p_j \approx \frac{1}{n} \sum_{t=1}^{n} O(t). \tag{2.112}$$

The Monte Carlo procedure is effective, because we can obtain a good estimate of $\langle O \rangle_{eq}$ also if $n \ll N$. In fact, because of ergodicity, we know (see Section 2.3.4) that the fraction of time spent by the stochastic process at state s_i is p_i for time going to infinity: The states s_k that are preferably visited by the trajectory of a stochastic process of finite length n are typically those with $p_k > 1/N$. Accordingly, the Monte Carlo method selects automatically, after some transient time depending on the initial state, the states corresponding to higher values of the equilibrium probability.

The main point is to estimate how the quality of the approximation depends on n. In fact, any practical implementation of the Monte Carlo procedure into a numerical algorithm is subject to various problems. For instance, the rule adopted to construct the "trajectory" of the stochastic process typically maintains some correlations among the sampled values $O(1), O(2), \ldots$. In particular, we can estimate the correlation time of the Monte Carlo process by measuring how the time autocorrelation function of the observable $O(t)$ decays in time, namely $\langle O(t)O(0) \rangle - \langle O(t) \rangle^2 \sim \exp(-t/\tau)$. If the process lasts over a time n, we can assume that the number of statistically independent samples is of the order n/τ. According to the central limit theorem (see Section 3.2), the error we make in the approximation (2.112) is $O(\sqrt{\tau/n})$: In practice, the Monte Carlo procedure is effective only if $n \gg \tau$.

Now, let us describe how the reversible Markov chain we are looking for can be explicitly defined for an Ising-type spin model, according to the most popular algorithm, called Metropolis. The steps of the algorithm are as follows:

1. We select a suitable initial state.
2. We select at random with uniform probability $1/L$ a spin variable, say the one at node n_k, and compute its local interaction energy with the nearby spins, E_{n_k}.
3. We flip the spin variable, that is, $\sigma_{n_k} \rightarrow -\sigma_{n_k}$, and we compute its local interaction energy with the nearby spins in this new configuration, E'_{n_k}.
4. If $E'_{n_k} < E_{n_k}$, the next state in the Markov process is the flipped configuration.
5. If $E'_{n_k} \geq E_{n_k}$, we "accept" the new configuration with probability

$$p^* = \exp(-\beta(E'_{n_k} - E_{n_k})).$$

This means we extract a random number r, uniformly distributed in the interval $[0, 1]$: If $r < p^*$, the next state is the flipped configuration; otherwise, it is equal to the old state.

---

[13] In statistical mechanics, there are few exactly solvable models, for example, the Ising model in $d = 1, 2$, where Z can be computed analytically in the thermodynamic limit $N \to \infty$.

6. We iterate the process starting again from step 2.

In general terms, in equilibrium statistical mechanics, the detailed balance condition is satisfied if

$$W_{ji}e^{-\beta E_i} = W_{ij}e^{-\beta E_j}, \tag{2.113}$$

where we have made explicit the equilibrium probability of occupancy of a microscopic state, given by the Gibbs weight. Therefore, the transition rates must satisfy the relation

$$\frac{W_{ji}}{W_{ij}} = e^{-\beta(E_j - E_i)}. \tag{2.114}$$

In general, if the transition rates are a function of the energy difference between the initial and final state,

$$W_{ji} = \omega\big(\beta(E_j - E_i)\big), \tag{2.115}$$

detailed balance is satisfied if

$$\frac{\omega(x)}{\omega(-x)} = e^{-x}. \tag{2.116}$$

The Metropolis algorithm is defined by

$$\omega(x) = \min\{1, e^{-x}\}, \tag{2.117}$$

which satisfies Eq. (2.116). We can mention another algorithm that satisfies Eq. (2.116) and it is constrained by the further condition $W_{ij} + W_{ji} = 1$, that is,

$$\omega(x) + \omega(-x) = 1, \tag{2.118}$$

which allows to obtain

$$\omega(x) = \frac{1}{1 + e^x}. \tag{2.119}$$

This algorithm was introduced by the American Roy Glauber in the context of the Ising model.[14]

2.4 Continuous Time Stochastic Processes

In this section, we illustrate the formulation of a general theory of stochastic processes, evolving in continuous state-space and time. In fact, many physical processes are better represented in this framework. For instance, if we want to describe a fluid, it is simpler to attribute continuous position $\mathbf{x}(t)$ and momentum $\mathbf{p}(t)$ coordinates to each fluid element (i.e., an infinitesimal portion of the fluid), rather than considering the same quantities for each particle in the fluid. Moreover, a continuous time variable, on its side, allows us to take advantage of the powerful machinery of differential equations.

---

[14] You should not make confusion between the Glauber transition rate (2.119) and the Glauber dynamics, also named spin-flip dynamics, which is used to model the nonconserved kinetics of an Ising-type model; see Appendix H.

2.4.1 Continuous Time Master Equation

As a first step we derive the time-continuous version of the master equation (2.107), while keeping the discrete nature of the state-space. For this purpose we assume that the Chapman–Kolmogorov relation (2.66) can be extended to a continuous time t,[15] as follows:

$$W_{ij}^{t+\Delta t} = \sum_k W_{ik}^{\Delta t} W_{kj}^{t}. \tag{2.120}$$

If Δt is an infinitesimal increment of time, we can imagine that $W_{ik}^{\Delta t}$ vanishes with Δt if $i \neq k$ and that $W_{ii}^{\Delta t} \to 1$ in the same limit. We can therefore write, up to terms $O((\Delta t)^2)$,

$$W_{ik}^{\Delta t} = \begin{cases} \mathcal{R}_{ik}\Delta t, & k \neq i \\ 1 - \mathcal{R}_{ii}\Delta t, & k = i. \end{cases} \tag{2.121}$$

The rates $\mathcal{R}_{ik}$ are not independent, because the normalization condition (2.57),

$$1 = \sum_i W_{ik}^{\Delta t} = 1 + \Delta t \left(-\mathcal{R}_{kk} + \sum_{i \neq k} \mathcal{R}_{ik} \right), \tag{2.122}$$

implies that

$$\mathcal{R}_{kk} = \sum_{i \neq k} \mathcal{R}_{ik}. \tag{2.123}$$

By substituting Eq. (2.121) into Eq. (2.120) and making use of Eq. (2.123), we can write

$$W_{ij}^{t+\Delta t} - W_{ij}^{t} = \Delta t \left(\sum_{k \neq i} W_{kj}^{t}\mathcal{R}_{ik} - W_{ij}^{t}\mathcal{R}_{ii} \right)$$

$$= \Delta t \left(\sum_{k \neq i} W_{kj}^{t}\mathcal{R}_{ik} - \sum_{k \neq i} W_{ij}^{t}\mathcal{R}_{ki} \right). \tag{2.124}$$

By dividing both members by Δt and performing the limit $\Delta t \to 0$, we obtain the continuous time master equation

$$\frac{dW_{ij}^{t}}{dt} = \sum_{k \neq i} (W_{kj}^{t}\mathcal{R}_{ik} - W_{ij}^{t}\mathcal{R}_{ki}). \tag{2.125}$$

We can now write it in terms of the probability $p_i(t)$, the system is in state i at time t, using the relation between $p_i(t)$ and W_{ij}^{t},

$$p_i(t) = \sum_j W_{ij}^{t} p_j(0). \tag{2.126}$$

In fact, it is sufficient to derive Eq. (2.126) with respect to time,

$$\frac{dp_i(t)}{dt} = \sum_j \frac{d W_{ij}^{t}}{d t} p_j(0) \tag{2.127}$$

---

[15] With respect to the case of discrete time, here W^t is not the elementary stochastic matrix raised to the power t. Rather, W^t should be understood as a transition matrix depending on the continuous parameter t.

and replacing (2.125) here earlier, we obtain

$$\frac{dp_i(t)}{dt} = \sum_j \sum_{k \neq i} (\mathcal{R}_{ik} W_{kj}^t - \mathcal{R}_{ki} W_{ij}^t) p_j(0). \tag{2.128}$$

Finally, we can write the continuous time master equation

$$\frac{dp_i(t)}{dt} = \sum_{k \neq i} (\mathcal{R}_{ik} p_k(t) - \mathcal{R}_{ki} p_i(t)). \tag{2.129}$$

Similar to its time-discrete version (2.107), this equation tells us that the variation in time of the probability of being in state s_i is obtained from the positive contribution of all transition processes from any state s_k to state s_i and from the negative contribution of all transition processes from state s_i to any other state s_k.

2.4.2 Equilibrium States versus Stationary States

Both equilibrium states and NESS are time-independent, macroscopic states, whose difference is easily highlighted by the continuous time master equation, Eq. (2.129).

Making use of the same notation adopted in Chapter 5, where we have to distinguish phase transitions between equilibrium states and NESS, this equation can be rewritten as follows:

$$\frac{dp(s,t)}{dt} = \sum_{s'} \left[p(s',t) w_{s,s'} - p(s,t) w_{s',s} \right], \tag{2.130}$$

where $p(s,t)$ is the probability that the system is in the microstate s at time t, while $w_{s,s'}$ is the transition rate between microstate s' and microstate s.

In general, a stationary (or steady) state is a macrostate characterized by occupation probabilities p that are independent of time, $p(s,t) = p(s)$, which implies

$$\sum_{s'} \left[p(s') w_{s,s'} - p(s) w_{s',s} \right] = 0 \quad \forall s \qquad \Longleftrightarrow \qquad \text{steady state.} \tag{2.131}$$

This relation means that the net flow of probability between each microstate s and all other microstates vanishes. As discussed in Sec. 2.3.5, an equilibrium state satisfies the condition of detailed balance, which corresponds to separately vanishing each term of the earlier summation,

$$p^{\text{eq}}(s') w_{s,s'} = p^{\text{eq}}(s) w_{s',s} \quad \forall s, s' \qquad \Longleftrightarrow \qquad \text{equilibrium state;} \tag{2.132}$$

that is, it corresponds to vanishing the net flow of probability between each pair of states, s, s'.

A NESS satisfies Eq. (2.131), but it does not satisfy Eq. (2.132). The two equations differ, not only because (2.132) is a stronger condition than (2.131); they also have a different use. The ensemble theory of equilibrium statistical mechanics provides the expression of $p^{\text{eq}}(s)$. For instance, the occupation probability of each microstate s in a system described by the canonical ensemble is given by the Gibbs distribution,

$$p^{\text{eq}}(s) = \frac{e^{-\beta E(s)}}{Z} \qquad \text{(canonical ensemble),} \tag{2.133}$$

where Z, the partition function, is the normalization factor. Therefore, Eq. (2.132) can be used to impose a relation that must be satisfied by transition rates,

$$\frac{w_{s',s}}{w_{s,s'}} = e^{-\beta(E(s')-E(s))}. \tag{2.134}$$

We might say that $p^{\text{eq}}(s)$ is the building block of equilibrium statistical mechanics and $w_{s',s}$ are derived quantities (in such a way as to satisfy detailed balance). In nonequilibrium theory, there is no general principle such as the Boltzmann weight and we do not know a priori the probability distribution for the nonequilibrium steady states. Rather, the model is defined by assigning the rates $w_{s',s}$. In this sense, Eqs. (2.132) and (2.131) have different uses: For an equilibrium state, we know $p^{\text{eq}}(s)$ and Eq. (2.132) provide a relation that must be satisfied by the transition rates; for a NESS, we define the transitions $w_{s',s}$ and Eq. (2.131) is a relation that must be satisfied by the probabilities of state occupation.

Now imagine that someone provides us with a set of transition rates, $w^*_{s',s}$, asking if they would lead the system to an equilibrium state or not. The question is equivalent to asking if they satisfy detailed balance or not, but we do not know what (if any) is the Hamiltonian governing the system. In other words, can we test detailed balance from the plain knowledge of the transition rates? The answer is positive and it is proved in the following theorem (see also Fig. 2.8).

Theorem 2.5   The transition rates $w_{s',s}$ satisfy the detailed balance condition

$$\frac{w_{s',s}}{w_{s,s'}} = \frac{p^{\text{eq}}(s')}{p^{\text{eq}}(s)} \qquad \forall s, s', \tag{2.135}$$

if and only if they satisfy the condition

$$\prod_{i=1}^{N} w_{s_{i+1},s_i} = \prod_{i=1}^{N} w_{s_{i-1},s_i} \tag{2.136}$$

for any set $(s_1, \ldots, s_N)$ of N microstates (with the definition $s_0 = s_N$ and $s_{N+1} = s_1$).

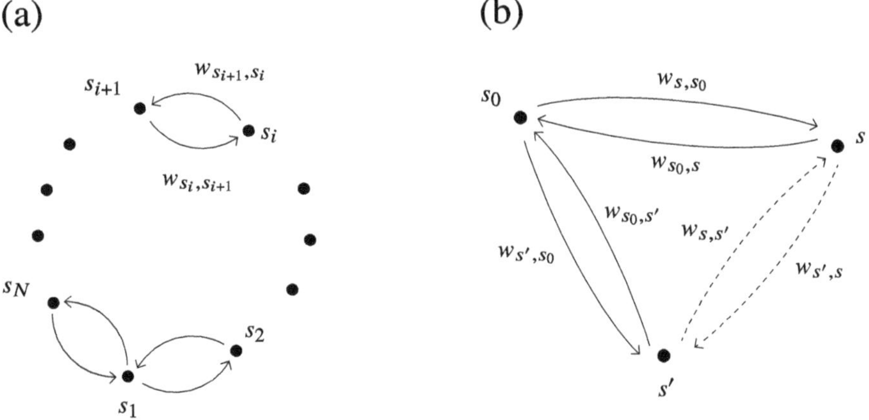

Fig. 2.8   Graphical representation of the transition rates appearing in the proof of Theorem 2.5: (a) Eq. (2.136); (b) Eqs. (2.138) and (2.140).

The first part of the theorem, detailed balance Eq. (2.135) implies Eq. (2.136) is trivial. As a matter of fact (see Fig. 2.8(a)),

$$\prod_{i=1}^{N} \frac{w_{s_{i+1},s_i}}{w_{s_{i-1},s_i}} = \prod_{i=1}^{N} \frac{p^{\mathrm{eq}}(s_{i+1})}{p^{\mathrm{eq}}(s_i)} = 1, \tag{2.137}$$

which is equivalent to Eq. (2.136).

The second part of the theorem requires us to define $p^{\mathrm{eq}}(s)$ from the transition rates. For this purpose, let us take a microstate as a reference state and label it s_0 (see Fig. 2.8(b)). For any state s, we define

$$p^{\mathrm{eq}}(s) \doteq p^{\mathrm{eq}}(s_0) \frac{w_{s,s_0}}{w_{s_0,s}}, \qquad \forall s, \tag{2.138}$$

where $p^{\mathrm{eq}}(s_0)$ has to be determined by the normalization of occupation probabilities,

$$\sum_s p^{\mathrm{eq}}(s) = 1. \tag{2.139}$$

In simple words, we define $p^{\mathrm{eq}}(s)$ in such a way that any state satisfies detailed balance with respect to a specified microstate s_0. This does not guarantee that detailed balance is satisfied for any pair of states s, s', and in general it is not. However, if Eq. (2.136) is valid, then it does, as we are going to prove.

Let us start from the relation $w_{s,s_0} w_{s',s} w_{s_0,s'} = w_{s',s_0} w_{s,s'} w_{s_0,s}$, which must be satisfied, because Eq. (2.136) is assumed to be true. We can rewrite it as

$$\left(\frac{w_{s,s_0}}{w_{s_0,s}} \right) w_{s',s} = \left(\frac{w_{s',s_0}}{w_{s_0,s'}} \right) w_{s,s'}, \tag{2.140}$$

and let us evaluate the two fractions between parentheses using Eq. (2.138):

$$\frac{p^{\mathrm{eq}}(s)}{p^{\mathrm{eq}}(s_0)} w_{s',s} = \frac{p^{\mathrm{eq}}(s')}{p^{\mathrm{eq}}(s_0)} w_{s,s'}. \tag{2.141}$$

Once we cancel $p^{\mathrm{eq}}(s_0)$ from both sides, the earlier equation proves that detailed balance between s and s' is valid.

In conclusion, we have given an alternative formulation of detailed balance that involves transition rates only, so we can test if detailed balance is broken or not by checking Eq. (2.136).

2.4.3 Stochastic Differential Equations

In Section 2.2.2, we introduced the Langevin equation for a Brownian particle (see Eq. (2.22)), on the basis of some heuristic arguments. The main conceptual weakness of this approach is the assumption that the components of the stochastic force, $\eta_i(t)$, acting on the Brownian particle are completely uncorrelated:

$$\langle \eta_i(t)\eta_j(t') \rangle = \Gamma \delta_{ij} \delta(t - t'), \tag{2.142}$$

where Γ is given by the fluctuation–dissipation relation (2.28). In a physical perspective, this is justified by assuming that the components of the stochastic force are independent of each other and their correlation time is practically negligible with respect to the typical time scale, $t_d = m/\tilde{\gamma} = 1/\gamma$, associated with the dynamics of the deterministic part, where m is the mass of the Brownian particle and $\tilde{\gamma}$ is the viscous friction.

Despite this assumption seeming quite plausible in physical phenomena (e.g., Brownian motion) usually modeled by stochastic processes, the mathematical structure of Eq. (2.22) is quite weird: The deterministic part is made of differentiable functions with respect to time, while $\eta_i(t)$ is intrinsically discontinuous at any time, due to Eq. (2.142). A way out of this problem is the introduction of Wiener processes, which can be considered the continuous time version of a random walk. The basic idea is quite simple: We use the integrated stochastic force to define a new stochastic process that can be made continuous in time,

$$W_i(t) = \int_0^t dt'\, \eta_i(t'). \tag{2.143}$$

The statistical average introduced in Section 2.2.2 has to be thought of as the average over the realizations of the stochastic process. If $\langle \eta_i(t) \rangle = 0$, then also $\langle W_i(t) \rangle = 0$, while (2.142) yields

$$\langle W_i(t) W_j(t') \rangle = \delta_{ij} \Gamma \min\{t, t'\}. \tag{2.144}$$

In particular,

$$\langle W_i^2(t) \rangle = \Gamma t, \tag{2.145}$$

meaning that the average squared amplitude of the Wiener process diffuses in time, with a generalized diffusion constant Γ, whose physical dimensions are determined by those of η, namely $[\Gamma] = [\eta]^2 [t]$.

The Wiener process defined in (2.143) obeys the following properties:

- $W_i(t)$ is a stochastic process continuous in time and with zero average, $\langle W_i(t) \rangle = 0$.
- For any $t_1 < t_2 < t_3$, the increments $(W_i(t_2) - W_i(t_1))$ and $(W_i(t_3) - W_i(t_2))$ are independent quantities, following the same distribution.
- For any $t_1 < t_2$, the probability distribution of the increments $(W_i(t_2) - W_i(t_1))$ is a Gaussian with zero average and variance $\Gamma(t_2 - t_1)$, which is a consequence of the central limit theorem; see Section 3.2.

Notice that $W_i(t)$ is not differentiable, meaning that we cannot define its time derivative, but we can define its infinitesimal increment $dW_i(t)$ for an arbitrarily small time interval dt, namely,

$$dW_i(t) = W_i(t + dt) - W_i(t) = \int_t^{t+dt} dt'\, \eta_i(t'), \tag{2.146}$$

which implies $\langle dW_i(t) \rangle = 0$ and $\langle (dW_i(t))^2 \rangle = \Gamma dt$.

In order to simplify the notation, the infinitesimal increment of the Wiener process is usually redefined as

$$dW_i(t) \rightarrow \sqrt{\Gamma}\, \Omega(t)\, \sqrt{dt}, \tag{2.147}$$

where $\Omega(t)$ represents a dimensionless stochastic process with zero average and unit variance. This relation is quite useful, because it attributes to the amplitude of the infinitesimal increment of the Wiener process a time scale, given by the square root of the infinitesimal time increment dt.

With these ingredients we can write the general form of a well-defined stochastic differential equation,

$$dX_i(t) = a_i(X(t), t)dt + b_i(X(t), t)dW_i(t), \quad i = 1, 2 \ldots, n, \tag{2.148}$$

where $X_i(t)$, $a_i(X(t), t)$, and $b_i(X(t), t)$ are the components of generic vectors of functions in $\mathbb{R}^n$: Each of these functions is differentiable with respect to its arguments. The physical interpretation of this general equation depends on the physical meaning attributed to $X_i(t)$.

It is worthwhile to rescale first $dW_i(t)$ and $b_i(X(t), t)$ as follows:

$$dW_i(t) \to \Omega(t)\sqrt{dt} \quad \Rightarrow \quad \langle (dW_i(t))^2 \rangle = dt \tag{2.149}$$

$$b_i(X(t), t) \to \sqrt{\Gamma}b_i(X(t), t), \tag{2.150}$$

in such a way that $b_i(X(t), t)$ acquires the physical dimensions of $\sqrt{\Gamma}$. For instance, Eq. (2.148) reproduces the problem of the Brownian particle in $\mathbb{R}^3$ (see Section 2.2.2) with $X_i(t) = v_i(t)$ (i.e., it is the i-th component of the velocity of the Brownian particle), $a_i(X(t), t) = -\gamma v_i(t)$ and $b_i(X(t), t) = \sqrt{\Gamma} = \sqrt{2D\gamma^2}$, with $D = T/(m\gamma)$, due to the Einstein formula (2.9).

In the case where $X_i(t))$ represents a space variable, $a_i(X(t), t)$ has the physical dimensions of a velocity, while $b_i^2(X(t), t)$ has the dimensions of a diffusion coefficient, because Eq. (2.149) implies that $[dW_i(t)] = [t]^{\frac{1}{2}}$. For instance, in the case of the stochastic particle subject to an elastic force (see Eq. (2.198)), $a_i(X(t), t) = -k X(t)$, where k is the Hook elastic constant divided by the friction coefficient $\tilde{\gamma}$ (i.e., $[k] = [t]^{-1}$) and $b_i^2(X(t), t) = 2D$, where D is the diffusion constant. More generally, $a_i(X(t), t)$ is called a generalized drift-function, while $b_i^2(X(t), t)$ is associated with a generalized diffusion coefficient.

Let us further simplify the notation by overlooking the subscript i

$$dX(t) = a(X(t), t) \, dt + b(X(t), t) \, dW(t), \tag{2.151}$$

and integrate this general stochastic differential equation, continuous in both space and time. On a formal ground this task can be immediately accomplished, writing the solution of Eq. (2.151) as

$$X(t) = X(0) + \int_0^t a(X(t'), t') \, dt' + \int_0^t b(X(t'), t') \, dW(t'). \tag{2.152}$$

If we want to extract any useful information from this formal solution, we have to perform its statistical average with respect to the Wiener process, as we have done for Brownian motion when passing from Eq. (2.23) to Eqs. (2.24) and (2.25) by averaging with respect to the components of the stochastic force $\eta_i(t)$.

Here, we have to face the further problem that the last integral is not uniquely defined when averaged over the stochastic process. In fact, according to basic mathematical analysis, the integral of a standard (Riemann) function $f(t)$ can be estimated by the Euler approximations

$$\int_0^t f(t')dt' \approx \sum_{i=0}^{N-1} f(t_i)\Delta t_i \approx \sum_{i=0}^{N-1} f(t_{i+1})\Delta t_i \tag{2.153}$$

where the support of t is subdivided into an ordered sequence of $N+1$ sampling times $(t_0 = 0, t_1, t_2 \ldots, t_{N-1}, t_N = t)$, with $\Delta t_i = t_{i+1} - t_i$. For the sake of simplicity, let us assume a uniform sampling, that is, $\Delta t_i = t/N$, $\forall i$. A theorem guarantees that in the limit $\Delta t \to 0$, both Euler approximations in Eq. (2.153) converge to the integral. Such a theorem does not hold for a Wiener process. The analogs of the Euler approximations for the stochastic integral are:

$$\mathcal{I}_I = \sum_{i=0}^{N-1} f(t_i)\Delta W(t_i), \tag{2.154}$$

$$\mathcal{I}_S = \sum_{i=0}^{N-1} f(t_{i+1})\Delta W(t_i). \tag{2.155}$$

Equations (2.154) and (2.155) are called the Itô and the Stratonovich[16] discretizations, respectively.

For very small $\Delta W(t_i)$, Eq. (2.155) can be approximated as

$$\mathcal{I}_S = \sum_{i=0}^{N-1} \left[f(t_i) + \alpha \Delta W'(t_i) + O\left(\Delta W'^2(t_i)\right) \right] \Delta W(t_i), \tag{2.156}$$

where $\Delta W'$ is, in general, a different realization of the Wiener process with respect to ΔW, and $\alpha = \frac{\delta f}{\delta W'}(t_i)$ is the functional derivative of $f(t)$ with respect to W' at t_i (see Appendix I.1).

By performing the limit $\Delta W \to 0$ and averaging over the Wiener process, we obtain

$$\left\langle \lim_{\Delta W \to 0} \mathcal{I}_I \right\rangle = \left\langle \int_0^t f(t') \, dW(t') \right\rangle = 0 \tag{2.157}$$

while, using Eq. (2.156),

$$\left\langle \lim_{\Delta W \to 0} \mathcal{I}_S \right\rangle = \left\langle \int_0^t f(t') \, dW(t') \right\rangle + \int_0^t \int_0^t \alpha(t'') \langle dW(t')dW(t'') \rangle = \int_0^t \alpha(t')dt' \tag{2.158}$$

where we have used the relation $\langle dW(t)dW(t') \rangle = \delta(t - t') \, dt$, which is a consequence of Eq. (2.149).

Notice that, at variance with Eq. (2.157), the last expression in Eq. (2.158) does not vanish: The Itô and the Stratonovich formulations are not equivalent. For the sake of completeness, here we just summarize this inequivalence by reporting some basic relations that stem from the very definitions of the Itô and Stratonovich stochastic integrals, Eqs. (2.154) and (2.155), respectively.

---

[16] After Kiyoshi Itô and Ruslan L. Stratonovich.

If we interpret Eq. (2.151) as an Itô stochastic differential equation, it can be shown that it is equivalent to the Stratonovich stochastic differential equation

$$dX(t) = \left[a(X(t),t) - \tfrac{1}{2} b(X(t),t)\partial_X b(X(t),t) \right] dt + b(X(t),t)\, dW(t). \tag{2.159}$$

Conversely, if Eq. (2.151) is interpreted as a Stratonovich stochastic differential equation, it can be shown that it is equivalent to the Itô stochastic differential equation

$$dX(t) = \left[a(X(t),t) + \tfrac{1}{2} b(X(t),t)\partial_X b(X(t),t) \right] dt + b(X(t),t) dW(t). \tag{2.160}$$

Note that only when $b(X(t),t)$ is a constant the two formulations are equivalent. This is why, in general, when dealing with models represented by stochastic differential equations, we have to choose and explicitly declare which formulation we are adopting to integrate these equations. If there are no specific reasons, in many applications it is advisable using the Itô formulation, because property (2.157) makes calculations simpler. For instance, this is the typical choice for models that assume the presence of a white (i.e., δ-correlated) noise. However, when noise is *colored*, as in many models of physical interest, the suitable choice is typically the Stratonovich one. For instance, this is the case when we deal with stochastic differential equations with multiplicative noise. In Appendix J we illustrate the example of the stochastic differential equation for the energy of a Brownian particle.

2.4.4 The Langevin Equation and Detailed Balance

Here we want to reconsider the physical interpretation of the Einstein relation (2.29), $\tilde{\Gamma} = 2\tilde{\gamma}T$, showing that its validity is equivalent to imposing detailed balance conditions for the Langevin equation of a Brownian particle subject to a force, generated by a conservative potential $U(x)$ (see Eq. (2.37)). This equation reads

$$m\ddot{x}(t) = -U'(x) - \tilde{\gamma}\dot{x}(t) + \tilde{\eta}(t), \tag{2.161}$$

where $\langle \tilde{\eta}(t) \rangle = 0$ and $\langle \tilde{\eta}(t)\tilde{\eta}(t') \rangle = \tilde{\Gamma}\delta(t - t')$.

In order to pursue our task, it is better to recast Eq. (2.161) in the more appropriate formalism of stochastic differential equations, namely

$$x(t + dt) = x(t) + \frac{p(t)}{m}dt \tag{2.162}$$

$$p(t + dt) = p(t) - U'(x(t))dt - \tilde{\gamma}\frac{p(t)}{m}dt + dW, \tag{2.163}$$

where dt is an infinitesimal time increment and dW is the infinitesimal increment of a Wiener process, as defined in Eq. (2.146). In practice, it is a Gaussian-distributed, stochastic variable with zero average and variance $\langle (dW)^2 \rangle = \int_0^{dt} d\tau' \int_0^{dt} d\tau'' \langle \tilde{\eta}(\tau')\tilde{\eta}(\tau'') \rangle = \tilde{\Gamma}dt$.[17]

Here, the delicate point is the correct formulation of detailed balance when the microscopic state is labeled by a quantity, the momentum $p = m\dot{x}$, which changes sign with time reversal. In fact, until now we have tacitly assumed that detailed balance applies

---

[17] For the sake of clarity, here we keep the Langevin formulation, rather than adopting the general formalism of Eq. (2.148), with the rescaling condition (2.149).

to systems, whose microscopic states are invariant under time reversal. For such systems, the formulation of detailed balance is simply $p_\alpha W_{\beta\alpha} = p_\beta W_{\alpha\beta}$ (see Eq. (2.109)), where $W_{\beta\alpha}$ is the transition rate between state α and state β and p_α is the probability to be in state α. Once we introduce the momentum variable, we should rather prove that

$$p_\alpha W_{\beta\alpha} = p_{\beta^*} W_{\alpha^*\beta^*}, \tag{2.164}$$

where the asterisk means the time-reversed state. Making reference to Eq. (2.163), labels read $\alpha = (x, p)$, $\beta = (x', p')$, $\alpha^* = (x, -p)$, and $\beta^* = (x', -p')$.

We can use the coupled stochastic equations (2.162) and (2.163) to obtain an explicit expression for the transition rates in Eq. (2.164) performed in the infinitesimal time increment dt. We can write

$$W_{\beta\alpha} = \delta\left(x' - x - \frac{p}{m}dt\right) \frac{1}{\sqrt{2\pi\tilde{\Gamma}dt}} \exp\left\{-\frac{\left(p' - p + U'(x)dt + \tilde{\gamma}\frac{p}{m}dt\right)^2}{2\tilde{\Gamma}dt}\right\} \tag{2.165}$$

$$W_{\alpha^*\beta^*} = \delta\left(x - x' + \frac{p'}{m}dt\right) \frac{1}{\sqrt{2\pi\tilde{\Gamma}dt}} \exp\left\{-\frac{\left(p' - p + U'(x')dt - \tilde{\gamma}\frac{p'}{m}dt\right)^2}{2\tilde{\Gamma}dt}\right\}. \tag{2.166}$$

where we have introduced the shorthand notation $x' \equiv x(t + dt)$ and $p' \equiv p(t + dt)$.

The first of these equations represents the transition rate $W_{\beta\alpha}$ in the infinitesimal time dt from (x, p) to (x', p'), where the Dirac-δ is a straightforward consequence of the deterministic equation (2.162), while the following factor expresses explicitly the assumptions that dW is a Gaussian-distributed, stochastic variable with zero average and variance $\langle (dW)^2 \rangle = \tilde{\Gamma} dt$. The equation of the transition rate $W_{\alpha^*\beta^*}$ from $(x', -p')$ to $(x, -p)$ is obtained by applying explicitly to the former equation the time reversal operation, that is, by exchanging x with x', p' with $-p$ and p with $-p'$.

It is important to point out that Eq. (2.162) shows that $(x' - x)$ is of order dt, while Eq. (2.163) indicates that $(p' - p)$, being proportional to dW, is at leading order $\sqrt{dt}$. Now, we can evaluate at leading order in dt the ratio of the transition rates (2.165) and (2.166) as follows:

$$\frac{W_{\beta\alpha}}{W_{\alpha^*\beta^*}} = \exp\left\{\frac{\tilde{\gamma}}{\tilde{\Gamma}}\left(\frac{p^2 - p'^2}{m} - \frac{2p}{m}U'(x)dt\right) + o(dt)\right\} \tag{2.167}$$

In fact, all the squared addenda in both of the exponentials cancel or can be neglected, yielding contributions of $o(dt)$. For what concerns the double products: The term $2(p - p')(U'(x) - U'(x')) \approx 2(p - p')U''(x)(x' - x) \sim o(dt^{\frac{3}{2}})$ can be neglected as well; the term $\frac{2\tilde{\gamma}}{m}(p - p')(p + p')$ yields $\frac{2\tilde{\gamma}}{m}(p^2 - p'^2)$; the term $-2\tilde{\gamma}(\frac{p}{m}U'(x) + \frac{p'}{m}U'(x'))dt$ at leading order can be approximated as $-4\tilde{\gamma}\frac{p}{m}U'(x)dt$. Note also that $\frac{p}{m}U'(x)dt = U'(x)(x' - x) \approx U(x') - U(x)$ and we can finally rewrite Eq. (2.167) as follows:

$$\frac{W_{\beta\alpha}}{W_{\alpha^*\beta^*}} = \exp\left\{\frac{2\tilde{\gamma}}{\tilde{\Gamma}}\left(E(\alpha) - E(\beta^*)\right) + o(dt)\right\}, \tag{2.168}$$

where

$$E(\alpha) \equiv E(x, p) = \frac{p^2}{2m} + U(x), \quad E(\beta^*) \equiv E(x', -p') = \frac{p'^2}{2m} + U(x'). \tag{2.169}$$

Equation (2.168) can be written as the ratio p_{β^*}/p_α, therefore attesting to the validity of the detailed balance, if and only if $2\tilde{\gamma}/\tilde{\Gamma}$ is the inverse absolute temperature, that is,

$$\tilde{\Gamma} = 2\tilde{\gamma}T. \tag{2.170}$$

We might repeat the same procedure in the simpler case of the overdamped Langevin equation for the Brownian particle subject to the force generated by a conservative potential $U(x)$, so that the inertial term $m\ddot{x}(t)$ is negligible. Rather than having the two coupled Eqs. (2.162) and (2.163), we simply have

$$x(t + dt) = x(t) - \frac{U'(x(t))}{\tilde{\gamma}} dt + \frac{1}{\tilde{\gamma}} dW, \tag{2.171}$$

and similar calculations lead to the same result (2.170). More precisely, the ratio between the transition rates turns out to have the approximate expression

$$\frac{W_{x'\,x}}{W_{x\,x'}} \approx \exp\left\{ \frac{2\tilde{\gamma}}{\tilde{\Gamma}} \Big(U(x) - U(x') \Big) + o(dt) \right\}, \tag{2.172}$$

Note that in the overdamped case, the energy of different states is given by the difference of the potential energies of the particle, since the Wiener process is just the result of the competition between a friction term and a conservative force, similar to the situation of the Ornstein-Uhlenbeck process (see Eq. (2.198)).

2.4.5 General Fokker–Planck Equation

We can use the general stochastic differential equation (2.151) to derive the general Fokker–Planck equation. Let us consider a function $f(X(t), t)$ that is at least twice differentiable with respect to X. We can write its Taylor series expansion as

$$df = \frac{\partial f}{\partial t} dt + \frac{\partial f}{\partial X} dX + \frac{1}{2} \frac{\partial^2 f}{\partial X^2} dX^2 + O(dt^2, dX^3). \tag{2.173}$$

Now we also assume that $X(t)$ obeys the general stochastic differential equation (2.151), which means that $dX = a\,dt + b\,dW$ (the arguments of the functions $a(X(t), t)$ and $b(X(t), t)$ have been overlooked). By substituting into Eq. (2.173) and discarding higher-order terms, we obtain the so-called Itô formula,

$$df = \left(\frac{\partial f}{\partial t} + a \frac{\partial f}{\partial X} \right) dt + \frac{1}{2} b^2 \frac{\partial^2 f}{\partial X^2} (dW)^2 + b \frac{\partial f}{\partial X} dW(t). \tag{2.174}$$

On the r.h.s. of this equation we have kept the contribution proportional to $(dW)^2$, because in order to solve this equation, one has to integrate and then average over the Wiener process. Accordingly, as a consequence of Eq. (2.149), we have $\langle (dW(t))^2 \rangle = dt$, that is, it contributes as the other two addenda that are proportional to dt.

The important result is that this new stochastic equation is formally equivalent to Eq. (2.151), as long as the functions $a(X(t), t)$ and $b(X(t), t)$ are appropriately redefined. We can conclude that any function $f(X(t), t)$, at least twice differentiable, obeys the same kind of stochastic differential equation obeyed by $X(t)$. The Itô formula is quite useful, because it allows us to transform a stochastic differential equation into another one, whose solution is known by a suitable guess of the function f.

Now, let us assume that f does not depend explicitly on t, that is, $\frac{\partial f}{\partial t} = 0$; we average Eq. (2.174) over the Wiener process and obtain

$$\frac{d}{dt} \langle f(X(t)) \rangle = \left\langle a \frac{\partial f}{\partial X} + \frac{1}{2} b^2 \frac{\partial^2 f}{\partial X^2} \right\rangle, \tag{2.175}$$

where we have used the conditions $\langle dW(t) \rangle = 0$ and $\langle (dW)^2 \rangle = dt$.

In order to specify the averaging operation, we introduce explicitly the probability density $P(X, t)$, defined on the state-space S of $X(t)$, where this stochastic variable takes the value X at time t. Given a function $f(X)$, its expectation value is

$$\langle f(X) \rangle = \int_S dX \, P(X, t) \, f(X), \tag{2.176}$$

and we can rewrite (2.175) as

$$\begin{aligned}
\frac{d}{dt} \langle f(X) \rangle &= \int_S dX f(X) \frac{\partial P(X, t)}{\partial t} \\
&= \int_S dX P(X, t) a \frac{\partial f}{\partial X} + \frac{1}{2} \int_S dX P(X, t) b^2 \frac{\partial^2 f}{\partial X^2}.
\end{aligned} \tag{2.177}$$

Integrating by parts both integrals in the last line, we obtain

$$\begin{aligned}
\int_S dX f(X) \frac{\partial P(X, t)}{\partial t} &= - \int_S dX f(X) \frac{\partial \, (a(X, t) P(X, t))}{\partial X} \\
&+ \frac{1}{2} \int_S dX f(X) \frac{\partial^2 (b^2(X, t) \, P(X, t))}{\partial X^2},
\end{aligned} \tag{2.178}$$

where it has been assumed that the probability density $P(X, t)$ vanishes at the boundary of S. The analogous assumption for the Brownian particle is that the distribution function of its position in space vanishes at infinity.

Since Eq. (2.178) holds for an arbitrary function $f(X)$, we can write the general Fokker–Planck equation for the probability density $P(X, t)$ as

$$\frac{\partial P(X, t)}{\partial t} = -\frac{\partial}{\partial X} (a(X, t) P(X, t)) + \frac{1}{2} \frac{\partial^2}{\partial X^2} (b^2(X, t) \, P(X, t)). \tag{2.179}$$

We point out that the previous equation has been obtained from the stochastic differential equation (2.151), which describes the evolution of each component of a vector of variables (e.g., coordinates) $X_i(t)$; see Eq. (2.148). In practice, in Section 2.4.3 we have assumed that all $X_i(t)$ are independent stochastic processes, so that the general Fokker–Planck equation has the same form for any $P(X_i, t)$. On the other hand, in principle one should deal with the general case, where $a_i(X, t)$ and $b_i(X, t)$ are different for different values of i. By assuming spatial isotropy for these physical quantities, one can overlook any dependence on i in the general Fokker–Planck equation. In practice, with these assumptions, all the components of the vector variable $X_i(t)$ are described by a single general Fokker–Planck equation, where $X(t)$ is a scalar quantity.

In a mathematical perspective, finding an explicit solution of Eq. (2.179) depends on the functional forms of the generalized drift coefficient $a(X, t)$ and of the generalized diffusion coefficient[18] $b^2/2$. On the other hand, as discussed in Section 2.4.6, interesting physical examples correspond to simple forms of these quantities.

2.4.6 Physical Applications of the Fokker–Planck Equation

Stationary Diffusion with Absorbing Barriers

Equation (2.179) can be expressed in the form of a conservation law, or continuity equation,

$$\frac{\partial P(X,t)}{\partial t} + \frac{\partial J(X,t)}{\partial X} = 0, \tag{2.180}$$

where

$$J(X,t) = a(X,t)P(X,t) - \frac{1}{2}\frac{\partial}{\partial X}(b^2(X,t)\,P(X,t)) \tag{2.181}$$

can be read as a probability current.

Since $X \in \mathbb{R}$, we can consider the interval $I = [X_1, X_2]$ and define the probability, $\mathcal{P}(t)$, that at time t the stochastic process described by (2.180) is in I,

$$\mathcal{P}(t) = \int_I P(X,t)dX. \tag{2.182}$$

By integrating both sides of Eq. (2.180) over the interval I, one obtains

$$\frac{\partial \mathcal{P}(t)}{\partial t} = J(X_1,t) - J(X_2,t). \tag{2.183}$$

The current is positive if the probability flows from smaller to larger values of X, and various boundary conditions can be imposed on I. For instance, the condition of reflecting barriers, that is, no flux of probability through the boundaries of I, amounts to $J(X_1,t) = J(X_2,t) = 0$ at any time t. Accordingly, the probability of finding the walker inside I is conserved, as a straightforward consequence of Eq. (2.183). Conversely, the condition of absorbing barriers implies that the walker reaching X_1 or X_2 will never come back to I, that is, $P(X_1,t) = P(X_2,t) = 0$ at any time t. When dealing with stochastic systems defined on a finite interval (a typical situation of numerical simulations), it may be useful to impose periodic boundary conditions that correspond to $P(X_1,t) = P(X_2,t)$ and $J(X_1,t) = J(X_2,t)$. In this case the probability $\mathcal{P}$ is conserved not because the flux vanishes at the borders, but because the inner and outer fluxes compensate.

---

[18] The mathematical problem of establishing conditions for the existence and the uniqueness of the solution of Eq. (2.179) demands specific assumptions on the space and time dependence of $a(X, t)$ and $b(X, t)$. They must be differentiable functions with respect to their arguments and they must be compatible with the physical assumption that $P(X, t)$ must rapidly vanish in the limit $X \to \pm\infty$ in such a way that the normalization condition $\int_{\mathbb{R}} P(X, t)dX = 1$ holds at any time t.

Stationary solutions $P^*(X)$ of Eq. (2.180) must fulfill the condition[19]

$$\frac{d}{dX}(a(X)P^*(X)) - \frac{1}{2}\frac{d^2}{dX^2}(b^2(X)P^*(X)) = 0, \tag{2.184}$$

where we have to assume that also the functions $a(X)$ and $b(X)$ do not depend on t. For instance, in the purely diffusive case, that is, $a(X) = 0$, $b^2(X) = 2D$, the general solution of Eq. (2.184) is

$$P^*(X) = C_1 X + C_2, \tag{2.185}$$

where C_1 and C_2 are constants to be determined by boundary and normalization conditions.

In the case of reflecting barriers at X_1 and X_2 in the interval I, there is no flux through the boundaries, so that the stationary solution must correspond to no flux in I, that is,

$$J^*(X_{1,2}) = -D \left.\frac{dP^*(X)}{dX}\right|_{X_{1,2}} = 0. \tag{2.186}$$

This condition implies $C_1 = 0$ and $P^*(X) = C_2 = |X_2 - X_1|^{-1}$, where the last expression is a straightforward consequence of the normalization condition $\int_{\mathbb{R}} P^*(X)dX = 1$. We can conclude that in this case, the stationary probability of finding a walker in I is a constant, given by the inverse of the length of I.

In the case of absorbing barriers, we obtain the trivial solution $P^*(X) = 0$ for $X \in I$, because for $t \to \infty$, the walker will eventually reach one of the absorbing barriers and disappear.

A more interesting case with absorbing barriers can be analyzed by assuming that at each time unit a new walker starts at site $X_0 \in I$; that is, we consider a stationary situation where a constant flux of walkers is injected in I at X_0 (see Fig. 2.9). Due to the presence of absorbing barriers, the stationary solution of Eq. (2.184) with a source generating walkers at unit rate in X_0 is

$$P^*(X) = \begin{cases} C_1(X - X_1), & \text{for } X < X_0 \\ C_2(X_2 - X), & \text{for } X > X_0, \end{cases} \tag{2.187}$$

which fulfills the condition of absorbing barriers $P^*(X_1) = P^*(X_2) = 0$. Moreover, we have to impose the continuity of $P^*(X)$ at the source point X_0. This condition yields the relation

$$C_1(X_0 - X_1) = C_2(X_2 - X_0). \tag{2.188}$$

Because of the presence of the flux source, Eq. (2.180) should be written, more correctly, as

$$\partial_t P + \partial_x J = F\delta(X - X_0). \tag{2.189}$$

---

[19] Partial derivatives turn into standard derivatives, because the stationary solution is, by definition, independent of t.

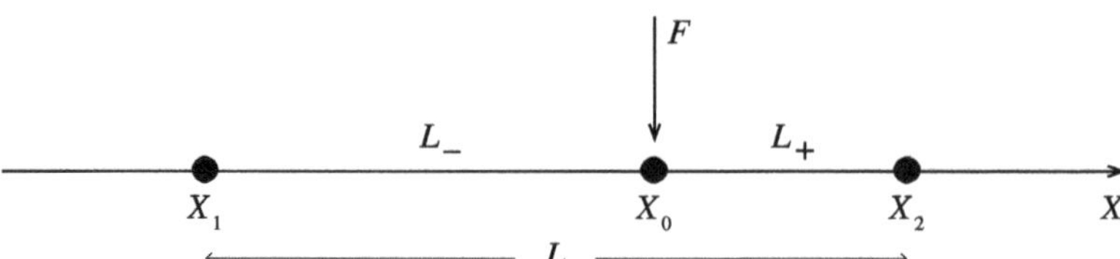

Fig. 2.9 Isotropic diffusion in a segment (X_1, X_2) of length L with absorbing boundaries at the interval extrema and a continuous injection of particles in $X = X_0$. The system relaxes to a steady state, where the stationary distribution of particle density varies linearly in each segment, because $P^*(X)$ is the solution of the equation $d^2 P^*/dX^2 = 0$. Absorbing boundaries imply $P^*(X_1) = P^*(X_2) = 0$. In $X = X_0$, $P^*(X)$ is continuous, but its derivative has a jump depending on the flux F (see Eq. (2.189)).

Therefore, the stationary solution must satisfy the relation

$$F = \int_{X_0^-}^{X_0^+} dX \, \partial_x J \tag{2.190}$$

$$= -D \left[\frac{dP^*}{dX}\bigg|_{X_0^+} - \frac{dP^*}{dX}\bigg|_{X_0^-} \right] \tag{2.191}$$

$$= D(C_2 + C_1). \tag{2.192}$$

The last equation, along with Eq. (2.188) gives

$$C_1 = \frac{F}{D}\frac{L_+}{L}, \qquad C_2 = \frac{F}{D}\frac{L_-}{L}, \tag{2.193}$$

where we have used the notations $L = X_2 - X_1$ for the total length of the interval and $L_- = X_0 - X_1$, $L_+ = X_2 - X_0$, for the left and right parts of the interval; see Fig. 2.9. The probability distributions and the currents are

$$P^*(X) = \begin{cases} \dfrac{F}{D}\dfrac{L_+}{L}(X - X_1) \\[2mm] \dfrac{F}{D}\dfrac{L_-}{L}(X_2 - X), \end{cases} \qquad J^*(X) = \begin{cases} -F\dfrac{L_+}{L} \equiv -J_-, & \text{for } X < X_0 \\[2mm] F\dfrac{L_-}{L} \equiv J_+, & \text{for } X > X_0, \end{cases} \tag{2.194}$$

where J_- and J_+ are the net fluxes of particles flowing in X_1 and X_2, respectively (and $J_- + J_+ = F$, because of matter conservation). Therefore, a particle deposited in X_0 has a probability

$$\Pi(X_1|X_0) = \frac{J_-}{F} = \frac{L_+}{L} = \frac{X_2 - X_0}{X_2 - X_1} \tag{2.195}$$

to be absorbed in X_1.

The same walker has an average exit time from I, $\mathcal{T}_I(X_0)$, which is equal to the number of walkers, $\mathcal{N}_w$, in the interval I divided by the total current flowing out (equal to F; see above). Since

$$\mathcal{N}_w = \int_{X_1}^{X_2} P^*(X)dX = \frac{F}{2D}L_- L_+, \tag{2.196}$$

we have

$$\mathcal{T}_I(X_0) = \frac{\mathcal{N}_w}{F} = \frac{1}{2D}(X_0 - X_1)(X_2 - X_0). \tag{2.197}$$

The reader can easily realize that we have just reconsidered in the continuous approach of the Fokker–Planck equation the problem of the random walk with absorbing barriers discussed in Section 2.3.2.

A Stochastic Particle Subject to an Elastic Force

We want to describe in the Itô formulation the motion of a stochastic particle in the presence of a conservative force and thermal fluctuations. In the case of an elastic force and of a constant diffusion coefficient, $b(X(t),t) = \sqrt{2D}$, this is known as the Ornstein–Uhlenbeck process,

$$dX(t) = -kX(t)\,dt + \sqrt{2D}\,dW(t), \tag{2.198}$$

where $X(t)$ is the position of the stochastic particle at time t and k is the Hook elastic constant divided by the friction coefficient $\tilde{\gamma}$. The physical interpretation of this equation is quite different from Newtonian mechanics: The presence of the Wiener process on the right-hand side points out that the overall process is the result of a balance among a friction term on the left-hand side (a force proportional to a velocity) and a conservative elastic force in the presence of thermal fluctuations. The inertial term, that is, the one proportional to the second derivative with respect to time of $X(t)$, is negligible on time scales larger than t_d, that is, we are in the so-called overdamped case.

This equation can be solved by introducing the function $Y(t) = X(t)\,e^{kt}$ that obeys the stochastic equation

$$dY(t) = e^{kt}(dX(t) + k\,X dt) = \sqrt{2D}\,e^{kt}\,dW(t), \tag{2.199}$$

which can be integrated between time 0 and time t, thus yielding

$$Y(t) = Y(0) + \sqrt{2D}\int_0^t e^{ks}\,dW(s). \tag{2.200}$$

Coming back to the variable $X(t)$, this equation becomes

$$X(t) = X(0)\,e^{-kt} + \sqrt{2D}\int_0^t e^{-k(t-s)}\,dW(s), \tag{2.201}$$

where $X(0)$ is the initial position of the particle. By averaging over the Wiener process, we finally obtain

$$\langle X(t)\rangle = X(0)\,e^{-kt} \tag{2.202}$$

and

$$\langle X^2(t)\rangle = \langle X(t)\rangle^2 + 2X(0)e^{-kt}\sqrt{2D}\int_0^t e^{-k(t-s)}\langle dW(s)\rangle \tag{2.203}$$

$$+ 2D\int_0^t e^{-k(t-s)}\int_0^t e^{-k(t-s')}\langle dW(s)dW(s')\rangle. \tag{2.204}$$

Using the relations $\langle dW(s)\rangle = 0$ and $\langle dW(s)dW(s')\rangle = \delta(s - s')ds$, we find

$$\langle X^2(t)\rangle - \langle X(t)\rangle^2 = \frac{D}{k}\left(1 - e^{-2kt}\right). \tag{2.205}$$

In the limit $t \to +\infty$ the average displacement vanishes, irrespective of the initial condition, while the variance of the stochastic process converges to D/k. The stochastic particle diffuses around the origin and its average squared displacement is distributed according to a Gaussian with zero average, while its variance is inversely proportional to the Hook constant k and proportional to the amplitude of fluctuations, that is, to the diffusion constant D. If we want to recover the result for the Brownian particle, we should take the limit $k \to 0$ in Eq. (2.205) before the limit $t \to \infty$, obtaining $\langle X^2(t)\rangle - \langle X(t)\rangle^2 = 2Dt$, as expected.

The Fokker–Planck equation of the Ornstein–Uhlenbeck process is

$$\frac{\partial P(X,t)}{\partial t} = k\frac{\partial}{\partial X}(X\,P(X,t)) + D\frac{\partial^2}{\partial X^2}P(X,t). \tag{2.206}$$

This equation can be solved via Fourier transformation and then using the method of characteristics. Here we limit to give the solution,

$$P(X,t) = \frac{1}{\sqrt{2\pi\sigma^2}}\exp\left(-\frac{(X - \langle X\rangle)^2}{2\sigma^2}\right) \tag{2.207}$$

with

$$\langle X\rangle = X_0\exp(-kt) \tag{2.208}$$

$$\sigma^2 = \frac{D}{k}[1 - \exp(-2kt)], \tag{2.209}$$

in accordance with what was found in Eqs. (2.202) and (2.205).

The Fokker–Planck Equation in the Presence of a Mechanical Force

Let us consider the Fokker–Planck equation for a particle subject to an external mechanical force $\mathbf{F}(\mathbf{x})$ generated by a conservative potential $U(\mathbf{x})$, that is,

$$\mathbf{F}(\mathbf{x}) = -\nabla U(\mathbf{x}). \tag{2.210}$$

Making use of the general formalism introduced in this section (see Eqs. (2.180) and (2.181)), we can write

$$\frac{\partial P(\mathbf{x},t)}{\partial t} = -\nabla\cdot\mathbf{J}, \tag{2.211}$$

where

$$\mathbf{J} = \frac{\mathbf{F}}{\tilde{\gamma}}P(\mathbf{x},t) - D\nabla P(\mathbf{x},t). \tag{2.212}$$

Here D is the usual diffusion coefficient and the parameter $\tilde{\gamma}$ is the same phenomenological quantity introduced in Section 2.2.2.

Here we avoid reporting the derivation of the solution $P(\mathbf{x}, t)$ of Eq. (2.211) and we limit our analysis to obtaining the explicit expression for the equilibrium probability distribution $P^*(\mathbf{x})$. This problem can be easily solved by considering that it is, by definition, independent of time, and Eq. (2.211) yields the relation

$$\nabla \cdot \mathbf{J} = 0. \tag{2.213}$$

On the other hand, for physical reasons, any (macroscopic) current must vanish at equilibrium, and the only acceptable solution of this equation is $\mathbf{J} = 0$. Under this condition, Eq. (2.212) reduces to

$$\frac{\nabla P^*(\mathbf{x})}{P^*(\mathbf{x})} = -\frac{\nabla U(\mathbf{x})}{\tilde{\gamma} D}, \tag{2.214}$$

or, equivalently, to

$$\nabla \ln P^* = -\frac{\nabla U}{\tilde{\gamma} D}. \tag{2.215}$$

This partial differential equation can be integrated if $U(\mathbf{x})$ diverges (and therefore $P(\mathbf{x})$ vanishes) for large $|\mathbf{x}|$,

$$P^*(\mathbf{x}) = A \exp\left(-\frac{U(\mathbf{x})}{T}\right), \tag{2.216}$$

where A is a suitable normalization constant and we have used the Einstein relation $D = T/\tilde{\gamma}$. As one should have expected on the basis of general arguments, we have obtained the equilibrium Gibbs distribution.

2.5 A Different Pathway to the Fokker–Planck Equation

In Section 2.4.5 we have derived the Fokker–Planck equation by the Itô formula (2.174) for the stochastic differential equation (2.151). A useful alternative formulation of the Fokker–Planck equation can be obtained from the fully continuous version of the Chapman–Kolmogorov equation (2.120), which is defined by the integral equation[20]:

$$W(X_0, t_0 | X, t + \Delta t) = \int_{\mathbb{R}} dY \, W(X_0, t_0 | Y, t) \, W(Y, t | X, t + \Delta t), \tag{2.217}$$

where the transition probability $W(X, t | X', t')$ from state X at time t to state X' at time t' exhibits a continuous dependence on both state-space and time variables.[21] As in Section 2.4.5, we can consider an arbitrary function $f(X(t))$, which has to vanish sufficiently rapidly

---

[20] Until now, with a discrete set of states, it has been useful to define W_{ij}^t as the transition rate between state j and state i, so as to use a matrix notation. Now, with a continuous set of states, it is simpler to use the notation given in Eq. (2.217), so that time flows from left to right and it is easier to follow the calculations.

[21] As discussed at the end of Section 2.4.5, we can assume the condition of isotropy, so that Eq. (2.217) holds for any spatial component X_i $(i = 1, \ldots, N)$ of the vector state-variable $\mathbf{X} \in \mathbb{R}^N$.

for $X \to \pm\infty$, in such a way that

$$\int_{\mathbb{R}} dX \, f(X) \, OW(X_0, t_0|X, t) < \infty, \tag{2.218}$$

for any operator $O = \mathbb{I}, \partial_t, \partial_X, \partial_{XX}, \ldots, (\partial_X)^n, \ldots$, the last symbol being a shorthand notation for the nth-order derivative with respect to the variable X. In particular, we can define the partial derivative of $W(X_0, t_0|X, t)$ with respect to time through the relation

$$\int_{\mathbb{R}} dX f(X) \frac{\partial W(X_0, t_0|X, t)}{\partial t} = \lim_{\Delta t \to 0} \frac{1}{\Delta t} \left\{ \int_{\mathbb{R}} dX f(X) [W(X_0, t_0|X, t + \Delta t) - W(X_0, t_0|X, t)] \right\}. \tag{2.219}$$

Making use of Eq. (2.217), we can write the first integral on the right-hand side in the form

$$\int_{\mathbb{R}} dX \, f(X) W(X_0, t_0|X, t + \Delta t) = \int_{\mathbb{R}} dX \, f(X) \int_{\mathbb{R}} dY \, W(X_0, t_0|Y, t) W(Y, t|X, t + \Delta t). \tag{2.220}$$

In the limit $\Delta t \to 0$, the transition probability $W(Y, t|X, t + \Delta t)$ vanishes unless X is sufficiently close to Y, so that we can expand $f(X)$ in a Taylor series,

$$f(X) = f(Y) + f'(Y)(X - Y) + \frac{1}{2} f''(Y)(X - Y)^2 + O(X - Y)^3.$$

By substituting into Eq. (2.219) and neglecting higher-order terms, we obtain

$$\int_{\mathbb{R}} dX \, f(X) \frac{\partial W(X_0, t_0|X, t)}{\partial t} = \lim_{\Delta t \to 0} \frac{1}{\Delta t} \left\{ \left[\int_{\mathbb{R}} dX \int_{\mathbb{R}} dY \left(f(Y) + f'(Y)(X - Y) \right. \right. \right.$$
$$\left. \left. + \frac{1}{2} f''(Y)(X - Y)^2 \right) \times W(X_0, t_0|Y, t) W(Y, t|X, t + \Delta t) \right]$$
$$\left. - \int_{\mathbb{R}} dX \, f(X) W(X_0, t_0|X, t) \right\}. \tag{2.221}$$

Because of the normalization condition

$$\int_{\mathbb{R}} dX \, W(Y, t|X, t') = 1, \tag{2.222}$$

the first term of the Taylor series expansion cancels with the last integral in Eq. (2.221), and we can finally rewrite Eq. (2.219) as

$$\int_{\mathbb{R}} dX \, f(X) \frac{\partial W(X_0, t_0|X, t)}{\partial t} = \lim_{\Delta t \to 0} \frac{1}{\Delta t} \left\{ \int_{\mathbb{R}} dY \int_{\mathbb{R}} dX \left[f'(Y)(X - Y) + \frac{1}{2} f''(Y)(X - Y)^2 \right] \right.$$
$$\left. \times W(X_0, t_0|Y, t) W(Y, t|X, t + \Delta t) \right\}. \tag{2.223}$$

We can now define the quantities

$$a(Y, t) = \lim_{\Delta t \to 0} \frac{1}{\Delta t} \int_{\mathbb{R}} dX (X - Y) \, W(Y, t|X, t + \Delta t) \tag{2.224}$$

$$b^2(Y, t) = \lim_{\Delta t \to 0} \frac{1}{\Delta t} \int_{\mathbb{R}} dX (X - Y)^2 \, W(Y, t|X, t + \Delta t) \tag{2.225}$$

and rewrite Eq. (2.223) as

$$\int_{\mathbb{R}} dX \, f(X) \, \frac{\partial W(X_0, t_0 | X, t)}{\partial t} = \int_{\mathbb{R}} dY \left[a(Y,t) f'(Y) + \frac{1}{2} b^2(Y,t) f''(Y) \right] W(X_0, t_0 | Y, t).$$

$$(2.226)$$

The right-hand side of this integral equation can be integrated by parts, and by renaming the integration variable $Y \rightarrow X$, we obtain

$$\int_{\mathbb{R}} dX \, f(X) \left[\frac{\partial W}{\partial t} + \frac{\partial}{\partial X} (a(X,t)W) - \frac{1}{2} \frac{\partial^2}{\partial X^2} (b^2(X,t)W) \right] = 0, \qquad (2.227)$$

where $W \equiv W(X_0, t_0 | X, t)$. Due to the arbitrariness of $f(X)$, we can finally write the Fokker–Planck equation for the transition probability $W(X_0, t_0 | X, t)$,

$$\frac{\partial W(X_0, t_0 | X, t)}{\partial t} = -\frac{\partial}{\partial X} (a(X,t) W(X_0, t_0 | X, t)) + \frac{1}{2} \frac{\partial^2}{\partial X^2} (b^2(X,t) W(X_0, t_0 | X, t)).$$

$$(2.228)$$

This equation is also known as the forward Kolmogorov equation. The pedagogical way this equation has been derived can be made more precise using the Kramers–Moyal expansion: In Appendix K we illustrate this technique for deriving the backward Kolmogorov equation, which describes the evolution with respect to the initial time t_0, rather than to the final time t. The backward equation reads

$$\frac{\partial W(X_0, t_0 | X, t)}{\partial t_0} = -a(X_0, t_0) \frac{\partial}{\partial X_0} W(X_0, t_0 | X, t) - \frac{1}{2} b^2(X_0, t_0) \frac{\partial^2}{\partial X_0^2} W(X_0, t_0 | X, t).$$

$$(2.229)$$

If there is time translational invariance, $W(X_0, t_0 | X, t)$ depends on $(t - t_0)$ only, so that the previous equation can be rewritten as

$$\frac{\partial W(X_0, t_0 | X, t)}{\partial t} = a(X_0, t_0) \frac{\partial}{\partial X_0} W(X_0, t_0 | X, t) + \frac{1}{2} b^2(X_0, t_0) \frac{\partial^2}{\partial X_0^2} W(X_0, t_0 | X, t).$$

$$(2.230)$$

Notice that this equation is not symmetric to Eq. (2.228), thus showing that the forward and backward formulations of the Kolmogorov equation are not straightforwardly related to each other. Eq. (2.229) is useful for some applications, such as the one discussed in Section 2.5.1.

We want also to point out that Eq. (2.228) tells us that the transition probability $W(X_0, t_0 | X, t)$ obeys a Fokker–Planck equation that is equivalent to the general Fokker–Planck equation (2.179) for the probability density $P(X, t)$. This is a consequence of the relation

$$P(X,t) = \int_{\mathbb{R}} dY \, P(Y, t') \, W(Y, t', | X, t), \qquad (2.231)$$

which is the space–time continuous version of Eq. (2.58). In fact, by deriving the previous equation with respect to t and using Eq. (2.228), one recovers Eq. (2.179). It is worth stressing that these equivalent formulations of the Fokker–Planck equation hold because we have assumed that $W(X_0, t_0 | X, t)$ vanishes in the limit $t \rightarrow t_0$ if $|X - X_0|$ remains finite. If this is not true, Eqs. (2.228) and (2.229) are not applicable, while the Chapman–Kolmogorov equation (2.217) still holds.

2.5.1 First Exit Time and the Arrhenius Formula

As an interesting application of what we have discussed in Section 2.4.5, we want to reconsider the problem of the escape time from an interval $I = [X_1, X_2]$. At page 78 we considered a symmetric, homogeneous diffusion process with absorbing barriers in $X_{1,2}$. Here we will consider a more complicated process and different boundary conditions. Let us start by setting the problem in general terms.

We can use the transition probability $W(X_0, t_0|X, t)$ to define the probability $\mathbb{P}_{X_0}(t)$ of being still in I at time t after having started from $X_0 \in I$ at time t_0:

$$\mathbb{P}_{X_0}(t) = \int_{X_1}^{X_2} dY\, W(X_0, t_0|Y, t). \tag{2.232}$$

In other words, $\mathbb{P}_{X_0}(t)$ is the probability that the exit time $\mathcal{T}_I(X_0)$ is larger than t. It should be stressed that Eq. (2.232) is not valid for any boundary conditions, because if the "particle" is allowed to exit the interval and reenter, such a relation between $\mathbb{P}_{X_0}(t)$ and $W(X_0, t_0|Y, t)$ is no longer true. In the following we are considering either reflecting boundary conditions or absorbing boundary conditions: In both cases, reentry is not allowed and Eq. (2.232) is valid.

Let us indicate with $\Pi(\mathcal{T})$ the probability density of the first exit time from I, starting at X_0 (to simplify the notation in the argument of $\Pi(\mathcal{T})$ we have overlooked the explicit dependence on I and X_0). By definition we have

$$\mathbb{P}_{X_0}(t) = \int_t^{+\infty} d\tau\, \Pi(\tau) \quad \rightarrow \quad \Pi(t) = -\frac{\partial \mathbb{P}_{X_0}(t)}{\partial t}. \tag{2.233}$$

The average exit time is given by the expression

$$\langle \mathcal{T}_I \rangle = \int_0^{+\infty} d\tau\, \Pi(\tau)\, \tau = -\int_0^{+\infty} d\tau \frac{\partial \mathbb{P}_{X_0}(\tau)}{\partial \tau}\, \tau = \int_0^{+\infty} d\tau \mathbb{P}_{X_0}(\tau), \tag{2.234}$$

where the last expression has been obtained by integrating by parts and assuming that $\mathbb{P}_{X_0}(t)$ vanishes sufficiently rapidly for $t \to +\infty$.

Now, we consider the backward Kolmogorov equation for the transition probability $W(X_0, t_0|Y, t)$ (2.230), specialized to the diffusive case, $b^2(X, t) = 2D$, with a time-independent drift term $a(X)$, namely,

$$\frac{\partial W(X_0, t_0|Y, t)}{\partial t} = a(X_0)\frac{\partial}{\partial X_0} W(X_0, t_0|Y, t) + D\frac{\partial^2}{\partial X_0^2} W(X_0, t_0|Y, t). \tag{2.235}$$

We can obtain an equation for $\mathbb{P}_{X_0}(t)$ by integrating both sides of Eq. (2.235) over Y, varying in the interval I, see Eq. (2.232):

$$\frac{\partial \mathbb{P}_{X_0}(t)}{\partial t} = a(X_0)\frac{\partial}{\partial X_0}\mathbb{P}_{X_0}(t) + D\frac{\partial^2}{\partial X_0^2}\mathbb{P}_{X_0}(t). \tag{2.236}$$

It is worth stressing that starting from the forward Kolmogorov equation (2.228), it would not have been possible to obtain such an equation. In the following we are going to replace the starting point X_0 with X, to make notations less cumbersome. Integrating further both sides of this equation over t and assuming $t_0 = 0$, we finally obtain an equation for $\langle \mathcal{T}_I(X) \rangle$:

$$-1 = a(X)\frac{\partial}{\partial X}\langle \mathcal{T}_I(X)\rangle + D\frac{\partial^2}{\partial X^2}\langle \mathcal{T}_I(X)\rangle, \tag{2.237}$$

where the left-hand side comes from the conditions $\mathbb{P}_X(0) = 1$ and $\mathbb{P}_X(+\infty) = 0$.

Equation (2.237) can be applied to the example discussed on page 79 by setting $a(X) = 0$ (no drift) and $\langle \mathcal{T}_I(X_1)\rangle = \langle \mathcal{T}_I(X_2)\rangle = 0$ (absorbing boundaries), yielding the solution

$$\langle \mathcal{T}_I(X)\rangle = \frac{1}{2D}(X - X_1)(X_2 - X), \tag{2.238}$$

which is the same result reported in Eq. (2.197).

It is possible to obtain the general solution of Eq. (2.237), once we observe that defining

$$\Phi(X) = \exp\left(\frac{1}{D}\int a(X)dX\right), \tag{2.239}$$

such equation can be rewritten as

$$\frac{d}{dX}\left(\Phi(X)\frac{d\langle \mathcal{T}(X)\rangle}{dX}\right) = -\frac{1}{D}\Phi(X), \tag{2.240}$$

which can be integrated twice, first giving

$$\frac{d\langle \mathcal{T}(X)\rangle}{dX} = -\frac{1}{D\,\Phi(X)}\int^X dY\,\Phi(Y) + \frac{C_1}{\Phi(X)}, \tag{2.241}$$

then

$$\langle \mathcal{T}(X)\rangle = C_1\int^X dY\,\frac{1}{\Phi(Y)} - \frac{1}{D}\int^X dY\,\frac{1}{\Phi(Y)}\int^Y dZ\Phi(Z) + C_2, \tag{2.242}$$

with the integration constants to be determined using the appropriate boundary conditions in $X = X_{1,2}$.

Let us now suppose absorbing conditions at both extrema, $\langle \mathcal{T}(X_1)\rangle = \langle \mathcal{T}(X_2)\rangle = 0$. The former condition can be implemented by taking the lower limit of the two integrals $\int^X dY\cdots$ equal to X_1, which automatically implies $C_2 = 0$. As for the lower limit of the integral $\int^Y dZ$, it is irrelevant, because its variation leads to redefine the constant C_1, so we can take it equal to X_1 as well. Now, Eq. (2.242) is written as

$$\langle \mathcal{T}(X)\rangle = C_1\int_{X_1}^X dY\,\frac{1}{\Phi(Y)} - \frac{1}{D}\int_{X_1}^X dY\,\frac{1}{\Phi(Y)}\int_{X_1}^Y dZ\Phi(Z), \tag{2.243}$$

with the right boundary condition, $\langle \mathcal{T}(X_2)\rangle = 0$, implying

$$C_1\int_{X_1}^{X_2} dY\,\frac{1}{\Phi(Y)} - \frac{1}{D}\int_{X_1}^{X_2} dY\,\frac{1}{\Phi(Y)}\int_{X_1}^Y dZ\Phi(Z) = 0. \tag{2.244}$$

We can finally write

$$\langle \mathcal{T}(X)\rangle = \frac{A(X; X_1, X_2) - B(X; X_1, X_2)}{D\int_{X_1}^{X_2}\frac{dY}{\Phi(Y)}}, \tag{2.245}$$

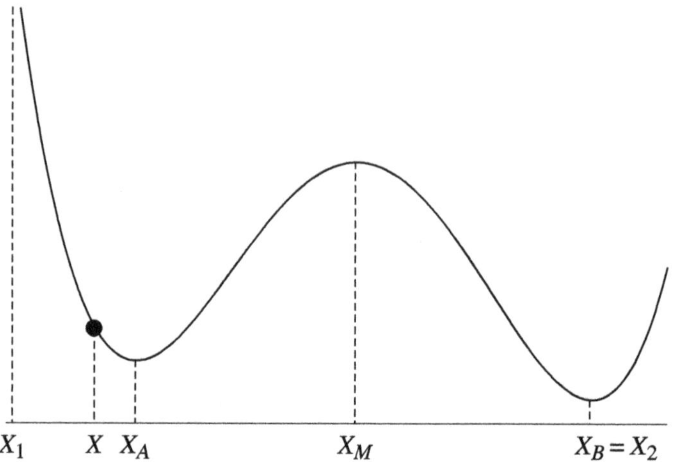

Fig. 2.10 Double well potential $U(X)$. The value X_1 can be finite or can diverge to $-\infty$.

where

$$A(X; X_1, X_2) = \int_{X_1}^{X_2} \frac{dY}{\Phi(Y)} \int_{X_1}^{Y} dZ \Phi(Z) \int_{X_1}^{X} \frac{dY'}{\Phi(Y')} \tag{2.246}$$

$$B(X; X_1, X_2) = \int_{X_1}^{X} \frac{dY}{\Phi(Y)} \int_{X_1}^{Y} dZ \Phi(Z) \int_{X_1}^{X_2} \frac{dY'}{\Phi(Y')}. \tag{2.247}$$

We conclude this section with an application of great physical relevance, the Arrhenius formula,[22] which describes the problem of the escape time of a stochastic diffusive process from an asymmetric double-well potential $U(X)$; see Fig. 2.10. In this case the generalized drift term is given by the expression $a(X) = -\frac{1}{\tilde{\gamma}} \frac{dU(X)}{dX}$, where $\tilde{\gamma}$ is the friction coefficient.

We want to ask the following question: If a diffusive walker starts at $t = 0$ close to the relative minimum X_A, what is the average time needed to overtake the barrier in X_M and to reach the absolute minimum in X_B? The question can be answered by computing the average time the particle takes to leave the interval $I = [X_1, X_2]$, with a reflecting barrier at X_1 and an absorbing barrier at $X_2 = X_B$. The left boundary condition yields

$$\lim_{X \to X_1^+} \frac{d\langle \mathcal{T}(X) \rangle}{dX} = 0 \tag{2.248}$$

and can be implemented (see Eq. (2.241)) taking $C_1 = 0$ and the lower limit of the integral equal to X_1, that is,

$$\frac{d\langle \mathcal{T}(X) \rangle}{dX} = -\frac{1}{D \, \Phi(X)} \int_{X_1}^{X} dY \, \Phi(Y), \tag{2.249}$$

where $\Phi(X)$, see Eq. (2.239), is now given by

$$\Phi(X) = e^{-\frac{1}{D\tilde{\gamma}} \int dx \frac{dU}{dx}} = e^{-\frac{1}{T} U(x)}.$$

---

[22] This formula, after the Swedish physicist Svante Arrhenius, has a great historical importance because it was originally formulated on the basis of heuristic arguments and successfully applied to the study of chemical reactions.

By applying the right boundary condition, $\langle \mathcal{T}(X_B) \rangle = 0$, we obtain the final, explicit result,

$$\langle \mathcal{T}(X) \rangle = \frac{1}{D} \int_X^{X_B} dY\, e^{\frac{U(Y)}{T}} \int_{X_1}^{Y} dZ\, e^{-\frac{U(Z)}{T}}. \tag{2.250}$$

The first integral is dominated by values of the variable Y close to the maximum of $U(X)$ in X_M, while the second integral is weakly dependent on Y for $Y \sim X_M$. Therefore, we can disregard such dependence and approximate the average exit time as follows:

$$\langle \mathcal{T}(X) \rangle \approx \frac{1}{D} \int_{X_1}^{X_M} dZ\, e^{-\frac{U(Z)}{T}} \int_X^{X_B} dY\, e^{\frac{U(Y)}{T}}. \tag{2.251}$$

The integral on Z is now a constant and is dominated by values of $U(Z)$ close to its minimum X_A, so we can use the Taylor series expansion ($\alpha_1 = U''(X_A)$)

$$U(Z) \simeq U(X_A) + \frac{1}{2}\alpha_1 (Z - X_A)^2 \tag{2.252}$$

and obtain the approximate estimate

$$\begin{aligned}
\int_{X_1}^{X_M} dZ\, e^{-\frac{U(Z)}{T}} &\approx e^{-\frac{U(X_A)}{T}} \int_{X_1}^{X_M} dZ\, e^{-\frac{\alpha_1}{2T}(Z-X_A)^2} \\
&\approx e^{-\frac{U(X_A)}{T}} \int_{-\infty}^{+\infty} dZ\, e^{-\frac{\alpha_1}{2T}(Z-X_A)^2} \\
&= \sqrt{\frac{2\pi T}{\alpha_1}}\, e^{-\frac{U(X_A)}{T}}.
\end{aligned} \tag{2.253}$$

In the second passage, we are allowed to extend the limits of the integral to $\pm\infty$ because we make an exponentially small error.

By applying a similar argument, we can observe that the integral over Y in Eq. (2.251) is dominated by values of Y close to X_M, and we can use the Taylor series expansion

$$U(Y) \simeq U(X_M) - \frac{1}{2}\alpha_2 (Y - X_M)^2, \tag{2.254}$$

with $\alpha_2 = -U''(X_M) > 0$, to obtain the approximate estimate

$$\begin{aligned}
\int_X^{X_B} dY\, e^{\frac{U(Y)}{T}} &\approx e^{\frac{U(X_M)}{T}} \int_X^{X_B} dY\, e^{-\frac{\alpha_2}{2T}(Y-X_M)^2} \\
&\approx e^{\frac{U(X_M)}{T}} \int_{-\infty}^{+\infty} dY\, e^{-\frac{\alpha_2}{2T}(Y-X_M)^2} \\
&= \sqrt{\frac{2\pi T}{\alpha_2}}\, e^{\frac{U(X_M)}{T}}.
\end{aligned} \tag{2.255}$$

In conclusion, the average exit time from a metastable potential well[23] is essentially independent of the starting point X, assuming it is close to X_A. The Kramers approximation

---

[23] The absorbing boundary condition on the right means that on the time scale of the exit process the particle has a vanishing probability to come back. Physically, this means that the new minimum in X_B must be lower than the minimum in X_A.

for the exit time can be finally written as

$$\langle \mathcal{T} \rangle = \frac{2\pi}{\sqrt{\alpha_1 \alpha_2}} \frac{T}{D} e^{\frac{\Delta U}{T}}, \tag{2.256}$$

where $\Delta U = U(X_M) - U(X_A)$, meaning that this quantity essentially depends on the height of the barrier separating X_A from X_B: The higher the barrier, the longer the average exit time from X_A.

The Arrhenius formula is essentially equivalent to Eq. (2.256) and provides an estimate of the reaction rate $\mathcal{R}$ of two diffusing chemical species, separated by a potential barrier ΔE, corresponding to the energy involved in the reaction process:

$$\langle \mathcal{T}(X) \rangle^{-1} \sim \mathcal{R} \propto \exp\left(-\frac{\Delta E}{T}\right). \tag{2.257}$$

2.6 Bibliographic Notes

The book by A. Einstein, *Investigations on the Theory of the Brownian Movement* (Dover, 1956), collects the five papers written by the great scientist between 1905 and 1908 about this central topic for the future developments of statistical mechanics. The book contains also very interesting notes by the editor, R. Fürth.

An interesting book in support of the atomic theory of matter was published in French by J. B. Perrin in 1913, and it is now available in several editions, with the English title *Atoms*.

An interesting article about the theory of Brownian motion is the original paper by G. E. Uhlenbeck and L. S. Ornstein, On the Theory of the Brownian Motion, *Physical Review*, **36** (1930) 823–841.

A mathematical approach to the theory of Brownian motion is contained in the book by E. Nelson, *Dynamical Theories of Brownian Motion*, 2nd ed. (Princeton University Press, 2001).

A formal approach to the perspective of probability theory and Markov chains can be found in W. Feller, *An Introduction to Probability Theory and Its Applications*, vol. 1, 3rd ed. (John Wiley & Sons, 1968).

A detailed and complete account about Monte Carlo methods and their applications can be found in K. Binder and D. W. Heermann, *Monte Carlo Simulation in Statistical Physics: An Introduction*, 6th ed. (Springer, 2019).

A clear, introductory textbook on stochastic processes is D. S. Lemons, *An Introduction to Stochastic Processes in Physics* (Johns Hopkins University Press, 2002).

A general and complete book about stochastic processes and their applications to stochastic differential equations and to Fokker–Planck equations is C. W. Gardiner, *Handbook of Stochastic Methods*, 4th ed. (Springer, 2009).

The Fokker–Planck equation in most of its facets and applications is discussed in H. Risken, *The Fokker–Planck Equation: Methods of Solution and Applications* (Springer-Verlag, 1984).

3

Fluctuations and Their Probability

This chapter includes classical arguments as well as modern topics, the latter covered more easily in reviews and monographs than in a textbook. First, we provide the reader an overview of the basic elements of probability theory, elaborated for tackling this problem. More precisely, in Section 3.2 we show how the Central Limit Theorem (CLT) allows us to establish under which conditions a normal statistics applies to a set of random variables. We also discuss how the statistics may deviate from a Gaussian distribution when the basic hypotheses of the CLT are relaxed. In Section 3.3 we introduce large deviations and extreme value statistics. It is important to point out that these concepts and the related mathematical tools are of great relevance in many fields of knowledge, because they allow us to extend our ability of statistical inference to the prediction of rare events. As a physical application of interest for this textbook, in Section 3.5 we illustrate how non-normal statistics, associated with heavy-tailed distribution functions, characterize anomalous diffusion processes. It is worth recalling that in Chapter 2 we have shown that Gaussian fluctuations are typically associated with normal diffusion processes. Relying upon a general model, the so-called continuous time random walk (CTRW), in this section we discuss how subdiffusive, diffusive, and superdiffusive processes, up to the ballistic regime, can be framed into a general mathematical scenario.

After these introductory sections dealing with a basic probabilistic approach to fluctuations, we turn our attention to the importance of fluctuations for statistical physics and thermodynamics, both in equilibrium and out-of equilibrium conditions. Some preliminary considerations are contained in Section 3.6. The fundamental approach proposed by Einstein to deal with these problems is reported in Section 3.7, where we show that a probabilistic interpretation can be attributed to the entropy of a fluctuating thermodynamic system. We point out that this theory also establishes a first, basic connection between equilibrium and nonequilibrium statistical mechanics, which will be definitely worked out in the context of the linear response theory (see Chapter 4).

It is worth anticipating here that linear response theory applies when strong simplifying assumptions, such as the local equilibrium hypothesis, hold. In this context one is dealing with nonequilibrium conditions of very large systems, evolving by a very slow macroscopic dynamics, that essentially amounts to a hydrodynamic behavior. One can argue that such conditions correspond to an adiabatic limit, where a thermodynamic transformation driving the system out of its equilibrium state can be performed reversibly, that is, by optimizing the efficiency of the transformation, as stated by Carnot's theorem.

If these simplifying assumptions do not apply, nonetheless some thermodynamically relevant information can be inferred through fluctuation relations associated with nonperturbative out-of-equilibrium conditions. In Section 3.8 we illustrate how the so-called

Jarzinski equality and Crooks theorem can be obtained from general measurement protocols in such out-of-equilibrium conditions in the presence of a single thermal reservoir. This scenario can be generalized to the case where a system is driven out of equilibrium by the presence of a thermal gradient imposed by two thermal reservoirs at different temperatures, yielding the formulation of a fluctuation theorem, which allows for a thermodynamic description in agreement with the second principle of thermodynamics.

3.1 Links between Statistics and Physics

In statistical physics the value of a given quantity, be it the position of a Brownian particle or the current flowing in a heat conductor, is subject to fluctuations, even if the measuring instrument is accurate. The focus of this chapter is on the statistics of such fluctuations, which may vanish or not in the thermodynamic limit. In the former case, our interest is about finite-size or finite-time effects. For example, what is the probability that the second law of thermodynamics is violated? If the number of degrees of freedom of our system is of the order of the Avogadro number, the answer is that we never observe such a violation, but for small systems (in which our interest has increased in recent times), the violation can be observed.

In some cases fluctuations are always relevant; they do not *compensate*. This is the case, for example, for anomalous diffusion, which may appear if the size distribution of the elementary hops of a random walker has a divergent variance: In this case the celebrated CLT does not apply. However, even if it does, we might be interested in the tail of the distribution rather than in its mean, or we might even be interested in the *extreme* event: What is the probability of the largest value of a fluctuating quantity?

Answering the earlier and similar questions needs to translate the problem of fluctuations in physical phenomena into the mathematical language of probability theory. Following his pioneering contributions to the theory of probability applied to physical problems, in particular to cosmology, the French mathematician Pierre-Simon de Laplace provided the first formulation of the CLT,[1] which was further refined by another French mathematician, Siméon-Denis Poisson. In fact, at the beginning of the nineteenth century, some outstanding mathematicians were primarily interested to establish the foundations of the theory of probability, because of its great impact for many applications in a wide spectrum of disciplines, including not only physics, but also biology, sociology, economy, and technology. The formulation of the theorem relied upon the assumption that one deals with an infinite sequence of independent, identically distributed variables. In this case one finds that any linear combination of such variables is normally distributed, that is, it obeys a universal Gaussian statistics, independent of the way the single variables are distributed. The theorem allows one to evaluate the average value and to estimate also the

---

[1] It seems that this wording ("zentraler Grenzwertsatz") was first used by the Hungarian mathematician George Pólya in the title of a scientific publication in the German journal *Mathematische Zeitschrift*. Pólya used such expression because the theorem "plays a central role in the calculus of probability." Most people think that central refers to the center of the distribution, which is also true.

error, or deviation, from the average. A more recent formulation of the CLT, which dates back to the beginning of the twentieth century, is due to the Finnish mathematician Jarl Waldemar Lindeberg and to the French mathematician Paul Lévy. This formulation extends the theorem to the case of a finite sequence of random variables, thus providing a more refined and complete account about deviations from the average.

A problem of great practical interest is the estimate of large deviations and of the extreme value statistics. In order to cope with this problem, one has to consider the statistics of events that are out of the standard deviation of the CLT. More precisely, large deviations theory goes beyond the Gaussian approximation, which is valid around the maximum of the probability distribution, while in extreme value statistics one is interested in knowing the statistics associated with the extremal, that is, minimal as well as maximal, values in a set of random variables. This is an important problem for analyzing any time series and it has applications to climate and finance, up to the physics of disordered systems.

As discussed in Chapter 2, a normal (Gaussian) statistics characterizes the fluctuations of diffusion processes and the sequence of *jumps* of a Brownian particle can be viewed as a dynamical manifestation of the universality of the CLT, which emerges also in the problem of the diffusive random walker. On the other hand, the complexity of natural phenomena indicates that there are stochastic processes that may escape a description based on a normal statistics. In order to deal with them in the sixties of the last century, the american mathematicians Elliott Waters Montroll and George Herbert Weiss introduced a suitable model, known as the continuous time random walk (CTRW), for extending the random walker problem to the case of anomalous diffusion. The basic ingredient of this model is the use of a non-normal statistics, that is, *heavy-tailed* probability distributions, for describing the jumps of a random walker. There are two formulations of this model: One is known as Lévy flights and it allows for independent statistics for the space run and for the time spent in a single *jump* of the random walker; the other, more physical one, known as Lévy walks (LWs), fixes a constant speed for each jump in such a way that the space run and the time spent in a single *jump* follow the same statistics. The main merit of the CTRW model is the identification of the general statistical conditions for reproducing normal diffusion as well as anomalous diffusion, either for subdiffusive or for superdiffusive processes, up to the ballistic limit.

The practical importance for thermodynamics of the fluctuations inherent in the approach based on statistical mechanics was first recognized by Einstein. His main contributions appeared in between 1902 and 1904, just before his celebrated paper on Brownian motion of 1905. In Einstein's original approach, the entropy can be associated with the probability of deviations from the equilibrium value of fluctuating macroscopic observables. Einstein's theory of thermodynamic fluctuations can be viewed as the building block of later elaborations, including the linear response theory of transport phenomena illustrated in Chapter 4.

More recent developments have been devoted to explore the role of fluctuations for establishing thermodynamic relations in out-of-equilibrium conditions in nonperturbative regimes (i.e., far from the linear response regime), for small systems, typically made by a number of microscopic elements, atoms, or molecules, much smaller than the Avogadro number, or even with fast thermodynamic transformations. In these situations the role of fluctuations of thermodynamic observables, such as energy or entropy, has to be reconsidered. Quite recently this kind of problem has become more and more important, because the thermodynamic properties of mesoscopic and microscopic systems have been widely investigated experimentally, while an increasing number of theoretical studies have been devoted to establish a solid ground for this kind of phenomena, nowadays interpreted in the context of a new research branch named stochastic thermodynamics.

3.2 The Central Limit Theorem

In this section we give a derivation of the CLT based on the characteristic function. We will emphasize its hypotheses and its limits, discussing which assumptions can be relaxed.

In simple terms, CLT shows that under certain conditions the linear combination of a diverging number N of random variables x_i, $X = \sum_{i=1}^{N} c_i x_i$, satisfies a Gaussian distribution, which explains its universality. The two main hypotheses are that x_i are independent and identically distributed (iid) variables. Since these assumptions are preserved if we pass from x_i to $y_i = c_i x_i$, the linear combination can be reduced to the sum.

Similarly to other derivations, we will introduce the variable

$$X = \frac{1}{\sqrt{N}} \sum_{i=1}^{N} (x_i - \langle x \rangle), \tag{3.1}$$

where each x_i follows the same distribution $p(x)$, with average value $\langle x \rangle = \int dx\, p(x) x$. The choice of the prefactor is easily explained by the observation that the variance of X, σ_X^2, is equal to the variance of each single random variable, σ^2,

$$\sigma_X^2 = \frac{1}{N} \sum_{i,j=1}^{N} \left(\langle x_i x_j \rangle - \langle x_i \rangle \langle x_j \rangle \right) = \frac{1}{N} \sum_{i=1}^{N} \left(\langle x_i^2 \rangle - \langle x_i \rangle^2 \right) = \sigma^2, \tag{3.2}$$

where we have used the property of independence, $\langle x_i x_j \rangle = \langle x_i \rangle \langle x_j \rangle$ if $i \neq j$. This result clearly depends on the finiteness of σ^2, that is, on the existence of the second moment of $p(x)$: This is the third hypothesis (in addition to independence and equal distribution) because CLT applies.

We stress that earlier result means that the variance of the arithmetic mean of x_i, that is, $s = \frac{1}{N} \sum_{i=1}^{N} x_i$, decays as $1/N$, which is a "proof" of the law of large numbers: The arithmetic mean tends to the average $\langle x \rangle$, evaluated according to $p(x)$.

Our proof of CLT goes through the characteristic function of a probability distribution, $\Gamma(k)$, that is, the average value of e^{ikx},

$$\Gamma(k) = \langle e^{ikx} \rangle = \int_{-\infty}^{+\infty} dx\, e^{ikx} p(x), \tag{3.3}$$

which is essentially the Fourier transform of $p(x)$. This also proves the existence of $\Gamma(k)$, because the finiteness of the integral $\int dx |p(x)|$ is certainly ensured by the normalizability of the distribution. In particular, for a Gaussian distribution,

$$p_G(x) = \frac{1}{\sqrt{2\pi\sigma^2}} e^{-\frac{x^2}{2\sigma^2}}, \tag{3.4}$$

its characteristic function is a Gaussian as well,

$$\Gamma_G(k) = \frac{1}{\sqrt{2\pi\sigma^2}} \int_{-\infty}^{+\infty} dx\, e^{ikx} e^{-\frac{x^2}{2\sigma^2}} = e^{-\frac{k^2\sigma^2}{2}}. \tag{3.5}$$

Let us now evaluate the characteristic function for the distribution $p_N(X)$ of X. We easily obtain

$$\Gamma_X(k) = \left\langle e^{ik\frac{1}{\sqrt{N}}\sum_i(x_i-\langle x\rangle)}\right\rangle \tag{3.6}$$

$$= \left\langle e^{i\frac{k}{\sqrt{N}}(x-\langle x\rangle)}\right\rangle^N \tag{3.7}$$

$$= \left(1 - \frac{k^2}{2N}\sigma^2 + O\left(\frac{k^3}{N^{3/2}}\right)\right)^N \tag{3.8}$$

$$= \exp\left(-\frac{k^2\sigma^2}{2}\right) + \text{correction terms}, \tag{3.9}$$

where we have used the formula

$$\lim_{N\to\infty}\left(1 + \frac{z}{N}\right)^N = e^z. \tag{3.10}$$

Comparing the final Eq. (3.9) with $\Gamma_G(k)$, see Eq. (3.5), we conclude that the random variable $X = \sum_i(x_i - \langle x\rangle)/\sqrt{N}$ is asymptotically ($N \to \infty$) distributed as a Gaussian, with zero average and with variance $\sigma_X^2 = \sigma^2$. Alternatively, we can say that the arithmetic mean $s = (\sum_i x_i)/N$ is asymptotically distributed as a Gaussian, with average $\langle x\rangle$ and with variance $\sigma_s^2 = \sigma^2/N$.

In the earlier derivation, we have assumed that X is the sum of a *diverging* number N of variables. In any real case, N is finite, which implies the existence of correction terms to the Gaussian distribution. Deriving such terms is not straightforward, also because the convergence of $p_N(X)$ to the Gaussian distribution $p_G(X)$ is not uniform; therefore, the validity of the approximation at a fixed N depends on X. It also depends on the specific expression of $p(x)$, of course. In fact, the probability distribution of the single variables x_i enters the goodness of the Gaussian approximation through two main features: the finiteness or not of all its moments, $\langle x^k\rangle$, and the vanishing or not of its third cumulant,[2] $Q_3 \equiv (x - \langle x\rangle)^3$ (which certainly vanishes if $p(x)$ is symmetric with respect to $\langle x\rangle$).

Let us consider a distribution $p(x)$ whose moments are all finite, which requires that for large x, it vanishes as a stretched exponential at least. Bearing in mind that X, see Eq. (3.1), is rescaled in such a way that $\langle X\rangle = 0$ and $\sigma_X = \sigma$, the Gaussian approximation is seen to be valid if $|X| < N^\gamma$, where the exponent $\gamma = 1/6$ if $Q_3 \neq 0$ and $\gamma = 1/4$ if $Q_3 = 0$. If instead $p(x)$ decays at large x as a power law, the goodness of the Gaussian approximation reduces until $\gamma = 0$, and the extension of the X variable over which CLT is applicable only increases logarithmically with N. Generally speaking, the determination of the tails of the

---

[2] All cumulants of a Gaussian distribution vanish from the third onward, see Appendix L.

distribution $p_N(X)$ is the subject of the large deviation theory, which will be discussed in Section 3.3. Such an approach will also allow a simple derivation of the exponent γ.

Let us now discuss the role of the different assumptions underlying the CLT and listed above Eq. (3.3), starting with the less relevant one: All variables x_i follow the same distribution $p(x)$. If this hypothesis is relaxed, the variable x_i has average μ_i and variance σ_i^2. Basically, we can continue to use the derivation done in Eqs. (3.6) to (3.9) if we replace $\langle x \rangle$ with μ_i in the definition (3.1) of X. The final result Eq. (3.9) is still valid, with $\sigma^2 = (\sum_{i=1}^{N} \sigma_i^2)/N \equiv D_N^2/N$, provided that σ_i^2/D_N^2 vanishes $\forall i$ in the limit $N \to \infty$. This means that the contribution of any random variable to the quantity D_N^2 is arbitrarily small for large N. It is called the Lindeberg condition.

The second and more important hypothesis to examine is the statistical independence of the random variables x_i. If this assumption does not apply, the passage from Eq. (3.6) to Eq. (3.7) is not allowed, because correlations between x_1 and x_2 imply that $\Gamma_{x_1+x_2}(k) \neq \Gamma_{x_1}(k)\Gamma_{x_2}(k)$. However, the CLT is still applicable if variables x_i are loosely coupled. Assuming that the correlation between x_i and x_{i+k} only depends on k, weak coupling means that

$$\sum_{k=1}^{\infty} C(k) \equiv \sum_{k=1}^{\infty} (\langle x_i x_{i+k} \rangle - \langle x_i \rangle \langle x_{i+k} \rangle) < \infty. \tag{3.11}$$

In practice the finiteness of correlations allows the following procedure. We group the N variables x_i in blocks of size L and we combine the L variables of each block in a single variable y_k, whose distribution is not expected to be Gaussian but whose variance is finite. The key point is that for sufficiently large L the new variables y_k are independent and CLT can be applied to them! More precisely, let us define the block variables as follows:

$$y_k = \frac{1}{\sqrt{L}} \sum_{i=(k-1)L+1}^{kL} (x_i - \langle x \rangle) \qquad k = 1, \ldots, \frac{N}{L} \tag{3.12}$$

and let us determine their variance and correlation:

$$\sigma_y^2 \equiv \langle y_k^2 \rangle - \langle y_k \rangle^2 \tag{3.13}$$

$$= \sigma^2 + \frac{2}{L} \sum_{1 \leq i < j \leq L} C(j-i) \tag{3.14}$$

$$\simeq \sigma^2 + 2 \sum_{k=1}^{\infty} C(k) \tag{3.15}$$

and

$$\langle y_k y_{k+1} \rangle - \langle y_k \rangle \langle y_{k+1} \rangle = \frac{1}{L} \sum_{i=1}^{L} \sum_{j=L+1}^{2L} C(j-1) \tag{3.16}$$

$$= \frac{1}{L} \left[\sum_{k=1}^{L} kC(k) + \sum_{k=L+1}^{2L-1} (2L-k)C(k) \right]. \tag{3.17}$$

If Eq. (3.11) is satisfied (i.e., if correlations are weak) for large (diverging) L the variance (3.15) is finite and the correlator (3.17) vanishes. These conditions are sufficient to apply the CLT to variables y_k, concluding that

$$X = \frac{1}{\sqrt{N}} \sum_{i=1}^{N} (x_i - \langle x \rangle) = \frac{1}{\sqrt{N/L}} \sum_{k=1}^{N/L} y_k \tag{3.18}$$

is Gaussianly distributed.

A simple but important example of the applicability of CLT to the case of weakly correlated random variables is provided by the numerical simulation of a physical system at equilibrium, when we follow its trajectory via molecular dynamics or Monte Carlo simulation: If we sample the trajectory at time intervals Δt, the different values of a given observable are correlated unless Δt is much larger than its correlation time τ. If $\Delta t \gg \tau$, the variables are uncorrelated and we can evaluate their arithmetic mean. According to CLT we know that such mean follows a Gaussian distribution with a variance vanishing as $1/N$. Instead, if $\Delta t \ll \tau$, the number of independent variables is of order $N^* \simeq N(\Delta t/\tau)$ and the variance vanishes as $1/N^*$.

Finally, let us discuss the hypothesis that the variance σ^2 of the single variable distribution $p(x)$ is finite. It is obvious that our derivation completely fails if σ diverges, because X and x have the same variance, see Eq. (3.2); therefore, σ_X diverges as well and X cannot be Gaussianly distributed. We can work out a specific example where $p(x)$ has a diverging variance, that of the Cauchy distribution, also called the Lorentzian distribution,

$$p(x) = \frac{1}{\pi(1 + x^2)}, \tag{3.19}$$

because its characteristic function can be easily calculated using the residue theorem,

$$\Gamma(k) = \langle e^{ikx} \rangle = \int_{-\infty}^{\infty} dx \frac{e^{ikx}}{\pi(1 + x^2)} = e^{-|k|}. \tag{3.20}$$

Therefore, the characteristic function of the arithmetic mean $s = (\sum_i x_i)/N$ is the same,

$$\Gamma_s(k) = (\Gamma(k/N))^N = \Gamma(k), \tag{3.21}$$

and $p_N(x) = p(x)$. In conclusion, if the variables x_i follow the Cauchy distribution, their mean follows the same distribution.

The sum of independent variables, each one following the same power-law distribution $p(x) \sim 1/|x|^{1+\alpha}$, will be found in other parts of this chapter. In Section 3.3.4 we will work out the specific case $\alpha = 3$ to show that the CLT has a restricted range of validity, because as soon as X is larger than some standard deviation, $P(X)$ follows the same power-law distribution. More generally, in Section 3.5.1 we will consider the Levy flights, a generalization of the standard random walk that has many applications and that is connected to $p(x) \sim 1/|x|^{1+\alpha}$ exactly as normal diffusion is connected to the CLT.

3.3 Large Deviations

In Section 3.2 we have shown that even if CLT is applicable, the Gaussian distribution may be valid only within a few σ (the standard deviation) from the average value. But rare events can be very important, so having information about their statistics is extremely valuable. The goal of this section is precisely to give some hints about the topic of large deviations and we start with a couple of examples, where we are able to derive analytically the expression for the full probability distribution of the arithmetic mean of iid variables x_i,[3]

$$s = \frac{1}{N} \sum_{i=1}^{N} x_i.$$
(3.22)

In the first example, x_i has a bimodal distribution; in the second example, x_i has an exponential distribution. Then, we will provide a more general treatment.

3.3.1 The Bimodal Distribution

We want to determine the distribution of the variable s defined here earlier knowing that each term of the sum takes the value $x_i = +1$ with probability p and the value $x_i = -1$ with probability $(1 - p)$. It is straightforward to evaluate the mean $\langle x \rangle = 2p - 1$ and the variance $\sigma^2 = 4p(1 - p)$ for the single variable.

If $N_{\pm}$ are the number of occurrences of $x_i = \pm 1$, the arithmetic mean reads $s = (N_+ - N_-)/N = 2(N_+/N) - 1$. The probability that there are exactly N_+ occurrences (in any order) is

$$P(N_+) = p^{N_+}(1 - p)^{N_-} \frac{N!}{N_+! N_-!},$$
(3.23)

which is correctly normalized because according to the binomial expansion we have

$$\sum_{N_+=0}^{N} P(N_+) = (p + (1 - p))^N = 1.$$
(3.24)

Using the Stirling approximation, $n! \simeq \sqrt{2\pi n}(n/e)^n$, we obtain

$$P(N_+) \simeq \sqrt{\frac{N}{2\pi N_+ N_-}}\, I, \qquad \text{with} \qquad I \equiv \left(\frac{pN}{N_+}\right)^{N_+} \left(\frac{(1-p)N}{N_-}\right)^{N_-}.$$
(3.25)

Let us now pass to the variable s, $P(s)ds = P(N_+)dN_+$, that is, $P(s) = (N/2)P(N_+)$, from which, using Eq. (3.25), we can write

$$P(s, N) = f(s, N)e^{-NC(s)},$$
(3.26)

where we have written I as an exponential and where the prefactor

$$f(s, N) \equiv \frac{N}{2}\sqrt{\frac{N}{2\pi N_+ N_-}} = \sqrt{\frac{N}{2\pi(1 + s)(1 - s)}}$$
(3.27)

has a subleading dependence on N.

---

[3] In the discussion of the CLT, we introduced the variable $X = \sum_i (x_i - \langle x \rangle)/\sqrt{N}$ because it was useful that the variance of X was of order one. In this context it is more useful to consider the arithmetic mean s.

The function $C(s)$, called the Cramer function or rate function, can be derived from the relation $I = e^{-NC(s)}$, giving

$$C(s) = \frac{1}{2}\left[(1+s)\ln\left(\frac{1+s}{2p}\right) + (1-s)\ln\left(\frac{1-s}{2(1-p)}\right)\right]. \tag{3.28}$$

The function $C(s)$ is plotted in Fig. 3.1.

3.3.2 The Exponential Distribution

In this case, each variable x_i entering the arithmetic mean $s = (\sum_i x_i)/N$ follows the exponential distribution

$$p(x) = \frac{1}{\mu}e^{-x/\mu}, \tag{3.29}$$

with mean $\langle x \rangle = \mu$ and variance $\sigma^2 = \mu^2$. The exact distribution followed by the variable s can be calculated as follows:

$$P(s, N) = \int_0^\infty dx_1 \cdots \int_0^\infty dx_N \frac{e^{-\frac{1}{\mu}(x_1+\cdots+x_N)}}{\mu^N} \delta\left(\frac{x_1 + \cdots + x_N}{N} - s\right) \tag{3.30}$$

$$= \left(\frac{N}{\mu}\right)^N s^{N-1} e^{-N\frac{s}{\mu}} \int_0^\infty dy_1 \cdots \int_0^\infty dy_N \delta\left(y_1 + \cdots + y_N - 1\right) \tag{3.31}$$

$$= \left(\frac{N}{\mu}\right)^N \frac{1}{(N-1)!} s^{N-1} e^{-N\frac{s}{\mu}}, \tag{3.32}$$

where the passage from the first to the second line is just due to the change of variables $x_i = (Ns)y_i$, and the N-dimensional integral appearing in the second line is equal to $(1/(N-1)!)$.[4]

Using the Stirling approximation, we find the approximate but more manageable expression,

$$P(s, N) \simeq \sqrt{\frac{N}{2\pi}} \left(\frac{e}{\mu}\right)^N s^{N-1} e^{-N\frac{s}{\mu}}, \tag{3.33}$$

which can be written in the form of Eq. (3.26), with

$$f(s, N) = \frac{1}{s}\sqrt{\frac{N}{2\pi}} \tag{3.34}$$

$$C(s) = \frac{s}{\mu} - 1 - \ln\left(\frac{s}{\mu}\right). \tag{3.35}$$

The function $C(s)$ is plotted in Fig. 3.1.

---

[4] Once that the s-dependence has been made explicit, the dependence on N and μ of the prefactor can simply be derived by the normalization condition $\int_0^\infty ds\, P(s, N) = 1$.

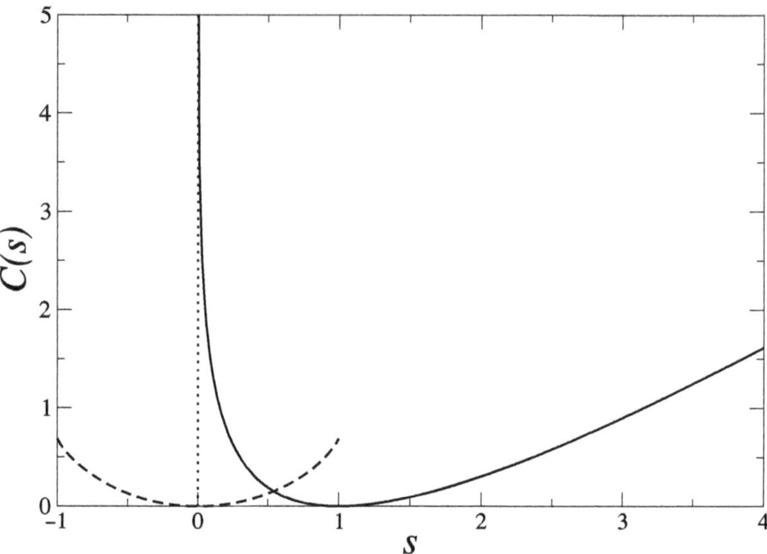

Fig. 3.1   The Cramer functions for the symmetric bimodal distribution (dashed line), see Eq. (3.28) with $p = \frac{1}{2}$, and for the exponential distribution (full line), see Eq. (3.35) with $\mu = 1$.

3.3.3 The Cramer Function

In the two previous examples, the probability distribution $P(s, N)$ of the arithmetic mean s of N iid variables has the form, $P(s, N) \simeq f(s, N)e^{-NC(s)}$. In general, this type of distribution allows to introduce the Cramer (or rate) function $C(s)$, which can be defined as a limit,

$$C(s) = \lim_{N \to \infty} -\frac{1}{N} \ln P(s, N), \tag{3.36}$$

so that any nonexponential dependence on N is washed out, as is the case for the prefactor $f(s, N)$. It is useful to remark some general properties of the Cramer function:

(i) $C(s)$ is always positive and it vanishes for $s = \langle x \rangle$, where it has a minimum.
(ii) In proximity of $\langle x \rangle$, we can write

$$C(s) \simeq C(\langle x \rangle) + \frac{C''(\langle x \rangle)}{2}(s - \langle x \rangle)^2 \simeq \frac{(s - \langle x \rangle)^2}{2\sigma^2}. \tag{3.37}$$

(iii) The function $C(s)$ is convex, $C''(s) > 0$.

The first property is satisfied by the bimodal distribution with $\langle x \rangle = 2p - 1$ and by the exponential distribution with $\langle x \rangle = \mu$. Therefore, for diverging N, $P(s, N) \to \delta(s - \langle x \rangle)$, which is the law of large numbers. The second property is equivalent to the CLT (see Section 3.2). The third property is satisfied by the bimodal distribution because $C''(s) = 1/(1 - s^2) > 0$ and by the exponential distribution because $C''(s) = 1/s^2 > 0$. The meaning of this property is related to the possibility to derive the Cramer function through its Legendre transformation, as it will be discussed here later. Before we deal with this task, let us consider the effects of the terms beyond the quadratic one in the expansion (3.37) around the minimum, because this allows us to reconsider the goodness of the CLT.

If we neglect the s-dependence of the prefactor, we can write

$$P(s, N) \sim e^{-N\left(\frac{z^2}{2\sigma^2} + c_3 z^3 + c_4 z^4 + o(z^4)\right)} \tag{3.38}$$

$$\sim P_G(s, N) e^{-N\left(c_3 z^3 + c_4 z^4\right)} \tag{3.39}$$

where $z = s - \langle x \rangle$, $c_k = (1/k!) d^k C(s)/ds^k |_{s=\langle x \rangle}$, and $P_G(s, N)$ is the Gaussian distribution resulting from the CLT. This is a good approximation to the extent that $Nz^3 \ll 1$ (if $c_3 \neq 0$) or that $Nz^4 \ll 1$ (if $c_3 = 0$). Since $s = (\sum_i x_i)/N$ while the variable X used in the context of the CLT, see Eq. (3.1), was defined as $X = \sum_i (x_i - \langle x \rangle)\sqrt{N}$, we have that $z = X/\sqrt{N}$ and the condition $Nz^3 \ll 1$ is equivalent to write $|X| \ll N^{1/6}$, while the condition $Nz^4 \ll 1$ is equivalent to write $|X| \ll N^{1/4}$. We have therefore found the validity conditions of the CLT discussed below Eq. (3.10), provided all moments are finite (see Section 3.3.4).

It is manifest that in most cases, we are unable to explicitly determine the cumulative distribution $P(s, N)$, and so we cannot use the definition (3.36) to find the Cramer function. In the general case we actually do not even know if such a function exists. The starting point to gain some insights into this problem is the evaluation of a quantity strictly related to the characteristic function of the cumulative variable s, that is, the average value of e^{qNs},

$$\langle e^{qNs} \rangle = \int ds P(s, N) e^{qNs} \tag{3.40}$$

$$\sim \int ds e^{N(qs - C(s))} \tag{3.41}$$

$$= e^{NL(q)}, \tag{3.42}$$

where

$$L(q) = \max_s (qs - C(s)) \tag{3.43}$$

is the Legendre transformation of the Cramer function.

If we limit here, as earlier, to a random variable s, which is the arithmetic mean of N iid variables x_i, the average value of e^{qNs} can be written in terms of the distribution $p(x)$ followed by each x_i, because

$$\langle e^{qNs} \rangle = \langle e^{qx_i} \rangle^N \tag{3.44}$$

and from Eqs. (3.42) and (3.44), we obtain

$$L(q) = \ln \langle e^{qx} \rangle. \tag{3.45}$$

Therefore, the Cramer function for the cumulative variable s is the Legendre transformation of the cumulant-generating function of the distribution $p(x)$ of the single variable x_i, see Appendix L. In Section 3.8.3, where we will discuss the Fluctuation theorem, we will see the use of the Cramer function in the context of the heat current distribution for a nonequilibrium steady state, see Eq. (3.221).

3.3.4 Power Law Distributions

It is interesting to consider a sum of iid variables whose distribution $p(x)$ decays at large x as a power law, $p(x) \sim 1/|x|^{1+\alpha}$, but with $\alpha > 2$ so that its variance is finite and CLT is applicable, but whose moments of order α or greater diverge. In this case the large deviation theorem does not apply because the distribution of the sum as well has a power law behavior.

Here we consider a specific distribution corresponding to $\alpha = 3$,

$$p(x) = \frac{2\sigma^3}{\pi(x^2 + \sigma^2)^2}, \tag{3.46}$$

whose variance is equal to σ^2 and whose characteristic function can be calculated explicitly,

$$\Gamma_x(k) = \left\langle e^{ikx} \right\rangle = (1 + \sigma|k|)e^{-\sigma|k|} \tag{3.47}$$

$$= 1 - \frac{\sigma^2 k^2}{2} + \frac{\sigma^3|k|^3}{3} + O(k^4). \tag{3.48}$$

The cubic term is singular in $k = 0$, as expected by the divergence of the third moment, see Eq. (L.5).

The characteristic function $\Gamma_X(k)$ of the collective variable[5] $X = (\sum_{i=1}^{N} x_i)/\sqrt{N}$ is simply derivable from $\Gamma_x(k)$, because

$$\Gamma_s(k) = \left(\Gamma_x\left(\frac{k}{\sqrt{N}}\right)\right)^N = \left(1 + \sigma\frac{|k|}{\sqrt{N}}\right)^N e^{-\sigma\sqrt{N}|k|}, \tag{3.49}$$

whose small$-k$ expansion writes[6]

$$\Gamma_s(k) = 1 - \frac{\sigma^2 k^2}{2} + \frac{\sigma^3|k|^3}{3\sqrt{N}} + O(k^4). \tag{3.50}$$

The singular cubic term is maintained, which means that $P(X)$ as well has diverging moments $\langle X^k \rangle$, with $k \geq 3$. We can even say more: Since the coefficient of the cubic term in Eq. (3.50) is equal to the same coefficient in Eq. (3.48) divided by $\sqrt{N}$, we can presume the same for the large $-X$ term of $P(X)$ respect to the large $-x$ term of $p(x)$,

$$p(x) \simeq \frac{2\sigma^3}{\pi x^4} \quad \text{(large } x\text{)}. \tag{3.51}$$

Therefore, for large deviations, we expect that

$$P(X) \simeq \frac{2\sigma^3}{\pi N^{1/2} X^4} \quad \text{(large } X\text{)}, \tag{3.52}$$

while for small deviations, CLT applies and

$$P(X) = \frac{1}{\sqrt{2\pi\sigma^2}} e^{-X^2/2\sigma^2}. \tag{3.53}$$

---

[5] We use again X rather than the arithmetic mean s because here we want to evaluate the limits of validity of the CLT, and the variance of X does not scale with N, unlike the variance of s.

[6] It is convenient to Taylor expand $\ln \Gamma_s(k)$, take the exponential, and expand again.

The comparison between the two expressions allows to determine the value X_0 beyond which the Gaussian approximation (3.53) is no more valid,

$$\frac{1}{\sqrt{2\pi\sigma^2}}e^{-X_0^2/2\sigma^2} \simeq \frac{2\sigma^3}{\pi N^{1/2}X_0^4},$$ (3.54)

that is, $X_0 \sim \sigma\sqrt{\ln N}$. This result is very different from what we have obtained below Eq. (3.39) using the large deviation function, therefore assuming that all moments $\langle X^k \rangle$ are finite: In that case X_0 grows as a power law of N, here it increases much more slowly.

3.3.5 Extreme Value Statistics

After having analyzed the probability distribution of the sum of N variables, either close to the maximum (CLT) or far from it (large deviations), it is useful to consider the distribution $\tilde{P}_N(x)$ of the *maximal* value among N i.i.d. random variables x_i, with $N \gg 1$. If we introduce the probability $P_>(x)$ that the single variable is larger than x,

$$P_>(x) = \int_x^\infty dy\, p(y),$$ (3.55)

then the probability $Q_<(x)$ that the maximum is smaller than x is

$$Q_<(x) = [1 - P_>(x)]^N = e^{N\ln(1-P_>(x))} \simeq e^{-NP_>(x)},$$ (3.56)

where we have used the vanishing of $P_>(x)$ for large x.

Therefore, $\tilde{P}_N(x) = dQ_</dx$ while the median value, x_m, satisfies the relation $Q_<(x_m) = 1/2$, that is,

$$P_>(x_m) \simeq \frac{\ln 2}{N}.$$ (3.57)

This result can be easily understood in terms of the law of large numbers. In fact, according to it, an event of probability p typically occurs Np times, so that we expect that the typical maximum occurs once.

It is useful to apply Eq. (3.57) to some distributions $p(x)$, keeping in mind that the key quantity to be determined is the probability $P_>(x)$ defined in Eq. (3.55). If $p(x) = Ae^{-cx^\alpha}$, we find[7]

$$P_>(x) = \frac{A}{c^{1/\alpha}}e^{-cx^\alpha}F(c^{1/\alpha}x),$$ (3.58)

where for large x, the x-dependence of $F(x)$ is weaker than the exponential factor. From earlier expression, we determine that the median value of the maximum has the following dependence on the number N of variables,

$$x_m \simeq (\ln N)^{1/\alpha},$$ (3.59)

which means that x_m grows slowly with N.

---

[7] The following relation can be easily proved for $\alpha = \frac{1}{2}, 1, 2$.

Instead, if $p(x)$ has a fat tail because it decays as a power law, $p(x) = A/x^{1+\alpha}$, then $P_>(x) = (A/\alpha)x^{-\alpha}$, and

$$x_m \sim N^{1/\alpha}. \tag{3.60}$$

It is useful to reconsider the initial hypotheses: The variables x_i must be uncorrelated and are distributed according to the same pdf. The first hypothesis can be easily relaxed if the random variables x_i are *weakly* correlated, that is, if correlations between x_i and $x_{i+\Delta}$ can be neglected as soon as Δ is larger than some correlation length ξ. If this is the case, previous theory can be applied provided that $N \gg \xi$, much in the same way we could relax the uncorrelation hypothesis to derive the CLT. In fact, for large N, a finite ξ allows to group the N variables in $N' = N/\xi$ subgroups C_j (with $j = 1, \ldots, N'$), each one characterized by the variable $y_j = \max_{i \in C_j} x_i$. Since $\max_j y_j = \max_i x_i$, the distribution of the maximum among the weakly correlated variables x_i can be traced back to the search of the maximum among the *uncorrelated* random variables y_j. However, we should keep in mind that we need to know (at least) the asymptotic behavior of the distribution of the variable $y = \max_{i=1,\xi} x_i$, where ξ is a finite, fixed length. Therefore we must be able to solve this problem either analytically (even approximately) or numerically.

Our approach to find x_m is also based on the hypothesis that the variables x_i are distributed according to the same pdf. If it is not so (but x_i are independent), the probability that the single variable is larger than x becomes dependent on i, $P_>^{(i)}(x) = \int_x^\infty dy\, p_i(y)$, and the distribution of the maximum is the derivative of $Q_<(x)$, where now $Q_<(x) \simeq e^{-\sum_{i=1}^N P_>^{(i)}(x)}$. Finally, Eq. (3.57) is replaced by

$$\sum_{i=1}^N P_>^{(i)}(x_m) \simeq \ln 2. \tag{3.61}$$

3.4 Random Matrices

The most interesting cases of the failing of the CLT arise when random variables are strongly interacting. If variables x_i can be understood as a sort of "time series" where correlation between x_i and x_{i+k} only depends on the "distance" k, we have seen that if correlations decay quickly enough, then CLT still applies. Here we mention an important example where all variables are correlated with each other: It is the theory of random matrices, which dates back to the 1950s and which has since shown its relevance in several areas of statistical physics. In those years, the Hungarian physicist Eugene Wigner investigated the energy levels of heavy nuclei and interpreted them as the eigenvalues of a suitable $N \times N$ real, symmetric matrix A whose entries A_{ij} are i.i.d. variables. If A_{ij} are normalized with zero mean and unit variance, and under some hypotheses on the finiteness of all moments, then the probability distribution of eigenvalues, $\rho(\lambda)$, converges for large N to the famous Wigner semicircle distribution,

$$\rho(\lambda) = \frac{1}{\pi N}\sqrt{2N - \lambda^2}. \tag{3.62}$$

In a sense, the Wigner law can be interpreted as a kind of equivalent of the CLT for random matrices. In CLT we start from i.i.d. variables x_i, we combine them linearly, and the result X is Gaussianly distributed.[8] Here, we start from i.i.d. variables A_{ij} and we combine them in a highly nonlinear manner to go from the elements of a matrix to its eigenvalues λ, which are distributed according to Eq. (3.62). It is interesting to remark that this problem can be mapped to a gas of one-dimensional interacting particles, which are subject to a potential well plus a logarithmic-like Coulomb repulsion[9]: The positions of the particles are the sought-after eigenvalues.

Furthermore, just as we are interested in the extreme statistics of X, we may be interested in the distribution of the maximal eigenvalue λ_M (i.e., in the earlier mapping, to the position of the rightmost particle). The median value of the maximum, λ_m, can be easily read off from the Wigner law, since distribution (3.62) has an upper cutoff $\lambda_c = \sqrt{2N}$. Therefore, $P_>(\lambda)$ (see Eq. (3.55)) vanishes for $\lambda \to \lambda_c$ and, according to Eq. (3.57), with increasing N, $\lambda_m \to \lambda_c$. It is less simple to derive the standard deviation of the distribution, which scales as $N^{-1/6}$, meaning that the median and mean are practically the same.

Deriving the full distribution is quite a technical task. Despite this, we are devoting the rest of this section to the resulting function, called Tracy–Widom (TW) distribution, because it plays an important role in the context of the Kardar–Parisi–Zhang (KPZ) universality class (see Section 7.7) and because it appears in different physical and mathematical areas. In particular, when studying the stochastic dynamics of a driven interface, TW distribution does not appear as an extreme value statistics: Instead, it describes the whole spectrum of fluctuations of a KPZ interface, which is not Gaussian nor symmetric. The fact that the TW distribution is not symmetric easily emerges from the abovementioned mapping with the model of interacting particles. Let's take the rightmost particle, which is repelled by the external potential from the right and by other particles from the left: Their combination results in a distribution that is clearly not symmetric.

There are actually three TW distributions, depending on how many real components there are for each element of the random matrix: If the matrix is real and symmetric (one component), we speak of Gaussian Orthogonal Ensemble (GOE); if the matrix is complex (two components), we speak of Gaussian Unitary Ensemble (GUE); if the matrix is quaternionic (four components), we speak of Gaussian Symplectic Ensemble (GSE). The distributions are called GOE/GUE/GSE TW distributions. The GUE-TW distribution, $F_2(s)$, is plotted and compared with the Gaussian curve in Fig. 3.2. The asymmetry is quantified by the skewness, see Appendix L, which vanishes for the Gaussian and it is approximately equal to 0.22 for $F_2(s)$. As for the tails of the TW distributions, they are strongly asymmetric: $F_2(s) \sim \exp(-|s|^3/12)$ for $s \to -\infty$ and $F_2(s) \sim \exp(-\frac{4}{3}s^{3/2})$ for $s \to \infty$.

---

[8] Let us remember that the tails of the distributions are not Gaussian.

[9] The classical model of particles in a potential well $U(x)$ and repelling each other with a logarithmic potential corresponds to a model of free fermions in a potential well $V(x)$, where $V(x)$ is parabolic if $U(x)$ is.

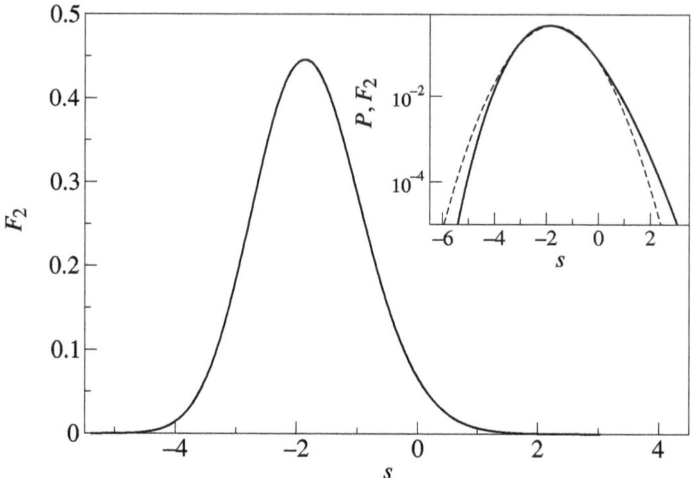

Fig. 3.2   One of the Tracy–Widom distributions, $F_2(s)$, called the Gaussian unitary ensemble. In the inset we plot it on a lin-log scale in order to highlight the difference between the TW distribution (solid line) with the Gaussian distribution (dashed line), plotted for the same average value $\bar{s} = -1.771$ and the same standard deviation, $\sigma = 0.9017$.

3.5 Generalized Random Walks

In Chapter 2, we used the mathematical language of discrete as well as of continuous stochastic processes to describe diffusive phenomena ruled by Gaussian probability distributions of observables. These processes are strictly related to the CLT because only its validity ensures that $X(N) = \sum_{i=1}^{N} x_i$ follows a Gaussian distribution: In this case the variance of $X(N)$ increases linearly with N, as expected for *normal* diffusion.

If the distribution of the single variables x_i has a diverging second moment, CLT no more applies and the resulting diffusion process is *anomalous*. In fact, there is a broad class of phenomena that escape standard diffusive processes and exhibit quite a different statistical description. As physical examples, we can mention turbulent fluid regimes, the dynamics of ions in an optical lattice, the diffusion of light in heterogeneous media, hopping processes of molecules along polymers, etc. Instances from other fields of science are the spreading of epidemics, the foraging behavior of bacteria and animals, the statistics of earthquakes and air traffic, and the evolution of the stock market. This incomplete list is enough to point out the importance of having suitable mathematical tools for describing the statistics of such a widespread class of phenomena that have in common an anomalous diffusive behavior.

The French mathematician Paul Lévy pioneered the statistical description of anomalous diffusive stochastic processes that are usually called Lévy processes; see Fig. 3.3 for an experimental example. Just to fix the ideas from the very beginning, a symmetric stochastic process described by the variable $x(t)$ (without prejudice of generality one can take

(a)   (b)

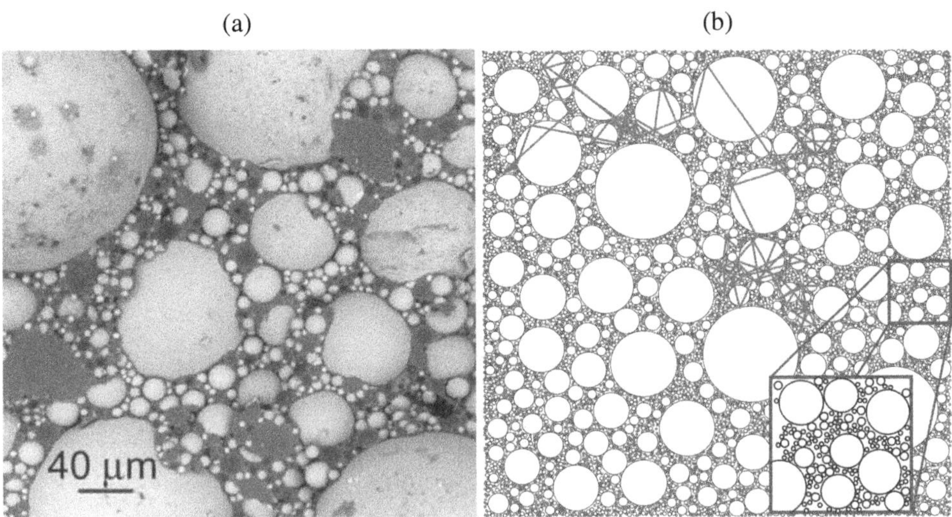

Fig. 3.3 (a) Electron micrograph of a Lévy glass. From M. Burresi et al., Weak Localization of Light in Superdiffusive Random Systems, *Physical Review Letters*, **108** (2012) 110604. (b) Simulated photon walk in a two-dimensional Lévy glass, with the inset showing the scale invariance of the glass. From P. Barthelemy, J. Bertolotti and D. S. Wiersma, Lévy Walk in an Inhomogeneous Medium, *Nature*, **453** (2008) 495–498.

$\langle x(t) \rangle = 0$) can be described by the time dependence of its variance,

$$\langle x^2(t) \rangle \propto t^\alpha, \quad 0 < \alpha < 2. \tag{3.63}$$

For $\alpha = 1$, we recover the standard diffusive process, while for $\alpha \neq 1$ we have anomalous diffusion and the cases $0 < \alpha < 1$ and $1 < \alpha < 2$ correspond to subdiffusive and superdiffusive stochastic processes, respectively. The limiting case $\alpha = 0$ corresponds, for example, to the Ornstein–Uhlenbeck process (see page 81), while the limiting case $\alpha = 2$ corresponds to ballistic motion.

3.5.1 Continuous Time Random Walk

The examples discussed in Section 2.3.2 describe fully discretized random walks: The walker performs a random sequence of equal-length space steps, each one in a unit time step, that is, the walker's steps are performed at a velocity with constant modulus. This simple way of representing a random walk can be generalized by assuming that the space steps, x, are continuous, independent, identically distributed random variables, according to a probability distribution $\eta(x)$ (as we have already discussed in Section 2.2.1 for an exponential distribution of step lengths). Similarly, we can assume that also the time intervals, t, needed to perform a step are continuous, independent, identically distributed random variables, according to a probability distribution $\omega(t)$. This implies that in this CTRW model, the walker performs elementary steps with different velocities. On the other hand, the very concept of velocity is still ill-defined, as well as for discrete random walk models, where the macroscopic motion is of diffusive type, despite the unit step being

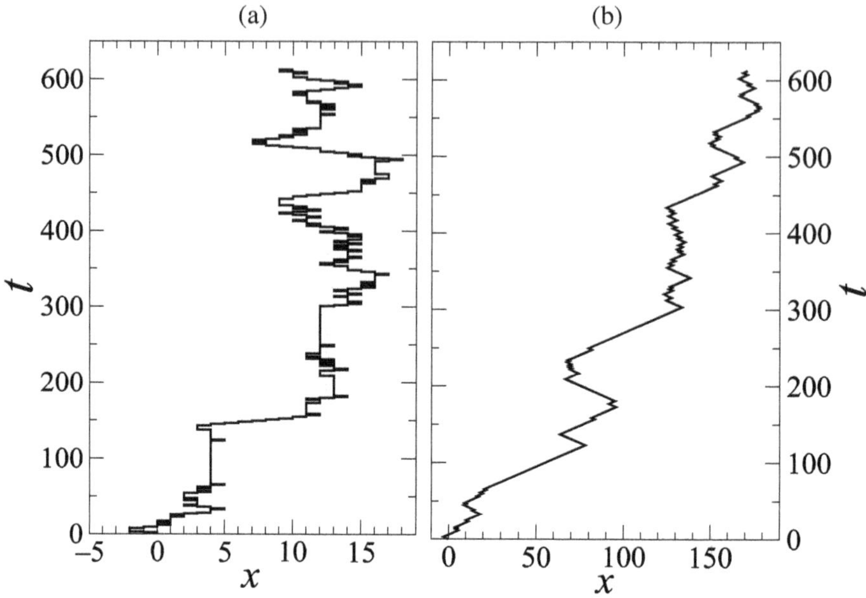

Fig. 3.4 (a) CTRW with step increment ± 1. (b) Lévy walk with velocity ± 1. Both evolutions are generated using the same realizations of time distribution $\omega(t)$ and of the $+/-$ signs for the direction of motion. We used $\omega(t) = \alpha/t^{1+\alpha}$ with $\alpha = 1.5$ for $t \geq 1$ and $\omega(t) = 0$ for $t < 1$.

performed at constant modulus of the velocity. In this perspective, it is more appropriate to interpret the continuous random variable t as a residence time of the walker at a given position in space, before performing a step in a vanishing time; see Fig. 3.4.

Let us now translate the CTRW model into a suitable mathematical language. For the sake of simplicity, we discuss here the one-dimensional case, whose extension to higher dimensions does not present additional difficulties. If we assume that the space where the random walker moves is isotropic, we have to assume that a jump of length x is equally probable to a jump of length $-x$; that is, we have to take into account a normalized symmetric distribution, $\eta(x) = \eta(-x)$ and $\int_{-\infty}^{+\infty} dx\, \eta(x) = 1$. The normalized probability distribution for the residence time $\omega(t)$ is defined for $0 \leq t \leq \infty$, with $\int_0^\infty \omega(t)dt = 1$.

It is useful to introduce the quantity

$$\psi(t) = \int_t^{+\infty} \omega(\tau)d\tau = 1 - \int_0^t \omega(\tau)d\tau, \tag{3.64}$$

which represents the probability of the walker not to make a step until time t (called survival probability). Now we want to define the random dynamics of the walker by introducing the probability $p(x,t)$ that a continuous-time random walker arrives at position x at time t, starting from the origin $x = 0$ at time $t = 0$. This probability can be written as[10]

$$p(x,t) = \int_{-\infty}^{+\infty} dy \int_0^t d\tau \eta(y)\, \omega(\tau)\, p(x-y, t-\tau) + \delta(x)\delta(t), \tag{3.65}$$

[10] The reader should note that $p(x,t)$ is the probability the walker *arrives* at position x at time t. The probability *to be* at position x at time t is $\rho(x,t)$, see Eq. (3.66).

which means that the position x can be reached at time t by any walker that arrived at any other position $x - y$ at time $t - \tau$, by making a jump of length y ($-\infty < y < +\infty$) after a waiting time τ ($0 \leq \tau \leq t$). Notice that causality is preserved, meaning that the evolution of the walker at time t depends only on what happened at previous times. We can use $p(x, t)$ and the survival probability $\psi(t)$ to obtain the probability that a CTRW is at position x at time t,

$$\rho(x, t) = \int_0^t d\tau\, p(x, t - \tau)\psi(\tau). \tag{3.66}$$

In fact, $\rho(x, t)$ is the sum over the time intervals τ during which a walker, which arrived at x at any previous time $t - \tau$, survived until time t.

We can use the Fourier–Laplace transformation to solve the integral Equations (3.65) and (3.66). Let us denote with k and s the dual variables of x and t, respectively. Eq. (3.65) is a convolution product for both variables x and t; by Fourier-transforming on x and by Laplace-transforming on t, this equation takes the simple form[11]

$$p(k, s) = \eta(k)\, \omega(s)\, p(k, s) + 1, \tag{3.67}$$

thus yielding

$$p(k, s) = \frac{1}{1 - \eta(k)\, \omega(s)}, \tag{3.68}$$

where

$$\eta(k) = \int_{-\infty}^{+\infty} dx\, e^{-ikx} \eta(x), \tag{3.69}$$

$$\omega(s) = \int_0^{+\infty} dt\, e^{-st} \omega(t). \tag{3.70}$$

Before performing the Fourier–Laplace transform of $\rho(x, t)$, it is useful to evaluate the Laplace transform of the survival probability, $\psi(t)$. Since its derivative is minus the residence time probability (see Eq. (3.64)), we can write

$$\omega(s) = \int_0^\infty dt e^{-st} \omega(t) \tag{3.71}$$

$$= -\int_0^\infty dt e^{-st} \psi'(t) \tag{3.72}$$

$$= 1 - s\psi(s), \tag{3.73}$$

where we have integrated by parts and used the condition $\psi(0) = 1$. We can use this result in the form

$$\psi(s) = \frac{1 - \omega(s)}{s} \tag{3.74}$$

to perform the Fourier–Laplace transform of $\rho(x, t)$, thus obtaining (see Eq. (3.66))

$$\rho(k, s) = \frac{1}{1 - \eta(k)\, \omega(s)} \frac{1 - \omega(s)}{s}. \tag{3.75}$$

---

[11] $p(x, t)$ can be extended to $t < 0$ by assuming $p(x, t) \equiv 0$ at negative times, so that the integration of $\delta(t)$ in Eq. (3.65) gives 1.

This is quite a useful equation, since the Fourier–Laplace transform of the density of CTRWs is found to depend in a simple way on $\eta(k)$ and $\omega(s)$. Moreover, the relation between the moments of the distribution of $\eta(x)$ and the derivatives of its Fourier transform $\eta(k)$, which is reported in Appendix M.3, applies also in this case, where discrete time steps have been substituted with continuous time steps, ruled by the probability density function $\omega(t)$. In analogy with Eq. (M.33), we can define the moments of the time-dependent distribution $\rho(x, t)$ as follows:

$$\langle x^m(t)\rangle = i^m \frac{d^m}{dk^m}\rho(k, t)\Big|_{k=0}. \tag{3.76}$$

For the Laplace-transformed moments, we can write

$$\langle x^m(s)\rangle = \int_0^{+\infty} dt \langle x^m(t)\rangle\, e^{-st} = i^m \frac{d^m}{dk^m}\rho(k, s)\Big|_{k=0}. \tag{3.77}$$

In particular, for $m = 2$, we have

$$\begin{aligned}
\langle x^2(s)\rangle &= -\frac{d^2}{dk^2}\rho(k, s)\Big|_{k=0}\\
&= -\frac{1-\omega(s)}{s}\frac{d^2}{d k^2}\frac{1}{1-\eta(k)\,\omega(s)}\Big|_{k=0}\\
&= -\frac{\omega(s)}{s\,[1-\omega(s)]}\eta''(k=0) - \frac{2\omega^2(s)}{s\,[1-\omega(s)]^2}(\eta'(k=0))^2,
\end{aligned} \tag{3.78}$$

where we have used that $\eta(k = 0) = \int_{-\infty}^{+\infty} dx\eta(x) = 1$. The two derivatives appearing in Eq. (3.78) encapsulate relevant information about the distribution $\eta(x)$.

Normal Diffusion

Let us start by investigating the case of normal diffusion, corresponding to $\omega(t)$ having a finite first moment and to $\eta(x)$ having a finite second moment. It is useful to investigate the limits $k \to 0$ and $s \to 0$, which provide information on the asymptotic behavior of $\rho(x, t)$ at large distances and for long times. In fact, in these limits we can use the approximate expressions

$$\eta(k) \approx \int_{-\infty}^{+\infty} dx \left(1 - ikx - \frac{k^2}{2}x^2 + \cdots\right)\eta(x) = 1 - ik\langle x\rangle - \frac{k^2}{2}\langle x^2\rangle + \cdots \tag{3.79}$$

$$\omega(s) \approx \int_0^{+\infty} dt\,(1 - st + \cdots)\omega(t) = 1 - \theta s + \cdots, \tag{3.80}$$

where we have considered the leading terms of the Taylor series expansion of the exponentials in Eqs. (3.69) and (3.70) and have introduced the quantities

$$\langle x \rangle = \int_{-\infty}^{+\infty} x\eta(x)dx \tag{3.81}$$

$$\langle x^2 \rangle = \int_{-\infty}^{+\infty} x^2\eta(x)dx \tag{3.82}$$

$$\theta = \langle t \rangle = \int_{0}^{+\infty} t\,\omega(t)\,dt. \tag{3.83}$$

For a symmetric distribution of spatial jumps, $\langle x \rangle = 0$ and $\langle x^2 \rangle = \sigma^2$. Coming back to Eq. (3.78), we can replace $\eta'(k = 0) = -i\langle x \rangle = 0$ and $\eta''(k = 0) = -\langle x^2 \rangle = -\sigma^2$, obtaining

$$\langle x^2(s) \rangle = \frac{\omega(s)}{s\,[1 - \omega(s)]}\sigma^2. \tag{3.84}$$

If we also replace the $s \to 0$ expansion of $\omega(s)$, we find

$$\langle x^2(s) \rangle = \frac{\sigma^2}{\theta s^2}, \tag{3.85}$$

whose Laplace antitransform is

$$\langle x^2(t) \rangle = \frac{\sigma^2}{\theta}t. \tag{3.86}$$

The last equation is proved observing that $\int_0^\infty dt\,e^{-st}t = 1/s^2$.

In conclusion, if θ and σ^2 are finite, we obtain standard diffusion, with

$$D = \frac{\sigma^2}{2\theta}. \tag{3.87}$$

This result can also be found by using expansions (3.79) and (3.80) in Eq. (3.75). Then, at leading order we can write $\left(1 - \eta(k)\omega(s)\right) \simeq \sigma^2 k^2/2 + \theta s$ and $\left(1 - \omega(s)\right) \simeq \theta s$, so that

$$\rho(k, s) \simeq \frac{1}{\left(\theta s + \frac{\sigma^2}{2}k^2\right)}\frac{\theta s}{s} \simeq \left(s + \frac{\sigma^2}{2\theta}k^2\right)^{-1}. \tag{3.88}$$

As shown in Appendix M.2, this is the Fourier–Laplace transformation of the solution of the diffusive equation

$$\frac{\partial \rho(x, t)}{\partial t} = D\frac{\partial^2 \rho(x, t)}{\partial x^2}, \tag{3.89}$$

with D given by Eq. (3.87).

Subdiffusion

The CTRW model also provides interesting examples of anomalous diffusion. For instance, let us consider the case where σ^2 is finite while θ diverges. This means that $\omega(t)$ is dominated by a power law for $t \to +\infty$,

$$\omega(t) \sim \frac{t_0^\alpha}{t^{\alpha+1}}, \tag{3.90}$$

with $0 < \alpha < 1$ and t_0 is now a suitable parameter with the dimension of a time, in such a way that ω keeps the correct dimension of an inverse time.

Notice that, even in the presence of such an asymptotic behavior, $\omega(t)$ is a well-defined probability distribution provided it can be normalized, which requires $\alpha > 0$ and that $\omega(t)$ is a smooth function for $t \to 0^+$.[12] On the other hand, the average residence time (see Eq. (3.83)) is found to diverge as $t^{1-\alpha}$ and the CTRW describes a much slower evolution than standard diffusion, that is, a subdiffusive process. In this case the Fourier–Laplace transform of $\omega(t)$ reads

$$\omega(s) = 1 - t_0^\alpha\, s^\alpha. \tag{3.91}$$

This result is a generalization of Eq. (3.80), which is not applicable here because $\alpha < 1$ and $\langle t \rangle$ diverges. However, $\omega(t)$ is still normalizable, so it must be

$$\lim_{s \to 0} \omega(s) = 1, \tag{3.92}$$

which allows us to write $\omega(s) - \omega(0) = \int_0^\infty dt(e^{-st} - 1)\omega(t)$. Since we are interested in obtaining an estimate for small values of s and since $0 < \alpha < 1$, we can write

$$\omega(s) - \omega(0) = \left(\int_0^1 + \int_1^\infty \right) dt(e^{-st} - 1)\omega(t)$$

$$\approx -s \int_0^1 dt\, t\omega(t) - t_0^\alpha \int_1^\infty dt\, \frac{1 - e^{-st}}{t^{1+\alpha}}$$

$$\approx -t_0^\alpha s^\alpha \int_0^\infty d\tau\, \frac{1 - e^{-\tau}}{\tau^{1+\alpha}} + O(s) \tag{3.93}$$

where we have approximated $(e^{-st} - 1)$ with $-st$ in the first integral, and we have used Eq. (3.90) in the second integral, whose lower limit can be shifted from $\tau = s$ to $\tau = 0$. In conclusion, since the definite integral in the last line is a number and t_0^α has just a dimensional role, we obtain Eq. (3.91).

We can now follow the same procedure adopted for the standard diffusive case, so in the limits $k \to 0$ and $s \to 0$, we can use the approximations $(1 - \eta(k)\omega(s)) \simeq \sigma^2 k^2/2 + (t_0\, s)^\alpha$ and $(1 - \omega(s)) \simeq (t_0\, s)^\alpha$, thus obtaining at leading order the expression

$$\rho(k, s) \simeq \frac{t_0^\alpha s^{\alpha-1}}{\frac{\sigma^2 k^2}{2} + (t_0 s)^\alpha} \simeq \left(s + \frac{\sigma^2 k^2}{2t_0^\alpha}\, s^{1-\alpha} \right)^{-1}, \tag{3.94}$$

which reduces to Eq. (3.88) for $\alpha = 1$, with $t_0 = 0$. Using this expression of $\rho(k, s)$, we can first compute the probability of finding the walker at the origin. This is given by the expression

$$\rho(x = 0, s) = \frac{1}{2\pi} \int_{-\infty}^{+\infty} dk\, \rho(k, s) \simeq \frac{t_0^\alpha s^{\alpha-1}}{2\pi} \int_{-\infty}^{+\infty} dk\, \left(k^2 \frac{\sigma^2}{2} + t_0^\alpha s^\alpha \right)^{-1}. \tag{3.95}$$

---

[12] Heuristic arguments suggest that physically interesting processes demand a vanishing probability distribution for a zero residence time of the CTRW.

With the change of variable $y = k\sqrt{\sigma^2/(2\,t_0^\alpha\,s^\alpha)}$, we can write

$$\rho(x = 0, s) \simeq \frac{1}{\pi\sqrt{2}}\,\frac{t_0^{\alpha/2}}{\sigma\,s^{1-\alpha/2}} \int_{-\infty}^{+\infty} dy\,\frac{1}{1+y^2} = \frac{1}{\sqrt{2}}\,\frac{t_0^{\alpha/2}}{\sigma\,s^{1-\alpha/2}}. \tag{3.96}$$

Since $\rho(x = 0, s)$ has not a finite limit for $s \to 0$, the line of reasoning to obtain (3.93) cannot be used and a more rigorous approach, making use of the so-called Tauberian theorems, should be adopted.[13] Using such theorems, the result for the probability of finding the walker at the origin is given by

$$\rho(0, t) = \frac{1}{\sqrt{2}}\,\frac{t_0^{\alpha/2}}{\sigma\,\Gamma(1 - \alpha/2)}\,t^{-\alpha/2}. \tag{3.97}$$

We can therefore conclude that this probability decays algebraically in time with an exponent $\alpha/2$, which for normal diffusion is $1/2$, with $t_0 = \theta$. It is important to stress that this expression provides just the asymptotic (i.e., large t) behavior of the probability density function of finding the walker at the origin.

From (3.94) we can compute all moments of the distribution. Here we just compute the time-dependent average square displacement, using the general formula (3.78), with $\langle x \rangle = 0$, and $\omega(s)$ given by (3.91), thus yielding up to leading order in k

$$\langle x^2(s) \rangle = -\frac{\omega(s)}{s\,[1 - \omega(s)]}\,\sigma^2 \approx \frac{1}{t_0^\alpha\,s^{\alpha+1}}\,\sigma^2. \tag{3.98}$$

By applying the results of the Tauberian theorems, we obtain

$$\langle x^2(t) \rangle = \frac{1}{\Gamma(1 + \alpha)\,t_0^\alpha}\,\frac{\sigma^2}{}\,t^\alpha = 2\,D_\alpha\,t^\alpha, \tag{3.99}$$

where D_α is a generalized diffusion coefficient.

For the sake of completeness, we want to mention that Eq. (3.94) can be read as the Fourier–Laplace transform of the following fractional partial differential equation

$$\frac{\partial^\alpha \rho(x, t)}{\partial t^\alpha} = D_\alpha\,\frac{\partial^2 \rho(x, t)}{\partial x^2}. \tag{3.100}$$

Although the important topic of fractional partial differential equations cannot be addressed here, we can nevertheless observe on an intuitive ground that Eq. (3.99) follows by a simple dimensional analysis of Eq. (3.100). In conclusion, Eq. (3.99) is the basic physical relation describing a subdiffusive random walk (i.e., $0 < \alpha < 1$) in continuous space and time.

---

[13] The Tauberian theorems allow us to determine the Laplace transform of a function $f(t)$ whose behavior for large t is of the form

$$f(t) \simeq t^{\gamma-1}\,\Phi(t) \quad \text{with} \quad 0 < \gamma < \infty,$$

where $\Phi(t)$ is a slowly varying function of t (e.g., $\lim_{t\to+\infty}\Phi(t) = c$, where c is a constant, or even $\phi(t) = [\log(t)]^n$). The Tauberian theorems tell us that the Laplace transform of $f(t)$ is

$$f(s) = \Gamma(\gamma)\,s^{-\gamma}\,\Phi(1/s).$$

Superdiffusion: Lévy Flights

Another interesting example contained in the CTRW model is the case where the average waiting time $\langle t \rangle = \theta$ is finite, while $\eta(x)$ is a normalized symmetric probability density function, whose asymptotic behavior obeys the power law

$$\eta(x) \sim |x|^{-(1+\alpha)}, \tag{3.101}$$

with $0 < \alpha < 2$. These processes are known as Lévy flights. For symmetry reasons, the first moment of $\eta(x)$ is zero, while the second moment is found to diverge as

$$\langle x^2 \rangle = \lim_{R \to \infty} \int^R dx\, |x|^{1-\alpha} \approx \lim_{R \to \infty} R^{2-\alpha}. \tag{3.102}$$

The Fourier transform of $\eta(x)$ can be evaluated with the spirit of Eq. (3.93), using the fact that the distribution probability is normalizable, so $\eta(k = 0) = 1$:

$$\eta(k = 0) - \eta(k) = \int_{-\infty}^{+\infty} dx \left(1 - e^{-ikx}\right) \eta(x) \tag{3.103}$$

$$\approx 2 \int_{1/|k|}^{\infty} dx \frac{1 - \cos(|k|x)}{x^{1+\alpha}} \tag{3.104}$$

$$= 2|k|^\alpha \int_1^\infty dy \frac{1 - \cos y}{y^{1+\alpha}}. \tag{3.105}$$

Finally, we find

$$\eta(k) \sim 1 - (\sigma\,|k|)^\alpha, \tag{3.106}$$

where σ is a suitable physical length scale.

In the limits $k \to 0$ and $s \to 0$, from Eq. (3.75) we obtain the approximate expression

$$\rho(k, s) \simeq \left(s + \frac{\sigma^\alpha}{\theta} |k|^\alpha\right)^{-1}. \tag{3.107}$$

By introducing the quantity $\mathcal{D}_\alpha = \sigma^\alpha/\theta$, we obtain the characteristic function (see Appendix L) of the generalized Levy law,

$$\rho(k, t) \simeq e^{-\mathcal{D}_\alpha t\,|k|^\alpha}, \tag{3.108}$$

where $\rho(k, t)$ is the inverse Laplace transform of $\rho(k, s)$ with respect to s, because $\int_0^\infty dt\, e^{-st}\, e^{-\mathcal{D}_\alpha t |k|^\alpha} = 1/(s + \mathcal{D}_\alpha |k|^\alpha)$. The probability distribution $p(x, t)$ can be obtained by antitransforming $\rho(k, t)$ in the Fourier space of k, but this is not an easy task, because the exponent α in Eq. (3.108) is not an integer. Nonetheless, it is possible to derive the fractional moments of order q of $\rho(x, t)$, as

$$\langle |x(t)|^q \rangle = \int_{-\infty}^{+\infty} \rho(x, t) |x(t)|^q dx \tag{3.109}$$

$$= \frac{1}{2\pi} \int_{-\infty}^{+\infty} dx |x|^q \int_{-\infty}^{+\infty} dk\, e^{ikx} e^{-\mathcal{D}_\alpha t\,|k|^\alpha} \tag{3.110}$$

$$= (\mathcal{D}_\alpha t)^{q/\alpha} \frac{1}{2\pi} \int_{-\infty}^{+\infty} dy |y|^q \int_{-\infty}^{+\infty} dp\, e^{ipy} e^{-|p|^\alpha} \tag{3.111}$$

$$\sim (\mathcal{D}_\alpha t)^{q/\alpha}, \tag{3.112}$$

where we have made two changes of variable, $p = (\mathcal{D}_\alpha t)^{1/\alpha} k$ and $y = x/(\mathcal{D}_\alpha t)^{1/\alpha}$, and the double integral in Eq. (3.111) is just a number. In particular, for $q = 2$ we have[14]

$$\langle x^2(t) \rangle \sim (\mathcal{D}_\alpha t)^{2/\alpha}. \tag{3.113}$$

As expected, for $\alpha = 2$ we recover the standard diffusive behavior, while for $\alpha < 2$ the mean squared displacement grows in time more than linearly. Notice that $\alpha = 1$ corresponds to a ballistic propagation and it is representative of the Cauchy distribution, see Eq. (3.19).

As we are going to discuss in Section 3.5.2, a suitable description of superdiffusion in the CTRW requires consideration of finite velocities by the introduction of the Lévy walk process.

3.5.2 Lévy Walks

Lévy processes are a general class of stochastic Markov processes with independent stationary time and space increments. For example, earlier we studied Lévy flights, characterized by a narrow distribution of time steps and a distribution of space increments given by Eq. (3.101). The main advantage of studying Lévy flights is that they are quite simple mathematical models of anomalous (super) diffusion. Their main drawback is that we are implicitly assuming that each step lasts over a fixed time interval, independent of the step length. Since this can be arbitrarily large, the random walker seems to move with an arbitrarily large velocity, a manifest drawback to giving a physical interpretation to such processes. Such physical interpretation can be restored by introducing Lévy walks (LWs), where the distribution of lengths is still given by

$$\eta(x) \sim \frac{1}{|x|^{1+\alpha}}, \tag{3.114}$$

but the path between the starting and the end points of a jump is assumed to be run by the walker at constant velocity $\mathbf{v}$; that is, the time t spent in a jump is proportional to its length x, with $t = x/|\mathbf{v}|$ (see Fig. 3.4(b)). It can be argued that a LW can be equally defined as a process where a walker moves at constant velocity in a time step t, whose asymptotic distribution is $\omega(t) \sim t^{-(1+\alpha)}$. In practice, we are assuming that time and length of a jump, being proportional to each other in each single jump, are defined by the same probability distribution. If $\alpha < 2$, the variance of $\omega(t)$ diverges and the distances of the jumps run by the walker obey the probability distribution (3.114). In the context of LW, one can describe superdiffusive behavior.

In general terms, we can formulate the LW process as a suitable modification of the CTRW model discussed in Section 3.5.1. The probability $p(x,t)$ that a LW arrives at position x at time t, starting from the origin $x = 0$ at time $t = 0$, still obeys an equation like (3.65), namely,

$$p(x,t) = \int_{-\infty}^{+\infty} dy \int_0^t d\tau \Omega(y,\tau)\, p(x-y, t-\tau) + \delta(x)\delta(t), \tag{3.115}$$

---

[14] In this context, the case $0 < \alpha < 1$, while formally included in the mathematical model of Levy flights, yields an average square displacement growing faster than t^2, a situation that does not correspond to any interpretation of physical interest.

where $\Omega(y, \tau)$ is the joint probability distribution of performing a step of length y in a time τ. Since a LW runs at constant velocity v, this quantity can be expressed as

$$\Omega(x, t) = \frac{1}{2}\delta(|x| - vt)\,\omega(t), \tag{3.116}$$

where

$$\omega(t) \sim t^{-(1+\alpha)} \tag{3.117}$$

for large values of t. By Fourier-transforming on x and by Laplace-transforming on t, Eq. (3.115) yields

$$p(k, s) = \frac{1}{1 - \Omega(k, s)}. \tag{3.118}$$

If $\Omega(x, t) = \eta(x)\omega(t)$ as in Eq. (3.65), we have $\Omega(k, s) = \eta(k)\omega(s)$ and we recover Eq. (3.68). In the present case instead, the Fourier–Laplace transform of $\Omega(x, t)$ (see Eq. (3.116)) is

$$\begin{aligned}
\Omega(k, s) &= \frac{1}{2}\int_{-\infty}^{+\infty} dx \int_{0}^{+\infty} dt\, e^{-ikx - st}\Big(\delta(-x - vt) + \delta(x - vt)\Big)\omega(t) \\
&= \frac{1}{2}\left[\int_{0}^{+\infty} dt\, e^{-(s-ivk)t}\,\omega(t) + \int_{0}^{+\infty} dt\, e^{-(s+ivk)t}\,\omega(t)\right] \\
&= \frac{1}{2}\big[\omega(s - ivk) + \omega(s + ivk)\big] = \mathrm{Re}\,\omega(s + ivk). \tag{3.119}
\end{aligned}$$

$\alpha < 1$, Ballistic Behavior

For $\alpha < 1$, we can use the result $\omega(s) = 1 - t_0^\alpha s^\alpha$, obtained for the CTRW model (see Eq. (3.91)), to write

$$\Omega(k, s) \simeq 1 - \frac{1}{2}t_0^\alpha\big[(s - ivk)^\alpha + (s + ivk)^\alpha\big]. \tag{3.120}$$

This equation must be handled with care, because we have noninteger powers of binomials. We must first take the limit of large distances ($k \to 0$), then the limit of long times ($s \to 0$); otherwise, asymptotic dynamics are limited by the artificially imposed finite size of the system. Therefore, we obtain

$$\Omega(k, s) \approx 1 - t_0^\alpha s^\alpha - \frac{t_0^\alpha v^2}{2}\alpha(1 - \alpha)k^2\, s^{\alpha-2}. \tag{3.121}$$

In analogy with definition (3.66), we want to use these quantities to obtain an expression for the probability that a LW is at position x at time t,

$$p(x, t) = \int_{-\infty}^{+\infty} dx' \int d\tau\, p(x - x', t - \tau)\Psi(x', \tau), \tag{3.122}$$

where now $\Psi(x', \tau)$ is the probability that a LW moves exactly by a distance x' in a time τ. Thus

$$\Psi(x, t) = \delta(|x| - vt)\int_{t}^{+\infty}\omega(t')dt', \tag{3.123}$$

which corresponds to the motion of a walker that proceeds at constant speed v under the condition that no scattering event occurs before time t. The same kind of calculation performed to obtain $\Omega(k, s)$ yields the expression of the Fourier–Laplace transform $\Psi(k, s)$,

$$\Psi(k, s) = \mathrm{Re}\,\psi(s + ivk), \tag{3.124}$$

where we recall that $\psi(s)$ is the Laplace transform of $\psi(t) = \int_t^\infty \omega(t')dt'$ and it is given by Eq. (3.74), which was obtained for CTRW, but still holds for LW.

By combining these results, the expression of the Fourier–Laplace transform of (3.122) is

$$\rho(k, s) = p(k, s)\,\Psi(k, s) = \frac{\Psi(k, s)}{1 - \Omega(k, s)}, \tag{3.125}$$

where the denominator can be found by Eq. (3.121). Making use of Eqs. (3.124) and (3.74), the numerator can be evaluated in a similar manner, giving

$$\Psi(k, s) = t_0^\alpha s^{\alpha-1} + \frac{t_0^\alpha v^2}{2}(\alpha - 1)(2 - \alpha)k^2\, s^{\alpha-3}. \tag{3.126}$$

We finally obtain

$$\rho(k, s) \simeq \frac{1}{s}\frac{s^\alpha - c^2 k^2 s^{\alpha-2}}{s^\alpha + b^2 k^2 s^{\alpha-2}}, \tag{3.127}$$

where $b^2 = v^2\alpha(1 - \alpha)/2$ and $c^2 = v^2(1 - \alpha)(2 - \alpha)/2$.

By applying Eq. (3.77) to this result, after some lengthy algebra and then by Laplace antitransforming in the s variable, we finally obtain

$$\langle x^2(t)\rangle = V^2 t^2, \tag{3.128}$$

that is, a ballistic behavior, where the characteristic spread velocity is given by $V = v\sqrt{1 - \alpha} < v$. This result is easily understandable, because for $\alpha < 1$ it is the longest jump to dominate, and it is ballistic. In the special case $\alpha = 1$, V vanishes and logarithmic corrections of the form $t^2/\ln(t)$ are expected for the mean square displacement.

$\alpha > 1$, Superdiffusion

For $1 < \alpha < 2$ we face quite a different situation because $\omega(t) \sim t_0^\alpha/t^{\alpha+1}$ has a finite first moment, that is, we are dealing with a finite characteristic time $\theta = \langle t \rangle$, while the second moment diverges. This suggests to evaluate the small-s behavior of the Laplace transform writing

$$\omega(s) - 1 + \theta s = \int_0^\infty dt\,\omega(t)\left(e^{-st} - 1 + st\right) \tag{3.129}$$

$$\simeq \int_0^{1/s} dt\,\frac{t_0^\alpha}{t^{\alpha+1}}\frac{s^2 t^2}{2}, \tag{3.130}$$

from which we obtain

$$\omega(s) \simeq 1 - \theta s + A s^\alpha, \tag{3.131}$$

where the quantity A is proportional to t_0^α.

Substituting this expression into Eq. (3.119), we obtain

$$\Omega(k, s) \simeq 1 - \theta s + \frac{1}{2} A [(s - ivk)^{\alpha} + (s + ivk)^{\alpha}]$$

$$\simeq 1 - \theta s + A s^{\alpha} - \frac{A v^2}{2} \alpha(\alpha - 1) k^2 s^{\alpha-2}$$

$$\simeq 1 - \theta s + \frac{A v^2}{2} \alpha(\alpha - 1) k^2 s^{\alpha-2}, \tag{3.132}$$

where the term s^{α} has been neglected with respect to the linear term s. As for the passage from the first to the second line, we first consider (as before) the limit $k \to 0$, then $s \to 0$. Making use again of Eqs. (3.124) and (3.74), we can also compute

$$\Psi(k, s) \simeq \theta - \frac{1}{2} A [(s - ivk)^{\alpha-1} + (s + ivk)^{\alpha-1}]$$

$$\simeq \theta - A s^{\alpha-1} + \frac{A v^2}{2} (\alpha - 1)(\alpha - 2) k^2 s^{\alpha-3}$$

$$\simeq \theta + \frac{A v^2}{2} (\alpha - 1)(\alpha - 2) k^2 s^{\alpha-3}, \tag{3.133}$$

where $s^{\alpha-1}$ has been neglected with respect to the constant term. By substituting Eqs. (3.132) and (3.133) into Eq. (3.125), we eventually obtain

$$\rho(k, s) \simeq \frac{1}{s} \frac{\theta - c_1^2 \, k^2 \, s^{\alpha-3}}{\theta + b_1^2 \, k^2 \, s^{\alpha-3}}, \tag{3.134}$$

where $c_1^2 = Av^2(\alpha - 1)(2 - \alpha)/2$ and $b_1^2 = Av^2\alpha(\alpha - 1)/2$. By applying Eq. (3.77) to this result, after (again) some lengthy algebra and then by Laplace antitransforming in the s variable (using the Tauberian theorems; see note 13 earlier), we obtain an expression for the mean square displacement,

$$\langle x^2(t) \rangle \simeq Ct^{3-\alpha}, \tag{3.135}$$

where $C \propto A(\alpha - 1)/\theta$. In this case we find a superdiffusive behavior that interpolates between the ballistic case ($\alpha = 1$) and standard diffusion ($\alpha = 2$). Also for $\alpha = 2$, logarithmic corrections of the form $t \ln(t)$ have to be expected.

For $\alpha > 2$, also terms of order $s^{\alpha-2}$ become negligible and $\rho(k, s)$, at leading order, recovers the form (3.88); that is, the LW boils down to a standard diffusive process. In fact, for $\alpha > 2$, the probability distribution $\omega(t)$ has finite average and variance and the asymptotic behavior of the corresponding LW can be argued to be equivalent to a standard Gaussian process equipped with the same average and variance.

3.5.3 Anomalous Diffusion: A Summary

The many mathematical details made explicit in this section may not help the reader in maintaining an overview of the main achievements therein contained. This is why it is worth providing here a summary. We have generalized the standard random walk (random space steps of fixed length at constant time steps) by introducing the distribution $\eta(x)$ of the jump length x and the distribution $\omega(t)$ of the time t between a jump and the next one. In the

CTRW, each jump is instantaneous, so t should be understood as the residence time rather than the flight time. If $\eta(x)$ has a finite second moment $\langle x^2 \rangle_\eta = \sigma^2$ and $\omega(t)$ has a finite mean $\langle t \rangle_\omega = \theta$ then standard RW is recovered, described by the diffusion equation with a diffusion coefficient $D = \sigma^2/2\theta$.

If either $\langle x^2 \rangle_\eta$ or $\langle t \rangle_\omega$ diverges, anomalous diffusion appears. For example, if σ^2 is finite but $\omega(t) \sim 1/t^{\alpha+1}$ (with $0 < \alpha < 1$), we obtain subdiffusion because $\langle x^2(t) \rangle \sim t^\alpha$, see Eq. (3.99). The reverse case, $\langle t \rangle_\omega$ is finite while $\langle x^2 \rangle_\eta$ diverges, is quintessential to Lévy flights and produces superdiffusion: If $\eta(x) \sim 1/|x|^{\alpha+1}$ (with $\alpha < 2$), then $\langle x^2(t) \rangle \sim t^{2/\alpha}$.

Lévy walks arise from the observation that the length x of a jump and the time t you have to wait before the next jump are not independent. In fact, jumps are not instantaneous and the bigger the jump, the longer the time it takes. LWs are characterized by ballistic jumps, that is, the walker travels at constant velocity v during the single jump, but different jumps are in randomly oriented directions. If the jump lengths follow the usual distribution $\eta(x) \sim 1/|x|^{1+\alpha}$, the global behavior of the walker depends on α: If $\alpha < 1$, the global behavior is also ballistic, see Eq. (3.128), with a velocity $V = v\sqrt{1 - \alpha} < v$; if $1 < \alpha < 2$, the average flight time is finite and the global behavior is superdiffusive, $\langle x^2(t) \rangle \sim t^{3-\alpha}$; if $\alpha > 2$, also the typical jump length is finite and the global motion is normal diffusion.

3.6 Statistical Fluctuations between Equilibrium and Out-of-equilibrium

Before illustrating the mathematical aspects of Einstein's theory of fluctuations, it is worth going through a few simple physical considerations. Fluctuations in thermodynamics are typically intended as small deviations from the equilibrium state. Such deviations may be imposed by measurement protocols or may emerge from the interaction with a (heat, particle, etc.) reservoir. For instance, on the basis of the principles of thermodynamics, we know that the specific heat at constant volume is given by the expression

$$C_V = \left(\frac{\partial \langle E \rangle}{\partial T} \right)_V, \tag{3.136}$$

where $\langle E \rangle \equiv U$ is the internal energy of the thermodynamic system under scrutiny. Thus, measuring C_V implies that one has to mildly drive a system out of equilibrium, changing its temperature T by a small amount dT and to evaluate the corresponding variation of the internal energy dU, after the system has relaxed to its equilibrium state at temperature $T + dT$, while keeping constant the volume V. This measurement protocol can be worked out by putting the physical systems in contact with a thermostat (reservoir), whose temperature can be varied.

Moreover, any thermodynamic (i.e., macroscopic) observable $\mathcal{A}$, which is not a conserved quantity, is subject to spontaneous fluctuations at equilibrium, also due to the continuously varying evolution of the underlying microscopic dynamics. This notwithstanding, its average value at equilibrium, $\langle \mathcal{A} \rangle$, is a well-defined quantity. Statistical mechanics allows one to compute such averages and, at the same time, to obtain an estimate of the fluctuations of $\mathcal{A}$. In a macroscopic system, the former are practically unaffected by the latter, because these are found to typically vanish for large values of the number of particles N. For instance, this argument plays a very important role in establishing the physical equivalence between statistical ensembles (at least for thermodynamic systems with short-range interactions) in the thermodynamic limit $N \to \infty$. Let us consider a system in contact with a heat reservoir at temperature $T = 1/\beta$: Its thermodynamics is described in terms of the canonical ensemble, that is, the probability of a microscopic state i with energy E_i is proportional to $\exp(-\beta E_i)$. Accordingly, the internal energy has the expression

$$\langle E \rangle = \frac{\sum_i E_i\, e^{-\beta E_i}}{\sum_i e^{-\beta E_i}} \tag{3.137}$$

and its fluctuations are

$$\Delta\langle E \rangle = \sqrt{\langle (E - \langle E \rangle)^2 \rangle} = \sqrt{\langle E^2 \rangle - \langle E \rangle^2}, \tag{3.138}$$

where

$$\langle E^2 \rangle = \frac{\sum_i E_i^2\, e^{-\beta E_i}}{\sum_i e^{-\beta E_i}}. \tag{3.139}$$

By deriving Eq. (3.137) with respect to β, one obtains

$$\frac{\partial \langle E \rangle}{\partial \beta} = \langle E \rangle^2 - \langle E^2 \rangle, \tag{3.140}$$

which allows us to rewrite (3.138) as follows

$$\Delta\langle E \rangle = \sqrt{-\frac{\partial \langle E \rangle}{\partial \beta}} = \sqrt{T^2 \frac{\partial \langle E \rangle}{\partial T}}. \tag{3.141}$$

Since for many thermodynamic systems the average energy is an extensive quanitity, that is, $\langle E \rangle \sim N$, Eq. (3.141) yields

$$\frac{\Delta\langle E \rangle}{\langle E \rangle} \sim \sqrt{\frac{1}{N}}. \tag{3.142}$$

This relation implies that for a macroscopic portion of matter ($N \sim N_A$), the relative fluctuations of the energy are practically undetectable and the system in contact with a thermal reservoir seems to exhibit a constant energy, as it happens for an isolated system, whose thermodynamics is described by the microcanonical statistical ensemble. This result also explains the reason why the interest of the founders of statistical mechanics was essentially focused on equilibrium averages, thus overlooking the role of vanishingly small fluctuations around equilibrium.

The alternative point of view of Einstein on the relevance of fluctuations can be understood if one rewrites Eq. (3.140) as follows:

$$\frac{\partial \langle E \rangle}{\partial T} = C_V = \frac{1}{T^2}\left(\langle E^2 \rangle - \langle E \rangle^2 \right). \tag{3.143}$$

This equation provides us a different way of estimating C_V, by measuring the fluctuations of the energy at a given temperature, thus showing that such fluctuations are directly related to measurable thermodynamic quantities. As a matter of proof that Einstein was deeply inspired by the role of fluctuations in thermodynamics, we can also recall that in his theory of Brownian motion (published in 1905), the diffusion coefficient D of a particle at position $\mathbf{r}$ and in equilibrium with a solvent at temperature T is given by the expression (see Eqs. (2.9) and (4.6))

$$D = \lim_{t \to \infty} \frac{\langle \mathbf{r}^2(t) \rangle - \langle \mathbf{r}(t) \rangle^2}{6t} = \frac{T}{6\pi \eta R} . \tag{3.144}$$

This tells us that by experimental measurements of the fluctuations of the position of a Brownian particle of radius R in thermodynamic equilibrium at temperature T with a solvent of viscosity η one can obtain an estimate (as better, as larger is t) of the temperature T. Conversely, using a thermometer for measuring T, these fluctuations can provide us an estimate of η, that is, a transport coefficient, characterizing a nonequilibrium stationary state of the solvent in the presence of a velocity gradient. This reminds us (see Section 1.2) that the heat conductivity κ of an ideal gas, a basic transport coefficient characterizing nonequilibrium properties induced by the presence of a temperature gradient, is proportional to its specific heat at constant volume C_V (see Eq. (1.18)). In view of Eq. (3.143), both of these remarks indicate that fluctuations also bridge equilibrium properties with nonequilibrium ones.

Similar arguments are further elaborated in Chapter 4, where we discuss the linear response theory and the Onsager reciprocity relations. At this point the reader can also appreciate the basic conceptual difference between thermodynamics, as a phenomenological theory, and statistical mechanics, as a fundamental predictive theory, which provides us, through fluctuations, an unified approach to equilibrium as well as to nonequilibrium thermodynamic phenomena.

3.7 Einstein's Approach

The basic idea of Einstein to deal with fluctuations amounts to obtain a probabilistic interpretation of the entropy S of a thermodynamic system. At first sight this could appear as a hopeless or ill-defined task, because the entropy is a thermodynamic state function, which is maximized at equilibrium. The physical considerations illustrated in Section 3.6 should help the reader to overcome this naive prejudice, because statistical mechanics imposes that fluctuations unavoidably affect any thermodynamic quantity, including S, which is expected to fluctuate around its maximum value at thermodynamic equilibrium.

In order to elaborate a suitable mathematical language for describing thermodynamic fluctuations, we can begin by considering first an isolated system with energy E, made of N particles contained in a volume V. This situation is described by the microcanonical statistical ensemble, where E is the value taken by the Hamiltonian functional $\mathcal{H}(q_i, p_i)$, where the position and momentum coordinates of all particles in the physical space are denoted by q_i and p_i, respectively, with $i = 1, \cdots, 3N$. According to Boltzmann, the entropy of this isolated system is given by the expression

$$S(E) = \ln(\Gamma(E)), \tag{3.145}$$

where $\Gamma(E)$ is the hypervolume of the phase space corresponding to a given value of E.[15] The thermodynamics of this isolated system is described by a set of macroscopic, fluctuating observables, such as the pressure P, the temperature T, and so on, that are functionals of the microstate $s = (q_1, \cdots, q_{3N}, p_1, \cdots, p_{3N})$. In full generality, we denote with $\mathcal{A}_l(s)$ with $l = 1, \cdots, m$ the set $\{\mathcal{A}_l(s)\}$ of all of these observables. At equilibrium, each one of these observables takes the value

$$\mathcal{A}_l^0 = \frac{1}{\Gamma(E)} \int ds \, \mathcal{A}_l(s) \, \delta(\mathcal{H}(s) - E), \tag{3.146}$$

where we have introduced the shorthand notation $ds = \prod_{i=1}^{3N} dq_i \, dp_i$.

Let $\Gamma(E, \{\mathcal{A}_l\})$ be the volume of the phase space where the set of fluctuating observables $\{\mathcal{A}_l(s)\}$ take specific values $\{\mathcal{A}_l\}$. By assuming that the system is ergodic, that is, that all microstates are statistically equivalent, we can define the probability of the system visiting any one of these microstates as follows:

$$P(E, \{\mathcal{A}_l\}) = \frac{\Gamma(E, \{\mathcal{A}_l\})}{\Gamma(E)}. \tag{3.147}$$

Accordingly, we can define the corresponding "conditional" entropy as

$$S(E, \{\mathcal{A}_l\}) = \ln\left[\Gamma(E, \{\mathcal{A}_l\})\right] \tag{3.148}$$

and rewrite

$$P(E, \{\mathcal{A}_l\}) = \frac{1}{\Gamma(E)} \exp\left[S(E, \{\mathcal{A}_l\})\right]. \tag{3.149}$$

Let us denote the set of deviations, that is, fluctuations, of the thermodynamic observables with respect to their equilibrium value as

$$\alpha_l = \mathcal{A}_l - \mathcal{A}_l^0. \tag{3.150}$$

By recalling that the entropy at thermodynamic equilibrium takes its maximum value with respect to any fluctuating observable, we can express this condition as follows:

$$\left(\frac{\partial S(E, \{\mathcal{A}_l\})}{\partial \mathcal{A}_k}\right)_0 = 0 \quad \forall k, \tag{3.151}$$

where the subscript 0 means that the derivative is evaluated for $\alpha_k = 0$.

When dealing with macroscopic systems, one expects that typical fluctuations are small, that is, $(\alpha_l / \mathcal{A}_l^0) \ll 1$, and the entropy can be expanded close to equilibrium as a Taylor series up to second order, as follows:

$$S(E, \{\mathcal{A}_l\}) \simeq S(E, \{\mathcal{A}_l^0\}) + \frac{1}{2} \sum_{j,k} \left(\frac{\partial^2 S(E, \{\mathcal{A}_l\})}{\partial \mathcal{A}_j \, \partial \mathcal{A}_k}\right)_0 \alpha_j \, \alpha_k. \tag{3.152}$$

---

[15] The Gibbs entropy is defined as $S(E) = \ln(\Sigma(E))$, where $\Sigma(E)$ is the hypervolume of the phase space corresponding to energies *smaller or equal* to E: $\Gamma(E) = \Sigma'(E)$. For most ordinary physical systems, the two definitions are equivalent, but in some cases, for example, when the energy spectrum is limited from above, it is recommended to use the Boltzmann definition.

Considering that, in virtue of the second principle of thermodynamics, $S(E, \{\mathcal{A}_l^0\})$ takes its maximum value, the symmetric matrix $\hat{\mathcal{D}}$, whose elements are

$$\mathcal{D}_{jk} = -\left(\frac{\partial^2 S(E, \{\mathcal{A}_l\})}{\partial \mathcal{A}_j \, \partial \mathcal{A}_k}\right)_0, \tag{3.153}$$

has to be a positive definite quadratic form, and we can approximate the probability in (3.149) as follows:

$$P(E, \{\alpha_l\}) \simeq C \, \exp\left[-\frac{1}{2} \sum_{jk} \mathcal{D}_{jk} \alpha_j \, \alpha_k\right], \tag{3.154}$$

which represents a multivariate Gaussian distribution. The normalization constant D is fixed by the condition

$$\int_{-\infty}^{+\infty} \prod_{l=1}^{m} d\alpha_l \, P(E, \{\alpha_l\}) = 1, \tag{3.155}$$

thus yielding

$$C = \sqrt{\frac{\det \hat{\mathcal{D}}}{(2\pi)^m}}. \tag{3.156}$$

Now, we have at our disposal an approximate expression of the probability distribution for (small) fluctuations of the thermodynamic observables in the microcanonical ensemble and we are in the position of computing its first two moments. This task can be accomplished by introducing the moment-generating function, see Eq. (L.2),

$$\mathcal{I}(\{h_l\}) \equiv \left\langle e^{\sum_l h_l \alpha_l} \right\rangle = \int_{-\infty}^{+\infty} \prod_{l=1}^{m} d\alpha_l \, e^{\sum_l h_l \alpha_l} \, P(E, \{\alpha_l\}) \simeq \exp\left[\frac{1}{2} \sum_{jk} h_j \, \mathcal{D}_{jk}^{-1} \, h_k\right], \tag{3.157}$$

which allows to easily obtain the moments of the distribution (3.154),

$$
\begin{aligned}
\langle \alpha_m \rangle &= \left. \frac{\partial \mathcal{I}(\{h_l\})}{\partial h_m} \right|_{\{h_l = 0, \, \forall l\}} \\
&= \left. \frac{\partial}{\partial h_m} \exp\left[\frac{1}{2} \sum_{jk} h_j \, \mathcal{D}_{jk}^{-1} \, h_k\right] \right|_{\{h_l = 0, \, \forall l\}} \\
&= \left. \frac{1}{2} \sum_{jk} (\mathcal{D}_{mk}^{-1} h_k + h_j \, \mathcal{D}_{jm}^{-1}) \, \mathcal{I}(\{h_l\}) \right|_{\{h_l = 0, \, \forall l\}} \\
&= \left. \sum_{k} \mathcal{D}_{mk}^{-1} \, h_k \, \mathcal{I}(\{h_l\}) \right|_{\{h_l = 0, \, \forall l\}} = 0
\end{aligned}
\tag{3.158}
$$

and

$$\begin{aligned}
\langle \alpha_m \, \alpha_n \rangle &= \frac{\partial^2 \mathcal{I}(\{h_l\})}{\partial h_m \, \partial h_n}\bigg|_{\{h_l=0,\,\forall l\}} \\
&= \frac{\partial}{\partial h_m} \sum_k (\mathcal{D}_{nk}^{-1} \, h_k \, \mathcal{I}(\{h_l\}))\bigg|_{\{h_l=0,\,\forall l\}} \\
&= \left[\mathcal{D}_{nm}^{-1} + \left(\sum_k \mathcal{D}_{nk}^{-1} h_k\right)\left(\sum_j \mathcal{D}_{mj}^{-1} h_j\right)\right] \mathcal{I}(\{h_l\})\bigg|_{\{h_l=0,\,\forall l\}} \\
&= \mathcal{D}_{mn}^{-1}.
\end{aligned} \tag{3.159}$$

As expected, the first moments of the multivariate Gaussian distribution (3.154) vanish, while the statistics of fluctuations is fully accounted for by the elements of the symmetric matrix (3.153). In particular, the variance of a fluctuating observable $\mathcal{A}_m$ is given by the expression

$$\langle \alpha_m^2 \rangle = \langle \mathcal{A}_m^2 \rangle - (\mathcal{A}_m^0)^2 = \mathcal{D}_{mm}^{-1} = \left[-\left(\frac{\partial^2 S(E,\{\mathcal{A}_l\})}{\partial \mathcal{A}_m^2}\right)_0\right]^{-1} \tag{3.160}$$

This tells us that we can obtain the information about the statistics of fluctuations of the observables of an isolated system from its entropy.

Einstein's theory of fluctuations can be extended to the case of a thermodynamic system contained in nonadiabatic walls and in equilibrium with the external universe. In full generality, such a situation is described by the so-called generalized statistical ensemble, where, at variance with the microcanonical ensemble, also the energy E and the number of particles N are fluctuating observables. The probability of observing the thermodynamic system in a microstate s is now given by the generalized Gibbs weight,

$$P(s,\{h_l\}) = \frac{1}{Z(T,\{h_l\})} e^{-\beta\,[\mathcal{H}(s) - \sum_l h_l \mathcal{A}_l(s)]}, \tag{3.161}$$

where

$$Z(T,\{h_l\}) = \int ds\, e^{-\beta\,[\mathcal{H}(s) - \sum_l h_l \mathcal{A}_l(s)]}. \tag{3.162}$$

As in the microcanonical case, $\{\mathcal{A}_l\}(s)$ is the set of fluctuating thermodynamic observables, while $\{h_l\}$ is the set of thermodynamic intensive parameters identifying the equilibrium state at temperature $T = \beta^{-1}$. For instance, h_l represents the chemical potential μ if $\mathcal{A}_l(s)$ is the number of particles $N(s)$, or the external magnetic field h if $\mathcal{A}_l(s)$ is the magnetization $M(s)$. The generalized free energy, whose minimum (constrained to the values of the intensive parameters $\{h_l\}$) identifies the equilibrium state at temperature T, is given by the expression

$$F(T,\{h_l\}) = -T \ln Z(T,\{h_l\}). \tag{3.163}$$

We can follow a procedure analogous to the one adopted for describing fluctuations in the microcanonical ensemble by defining the probability of observing values E and $\mathcal{A}_l$, $P(E,\{\mathcal{A}_l\})$, which are different from the equilibrium ones,

$$E^0 \equiv U = \frac{1}{Z(T,\{h_l\})} \int ds\, \mathcal{H}(s)\, e^{-\beta\,[\mathcal{H}(s) - \sum_l h_l \mathcal{A}_l(s)]} \tag{3.164}$$

and

$$\mathcal{A}_l^0 = \frac{1}{Z(T, \{h_l\})} \int ds\, \mathcal{A}_l(s)\, e^{-\beta\,[\mathcal{H}(s) - \sum_l h_l \mathcal{A}_l(s)]}. \tag{3.165}$$

Starting from the expression

$$P(E, \{\mathcal{A}_l\}) = \int ds P(s, \{h_l\})\, \delta(E - \mathcal{H}(s)) \prod_l \delta(\mathcal{A}_l - \mathcal{A}_l(s)) \tag{3.166}$$

with $P(s, \{h_l\})$ given in Eq. (3.161), we obtain

$$P(E, \{\mathcal{A}_l\}) = e^{\beta\, F(T, \{h_l\})}\, e^{S(E, \{\mathcal{A}_l\})}\, e^{-\beta(E - \sum_l h_l \mathcal{A}_l)} \equiv e^{-\beta\, \Delta\mathcal{F}(E, \{\mathcal{A}_l\})}, \tag{3.167}$$

where

$$\Delta\mathcal{F}(E, \{\mathcal{A}_l\}) = E - \sum_l h_l\, \mathcal{A}_l - T\, S(E, \{\mathcal{A}_l\}) - F(T, \{h_l\}) \tag{3.168}$$

is the so-called *availability*, that is, the difference between the free energy constrained to the nonequilibrium values E and $\{\mathcal{A}_l\}$ of the fluctuating observables and the equilibrium free energy (3.163). In the earlier expression $S(E, \{\mathcal{A}_l\})$ is the analogous of the conditional entropy defined in Eq. (3.148): Note that, at variance with the microcanonical case, the energy here can fluctuate and E represents a value of the internal energy different from the equilibrium one, E^0.

Since by definition $\Delta\mathcal{F}(E^0, \{\mathcal{A}_l^0\}) = 0$, we can expand also $S(E, \{\mathcal{A}_l\})$ as a Taylor series around the equilibrium values E^0 and $\{\mathcal{A}_l^0\}$ up to the second order. In order to simplify the notation, we define the $(m+1)$-th variable $\mathcal{A}_{m+1} = E$, so that $S(E, \{\mathcal{A}_l\})$ with $l = 1, \ldots, m$ can be written as $S(\{\mathcal{A}_l\})$ with $l = 1, \ldots, m+1$. In this way, we obtain

$$\frac{\Delta\mathcal{F}(E, \{\mathcal{A}_l\})}{T} \simeq -\frac{1}{2} \sum_{j,k=1}^{m+1} \left.\frac{\partial^2 S(\{\mathcal{A}_l\})}{\partial \mathcal{A}_j\, \partial \mathcal{A}_k}\right|_0 \alpha_j \alpha_k, \tag{3.169}$$

where the linear terms of this Taylor series expansion are null because the thermodynamic equilibrium state corresponds to a minimum of the free energy $F(T, \{h_l\})$. Therefore, we finally obtain the approximate expression

$$P(E, \{\alpha_l\}) \simeq C \exp\left[-\frac{1}{2} \sum_{j,k=1}^{m+1} \mathcal{D}_{jk} \alpha_j \alpha_k\right]. \tag{3.170}$$

where α_l are defined in Eq. (3.150), C in Eq. (3.156) and where the matrix elements $\mathcal{D}_{jk}$

$$\mathcal{D}_{jk} = -\left(\frac{\partial^2 S(E, \{\mathcal{A}_l\})}{\partial \mathcal{A}_j\, \partial \mathcal{A}_k}\right)_0, \tag{3.171}$$

belong to the symmetric matrix $\hat{\mathcal{D}}$, which is a positive definite quadratic form. It is worth recalling that $S(E, \{\mathcal{A}_l\})$ in the previous formula is the entropy of a system in equilibrium at temperature T, so that E is a fluctuating observable, as well as any other $\mathcal{A}_l$, while in the entropy of an isolated system, defined in (3.148), E is a physical parameter, fixing the value of the energy of the system.

Anyway, we have recovered the same formalism worked out for describing fluctuations in the microcanonical ensemble, and from this point on any explicit calculation proceeds accordingly. In particular, we can specialize the calculation to the case where the only fluctuating observable is the energy, that is, to $\alpha = E - E^0$, while all the other fluctuating observables take their equilibrium value, that is, $\mathcal{A}_l = \mathcal{A}_l^0$.

This scenario corresponds to the canonical equilibrium statistical ensemble, where the free energy of the thermodynamic system made of N particles contained in a volume V is also a function of the temperature T, namely $F(T; V, N)$. Specifically, we can use Eq. (3.159) for computing the fluctuations of the energy as follows:

$$\langle (E - E^0)^2 \rangle = \langle E^2 \rangle - \langle E \rangle^2$$
$$= -\left[\left(\frac{\partial^2 S(E)}{\partial E^2} \right)_0 \right]^{-1} = -\left[\left(\frac{\partial}{\partial E} \frac{1}{T} \right)_0 \right]^{-1}$$
$$= \left[\frac{1}{T^2} \left(\frac{\partial T}{\partial E} \right)_0 \right]^{-1} = T^2 C_V, \tag{3.172}$$

thus recovering Eq. (3.143). Note that in this calculation we have made use of the well-known thermodynamic relation $\left(\frac{\partial S}{\partial E} \right)_0 = \frac{1}{T}$ and of the definition of the specific heat at constant volume, given in Eq. (3.136).

We conclude this section by pointing out that the formal equivalence of Einstein's theory of fluctuations in the microcanonical, canonical, and generalized equilibrium statistical ensembles applies as a consequence of the equivalence among these equilibrium ensembles in the thermodynamic limit. When such an equivalence does not hold, for instance, in thermodynamic systems made of particles interacting by long-range forces, this theory of fluctuations has to be reconsidered in the light of suitable physical arguments.

3.8 Stochastic Thermodynamics: An Introduction

When one has to deal with nonperturbative out-of-equilibrium regimes, or with small systems (i.e., systems whose number of degrees of freedom N is much smaller than the Avogadro number N_A), or even with fast thermodynamic transformations, the role of fluctuations has to be reconsidered in the framework of a new approach, known as stochastic thermodynamics.

The practical interest of the fluctuation relations hereafter discussed is associated with the possibility of obtaining a thermodynamic description of such processes. For instance, the variations of free energy in a system can be measured experimentally by driving it in out-of-equilibrium conditions, which are certainly easier to control, than keeping it at thermal equilibrium, where fluctuations typically affect the accuracy of any measurement, in particular when dealing with small systems. In fact, fluctuation relations and theorems have provided deep insight into various physical processes, such as current fluctuations in small electric circuits, heat current fluctuations in nanomaterials, and manipulation of polymers and biomolecules by optical traps.

3.8.1 Jarzynski Equality

This section aims at illustrating some physically relevant relations when the role of thermodynamic fluctuations is explicitly taken into account in nonequilibrium processes. For the sake of simplicity, let us make reference to an ideal experiment, where a thermodynamic system is initially prepared in an equilibrium thermal state at a temperature T, fixed by a thermal reservoir (universe). Then the system is isolated from the reservoir and it is made to evolve by a suitable protocol to a new state and again it is thermalized at the same temperature T, imposed by the thermal reservoir. For instance, this could be the case of an elastic bid or a polymer stretched by mechanical (conservative) forces, that is, during the stretching process, the polymer evolves deterministically in the absence of any contact with the thermal reservoir, which is put in thermal contact with the polymer only in its initial and final states. Accordingly, both of these states are genuine equilibrium thermodynamic states, while the mechanical stretching is by construction a nonequilibrium thermodynamic process. How can one approach a quantitative description of such an experiment? One can easily realize that no reproducible predictions are available for a single realization of this protocol, because of the indeterminacy of inherent thermal fluctuations from the reservoir in the initial state. On the other hand, one can repeat the experiment several times and obtain the probability distribution $\rho(W)$ of the different values of the work W measured in each realization. If the system we are dealing with were truly macroscopic, fluctuations would essentially become immaterial and we could reasonably assume that $\rho(W)$ boils down to a Dirac distribution $\delta(W - \overline{W})$, where the result of work measurements $\overline{W}$ is at most affected by the accuracy of instruments. In these conditions, we can say that Clausius' inequality should hold, namely

$$\overline{W} \geq \Delta F \equiv F_1 - F_0 \,, \tag{3.173}$$

where F_0 and F_1 are the Helmholtz free energies in the initial and final equilibrium states, respectively. Note that in this case the equality in Eq. (3.173) holds only for reversible processes. Moreover, Eq. (3.173) contains all that we know about the thermodynamics, concerning the above-sketched experimental protocol for a macroscopic system.

On the other hand, if the system is not macroscopic, that is, it is made by a number of degrees of freedom $N \ll N_A$, $\rho(W)$ is expected to be represented qualitatively as in Fig. 3.5. Note that the range of values of W may extend to the left of ΔF: This implies that the system can be stretched extracting work from the reservoir and this may happen due to large, rare thermal fluctuations that may induce special initial conditions for which stretching is accompanied by a negative ΔF.

In a single process of a "small" system, we can therefore observe a violation of the second principle of thermodynamics, because the mechanical work W may be extracted in an isothermal transformation, being T unchanged in the initial and final equilibrium states of the system. Accordingly, we can assume that Clausius' inequality applies in a statistical sense, namely

$$\langle W \rangle = \int dW \, \rho(W) W \geq \Delta F \equiv F_1 - F_0 \,. \tag{3.174}$$

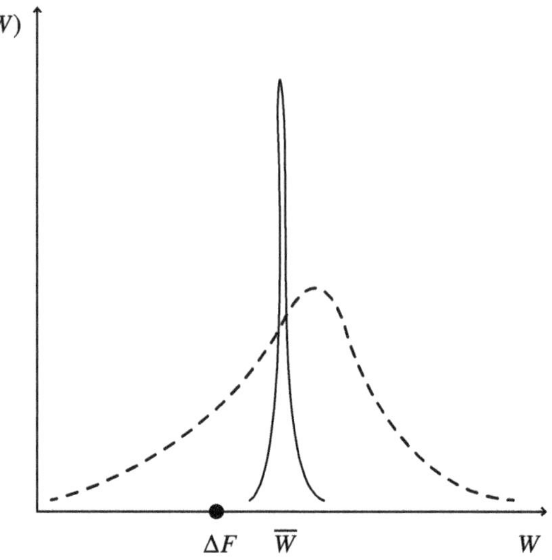

Fig. 3.5 The probability distribution $\rho(W)$ of the values of the work W for a macroscopic system (full line) and for a mesoscopic system (dashed line), subject to an irreversible process.

The averaging operation $\langle \bullet \rangle$ is performed over realizations of the experimental protocol, where the second principle of thermodynamics is typically respected and rarely violated, but, on average, never contradicted.

We can say more about thermodynamic relations in the experimental protocol, taking advantage of a Hamiltonian description of the physical system under scrutiny, made of N particles and whose degrees of freedom are represented in d space dimensions by the canonically conjugated variables of position $\vec{q}_i(t)$ and momentum $\vec{p}_i(t)$ (with $i = 1, \cdots, dN$): They identify univocally a microstate $s(t) = \left(\vec{q}_i(t), \vec{p}_i(t)\right)$ at time t. If the system is in thermal equilibrium with a heat reservoir at temperature $T = 1/\beta$, the equilibrium probability that it can be found in a microstate $s(t)$ at time t is given by the Boltzmann–Gibbs statistical formula

$$p_e[s(t)] = \frac{1}{Z_t} \exp\left[-\beta \mathcal{H}[s(t)]\right], \tag{3.175}$$

where $\mathcal{H}[s(t)]$ represents the internal energy of the system in the state $s(t)$,

$$Z_t = e^{-\beta F_t} = \int_{\mathcal{V}} d\mathcal{V} \ \exp\left[-\beta \mathcal{H}(s(t))\right] \tag{3.176}$$

is the partition function and F_t is the Helmholtz free energy. In the former equation we have also introduced the shorthand notation $\mathcal{V}$ for the phase space, whose infinitesimal volume element is[16]

---

[16] The factor $1/N!$ stems from the hypothesis that identical particles are undistinguishable: It was originally introduced by Otto Sackur and Hugo Martin Tetrode for recovering the additivity of the entropy of the ideal gas.

$$d\mathcal{V} \equiv \frac{1}{N!} \frac{\prod_{i=1}^{dN} dq_i \, dp_i}{(2\pi\hbar)^{dN}},$$

(3.177)

where $\hbar = 1.05457 \times 10^{-34}$ Js denotes the reduced Planck constant.

We want to point out that we consider here an explicit dependence of the partition function Z_t on time t, because we aim at describing an ideal experimental protocol where we modify the macroscopic parameters defining the equilibrium state of the system, when passing from the initial to the final state. For instance, in the case of an experiment where we stretch a polymer, we have $Z = Z(T, L)$, where the length L of the polymer is a time-dependent quantity, with the constraint it has to take the fixed values L_0 and L_1 in the initial and final states, respectively.

For the sake of simplicity, the adopted ideal protocol assumes that the evolution of the system in between its initial and final states, that is, when the system is isolated from the thermal bath, is made by the deterministic Hamilton equations. Let us denote with $s(0)$ and $s(1)$ the initial and final microscopic states, respectively. The work done on a single realization of the measurement protocol is given by the difference of the energies in $s(0)$ and in $s(1)$,

$$W = \mathcal{H}[s(1)] - \mathcal{H}[s(0)].$$

(3.178)

Now, we can easily compute

$$\langle \exp(-\beta W) \rangle = \int ds(0) \, p_e[s(0)] \, \exp(-\beta W)$$

$$= \frac{1}{Z_0} \int ds(0) \, \exp(-\beta \mathcal{H}[s(1)]).$$

(3.179)

On the other hand, $s(1)$ is univocally determined by $s(0)$ through the deterministic and time-reversible evolution described by the Hamilton equations, therefore

$$\int ds(0) \, \cdots = \int ds(1) \left| \frac{\partial s(1)}{\partial s(0)} \right|^{-1} \cdots .$$

(3.180)

Due to the Liouville theorem, which implies the conservation of volumes in Hamiltonian deterministic evolution, the Jacobian determinant $\left| \frac{\partial s(1)}{\partial s(0)} \right| = 1$ and we finally obtain

$$\langle \exp(-\beta W) \rangle = \frac{1}{Z_0} \int ds(1) \, \exp(-\beta \mathcal{H}[s(1)]) = \frac{Z_1}{Z_0} = e^{-\beta(F_1 - F_0)} .$$

(3.181)

In summary, we have proved the nonequilibrium relation

$$\langle e^{-\beta W} \rangle = e^{-\beta \Delta F} ,$$

(3.182)

which is known as Jarzinski equality.

Some comments are in order. Taking advantage of the well-known Jensen's inequality[17] $\langle \exp(-\beta W) \rangle \geq \exp\langle(-\beta W)\rangle$, from Eq. (3.182), one obtains

$$e^{-\beta \Delta F} \geq e^{\langle -\beta W \rangle} \quad \rightarrow \quad \langle W \rangle \geq \Delta F = F_1 - F_0 ,$$

(3.183)

[17] For any convex function $g(x)$, $\langle g(x) \rangle \geq g(\langle x \rangle)$.

where the second relation confirms that Jarzinski equality is consistent with the second principle of thermodynamics, expressed by Eq. (3.174). Moreover, Eq. (3.182) provides a precise quantitative prediction for a nonequilibrium experiment, where the average is over realizations of the experimental protocol.

If we are considering a situation where we do work on the system, for example, we stretch the polymer, then $F_1 > F_0$ and $\langle W \rangle > 0$. On the other hand, if we perform the inverse protocol (e.g., relaxing the polymer), then $F_1 < F_0$ and we can obtain work from the system, that is, $\langle W \rangle < 0$, and Eq. (3.183) could be better expressed as $\langle -W \rangle \leq F_0 - F_1$. The latter expression is the standard relation one can find in thermodynamics textbooks, as one of the possible formulations of the second principle of thermodynamics: In an isothermal transformation, the work extracted from a thermodynamic macroscopic system, that is, $|W|$, is always smaller than the (positive) difference between the free energies of the initial and final states.

In Section 3.8.2, we discuss a more general fluctuation relation, namely Crooks' fluctuation theorem, where the experimental protocol is performed while keeping the system in thermal equilibrium with a heat reservoir at temperature T. We also show that in this case Jarzinski equality is recovered as a special case of this theorem.

3.8.2 Crooks Fluctuation Theorem

In Section 3.8.1, we have proved a first interesting and useful fluctuation relation, namely Jarzinski equality (3.182), for a nonequlibrium process, making reference to an experimental protocol where equilibrium thermodynamics enters only in the initial and final states of the system, while the nonequilibrium process is represented by the time-reversible deterministic Hamiltonian dynamics of an isolated system. This is quite a mathematical idealization of a true experimental situation, where one typically has to deal with a physical system which evolves in thermal contact with the environment (thermostat). As we are going to show, the mathematical simplification of assuming a deterministic evolution does not affect the significance of the obtained result, because it depends uniquely on the assumption that the initial and final states are equilibrium ones.

In this section we want to illustrate a more general fluctuation relation, known as Crooks' theorem. We still make reference to an experimental protocol where we perform a nonequilibrium transformation of a mesoscopic physical system, but with two differences. First, we now consider an evolution of the system that is maintained in thermal contact with a heat reservoir at a fixed temperature T. Accordingly, we have to describe the nonequilibrium evolution of the system making use of a stochastic dynamics, that is, an evolution subject to thermal fluctuations, rather than the Hamiltonian deterministic dynamics employed in Section 3.8.1. Second, we want to consider an experimental protocol where at any realization we perform a cycle, that is, we drive the system from the initial equilibrium state $s(0)$ (with free energy F_0) to the equilibrium state $s(1)$ (with free energy F_1) and then backward from $s(1)$ to $s(0)$. In each realization of this nonequilibrium cyclic transformation, we measure the work associated with the forward process, $s(0) \to s(1)$, and the work associated with the backward process, $s(1) \to s(0)$. Without prejudice to generality, we can assume $\Delta F = F_1 - F_0 > 0$. For instance, the forward process could be the stretching of a polymer,

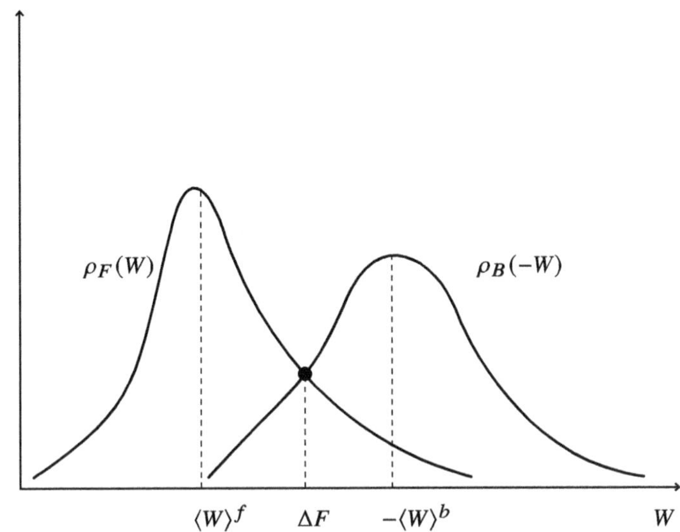

Fig. 3.6 A microscopic system is subject to an irreversible cyclic transformation, in contact with a thermal reservoir. The function $\rho^f(W)$ is the probability distribution of the work W made in the forward protocol; the function $\rho^b(-W)$ is the same for the backward protocol, with the work W changed of sign.

where typically W is positive and, accordingly, the backward process is the shrinking of the polymer, where typically W is negative, that is, the system makes work on the environment. By repeating this nonequilibrium cyclic transformation several times, we can obtain the normalized probability distributions $\rho^f(W)$ and $\rho^b(-W)$, which are qualitatively represented in Fig. 3.6: Note that in this representation both probability distributions are peaked at positive values of W.

As argued in Section 3.8.1, we are dealing with a situation where in a single realization of the experimental protocol, one can observe, very rarely, the violation of the second principle of thermodynamics, because a cycle may produce work, that is, the total work in a single cycle may be negative, due to the randomness of thermal fluctuations. On the other hand, we can assume that the second principle of thermodynamics still applies in the following statistical form

$$\int dW \, W(\rho^f(W) + \rho^b(W)) = \langle W \rangle^f + \langle W \rangle^b \geq 0, \tag{3.184}$$

thus stating that an isothermal nonequilibrium (i.e., irreversible) cyclic transformation can be performed only by making work on the system. In the earlier equation, $\langle W \rangle^f$ and $-\langle W \rangle^b$ are the positive means of the distributions $\rho^f(W)$ and $\rho^b(-W)$, respectively (see Fig. 3.6).

It is worth recalling that the work done on the system during a cycle is necessarily positive only for a macroscopic system, where fluctuations are essentially immaterial and the distributions are Dirac deltas, apart from instrumental inaccuracy: $\rho^f(W) \rightarrow \delta(W - \langle W \rangle^f)$ and $\rho^b(-W) \rightarrow \delta(W + \langle W \rangle^b)$.

For a mesoscopic system, however, we can say more; in particular, we can prove the following fluctuation relation

$$\frac{\rho^f(W)}{\rho^b(-W)} = e^{\beta(W-\Delta F)}, \tag{3.185}$$

which is known as Crooks fluctuation theorem.[18]

Before providing a proof of this theorem, we want to anticipate some comments about its implications. First, one can observe that the distributions $\rho^f(W)$ and $\rho^b(-W)$ intersect in $W = \Delta F > 0$. Moreover, we can prove that Crooks fluctuation theorem implies Jarzinski equality (3.182). In fact, rewriting Eq. (3.185) as

$$\rho^f(W)e^{-\beta W} = \rho^b(-W)e^{-\beta\Delta F} \tag{3.186}$$

and integrating both sides of this equation over W, we obtain

$$\langle e^{-\beta W}\rangle = e^{-\beta\Delta F}. \tag{3.187}$$

Finally, we can discuss a large deviation argument by computing the probability of observing, during the nonequilibrium experimental protocol, a violation of the second law of thermodynamics, relying upon the nonequilibrium, fluctuation relations. This task can be easily accomplished by considering that such a violation occurs if in a single realization of the forward protocol one measures a work $W < \Delta F - \mathcal{W}$, where $\mathcal{W} > 0$ determines the amplitude of the deviation. The probability of this event is given by the expression

$$P[W < \Delta F - \mathcal{W}] \equiv \int_{-\infty}^{\Delta F - \mathcal{W}} dW \rho^f(W). \tag{3.188}$$

We can obtain an upper bound of this expression by the sequence of inequalities

$$\int_{-\infty}^{\Delta F - \mathcal{W}} dW \rho^f(W) < \int_{-\infty}^{\Delta F - \mathcal{W}} dW \rho^f(W)\, e^{\beta(\Delta F - \mathcal{W} - W)}$$

$$< e^{-\beta\mathcal{W}} \int_{-\infty}^{+\infty} dW \rho^f(W)\, e^{\beta(\Delta F - W)}. \tag{3.189}$$

Due to Jarzinski equality (3.182), the last integral is equal to one and we can conclude that

$$P[W < \Delta F - \mathcal{W}] < e^{-\beta\mathcal{W}}, \tag{3.190}$$

which shows that the probability of observing a large deviation $\mathcal{W} \gg T$ from the second law of thermodynamics is less than exponentially small.

The mathematical discussion contained in the next part also allows the reader to appreciate explicitly that Crooks theorem contains more information than Jarzinski equality, since it makes reference to a cyclic transformation, where the forward and backward stochastic trajectories are related by the time-reversal symmetry.

---

[18]   In the thermodynamic limit, ρ^f and ρ^b are Dirac deltas and Eq. (3.185) reduces to: $\langle W\rangle^f = -\langle W\rangle^b = \Delta F$.

Stochastic Dynamical Model for a Cyclic Transformation

In order to describe the dynamics of the cyclic isothermal transformation, we can assume that both the forward and backward evolutions are stochastic processes, performed in n discrete time steps $t_1, t_2, \cdots, t_n$. We assume that at time t_i, the system is in an equilibrium state s_i, characterized by the value of a control parameter λ_i. For instance, in the example of the polymer λ_i is the value of its length at time t_i, which is fixed by the experimental protocol. For what concerns the forward stochastic evolution, we assume that it is composed by a two-step sequence,

$$(s_i, \lambda_i) \rightarrow (s_i, \lambda_{i+1}) \rightarrow (s_{i+1}, \lambda_{i+1}), \tag{3.191}$$

that is, we start at time t_i with a system in s_i, where the control parameter takes the value λ_i and we first modify its value to λ_{i+1} (e.g., we stretch the polymer), and then we let the system evolve to its new state s_{i+1} at time $t + 1$. We can assume that the first step is instantaneous, while the second one corresponds to a Markov process, whose transition rate $\mathcal{R}(s_i \rightarrow s_{i+1}; \lambda_{i+1})$ depends only on the initial state s_i and on λ_{i+1}.[19]

The backward stochastic evolution is the time-reversed version of the forward one, that is, it is composed by the two-step sequence

$$(s_i, \lambda_i) \rightarrow (s_{i-1}, \lambda_i) \rightarrow (s_{i-1}, \lambda_{i-1}), \tag{3.192}$$

where we first make the system evolve by a Markov process to its new state s_{i-1} and then we let the control parameter change instantaneously to the new value λ_{i-1}. Accordingly, the backward transition rate $\mathcal{R}(s_i \rightarrow s_{i-1}; \lambda_i)$ is assumed to depend only on s_i and λ_i.

Before passing to an explicit description of the stochastic dynamics, let us first observe that in the forward process the overall variation of energy during the experimental protocol can be expressed as the sum of two contributions, namely

$$\Delta E = E(s_n, \lambda_n) - E(s_1, \lambda_1) \equiv W^f + Q^f, \tag{3.193}$$

where

$$W^f = \sum_{i=1}^{n-1} \left[E(s_i; \lambda_{i+1}) - E(s_i; \lambda_i) \right], \tag{3.194}$$

$$Q^f = \sum_{i=1}^{n-1} \left[E(s_{i+1}; \lambda_{i+1}) - E(s_i; \lambda_{i+1}) \right]. \tag{3.195}$$

W^f represents the overall work done on the system, which is associated with the changes of the control parameter (step one in Eq. (3.191)) along the forward trajectory. Q^f epitomizes the total heat absorbed from the environment (heat reservoir), which is associated with the passages between equilibrium states (step two in Eq. (3.191)) along the forward trajectory.

---

[19] Here we adopt a notation for the transition rates, $\mathcal{R}(s_i \rightarrow s_{i+1}; \lambda_{i+1})$, which is more explicit than the one used in Section 2.4. We do this for greater clarity and because we want to point out the dependence of the transition rates on the control parameter of the experimental protocol.

Now, let us come back to the cyclic stochastic evolution. The probability of observing a forward trajectory $\sigma \equiv (s_1 \to s_2 \cdots s_{n-1} \to s_n)$ can be written as

$$P^f[\sigma] = p_e[s_1]\, \mathcal{R}(s_1 \to s_2; \lambda_2) \cdots \mathcal{R}(s_{n-1} \to s_n; \lambda_n), \tag{3.196}$$

that is, it is given by the product of the equilibrium probability of finding the system at time t_1 in s_1 (see Eq. (3.175)) times the forward transition rates up to s_n. Analogously, the probability of observing a backward trajectory $\overline{\sigma}$ starting at state s_n and eventually reaching the state s_1, after having visited the states $s_{n-1}, \cdots, s_2$, can be written as

$$P^b[\overline{\sigma}] = p_e[s_n]\, \mathcal{R}(s_n \to s_{n-1}; \lambda_n) \cdots \mathcal{R}(s_2 \to s_1; \lambda_2). \tag{3.197}$$

Moreover, since the overall cyclic protocol is isothermal, we can assume that the detailed balance relation holds, namely

$$\frac{\mathcal{R}(s_i \to s_{i+1}; \lambda_{i+1})}{\mathcal{R}(s_{i+1} \to s_i; \lambda_{i+1})} = \frac{e^{-\beta E(s_{i+1}; \lambda_{i+1})}}{e^{-\beta E(s_i; \lambda_{i+1})}}. \tag{3.198}$$

Making use of the last two assumptions, we obtain the relation

$$\frac{\mathcal{R}(s_1 \to s_2; \lambda_2) \cdots \mathcal{R}(s_{n-1} \to s_n; \lambda_n)}{\mathcal{R}(s_n \to s_{n-1}; \lambda_n) \cdots \mathcal{R}(s_2 \to s_1; \lambda_2)} = \exp\left(-\beta \sum_{i=1}^{n-1} \left[E(s_{i+1}; \lambda_{i+1}) - E(s_i; \lambda_{i+1}) \right] \right)$$
$$= e^{-\beta Q^f}. \tag{3.199}$$

Finally, using Eqs. (3.196), (3.197), and (3.199), and exploiting the equilibrium relation

$$\frac{p_e[s_1]}{p_e[s_n]} = \frac{Z_n}{Z_1} e^{\beta \Delta E} = e^{\beta(\Delta E - \Delta F)}, \tag{3.200}$$

we obtain

$$\frac{P^f[\sigma]}{P^b[\overline{\sigma}]} = \frac{p_e[s_1]}{p_e[s_n]} e^{-\beta Q^f} = e^{\beta(\Delta E - \Delta F - Q^f)} = e^{\beta(W^f - \Delta F)}. \tag{3.201}$$

In order to prove the Crooks fluctuation theorem, see Eq. (3.185), it is necessary to write the probability distributions $\rho^f(W)$ and $\rho^b(-W)$ in terms of integrals over all possible trajectories,

$$\rho^f(W) = \int d\sigma\, P^f[\sigma]\, \delta(W - W^f[\sigma]),$$
$$\rho^b(-W) = \int d\overline{\sigma}\, P^b[\overline{\sigma}]\, \delta(W + W^b[\overline{\sigma}]), \tag{3.202}$$

and to remark, see Eq. (3.194), that

$$W^b[\overline{\sigma}] = -W^f[\sigma]. \tag{3.203}$$

Now, if we write Eq. (3.201) in the form $P^f[\sigma] = P^b[\overline{\sigma}]e^{\beta(W^f-\Delta F)}$, we obtain

$$
\begin{aligned}
\rho^f(W) &= \int d\sigma\, P^f[\sigma]\, \delta(W - W^f[\sigma]) \\
&= \int d\sigma\, P^b[\overline{\sigma}]e^{\beta(W^f[\sigma]-\Delta F)}\delta(W - W^f[\sigma]) \qquad (3.204) \\
&= \int d\overline{\sigma}\, P^b[\overline{\sigma}]e^{\beta(-W^b[\overline{\sigma}]-\Delta F)}\delta(W + W^b[\overline{\sigma}]) \qquad (3.205) \\
&= \rho^b(-W)\, e^{\beta(W-\Delta F)},
\end{aligned}
$$

where the passage from Eq. (3.204) to Eq. (3.205) makes use of Eq. (3.203) and requires the time reversal symmetry in the cyclic protocol, $\int d\sigma = \int d\overline{\sigma}$.

It is worth noting that Crooks relation, $\rho^f(W)/\rho^b(-W) = e^{\beta(W-\Delta F)}$, looks like a generalized detailed balance relation. This similarity is even stronger if we remark that the multiple-steps trajectory $s_1 \to s_2 \cdots s_{n-1} \to s_n$ might be replaced by a single step allowing to pass from an initial state s_1 to a final state s_n, so that such relation can be understood to relate any pair of states corresponding to different values of some control parameter λ. Doing so, Equations from (3.196) to (3.199) would be simplified, but the multiple-steps strategy adopted here has some similarities with Section 3.8.3, which justifies its usage.[20]

3.8.3 Fluctuation Theorem for a NESS

Up to now, we have considered processes where the system is driven out-of-equilibrium while interacting with a single thermal reservoir, either in the initial and final states (Section 3.8.1), or all over a cyclic transformation (Section 3.8.2). Another situation of physical interest amounts to keeping a system in a stationary out-of-equilibrium state, where mass, momentum, and energy currents flow through the system, driven by a gradient of density, velocity, or temperature imposed on the system. Such situations are those illustrated in Section 1.2.2, where we discuss the phenomenological approach to transport phenomena according to basic kinetic theory.

Here we reconsider such situations in the light of a more refined theoretical approach, which takes explicitly into account the role of thermal fluctuations. More precisely, we aim at showing that a fluctuation theorem, analogous to the Crooks' one, holds for fluctuations in stationary out-of-equilibrium conditions. In summary, this theorem can be expressed as follows:

$$
\lim_{t\to\infty}\frac{1}{t}\ln\frac{P_t(J)}{P_t(-J)} = J\left(\frac{1}{T_1} - \frac{1}{T_2}\right) \equiv \Sigma, \qquad (3.206)
$$

where $P_t(\pm J)$ is the probability that in a time t one measures the value $\pm J$ of the heat current between the two thermostats at temperatures T_1 and T_2, and Σ is the entropy production rate of an out-of-equilibrium thermodynamic system, which dissipates the heat $Q = Jt$ in a time t. As made evident by the limit $t \to \infty$, this fluctuation theorem holds asymptotically, that is, for time scales much larger than those associated with fluctuations. For pedagogical

---

[20] In addition to that, it is reasonable to think that an experimental protocol is more similar to a multiple-steps transition than to a single-step one.

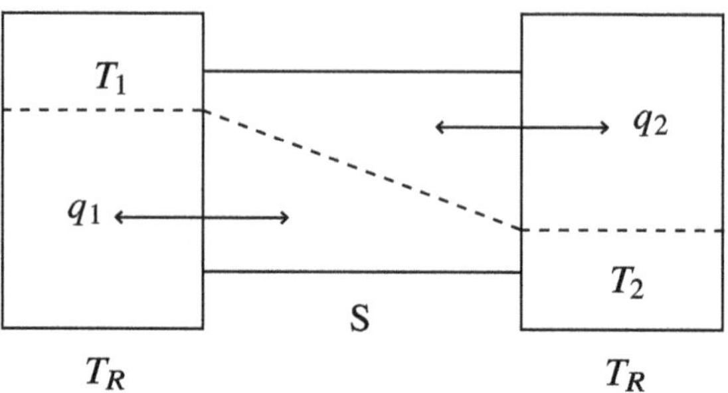

Fig. 3.7   A thermodynamic system S in contact with two thermal reservoirs R at different temperatures $T_1 > T_2$. The dashed line shows the temperature profile that, according to Fourier's law, is linear inside S. Thermal reservoirs exchange heat q_1, q_2 with the system: If $q_1 > 0$, the left reservoir releases heat to the system; if $q_2 > 0$, the right reservoir releases heat to the system.

reasons, it is worth illustrating this fluctuation theorem making reference to the specific example of a thermodynamic system S driven in a stationary nonequilibrium state, imposed by the contacts with two thermal reservoirs at different temperature T_1 and T_2, see Fig. 3.7. We need some preliminary mathematical considerations for describing properly such a setup. We assume that the evolution in continuous time of our physical system is given by the master Equation (2.129),

$$\frac{dp_i(t)}{dt} = \sum_j \left(\mathcal{R}_{ij} p_j(t) - \mathcal{R}_{ji} p_i(t) \right), \tag{3.207}$$

where $p_i(t)$ is the probability of observing the system in state i at time t, $\mathcal{R}_{ij}$ is the transition rate from state j to state i. Note that at any time t the normalization condition $\sum_i p_i(t) = 1$ is preserved. Moreover, the physical dimension of $\mathcal{R}_{ij}$ is $[t^{-1}]$ and also these transition rates have to satisfy the normalization condition (2.123),

$$\mathcal{R}_{jj} = \sum_{i \neq j} \mathcal{R}_{ij}. \tag{3.208}$$

In full analogy to what we have discussed in Section 2.3.5 for the time-discrete version of the master equation, Eq. (3.207) makes $p_i(t)$ evolve asymptotically to a stationary distribution probability $\bar{p}_i$, satisfying the condition $\frac{d\bar{p}_i}{dt} = 0$. If the system is at equilibrium, $\bar{p}_i$ represents the equilibrium probability and the detailed balance condition

$$\mathcal{R}_{ij} \bar{p}_j - \mathcal{R}_{ji} \bar{p}_j = 0 \quad , \quad \forall i, j \tag{3.209}$$

holds for any pair of states (i, j). Let us recall that the detailed balance conditions amount to impose a microreversibility in time of the stochastic process engendered by the master Equation (3.207).

When S is put in contact with a thermal reservoir at temperature T, the transition from a state j to a state i is associated also with the transfer of an amount of heat q from the reservoir, and the detailed balance condition becomes

$$\mathcal{R}_{ij} e^{-\beta E_j} - \mathcal{R}_{ji} e^{-\beta E_i} = 0 \quad , \quad \forall i, j. \tag{3.210}$$

Here we have made explicit the dependence of the transition rates on the amount of heat transferred by the reservoir. Actually, if an amount of heat q is transferred[21] for the change of state from j to i, an amount of heat $-q$ has to be transferred for the inverse change of state, from i to j. Moreover, we have expressed the equilibrium probability by the Boltzmann–Gibbs formula, that is, $\bar{p}_i = \exp\left(-\beta E_i\right)/Z$.

By definition, we have that $q = E_i - E_j$ and we can rewrite Eq. (3.210) as follows:

$$\mathcal{R}_{ij} = \mathcal{R}_{ji} e^{-\beta q} \quad , \quad \forall i, j. \tag{3.211}$$

Note that in a cyclic transformation of a system in contact with a single thermal reservoir, the total amount of heat exchanged with S is null.

Now, let us suppose that S is in contact with two thermal reservoirs at different temperatures T_1 and T_2. In this case, Eq. (3.211) must be generalized to the following balance condition:

$$\mathcal{R}_{ij} = \mathcal{R}_{ji} e^{-\beta_1 q_1} e^{-\beta_2 q_2} \quad , \quad \forall i, j, \tag{3.212}$$

where q_1 and q_2 are the amounts of heat exchanged with the reservoir at temperature $T_1 = \frac{1}{\beta_1}$ and $T_2 = \frac{1}{\beta_2}$, respectively, for the change of state from j to i (see Fig. 3.7). Relying upon this generalized balance condition, we can follow a procedure analogous to the one adopted for proving Crooks fluctuation theorem by considering the stochastic evolution of system S along a stochastic trajectory σ lasting over a time t, going through the sequence of states $1, 2, \cdots, j, \cdots, n$ at constant pace $\tau = \frac{t}{n}$. The probability $P[\sigma]$ of observing this trajectory can be expressed as follows:

$$P[\sigma] = \mathcal{R}_{n\,n-1} \cdots \mathcal{R}_{j\,j-1} \cdots \mathcal{R}_{21} P_s(1), \tag{3.213}$$

where $q_1(j)$ and $q_2(j)$ are the amounts of heat exchanged with the two reservoirs for the change of state from $j - 1$ to j, while $P_s(1)$ is the stationary probability of having S in the initial state of σ. Note that this stationary probability is different from the equilibrium probability, because we are dealing with a stochastic trajectory associated with a stationary nonequilibrium process, a quantity that is generally unknown in the absence of a suitable physical model of S and of the thermal reservoirs.

The probability $P[\overline{\sigma}]$ of observing the reversed trajectory $\overline{\sigma}$ going through the sequence of states $n, n - 1, \cdots, j, \cdots, 1$ at constant pace τ is given by the expression

$$P(\overline{\sigma}) = P_s(n)\mathcal{R}_{n-1\,n} \cdots \mathcal{R}_{j-1\,j} \cdots \mathcal{R}_{12}. \tag{3.214}$$

---

[21] A positive q means that heat is absorbed from the thermostat. A negative q means that heat is released to the thermostat.

Making use of Eq. (3.212), we obtain a relation between the probabilities of the trajectories σ and $\overline{\sigma}$, namely

$$P[\sigma] = \frac{P_s(1)}{P_s(n)}\, e^{-\beta_1 Q_1}\, e^{-\beta_2 Q_2}\, P[\overline{\sigma}], \tag{3.215}$$

where

$$Q_1 = q_1(1) + q_1(2) + \cdots + q_1(n) \quad , \quad Q_2 = q_2(1) + q_2(2) + \cdots + q_2(n) \tag{3.216}$$

are the total amounts of heat exchanged by S with the reservoirs along the trajectory σ.

In order to prove the fluctuation theorem (3.206), we have to impose two physical conditions. The first one is that the system cannot accumulate an infinite amount of energy along the trajectory σ. Since straightforward considerations about energy balance impose the equality $E_n - E_1 = Q_1 + Q_2$, this condition amounts to assuming that there exists a constant $\mathcal{E}$ such that

$$|Q_1 + Q_2| < \mathcal{E}. \tag{3.217}$$

The second condition is that the rate $\frac{P_s(i)}{P_s(j)}$ is a bounded quantity for any pair of states i and j, that is, it exists a constant $\rho > 1$ such that

$$\frac{1}{\rho} < \frac{P_s(i)}{P_s(j)} < \rho. \tag{3.218}$$

By combining these two conditions with Eq. (3.215), we can write

$$\frac{1}{\rho\, e^{\beta_2\, \mathcal{E}}}\, e^{(\beta_2-\beta_1)\, Q_1}\, P[\overline{\sigma}] \; < P[\sigma] < \; \rho\, e^{\beta_2\, \mathcal{E}}\, e^{(\beta_2-\beta_1)\, Q_1}\, P[\overline{\sigma}]. \tag{3.219}$$

By summing over all the stochastic trajectories corresponding to a given value of the stationary heat current $J = \frac{Q_1}{t}$, this equation yields the inequalities

$$\frac{1}{\rho\, e^{\beta_2\, \mathcal{E}}}\, e^{(\beta_2-\beta_1)\, Jt}\, P(-J) \; < P(J) < \; \rho\, e^{\beta_2\, \mathcal{E}}\, e^{(\beta_2-\beta_1)\, Jt}\, P(-J). \tag{3.220}$$

Before proceeding, let us recall that our system is in a nonequilibrium steady state because its ends are attached to two thermal reservoirs at different temperatures. The quantity $P(J)$ is the probability that during a large time interval t, the system exchanges a quantity of heat $Q_1 = Jt$ with the reservoir at temperature T_1. The condition of stationarity requires that the current "entering" the system must equal the current "leaving" the system, which implies that $\lim_{t\to\infty}(Q_1 + Q_2)/t = 0$.

The total exchange of heat with a thermal reservoir is the sum of a very large number of random, microscopic exchanges, and J plays the role of the average value of such exchanges. Therefore, the analysis of the distribution $P(J)$ is conceptually similar to the problem of large deviations studied in Section 3.3. In particular, it is useful to refer to the definition of the Cramer function, see Eq. (3.36), where N should now be replaced by time t. In poor terms, the Cramer function, also called the large deviation functional $C(J)$, is defined through the relation

$$P(J) \sim e^{-t\, C(J)}, \tag{3.221}$$

where the symbol "of the order of" means that all nonexponential prefactors can be neglected. Therefore, the inequalities of Eq. (3.220) in the large time limit, $t \to \infty$, boil down to the fluctuation theorem

$$C(J) = C(-J) - (\beta_2 - \beta_1)J, \tag{3.222}$$

or, equivalently

$$\lim_{t \to \infty} \frac{1}{t} \ln \frac{P(J)}{P(-J)} = J \left(\frac{1}{T_2} - \frac{1}{T_1} \right), \tag{3.223}$$

which corresponds to the general formulation of the fluctuation theorem reported in Eq. (3.206).

We conclude this section with a few remarks of physical interest. Condition (3.217) can be rephrased by assuming that in stationary nonequilibrium conditions, the total amount of heat Q_1, provided to S by the thermal reservoir at temperature T_1, is eventually transferred to the heat reservoir at temperature T_2, that is, $Q_2 \simeq -Q_1$. As a consequence, despite both Q_1 and Q_2 being proportional to t, the total amount of heat, $Q_1 + Q_2$, stored by the system S remains finite, and the variation of entropy ΔS_r in the two reservoirs is given by the expression

$$\Delta S_r = -\left(\frac{Q_2}{T_2} + \frac{Q_1}{T_1} \right) \simeq Q_1 \left(\frac{1}{T_2} - \frac{1}{T_1} \right). \tag{3.224}$$

In stationary nonequilibrium conditions and for large values of t, we can say that ΔS_r amounts to the total variation of entropy ΔS_{tot}.

If we rewrite Eq. (3.222) as follows

$$C(J) - C(-J) = J \left(\frac{1}{T_1} - \frac{1}{T_2} \right), \tag{3.225}$$

we can observe that, despite the large deviation functional $C(J)$ being a model-dependent quantity, that is, it depends on the details of the physical system S and of the thermal reservoirs, its difference between reversed trajectories is fully unaffected by such details and it is linearly dependent on the heat current J, with a universal slope that depends on T_1 and T_2, only. This allows to design suitable experimental protocols for obtaining quantitative estimates of these differences between large deviation functionals (i.e., nonequilibrium thermodynamic state functions analogous to the free energy of equilibrium thermodynamics) by measuring the heat current J.

Finally, note that if $T_1 > T_2$ and $J > 0$, Eq. (3.225) implies that $C(J) < C(-J)$. This inequality yields the following result:

$$\frac{\langle Q_1 \rangle}{t} = \langle J \rangle = \int_0^{+\infty} J\Big(P(J) - P(-J)\Big)dJ > 0, \tag{3.226}$$

which indicates that on average the heat current flows from the hot reservoir to the cold one. This means that in such conditions the second principle of thermodynamics holds true on average, that is, there is a very small, but non-negligible probability that in a single trajectory the second principle of thermodynamics may be violated as an effect of fluctuations.

3.9 Bibliographic Notes

A. I. Khinchin, *Mathematical Foundations of Statistical Mechanics* (Dover Publications, 1999) offers a detailed proof of CLT and the explicit form of correction terms for finite N.

D. Sornette, *Critical Phenomena in Natural Science*, 2nd ed. (Springer, 2006) is a good source for CLT and large deviations.

The book *Introduction to Random Matrices* by G. Livan, M. Novaes, and P. Vivo (Springer, 2018) is a pedagogical account of the topic.

A useful reference for many questions faced in this chapter and beyond are the lecture notes by B. Derrida, *Fluctuations et grandes déviations autour du Seconde Principe* (Collège de France, 2015–2016).

A recent monograph on stochastic thermodynamics is the book by L. Peliti and S. Pigolotti, *Stochastic Thermodynamics: An Introduction* (Princeton University Press, 2021).

A formulation of the fluctuation theorem for the stationary dynamics of chaotic Hamiltonian systems is discussed in G. Gallavotti and E. G. D. Cohen, *Physical Review Letters*, **74**(14), 2694 (1995) and *Journal of Statistical Physics*, **80**(5–6), 931–970 (1995).

The main references about Jarzinski equality and Crooks' theorem are C. Jarzynski, *Physical Review Letters*, **78**, 2690 (1997) and *Physical Review E*, **56**, 5018 (1997); G. E. Crooks, *Physical Review E*, **60**, 2721 (1999) and *Physical Review E*, **61**, 2361 (2000).

An extended overview about recent achievements in the theory of random walks and Lévy processes is contained in J. Klafter and I. M. Sokolov, *First Steps in Random Walks: From Tools to Applications* (Oxford University Press, 2011). This book includes additional information about fractional partial differential equations.

A reference textbook of functional analysis is W. Rudin, *Functional Analysis* (McGraw-Hill, 1991), where details about Tauberian theorems can be found.

4

Linear Response Theory and Transport Phenomena

For pedagogical reasons, this chapter is introduced by showing how one can obtain the Kubo relation for a Brownian particle (Section 4.2). In fact, we aim at pointing out from the very beginning the importance of equilibrium time correlation functions, which result to be basic ingredients also for the perturbative approach to nonequilibrium phenomena, based on linear response theory.

In order to generalize a similar approach, we have first to rely on the assumption that any thermodynamic observable, even at equilibrium, has to be a fluctuating quantity. As we illustrate in Section 4.3, a preliminary step in the direction of a linear response theory amounts to assuming that thermodynamic observables obey a fluctuation-driven evolution, described by generalized Langevin and Fokker–Planck equations, in full analogy with a Brownian particle. The following step in this direction needs the explicit dynamical description of the physical mechanism of fluctuations of a thermodynamic observable as the result of the effect of an associated perturbation field, switched on in the past (Section 4.4). In fact, this approach allows us to establish a quantitative relation with the equilibrium time correlation function of the perturbed thermodynamic observable, thus yielding a fully general fluctuation–dissipation relation. This basic formalism is applied for computing several quantities of physical interest, such as the work done in the presence of a time-dependent field, the susceptibility associated with the perturbed observable, together with the classical example of the damped harmonic oscillator in the presence of a time-dependent driving force.

Moreover, by exploiting the formal equivalence of the Fokker–Planck equation with a continuity equation, linear response theory also provides a basic hydrodynamic approach to transport phenomena. As shown in Section 4.5, it allows to obtain the generalized Green–Kubo relations, by which one can compute the transport coefficients, such as diffusivity, viscosity, and thermal conductivity of a thermodynamic system in terms of the equilibrium time correlation functions of the mass, momentum, and energy currents, respectively.

This important result can be extended also to coupled transport processes. As a preliminary step in Section 4.6, we discuss how linear response theory can be generalized in the presence of various fluctuating observables. Then, the bridge with the formulation of the thermodynamics of irreversible processes can be established by considering entropy as a genuine dynamical variable (Section 4.7). The key ingredient is its production rate, which can be expressed as a combination of generalized thermodynamic forces (affinities) and the corresponding fluxes (currents). By combining this result with the Onsager reciprocity relations and the Onsager theorem for charged transport, we can formulate a phenomenological theory of coupled transport processes, which is summarized by the Onsager matrix of generalized transport coefficients: Its properties and symmetries again originate from those of equilibrium time correlation functions of pairs of thermodynamic observables. We point out that, in the linear response regime, entropy (despite being considered a dynamical variable) maintains the same relation with basic thermodynamic quantities (internal energy, pressure, and chemical potentials) as in the equilibrium case. This amounts to asserting that also for stationary non-equilibrium processes associated with coupled transport phenomena, local equilibrium conditions hold. Several examples of coupled transport are reported in Section 4.8.

For the sake of completeness, a short illustration of linear response theory for quantum systems is given in Appendix N. In fact, despite its formal analogy with the classical case, its application to specific examples also requires the knowledge of quantum field theory. Therefore, it can be easily accessible only for readers familiar with this topic.

4.1 A Theoretical Basis for Transport Processes

Linear response theory is a perturbative approach to nonequilibrium processes that allows us to establish a relation between equilibrium time correlation functions of fluctuating observables and measurable quantities of physical interest, such as susceptibilities, transport coefficients, relaxation rates, and so on. For instance, the diffusion coefficient of a Brownian particle can be related to its velocity autocorrelation function by the Kubo relation. The basic idea is that we deal with systems whose evolution remains always close to an equilibrium state, also when some perturbation drives them out of it. At first sight, all of that may appear contradictory: How can it be that equilibrium properties are associated with nonequilibrium ones? A preliminary answer to this question is given at the end of Section 1.2.2 in the framework of elementary kinetic theory of transport phenomena: In the presence of a macroscopic gradient of physical quantities, equilibrium conditions set in locally. Such an assumption may appear reasonable if the gradient is moderate, that is, if we gently drive the system out of equilibrium so that, on local space and time scales, significantly larger than the mean free path of particles and of the equilibrium relaxation time, the system always remains very close to equilibrium conditions. In the same spirit, Einstein formulated a theory of fluctuations close to equilibrium, described in Section 3.7.

Another important consideration for proceeding in the direction of linear response theory comes from the general Fokker–Planck formulation of stochastic processes discussed in Chapter 2. This dynamical description can be applied to the fluctuation-driven evolution of any thermodynamic, that is, macroscopic, observable. We should admit that this is not a straightforward assumption and it demanded some efforts before being fully understood. Once established that close to or at equilibrium also thermodynamic observables fluctuate, we need to identify the mechanism at the origin of these fluctuations and describe it in a suitable mathematical language. This task can be accomplished by making use of a Hamiltonian description, where the presence of a perturbation field induces the fluctuation of a thermodynamic observable. In this framework, the equilibrium side of linear response theory is associated with the autocorrelation function of the thermodynamic observable averaged over the Boltzmann–Gibbs equilibrium statistical measure, which implies that the system is in contact with a thermal reservoir at a given temperature. A mathematical description of the linear response theory is based on the fluctuation–dissipation theorem: It concerns the analytic properties of the linear response function, which is proportional to the time derivative of the equilibrium autocorrelation function. This theorem has a clear physical importance, because it allows us to establish a rigorous relation between the fluctuation and the dissipation mechanisms, which are the key ingredients of linear response theory.

Relying on such a general approach, one can also elaborate a hydrodynamic approach to transport processes. As we have seen in Section 2.4.6, the Fokker–Planck equation can be cast into the form of a hydrodynamic equation for a fluid system. Making use of constitutive relations that allow us to express density currents in terms of density gradients (as in elementary kinetic theory of transport; see Section 1.2.2), we can obtain the general Green–Kubo relations for diffusivity, viscosity, and heat conductivity of any thermodynamic system, whose transport properties are formally equivalent to those of a fluid. As in the case of the Brownian particle, such relations allow us to calculate these transport coefficients from the integral of the equilibrium correlation function of the total mass, momentum, and energy currents. The practical interest of the Green–Kubo relations also relies on the possibility of obtaining numerical estimates of the equilibrium correlation functions by numerical calculations performed by suitable molecular dynamics algorithms.

Linear response theory can be generalized by considering the possibility that a perturbation field acting on a given thermodynamic observable may have influence over other thermodynamic observables. This is quite a natural idea, because the presence of a perturbation field modifies the evolution of microscopic degrees of freedom of the systems, and this change, in principle, concerns any other macroscopic observable, because it is a function of the same degrees of freedom. The main reason for this approach is summarized by the Onsager regression hypothesis: When studying the evolution of a physical system around equilibrium, we cannot distinguish if an instantaneous value of an observable, different from its equilibrium value, is due to a spontaneous fluctuation or to a perturbation. As a consequence, the regression of the instantaneous value of the observable toward its equilibrium value is related to the decay of the time correlation functions. It is important to point out that this statement applies to the correlation function of any pair of fluctuating observables. Their symmetry with respect to time reversal is a crucial ingredient for characterizing the properties of the Onsager matrix of generalized transport coefficients. This is the main tool for describing coupled transport phenomena, such as thermomechanical, thermoelectric, and galvanomagnetic effects in continuous systems. In fact, a phenomenological theory of coupled transport can be constructed by combining linear response theory with a thermodynamic formulation of irreversible processes. This is based on the idea that nonequilibrium states of a thermodynamic system can be characterized by the entropy production rate, see Section 3.8.3, expressed as the product of generalized thermodynamic forces, usually called affinities, with the fluxes of the corresponding observable. In a phenomenological approach, we can assume that fluxes and affinities are proportional to each other, and we conclude that the entries of the Onsager matrix can be expressed again in terms of correlation functions of pairs of fluctuating observables.

Linear response theory can also be formulated for quantum systems. On a conceptual ground, quantum processes, at variance with classical ones, are characterized by the presence of intrinsic fluctuations, whose practical manifestation is that we expect to obtain statistical inferences rather than deterministic ones even at the microscopic level. In the framework of a thermodynamic description such a difference is much less relevant, although extending linear response theory to quantum systems demands the introduction of suitable tools of quantum theory, such as the density matrix formalism and time ordering, when dealing with noncommuting quantum observables (see Appendix N).

4.2 The Kubo Formula for the Brownian Particle

In Section 2.4.5 we introduced a general Fokker–Planck equation for a generic stochastic process $X(t)$. In order to explore in more detail, the dynamical content of this equation, let us first consider the easy example of the diffusion equation (2.49) for a Brownian particle. We assume space isotropy and that $x(t)$, $y(t)$, and $z(t)$ are described by independent stochastic processes, so that we can treat the diffusion equation for just one of these variables, say $x(t)$,

$$\frac{\partial P(x,t)}{\partial t} = D\frac{\partial^2}{\partial x^2}P(x,t). \tag{4.1}$$

The explicit solution for $P(x,t)$ (see Eq. (2.50)) can be used to compute the average square displacement, given in Eq. (2.51),

$$\langle x^2\rangle - \langle x\rangle^2 = 2Dt. \tag{4.2}$$

If we assume the initial position of the Brownian particle at the origin, that is, $\mathbf{r}(0) = 0$,

$$\langle \mathbf{r}^2\rangle = \langle x^2\rangle + \langle y^2\rangle + \langle z^2\rangle = 6Dt. \tag{4.3}$$

It is noteworthy that the same result can be obtained by multiplying both sides of Eq. (4.1) by x^2 and integrating over x,

$$\frac{\partial}{\partial t}\int_{-\infty}^{+\infty} dx\, x^2\, P(x,t) = D\int_{-\infty}^{+\infty} dx\, x^2\frac{\partial^2}{\partial x^2}P(x,t). \tag{4.4}$$

The right-hand side of this equation can be integrated twice by parts over x: Since $P(x,t)$ and its derivatives with respect to x vanish for $x \to \pm\infty$ more rapidly than any power of x, one finally obtains

$$\frac{\partial\langle x^2(t)\rangle}{\partial t} = 2D, \tag{4.5}$$

whose time integration gives, again, Eq. (4.2). We should stress that the Fokker–Planck equation does not cover time scales smaller than or of the order of $t_d = m/\tilde{\gamma}$, which is the time necessary to attain equilibrium. This is the reason why Eq. (4.2) differs from Eq. (2.33), whose limiting behaviors are given in Eq. (2.35). It is therefore more correct to write[1]

$$\lim_{t \to \infty} \frac{\langle x^2(t) \rangle}{t} = 2\,D. \tag{4.6}$$

We can obtain a useful relation between the diffusion constant D of the Brownian particle and its velocity autocorrelation function as follows. We start from the relation $x(t) = \int_0^t dt'\, v_x(t')$ that defines the instantaneous position of the Brownian particle $x(t)$ in terms of its velocity component in the x-direction $v_x(t)$, so we can write

$$x^2(t) = \int_0^t dt' \int_0^t dt'' v_x(t')\, v_x(t''). \tag{4.7}$$

By deriving both sides with respect to time, we obtain

$$\frac{\partial x^2(t)}{\partial t} = 2\, v_x(t) \int_0^t dt'\, v_x(t') = 2 \int_0^t dt'\, v_x(t)v_x(t'). \tag{4.8}$$

We can average both sides of this equation and we obtain

$$\frac{\partial \langle x^2(t) \rangle}{\partial t} = 2 \int_0^t d\,t'\, \langle v_x(t)v_x(t') \rangle, \tag{4.9}$$

where the integral on the right-hand side contains the velocity autocorrelation function of the Brownian particle. Since the process is stationary, that is, invariant under time translation, we can shift the origin of time as follows:

$$\langle v_x(t)\, v_x(t') \rangle = \langle v_x(t - t')\, v_x(0) \rangle. \tag{4.10}$$

By redefining the integration variable as $\tau = t - t'$, we obtain

$$\frac{\partial \langle x^2(t) \rangle}{\partial t} = 2 \int_0^t d\,\tau\, \langle v_x(\tau)v_x(0) \rangle \tag{4.11}$$

and using Eq. (4.6), we can finally write

$$D = \lim_{t \to \infty} \int_0^t d\,\tau \langle v_x(\tau)\, v_x(0) \rangle. \tag{4.12}$$

Equation (4.12) is the simplest form of the so-called Kubo relation. It tells us that the diffusion coefficient D of the Brownian particle can be measured by estimating the asymptotic (equilibrium) behavior of its velocity correlation function. It is important to note that we are facing the first example of a more general situation, where transport coefficients, that is, physical quantities typically associated with nonequilibrium processes, can be obtained from equilibrium averages of correlation functions of suitable current-like observables.

In order to extend the Kubo relation to other transport coefficients of a generic thermodynamic system, such as viscosity or thermal conductivity, we must first work out the linear response theory. As we show in Section 4.5, the results of linear response theory can be used to formulate a hydrodynamic description based on the Fokker–Planck equation for fluctuating observables associated with conserved quantities.

---

[1] If the average value $\langle x^2(t) \rangle$ is intended as a time average of the square displacement of a single walker, the limit of diverging time is necessary. Instead, if it is an ensemble average over many different walkers, Eq. (4.6) is correct at any time.

4.3 Generalized Brownian Motion

In Chapter 2 we extensively discussed the theory of fluctuations for a Brownian particle via the Langevin and the Fokker–Planck equations, while providing in Section 2.4 a generalization of these equations to any stochastic quantity, subject to uncorrelated fluctuations, such as the thermal fluctuations of a thermodynamic system at or close to its equilibrium state at a given temperature T. Here we want to proceed along this direction starting from a basic argument: In principle, there is no limitation to applying this mathematical description to the fluctuation-driven evolution of any thermodynamic, that is, macroscopic, observable. However, the introductory level of this book suggests we avoid entering the awkward pathway of a mathematically rigorous justification of such an assumption. Therefore, we limit ourselves to a few simple explanatory remarks.

As well as the position and the velocity of a Brownian particle, also a thermodynamic observable $X(t)$ fluctuates in time at thermal equilibrium. Such fluctuations, as it happens for the Brownian particle, are originated by the interactions with an exceedingly large number (typically, the Avogadro number $N_A \approx 10^{23}$) of microscopic degrees of freedom present in the system. In practice, we assume that at thermal equilibrium all of these microscopic degrees of freedom evolve as independent stochastic variables and also independent of the actual value of $X(t)$. Due to their large number, relative fluctuations are expected to be quite small and eventually vanish in the thermodynamic limit for any macroscopic observable. This statement is a consequence of the central limit theorem discussed in Section 3.2. Since fluctuations are irrelevant in the thermodynamic limit, the finiteness of the system we are dealing with is a necessary condition for the theory of generalized Brownian motion.

In complete analogy with this, we write the Langevin-like equation[2]

$$\tilde{\gamma}\frac{d\,X(t)}{d\,t} = \mathcal{F}(X) + \tilde{\eta}(t), \tag{4.13}$$

where $\mathcal{F}$ is a generalized force, $\tilde{\gamma}$ is a generalized friction coefficient, and $\tilde{\eta}(t)$ is the stochastic term, epitomizing the fluctuations originated by the interaction of the thermodynamic observable X with the microscopic degrees of freedom.

With the earlier assumptions on the origin of fluctuations, the stochastic term has to fulfill the properties

$$\langle \tilde{\eta}(t) \rangle = 0 \tag{4.14}$$

$$\langle \tilde{\eta}(t)\,\tilde{\eta}(t') \rangle = \tilde{\Gamma}\,\delta(t - t'), \tag{4.15}$$

where $\langle \cdot \rangle$ are equilibrium averages. The latter condition of fully time-uncorrelated stochastic fluctuations should be interpreted, in a physical perspective, as the assumption that $X(t)$ varies over much longer time scales than those typical of $\tilde{\eta}(t)$.

This problem is formally analogous to the example of a Brownian particle subject to an external force, originated by a conservative potential (see Section 2.4.6). In fact, we can

---

[2] We use a notation that is in full agreement with Section 2.2.2.

formally write the Fokker–Planck equation corresponding to Eq. (4.13) as

$$\frac{\partial P(X,t)}{\partial t} = -\frac{\partial \mathcal{J}(X,t)}{\partial X},\qquad(4.16)$$

where

$$\mathcal{J} = v(X)P(X,t) - \frac{\tilde{\Gamma}}{2\tilde{\gamma}^2}\frac{\partial}{\partial X}P(X,t).\qquad(4.17)$$

Here

$$v(X) = \frac{\mathcal{F}}{\tilde{\gamma}}\qquad(4.18)$$

is the analog of the limiting velocity of the Brownian particle, and $\tilde{\Gamma}/\tilde{\gamma}^2$ is the strength of the correlation of the noise $\tilde{f} \equiv \tilde{\eta}/\tilde{\gamma}$.

Close to thermodynamic equilibrium, the expression of $\mathcal{F}$ comes from Einstein's theory of equilibrium fluctuations (see Section 3.7), which provides us with an expression of the equilibrium probability density of the values taken by X due to fluctuations,

$$P(X) = C \exp\left(-\frac{\mathcal{U}(X)}{T}\right),\qquad(4.19)$$

where $\mathcal{U}$ is the thermodynamic potential expressed as a function of X. In particular, one has

$$\mathcal{U}(X^*) = U(V,T),\qquad(4.20)$$

where X^* is the equilibrium value of X and $U(V,T)$ is the internal energy of the thermodynamic system.

By assuming stationarity conditions for $P(X)$ and making use of the fluctuation–dissipation relation (2.29), $\tilde{\Gamma} = 2\tilde{\gamma}T$, we can rewrite (4.18) as

$$v(X) = \frac{\mathcal{F}}{\tilde{\gamma}} = -\frac{2T}{\tilde{\Gamma}}\frac{\partial \mathcal{U}(X)}{\partial X},\qquad(4.21)$$

where the generalized thermodynamic force $\mathcal{F}$ has been written as the derivative of the potential $\mathcal{U}$ with respect to the perturbed thermodynamic observable X. We can therefore rewrite Eq. (4.13) in the form

$$\frac{dX}{dt} = -\frac{2T}{\tilde{\Gamma}}\frac{\partial \mathcal{U}(X)}{\partial X} + \tilde{f}(t).\qquad(4.22)$$

We can conclude that making use of Einstein's theory of fluctuations and of the analogy with the problem of a Brownian particle subject to a conservative force, we have derived a heuristic expression for the generalized Brownian motion in terms of the nonlinear Langevin-like equation (4.22).

We can proceed along this direction by considering that in thermodynamic equilibrium conditions $\mathcal{U}(X^*)$ is a minimum of the generalized potential $\mathcal{U}(X)$ with respect to X, that is,

$$\left(\frac{\partial \mathcal{U}(X)}{\partial X}\right)_{X=X^*} = 0,\qquad(4.23)$$

so we can write

$$\mathcal{U}(X) = U(V,T) + \frac{1}{2}\left(\frac{\partial^2 \mathcal{U}(X)}{\partial X^2}\right)_{X=X^*}(X - X^*)^2 + \cdots. \tag{4.24}$$

Notice that in a thermodynamic system made of N particles, relative fluctuations at equilibrium of any macroscopic observable, that is, $|X - X^*|$, are of the order of $N^{-\frac{1}{2}}$, as predicted by the central limit theorem (see Section 3.2): This is why higher powers can be neglected with respect to the quadratic term. Since we are interested in studying the fluctuations of X close to its equilibrium value X^*, in Eq. (4.22) we can evaluate the derivative of the potential making use of Eq. (4.24), obtaining the linearized form

$$\frac{dX}{dt} = -\frac{2T}{\tilde{\Gamma}}\,\chi_X^{-1}\,(X - X^*) + \tilde{f}(t). \tag{4.25}$$

Here the coefficient

$$\chi_X^{-1} = \left(\frac{\partial^2 \mathcal{U}(X)}{\partial X^2}\right)_{X=X^*} \tag{4.26}$$

represents the inverse of the equilibrium thermodynamic susceptibility, associated with the thermodynamic observable X, that is, the same quantity that in a magnetic system defines how the magnetization responds to the variation of an external magnetic field. In Section 4.4, we determine the susceptibility within a linear response theory.

4.4 Linear Response to an External Force

In this section we discuss how a physical system responds to an external time-dependent force that is assumed to gently perturb it. We can imagine we are observers trying to get some inference about the effects of a spontaneous fluctuation of the physical system under examination, but we could equally think that we are applying some perturbation to the system in order to force it from the outside or to perform a measurement on it. All of these situations correspond to the same mathematical description, whose basic ingredients are now presented in the language of classical mechanics. As we shall shortly discuss in Appendix N, they can be adapted by suitable modifications also to quantum mechanical systems and even to quantum field theory.

First, we introduce the basic mathematical ingredients. We assume that the physical system under examination is made of N particles, obeying the laws of classical mechanics. Their position and momentum coordinates in a physical space of dimension d are usually denoted as $q_i(t)$ and $p_i(t)$, with $i = 1, 2, \ldots, dN$. If the system has energy E, the Hamilton functional, $\mathcal{H}(q_i(t), p_i(t)) = E$, provides all information about the deterministic dynamics of the system through the Hamilton equations

$$\frac{dq_i(t)}{dt} = \frac{\partial \mathcal{H}}{\partial p_i(t)}, \tag{4.27a}$$

$$\frac{d\,p_i(t)}{d\,t} = -\frac{\partial\,\mathcal{H}}{\partial q_i(t)}. \tag{4.27b}$$

This means that, due to the time reversal symmetry of Hamilton's equations, the forward and backward time evolution of the system is uniquely determined by the initial conditions $\{q_i(0), p_i(0)\}$.

On the other hand, here we are interested in describing our system in a thermodynamic language, rather than in a mechanical one. Equilibrium statistical mechanics provides us with the basic ingredient via the Boltzmann–Gibbs formula,

$$\Pi(q_i, p_i) = \frac{1}{Z}\, e^{-\beta\,\mathcal{H}(q_i, p_i)}, \tag{4.28}$$

where $\Pi(q_i, p_i)$ is the probability of observing the system in the microstate (q_i, p_i) and $\beta = 1/T$, with T the temperature of the thermal bath, which keeps our system at equilibrium. The partition function

$$Z = \int_{\mathcal{V}} d\mathcal{V}\; e^{-\beta\,\mathcal{H}(q_i, p_i)} = e^{-\beta F} \tag{4.29}$$

provides the relation with equilibrium thermodynamics, because F is the free energy, which is a function of temperature T and of thermodynamic observables, $F = F(T, \{X\})$, where X can be the volume V occupied by the system, the number of particles N present in the system, its magnetization M, and so on.

In Eq. (4.29) we have adopted the same notation of Eq. (3.176), where $\mathcal{V}$ denotes the phase space, whose infinitesimal volume element is[3]

$$d\mathcal{V} \equiv \frac{1}{N!}\,\frac{\prod_{i=1}^{dN} dq_i\, dp_i}{(2\pi\hbar)^{dN}}, \tag{4.30}$$

where $\hbar = 1.05457 \times 10^{-34}$ Js denotes the reduced Planck constant.[4]

Now, we want to consider the situation in which we disturb the system, driving it out of its thermodynamic equilibrium state to a new state described by a perturbed Hamiltonian $\mathcal{H}'$. For a quantitative description of the response of the system, we have to perform a measurement over a suitable macroscopic, that is, thermodynamic, observable $X(q_i, p_i)$, that is also a functional defined onto phase space $\mathcal{V}$, and we assume that it is not a conserved quantity, namely, $dX/dt \neq 0$.

The perturbed Hamiltonian can be written as

$$\mathcal{H}' = \mathcal{H} - h\,X, \tag{4.31}$$

where h is the perturbation field, or generalized thermodynamic force, that allows perturbation of the system by modifying the value of X. More generally, h and X are conjugate thermodynamic variables, like pressure and volume or chemical potential and number of particles, which are related, at thermodynamic equilibrium, through basic

---

[3] The factor $1/N!$ stems from the hypothesis that identical particles are undistinguishable: It was originally introduced by Otto Sackur and Hugo Martin Tetrode for recovering the additivity of the entropy of the ideal gas.

[4] We write $2\pi\hbar$ rather than h because, here later, this symbol is used for denoting a generalized field.

thermodynamic relations. In fact, by replacing $\mathcal{H}$ with $\mathcal{H}'$ in Eq. (4.29), we obtain

$$Z'_h = \int_{\mathcal{V}} d\mathcal{V}\ e^{-\beta(\mathcal{H}-hX(q_i,p_i))} = e^{-\beta F_h}. \tag{4.32}$$

Deriving with respect to h, we obtain

$$\beta \int_{\mathcal{V}} d\mathcal{V} X e^{-\beta(\mathcal{H}-hX(q_i,p_i))} = -\beta Z_h \frac{\partial F_h}{\partial h}, \tag{4.33}$$

that is to say,

$$\frac{\partial F_h}{\partial h} = -\frac{1}{Z_h} \int_{\mathcal{V}} d\mathcal{V} X e^{-\beta(\mathcal{H}-hX(q_i,p_i))} \tag{4.34}$$

$$= -\langle X \rangle. \tag{4.35}$$

The earlier relation applies to any pair of conjugated thermodynamic variables: magnetic field and magnetization, temperature and entropy, pressure and volume, chemical potential, and number of particles. In particular, this type of relation allows us to determine the static or equilibrium susceptibility χ, that is, the response of an extensive property under variation of the conjugated intensive variable h, in the limit $h \to 0$:

$$\chi = \lim_{h \to 0} \frac{\partial \langle X \rangle}{\partial h}$$

$$= \lim_{h \to 0} \frac{\partial}{\partial h} \left[\frac{1}{Z_h} \int_{\mathcal{V}} d\mathcal{V} X e^{-\beta(\mathcal{H}-hX(q_i,p_i))} \right]$$

$$= \lim_{h \to 0} \left\{ -\frac{\beta}{Z_h^2} \left[\int_{\mathcal{V}} d\mathcal{V} X e^{-\beta(\mathcal{H}-hX(q_i,p_i))} \right]^2 + \frac{\beta}{Z_h} \int_{\mathcal{V}} d\mathcal{V} X^2 e^{-\beta(\mathcal{H}-hX(q_i,p_i))} \right\}$$

$$= \beta[\langle X^2 \rangle_0 - \langle X \rangle_0^2]. \tag{4.36}$$

For instance, the static magnetic susceptibility of a magnetic system, whose perturbed Hamiltonian has the form $\mathcal{H} - HM$, where M is the magnetization (the extensive quantity) conjugated to the magnetic field H (the intensive quantity), is given by

$$\chi = \lim_{h \to 0} \frac{\partial M}{\partial H} = \beta[\langle M^2 \rangle_0 - \langle M \rangle_0^2]. \tag{4.37}$$

While this section has dealt with a constant perturbation that just shifts the equilibrium, Section 4.4.1 concerns a time-dependent perturbation field, $h(t)$. The study of how the system responds to such perturbations leads to the celebrated fluctuation–dissipation relations.

4.4.1 Linear Response and Fluctuation–Dissipation Relation

In order to provide a simple mathematical description, it is worth posing the problem as follows: We can assume that the system was prepared, very far in the past, in an equilibrium state of the perturbed Hamiltonian $\mathcal{H}' = \mathcal{H} - h(t)X$, where $h(t) = h$ for $-\infty < t < 0$. At $t = 0$ the perturbation is switched off, $h(t) = 0$ for $t > 0$, and we let the system relax to the equilibrium state of the unperturbed Hamiltonian $\mathcal{H}$. The relaxation process is guaranteed by the presence of the thermal bath. Up to $t = 0$, the equilibrium value of X is defined by the expression

$$X^* \equiv \langle X \rangle_{eq} = \frac{1}{Z'} \int_{\mathcal{V}} d\mathcal{V} \; e^{-\beta \mathcal{H}'(q_i, p_i)} X(q_i, p_i), \tag{4.38}$$

where Z' is given by Eq. (4.32). In what follows we adopt the standard notation $\langle \cdot \rangle_{eq}$ for the equilibrium average of any observable. When $h(t)$ is switched off, the evolution of the microscopic variables $\{q_i(t), p_i(t)\}$ starts to be ruled by Eq. (4.27) (while for $t < 0$ their evolution was ruled by the same equations with $\mathcal{H}'$ replacing $\mathcal{H}$), so that, for $t > 0$, $X(t)$ is an explicit function of time: Similarly to a Brownian particle, it will relax, starting from X^*, to its equilibrium value for the unperturbed Hamiltonian $\mathcal{H}$, that is, the Eq. (4.38) where $\mathcal{H}'$ is replaced by $\mathcal{H}$.

In order to obtain a suitable expression for the time evolution of $X(t)$ during its relaxation, that is, for $t > 0$, we can write

$$\langle X(t) \rangle = \frac{\int_{\mathcal{V}} d\mathcal{V} \; e^{-\beta(\mathcal{H}(q_i, p_i) - hX(0))} X(t)}{\int_{\mathcal{V}} d\mathcal{V} \; e^{-\beta(\mathcal{H}(q_i, p_i) - hX(0))}}, \tag{4.39}$$

where $X(t)$ at the numerator of the right-hand side is formally obtained by integrating Eq. (4.27), starting from initial conditions $\{q_i, p_i\}$, which must be weighted according to the equilibrium distribution of the perturbed Hamiltonian. This is the reason why in both the numerator and denominator is $\mathcal{H} - hX(0)$, where $X(0) = X(q_i(0), p_i(0))$ denotes the value of X when the perturbation is switched off.

Assuming that h is small, we can expand $e^{\beta h X(0)}$ at the first order in h and therefore we can approximate the right-hand side as

$$\langle X(t) \rangle \approx \frac{\int_{\mathcal{V}} d\mathcal{V} \; e^{-\beta \mathcal{H}(q_i, p_i)} \, (1 + \beta h X(0)) X(t)}{\int_{\mathcal{V}} d\mathcal{V} \; e^{-\beta \mathcal{H}(q_i, p_i)} \, (1 + \beta h X(0))}. \tag{4.40}$$

Dividing numerator and denominator by the unperturbed partition function $Z_0 = \int_{\mathcal{V}} d\mathcal{V} \; e^{-\beta \mathcal{H}(q_i, p_i)}$, we obtain

$$\langle X(t) \rangle = \frac{\langle (1 + \beta h X(0)) \, X(t) \rangle_0}{\langle 1 + \beta h X(0) \rangle_0}, \tag{4.41}$$

where $\langle \cdots \rangle_0$ represents the statistical average for the unperturbed system. At first order in h we finally obtain the fluctuation–dissipation relation

$$\langle X(t) \rangle - \langle X \rangle_0 = \beta h \left(\langle X(t) X(0) \rangle_0 - \langle X \rangle_0^2 \right), \tag{4.42}$$

where we have replaced $\langle X(t) \rangle_0$ with $\langle X \rangle_0$, because the equilibrium expectation value of X, as well as that of any other thermodynamic quantity, is independent of time. It is worth noting that in the limit $t, h \rightarrow 0$ we obtain Eq. (4.37).

We can conclude that the average deviation of a perturbed macroscopic observable from its unperturbed average evolution is linear in the amplitude of the perturbation field h and proportional to the equilibrium average of its connected time autocorrelation function. The latter quantity measures the equilibrium fluctuations of X, while the former describes the relaxation process of the perturbed variable. This is why Eq. (4.42) is called the fluctuation–dissipation relation, which is the cornerstone of linear response theory. This relation has important conceptual implications for the physics of nonequilibrium processes. In practice, it tells us that the nonequilibrium evolution of a gently perturbed macroscopic thermodynamic system is essentially indistinguishable from its fluctuations around equilibrium, so that in the linear regime the nonequilibrium dynamics of a macroscopic observable can be measured by its time autocorrelation function. Without prejudice to generality, we can set to zero the equilibrium average values of X, namely $\langle X \rangle_0 = 0$, and rewrite the fluctuation–dissipation relation (4.42) in a simpler form,

$$\langle X(t) \rangle = \beta h \langle X(t) X(0) \rangle_0 . \tag{4.43}$$

A formulation of the fluctuation–dissipation relation in the presence of a general time-dependent perturbation $h(t)$ introduces the response function $\Xi(t, t')$,

$$\langle X(t) \rangle = \int d t' \, \Xi(t, t') \, h(t'), \tag{4.44}$$

which is based on the consideration that the response at time t of a perturbed observable depends on the value of the perturbation field at any other time $t' < t$. On the other hand, if we want to preserve the physical assumption of causality, we have to impose that $\Xi(t, t')$ cannot depend on values of the perturbation field taken at time $t' > t$. Moreover, consistently with Eq. (4.43), $\Xi(t, t')$ has to be an equilibrium quantity; therefore, it has to be invariant under time translations, that is, it should depend on $(t - t')$. Thus

$$\Xi(t, t') = \begin{cases} \Xi(t - t'), & \text{if } t > t' \\ 0, & \text{otherwise.} \end{cases} \tag{4.45}$$

Important properties of the response function $\Xi(t - t')$, including the Kramers–Krönig relations, are reported in Appendix O.

We can now determine the explicit expression of the response function, comparing Eqs. (4.43) and (4.44). Since the former has been derived assuming

$$h(t) = \begin{cases} h, & \text{if } t < 0 \\ 0, & \text{otherwise,} \end{cases} \tag{4.46}$$

it must be

$$\int_{-\infty}^{0} dt' \, \Xi(t - t') = \beta \langle X(t) X(0) \rangle_0 , \tag{4.47}$$

or, after introducing the change of variable $\tau = (t - t')$,

$$\int_{t}^{+\infty} d\tau \, \Xi(\tau) = \beta \langle X(t) X(0) \rangle_0 . \tag{4.48}$$

Taking the time derivative of both sides and using causality, we finally obtain

$$\Xi(t) = -\beta \Theta(t) \frac{d}{dt} \langle X(t) X(0) \rangle_0 , \tag{4.49}$$

where $\Theta(t)$ is the Heaviside step distribution, which is equal to one for $t > 0$ and vanishes otherwise.

Equation (4.49) is a more general form of the fluctuation–dissipation relation in the linear response limit, because any explicit dependence on h has disappeared and the response to any (small) perturbation $h(t)$ can be derived from Eq. (4.44). A further way of expressing the same content of Eq. (4.49) can be obtained by considering its Fourier transform,

$$\Xi(\omega) = \int_{-\infty}^{+\infty} dt\, e^{-i\omega t}\, \Xi(t)$$

$$= -\beta \int_{-\infty}^{+\infty} dt\, e^{-i\omega t}\, \Theta(t)\, \frac{d}{dt} \langle\, X(t)\, X(0)\,\rangle_0. \tag{4.50}$$

The last expression is the Fourier transform of a product of functions, which is known to be equal to the convolution product ($\star$) of the single Fourier-transformed functions, according to the general relation

$$\int_{-\infty}^{+\infty} dt\, e^{-i\omega t}\, f(t)\, g(t) = \frac{1}{2\pi} \int_{-\infty}^{+\infty} d\omega'\, f(\omega - \omega')\, g(\omega') \equiv \frac{1}{2\pi} f(\omega) \star g(\omega). \tag{4.51}$$

With reference to Eq. (4.50), we have

$$f(t) = \Theta(t)$$

$$g(t) = \frac{d}{dt} \langle\, X(t)\, X(0)\,\rangle_0 \equiv \frac{d}{dt} C(t), \tag{4.52}$$

where we have introduced the shorthand notation $C(t)$ for $\langle\, X(t)\, X(0)\,\rangle_0$. Since the Fourier transform of the time derivative of a function is the Fourier transform of the function multiplied by a factor of $i\omega$, we have $g(\omega) = i\omega C(\omega)$, where $C(\omega)$ is the Fourier transform of $C(t)$. Due to time translation invariance at equilibrium, this quantity must be an even function of t,[5] so that its Fourier transform must be real.[6]

In conclusion, $\Xi(\omega)$ can be written as

$$\Xi(\omega) = -\beta \int_{-\infty}^{+\infty} \frac{d\omega'}{2\pi}\, \Theta(\omega - \omega')(i\omega')C(\omega'). \tag{4.53}$$

This expression is well defined only in the sense of distributions, because we need to make convergent the integral expressing $\Theta(\omega)$ as follows (see also Eqs. (O.19) and (O.20)):

$$\Theta(\omega) = \lim_{\epsilon \to 0^+} \int_0^{+\infty} dt\, e^{-(\epsilon + i\omega)t} = \lim_{\epsilon \to 0^+} \frac{1}{\epsilon + i\omega} = \pi\delta(\omega) - i\mathrm{PV}\left(\frac{1}{\omega}\right), \tag{4.54}$$

where the PV denotes the principal value of the argument.

---

[5] In fact, at equilibrium $\langle\, X(t)\, X(0)\,\rangle_0$ does not change if we shift the time scale by a given amount τ: In particular, if we take $\tau = -t$, we obtain $\langle\, X(t)\, X(0)\,\rangle_0 = \langle\, X(0)\, X(-t)\,\rangle_0$.

[6] According to the Wiener–Khinchin theorem, the Fourier transform of the correlation function, $C(\omega)$, of a fluctuating observable $X(t)$ represents its power spectrum. More precisely, the theorem states that the following relation holds:

$$C(\omega) = \int_{-\infty}^{+\infty} dt\, e^{-i\omega t} \langle\, X(t)\, X(0)\,\rangle_0 = \lim_{T \to +\infty} \frac{1}{T} \langle |X_T(\omega)|^2 \rangle$$

where

$$X_T(\omega) = \int_{-T/2}^{T/2} dt\, X(t) e^{-i\omega t}.$$

Making use of this relation and considering that, according to our conventions (see Appendix O), $\Xi(\omega)$ is analytic in the lower half complex plane, we obtain

$$\Xi(\omega) = -\beta \left[\frac{(i\omega)}{2} C(\omega) + \mathrm{PV} \int_{-\infty}^{+\infty} \frac{d\omega'}{2\pi} \frac{\omega'}{\omega - \omega'} C(\omega') \right]. \tag{4.55}$$

This equation tells us that the imaginary part of the Fourier transform of the response function, $\Xi^{\mathrm{I}}(\omega)$, is related to the power spectral density of thermal fluctuations of the macroscopic variable $X(t)$, more precisely,

$$\Xi^{\mathrm{I}}(\omega) = -\beta \frac{\omega}{2} C(\omega). \tag{4.56}$$

The imaginary part is also related to the real part of the same Fourier transform, $\Xi^{\mathrm{R}}(\omega)$, by the Kramers–Krönig relation (see Eq. (O.22)) as

$$\Xi^{\mathrm{R}}(\omega) = \mathrm{PV} \int_{-\infty}^{+\infty} \frac{d\omega'}{\pi} \frac{1}{\omega - \omega'} \Xi^{\mathrm{I}}(\omega'). \tag{4.57}$$

The real and imaginary parts of the response function have a well-defined parity. Since $C(\omega) = C(-\omega)$ (see note 5 earlier), from Eq. (4.56) it is straightforward that $\Xi^{\mathrm{I}}(\omega)$ is an odd function. Using this result and changing the sign of the integration variable in Eq. (4.57), we find that $\Xi^{\mathrm{R}}(\omega)$ is an even function.

An alternative way of writing the Kramers–Krönig relation, also discussed in Appendix O, can be found coming back to Eq. (4.53), using the central expression of Eqs. (4.54) and (4.56),

$$\Xi(\omega) = \lim_{\epsilon \to 0^+} \int_{-\infty}^{+\infty} \frac{d\omega'}{\pi} \frac{\Xi^{\mathrm{I}}(\omega')}{\omega' - \omega - i\epsilon}. \tag{4.58}$$

Equations (4.56) to (4.58) are the most widely used formulations of the so-called fluctuation–dissipation theorem.

4.4.2 Work Done by a Time-Dependent Field

We want to consider the case where the time-dependent field $h(t)$, whose physical interpretation is that of a generalized thermodynamic force associated with the observable $X(t)$, is switched on and we aim at measuring the work W done by it on the system, which is given by the expression

$$W = \int dt \, h(t) \left\langle \frac{dX(t)}{dt} \right\rangle. \tag{4.59}$$

For physical reasons we assume that $h(t)$ is a sufficiently regular function of time t and, in particular, that its Fourier transform exists. For instance, we could deal with a field $h(t)$ that is a sufficiently rapidly vanishing function for $t \to \pm\infty$, or even a periodic function, as we discuss later in Section 4.4.3 for the damped harmonic oscillator. In the former case, W can be computed for $-\infty < t < +\infty$, provided the integral in Eq. (4.59) exists, while in the second case it makes sense to compute the work done on the system over a period. Let us consider here the former case, so that we can write

$$W = \int_{-\infty}^{+\infty} dt\, h(t) \left\langle \frac{dX(t)}{dt} \right\rangle. \tag{4.60}$$

We can use Eq. (4.44) and write

$$W = \int_{-\infty}^{+\infty} dt\, h(t) \int_{-\infty}^{+\infty} dt'\, \frac{\partial}{\partial t} \Xi(t - t')\, h(t'). \tag{4.61}$$

In the right-hand side of this equation, we can substitute $\Xi(t - t')$ with its definition in terms of its Fourier antitransform and write

$$W = \int_{-\infty}^{+\infty} dt\, h(t) \int_{-\infty}^{+\infty} dt'\, \frac{\partial}{\partial t} \int_{-\infty}^{+\infty} \frac{d\omega}{2\pi} \Xi(\omega) e^{i\,\omega(t-t')}\, h(t')$$

$$= \int_{-\infty}^{+\infty} \frac{d\omega}{2\pi}\, (i\omega)\, \Xi(\omega) \int_{-\infty}^{+\infty} dt\, e^{i\,\omega t}\, h(t) \int_{-\infty}^{+\infty} dt'\, e^{-i\,\omega t'}\, h(t').$$

Now we can take into account that $h(t)$ is a real quantity, that is, $h(-\omega) = h^*(\omega)$, so that $h(-\omega)h(\omega) = |h(\omega)|^2$ and we can finally write

$$W = \int_{-\infty}^{+\infty} \frac{d\omega}{2\pi}\, (i\omega)\Xi(\omega)|h(\omega)|^2. \tag{4.62}$$

Due to the presence of the linear factor $i\omega$, the Fourier transform of the response function contributes to the integral on the right-hand side with its odd component, that is, its imaginary part $\Xi^{\mathrm{I}}(\omega)$, and we can finally write

$$W = -\int_{-\infty}^{+\infty} \frac{d\omega}{2\pi}\, \omega\, \Xi^{\mathrm{I}}(\omega)|h(\omega)|^2 = \frac{\beta}{2} \int_{-\infty}^{+\infty} \frac{d\omega}{2\pi}\, \omega^2 C(\omega)|h(\omega)|^2, \tag{4.63}$$

where we have used Eq. (4.56) to relate $\Xi^{\mathrm{I}}(\omega)$ with $C(\omega)$.

We have obtained that the work done on the system by the field $h(t)$ depends on the imaginary part of the Fourier transform of the response function. This is the reason why Eqs. (4.56) to (4.58) are named fluctuation–dissipation relations. Moreover, Eq. (4.63) tells us that the work done on the system is related through $C(\omega)$ to the power spectrum of the fluctuating observable $X(t)$ (see also note 5 earlier), thus implying that W, that is, the work done on a system by a generalized thermodynamic force $h(t)$, is a positive quantity, as it should be.

4.4.3 Simple Applications of Linear Response Theory

Some basic physical examples can help in appreciating the main features of the linear response approach and the role of the response function $\Xi(t - t')$, whose properties are illustrated in Appendix O.

Susceptibility

As a first instance, let us start from Eq. (4.44). Using the property $\Xi(t, t') = \Xi(t - t')$, it can be written in Fourier space as

$$\langle X(\omega) \rangle = \Xi(\omega) h(\omega). \tag{4.64}$$

The susceptibility χ is the physical quantity measuring the response of the observable to the perturbation field (e.g., the magnetization emerging in a magnetic material when a small magnetic field is switched on). It can be defined as

$$\chi = \lim_{\omega \to 0} \Xi(\omega) = \frac{\partial \langle X \rangle(\omega)}{\partial h(\omega)}\bigg|_{\omega=0}. \tag{4.65}$$

This definition coincides with that of the static susceptibility given in Eq. (4.36). In fact, if we consider Eq. (4.50), we can write

$$\lim_{\omega \to 0} \Xi(\omega) = \lim_{\epsilon \to 0^+} -\beta \int_0^{+\infty} dt e^{-\epsilon t} \frac{d}{dt} \langle X(t) X(0) \rangle_0$$

$$= \lim_{\epsilon \to 0^+} \left\{ \left[-\beta e^{-\epsilon t} \langle X(t) X(0) \rangle_0 \right]_0^{+\infty} - \beta\epsilon \int_0^{+\infty} dt e^{-\epsilon t} \langle X(t) X(0) \rangle_0 \right\}$$

$$= \beta \langle X^2(0) \rangle_0 = \beta \langle X^2 \rangle_0, \tag{4.66}$$

where the last equality holds because the equilibrium average $\langle X^2 \rangle_0$ is independent of time. This proves that definition (4.65) coincides with Eq. (4.36) for $\langle X \rangle_0 = 0$.

An alternative way of expressing the static susceptibility stems from Eq. (4.58). In fact, we can also write

$$\chi = \lim_{\omega \to 0} \lim_{\epsilon \to 0^+} \int_{-\infty}^{+\infty} \frac{d\omega'}{\pi} \frac{\Xi^{\mathrm{I}}(\omega')}{\omega' - \omega - i\epsilon} \tag{4.67}$$

and commuting the two limits, we find

$$\chi = \lim_{\epsilon \to 0^+} \int_{-\infty}^{+\infty} \frac{d\omega'}{\pi} \frac{\Xi^{\mathrm{I}}(\omega')}{\omega' - i\epsilon}. \tag{4.68}$$

We can interpret this equation as follows: If we know how much a system dissipates at any frequency in the presence of a perturbation field, we can determine the response of the system at zero frequency, that is, its susceptibility, by summing over all of these contributions. This is called the thermodynamic sum rule.

The Damped Harmonic Oscillator

The dynamics of a damped harmonic oscillator under the action of a driving force $F(t)$ is described by the differential equation

$$m\ddot{x}(t) + \tilde{\gamma}\dot{x}(t) + Kx(t) = F(t). \tag{4.69}$$

Since it is a linear equation, it is not necessary that $F(t)$ is a weak perturbation and the response function $\Xi(t - t')$ is just the Green's function $G(t - t')$, which allows us to find the solution of Eq. (4.69) as a function of $F(t)$, namely,

$$x(t) = \int_{-\infty}^{+\infty} dt' \, G(t - t') \, F(t'). \tag{4.70}$$

The standard method of solving this problem amounts to expressing $G(t)$ by its Fourier antitransform $G(\omega)$

$$G(t) = \frac{1}{2\pi} \int_{-\infty}^{+\infty} d\omega e^{i\omega t} G(\omega) \tag{4.71}$$

and rewriting Eq. (4.70) as

$$x(t) = \int_{-\infty}^{+\infty} dt' \; \frac{1}{2\pi} \int_{-\infty}^{+\infty} d\omega \, e^{i\omega(t-t')} G(\omega) \, F(t').$$
(4.72)

We can substitute this formal solution into Eq. (4.69) and obtain

$$\int_{-\infty}^{+\infty} \frac{d\omega}{2\pi} \int_{-\infty}^{+\infty} dt' \left(-m\omega^2 + i\tilde{\gamma}\omega + K \right) e^{i\omega(t-t')} G(\omega) \, F(t') = F(t).$$
(4.73)

This equation is solved if

$$G(\omega) = -\frac{1}{m\omega^2 - i\tilde{\gamma}\omega - K},$$
(4.74)

because of the equality (in the sense of distributions)

$$\delta(t) = \frac{1}{2\pi} \int_{-\infty}^{+\infty} d\omega e^{i\omega t}.$$
(4.75)

It is common to introduce the natural frequency of the undamped oscillator, $\omega_0 = \sqrt{K/m}$, and write

$$G(\omega) = -\frac{1}{m(\omega^2 - i\gamma\omega - \omega_0^2)},$$
(4.76)

where $\gamma = \tilde{\gamma}/m$.

According to Eq. (4.65), the susceptibility of the damped and forced harmonic oscillator is

$$\chi = \lim_{\omega \to 0} G(\omega) = \frac{1}{m\omega_0^2}.$$
(4.77)

Thus, in this zero-frequency limit, the Fourier transform of Eq. (4.70) provides us with the response of the oscillator to a static (i.e., time-independent) force F

$$x = \frac{F}{m\omega_0^2},$$
(4.78)

that is, the result we expect on the basis of a straightforward visual inspection of Eq. (4.69). In general, we can rewrite (4.76) as

$$G(\omega) = -\frac{1}{m \, (\omega - \omega_1)(\omega - \omega_2)}$$
(4.79)

where

$$\omega_1 = \Omega + i\frac{\gamma}{2}$$

$$\omega_2 = -\Omega + i\frac{\gamma}{2}$$

are the poles of the response function in the complex plane, with

$$\Omega^2 = \omega_0^2 - \frac{\gamma^2}{4}.$$

The position of the poles $\omega_{1,2}$ in the complex plane is sketched in Fig. 4.1, in the underdamped case, that is, $\omega_0^2 > \gamma^2/4$, on the left and in the overdamped case, that is,

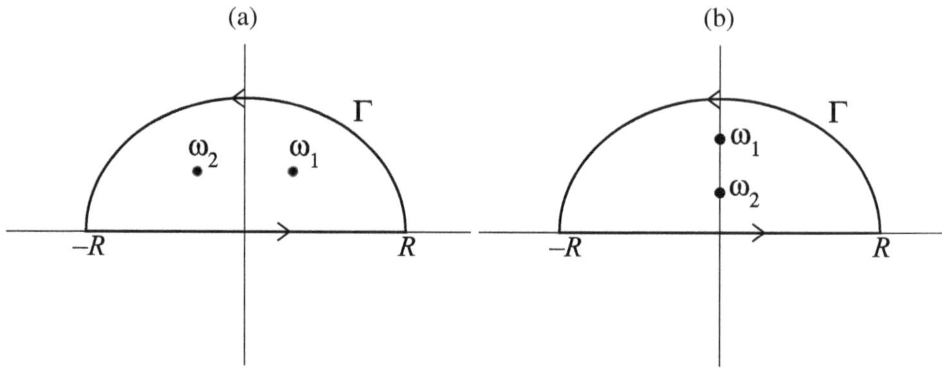

Fig. 4.1   The complex plane for calculating the integral Eq. (4.81) with the method of residues. The two poles have different locations in the underdamped case (a) and in the overdamped case (b), but they are always in the upper half complex plane, therefore preserving causality.

$\omega_0^2 < \gamma^2/4$, on the right. In both cases the poles lay in the upper half complex plane and causality is preserved, that is, $G(t) = 0$ for $t < 0$ (see Appendix O).

We can obtain an explicit expression of $G(t)$ for $t > 0$ by considering the inverse Fourier transform of $G(\omega)$

$$G(t) = -\frac{1}{2\pi} \int_{-\infty}^{+\infty} d\omega e^{i\omega t} \frac{1}{m(\omega - \omega_1)(\omega - \omega_2)}. \tag{4.80}$$

The right-hand side of this equation can be evaluated by considering the analytic extension to the complex plane of the function to be integrated ($\omega \to z$) and by evaluating its integral over a closed contour Γ, which runs along the real axis from $-R$ to $+R$ and is closed in the upper-half complex plane by a semicircle of radius R (see Fig. 4.1),

$$G(t) = -\frac{1}{2\pi} \oint_\Gamma dz \frac{e^{izt}}{m(z - \omega_1)(z - \omega_2)}. \tag{4.81}$$

By performing the limit $R \to \infty$, the part of the integral over the semicircle of radius R vanishes due to the Jordan lemma, and the theorem of residues yields the expression

$$G(t) = -\frac{1}{2\pi m} 2\pi i \left[\frac{e^{i\omega_1 t}}{\omega_1 - \omega_2} + \frac{e^{i\omega_2 t}}{\omega_2 - \omega_1} \right] \tag{4.82}$$

$$= \frac{\sin(\Omega t)}{m\Omega} e^{-\gamma t/2}. \tag{4.83}$$

In the strongly overdamped regime, the inertial term in Eq. (4.69) can be neglected and the Fourier transform of the response function (see Eq. (4.76)) can be approximated as

$$G(\omega) \approx \frac{1}{K + i\tilde{\gamma}\omega}. \tag{4.84}$$

The pole is in $\omega = iK/\tilde{\gamma}$ and causality is preserved, while the imaginary part of $G(\omega)$ is

$$G_I(\omega) = -\frac{\tilde{\gamma}\omega}{K^2 + \tilde{\gamma}^2\omega^2}, \tag{4.85}$$

which is an odd function of ω, as discussed in Appendix O.

Making use of Eq. (4.56), with $X(t) \to x(t)$, we find that

$$C(\omega) = -\frac{2}{\beta\omega}G_I(\omega) = 2T\frac{\tilde{\gamma}}{K^2 + \tilde{\gamma}^2\omega^2}, \tag{4.86}$$

where $C(\omega)$ is the Fourier transform of the correlation function

$$C(t) = \lim_{T\to\infty}\frac{1}{T}\int_{-T/2}^{T/2}dt'\,x(t'+t)\,x(t') \equiv \langle x(t'+t)\,x(t')\rangle_{t'}. \tag{4.87}$$

By the Wiener–Khinchin theorem (see note 5 earlier), we know that $C(\omega)$ is proportional to the power spectrum $|x(\omega)|^2$, which, in the overdamped regime, has the form of the Lorentzian function of ω, given in the right-hand side of Eq. (4.86).

We can also compute the energy dissipated by the damped harmonic oscillator, which is the work per unit time performed by the driving force $F(t)$:

$$\begin{aligned}
\frac{dW}{dt} &= F(t)\,\dot{x}(t) \\
&= F(t)\frac{d}{dt}\int_{-\infty}^{+\infty}dt'\,G(t-t')\,F(t') \\
&= F(t)\int_{-\infty}^{+\infty}dt'\int_{-\infty}^{+\infty}\frac{d\omega}{2\pi}(i\omega)G(\omega)e^{i\omega(t-t')}\,F(t') \\
&= F(t)\int_{-\infty}^{+\infty}d\omega\frac{e^{i\omega t}}{2\pi}(i\omega)G(\omega)\int_{-\infty}^{+\infty}dt'\,e^{-i\omega t'}\,F(t') \\
&= \frac{F(t)}{2\pi}\int_{-\infty}^{+\infty}d\omega(i\omega)G(\omega)e^{i\omega t}\,F(\omega).
\end{aligned} \tag{4.88}$$

By definition, the driving force $F(t)$ is a real quantity. We specialize our analysis to the case where it is a periodic function of time with frequency ν, namely, $F(t) = F_0\cos(\nu t)$. Its Fourier transform reads

$$F(\omega) = \frac{F_0}{2}\big[\delta(\omega-\nu) + \delta(\omega+\nu)\big]. \tag{4.89}$$

Substituting this expression into Eq. (4.88), we obtain

$$\frac{dW}{dt} = \frac{i}{4\pi}\nu\,F_0^2\cos(\nu t)\big[G(\nu)e^{i\nu t} - G(-\nu)e^{-i\nu t}\big]. \tag{4.90}$$

The work performed by the driving force over one period

$$\overline{W} = \int_0^{2\pi/\nu}dt\,\frac{dW}{dt} \tag{4.91}$$

is easily obtained by remarking that

$$\overline{\cos(\nu t)e^{\pm i\nu t}} = \frac{1}{2}, \tag{4.92}$$

thus yielding

$$\overline{W} = \frac{i}{4\pi}\frac{\nu F_0^2}{2}\big[G(\nu) - G(-\nu)\big]. \tag{4.93}$$

According to the parity properties of the Fourier transform of the response function, its real part is even and its imaginary part is odd. So, we can finally write

$$\overline{W} = -\frac{F_0^2}{4\pi} \nu \, G_I(\nu).$$
(4.94)

In the strong overdamped limit we can use Eq. (4.85) and write

$$\overline{W} = \frac{F_0^2}{4\pi} \frac{\tilde{\gamma}\nu^2}{K^2 + \tilde{\gamma}^2\nu^2},$$
(4.95)

which is, as expected, a positive quantity.

4.5 Hydrodynamics and the Green–Kubo Relation

In Section 2.4.6, we have seen that the Fokker–Planck equation for the probability of finding a stochastic particle at position $\mathbf{x}$ at time t in the presence of a conservative potential $U(\mathbf{x})$ has the formal structure of a continuity equation

$$\frac{\partial P(\mathbf{x}, t)}{\partial t} = -\nabla \cdot \mathbf{J}(\mathbf{x}, t).$$
(4.96)

In the absence of external forces, the current vector reads

$$\mathbf{J} = -D\nabla P(\mathbf{x}, t),$$
(4.97)

with D denoting the diffusion coefficient. If we multiply the probability $P(\mathbf{x}, t)$ by the total number of Brownian particles and by their mass m (we consider the simple case of identical noninteracting Brownian particles), the mass density of Brownian particles, $\rho_m(\mathbf{x}, t)$, obeys a similar equation,

$$\frac{\partial \rho_m(\mathbf{x}, t)}{\partial t} = -\nabla \cdot \mathbf{J}_m(\mathbf{x}, t),$$
(4.98)

where the current density is defined by the constitutive relation

$$\mathbf{J}_m(\mathbf{x}, t) = -D\nabla\rho_m(\mathbf{x}, t).$$
(4.99)

We want to point out that for an ensemble of identical noninteracting Brownian particles, the current is just proportional to their density gradient, consistent with Eq. (1.16).

A hydrodynamic description of a system where a set of macroscopic observables is conserved (e.g., mass, momentum, energy, charge, spin, etc.) is based on continuity equations of the same kind as Eq. (4.98). More precisely, for any conserved quantity a, we can write the continuity equation

$$\frac{\partial \rho_a(\mathbf{x}, t)}{\partial t} = -\nabla \cdot \mathbf{J}_a(\mathbf{x}, t),$$
(4.100)

where $\rho_a(\mathbf{x}, t)$ is the density of a conserved quantity a in the system and $\mathbf{J}_a(\mathbf{x}, t)$ is the correspondent density current. A constitutive relation like Eq. (4.99) can be obtained by going through the same kind of calculations performed in Section 1.2.2 to find an expression for the fluxes of mass, velocity, and energy as functions of the gradient of the corresponding quantities along a given direction. Equation (4.101) is just the generalization of these calculations for a density current vector:

$$\langle \mathbf{J}_a(\mathbf{x}, t) \rangle = -D_a \nabla \langle \rho_a(\mathbf{x}, t) \rangle. \tag{4.101}$$

It is important to observe that while Eq. (4.100) expresses an exact, microscopic conservation law, Eq. (4.101) is a phenomenological equation that relates nonequilibrium averages of fluctuating quantities. In fact, the constitutive equation holds in the presence of an applied density gradient, a setup where the role of fluctuations is intrinsic to the transport process. Since Eqs. (4.100) and (4.101) hold for any conserved quantity a, in what follows it is worthwhile overlooking the subscript a and we can write a general Fokker–Planck equation for average quantities after having substituted (4.101) into (4.100):

$$\frac{\partial \langle \rho(\mathbf{x}, t) \rangle}{\partial t} = D \nabla^2 \langle \rho(\mathbf{x}, t) \rangle. \tag{4.102}$$

On the other hand, linear response theory tells us that, for moderate applied gradients, local density fluctuations are practically indistinguishable from spontaneous relaxation to equilibrium of the same quantity (e.g., see Eq. (4.43)); more precisely, they are proportional to each other. In this case we have to consider that we are dealing with densities and current densities, that is, quantities that depend on both space and time. Accordingly, we can formally replace in Eq. (4.102) the average nonequilibrium fluctuation[7] $\langle \rho(\mathbf{x}, t) \rangle$ with the equilibrium correlation function of the density,

$$C(\mathbf{x} - \mathbf{y}, t - t') = \langle \rho(\mathbf{x}, t) \, \rho(\mathbf{y}, t') \rangle_0, \tag{4.103}$$

where, for the sake of simplicity, we have assumed space translation invariance, while time translational invariance is granted at equilibrium. We can finally write

$$\frac{\partial C(\mathbf{x} - \mathbf{y}, t - t')}{\partial t} = D \nabla^2 C(\mathbf{x} - \mathbf{y}, t - t'). \tag{4.104}$$

It is worth rewriting this equation for the space Fourier transform of $C(\mathbf{x} - \mathbf{y}, t - t')$,

$$C(\mathbf{k}, t - t') = \int_V d\mathbf{x} \, e^{-i\mathbf{k} \cdot (\mathbf{x} - \mathbf{y})} \, C(\mathbf{x} - \mathbf{y}, t - t'), \tag{4.105}$$

where $\int_V$ denotes the integral over the volume V of the system. Due to translation invariance, this expression is independent of $\mathbf{y}$ and we can write

$$
\begin{aligned}
C(\mathbf{k}, t - t') &= \frac{1}{V} \int_V d\mathbf{y} \int_V d\mathbf{x} \, e^{-i\mathbf{k} \cdot (\mathbf{x} - \mathbf{y})} \, C(\mathbf{x} - \mathbf{y}, t - t') \\
&= \int \int \frac{d\mathbf{x} \, d\mathbf{y}}{V} \, e^{-i\mathbf{k} \cdot (\mathbf{x} - \mathbf{y})} \, \langle \rho(\mathbf{x}, t) \, \rho(\mathbf{y}, t') \rangle_0 \\
&= \frac{1}{V} \langle \rho(\mathbf{k}, t) \, \rho(-\mathbf{k}, t') \rangle_0.
\end{aligned}
\tag{4.106}
$$

---

[7] Let us recall that in Section 4.4.1 we have finally assumed that the equilibrium average of the quantity under study vanishes, so $\langle \rho \rangle_0 = 0$.

Because of time translation invariance of the correlation function, we can set $t > t' = 0$ and write the Fokker–Planck equation for the correlation function in the form

$$\frac{\partial C(\mathbf{k}, t)}{\partial t} = -D\,k^2 C(\mathbf{k}, t), \tag{4.107}$$

which can be derived by taking the spatial Fourier transform of Eq. (4.104) and applying the Laplacian to both sides of Eq. (4.105). Its solution is

$$C(\mathbf{k}, t) = \exp\left(-Dk^2 t\right) C(\mathbf{k}, 0), \tag{4.108}$$

where $k^2 = |\mathbf{k}|^2$ and $t > 0$. The solution can be extended to $t < 0$ by considering that at equilibrium the correlation function is independent of the choice of the origin of time, that is,

$$C(\mathbf{k}, -t) = \frac{1}{V}\langle \rho(\mathbf{k}, -t)\rho(-\mathbf{k}, 0)\rangle_0 \tag{4.109}$$

$$= \frac{1}{V}\langle \rho(\mathbf{k}, 0)\rho(-\mathbf{k}, t)\rangle_0 \tag{4.110}$$

$$= C(-\mathbf{k}, t) \tag{4.111}$$

$$= C(\mathbf{k}, t), \tag{4.112}$$

where the last equality is derived from the spatial parity of the correlation function, $C(\mathbf{r}, t) = C(-\mathbf{r}, t)$. In conclusion, we can write for any t,

$$C(\mathbf{k}, t) = \exp\left(-D\,k^2\,|t|\right) C(\mathbf{k}, 0). \tag{4.113}$$

The time Fourier transform of $C(\mathbf{k}, t)$ is

$$C(\mathbf{k}, \omega) = \int_{-\infty}^{+\infty} dt\, e^{-i\omega t} C(\mathbf{k}, t) \tag{4.114}$$

$$= C(\mathbf{k}, 0) \int_{-\infty}^{+\infty} dt\, e^{-i\omega t} e^{-Dk^2 |t|} \tag{4.115}$$

$$= C(\mathbf{k}, 0) \int_{0}^{+\infty} dt\, e^{-Dk^2 t} \left(e^{-i\omega t} + e^{i\omega t}\right) \tag{4.116}$$

$$= C(\mathbf{k}, 0) \left[\frac{1}{Dk^2 + i\omega} + \frac{1}{Dk^2 - i\omega}\right] \tag{4.117}$$

$$= C(\mathbf{k}, 0)\,\frac{2D\,k^2}{\omega^2 + (Dk^2)^2}. \tag{4.118}$$

Therefore, $C(\mathbf{k}, \omega)$ is a Lorentzian function centered in $\omega = 0$ and of width Dk^2. If we divide both sides by k^2 and take the limit $k \to 0$, we find

$$\lim_{k \to 0} \frac{1}{k^2} C(\mathbf{k}, \omega) = C\,\frac{2D}{\omega^2}, \tag{4.119}$$

where $C = C(\mathbf{k} = 0, t = 0)$. So, we can obtain an expression for the generalized diffusion coefficient as

$$D = \frac{1}{2C} \lim_{\omega \to 0} \lim_{k \to 0} \frac{\omega^2}{k^2} C(\mathbf{k}, \omega). \tag{4.120}$$

In fact, D is a quantity independent of $\mathbf{k}$ and ω and its value is obtained by taking the infinite space ($\lim_{\mathbf{k} \to 0}$) and the infinite time ($\lim_{\omega \to 0}$) limits of the density correlation function. In order to obtain the Green–Kubo relation, we want to write D as a function of the density current that appears in Eq. (4.100). This equation, for space Fourier transformed variables, is

$$\frac{\partial \rho(\mathbf{k}, t)}{\partial t} + i\mathbf{k} \cdot \mathbf{J}(\mathbf{k}, t) = 0, \tag{4.121}$$

where

$$J_j(\mathbf{k}, t) = \int d\mathbf{x}\, e^{-i\mathbf{k}\cdot\mathbf{x}} J_j(\mathbf{x}, t), \quad j = x, y, z. \tag{4.122}$$

Then, using Eq. (4.106), we can write

$$\frac{\partial}{\partial t}\frac{\partial}{\partial t'} C(\mathbf{k}, t - t') = \left\langle \frac{1}{V} \frac{\partial \rho(\mathbf{k}, t)}{\partial t} \frac{\partial \rho(-\mathbf{k}, t')}{\partial t'} \right\rangle$$

$$= \frac{1}{V} \sum_{i,j} k_i\, k_j\, \langle J_i(\mathbf{k}, t) J_j(-\mathbf{k}, t')\rangle_0 . \tag{4.123}$$

We can now perform the time Fourier transform of both sides. The left-hand side gives

$$\int_{-\infty}^{+\infty} d(t - t') e^{-i\omega(t-t')} \frac{\partial}{\partial t}\frac{\partial}{\partial t'} C(\mathbf{k}, t - t') = \omega^2 C(\mathbf{k}, \omega). \tag{4.124}$$

If we divide by k^2, for small values of $\mathbf{k}$ we can write

$$\lim_{\omega\to 0}\lim_{\mathbf{k}\to 0} \frac{\omega^2}{k^2} C(\mathbf{k}, \omega) = \lim_{\omega\to 0} \int_{-\infty}^{+\infty} d(t - t') e^{-i\omega(t-t')} \lim_{\mathbf{k}\to 0} \sum_{i,j} \frac{k_i\, k_j}{k^2} \frac{1}{V} \langle J_i(\mathbf{k}, t) J_j(-\mathbf{k}, t')\rangle_0$$

$$= \lim_{\omega\to 0} \int_{-\infty}^{+\infty} d(t - t') e^{-i\omega(t-t')} \sum_{i,j} \frac{k_i\, k_j}{k^2} \frac{1}{V} \langle J_i^T(t) J_j^T(t')\rangle_0,$$

where $J_i^T(t)$ denotes the ith component of the total current, that is,

$$J_i^T(t) \equiv \int_V d\mathbf{x} J_i(\mathbf{x}, t) = \lim_{\mathbf{k}\to 0} J_i(\mathbf{k}, t). \tag{4.125}$$

For an isotropic system in d dimensions, we have

$$\langle J_i^T(t) J_j^T(t')\rangle = \frac{1}{d} \delta_{ij}\, \langle \mathbf{J}^T(t) \cdot \mathbf{J}^T(t')\rangle_0 . \tag{4.126}$$

We can finally rewrite Eq. (4.120) as

$$D = \frac{1}{2C} \lim_{\omega\to 0} \int_{-\infty}^{+\infty} d(t - t') e^{-i\omega(t-t')} \sum_{i,j} \frac{k_i\, k_j}{k^2} \frac{1}{V}\frac{1}{d}\delta_{ij}\langle \mathbf{J}^T(t) \cdot \mathbf{J}^T(t')\rangle_0$$

$$= \frac{1}{dVC}\frac{1}{2} \lim_{\omega\to 0} \int_{-\infty}^{+\infty} d(t - t') e^{-i\omega(t-t')} \langle \mathbf{J}^T(t) \cdot \mathbf{J}^T(t')\rangle_0$$

$$= \frac{1}{dVC} \int_0^{+\infty} dt\, \langle \mathbf{J}^T(t) \cdot \mathbf{J}^T(0)\rangle_0, \tag{4.127}$$

where we have used the fact that the autocorrelation function is even under time reversal $t \to -t$. The previous equation can be rewritten in a mathematically more appropriate form, because the integral should be regularized as follows:

$$D = \frac{1}{dVC} \lim_{\varepsilon \to 0^+} \int_0^{+\infty} dt \, e^{-\varepsilon t} \langle \mathbf{J}^T(t) \cdot \mathbf{J}^T(0) \rangle_0. \tag{4.128}$$

This equation is known as the conventional Green–Kubo relation, which expresses a transport coefficient D, associated with a density $\rho(\mathbf{x}, t)$, in terms of the time integral of the autocorrelation function of the total current $\mathbf{J}^T(t)$. Note that in order to determine D we also have to compute

$$C = \lim_{\mathbf{k} \to 0} C(\mathbf{k}, t = 0), \tag{4.129}$$

which is proportional to the value of the static autocorrelation function of $\rho(\mathbf{k}, t)$ (see Eq. (4.106)).

The Green–Kubo relation is a very interesting result of linear response theory since it allows computation of transport coefficients in terms of equilibrium averages of total current correlation functions. This is also of great practical interest, because we can compute transport coefficients by numerical simulations, which provide reliable numerical estimates of $\langle \mathbf{J}^T(t) \cdot \mathbf{J}^T(0) \rangle_0$. For instance, this method is quite useful for the progress of the theory of liquids and also for investigating anomalous transport properties in low-dimensional systems, such as carbon nanotubes and suspended graphene layers.

A widely used method to perform numerical simulations of physical models is classical molecular dynamics. Let us just sketch the basic ingredients of this method. In the simple case of a homogeneous system, we deal with N equal particles of mass m, interacting among themselves, whose energy is given by a Hamiltonian $\mathcal{H}(\{\mathbf{r}_i(t), \mathbf{v}_i(t)\}_{i=1}^N)$, where $\mathbf{r}_i(t)$ and $\mathbf{v}_i(t)$ are the position and velocity coordinates of the ith particle. The equations of motion of the particles are given by the Hamilton equations and they can be integrated by suitable symplectic algorithms.[8] The system can be put in contact with reservoirs, which can be represented in the simulations by additional dynamical variables generated by deterministic or stochastic rules. The dynamical equations of the system plus reservoirs are integrated over a sufficiently long time interval to obtain reliable equilibrium averages of the total current correlation function appearing in Eq. (4.128). In general, we can invoke ergodicity for the system under consideration, that is, the equivalence between time and ensemble averages. In a strictly rigorous mathematical sense, there are few cases where ergodicity can be proved. On the other hand, weak ergodicity[9] can be assumed to be valid for many models, where relaxation processes emerge spontaneously as a result of the interactions

---

[8] Symplectic integration algorithms are typically employed for approximating Hamiltonian dynamics into the form of a map evolving by finite time steps. In fact, the map is written in such a way to preserve any volume in phase space, that is, to satisfy Liouville's theorem. The main practical consequence is that the approximate symplectic algorithm conserves all the invariant quantities of the original Hamiltonian.

[9] This concept was introduced by the Russian mathematician Aleksandr Khinchin as a heuristic substitute of true ergodicity. When, starting from generic initial conditions, all quantities of physical interest typically exhibit a fast relaxation to their equilibrium values, we can assume that in practice ergodicity holds, although a true rigorous proof is unavailable. The typical test of this hypothesis amounts to checking that the time autocorrelation function of the observable $X(t)$ decays exponentially fast in time,

$$\langle X(t) X(0) \rangle \sim e^{-t/t_0}.$$

In general, the relaxation time scale t_0 is expected to be weakly dependent on the observable $X(t)$.

among particles and between particles and reservoirs.[10] The quality of numerical estimates of time correlation functions appearing in the Green–Kubo relations can be improved by averaging over many different initial conditions and realizations of the stochastic processes associated with the presence of reservoirs.

We want to conclude this section by reporting, as an example, the expressions of transport coefficients for a one-dimensional system of N particles of mass m, interacting through a nearest-neighbor confining potential $V(x_{n+1} - x_n)$, where the subscript n labels the lattice site and the variable x_n can be read as the displacement of the nth particle with respect to its equilibrium position. We assume periodic boundary conditions. In this case the Green–Kubo relations take a particularly simple form and physical interpretation. The dynamics of particles is ruled by Newton's dynamics,

$$m\ddot{x}_n = -F_n + F_{n-1}$$

$$F_n = -V'(x_{n+1} - x_n),$$

where $V'(x)$ is a shorthand notation for the derivative of $V(x)$ with respect to its argument. By assuming that the velocity of the center of mass is zero,[11] one obtains the following expressions for the bulk viscosity η_B and the thermal conductivity κ,

$$\eta_B = \frac{1}{NT} \int_0^{+\infty} dt \, \langle J_p(t) J_p(0) \rangle, \tag{4.130}$$

where

$$J_p(t) = \sum_{n=1}^{N} F_n(t) \tag{4.131}$$

is the total momentum flux, and

$$\kappa = \frac{1}{NT^2} \int_0^{+\infty} dt \, \langle J_E(t) J_E(0) \rangle, \tag{4.132}$$

where

$$J_E(t) = \frac{1}{2} \sum_{n=1}^{N} F_n(t)(\dot{x}_{n+1} + \dot{x}_n) \tag{4.133}$$

is the total heat flux.

4.6 Generalized Linear Response Function

We have to consider that if $h(t)$ is able to perturb its conjugated macroscopic observable $X(t)$ to which it is directly coupled, it also can influence the dynamical state of other macroscopic

---

[10] When stochastic rules are at work in molecular dynamics simulations, they are usually expected to facilitate the relaxation processes by introducing in the dynamics incoherent fluctuations in the form of noise. In fact, it is well known that a purely deterministic dynamics, even in the case of large systems, may exhibit dynamical regimes where ergodic conditions cannot be achieved or might eventually show up over practically inaccessible time scales.

[11] This condition can be set in the initial condition and automatically preserved, because periodic boundary conditions allow for symmetry by translation invariance in the lattice, that is, the conservation of momentum.

observables. More precisely, we can generalize what was discussed in Section 4.4 by introducing a perturbed Hamiltonian that depends on a set of thermodynamic observables $X_k(t)$ and their conjugate perturbation fields $h_k(t)$,

$$\mathcal{H}' = \mathcal{H} - \sum_k h_k(t)\, X_k, \tag{4.134}$$

where, without prejudice to generality, we can assume $\langle X_k \rangle_0 = 0 \; \forall k$.

In the following we focus on the macroscopic observable $X_i(t)$ and we evaluate its response to the switching of the perturbation field $h_j(t)$, conjugate to $X_j(t)$. In analogy with (4.39), the starting point is the equation

$$\langle X_i(t) \rangle = \frac{1}{Z'} \int_{\mathcal{V}} d\mathcal{V} \; e^{-\beta \mathcal{H}'(q_i, p_i)} X_i(t). \tag{4.135}$$

In the linear approximation, the response function can be evaluated by calculating the derivative of Eq. (4.135) with respect to h_j,

$$\begin{aligned}
\frac{\partial \langle X_i(t) \rangle}{\partial h_j} &= \frac{\beta}{Z'} \int_{\mathcal{V}} d\mathcal{V} \; e^{-\beta \mathcal{H}'(q_i, p_i)} X_i(t) X_j(0) \\
&\quad - \frac{\beta}{(Z')^2} \int_{\mathcal{V}} d\mathcal{V} \; e^{-\beta \mathcal{H}'(q_i, p_i)} X_i(t) \int_{\mathcal{V}} d\mathcal{V} \; e^{-\beta \mathcal{H}'(q_i, p_i)} X_j(0),
\end{aligned} \tag{4.136}$$

and by evaluating this expression for $h_j = 0$, so that (4.136) yields the relation (compare with Eq. (4.42))

$$\langle X_i(t) \rangle - \langle X_i(t) \rangle_0 = \beta h_j \Big(\langle X_j(0) X_i(t) \rangle_0 - \langle X_j(0) \rangle_0 \langle X_i(t) \rangle_0 \Big) + O(h_j^2). \tag{4.137}$$

Following the same procedure sketched in Section 4.4.1, we can finally obtain, in the linear approximation, an expression for the generalized response function $\Xi_{ij}(t)$,

$$\Xi_{ij}(t) = -\beta\, \Theta(t) \frac{d}{dt} \langle X_i(t)\, X_j(0) \rangle_0 , \tag{4.138}$$

to be compared with Eq. (4.49). It is worth recalling that also in this general case we have assumed $\langle X_i \rangle_0 = \langle X_j \rangle_0 = 0$. This generalized response function satisfies the fluctuation–dissipation relation

$$\langle X_i(t) \rangle = \int dt'\, \Xi_{ij}(t - t') h_j(t'), \tag{4.139}$$

whose version in the Fourier space reads

$$\langle X_i(\omega) \rangle = \Xi_{ij}(\omega)\, h_j(\omega). \tag{4.140}$$

We can also rewrite Eq. (4.138) in the Fourier space as

$$\Xi_{ij}(\omega) = -\beta \int_{-\infty}^{+\infty} \frac{d\omega'}{2\pi} \Theta(\omega - \omega')(i\omega')\, C_{ij}(\omega'), \tag{4.141}$$

where $C_{ij}(\omega) = \langle X_i^*(\omega) X_j(\omega) \rangle_0$ is the Fourier transform of the time correlation function $\langle X_i(t) X_j(0) \rangle_0$. The previous equation generalizes Eq. (4.53).

All the properties of the response function discussed in Section 4.4.1 straightforwardly hold for $\Xi_{ij}(\omega)$. In particular, the fluctuation–dissipation theorem (generalization of Eq. (4.56)) is

$$\Xi_{ij}^I(\omega) = -\frac{\beta\omega}{2}C_{ij}(\omega),\tag{4.142}$$

while the Kramers–Krönig relation (generalization of Eq. (4.57)) now is

$$\Xi_{ij}^R(\omega) = \text{PV}\int_{-\infty}^{+\infty}\frac{d\omega'}{\pi}\frac{1}{\omega'-\omega}\,\Xi_{ij}^I(\omega'),\tag{4.143}$$

where $\Xi_{ij}^R(\omega)$ and $\Xi_{ij}^I(\omega)$ denote the real and the imaginary part of the response function $\Xi_{ij}(\omega)$, respectively.

Eq. (4.142) shows that the imaginary part of the Fourier transform of the response function is related to the equilibrium fluctuations of the same pair of observables through the Fourier transform of their time correlation function. In Appendix N, we illustrate a similar linear-response approach for quantum systems.

4.6.1 Onsager Regression Relation and Time Reversal

In this section we want to illustrate the Onsager regression relation in the framework of the general linear response approach discussed in Section 4.5. Taking advantage of time–translation invariance at equilibrium, we can rewrite Eq. (4.138) as

$$\Xi_{ij}(t-t') = \beta\,\Theta(t-t')\frac{d}{dt'}\langle\, X_i(t)\,X_j(t')\,\rangle_0 .\tag{4.144}$$

Let us now consider perturbation fields whose time dependence is given by

$$h_j(t) = h\,\Theta(-t)\,e^{\epsilon t},\tag{4.145}$$

where $0 < \epsilon \ll 1$ is a small parameter, such that $h_j(t)$ slowly increases from $-\infty$ to its maximum amplitude h for $t \to 0$, just before being switched off at $t = 0$. Then, for $t > 0$, Eq. (4.139) becomes

$$\langle X_i(t)\rangle = \beta\,h\int_{-\infty}^{0}dt'\,e^{\epsilon t'}\frac{d}{dt'}\langle\, X_i(t)\,X_j(t')\,\rangle_0 .\tag{4.146}$$

We can integrate by parts the right-hand side and finally obtain[12]

$$\langle X_i(t)\rangle = \beta\,h\left\{\left[e^{\epsilon t'}\langle\, X_i(t)\,X_j(t')\,\rangle_0\right]_{t'=-\infty}^{t'=0} - \epsilon\int_{-\infty}^{0}dt'\,e^{\epsilon t'}\langle\, X_i(t)\,X_j(t')\,\rangle_0\right\}$$

$$= \beta\,h\,\langle\, X_i(t)\,X_j(0)\,\rangle_0 + O(\epsilon).\tag{4.147}$$

This result is a generalization of Eq. (4.137) because in some sense it takes into account the switching process of the perturbation, which has now been shown to be of order ϵ. Eq. (4.147) is known as the Onsager regression relation, which expresses quite an important concept pointed out by the Norwegian physicist Lars Onsager. He argued that if we are studying a physical system during its evolution around equilibrium, we cannot distinguish if

---

[12] We must first take the appropriate limits for t', then take the limit $\epsilon \to 0$.

an instantaneous value of an observable $\langle X_i(t) \rangle$, different from its equilibrium value $\langle X_i \rangle_0$, is due to a spontaneous fluctuation or to a perturbation switched off in the past.[13] This statement implies that the way a fluctuating observable spontaneously decays toward its equilibrium value cannot be distinguished by the way its time correlation function decays at equilibrium.

It is important to point out that the fluctuating macroscopic observables $X_i(t)$ that we consider in the linear response theory are functions of time t through the canonical coordinates $(q_k(t), p_k(t))$, because we have assumed that the physical system they belong to is represented by a Hamiltonian $\mathcal{H}(q_k(t), p_k(t))$ (see Section 4.4). As we have discussed earlier, equilibrium quantities, like time correlation functions of fluctuating observables, also characterize nonequilibrium properties in the linear response theory. In a statistical sense, the equilibrium Gibbs measure is expressed in terms of the Hamiltonian, together with its symmetries. In particular, the basic one is the time-translational symmetry that is associated with the conservation of energy in the isolated system. In particular, this implies that the Hamiltonian is invariant under time reversal. For consistency reasons we also have to assume that all observables $X_i(t) \equiv X_i(q_k(t), p_k(t))$ have a definite parity with respect to time reversal, according to the way they depend on microscopic canonical coordinates.[14] For instance, observables such as particle and energy densities have even parity with respect to time reversal, because they are even functions of the particle velocities, while momentum density has odd parity, because it depends linearly on the particle velocities.

All this matter can be expressed in mathematical language by introducing a time-reversal operator $\mathcal{T}$ that applies to the phase space points as

$$\mathcal{T}(q_k(t), p_k(t)) = (q_k(t), -p_k(t)) \,, \quad \text{with} \quad \mathcal{T}\mathcal{H} = \mathcal{H}, \tag{4.148}$$

that is, $\mathcal{T}$ inverts the components of momenta of all particles in the system, and $\mathcal{H}$ has to be an even function of momenta. This implies that if the trajectory $(q_k(t), p_k(t))$ is a solution of the Hamilton Equation (4.27) with initial conditions $(q_k(0), p_k(0))$, the time-reversed trajectory $\mathcal{T}(q_k(t), p_k(t))$ is also a solution with initial condition $\mathcal{T}(q_k(0), p_k(0))$. We can assume that all observables $X_i(t)$, as well as $\mathcal{H}$, are eigenfunctions of $\mathcal{T}$, because they have a well-defined parity. Thus

$$X_i(-t) \equiv \mathcal{T}X_i(t) = X_i(\mathcal{T}(q_k(t), p_k(t))) = X_i((q_k(t), -p_k(t))) = \tau_i X_i(t), \tag{4.149}$$

where the eigenvalues $\tau_i = \pm 1$ determine the parity of observable $X_i(t)$ with respect to time reversal.

For what we are going to discuss later on, it is important to determine how $\mathcal{T}$ acts over the time correlation functions $\langle X_i(t) X_j(t') \rangle_0$. We can assume that we are in a given initial state $(q_k(0), p_k(0))$ at $t' = 0$ and apply to it the time reversal operator $\mathcal{T}(q_k(0), p_k(0)) = (q_k(0), -p_k(0))$. For instance, in a Hamiltonian system made of identical particles, this operation amounts to changing the initial conditions by keeping the same space coordinates, but reversing the velocities. According to Eq. (4.149), we

---

[13] The average $\langle \, \cdot \, \rangle$ is performed over a statistical ensemble of perturbed systems, subject to the same perturbation field $h(t)$ as the one defined in Eq. (4.145).

[14] We are considering Hamiltonian systems in the absence of magnetic fields.

have $\mathcal{T} X_j(0) = \tau_j X_j(0)$. We now let the system evolve for a time $-t$ to the state $(q_k(-t), p_k(-t))$ where the corresponding value of $X_i(-t)$ has to satisfy the eigenvalue relation $\mathcal{T} X_i(-t) = \tau_i X_i(-t)$. At thermodynamic equilibrium, the probability of the initial state $(q_k(0), p_k(0))$ and of the time-reversed state $(q_k(0), -p_k(0))$ are the same, because of the invariance of $\mathcal{H}$ under the action of $\mathcal{T}$. We can finally write[15]

$$\langle\, X_i(t)\, X_j(0)\, \rangle_0 = \langle\, \mathcal{T} X_i(-t)\, \mathcal{T} X_j(0)\, \rangle_0 = \tau_i \tau_j \langle\, X_i(-t)\, X_j(0)\, \rangle_0 . \tag{4.150}$$

We can conclude that the correlation functions of observables with the same parity with respect to time reversal are even functions of time, while those of observables with different parity are odd functions of time. This result has important consequences for the properties of the response function (4.144), that, for $t > 0$, can be rewritten

$$\Xi_{ij}(t) = -\beta \frac{d}{dt}\langle\, X_i(t)\, X_j(0)\, \rangle_0 = -\beta\, \tau_i\, \tau_j\, \frac{d}{dt}\langle\, X_i(-t)\, X_j(0)\, \rangle_0 . \tag{4.151}$$

Because of the invariance of the Hamiltonian dynamics under time shifts at equilibrium, we can also write

$$\Xi_{ij}(t) = -\beta\, \tau_i\, \tau_j\, \frac{d}{dt}\langle\, X_i(0)\, X_j(t)\, \rangle_0 = \tau_i\, \tau_j\, \Xi_{ji}(t). \tag{4.152}$$

This is an important relation that will be used in Section 4.7 to establish symmetries among kinetic coefficients in the Onsager theory of linear response.

4.7 Entropy Production, Fluxes, and Thermodynamic Forces

In this section we aim at providing a general description of nonequilibrium processes in the linear response regime based on the thermodynamics of irreversible processes. This is summarized by the second principle of thermodynamics: When a physical system is in nonequilibrium conditions, its entropy, S, must increase, namely, $dS \geq 0$.[16]

4.7.1 Nonequilibrium Conditions between Macroscopic Systems

For pedagogical reasons, we discuss the simple case where a physical system, in contact with an external reservoir, is in nonequilibrium conditions with it. We assume that in this universe the total variation of entropy $d\mathbb{S}$ can be written as

$$d\mathbb{S} = dS + dS_r, \tag{4.153}$$

that is, it is the sum of the variation of entropy in the system, dS, and in the reservoir, dS_r. Notice that dS_r is not necessarily positive, depending on the way the reservoir is coupled to the system. For an adiabatically insulated system, $dS_r = 0$ by definition and $d\mathbb{S} \equiv dS \geq 0$.

---

[15] We recall once again that the average, equilibrium values of the various quantities are assumed to vanish, without loss of generality.

[16] Such a statement is certainly valid for physical systems where the interaction forces among the particles are short range; when long-range forces are present the situation is more complicated because additivity is lost and Eq. (4.157) does not apply.

More generally, if we assume that the reservoir is at (absolute) temperature T, the variation of entropy in the reservoir is proportional to the total heat supplied to the system through the Carnot–Clausius formula,

$$dS_r = -\frac{dQ}{T}, \tag{4.154}$$

where $dQ > 0$ means that the system absorbs heat from the reservoir. Since $dS \geq 0$, in nonequilibrium conditions, when the system can exchange only heat with the reservoirs, we have

$$d\mathbb{S} \geq \frac{dQ}{T}. \tag{4.155}$$

The previous equation is a way of expressing the second law of thermodynamics.

The problem of nonequilibrium thermodynamics amounts to establishing a relation between the way entropy varies in time and the irreversible phenomena that generate such a variation. A first step in this direction can be accomplished by considering that the entropy depends on a set of extensive thermodynamic observables X_i, namely,

$$S \equiv S(\{X_i\}). \tag{4.156}$$

Due to their extensive character, we can assume that a given value $\mathbb{X}_i$ taken in our universe by one of these observables has to be the sum of the values taken by this observable in the system, X_i, and in the reservoir, $X_i^{(r)}$, that is,

$$\mathbb{X}_i = X_i + X_i^{(r)}. \tag{4.157}$$

If fluctuations allow both X_i and $X_i^{(r)}$ to be unconstrained, we can define the thermodynamic force $\mathbb{F}_i$ associated with the extensive observable X_i as

$$\mathbb{F}_i \equiv \left(\frac{\partial \mathbb{S}}{\partial \mathbb{X}_i}\right)\Big|_{\mathbb{X}_i=X_i} = \frac{\partial S}{\partial X_i} - \frac{\partial S_r}{\partial X_i^{(r)}} \equiv F_i - F_i^{(r)}, \tag{4.158}$$

where we have defined

$$F_i = \left(\frac{\partial S}{\partial X_i}\right) \tag{4.159}$$

$$F_i^{(r)} = \left(\frac{\partial S_r}{\partial X_i^{(r)}}\right) \tag{4.160}$$

and we have taken explicitly into account that, as a consequence of the closure condition (4.157), $dX_i = -dX_i^{(r)}$, since the value $\mathbb{X}_i$ is fixed in our universe, that is, $d\mathbb{X}_i = 0$. Notice that the partial derivatives in Eq. (4.158) are performed while keeping constant all the other extensive thermodynamic observables. These generalized thermodynamic forces $\mathbb{F}_i$, that are also called affinities, are responsible for driving the system in out-of-equilibrium conditions. At thermodynamic equilibrium, the total entropy in the universe $\mathbb{S}$ is by definition maximal, so that $\mathbb{F}_i = 0$. On the other hand, if $\mathbb{F}_i \neq 0$ an irreversible process sets in, spontaneously driving the system toward an equilibrium state.

We want to point out that an equivalent way of introducing the concept of affinities can be obtained by reinterpreting S and S_r as the entropies of two subsystems forming a closed, that is, isolated, system whose entropy is given by $\mathbb{S}$. At a classical level, the two interpretations are equivalent, because it is natural to assume that a physical system can be isolated, while the two subsystems, by definition, have the same physical properties. At a quantum level, the idealization of an isolated physical system seems quite an ill-defined concept. Moreover, the presence of a thermal reservoir, which represents the complement with respect to the universe of the system under scrutiny, implies that the reservoir is not necessarily made only of the same kind of physical units (particles, atoms, molecules, etc.) that are present in the system. In this perspective, the interpretation with the reservoir seems to be more general and appropriate. On the other hand, in this case we have to postulate the existence of a physical mechanism that allows the reservoir to interact with the system. Far from being a purely philosophical argument, this is a crucial point to be taken into account, if we aim at constructing a physical model of a suitable thermal reservoir, in both the classical and quantum cases. In particular, this is of primary importance if we have to deal with numerical simulations in the presence of a thermal reservoir, and various schemes can be adopted, including stochastic as well as deterministic formulations. Despite their intellectual and practical interest, these topics are not considered in this textbook.

We can exemplify our previous considerations about a system in out-of-equilibrium conditions by first recalling that typical examples of extensive thermodynamic observables are the internal energy U, the volume V, and the number of particles N_j of the species j contained in the system. The Gibbs relation

$$T\,dS = dU + P\,dV - \sum_j \mu_j\,dN_j \tag{4.161}$$

provides us with the functional dependence of S on these extensive observables, where T is the absolute temperature, P is the pressure, and μ_j is the chemical potential of particles of species j. Notice that this relation deals with entropy variations at close to equilibrium conditions. We assume that such a functional dependence is valid also for the above-described out-of-equilibrium conditions. For instance, we can still define the absolute temperature of the system and of the reservoir (or of the two subsystems) by the expression

$$\left(\frac{\partial S}{\partial U}\right)_{V,N_j} = \frac{1}{T} \tag{4.162}$$

and, analogously, the pressure and chemical potentials by the expressions

$$\left(\frac{\partial S}{\partial V}\right)_{U,N_j} = \frac{P}{T}, \qquad \left(\frac{\partial S}{\partial N_j}\right)_{U,V} = -\frac{\mu_j}{T}, \tag{4.163}$$

where we adopt the standard notation for partial derivatives and keep the subscript quantities constant.

With reference to Eq. (4.158), the affinities stemming from the Gibbs relation (4.161) are

$$\mathbb{F}_U = \frac{1}{T} - \frac{1}{T_r}, \qquad \mathbb{F}_V = \frac{P}{T} - \frac{P_r}{T_r}, \qquad \mathbb{F}_{N_j} = -\frac{\mu_j}{T} + \frac{(\mu_j)_r}{T_r}. \tag{4.164}$$

The vanishing of all affinities implies that the system and the reservoir share the same temperature, pressure, and chemical potentials; therefore, they are at equilibrium. On the other hand, if the difference of the inverse temperatures is nonzero, there is a heat flow between the system and the reservoir that aims at establishing global equilibrium conditions.

For instance, we can think that the system and the reservoir are in contact through a thin diathermal wall, which allows the heat to flow from the higher temperature to the lower one, until the system eventually takes the temperature of the reservoir (or the two subsystems eventually exhibit the same temperature, which results from a suitable average of the original temperatures). This is to point out that, in general, the response to an applied thermodynamic force, $\mathbb{F}_i$, should be related to the rate of change over time of the extensive observable X_i. More precisely, we can define the fluxes (rate of changes) as

$$J_i = \frac{dX_i}{dt}. \tag{4.165}$$

When the affinity vanishes, the corresponding flux also vanishes. The thermodynamics of irreversible processes aims at establishing the relation between fluxes and affinities. For this purpose it is useful to introduce an explicit dependence of the entropy on time, so we can define the entropy production rate as

$$\frac{d\mathbb{S}}{dt} = \sum_i \left(\frac{\partial S}{\partial X_i} - \frac{\partial S_r}{\partial X_i^{(r)}} \right) \frac{dX_i}{dt} = \sum_i \mathbb{F}_i \, J_i. \tag{4.166}$$

Since our universe is isolated, the second principle of thermodynamics states that in out-of-equilibrium conditions, the total entropy has to increase, namely,

$$\frac{d\mathbb{S}}{dt} = \sum_i \mathbb{F}_i \, J_i \geq 0, \tag{4.167}$$

and it vanishes only when all affinities and the corresponding fluxes vanish, that is, at thermodynamic equilibrium.[17]

4.7.2 Phenomenological Equations

After having identified the affinities, we can proceed to the main goal of the thermodynamics of irreversible processes, that is, finding a relation between affinities and fluxes. For this purpose, within a linear nonequilibrium regime, we suppose that fluctuating macroscopic observables $X_i(t)$ evolve in time according to linear equations of the form

$$J_i = \frac{d}{dt} X_i(t) = - \sum_k \mathcal{M}_{ik} X_k(t), \tag{4.168}$$

where $\mathcal{M}_{ik}$ are phenomenological coefficients, depending both on the nature of the specific fluctuating quantities and on the specific conditions leading the system out of equilibrium. If only one quantity fluctuates, this assumption leads to an exponential relaxation toward an asymptotic, vanishing value, in line with the linear response theory, where the exponential relaxation is the result of the equilibrium correlation function.

As we are going to show later in this section, such an assumption yields a linear relation between fluxes and affinities. This relation is consistent with the assumption of a linear

---

[17] There are peculiar situations in coupled transport processes (see Section 4.8), where, as a consequence of symmetries, a flux can be made to vanish by applying suitable, nonvanishing affinities, although the system is kept in stationary out-of-equilibrium conditions.

response to an applied thermodynamic force, meaning that, at leading order, the flux of an extensive thermodynamic observable is proportional to the affinities. The proportionality constants are the so-called (linear) kinetic coefficients.[18]

On the other hand, some physical arguments associated with the fluctuating nature of $X_i(t)$ suggest that we should reconsider the meaning we attribute to the set of differential Equations (4.168). The first consideration is that we expect that Eq. (4.168) describes confidently the linear response regime over time scales sufficiently longer than the correlation time, t_c, of spontaneous fluctuations in the system: In this sense t has to be intended as a coarse-grained variable. Moreover, the flux $J_i(t)$ can be more properly defined by substituting $X_i(t)$ with its average over realizations of the fluctuation–relaxation process,[19] namely,

$$J_i = \frac{d}{dt}\langle X_i(t)\rangle = -\sum_k \mathcal{M}_{ik}\langle X_k(t)\rangle, \tag{4.169}$$

where the phenomenological coefficients are assumed to be physical quantities independent of the average over realizations.

We can use the Onsager regression relation (4.147) and rewrite Eq. (4.169) as

$$\frac{d}{dt}\langle X_i(t)\, X_j(0)\rangle_0 = -\sum_k \mathcal{M}_{ik}\langle X_k(t)\, X_j(0)\rangle_0, \tag{4.170}$$

with the caveat that it is valid for $t > 0$. If X_i and X_j are both invariant under time reversal (i.e., $\tau_i = \tau_j = 1$), we have

$$\langle X_i(t)\, X_j(0)\rangle_0 = \langle X_j(t)\, X_i(0)\rangle_0. \tag{4.171}$$

By differentiating both sides of this equation with respect to time and using Eq. (4.170), we obtain

$$\sum_k \mathcal{M}_{ik}\langle X_k(t)\, X_j(0)\rangle_0 = \sum_k \mathcal{M}_{jk}\langle X_k(t)\, X_i(0)\rangle_0, \tag{4.172}$$

which can be evaluated for $t = 0$, yielding the relation

$$\sum_k \mathcal{M}_{ik}\langle X_k\, X_j\rangle_0 = \sum_k \mathcal{M}_{jk}\langle X_k\, X_i\rangle_0, \tag{4.173}$$

where $\langle X_k\, X_j\rangle_0$ is a shorthand notation for the equal-time equilibrium correlation. If we define the matrix L, whose elements are

$$L_{ij} = \sum_k \mathcal{M}_{ik}\langle X_k\, X_j\rangle_0, \tag{4.174}$$

Eq. (4.173) implies that L is symmetric,

$$L_{ij} = L_{ji}. \tag{4.175}$$

This symmetry property between phenomenological transport coefficients is known as the Onsager reciprocity relation.

---

[18] In general, one could also consider higher-order kinetic coefficients by assuming additional nonlinear dependences of fluxes on affinities: Such higher-order coefficients may become relevant in out-of-equilibrium regimes dominated by nonlinear effects.

[19] This assumption can be justified with the same arguments that we have used for replacing Eq. (4.100) with Eq. (4.102).

Now we want to show that by making use of Eqs. (4.170), we can obtain a linear relation between fluxes and affinities of the form

$$J_i = \frac{d}{dt} X_i(t) = \sum_j L_{ij} \mathbb{F}_j, \tag{4.176}$$

where the matrix elements L_{ij} are defined in Eq. (4.174). Before proving the validity of these phenomenological relations, we want to observe that they allow us to express the elements of the Onsager matrix as

$$L_{ij} = \left(\frac{\partial J_i}{\partial \mathbb{F}_j} \right)_{\mathbb{F}_k = 0}, \tag{4.177}$$

that is, the matrix L of the linear kinetic coefficients contains elements that are computed at equilibrium conditions, that is, when all affinities $\mathbb{F}_k$ vanish. In other words, the L_{ij} can be expressed in terms of quantities such as temperature, pressure, chemical potentials, and so on, that characterize the equilibrium state of the system.[20]

As we have obtained Eq. (4.170) from Eq. (4.168), in this case we should also consider the fluctuating nature of the quantities $X_i(t)$ and $\mathbb{F}_k(t)$ appearing in Eq. (4.176). We can proceed by formally multiplying both sides of Eq. (4.176) by $X_j(0)$ and considering the equilibrium averages of the products, thus obtaining

$$\frac{d}{dt} \langle X_i(t) X_j(0) \rangle_0 = \sum_k L_{ik} \langle \mathbb{F}_k(t) X_j(0) \rangle_0. \tag{4.178}$$

Taking into account Eq. (4.170), we can write

$$\sum_k \mathcal{M}_{ik} \langle X_k(t) X_j(0) \rangle_0 = - \sum_k L_{ik} \langle \mathbb{F}_k(t) X_j(0) \rangle_0. \tag{4.179}$$

In particular, this relation holds for equal-time equilibrium correlations

$$\sum_k \mathcal{M}_{ik} \langle X_k X_j \rangle_0 = - \sum_k L_{ik} \langle \mathbb{F}_k X_j \rangle_0. \tag{4.180}$$

If we are considering nonequilibrium conditions between macroscopic systems, the right-hand side of this equation can be written using Eq. (4.158) as

$$\langle \mathbb{F}_k X_j \rangle_0 = \left\langle \frac{\partial S}{\partial X_k} X_j \right\rangle_0 - \left\langle \frac{\partial S_r}{\partial X_k^{(r)}} X_j \right\rangle_0. \tag{4.181}$$

If the system is in contact with a reservoir, the second term on the right-hand side can be neglected compared to the first, because at equilibrium, correlations between the entropy of the reservoir and extensive observables are irrelevant. Therefore, using Eq. (4.159), we can eventually write

$$\langle \mathbb{F}_k X_j \rangle_0 = \left\langle F_k X_j \right\rangle_0. \tag{4.182}$$

---

[20] This can be viewed as a generalization of what we found in Section 1.2.2, where the transport coefficients derived from the kinetic theory have been expressed as functions of particle density, mean free path, average velocity, specific heat at constant volume, and so on.

The equilibrium correlation on the right-hand side of Eq. (4.182) can be evaluated by considering the fundamental postulate of equilibrium statistical mechanics, according to which

$$S(\{X_i\}) = \ln \Gamma, \tag{4.183}$$

where Γ is the volume in the phase space, which is compatible with the assigned values of the extensive thermodynamic variables X_i. The main consequence of this postulate is that the equilibrium value of any observable $O(\{X_i\})$ can be defined as

$$\langle O(\{X_i\})\rangle_0 = \frac{1}{\mathcal{N}} \int \prod_i dX_i O(\{X_i\}) e^{S(\{X_i\})}, \tag{4.184}$$

where the normalization factor is given by the expression

$$\mathcal{N} = \int \prod_i dX_i e^{S(\{X_i\})}. \tag{4.185}$$

If we take $O(\{X_i\}) = \frac{\partial S}{\partial X_k} X_j$, we can write

$$
\begin{aligned}
\left\langle \frac{\partial S}{\partial X_k} X_j \right\rangle_0 &= \frac{1}{\mathcal{N}} \int \prod_i dX_i \frac{\partial S}{\partial X_k} X_j e^{S(\{X_i\})} \\
&= \frac{1}{\mathcal{N}} \int \prod_i dX_i \frac{\partial}{\partial X_k} \left(e^{S(\{X_i\})} \right) X_j
\end{aligned}
\tag{4.186}
$$

and integrating by parts, we obtain

$$
\langle F_k X_j \rangle_0 = \left\langle \frac{\partial S}{\partial X_k} X_j \right\rangle_0 \tag{4.187}
$$

$$
= -\delta_{jk} + \frac{1}{\mathcal{N}} \int \prod_{i \neq k} dX_i X_j \, e^{S(\{X_i\})} \Big|_{X_k^{\min}}^{X_k^{\max}} \tag{4.188}
$$

$$
= -\delta_{jk}, \tag{4.189}
$$

where the second part of the integral, coming from integration by parts, can be neglected in the thermodynamic limit, while the other part reduces to the Kronecker delta δ_{jk}, because the extensive thermodynamic variables are assumed to be independent of each other.

Substituting this result into Eq. (4.180), we recover the definition (4.174), thus showing the validity of the phenomenological relations (4.176). We can conclude that such relations represent an effective approximation in a linear response regime, where local equilibrium conditions hold.

4.7.3 Variational Principle

Using Eq. (4.176), the entropy production rate, given in Eq. (4.166), can be expressed as a bilinear form of the affinities,

$$\frac{d\mathbb{S}}{dt} = \sum_{ij} L_{ij} \, \mathbb{F}_i \, \mathbb{F}_j. \tag{4.190}$$

In out-of-equilibrium conditions, the right-hand side of this equation has to be a positive quantity, which must vanish at equilibrium. In particular, if we switch off all currents other than J_i, we have $\frac{d\mathbb{S}}{dt} = L_{ii}\mathbb{F}_i^2$, which implies

$$L_{ii} > 0 \quad \forall i. \tag{4.191}$$

We can also compute the time derivative of the entropy production rate,

$$\frac{d^2\mathbb{S}}{dt^2} = 2\sum_{ij}\sum_{k}\frac{\partial \mathbb{F}_i}{\partial X_k}\frac{dX_k}{dt}L_{ij}\,\mathbb{F}_j \tag{4.192}$$

$$= 2\sum_{ik}\frac{\partial \mathbb{F}_i}{\partial X_k}J_k\,J_i$$

$$= 2\sum_{ik}\frac{\partial^2\mathbb{S}}{\partial X_k\partial X_i}J_k\,J_i, \tag{4.193}$$

where we have used, in order, the symmetry condition (4.175), the phenomenological relation (4.176), and the definition of generalized thermodynamic force in Eq. (4.158).

Since the entropy $\mathbb{S}(\{X_i\})$ is a concave function of its arguments, we can conclude that the right-hand side of Eq. (4.193) is a negative quadratic form and

$$\frac{d^2\mathbb{S}}{dt^2} \le 0. \tag{4.194}$$

Therefore, in a linear regime, the entropy production rate, equal to $d\mathbb{S}/dt$, spontaneously decreases during the system evolution and there is a natural tendency of the system to minimize the entropy production rate. If the out-of-equilibrium conditions are transient, the entropy production rate will eventually vanish, while it may approach a minimum finite value when stationary out-of-equilibrium conditions are imposed.

In order to illustrate the latter scenario, we take as an example a system that is maintained in nonequilibrium conditions by fixing the value of one affinity, say $\mathbb{F}_i$, while all other affinities, $\mathbb{F}_k$, are free to vary. Making use of the rate of entropy production Eq. (4.190) and of the symmetry of the matrix L, we can write

$$\left.\frac{\partial\,\dot{\mathbb{S}}}{\partial\,\mathbb{F}_k}\right|_{k\neq i} = 2\sum_{j}L_{kj}\,\mathbb{F}_j = 2\,J_k. \tag{4.195}$$

By this relation and Eq. (4.194), we can conclude that the stationary state, that is, the nonequilibrium state corresponding to the vanishing of all currents J_k other than J_i, corresponds to an extremal state with respect to the varying affinities. The evaluation of the second derivative,

$$\frac{\partial^2\,\dot{\mathbb{S}}}{\partial\,\mathbb{F}_k^2} = 2L_{kk} > 0, \tag{4.196}$$

tells us that we are dealing with a minimum entropy production rate.

Let us consider a system in contact with two reservoirs, which we name a and b. In this case, Eq. (4.166) becomes

$$\frac{d\mathbb{S}}{dt} = \sum_i \left(\frac{\partial S}{\partial X_i} J_i + \frac{\partial S_a}{\partial X_i^{(a)}} J_i^{(a)} + \frac{\partial S_b}{\partial X_i^{(b)}} J_i^{(b)} \right), \tag{4.197}$$

where we have denoted with $J_i^{(a)}$ and $J_i^{(b)}$ the fluxes of the observable X_i in the reservoirs. The closure condition analogous to (4.157) in this case reads

$$\mathbb{X}_i = X_i + X_i^{(a)} + X_i^{(b)} \tag{4.198}$$

and yields the relation

$$d\mathbb{X}_i = 0 = dX_i + dX_i^{(a)} + dX_i^{(b)} \tag{4.199}$$

because the value of $\mathbb{X}_i$ is fixed in the universe. Since the system can be considered much smaller than the two reservoirs, the variations of the observable X_i can be neglected with respect to the corresponding variations in the reservoirs, and we can rewrite Eq. (4.197) in the approximate form

$$\frac{d\mathbb{S}}{dt} \approx \sum_i \left(\frac{\partial S_a}{\partial X_i^{(a)}} - \frac{\partial S_b}{\partial X_i^{(b)}} \right) J_i^{(a)} \tag{4.200}$$

because $J_i^{(a)} = \frac{dX_i^{(a)}}{dt} = -\frac{dX_i^{(b)}}{dt} = -J_i^{(b)}$. We can conclude that, in stationary conditions, the flux J_i flowing through the system is approximately equal to the one flowing through the reservoirs $J_i \approx J_i^{(a)} = -J_i^{(b)}$, while the corresponding affinity is given by the approximate expression

$$\mathbb{F}_i \approx \frac{\partial S_a}{\partial X_i^{(a)}} - \frac{\partial S_b}{\partial X_i^{(b)}}, \tag{4.201}$$

which amounts to the difference of the affinities in the two reservoirs. As an explicit instance, we can think of a system in contact with two heat reservoirs at different temperatures, $T^{(a)}$ and $T^{(b)}$, with $T^{(a)} > T^{(b)}$. The only varying observable in the reservoirs is the internal energy U, and we have an average outgoing flux from the reservoir at temperature $T^{(a)}$ and an average incoming flux to the reservoir at temperature $T^{(b)}$; Eq. (4.200) becomes

$$\frac{d\mathbb{S}}{dt} \approx \left(\frac{1}{T^{(b)}} - \frac{1}{T^{(a)}} \right) |J_u^{(a)}|, \tag{4.202}$$

where $J_u^{(a)}$ denotes the energy current flowing through the system. This case can be generalized to an arbitrary number of reservoirs in contact with the system.

All together, we can state the following variational principle: The stationary state corresponds to a minimum value of the entropy production rate, compatible with the external constraints imposed by the presence of affinities yielding nonvanishing currents.

4.7.4 Nonequilibrium Conditions in a Continuous System

As we have seen in Section 4.7.1, when we deal with out-of-equilibrium conditions between a system and a reservoir or between two homogeneous subsystems, the identification of the affinities $\mathbb{F}_i$ (see Eq. (4.164)) is a straightforward consequence of Gibbs relation (4.161). On the other hand, it may happen that heat, rather than flowing through a thin diathermal wall separating the two units of a system, flows along a piece of matter, where the temperature varies continuously in space due to the presence of a temperature gradient. For instance, this is the typical situation when a heat conductor is put in contact at its ends with two thermal reservoirs at different temperatures. In cases like this, the identification of the affinity in the system is not straightforward.

The main problem is that we need a suitable definition of entropy in a continuous system driven out of equilibrium. We can overcome this conceptual obstacle by assuming that local equilibrium conditions hold in the system. This means that we can associate to a small region of the system a local entropy, whose functional dependence on the extensive observables is the same as the one given at equilibrium by the Gibbs relation. Thus,

$$dS = \sum_i F_i \, dX_i, \tag{4.203}$$

which is formally equivalent to Eq. (4.159).

This is quite a crucial assumption, which is expected to hold in a continuous thermodynamic system gently driven out of equilibrium, in such a way that its entropy is still a well-defined local quantity, which maintains the same functional dependence on the extensive thermodynamic quantities. Said differently, we are assuming that in a continuous system mildly driven out of equilibrium, local equilibrium conditions spontaneously emerge on a spatial scale of the order of a few mean-free paths of the particles in the system and on a time scale of the order of a few collision/interaction times between such particles. In this perspective, one can easily realize that we are considering out-of-equilibrium conditions, where the effect of perturbations from the surrounding world is sufficiently small to remain in a linear response regime, in such a way that the system can relax locally to equilibrium conditions or, more properly, to a stationary state that is locally indistinguishable from an equilibrium one. The validity of this hypothesis can be justified a posteriori by the reliability of the predictions of the linear response theory. When strong perturbations or driving forces are applied to an out-of-equilibrium system, the linear approach fails and different descriptions have to be adopted, as it happens for turbulent regimes or anomalous transport phenomena.

Now we aim at finding an explicit expression for the affinities in a continuous system. In this respect, it is important to point out that, at variance with the macroscopic case discussed in Section 4.7.1, all the quantities in Eq. (4.203) should exhibit their local nature by an explicit dependence on the space coordinates, which disappears in stationary conditions, that is, when a constant flux flows through the system. If the system was at equilibrium, these fluxes would vanish.

In order to specify the problem, we can consider an infinitely extended system in three space dimensions. In this case, it is more appropriate to assume that a relation such as Eq. (4.203) holds for the intensive quantities, that is, the corresponding functions per unit volume $s(\{x_i\})$:

$$ds = \sum_i F_i \, dx_i, \tag{4.204}$$

so that we can write

$$F_i = \left(\frac{\partial s}{\partial x_i}\right). \tag{4.205}$$

As a consequence, in this formulation based on intensive quantities, none of the x_i corresponds to the volume V, while all the local quantities F_i are the same functions of the extensive observables, as is the case at equilibrium. Moreover, the fluxes J_i have to be replaced by three-dimensional vectors, and Eq. (4.204) suggests the possibility of defining an entropy current density $\mathbf{j}_s$, by means of the current densities $\mathbf{j}_i$ associated with the intensive observables x_i, as

$$\mathbf{j}_s = \sum_i F_i \, \mathbf{j}_i. \tag{4.206}$$

In a region of a continuous system we can say that the rate of local production of entropy is given by the entropy entering/leaving this region (surface term) plus the rate of increase of entropy within this region (volume term), as

$$\frac{ds}{dt} = \frac{\partial s}{\partial t} + \nabla \cdot \mathbf{j}_s. \tag{4.207}$$

The first term on the right-hand side can be obtained from Eq. (4.204), namely,

$$\frac{\partial s}{\partial t} = \sum_i F_i \frac{\partial x_i}{\partial t}, \tag{4.208}$$

while the second one is given by the expression

$$\nabla \cdot \mathbf{j}_s = \nabla \cdot \left(\sum_i F_i \, \mathbf{j}_i\right) = \sum_i \left(\nabla F_i \cdot \mathbf{j}_i + F_i \, \nabla \cdot \mathbf{j}_i\right). \tag{4.209}$$

Therefore, we can rewrite Eq. (4.207) as

$$\frac{ds}{dt} = \sum_i F_i \frac{\partial x_i}{\partial t} + \sum_i \nabla F_i \cdot \mathbf{j}_i + \sum_i F_i \, \nabla \cdot \mathbf{j}_i \tag{4.210}$$

$$= \sum_i F_i \left(\frac{\partial x_i}{\partial t} + \nabla \cdot \mathbf{j}_i\right) + \sum_i \nabla F_i \cdot \mathbf{j}_i. \tag{4.211}$$

Here the quantities in parentheses vanish if the extensive observables are conserved quantities, like the energy (in the absence of chemical reactions) and the mole numbers, because in this case the continuity equations,

$$\frac{\partial x_i}{\partial t} + \nabla \cdot \mathbf{j}_i = 0, \tag{4.212}$$

must hold. In conclusion, we finally obtain

$$\frac{ds}{dt} = \sum_i \nabla F_i \cdot \mathbf{j}_i, \tag{4.213}$$

which should be compared with Eq. (4.166). This indicates that the affinities $\mathbb{F}_i$, introduced for macroscopic systems as the difference of thermodynamic quantities F_i (see Eq. (4.158)), in the continuous case transform into the gradient of the same quantities, that is,

$$\mathbb{F}_i = F_i - F_i^{(r)} \quad \rightarrow \quad \mathbf{F}_i = \nabla F_i(\mathbf{r}). \tag{4.214}$$

For instance, if we consider the x-component of the energy current density $(j_u)_x$ flowing in the continuous system, the quantity multiplying it in Eq. (4.213) is given by the gradient of the inverse temperature along the x-direction, namely $\partial_x(1/T)$. In fact, it is worth recalling that in the nonequilibrium case, the temperature T as well as the entropy density and the current densities are functions of the space coordinates. In a similar way, if we consider the x-component of the ith mole number current density $(j_{n_i})_x$, the associated affinity is given by $-\partial_x(\mu_i/T)$, where μ_i is also a function of the space coordinates.

4.8 Physical Applications of Onsager Reciprocity Relations

Up to now in this chapter we have illustrated the basic mathematical and theoretical aspects of the thermodynamics of irreversible processes, based on the linear response approach and on the local equilibrium hypothesis for continuous systems. In what follows we describe some examples and applications that should help the reader to appreciate the physical content of this approach and to understand how to use such theoretical tools for studying transport processes in nonequilibrium phenomena. Since the case of charged particles in the presence of a magnetic field requires a specific description, we first discuss some examples of coupled transport of neutral particles, then the case of charged particles.

4.8.1 Coupled Transport of Neutral Particles

Mechanothermal and Thermomechanical Effects

As a first pedagogical example, we consider a system made of two volumes containing the same gas that are in contact through a wall, where a small hole allows for a slow exchange of particles. The exchanged particles also carry their (internal) energy from one gas to the other. In this sense, this geometry represents the simplest example of coupled transport. Making reference to the conditions described in Section 4.7.1, we can consider one of the two gases as the reservoir and the other as the thermodynamic system. The affinities (see Eq. (4.164)) in this simple example are

$$\mathbb{F}_U = \frac{1}{T} - \frac{1}{T_r} \tag{4.215}$$

$$\mathbb{F}_N = -\frac{\mu}{T} + \frac{\mu_r}{T_r}, \tag{4.216}$$

where T, μ and T_r, μ_r are the temperatures and chemical potentials in the system and in the reservoir, respectively. If U is the internal energy and N is the number of particles of the system, the corresponding fluxes are

$$J_U = \frac{dU}{dt} \quad \text{and} \quad J_N = \frac{dN}{dt} \tag{4.217}$$

and the phenomenological Equations (4.176) are

$$J_U = L_{UU}\,\mathbb{F}_U + L_{UN}\,\mathbb{F}_N, \tag{4.218}$$

$$J_N = L_{NU}\,\mathbb{F}_U + L_{NN}\,\mathbb{F}_N. \tag{4.219}$$

Since both observables are even with respect to the time reversal operator $\mathcal{T}$, $\tau_U = \tau_N = +1$, the following conditions hold:

$$L_{UN} = L_{NU}, \quad L_{UU} > 0, \quad L_{NN} > 0, \quad L_{UU}L_{NN} - L_{UN}^2 = \det L > 0. \tag{4.220}$$

The first condition comes from the symmetry of the Onsager matrix (see Eq. (4.175)), and the following inequalities derive from the condition

$$\frac{d\mathbb{S}}{dt} = \sum_{ij} L_{ij}\,\mathbb{F}_i\,\mathbb{F}_j \geq 0, \tag{4.221}$$

which must hold for any $\mathbb{F}_i$ and $\mathbb{F}_j$.[21]

If the gases in the two volumes are at the same temperature T (thermal equilibrium), $\mathbb{F}_U = 0$, and the energy flux is given by

$$J_U = L_{UN}\left(-\frac{\mu}{T} + \frac{\mu_r}{T}\right), \tag{4.222}$$

while the particle flux is

$$J_N = L_{NN}\left(-\frac{\mu}{T} + \frac{\mu_r}{T}\right), \tag{4.223}$$

thus yielding the relation

$$\frac{J_U}{J_N} = \frac{L_{UN}}{L_{NN}}. \tag{4.224}$$

In this situation we have coupled the transport of particles and heat by a purely mechanical effect even in thermal equilibrium conditions (mechanothermal effect).

Another interesting situation is when the stationary flux of particles vanishes $J_N = 0$, and we let just the energy flux survive. In this case the coupled transport yields a thermomechanical effect. The phenomenological relation (4.219) reduces to

$$\frac{\mathbb{F}_N}{\mathbb{F}_U} = -\frac{L_{NU}}{L_{NN}} = -\frac{L_{UN}}{L_{NN}}, \tag{4.225}$$

where the last equality is a consequence of the Onsager reciprocity relation (4.175). We can also conclude that the thermomechanical effect is characterized by the presence of a stationary energy current that is a combined effect of both affinities, because temperatures and chemical potentials are different both in the system and in the reservoir. In particular, we find that

$$J_U = \frac{\det L}{L_{NN}}\,\mathbb{F}_U. \tag{4.226}$$

---

[21] More precisely, if we choose $\mathbb{F}_N = 0$ or $\mathbb{F}_U = 0$, we find the positivity of the diagonal terms, and if we choose the minima $\frac{\mathbb{F}_U}{\mathbb{F}_N} = -\frac{L_{UN}}{L_{UU}}$ or $\frac{\mathbb{F}_U}{\mathbb{F}_N} = -\frac{L_{NN}}{L_{UN}}$, we find the positivity of the determinant.

Notice also that the ratio between the fluxes in the first case (i.e., the mechanothermal effect) is equal to minus the ratio of affinities in the second case (i.e., the thermomechanical effect). This shows that Onsager relations establish interesting phenomenological similarities among quite different physical phenomena.

Coupled Transport in Linear Continuous Systems

As discussed in Section 4.7.2, the linear approach to transport phenomena, yielding a proportionality between fluxes and affinities (see Eq. (4.176)), stems from phenomenological considerations. For instance, in many thermal conductors, it has been found experimentally that, in the absence of a matter current density, the energy current density $\mathbf{j}_u$ is proportional to the applied temperature gradient, according to Fourier's law

$$\mathbf{j}_u = -\kappa \, \nabla T(\mathbf{r}) = \kappa \, T^2 \, \nabla \left(\frac{1}{T} \right), \tag{4.227}$$

where κ is the thermal conductivity (see Section 1.2.2). For the sake of simplicity, in the examples hereafter discussed, we assume we are dealing with homogeneous isotropic systems, so that all phenomenological transport coefficients do not depend on the space directions. In fact, the last expression in Eq. (4.227) indicates that the energy current density $\mathbf{j}_u$ depends linearly on the gradient of the affinity, $\nabla(1/T)$ (see Section 4.7.4), with a proportionality coefficient κT^2, where κ is a scalar quantity. On a physical ground, it is worth stressing that Fourier's law is expected to hold for small temperature gradients, while a dependence on higher-order terms, such as $[\nabla(1/T)]^2$ and $[\nabla(1/T)]^3$, is expected to emerge for values of the applied temperature gradient sufficiently large to yield strong nonlinear effects.[22]

Another well-known example of a linear phenomenological law is Fick's law of diffusion, which tells us that, in the absence of an energy current density, the matter current density can be written as

$$\mathbf{j}_\rho = -D \, \nabla \rho(\mathbf{r}), \tag{4.228}$$

where D is the diffusion coefficient and ρ is the matter density field.[23] In a linear regime at fixed temperature T, ρ is proportional to the chemical potential (per unit volume) μ and Eq. (4.228) can be rewritten as

$$\mathbf{j}_\rho = D \, \nabla \left(\frac{-\mu(\mathbf{r})}{T} \right). \tag{4.229}$$

Also in this case we are facing linear conditions, which correspond to sufficiently moderate matter density gradients, in such a way that the matter current density is found to be proportional to the gradient of the corresponding affinity.

Now we want to consider the situation of a continuous thermodynamic system made of a single species of particles contained in a fixed volume V, where stationary energy and

---

[22] Deviations from the phenomenological Fourier's law have been predicted and experimentally observed in low-dimensional models, such as carbon nanotubes and graphene layers. In such cases, the deviations are due to specific anomalous diffusive behaviors (like the superdiffusive regime described in Section 3.5), which are typical of low-dimensional fluids and nanostructures in the absence of on-site potentials.

[23] It is a number density, not a mass density.

matter current densities are present at the same time. In this case, the Gibbs relation for intensive observables Eq. (4.204) specializes to

$$du = Tds + \mu d\rho,$$
(4.230)

where u, s, and ρ are the densities of internal energy, entropy, and number of particles, respectively. Taking into account the previous phenomenological considerations, we can write the linear coupled transport equations (4.176) in the form

$$\mathbf{j}_u = L_{uu}\,\nabla\!\left(\frac{1}{T}\right) + L_{u\rho}\,\nabla\!\left(\frac{-\mu}{T}\right)$$
(4.231)

$$\mathbf{j}_\rho = L_{\rho u}\,\nabla\!\left(\frac{1}{T}\right) + L_{\rho\rho}\,\nabla\!\left(\frac{-\mu}{T}\right),$$
(4.232)

where $\mathbf{j}_u$ and $\mathbf{j}_\rho$ are the current densities. Note that, even if not explicitly indicated, all the physical quantities appearing in these equations are functions of the coordinate vector $\mathbf{r}$. The symmetric structure of Onsager matrix of linear coefficients (see Eq. (4.175)) yields the relation $L_{u\rho} = L_{\rho u}$, while, according to Eq. (4.214), the affinities are

$$\mathbf{F}_u = \nabla\!\left(\frac{1}{T}\right)$$

$$\mathbf{F}_\rho = \nabla\!\left(\frac{-\mu}{T}\right).$$
(4.233)

The entropy density production rate in the system is given by the expression

$$\frac{ds}{dt} = \mathbf{j}_u \cdot \nabla\!\left(\frac{1}{T}\right) + \mathbf{j}_\rho \cdot \nabla\!\left(\frac{-\mu}{T}\right),$$
(4.234)

which exemplifies Eqs. (4.213) and (4.214) for an out-of-equilibrium continuous system in the presence of irreversible processes, produced by thermodynamic forces (in the form of applied gradients) determining energy and mass current densities flowing through the system.

We want to point out that Eq. (4.234) is obtained as a consequence of the continuity equations,

$$\frac{\partial u}{\partial t} + \nabla \cdot \mathbf{j}_u = 0$$
(4.235)

$$\frac{\partial \rho}{\partial t} + \nabla \cdot \mathbf{j}_\rho = 0,$$
(4.236)

which express the conservation of energy and matter densities in the system. In general, stationary conditions imply divergence-free currents; that is, in this case we have $\nabla \cdot \mathbf{j}_u = \nabla \cdot \mathbf{j}_\rho = 0$. Notice that this is not the case for the entropy density current, because stationary conditions are associated with finite currents and with a finite entropy production rate. Combining Eq. (4.234) and the Onsager relations (4.231) and (4.232), we can also write

$$\frac{ds}{dt} = \sum_{\alpha,\beta} L_{\alpha\beta}\mathbf{F}_\alpha \cdot \mathbf{F}_\beta,$$
(4.237)

with indices α and β running over the subscripts u and ρ. The latter relation is the analog of Eq. (4.190) in the case of a continuous system.

4.8.2 Onsager Theorem and Transport of Charged Particles

Another phenomenological example of a linear transport phenomenon is Ohm's law, where the electric current (i.e., the flux) is found to be proportional to the electric potential difference (i.e., the affinity) applied at the ends of a standard electric conductor (e.g., a copper or a silver wire), whose ohmic resistance (the inverse of the transport coefficient named conductivity) is the proportionality constant.

On the other hand, when dealing with transport phenomena, including electromagnetic effects the Onsager reciprocity relation (4.175) has to be reconsidered in the light of Onsager theorem, which can be formulated as

$$L_{ij}(\mathbf{B}) = L_{ji}(-\mathbf{B}), \tag{4.238}$$

where we have indicated explicitly only the dependence of the linear kinetic coefficients on the applied external magnetic field $\mathbf{B}$, although they have to be intended as functions of the other intensive parameters also. This theorem states that the value of the linear kinetic coefficient L_{ij} measured in the presence of an applied external magnetic field $\mathbf{B}$ is equal to the value of the linear kinetic coefficient L_{ji} measured in the reversed magnetic field $-\mathbf{B}$. Said differently, in the presence of an external magnetic field $\mathbf{B}$, the Onsager reciprocity relation is maintained by changing the sign of $\mathbf{B}$. We do not report here the proof of this theorem, which is related to the time reversal properties of the magnetic field.

In what follows we illustrate the Onsager theory in the case of the transport of charged particles, in particular electrons in metals or semiconductor materials. This theory allows us to explain many interesting coupled transport phenomena that are quite common in our everyday life, for example, when we experience the cooling in a refrigerator or the heat produced by a lamp or by the electronic circuits of a computer. In particular, we first deal with thermoelectric effects in one-dimensional conductors. This analysis exemplifies how the Onsager theory can be used to establish relations between phenomenological coefficients as a result of the relations between Onsager kinetic coefficients. The situation is just more involved, but it works the same way when the same approach is applied to the study of thermomagnetic effects in three-dimensional conductors. This is the second set of cases that we discuss hereafter, as they are the standard framework for the application of the Onsager theorem, concerning the effects associated with the presence of an external magnetic field on charged particle currents.

Thermoelectric Effects

These physical phenomena concern the coupled transport of electric and heat currents in the absence of a magnetic field. They result from the mutual interference of heat flow and electric current. We can cast them in the language of the Onsager theory by considering, for the sake of simplicity, a conductor where the electric current and the heat current flow in one dimension and the electric current is carried by electrons only, as it happens in a metal wire.[24] We adopt here the description of the conductor as a continuous system and we specialize the Gibbs relation (4.204) to

$$ds = \frac{1}{T}\, du - \frac{\mu}{T} dn, \tag{4.239}$$

where T is the temperature, s is the local entropy density, u is the local energy density, n is the number of electrons per unit volume, and, accordingly, μ is the electrochemical potential per particle (not per mole, as usual). In principle, we should add to the right-hand side the contribution of other components, $- \sum_i (\mu_i/T)dn_i$, like the atomic nuclei, that form the structure of the conductor. However, their role in thermoelectric effects is immaterial, because in a linear regime, components other than electrons do not contribute to the electric current flowing through the conductor. Therefore, Eq. (4.206) in this case becomes

$$j_s = \frac{1}{T}\, j_u - \frac{\mu}{T}\, j_n, \tag{4.240}$$

where j_s, j_u, and j_n are the one-dimensional current densities of entropy, energy, and number of electrons, respectively. Since we are dealing with a continuous system, we can also write the entropy density production rate (see Eq. (4.213)) as

$$\frac{ds}{dt} = \partial_x \left(\frac{1}{T}\right) j_u - \partial_x \left(\frac{\mu}{T}\right) j_n, \tag{4.241}$$

where we have denoted with x the direction along the one-dimensional conductor.

Given the negative sign of the elementary charge, it is useful to consider the (positive) flux $-j_n$, which can be used, along with j_u, to write the linear coupled transport Equations (4.176) in the form

$$- j_n = L'_{nn}\, \partial_x \frac{\mu}{T} + L'_{nu} \partial_x \frac{1}{T} \tag{4.242}$$

$$j_u = L'_{un}\, \partial_x \frac{\mu}{T} + L'_{uu} \partial_x \frac{1}{T}, \tag{4.243}$$

where the use of the prime symbol for the Onsager coefficients is made clearer by the following considerations. In fact, for physical reasons, it is better to describe thermoelectric effects by replacing the current density of the total internal energy with the current density of heat. Making reference to the thermodynamic relation $dQ = T\, dS$, we can define a heat current density as

$$j_q = T\, j_s, \tag{4.244}$$

where we maintain the lowercase subscripts to denote the intensive quantities. Using Eq. (4.240), we can write

$$j_q = j_u - \mu\, j_n. \tag{4.245}$$

This result can be interpreted by considering that μ amounts to the potential energy per particle, so that $\mu\, j_n$ represents the potential energy current density. In this sense, subtracting

---

[24] Despite these limiting hypotheses, the theory, in principle, allows one to deal with problems where charge is transported also by other kinds of particles, such as protons and charged nuclei in plasmas or ions in chemical reactions.

μj_n from the total energy current density allows us to think of the heat current density j_q as a sort of kinetic energy current density. We can now rewrite Eq. (4.241) in the form

$$\frac{ds}{dt} = \left(\partial_x \frac{1}{T}\right) j_q - \frac{1}{T}\left(\partial_x \mu\right) j_n, \tag{4.246}$$

so that, in the heat representation, the affinities associated with j_q and $-j_n$ become $\partial_x(1/T)$ and $(1/T)\partial_x \mu$, respectively. Therefore, the linear coupled transport Equations (4.176) for the heat current and the number of electrons current can be rewritten as

$$- j_n = L_{nn} \frac{1}{T} \partial_x \mu + L_{nq} \partial_x \frac{1}{T} \tag{4.247}$$

$$j_q = L_{qn} \frac{1}{T} \partial_x \mu + L_{qq} \partial_x \frac{1}{T}. \tag{4.248}$$

These relations replace Eqs. (4.242) and (4.243) in the heat representation.

Now, we want to find a physical interpretation of the earlier linear kinetic coefficients L_{ij}. First, we observe that $L_{nq} = L_{qn}$, because no external magnetic field is applied. Then, we remark that the electrochemical potential μ is made of two contributions,

$$\mu = \mu_e + \mu_c, \tag{4.249}$$

which correspond to the electrical (μ_e) and chemical (μ_c) contributions to the potential, respectively. In particular, we can write $\mu_e = e\,\Phi$, where e is the (negative) electron charge and Φ is the electrostatic potential. On the other hand, μ_c is a function of T and of the electronic concentration. In other words, the electrochemical potential per unit charge $(1/e)\mu$ is such that its gradient $(1/e)\partial_x \mu$ is the sum of the electric field $(1/e)\partial_x \mu_e$ and of an effective driving force $(1/e)\partial_x \mu_c$ associated with the presence of a concentration gradient of electrons. If we assume that the conductor is made of a homogeneous isothermal material, $\partial_x \mu_c = 0$ and $\partial_x \mu = \partial_x \mu_e$, while the electric conductivity σ is defined as the electric current density $e j_n$ per unit potential gradient $(1/e)\partial_x \mu$, that is, the electromotive force. Thus, for $\partial_x T = 0$ (isothermal condition),

$$\sigma = -\frac{e^2 j_n}{\partial_x \mu}. \tag{4.250}$$

Eq. (4.247), for spatially constant T, yields the relation

$$L_{nn} = \frac{\sigma T}{e^2}, \tag{4.251}$$

which allows us to attribute an explicit physical interpretation to the Onsager coefficient L_{nn} in the case of thermoelectric transport.

The heat conductivity, on its side, when $j_n = 0$, that is, no electric current flows through the conductor, is given by Fourier's law (see Eq. (4.227)),

$$\kappa = -\frac{j_q}{\partial_x T}. \tag{4.252}$$

Using the condition $j_n = 0$, from Eq. (4.247), it is possible to deduce the equation

$$\partial_x \mu = -T \frac{L_{nq}}{L_{nn}} \partial_x \frac{1}{T} = \frac{L_{nq}}{T L_{nn}} \partial_x T \tag{4.253}$$

that, once replaced in (4.248), gives the result

$$\kappa = \frac{\det L}{T^2 \, L_{nn}}. \tag{4.254}$$

This equation, Eq. (4.251), and the symmetry relation $L_{nq} = L_{qn}$, put together, imply that only one of the Onsager coefficients remains to be determined. In the examples that we discuss hereafter, we describe, in specific thermoelectric geometries, how we can find an additional, physical relation allowing us to determine all the Onsager coefficients.

The Seebeck Effect

This effect was discovered by the Estonian physicist Thomas Johannes Seebeck in 1821. He observed that a circuit made by two different conductors, whose junctions are taken at different temperatures, produced an electric current that was signaled by the deflection of a magnetic needle. Later the Danish scientist Hans Christian Ørsted realized that such a setup points out the manifestation of a thermoelectric effect: The difference of temperature at the junctions and the different electronic mobility in the two materials produces an effective *thermoelectric power*, which amounts to an electromotive force (emf) making a current flow through the circuit. In order to interpret this phenomenon according to the Onsager kinetic theory, we can adopt a different setup, that is, a thermocouple (see Fig. 4.2): The junctions connecting the two conductors A and B (e.g., one copper and one iron wire or a couple of differently doped semiconductors) are kept at different temperatures, $T_1 < T_2$, and a voltmeter is inserted along one of the two arms of the thermocouple, allowing for the passage of the heat flow, but not of the electric current. In practice, the voltmeter measures the emf, which is produced by the heat flow, while $j_n = 0$. We assume that the voltmeter is at a point along the arm at some temperature intermediate between T_1 and T_2 and that it is sufficiently small and made by a good heat conductor, in such a way as to not perturb significantly the heat flow.

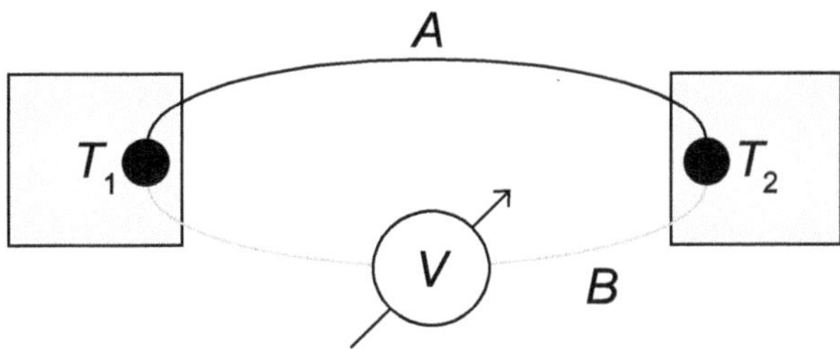

 Experimental apparatus to observe the Seebeck effect. Two different conducting (or semiconducting) wires, A and B, are joined at their ends (black dots), which are kept at different temperatures T_1, T_2. This temperature gradient induces a difference of electric potential (electromotive force) at the inputs of the voltmeter V.

Since $j_n = 0$, we can use Eq. (4.253) to determine the relation between $\partial_x \mu$ and $\partial_x T$, which can be written more transparently as

$$d\mu = \frac{L_{nq}}{T\,L_{nn}}\,dT, \tag{4.255}$$

that is, directly linking the variation of temperature with the variation of chemical potential. This equation allows us to compute the difference of electrochemical potential between the junctions along the two arms of the thermocouple as

$$\mu_2 - \mu_1 = \int_1^2 \frac{L_{nq}^A}{T\,L_{nn}^A}\,dT \tag{4.256}$$

$$\mu_2 - \mu_r = \int_r^2 \frac{L_{nq}^B}{T\,L_{nn}^B}\,dT \tag{4.257}$$

$$\mu_\ell - \mu_1 = \int_1^\ell \frac{L_{nq}^B}{T\,L_{nn}^B}\,dT, \tag{4.258}$$

where μ_r and μ_ℓ are the values of the electrochemical potential measured at the right and left electric contacts of the voltmeter, respectively. The previous equations yield the relation

$$\mu_r - \mu_\ell = \int_1^2 \left(\frac{L_{nq}^A}{T\,L_{nn}^A} - \frac{L_{nq}^B}{T\,L_{nn}^B} \right) dT. \tag{4.259}$$

We have assumed that the voltmeter is at a fixed value of the local temperature and the voltage that it measures is given by the expression

$$V = \frac{1}{e}(\mu_r - \mu_\ell) = \int_1^2 \left(\frac{L_{nq}^A}{e\,T\,L_{nn}^A} - \frac{L_{nq}^B}{e\,T\,L_{nn}^B} \right) dT. \tag{4.260}$$

The thermoelectric power of a thermocouple, ε_{AB}, is defined as the increment of voltage per unit temperature difference, while its sign is conventionally said to be positive if the increment of voltage drives the current from A to B at the hot junction, T_2. In practice, if $T_1 = T$ and $T_2 = T + \Delta T$, in the limit of small ΔT

$$\varepsilon_{AB} = \frac{V}{\Delta T} \equiv \varepsilon_B - \varepsilon_A, \tag{4.261}$$

where

$$\varepsilon_X = -\frac{L_{nq}^X}{e\,T\,L_{nn}^X}, \qquad X = A, B, \tag{4.262}$$

defines the absolute thermoelectric power of a single electric conductor.

As announced after Eq. (4.254), this is the first example of an additional relation between Onsager coefficients and a physical, measurable quantity, the thermoelectric power of a conductor. We can now express all coefficients in terms of σ, κ, and ε: L_{nn} is given by Eq. (4.251), then $L_{nq} = L_{qn} = -(T^2 \sigma \varepsilon)/e$ are derivable from Eq. (4.262), and finally $L_{qq} = (T^3 \sigma \varepsilon^2 + T^2 \kappa)$ can be found by Eq. (4.254). Inserting these expressions in Eqs. (4.247) to

(4.248), we obtain

$$- j_n = \left(\frac{\sigma}{e^2}\right)\partial_x \mu - \left(\frac{T^2 \sigma \varepsilon}{e}\right)\partial_x \frac{1}{T} \tag{4.263}$$

$$j_q = -\left(\frac{T\sigma \varepsilon}{e}\right)\partial_x \mu + (T^3 \sigma \varepsilon^2 + T^2 \kappa)\partial_x \frac{1}{T}. \tag{4.264}$$

By multiplying Eq. (4.263) by the factor $T\varepsilon e$ and summing it to Eq. (4.264), we find the relation

$$j_q = T \varepsilon e \, j_n + T^2 \kappa \, \partial_x \frac{1}{T}. \tag{4.265}$$

Recalling that $j_s = j_q/T$, we finally obtain

$$j_s = \varepsilon e \, j_n + T \kappa \, \partial_x \frac{1}{T}. \tag{4.266}$$

This formula tells us that each electron contributes to the entropy current as a charge carrier (first term on the right-hand side) and as a heat carrier (second term on the right-hand side). In particular, its contribution as a charge carrier amounts to an entropy εe, which allows us to interpret the thermoelectric power as the entropy transported per unit charge by the electronic current.

The Peltier Effect

The Peltier effect was discovered in 1834 by the French physicist Jean Peltier. He observed that when a current flows through the junction between two different conductors, heat is exchanged between the junction and the environment; see Fig. 4.3. One can easily realize that such a device may work as a heater or as a cooler, according to the direction of the electric current. The Peltier effect is therefore different from the Joule effect in two (related) respects: It is due to the presence of a junction and it may produce heating as well as cooling. The Peltier effect is usually described as the reverse of the Seebeck effect: In the former, a heat current is induced by an electric current; in the latter, it is the opposite.

The junction between the conductors A and B is assumed to be kept in isothermal conditions. The Onsager theory predicts that a certain amount of heat has to be supplied

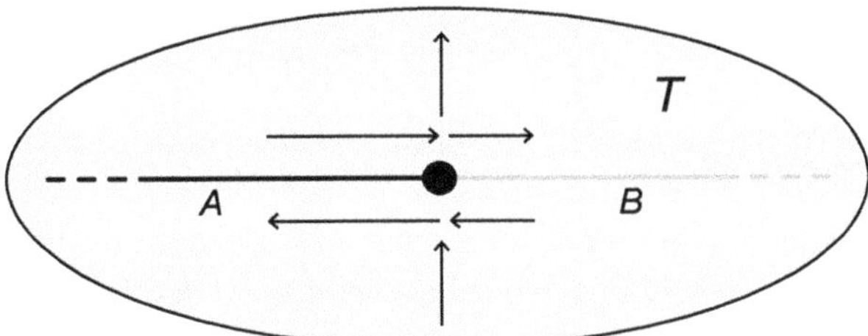

Fig. 4.3   The diagram of the Peltier effect. A junction (black dot) of two different conductors A and B, when kept at constant temperature T, must exchange heat with the reservoir in order to sustain a constant electric current. The horizontal arrows correspond to j_q^A and j_q^B: Their difference amounts to the heat exchanged with the environment at temperature T, which may be either positive or negative; see the two cases above and below the junction.

or removed from the junction in order to keep the electronic current stationary. The total energy current is discontinuous through the junction. In fact, in each conductor it is given by the expression (see Eq. (4.245))

$$j_u = j_q + \mu\, j_n. \tag{4.267}$$

On the other hand, at the isothermal junction, μ and j_n must be continuous in order to have a stationary current, so that the discontinuity in j_u amounts to

$$j_u^A - j_u^B = j_q^A - j_q^B. \tag{4.268}$$

The condition of maintaining the setup at constant temperature implies that no thermal gradient is present and Eqs. (4.263) and (4.264) simplify to

$$j_n = -\left(\frac{\sigma}{e^2}\right)\partial_x \mu \tag{4.269}$$

$$j_q = -\left(\frac{T\sigma\,\varepsilon}{e}\right)\partial_x \mu, \tag{4.270}$$

which can be combined to obtain the relation

$$j_q = T\,\varepsilon\,(e\,j_n). \tag{4.271}$$

From this equation we can write

$$j_q^B - j_q^A = T(\varepsilon_B - \varepsilon_A)(e\,j_n), \tag{4.272}$$

thus showing that the discontinuity of the heat current at the junction is due to the different thermoelectric power of the conductors and is proportional to the electric current, $e\,j_n$. The proportionality constant defines the Peltier coefficient of the junction,

$$\Pi_{AB} = \frac{(j_q^B - j_q^A)}{e\,j_n} = T(\varepsilon_B - \varepsilon_A), \tag{4.273}$$

which can be interpreted as the heat to be supplied to the junction per unit of electric current flowing from A to B. Equation (4.273) proves the relation between the Peltier and the Seebeck effects, showing that the two effects are simply proportional to one another.

The Thomson–Joule Effect

The Thomson–Joule effect appears when an electric current flows through a conductor with an applied temperature gradient. We consider first a conductor through which a heat current flows in the absence of an electric one. Because of the applied temperature gradient, the temperature field $T(x)$ along the conductor is determined by the dependence on temperature of its kinetic coefficients. We can now perform an idealized setup, where the conductor is put in contact at each point x with a heat reservoir at temperature $T(x)$. In such conditions, there is no heat exchanged between the conductor and the reservoirs. Then, we switch on a stationary electric current j_n flowing through the conductor, thus producing a heat exchange with the reservoirs, in such a way that any variation of the energy current through the conductor has to be supplied by the reservoirs. Before switching on the electric current, we know that the energy current must be conserved, that is, $\partial_x j_u = 0$. After the electric current is switched on, according to Eq. (4.267), we can write

$$\partial_x j_u = \partial_x j_q + j_n \partial_x \mu, \tag{4.274}$$

because j_n is a constant current. Using Eqs. (4.265) and (4.263), we can rewrite this equation as

$$\partial_x j_u = \partial_x \left(T \varepsilon e j_n + T^2 \kappa \partial_x \frac{1}{T} \right) + j_n \left(-\frac{e^2}{\sigma} j_n + T^2 \varepsilon e \partial_x \frac{1}{T} \right). \tag{4.275}$$

In the adopted setup, the only quantities[25] that depend on the space coordinate x are T and ε, and we can simplify Eq. (4.275), obtaining

$$\partial_x j_u = T \left(\partial_x \varepsilon \right) e j_n - \kappa \partial_{xx} T - \frac{e^2}{\sigma} j_n^2. \tag{4.276}$$

Since in the absence of an electric current, that is, $j_n = 0$, we must have $\partial_x j_u = 0$; the temperature field must be such that the second addendum on the right-hand side must vanish, that is,

$$\kappa \partial_{xx} T = 0. \tag{4.277}$$

This means that the temperature field $T(x)$ is expected to exhibit a linear dependence on the space coordinate x, consistent with Fourier's law. Thus, assuming as a first approximation that the temperature profile does not change when $j_n \neq 0$, we can simplify Eq. (4.276) as

$$\partial_x j_u = T \left(\partial_x \varepsilon \right) e j_n - \frac{e^2}{\sigma} j_n^2. \tag{4.278}$$

On the other hand, the thermoelectric power depends on x because it is a function of temperature, so we can write

$$\partial_x \varepsilon = \frac{d\varepsilon}{dT} \partial_x T, \tag{4.279}$$

from which

$$\partial_x j_u = T \frac{d\varepsilon}{dT} \partial_x T e j_n - \frac{e^2}{\sigma} j_n^2. \tag{4.280}$$

The second term on the right-hand side is the so-called Joule heat, which is produced even in the absence of a temperature gradient. The first term is the so-called Thomson heat, which has to be absorbed by the reservoirs to maintain the temperature gradient, $\partial_x T$, when the electric current flows through the conductor. We can define the Thomson coefficient, τ, as the amount of Thomson heat absorbed per unit electric current $(e j_n)$ and per unit temperature gradient $(\partial_x T)$, obtaining

$$\tau = T \frac{d\varepsilon}{dT}. \tag{4.281}$$

---

[25] In principle κ should also depend on T and, accordingly, on x, but for sufficiently small temperature gradients, it can be assumed to be constant.

Making use of this definition and of Eq. (4.273), we can establish a relation between the Peltier and Thomson coefficients with the thermoelectric power,

$$\frac{d\,\Pi_{AB}}{d\,T} = (\tau_B - \tau_A) + (\varepsilon_B - \varepsilon_A),\tag{4.282}$$

which can be interpreted as a consequence of the energy conservation. In fact, the thermoelectric power of a junction is the result of the contributions of the heat per unit temperature and per unit electric current supplied to the junction by the Peltier and by the Thomson effects.

Thermomagnetic and Galvanomagnetic Effects

We now consider a three-dimensional conductor in which heat current and electric current can flow simultaneously in the presence of an external magnetic field $\mathbf{B}$. For the sake of simplicity, we assume that $\mathbf{B}$ is directed along the z-axis, while currents and affinities (i.e., gradients) depend only on the coordinates x and y. In this configuration we are assuming a planar symmetry, meaning that the observed effects are independent of z.[26] The basic formulas are essentially the same employed for studying the thermoelectric effects, extended to the case $d = 2$. Eqs. (4.240), (4.244), and (4.245) become

$$\mathbf{j}_s = \frac{1}{T}\mathbf{j}_u - \frac{\mu}{T}\mathbf{j}_n\tag{4.283}$$

$$\mathbf{j}_q = T\mathbf{j}_s\tag{4.284}$$

$$\mathbf{j}_q = \mathbf{j}_u - \mu\mathbf{j}_n,\tag{4.285}$$

respectively, while Eq. (4.246) turns into

$$\begin{aligned}
\frac{ds}{dt} &= \nabla\frac{1}{T}\cdot\mathbf{j}_q - \frac{1}{T}\nabla\mu\cdot\mathbf{j}_n\\
&= \partial_x\frac{1}{T}\,(j_q)_x - \frac{1}{T}\partial_x\mu\,(j_n)_x + \partial_y\frac{1}{T}\,(j_q)_y - \frac{1}{T}\partial_y\mu\,(j_n)_y.
\end{aligned}\tag{4.286}$$

Following the same procedure adopted for thermoelectric effects, we can write the set of equations analogous to Eqs. (4.247) and (4.248) in the form

$$-(j_n)_x = L_{11}\frac{1}{T}\partial_x\mu + L_{12}\partial_x\frac{1}{T} + L_{13}\frac{1}{T}\partial_y\mu + L_{14}\partial_y\frac{1}{T}\tag{4.287}$$

$$(j_q)_x = L_{21}\frac{1}{T}\partial_x\mu + L_{22}\partial_x\frac{1}{T} + L_{23}\frac{1}{T}\partial_y\mu + L_{24}\partial_y\frac{1}{T}\tag{4.288}$$

$$-(j_n)_y = L_{31}\frac{1}{T}\partial_x\mu + L_{32}\partial_x\frac{1}{T} + L_{33}\frac{1}{T}\partial_y\mu + L_{34}\partial_y\frac{1}{T}\tag{4.289}$$

$$(j_q)_y = L_{41}\frac{1}{T}\partial_x\mu + L_{42}\partial_x\frac{1}{T} + L_{43}\frac{1}{T}\partial_y\mu + L_{44}\partial_y\frac{1}{T}\,.\tag{4.290}$$

Space isotropy in the (x, y)-plane implies, in particular, the invariance under a $(\pi/2)$-rotation, corresponding to the transformation $x \to y$ and $y \to -x$. Considering that under

---

[26] If the system extends over a finite size along z, we assume that surface corrections are negligible.

this transformation, $(j_{n,q})_x \to (j_{n,q})_y$, this invariance allows us to find the following relations among coefficients:

$$L_{31} = -L_{13} \quad L_{32} = -L_{14} \quad L_{33} = L_{11} \quad L_{34} = L_{12}$$
$$L_{41} = -L_{23} \quad L_{42} = -L_{24} \quad L_{43} = L_{21} \quad L_{44} = L_{22}.$$

We now consider a reflection of the x-axis, which must be accompanied by the inversion of the magnetic field. Invariance under this transformation implies that L_{11}, L_{12}, L_{21}, and L_{22} are even functions of $\mathbf{B}$, while L_{13}, L_{14}, L_{23}, and L_{24} are odd functions of $\mathbf{B}$. Finally, using the general relations (4.238) between Onsager coefficients when we reverse the sign of $\mathbf{B}$, we find the additional relations $L_{23} = L_{14}$ and $L_{21} = L_{12}$. In conclusion, the previous set of equations can be rewritten in such a way as to contain only six independent kinetic coefficients:

$$- (j_n)_x = L_{11} \frac{1}{T} \partial_x \mu + L_{12} \partial_x \frac{1}{T} + L_{13} \frac{1}{T} \partial_y \mu + L_{14} \partial_y \frac{1}{T} \tag{4.291}$$

$$(j_q)_x = L_{12} \frac{1}{T} \partial_x \mu + L_{22} \partial_x \frac{1}{T} + L_{14} \frac{1}{T} \partial_y \mu + L_{24} \partial_y \frac{1}{T} \tag{4.292}$$

$$-(j_n)_y = -L_{13} \frac{1}{T} \partial_x \mu - L_{14} \partial_x \frac{1}{T} + L_{11} \frac{1}{T} \partial_y \mu + L_{12} \partial_y \frac{1}{T} \tag{4.293}$$

$$(j_q)_y = -L_{14} \frac{1}{T} \partial_x \mu - L_{24} \partial_x \frac{1}{T} + L_{12} \frac{1}{T} \partial_y \mu + L_{22} \partial_y \frac{1}{T} . \tag{4.294}$$

With this result at hand, we can make use of the definitions of the phenomenological transport coefficients already introduced to describe the thermoelectric effects to provide an analogous description of magnetoelectric and galvanometric effects.

Isothermal and Adiabatic Electric Conductivities

For the sake of simplicity, we assume that an electric current flows in the x-direction, that is, $(j_n)_x \neq 0$ and $(j_n)_y = 0$. The electric conductivity of the material is defined by the relation

$$\sigma = -\frac{e^2 \, (j_n)_x}{\partial_x \mu}. \tag{4.295}$$

The isothermal transport conditions correspond to the absence of thermal gradients, that is, $\partial_x T = \partial_y T = 0$. We can use Eq. (4.293) to find $\partial_y \mu = (L_{13}/L_{11})\partial_x \mu$, then replace in Eq. (4.291) to obtain an expression of the isothermal conductivity σ_{I} in terms of the Onsager kinetic coefficients,

$$\sigma_{\mathrm{I}} = \frac{e^2}{T \, L_{11}} \left(L_{11}^2 + L_{13}^2 \right). \tag{4.296}$$

The adiabatic transport conditions correspond to a situation where a gradient of temperature is allowed only along the y-direction, transverse to the current $(j_n)_x$, while no transverse current can flow, that is, $\partial_x T = (j_n)_y = (j_q)_y = 0$. Now we need three equations, (4.291), (4.293), and (4.294), for expressing the adiabatic electric conductivity in terms of the Onsager kinetic coefficients, thus obtaining[27]

$$\sigma_{\mathrm{A}} = \frac{e^2 D_3}{T D_2},$$

(4.297)

where D_2 and D_3 are the determinants of the reduced matrices:

$$D_2 = \begin{vmatrix} L_{11} & L_{12} \\ L_{12} & L_{22} \end{vmatrix}, \qquad D_3 = \begin{vmatrix} L_{11} & L_{13} & L_{14} \\ -L_{13} & L_{11} & L_{12} \\ -L_{14} & L_{12} & L_{22} \end{vmatrix}.$$

(4.298)

Isothermal and Adiabatic Heat Conductivities

Similar to what we did for the electric conductivities, we now assume that a thermal gradient is present only in the x-direction, that is, $\partial_x T \neq 0$ and $\partial_y T = 0$. The heat conductivity of the material is defined by the relation

$$\kappa = -\frac{(j_q)_x}{\partial_x T}.$$

(4.299)

In this case, the isothermal transport conditions correspond to the absence of electric currents, namely, $(j_n)_x = (j_n)_y = 0$. In this case, we have to use Eqs. (4.291) to (4.293) for expressing the isothermal heat conductivity in terms of the Onsager kinetic ones, thus obtaining

$$\kappa_{\mathrm{I}} = \frac{1}{T^2} \frac{D_3}{L_{11}^2 + L_{13}^2}.$$

(4.300)

The appearance of the determinant D_3, the same introduced in the expression of the adiabatic electric conductivity Eq. (4.297), at the numerator of κ_I is again a manifestation of the symmetries of the Onsager matrix.

The conditions for adiabatic heat transport correspond to the absence of electronic current and to the confinement of the heat current to the x-direction, namely, $(j_n)_x = (j_n)_y = (j_q)_y = 0$. Imposing these conditions in Eqs. (4.291), (4.293), and (4.294), the fourth equation (4.292) yields

$$\kappa_{\mathrm{A}} = \frac{D_4}{T^2 D_3},$$

(4.301)

where

$$D_4 = \begin{vmatrix} L_{11} & L_{12} & L_{13} & L_{14} \\ L_{12} & L_{22} & L_{14} & L_{24} \\ -L_{13} & -L_{14} & L_{11} & L_{12} \\ -L_{14} & -L_{24} & L_{12} & L_{22} \end{vmatrix}$$

(4.302)

is the determinant of the complete matrix of Onsager kinetic coefficients.

---

[27] The mathematics behind these calculations is the standard procedure of solving linear systems of equations.

Isothermal Hall and Nernst Effects

This case corresponds to a situation where the presence of the external magnetic field $\mathbf{B}$ in the z-direction yields an electric current flowing along the x-direction, while a transverse emf is applied along the y-direction. The effect is essentially due to the Lorentz force acting on electrons (or, more generally, on charge carriers). The Hall coefficient is defined as

$$R_\mathrm{H} = \frac{\partial_y \mu}{e^2 |\mathbf{B}| \, (j_n)_x}. \tag{4.303}$$

The isothermal conditions correspond to a null transverse electric current and to the absence of thermal gradients, that is, $(j_n)_y = \partial_x T = \partial_y T = 0$. An expression of R_H as a function of Onsager kinetic coefficients can be obtained using Eqs. (4.291) and (4.293),

$$R_\mathrm{H} = -\frac{T}{e^2 \, |\mathbf{B}|} \frac{L_{11}}{L_{11}^2 + L_{13}^2}. \tag{4.304}$$

The adiabatic case corresponds to $\partial_y T \neq 0$ and we have no way to compute a similar expression for the Hall coefficient from Onsager relations. In fact, the presence of a transverse thermal gradient does not allow us to identify the expression $(1/e)\partial_y \mu$ as the actual emf acting on charge carriers, and the problem lacks a unique solution.

We face a similar situation with the Nernst effect, which amounts to producing a thermal gradient along the x-direction, while an emf is applied in the y-direction in the absence of electronic currents, that is, $(j_n)_x = (j_n)_y = 0$. The Nernst coefficient is defined as

$$N = -\frac{\partial_y \mu}{e|\mathbf{B}|\partial_x T}. \tag{4.305}$$

The isothermal condition refers just to the absence of a temperature gradient along the y-direction, that is, $\partial_y T = 0$: Making use again of Eqs. (4.291) and (4.293), we can express the isothermal Nernst coefficient in terms of the Onsager kinetic coefficients as

$$N_\mathrm{I} = \frac{1}{e|\mathbf{B}|T} \frac{L_{11}L_{14} - L_{12}L_{13}}{L_{11}^2 + L_{13}^2}. \tag{4.306}$$

Ettingshausen and Righi–Leduc Effects

In this case we deal with a situation where both currents in the y-direction are absent, that is, $(j_n)_y = (j_q)_y = 0$, while in the x-direction there is no thermal gradient, $\partial_x T = 0$ (Ettingshausen effect), or no electric current, $(j_n)_x = 0$ (Righi–Leduc effect). The Ettingshausen coefficient is defined as

$$P = \frac{\partial_y T}{e|\mathbf{B}|(j_n)_x}. \tag{4.307}$$

and making use of Eqs. (4.291), (4.293), and (4.294), it can be expressed as

$$P = \frac{T^2}{e|\mathbf{B}|} \frac{L_{11}L_{14} - L_{12}L_{13}}{D_3}. \tag{4.308}$$

The Righi–Leduc coefficient is defined as

$$A = \frac{\partial_y T}{|\mathbf{B}| \partial_x T} \tag{4.309}$$

and making use again of Eqs. (4.291), (4.293), and (4.294), it can be expressed as

$$A = \frac{D_3'}{|\mathbf{B}| D_3}, \tag{4.310}$$

where

$$D_3' = \begin{vmatrix} L_{11} & L_{13} & -L_{12} \\ -L_{13} & L_{11} & L_{14} \\ -L_{14} & L_{12} & L_{24} \end{vmatrix}. \tag{4.311}$$

Summarizing, we have obtained eight different phenomenological coefficients as a function of six independent Onsager kinetic coefficients. This means that there must be two relations among the phenomenological coefficients. One, pointed out by the American physicist Percy Williams Bridgman, is

$$T\, N_1 = \kappa_1 P, \tag{4.312}$$

which can be immediately checked. Obtaining the second relation is definitely more involved because of the presence of third- and fourth-order powers of the Onsager kinetic coefficients. It is important to point out that, as a consequence of the parity properties with respect to $\mathbf{B}$ of the Onsager coefficients, all the phenomenological coefficients of the various thermomagnetic and galvanomagnetic effects are even functions of $\mathbf{B}$, as we can argue by the physical consideration that the measurement of these coefficients cannot depend on the switching from $\mathbf{B}$ to $-\mathbf{B}$.

4.9 Bibliographic Notes

A classical textbook about linear response theory, the fluctuation–dissipation theorem, and their applications to nonequilibrium statistical mechanics is R. Kubo, M. Toda, and N. Hashitsume, *Statistical Physics II: Nonequilibrium Statistical Mechanics*, 3rd ed. (Springer, 1998). In particular, this book contains a rigorous extension of the formalism of stochastic differential equations to macroscopic observables by the so-called projection formalism.

A classical textbook on nonequilibrium thermodynamics containing a detailed presentation of the Onsager theory of transport processes is S. R. de Groot and P. Mazur, *Non-Equilibrium Thermodynamics* (Dover, 1984).

A pedagogical introduction to entropy is the book by D. S. Lemons, *A Student's Guide to Entropy* (Cambridge University Press, 2013).

An interesting and modern book about nonequilibrium statistical mechanics of turbulence, transport processes, and reaction–diffusion processes is J. Cardy, G. Falkovich, and K. Gawedzki, *Non-equilibrium Statistical Mechanics and Turbulence* (Cambridge University Press, 2008). This book contains the lectures given by the authors at a summer school. This notwithstanding, these lectures concern frontier topics in nonequilibrium statistical mechanics that may not be immediately accessible to undergraduate students.

A comprehensive account on anomalous transport phenomena in low-dimensional systems can be found in the book by S. Lepri (ed.), *Thermal Transport in Low Dimensions, Lecture Notes in Physics* (Springer, 2016). Some of the contributions contained in this book illustrate experimental results.

A mathematical textbook about analytic functions and their Fourier and Laplace transforms, together with functional analysis, is W. Rudin, *Functional Analysis* (McGraw-Hill, 1991).

Many interesting aspects concerning the applications of linear response theory and Kubo formalism to linear irreversible thermodynamics are contained in the book by D. J. Evans and G. P. Morris, *Statistical Mechanics of Nonequilibrium Liquids*, 2nd ed. (ANU Press, 2007). This book also contains an instructive part concerning the use of computer algorithms useful for performing numerical simulations of fluids.

A short survey about models of thermal reservoirs useful for numerical simulations is discussed in the review paper by S. Lepri, R. Livi, and A. Politi, Thermal Conduction in Classical Low-dimensional Lattices, *Physics Reports*, **377** (2003) 1–80.

5

From Equilibrium to Out-of-Equilibrium Phase Transitions: Driven Lattice Gases

This chapter is a bridge between equilibrium and nonequilibrium phase transitions. They have in common several basic concepts, such as control parameters, order parameters, and critical exponents, but they also have in common tools and methods, such as phenomenological scaling theory, mean-field theory, and renormalization group. Rather than taking their knowledge for granted, after a brief introduction to the importance of phase transitions, we dedicate the long Section 5.2 to provide the reader a not-so-short guide to equilibrium phase transitions, where all of these concepts and methods are illustrated, making reference to well-known models of physical phenomena, such as the liquid–gas and the ferro-paramagnetic equilibrium phase transitions. We want to point out that this reference overview is also introductory to Chapter 6, where we discuss another class of nonequilibrium phase transitions.

For pedagogical reasons, we have decided to approach nonequilibrium phase transitions by considering first the lattice gas "interpretation" of the Ising model, then driving it out of equilibrium through the application of a force favoring the hopping of particles in a given direction, therefore obtaining what is called a bulk-driven system. This is the so-called Katz–Lebowitz–Spohn (KLS) model, studied in Section 5.3, which shows the importance of boundary conditions. The effect of boundaries is analyzed in Section 5.4, where we consider the case of periodic boundary conditions and the case where the system is also boundary-driven. With periodic boundary conditions, the system is out of equilibrium because hopping is asymmetric, but the statistics of the steady state is the same as the equilibrium statistics when hopping is symmetric. If we attach the system to two reservoirs of particles, see Section 5.4, we obtain a much more interesting out-of-equilibrium system, whose dynamical state depends on the details of the reservoirs and which may be characterized by a low density of particles, by a high density of particles, or by a maximization of the current of particles.

The physical scenario can become even more interesting if we consider the combining effect of two different types of particles, flowing in opposite directions. If boundary conditions are symmetric (same injection rate, same extraction rate), we can ask ourselves if a spontaneous symmetry breaking can occur between the two classes of particles. We perform this analysis in Section 5.5, focusing on the high-density phase, which is more promising to find a symmetry breaking. In fact, such phase transition does occur, showing that in nonequilibrium conditions they can exist even in one spatial dimension.

5.1 The Importance of Phase Transitions

Phase transitions are physical phenomena that are present in our everyday life: liquid water cools into ice and water vapor condenses on the surface of a cold glass; mixtures of various components separate into different phases; a piece of magnetized iron loses its magnetization by heating; traffic flow collapses into a jam; and so on. Phase transitions provide us with the main example of how matter can go through changes of state and properties under the action of varying physical conditions. It is astonishing that phase transitions occur at the atomic and molecular levels as well as at cosmological scales. Moreover, independently of the actual physical scales, they exhibit universal properties, which stem from the basic symmetries contained in the interaction rules between the constituents of matter.

All of these features have been widely investigated in many different realms of experimental science over the last centuries. Statistical mechanics has provided us with the theoretical framework by which these phenomena can be interpreted and predicted. This is certainly one of its major breakthroughs, although we have to point out that our present knowledge about equilibrium phase transitions relies on a much more solid ground than the one we have about nonequilibrium ones. We can say that many concepts and methods that have been worked out for describing equilibrium phase transitions can be borrowed and adapted to nonequilibrium phase transitions, relying on the power of analogies in the physical world.

Anyway, in general this is not enough, because there are some crucial differences between equilibrium and nonequilibrium phenomena. For instance, at equilibrium we can deal with statistical properties associated with static conditions, while in out-of-equilibrium conditions we have to deal with an explicit dynamical description, where time enters as a basic ingredient and takes the form of an additional dimension: if in equilibrium statistical physics spatial fluctuations are important and let mean-field theory fail in low dimension, in out-of-equilibrium systems we must also consider temporal fluctuations. Moreover, nonequilibrium phase transitions occur between different stationary states, when tuning a suitable control parameter, such as the strength of a driving field or the gradient intensity applied at the boundaries of the system. Finally, another important difference is that the free energy, which has a very relevant role at equilibrium, generally has no out-of-equilibrium counterpart. This is why a standard thermodynamic interpretation cannot apply to nonequilibrium phase transitions.

This chapter is focused on the driven lattice gas, that is, a lattice gas that is in contact with different reservoirs and whose dynamics is forced to favor the hopping in a given direction: In this case, steady states are characterized by the presence of macroscopic currents, which is obviously prevented at equilibrium. Instead, in Chapter 6, we will study systems that are intrinsically out of equilibrium, not because we are imposing a drift with an external drive. Before passing to study nonequilibrium phase transitions, we are going to briefly explain the main concepts of equilibrium phase transitions.

5.2 Basic Concepts and Tools of Equilibrium Phase Transitions

Phase transitions in physical systems at thermodynamic equilibrium occur because some symmetry of the system at hand breaks when a control parameter, for example, temperature, pressure, or volume, is varied.

In first-order phase transitions, this mechanism is associated with the exchange of a finite amount of energy (heat) with the environment, as happens when passing from a gas to a liquid phase for $T < T_c$ (see the following section for the van der Waals model). In this case one observes a discontinuity in the fluid density $\rho = N/V$, when the pressure P or the temperature T are varied by an infinitesimal amount. The heat supplied by the environment when passing from the liquid to the gas phase is what is needed to overtake the interaction energy between the atoms in the liquid phase.

In continuous, that is, second-order, phase transitions, such as the one occurring along the critical isothermal line for a van der Waals gas, or as happens in the Ising model of a ferromagnet in the absence of a magnetic field (see Section 5.2.2), the symmetry breaks spontaneously, because of microscopic fluctuations. These continuous phase transitions are characterized by the presence of long-range correlations at the transition point, which impose the typical scale-invariant structure of critical behavior. The main consequence is the existence of critical exponents, associated with the power-law behavior of physical quantities, which may either vanish or diverge at the critical point. Dimensional analysis as well as renormalization group methods have usually been employed to predict these exponents approximately and, if possible, exactly.

5.2.1 Phase Transitions and Thermodynamics

In 1873 the Dutch scientist Johannes Diderik van der Waals realized that the changes of state observed experimentally in real gases could not be accounted for by the equation of state of the ideal gas,

$$PV = NRT, \tag{5.1}$$

where the product of pressure P and volume V is proportional to the temperature T and to the number of moles N contained in the gas, with a proportionality constant equal to the so-called constant of gases, $R = 8.314 \text{ J mol}^{-1} \text{ K}^{-1}$. The earlier equation can also be written as

$$Pv = T, \tag{5.2}$$

where $v = V/N$ is the volume per molecule.

In fact, van der Waals proposed to modify Eq. (5.2) by taking into account two basic facts. At variance with the ideal gas model, molecules in a real gas are not "point-like" particles but occupy a finite volume. One can easily realize that this is a crucial feature to be taken into account when the gas is compressed to make it change to a liquid or a solid. Accordingly, one should subtract from the volume of the container V a phenomenological quantity b per particle (the so-called covolume), so that $V - Nb$ represents the effective volume available to the molecules.

Moreover, molecules in a real gas are subject to an attractive interaction, which reduces the pressure exerted on the walls of the volume containing the real gas. By combining these two heuristic ingredients (see Appendix P for more details), van der Waals obtained the following equation of state:

$$P = \frac{T}{v - b} - \frac{a}{v^2},\tag{5.3}$$

which is characterized by a critical value $T_c = 8a/(27b)$ below which isothermal curves $P(v)$ are no longer a decreasing function of the volume. At $T = T_c$, there is an inflection point that allows us to define a critical volume (per particle) $v_c = 3b$ and a critical pressure $P_c = P(v_c, T_c) = a/(27b^2)$. If we rescale thermodynamical variables with respect to their critical values, we obtain the parameter-free equation of state given in Eq. (P.12),

$$P^* = \frac{8}{3} \frac{T^*}{v^* - \frac{1}{3}} - \frac{3}{(v^*)^2}.\tag{5.4}$$

Despite its heuristic character, Eq. (5.3) has the main merit of pointing out the concept of universality. In fact, any van der Waals gas, independent of its atomic or molecular constituents (i.e., independent of the specific values of a and b), is found to obey the same equation of state given in Eq. (5.4) and plotted for different values of T^* in Fig. 5.1. For $T^* > 1$ (upper curve), we are in the gas phase and $P(v)$ is a monotonously decreasing function. At $T^* = 1$ (the critical temperature, middle curve) comes an inflection point, and for $T^* < 1$ (lower curve and inset) is a region where P would increase with v. This unphysical piece of isothermal curve should be replaced by a horizontal line according to the Maxwell construction and joining the two phases in equilibrium, the liquid A and the gas B. The reason why the van der Waals equation cannot account for this process is that it cannot describe a nonhomogeneous system.

The passage from state A to state B along one of the isobaric and isothermal segments corresponds to a first-order phase transition, where a finite amount of heat has to be supplied to the fluid. Since temperature and pressure remain constant, the supplied heat serves to break the interaction energy between particles in the liquid phase.[1] At $T = T_c$ the phase transition from the liquid to the gas phase occurs continuously, without heat exchange with the environment. According to the classification of the Austrian Paul Ehrenfest, this scenario corresponds to a second-order phase transition.

The van der Waals equation points out another distinctive feature of equilibrium second-order phase transitions, that is, the singular behavior of thermodynamic observables at the critical temperature T_c. For instance, the isothermal compressibility χ of a van der Waals gas at the critical temperature diverges as

$$\chi = -\left(\frac{\partial V}{\partial P}\right)_T \sim |T - T_c|^{-\gamma},\tag{5.5}$$

---

[1] In the reversed transformation, $B \to A$, the same amount of (latent) heat is returned to the environment. This is what happens with the condensation process that comes before a rainfall.

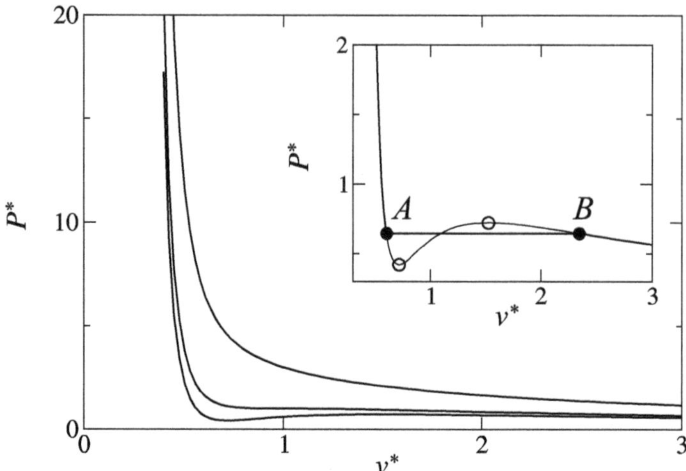

Fig. 5.1 Plot of the rescaled van der Waals Equation (5.4), for $T^* = 1.5$ (gas phase, upper curve), $T^* = 1$ (critical point, middle curve), and $T^* = 0.9$ (phase coexistence, lower curve). In the inset we focus on $T^* < 1$, where there is coexistence between the liquid phase A and the gas phase B. In between A and B, the equilibrium curve is the horizontal segment. The curved portions between solid and open circles correspond to metastable states, and the one between open circles corresponds to unstable states (with the pressure unphysically increasing with the volume at constant temperature).

where γ is the corresponding critical exponent. This result can be easily obtained by differentiating Eq. (5.4) with respect to P^* while keeping T^* constant, thus obtaining

$$-\left(\frac{\partial v^*}{\partial P^*}\right)_{T^*} = \left(\frac{8T^*}{3\left(v^* - \frac{1}{3}\right)^2} - \frac{6}{(v^*)^3}\right)^{-1}. \tag{5.6}$$

For $v = v_c$ (i.e., $v^* = 1$), the above equation simplifies to

$$-\left(\frac{\partial v^*}{\partial P^*}\right)_{T^*} = \frac{1}{6}(T^* - 1)^{-1}, \tag{5.7}$$

which gives[2] (see Eq. (5.5)) $\gamma = 1$. We should also point out that experimental measurements yield $\gamma \simeq 1.25$.

The singular behavior at the critical point of various observables other than the isothermal compressibility is the typical signature of critical phenomena. For instance, the specific heat at constant volume of real gases, c_V, has been found to diverge at T_c as

$$c_V \sim |T - T_c|^{-\alpha}, \tag{5.8}$$

where the experimental estimate of the critical exponent α is close to 0.12. Also in this case, the van der Waals theory fails, since it predicts just a discontinuity of c_V at $T = T_c$ (see Appendix P). These discrepancies are a consequence of the mean-field nature of the van der Waals phenomenological equation, which underestimates the strongly correlated

---

[2] Coming back to unreduced variables, we get $\chi = -\left(\frac{\partial V}{\partial P}\right)_T = \frac{N v_c T_c}{6 P_c}(T - T_c)^{-1} = 4b^2(T - T_c)^{-1}.$

nature of microscopic fluctuations at the critical point. This scenario can be understood in the light of the Landau theory of critical phenomena (see Section 5.2.3).

5.2.2 Phase Transitions and Statistical Mechanics

A microscopic theory of critical phenomena associated with second-order phase transitions has been successfully worked out in equilibrium statistical mechanics. This has been made possible by studying simple models that share with many physical systems two basic universal features, namely, symmetry and dimensionality. For instance, binary mixtures and lattice gas models have been found to be equivalent to the Ising model of ferromagnetism. Hereafter we report a short description of the latter model, which can be considered as the test bed for the main achievements in this field.

The Ising model of ferromagnetism is represented by the Hamiltonian

$$\mathcal{H}_I[\{\sigma_i\}] = -\sum_{\langle ij \rangle} J_{ij}\, \sigma_i\, \sigma_j - \sum_i H_i\, \sigma_i. \tag{5.9}$$

The first sum is performed over all distinct and unordered pairs of nearest-neighbor sites of a regular lattice, where the discrete spin-like variables $\sigma_i = \pm 1$ are located. The interaction coupling constant J_{ij}, called the exchange constant, is positive in order to favor spin alignment for minimizing the local energy, while H_i is a local magnetic field favoring the alignment of σ_i with its sign. If we assume homogeneity of the physical parameters, we can deal with the simplified version of the Ising model where $J_{ij} = J$ and $H_i = H$, independent of the lattice sites i and j:

$$\mathcal{H}_I[\{\sigma_i\}] = -J\sum_{\langle ij \rangle} \sigma_i\, \sigma_j - H\sum_i \sigma_i. \tag{5.10}$$

In the absence of the external magnetic field, that is, $H = 0$, the Ising Hamiltonian is symmetric with respect to the global "spin-flip" transformation $\sigma_i \rightarrow -\sigma_i$ (i.e., it exhibits the discrete $\mathbb{Z}_2$ symmetry).

In equilibrium statistical mechanics, we assume that the system is in contact with a thermostat at temperature T and that the probability $p(\{\sigma_i\})$ of a spin configuration $\{\sigma_i\}$ is given by the expression

$$p(\{\sigma_i\}) = \frac{1}{Z} \exp\left(-\frac{\mathcal{H}_I[\{\sigma_i\}]}{T}\right), \tag{5.11}$$

where

$$Z(V,T,H) = \sum_{\{\sigma_i\}} \exp\left(-\frac{\mathcal{H}_I[\{\sigma_i\}]}{T}\right) \tag{5.12}$$

is the partition function, obtained by summing the so-called Boltzmann weights over the collection of all possible spin configurations. The relation with thermodynamics can be established by the following equation:

$$F(V,T,H) = -T \ln Z, \tag{5.13}$$

where $F(V, T, H)$ is the thermodynamic free energy, which depends on the volume of the system V,[3] on the temperature T and on the external magnetic field H.

Equation (5.13) can be justified by rewriting it in the form

$$\sum_{\{\sigma_i\}} \exp\left(\frac{F(V, T, H) - \mathcal{H}_I[\{\sigma_i\}]}{T}\right) = 1 \tag{5.14}$$

and by differentiating both sides with respect to T, thus yielding the relation

$$\sum_{\{\sigma_i\}} \exp\left(\frac{F(V, T, H) - \mathcal{H}_I[\{\sigma_i\}]}{T}\right)\left[-\frac{1}{T^2}\left(F - \mathcal{H}_I[\{\sigma_i\}] - T\frac{\partial F}{\partial T}\right)\right] = 0. \tag{5.15}$$

Since the equilibrium value of any observable $O[\{\sigma_i\}]$ is given by the expression

$$\langle O \rangle = \frac{1}{Z} \sum_{\{\sigma_i\}} O[\{\sigma_i\}] \exp\left(-\frac{\mathcal{H}_I[\{\sigma_i\}]}{T}\right), \tag{5.16}$$

multiplying both sides of Eq. (5.15) by $-T^2$, we obtain

$$F - \langle \mathcal{H}_I \rangle - T\frac{\partial F}{\partial T} = F - U - T\frac{\partial F}{\partial T} = 0, \tag{5.17}$$

where $U \equiv \langle \mathcal{H}_I \rangle$ is the internal energy and the last equation can be written in the form

$$F = U - TS, \tag{5.18}$$

which is the thermodynamic definition of the free energy F, where $S = -\frac{\partial F}{\partial T}$ is the entropy.[4]

The ferromagnetic phase transition characterizing the homogeneous Ising model (5.10) can be described by studying the dependence of the magnetization M on the temperature T. These quantities are usually called the order parameter and the control parameter, respectively. While T is fixed by the thermostat, M is defined as the equilibrium average of the sum of all spin variables over the lattice volume V, namely,

$$M(V, T, H) = \left\langle \sum_i \sigma_i \right\rangle$$

$$= \frac{1}{Z} \sum_{\{\sigma_i\}} \left(\sum_i \sigma_i\right) \exp\left(-\frac{\mathcal{H}_I[\{\sigma_i\}]}{T}\right) \tag{5.19}$$

$$= T\frac{\partial \ln Z}{\partial H}.$$

In the absence of the external magnetic field, that is, $H = 0$, and for[5] $d \geq 2$, a qualitative description is summarized by the phase diagram shown in Fig. 5.2. In the high-temperature phase (point A) M vanishes, because entropic fluctuations cancel the ordering effect of

---

[3] This dependence stems from the summation over all spin configuration of a system that occupies a volume V, which is proportional to the total number of spins.

[4] Therefore, the relation (5.13) between the free energy and the partition function has been justified by showing that the well-known thermodynamic relation (5.18) can be derived from it. Alternatively, we might start from Eq. (5.18) and derive Eq. (5.13).

[5] For $d = 1$, a non vanishing magnetization only appears at $T = 0$; see Appendix H.1.

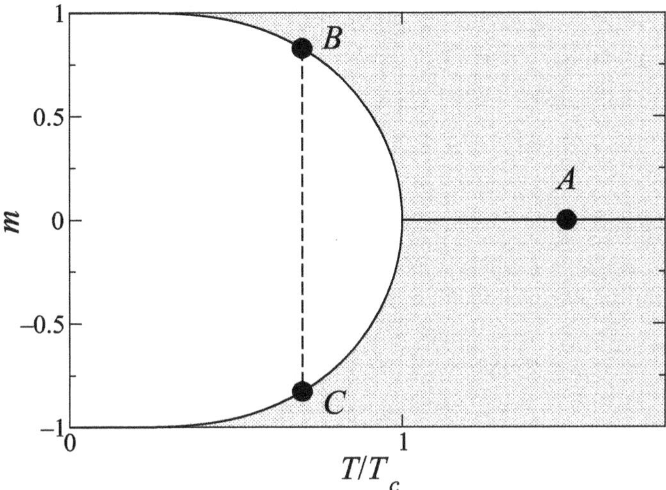

Fig. 5.2 Schematic plot of the temperature dependence of the order parameter density, $m = M/V$, in an Ising-like system. A, B, and C are equilibrium states at zero magnetic field. A is a disordered state at $T > T_c$, while B, C are a pair of equivalent, ordered states at $T < T_c$. The shaded region corresponds to equilibrium states for $H \neq 0$.

the ferromagnetic coupling J. By cooling the system through the critical temperature T_c, it acquires a spontaneous magnetization, either positive or negative, along one of the two branches bifurcating at T_c. This phenomenon occurs, because the magnetic susceptibility of the Ising ferromagnet,

$$\chi(V,T) = \frac{1}{V} \lim_{H \to 0} \frac{\partial M(V,T,H)}{\partial H}, \tag{5.20}$$

diverges at T_c, so that even a random microscopic fluctuation is able to produce a macroscopic effect. In a statistical perspective, we can say that if we repeat the cooling procedure several times through T_c, positive or negative spontaneous magnetization is eventually observed with equal probability, as in a toss-a-coin process. Let us assume that one cooling procedure has selected a positively magnetized state B, and now, we want to perform a thermodynamic transformation making the system pass from B, to its symmetric, negatively magnetized state C. Such a thermodynamic transformation is a first-order phase transition, because it can be performed by switching on a negative external magnetic field, that has to supply a sufficient amount of magnetic energy to make a macroscopic portion of spins flip. If, after flipping the magnetization, the external field H is switched off, we obtain state C. In general terms, the shaded region corresponds to equilibrium states in the presence of a nonvanishing field, either positive ($M > 0$) or negative ($M < 0$). The white region, corresponding to nonequilibrium states, will be studied in Chapter 8.

In order to obtain a quantitatively rigorous description of the phase diagram of the Ising ferromagnetic transition, one should compute explicitly the partition function (5.12), which is equivalent to computing the free energy $F(V,T,H)$ of the Ising model. A detailed account of this problem goes beyond the aim of this book; here we just mention that this calculation is almost straightforward for $d = 1$ (see Appendix H.1), while for $d = 2$ the solution was found by Onsager in the last century, making use of refined mathematical methods based on the transfer matrix approach. The exact solution in $d = 3$ is still unknown, while in the limit $d \to \infty$ it can be shown that a mean-field approach (see Section 5.2.3) allows one to compute $F(V, T, H)$. Numerical studies confirm that the qualitative scenario shown in Fig. 5.2 holds for $d \geq 2$. From a quantitative point of view, we should distinguish between (irrelevant) properties that depend on the microscopic details, such as the value of T_c, and (relevant) properties that depend on general features as the space dimension d, such as the critical exponents (see Section 5.2.4).

Before concluding this section, we want to comment on the crucial role played by the thermodynamic limit in the theory of equilibrium phase transitions. In fact, a thermodynamic system is assumed to be made up of an extremely large number of particles, such as the Avogadro number of atoms or molecules contained in a mole of a gas. This physical assumption can be translated into a rigorous mathematical formulation by a precise definition: The thermodynamic limit is obtained by making the number of microscopic variables N (e.g., the number of spins of the Ising model) go to infinity, while their density $\rho = N/V$ is kept constant.

We have mentioned earlier that the magnetic susceptibility $\chi(V, T)$ defined in Eq. (5.20) diverges at T_c. This indicates the presence of a singular (nonanalytic) behavior of χ, that can be traced back to M and Z through Eqs. (5.19) and (5.20). On the other hand, Z is defined as a sum of analytic functions (see Eq. (5.12)), and a nonanalyticity in one of its derivatives, such as χ, may appear only if this sum is made by an infinite number of nonvanishing contributions. This simple mathematical argument shows that, in order to reproduce in the theory the diverging behavior of χ at T_c, it is necessary to sum Z over an infinite volume V (i.e., over an infinite number of spin configurations), while keeping ρ constant. Since a rigorous mathematical theory of phase transitions has to take into account from the very beginning the thermodynamic limit as a basic ingredient, it is more appropriate to deal with intensive quantities rather than extensive ones. In particular, we have to redefine the magnetic susceptibility as

$$\chi(\rho, T) = \lim_{H \to 0} \frac{\partial m(\rho, T, H)}{\partial H}, \tag{5.21}$$

where

$$m(\rho, T, H) = \lim_{V \to \infty} \frac{M(V, T, H)}{V} \tag{5.22}$$

is the magnetization density. Consistently, we can define the free energy density as

$$f(\rho, T, H) = \lim_{V \to \infty} \frac{F(V, T, H)}{V}. \tag{5.23}$$

5.2.3 Landau Theory of Critical Phenomena

A general phenomenological theory of phase transitions was proposed in the last century by the Soviet Lev Davidovich Landau. The basic idea amounts to describing a phase transition close to the critical point by introducing an effective free energy density $f(m, T, H)$ that depends on the temperature T (or any equivalent thermodynamic quantity playing the role of a control parameter), on a coarse-grained order parameter density $m(\mathbf{x})$, and on its conjugate field $H(\mathbf{x})$, where $\mathbf{x}$ is the position vector in a space of dimension d. The attribute "coarse-grained" means that $m(\mathbf{x})$ is averaged over a suitable length scale, in such a way that it is insensitive to fluctuations on the atomic scale. For instance, in a magnetic system such as those described by the Ising model (see Section 5.2.2), $m(\mathbf{x})$ is the magnetization density and $H(\mathbf{x})$ is an external magnetic field, which may be dependent on position.

The basic assumption of the Landau theory is that, close to the transition point denoted by the critical value of the temperature T_c, $m(\mathbf{x})$ is small and $f(m, T, H)$ can be expanded at the lowest orders in powers of $m(\mathbf{x})$ and of its derivatives as

$$f(m, T, H) = \frac{1}{2}|\nabla m(\mathbf{x})|^2 - H(\mathbf{x})\, m(\mathbf{x}) + \frac{1}{2}a\, m^2(x) + b\, m^3(\mathbf{x}) + u\, m^4(\mathbf{x}) + \cdots . \quad (5.24)$$

In Appendix Q we provide an explicit derivation of the exact form of $f(m, T, 0)$ from an Ising-like Hamiltonian. In the spirit of the Landau approach, the earlier expression can be thought of as a general, small-m and small-H expansion, which is valid close to the critical point, which corresponds to $T = T_c$ and $H = 0$.

The first term on the right-hand side (whose prefactor can be set equal to $\frac{1}{2}$ by a suitable rescaling of the energy; see Eq. (Q.18)) accounts for spatial variations of the order parameter, which are not negligible at the critical point. The second term accounts for a linear coupling with $H(\mathbf{x})$, and the following terms represent a perturbative expansion with respect to $m(\mathbf{x})$, with a, b, and u being phenomenological parameters. In particular (see Appendix Q), a is assumed to be proportional to the so-called reduced temperature,

$$a = a_0\, t, \quad \text{with} \quad t = \frac{(T - T_c)}{T_c}, \quad (5.25)$$

where a_0 is a positive constant and also $u > 0$ in order to guarantee a meaningful free energy, because the thermodynamic equilibrium states correspond to the absolute minima of the free energy density $f(m, T, H)$. As for b, it should vanish in systems invariant under change of sign of $m(\mathbf{x})$, as in the case of the ferromagnetic Ising model, because we require that $f(-m, T, -H) = f(m, T, H)$. This is no longer true, and $b \neq 0$, if the system does not hold the Ising symmetry, as occurs in the first-order van der Waals transition; see Appendix P.

The homogeneous mean-field solution of the Landau theory of critical phenomena is obtained by neglecting spatial fluctuations, so we deal with the mean-field free energy density,

$$f(m, T, H) = \frac{1}{2}am^2 + um^4 - H\, m. \quad (5.26)$$

The thermodynamic equilibrium value of m is obtained by solving the equation $\partial f / \partial m = 0$, that is,

$$a\, m + 4u\, m^3 - H = 0, \quad (5.27)$$

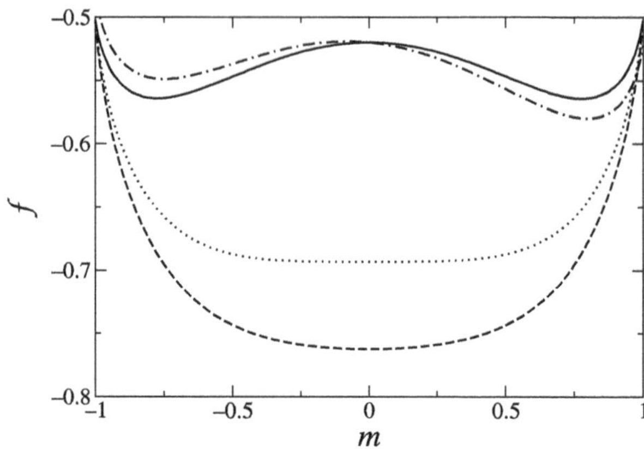

Fig. 5.3 The Ginzburg–Landau free energy (5.26) for $H = 0$, above T_c (dashed line), at T_c (dotted line), and below T_c (solid line). A field $H > 0$ modifies the double well potential as indicated by the dot-dashed line.

and by identifying the absolute minimum of $f(m, T, H)$. In the case of $H = 0$, the absolute minima of $f(m, T, H)$ are

$$m = \begin{cases} 0, & \text{for } T > T_c \\ \pm \left(\frac{a_0}{4u}\right)^{\frac{1}{2}} |t|^{\frac{1}{2}}, & \text{for } T < T_c. \end{cases} \tag{5.28}$$

Accordingly, the mean-field theory predicts that the magnetization vanishes when T approaches T_c from below as $m \sim |t|^\beta$, with $\beta = 1/2$. See Fig. 5.3 for a pictorial representation of $f(m, T, 0)$ above, at, and below the critical temperature T_c. This result provides a posteriori explanation of the definition (5.25) of the phenomenological parameter a. In fact, for $T > T_c$, a is positive and the minimum of $f(m, T, 0)$ is at $m = 0$, while for $T < T_c$, a is negative and the minima of $f(m, T, 0)$ correspond to a finite magnetization.

Beyond the magnetization, other thermodynamic observables exhibit power-law behavior close to or at the critical point. In particular, the equilibrium values of the magnetic susceptibility χ and of the specific heat at constant volume c_V are expected to diverge at T_c as

$$\chi = \left.\frac{\partial m}{\partial H}\right|_{H=0} \sim |t|^{-\gamma} \tag{5.29}$$

$$c_V = -T \left.\frac{\partial^2 f}{\partial T^2}\right|_{H=0} \sim |t|^{-\alpha}, \tag{5.30}$$

where both critical exponents α and γ are positive numbers. An explicit expression of the equilibrium value of χ in the mean-field approximation can be obtained by differentiating Eq. (5.27),

$$(a + 12\, u\, m^2)dm = dH, \tag{5.31}$$

and making use of Eqs. (5.25) and (5.28), thus yielding the expression

$$\chi = \begin{cases} \frac{1}{a_0}|t|^{-1}, & \text{for } T > T_c \\ \frac{1}{2a_0}|t|^{-1}, & \text{for } T < T_c \end{cases} \tag{5.32}$$

so that $\gamma = 1$. In order to obtain c_V, we can write the equilibrium value of the mean-field free energy for $H = 0$ by combining Eqs. (5.26), (5.27), and (5.28):

$$f(m, T, 0) = \begin{cases} 0, & \text{for } T > T_c \\ -\left(\frac{a_0^2}{16u}\right)t^2, & \text{for } T < T_c. \end{cases} \tag{5.33}$$

By applying definition (5.30), one obtains

$$c_V = \begin{cases} 0, & \text{for } T > T_c \\ \frac{T_c a_0^2}{8u}, & \text{for } T < T_c, \end{cases} \tag{5.34}$$

which predicts a finite discontinuity of c_V at T_c, so that $\alpha = 0$.

At the critical point $T = T_c$ (i.e., $a = 0$), the magnetization m is expected to exhibit a power-law dependence on the external magnetic field H of the form $m \sim H^{1/\delta}$. From Eq. (5.27), we obtain

$$m(T_c, H) = \left(\frac{H}{4u}\right)^{\frac{1}{3}}, \tag{5.35}$$

thus yielding $\delta = 3$.

The mean-field solution provides an approximate description of critical phenomena, since it completely disregards the role of fluctuations, which are known to play an important role close to the critical point, as mentioned in Section 5.2.2. We can take into account fluctuations by studying the connected autocorrelation function,

$$C(\mathbf{x}) = \langle m(\mathbf{x})m(0)\rangle - \langle m(\mathbf{x})\rangle\langle m(0)\rangle = \langle m(\mathbf{x})m(0)\rangle - \langle m(0)\rangle^2, \tag{5.36}$$

where the brackets $\langle \cdots \rangle$ denote the equilibrium average, and the second equality stems from the assumption of translation invariance. For simplicity, let us consider the case $H = 0$ and $T > T_c$, that is, $\langle m(0)\rangle = 0$. By introducing the Fourier transform of $m(\mathbf{x})$

$$m(\mathbf{k}) = \int d^d\mathbf{x}\, e^{-i\mathbf{k}\cdot\mathbf{x}}\, m(\mathbf{x}) \tag{5.37}$$

the Fourier transform of $C(\mathbf{x})$ takes the simple form

$$C(\mathbf{k}) = \frac{1}{(2\pi)^d}\langle |m(\mathbf{k})|^2\rangle. \tag{5.38}$$

We can now determine $C(\mathbf{k})$, because $|m(\mathbf{k})|^2$ is related to the quadratic part of the free energy so that its average value, which appears in Eq. (5.38), can be easily evaluated using the equipartition theorem. In the paramagnetic region and if $H = 0$, we can consider only the quadratic part of the free energy,

$$f(m(\mathbf{x}), T, 0) \approx \frac{1}{2}|\nabla m(\mathbf{x})|^2 + \frac{1}{2}a\, m^2(\mathbf{x}), \tag{5.39}$$

whose space integral can be evaluated in the Fourier space,

$$
\begin{aligned}
\int d\mathbf{x} f(m(\mathbf{x}), T, 0) &= \frac{1}{2} \int d\mathbf{x} \frac{1}{(2\pi)^{2d}} \int d\mathbf{k}_1 \int d\mathbf{k}_2 (a - \mathbf{k}_1 \cdot \mathbf{k}_2) e^{i\mathbf{k}_1 \cdot \mathbf{x}} e^{i\mathbf{k}_2 \cdot \mathbf{x}} m(\mathbf{k}_1) m(\mathbf{k}_2) \\
&= \frac{1}{2} \int d\mathbf{k}_1 \int d\mathbf{k}_2 (a - \mathbf{k}_1 \cdot \mathbf{k}_2) \frac{1}{(2\pi)^d} m(\mathbf{k}_1) m(\mathbf{k}_2) \delta(\mathbf{k}_1 + \mathbf{k}_2) \\
&= \frac{1}{2} \frac{1}{(2\pi)^d} \int d\mathbf{k} (a + \mathbf{k}^2) |m(\mathbf{k})|^2 \\
&\equiv \int d\mathbf{k} f(m(\mathbf{k}), T, 0).
\end{aligned}
$$

In the first line earlier we have expressed $m(\mathbf{x})$ and $\nabla m(\mathbf{x})$ in terms of $m(\mathbf{k})$; in the second line we have spatially integrated the exponents, obtaining a Dirac delta; and in the third line we have used it to pass to a single integral in the $\mathbf{k}$-space.

We can now use the equipartition theorem, writing

$$
T = \langle f(m(\mathbf{k}), T, 0) \rangle = \frac{1}{2} \frac{1}{(2\pi)^d} (\mathbf{k}^2 + a) \langle |m(\mathbf{k})|^2 \rangle \tag{5.40}
$$

and finally obtaining

$$
C(\mathbf{k}) = \frac{1}{(2\pi)^d} \langle |m(\mathbf{k})|^2 \rangle = \frac{2T}{\mathbf{k}^2 + a}. \tag{5.41}
$$

By antitransforming this expression, we can write the connected autocorrelation function in the so-called Ornstein–Zernike form,

$$
C(\mathbf{x}) \sim |\mathbf{x}|^{2-d} e^{-|\mathbf{x}|/\xi}, \tag{5.42}
$$

where $\xi = a^{-1/2}$ is the correlation length, that is, the range of fluctuation correlations for $T > T_c$. By recalling Eq. (5.25), we can see that ξ diverges for $T \to T_c^+$ as

$$
\xi \sim t^{-\nu}, \quad \text{with} \quad \nu = \frac{1}{2}. \tag{5.43}
$$

We conclude this part by evaluating the relevance of fluctuations, which have been disregarded in the previous mean-field approach. We can assess it through the ratio of the spatial fluctuations of the order parameter over a distance ξ (the only relevant length scale, if the system is close to criticality) and the average value of the order parameter itself. This means we are tacitly assuming to be in the ordered region, $T < T_c$. We must therefore consider the ratio

$$
\left. \frac{C(\xi)}{m^2} \right|_{T \to T_c^-} = \frac{\xi^{2-d}}{\frac{a_0}{4u} |t|}, \tag{5.44}
$$

where the numerator has been determined by Eq. (5.42) and the denominator by Eq. (5.28). Since $\xi = (a_0 |t|)^{-1/2}$, this ratio can be simplified as

$$
\left. \frac{C(\xi)}{m^2} \right|_{T \to T_c^-} \sim 4u (a_0 |t|)^{(d-4)/2}, \tag{5.45}
$$

and the requirement that fluctuations are negligible in the limit $t \to 0$ is equivalent to saying that such a ratio is much smaller than one, which is true, according to Eq. (5.45), if $d > 4$.

This is known as the Ginzburg criterion, which points out the crucial role played by space dimension d in critical phenomena, identifying the upper critical dimension $d_c^u = 4$. In fact, we can say that for $d > d_c^u$, the mean-field solution and its critical exponent are correct,[6] while for $d < d_c^u$, the mean-field solution fails. Along with d_c^u, it is possible to define a *lower* critical dimension, d_c^l, below which thermal fluctuations are so important that an ordered phase can exist only in their absence, that is, at $T = 0$. For the Ising universality class, $d_c^l = 1$.

5.2.4 Critical Exponents and Scaling Hypothesis

In Section 5.2.3, we have sketched the main elements of the Landau theory of phase transitions, which is particularly useful for describing critical phenomena. On the other hand, this theory is based on many approximations, which are unable to capture some crucial aspects of critical phenomena associated with the peculiar role of dimensionality and fluctuations. For instance, the critical exponents predicted by the Landau theory, called classical exponents, are different from those obtained in experiments or from exact/numerical solutions of specific models, thus proving the failure of this approximate theory. In Table 5.1, it is possible to compare the critical exponents of the Ising model for different spatial dimensions: $d = 2$ (exact values), $d = 3$ (numerical estimates), and $d \geq 4$ (classical, mean-field values). The exponent η is going to be defined below Eq. (5.46).

A better description of critical phenomena can be made by introducing the scaling hypothesis, which is partially inspired by heuristic arguments. In fact, it has been observed that the universality of critical phenomena stems from the property of scale invariance of thermodynamic observables at the critical point. The considerations that follow are based on two main ideas concerning the form of the correlation function close to the critical point and the "singular part" of the free energy density.

According to the Landau theory, the connected autocorrelation function has the Ornstein–Zernike form (5.42). If we generalize the exponent of the power term, this form has a validity that goes beyond the mean-field theory. Therefore, close to the critical point and in the absence of external sources (i.e., $H(\mathbf{x}) = 0$), the connected autocorrelation function

Table 5.1 Exact values of the critical exponents for the Ising model in $d = 2$, numerics in $d = 3$, and mean-field values

Exponent	Ising ($d = 2$)	Ising ($d = 3$)	Mean-Field
α	0	0.110	0
β	1/8	0.326	1/2
γ	7/4	1.237	1
δ	15	4.790	3
ν	1	0.630	1/2
η	1/4	0.0363	0

---

[6] The case $d = 4$ deserves a special attention, because subleading logarithmic corrections to the scaling behavior at the critical point should be taken into account.

$C(\mathbf{x})$, defined in (5.36), is assumed to have the universal functional form

$$C(\mathbf{x}) \sim \frac{e^{-|\mathbf{x}|/\xi}}{|\mathbf{x}|^p} \tag{5.46}$$

with

$$p = d - 2 + \eta \tag{5.47}$$

$$\xi \propto |t|^{-\nu}. \tag{5.48}$$

The exponent η has been found (in experiments and models) to take values different from zero, at variance with the Landau theory, which predicts $\eta = 0$; see Eq. (5.42). The divergence of the correlation length ξ when $|t| \to 0$ means that the only physical length scale that matters close to T_c is ξ itself. The functional form of the autocorrelation function implies that at $T = T_c$, $C(\mathbf{x})$ has a power-law behavior, which allows to define the exponent η.

The second idea concerns the free energy density f, which is nonanalytic at T_c, because quantities derived from it are singular. Since the nonanalyticity comes from the thermodynamic limit, the dependence of the singular component of f on the reduced temperature $t = (T - T_c)/T_c$ is assumed to be fixed by dimensional considerations: Since the free energy density has the dimension of the inverse volume $V^{-1} = L^{-d}$ (see Eq. (5.23)), we simply write

$$f \sim \xi^{-d} \sim |t|^{d\nu}. \tag{5.49}$$

In the following, we start from Eqs. (5.46) and (5.49) and using standard thermodynamic and statistical relations, we derive the singular behavior of various physical quantities, finding that all critical exponents introduced in Section 5.2.3 are not independent of each other, because some relations among them, called scaling laws, hold. In particular, once the functional form (5.46) of the autocorrelation function has been fixed, all critical exponents can be expressed by simple algebraic relations as functions of η and ν, together with the space dimension d.

A first relation can be obtained by observing that, close to T_c, the specific heat at constant volume c_V (see Eq. (5.30)) can be written as

$$c_V = -T \frac{\partial^2 f}{\partial T^2}\Big|_{H=0} \simeq -\frac{1}{T_c} \frac{\partial^2 f}{\partial t^2}\Big|_{H=0} \sim |t|^{d\nu-2}, \tag{5.50}$$

where the last relation is a consequence of Eq. (5.49). By comparing this result with the expected scaling relation reported in Eq. (5.30), we have

$$\boxed{d\nu = 2 - \alpha}, \tag{5.51}$$

which is known as the Josephson scaling law.

A second relation can be derived by first rewriting (5.46) as

$$C(\mathbf{x}, \xi) \sim \xi^{-p} C(\mathbf{y}), \tag{5.52}$$

where $\mathbf{y} = \mathbf{x}/\xi$ is a dimensionless variable, as well as the scaling function $C(\mathbf{y}) = |\mathbf{y}|^{-p} e^{-|\mathbf{y}|}$. Accordingly, the physical scale of C is determined only by the factor ξ^{-p}. Making reference to Eq. (5.36), dimensional analysis allows us to write

$$m \sim C^{1/2} \sim \xi^{-p/2} = |t|^{p\nu/2}.$$

(5.53)

Since the magnetization is expected to vanish as $m \sim |t|^{\beta}$, we can write the scaling relation

$$\beta = \nu(d - 2 + \eta)/2,$$

(5.54)

which will be used later.

A third scaling relation can be obtained by the thermodynamic relation between the magnetic susceptibility χ and the correlation function $C(\mathbf{x})$,[7]

$$\chi = \frac{1}{T} \int d\mathbf{x}\, C(\mathbf{x}).$$

(5.55)

Dimensional analysis of this relation implies that the scaling dependence of χ on the correlation length ξ close to T_c has to be of the form

$$\chi \sim \xi^d \, \xi^{-p} = \xi^{2-\eta} = |t|^{-\nu(2-\eta)},$$

(5.56)

where the factor ξ^d is just due to the measure of the space integral in d dimensions. Considering that χ is expected to diverge at T_c as in Eq. (5.29), we can obtain the scaling relation

$$\boxed{\gamma = \nu(2 - \eta)}\,,$$

(5.57)

which is known as the Fisher scaling law.

The fourth final relation can be derived by considering that the magnetization density m is defined by the relation

$$m = -\frac{\partial f}{\partial H}\bigg|_{H=0}.$$

(5.58)

Dimensional analysis of this equation (see Equations (5.49) and (5.53)) yields

$$H \sim \xi^{-d} \, \xi^{p/2} = \xi^{(\eta-d-2)/2} \sim |t|^{\nu(d+2-\eta)/2}.$$

(5.59)

Phenomenological considerations indicate that at T_c the equation of state of a ferromagnetic system is given by the relation

$$m \sim H^{1/\delta},$$

(5.60)

which establishes how the magnetization density m depends on the amplitude of the perturbation produced by an external magnetic field H. Since m is expected to scale at T_c as $|t|^{\beta}$, we can conclude that H has to scale as

$$H \sim |t|^{\beta\,\delta},$$

(5.61)

thus yielding the relation

$$\beta\,\delta = \nu(d + 2 - \eta)/2.$$

(5.62)

---

[7] This relation can be derived using Eqs. (5.20) and (5.19).

By subtracting Eq. (5.54) from Eq. (5.62) and taking into account Eq. (5.57), we can write

$$\boxed{\beta(\delta - 1) = \gamma}, \tag{5.63}$$

which is known as the Widom scaling law. Finally, by summing Eqs. (5.54) and (5.62) and taking into account Eqs. (5.51), we obtain $\beta(\delta + 1) = d\nu = 2 - \alpha$. Since, according to Eqs. (5.63), $\beta\delta = \beta + \gamma$, we can write

$$\boxed{\alpha + 2\beta + \gamma = 2} \tag{5.64}$$

which is known as the Rushbrooke scaling law.

In summary, the six phenomenological critical exponents α, β, γ, δ, η, and ν obey the four scaling laws, (5.51), (5.57), (5.63), and (5.64), thus proving that only two of them are actually independent and sufficient to describe the critical behavior associated with a continuous equilibrium phase transition. Notice that also the classical exponents obtained in Section 5.2.3 obey these four scaling laws,[8] because their validity is independent of the way one computes f.

5.2.5 Phenomenological Scaling Theory

The scaling analysis of Section 5.2.4 is based on the observation that close to the critical point, the correlation length ξ dominates over all other scales and the behavior of physical quantities is derived using thermodynamic and statistical relations. In this section we offer a more elegant and transparent way to formulate a phenomenological scaling theory of equilibrium phase transitions, which will be useful to study nonequilibrium phase transitions as well (see Section 6.3.2).

In fact, we can assume that all observables exhibit scaling properties close to the critical point; that is, they must be homogeneous functions of some scaling parameter, which have to be eventually related to the correlation length $\xi \sim |t|^{-\nu}$ or, equivalently, to t. In this respect, we point out that in the Landau theory of critical phenomena, the free energy density can be assumed to depend on two physical variables, namely, t and H. Accordingly, any other thermodynamic observable has to depend on the same variables. For instance, close to the critical point, that is, for $t, H \to 0$, the scaling hypothesis amounts to imposing the condition that the magnetization density $m(t, H)$ has to be invariant by suitably rescaling itself and its arguments, thus

$$m(t, H) = \Lambda_m m(\Lambda_t t, \Lambda_H H). \tag{5.65}$$

For $H = 0$ we know that the magnetization density vanishes as $m \sim |t|^\beta$. We are free to choose $\Lambda_t = \Lambda = |t|^{-1}$; this is quite a natural choice, because we are assuming that our reference scaling parameter Λ amounts to the inverse of the rescaled distance from the critical temperature T_c. We can specialize the previous formula to the relation

$$\Lambda_m m(1, 0) = m(t, 0) \sim |t|^\beta \quad \to \quad \Lambda_m = |t|^\beta = \Lambda^{-\beta}. \tag{5.66}$$

---

[8] The hyperscaling relation (5.51), depending on the dimension d, is satisfied by classical exponents if we replace d with d_c^u.

Applying the same argument to $m(0, H) = \Lambda_m m(0, \Lambda_H H)$ and taking into account the phenomenological scaling relation (5.60), we obtain $H^{1/\delta} = \Lambda_m (\Lambda_H H)^{1/\delta}$. We can therefore express the scaling parameter Λ_H in terms of Λ:

$$\Lambda_H = \Lambda_m^{-\delta} = \Lambda^{\beta\delta}, \tag{5.67}$$

obtaining

$$m(t, H) = \Lambda^{-\beta} m(\Lambda t, \Lambda^{\beta\delta} H). \tag{5.68}$$

A scaling relation analogous to Eq. (5.68) holds for the free energy density,

$$f(t, H) = \Lambda_f f(\Lambda t, \Lambda^{\beta\delta} H). \tag{5.69}$$

In order to obtain Λ_f, we can use the scaling law for c_V (see Eq. (5.30)),

$$c_V \sim \Lambda^{\alpha} \sim \left. \frac{\partial^2 f(t, H)}{\partial t^2} \right|_{H=0} = \Lambda_f \Lambda^2 \left. \frac{\partial^2 f(t, H)}{\partial t^2} \right|_{H=0, t=0}, \tag{5.70}$$

yielding $\Lambda_f = \Lambda^{\alpha-2}$, that is,

$$f(t, H) = \Lambda^{\alpha-2} f(\Lambda t, \Lambda^{\beta\delta} H). \tag{5.71}$$

This scaling formula can be used for establishing relations among the different scaling exponents. According to the arguments reported in Section 5.2.4 (see Eq. (5.49)), simple dimensional analysis imposes that the free energy density scales as

$$f(t, 0) \sim \Lambda^{-d\nu}. \tag{5.72}$$

Comparison of Eqs. (5.71) and (5.72) allows us to obtain the Josephson law, Eq. (5.51),

$$\boxed{d\nu = 2 - \alpha.} \tag{5.73}$$

Moreover, by deriving both sides of Eq. (5.71) with respect to H, and recalling that $m = -\frac{\partial f(t, H)}{\partial H}$, we obtain

$$m(t, H) = \Lambda^{\alpha-2} \Lambda^{\beta\delta} m(\Lambda t, \Lambda^{\beta\delta} H). \tag{5.74}$$

Comparing this relation with Eq. (5.65), we obtain $\Lambda_m = \Lambda^{\alpha-2+\beta\delta}$. Since $\Lambda_m = \Lambda^{-\beta}$, we find

$$\beta = 2 - \alpha - \beta\delta. \tag{5.75}$$

If we now derive both sides of Eq. (5.74) with respect to H and recall that $\chi = \frac{\partial m}{\partial H}$, we obtain

$$\chi(t, H) = \Lambda^{\alpha-2} \Lambda^{2\beta\delta} \chi(\Lambda t, \Lambda^{\beta\delta} H). \tag{5.76}$$

Making use of the phenomenological scaling relation $\chi(t, 0) \sim \Lambda^{\gamma}$ (see Eq. (5.29)), we can write

$$\gamma = 2\beta\delta + \alpha - 2. \tag{5.77}$$

Summing Eqs. (5.75) and (5.77), we obtain the Widom scaling law, Eq. (5.63),

$$\boxed{\beta(\delta - 1) = \gamma,} \tag{5.78}$$

while summing twice Eq. (5.75) to Eq. (5.77), we obtain the Rushbrooke scaling law, Eq. (5.64),

$$\boxed{\alpha + 2\beta + \gamma = 2.}$$
(5.79)

Other quantities that exhibit scale invariance (i.e., homogeneity) properties close to the critical point are the correlation length ξ (see Eqs. (5.42) and (5.43)),

$$\xi(t, H) = \Lambda^{\nu} \xi(\Lambda t, \Lambda^{\beta\delta} H),$$
(5.80)

and the autocorrelation function,

$$C(\mathbf{x}, t, H) = \Lambda_c C(\Lambda_{\mathbf{x}} \mathbf{x}, \Lambda t, \Lambda^{\beta\delta} H).$$
(5.81)

Notice that the dependence of C on the position variable $\mathbf{x}$ requires us to introduce a scaling parameter $\Lambda_{\mathbf{x}}$. As a direct consequence of Eq. (5.46), we can conclude that

$$\Lambda_{\mathbf{x}} = \Lambda^{\nu},$$
(5.82)

because the ratio $|\mathbf{x}|/\xi$ should not rescale, while

$$\Lambda_c = \Lambda_{\mathbf{x}}^{-p} = \Lambda^{-p\nu}.$$
(5.83)

Taking into account relation (5.55), we can write

$$\chi \sim \Lambda_{\mathbf{x}}^{d} \Lambda_c = \Lambda^{\nu(2-\eta)},$$
(5.84)

thus recovering the Fisher scaling law, Eq. (5.57),

$$\boxed{\gamma = \nu(2 - \eta).}$$
(5.85)

5.2.6 Scale Invariance and Renormalization Group

Taking inspiration from what we have discussed in Section 5.2.5, we can reformulate the scaling hypothesis in a more general form. In fact, if f describes a system that exhibits a continuous phase transition, it must be a homogeneous function of its variables, that is, the control parameter t and the source field H, in the vicinity of the critical point $t = 0$. Thus we can write

$$f(t, H) = \Lambda^{-d} f(\Lambda_t t, \Lambda_H H),$$
(5.86)

where Λ is an arbitrary length scale that close to the critical point coincides with the correlation length ξ (see Eq. (5.48)),

$$\Lambda \sim \xi \sim t^{-\nu},$$
(5.87)

and we have explicitly taken into account that $f(t, H)$ has the physical dimension of an inverse volume in a space of dimension d.

We can express the two scaling parameters, Λ_t and Λ_H as

$$\Lambda_t = \Lambda^{D_t}$$
(5.88)

$$\Lambda_H = \Lambda^{D_H},$$
(5.89)

where D_t and D_H are the scaling exponents, or dimensions, associated with the variables t and H, respectively. As discussed at the end of Section 5.2.4, the knowledge of these two exponents is enough to determine all the other phenomenological exponents by the four scaling laws therein derived. Since t is a dimensionless quantity, we can fix $D_t = 1/\nu$, so that $\Lambda_t = \xi^{1/\nu} = |t|^{-1}$, and $D_H = \beta\delta/\nu$, so that $\Lambda_H = \xi^{D_H} = |t|^{-\beta\delta}$ (see Eq. (5.61)). This choice corresponds to the phenomenological scaling theory described in Section 5.2.5.

On the other hand, it is possible to compute D_t and D_H explicitly by applying the scale invariance hypothesis directly to the Hamiltonian $\mathcal{H}(\phi(\mathbf{x}), t, H)$ of the model at hand, which depends on some state variable $\phi(\mathbf{x})$, $\mathbf{x}$ being the space coordinate in d dimensions, and on the physical parameters t and H. This is the basic idea of the renormalization group method. In order to illustrate this approach, we have to recall first the definition of the partition function and its relation with the free energy density (see Section 5.2.2):

$$Z(t, H, V) = e^{-F(t,H,V)} = \int_V d\phi(\mathbf{x}) e^{-\mathcal{H}(\phi(\mathbf{x}),t,H)} \tag{5.90}$$

$$f(t, H) = \frac{F(t, H, V)}{V}, \tag{5.91}$$

where both $F(t, H, V)$ and $\mathcal{H}(\phi(\mathbf{x}), t, H)$ are measured in units of T. This standard choice allows us to simplify the notation and is equivalent to rescaling the physical parameters of the model, for example, $\frac{J}{T} \to J$ and $\frac{H}{T} \to H$.

In order to apply the scale-invariance hypothesis to $\mathcal{H}(\phi(\mathbf{x}), t, H)$, we have to introduce a length scale explicitly. A way to accomplish this task amounts to modifying the dependence of the state variable ϕ from a continuous space coordinate vector $\mathbf{x}$ to a discrete lattice coordinate vector,

$$\phi(\mathbf{x}) \to \phi_i, \tag{5.92}$$

where i is a set of d integer coordinates $i_1, i_2, \ldots, i_d$, which identify a site on a d-dimensional cubic lattice at distance ℓ_0 from its nearest-neighbor sites. For instance, this is the case of the Ising model introduced in Section 5.2.2. Without prejudice to generality, we can set the lattice spacing $\ell_0 = 1$.

We can subdivide the lattice volume into blocks of size ℓ^d, $\ell > 1$, and we can associate with each block a new collective block variable

$$\phi'_j = \sum_{i \in j} \phi_i, \tag{5.93}$$

where j is the coordinate vector of the block and i those of the sites belonging to it. Here we adopt the rule of defining the collective block variable by summing over all the original variables in the block, but one can adopt different rules, provided the new block-transformed Hamiltonian preserves the symmetries of the original one. For instance, in the case of the Ising model, where $\phi_i = \sigma_i = \pm 1$, we can define the block transformation as

$$\sigma'_j = \begin{cases} +1, & \text{if } \sum_{i \in j} \sigma_i > 0 \\ -1, & \text{if } \sum_{i \in j} \sigma_i < 0 \\ \sigma_k, & \text{if } \sum_{i \in j} \sigma_i = 0, \end{cases} \tag{5.94}$$

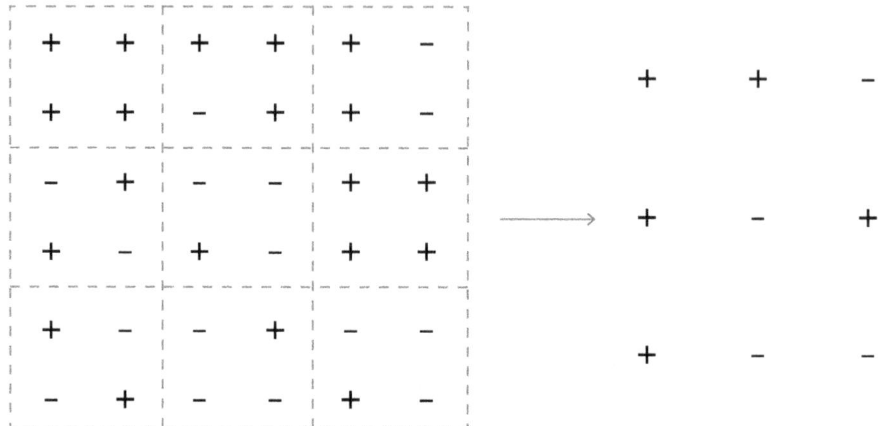

Fig. 5.4 Starting from an Ising model of unitary lattice constant, we perform a coarse graining by replacing each separate set of four spins with a single spin, following the majority rule. In case of a tie, the sign of the spin is chosen randomly. The new lattice constant is $\ell = 2$.

where σ_k is the value of a spin chosen at random in the block. This is known as the majority rule, illustrated in Fig. 5.4.

We have to assume that close to the critical point, this block transformation $\mathcal{R}$ (or any other transformation that amounts to rescaling the length of the lattice spacing) does not change the form of the original Hamiltonian, but it rescales the values of its variables and parameters.[9] The previous assumption implies that[10]

$$\mathcal{R}\left[\mathcal{H}(\phi_i, t, H)\right] = \mathcal{H}'(\phi_i', t', H') \tag{5.95}$$

and also

$$Z(t, H, V) = Z(t', H', V') \tag{5.96}$$

$$f(t, H) = \ell^{-d} f(t', H'), \tag{5.97}$$

where $Z(t', H', V')$ and $f(t', H')$ are functionals of the renormalized Hamiltonian $\mathcal{H}'(\phi_i', t', H')$.

For the moment we do not need to specify $\mathcal{R}$. We can observe that the homogeneity property (5.86) is recovered by assuming that

$$t' = \ell^{D_t} t \tag{5.98}$$

$$H' = \ell^{D_H} H. \tag{5.99}$$

Close to the critical point we can expect that both t and H are very small as well as the renormalized variables t' and H' in such a way that at leading order $\mathcal{R}$ is a linear

---

[9] The validity of the renormalization group method relies on the assumption that its results are independent of the adopted method for the lattice rescaling transformation.

[10] Since performing a renormalization transformation involves the trace over a subset of state variables ϕ_i (see the examples discussed in Appendices H.1 and H.2), $\mathcal{H}'(\phi_i', t', H')$ also contains some additional constant $C(t, H)$.

transformation of the form

$$t' = \tau t + O(t^2) \tag{5.100}$$

$$H' = \zeta H + O(H^2), \tag{5.101}$$

thus yielding the relations

$$\frac{1}{\nu} = D_t = \frac{\ln \tau}{\ln \ell} \tag{5.102}$$

$$\frac{\beta \delta}{\nu} = D_H = \frac{\ln \zeta}{\ln \ell}, \tag{5.103}$$

which could allow us to obtain the scaling exponents from the knowledge of τ and ζ.

Despite the heuristic value of this formulation of the renormalization group method worked out in real space, mainly due to the intuition of the American physicist Leo Philip Kadanoff, many of the simplifying assumptions that have been introduced need to be properly reconsidered if we want to translate the formal renormalization transformation $\mathcal{R}$ into an explicit algorithm. In order to make the reader appreciate the difficulties inherent in such a procedure, we discuss some examples in Appendixes H.1 and H.2. However, describing in full generality the renormalization group transformations goes beyond the aims of this book.

For what we are going to discuss in Section 7.7.2 and in Appendix R, it is more useful to sketch the renormalization group approach in Fourier space, which was originally proposed by the American physicist Kenneth Geddes Wilson. At variance with the real space renormalization group procedure, which typically needs to be worked out in an infinite-dimensional parameter space (see Appendix H.2), we can consider a Hamiltonian that contains the minimum number of terms necessary to determine a critical behavior. For pedagogical reasons, it is worth illustrating this method by an example, considering the case of the Landau theory at the lowest order, whose Hamiltonian, in d space dimensions, is[11]

$$\mathcal{H}[m(\mathbf{x})] = \frac{1}{2} \int d\mathbf{x} \left(|\nabla m(\mathbf{x})|^2 + a\, m^2(\mathbf{x}) \right). \tag{5.104}$$

Using Fourier-transformed variables,

$$m(\mathbf{k}) = \int_{-\infty}^{+\infty} d\mathbf{x}\, e^{-i\,\mathbf{k}\cdot\mathbf{x}}\, m(\mathbf{x}), \tag{5.105}$$

and following the calculations at page 210, we can rewrite Eq. (5.104) as

$$\mathcal{H}[m(\mathbf{k})] = \frac{1}{2} \frac{1}{(2\pi)^d} \int d\mathbf{k}\, (k^2 + a)|m(\mathbf{k})|^2, \tag{5.106}$$

where $k^2 = |\mathbf{k}|^2$. We can introduce a cutoff Λ in $\mathbf{k}$-space that fixes a wavelength, above which we assume a coarse-graining of the model; as in the case of renormalization transformations in the space of coordinates where fluctuations below a certain scale a_0 are ignored, here we should ignore fluctuations of wavelength larger than $\Lambda = 2\pi/a_0$.

---

[11] It is enough to fix a single parameter a to fix the scale of m.

We can write the partition function of the Landau model as

$$Z = \prod_{|\mathbf{k}|<\Lambda} \int dm(\mathbf{k})\, dm^*(\mathbf{k}) e^{-\beta\mathcal{H}[m(\mathbf{k})]}. \tag{5.107}$$

A renormalization transformation can be performed by defining a new energy $\mathcal{H}'[m(\mathbf{k})]$, that is obtained by summing over all the wavelengths whose modulus is between Λ/b and Λ, with $b > 1$,

$$e^{-\mathcal{H}'[m(\mathbf{k})]} \equiv e^{F_\Lambda} \prod_{\Lambda/b<|\mathbf{k}|<\Lambda} \int dm(\mathbf{k})\, dm^*(\mathbf{k}) e^{-\beta\mathcal{H}[m(\mathbf{k})]}, \tag{5.108}$$

where F_Λ is a function of the cutoff Λ and of the coupling constant a. Once we have integrated over all degrees of freedom, F_Λ converges to the free energy F.

By this definition we have obtained a new Hamiltonian that depends on $m(\mathbf{k})$ and is defined only for $|\mathbf{k}| < \Lambda/b$. On the other hand, $\mathcal{H}'[m(\mathbf{k})]$ takes the same form of $\mathcal{H}[m(\mathbf{k})]$, but it depends on a new coupling a',

$$\mathcal{H}'[m(\mathbf{k})] = \frac{1}{2}\frac{1}{(2\pi)^d} \int d\mathbf{k}\,(k^2 + a')|m(\mathbf{k})|^2. \tag{5.109}$$

In terms of the new Hamiltonian $\mathcal{H}'$, the partition function can be rewritten as

$$Z = e^{-F_\Lambda} \prod_{|\mathbf{k}|<\Lambda/b} \int dm(\mathbf{k})\, dm^*(\mathbf{k}) e^{-\beta\mathcal{H}'[m(\mathbf{k})]}. \tag{5.110}$$

We can restore the original cutoff Λ by performing a suitable change of variable in the definition of $\mathcal{H}'[m(\mathbf{k})]$, which amounts to rescaling $\mathbf{k}$ by the factor b, that is, $\mathbf{k}' = b\mathbf{k}$,

$$\mathcal{H}'[m(\mathbf{k}')] = \frac{1}{2}\frac{1}{(2\pi)^d} b^{-d} \int d\mathbf{k}'\left(\frac{1}{b^2}k'^2 + a\right)|m(\mathbf{k}'/b)|^2. \tag{5.111}$$

The last step of the renormalization procedure in $\mathbf{k}$-space amounts to restoring the original normalization of the order parameter; that is, we want the coefficient of the term k'^2 again to be 1. We can obtain this result by the replacement

$$m(\mathbf{k}'/b) = \sqrt{b^{d+2}}m'(\mathbf{k}'). \tag{5.112}$$

This operation is the analog of the Kadanoff block-spin transformation. The renormalized Hamiltonian eventually is

$$\mathcal{H}'[m(\mathbf{k}')] = \frac{1}{2}\frac{1}{(2\pi)^d} \int d\mathbf{k}'\left(k'^2 + a'\right)|m'(\mathbf{k}')|^2, \tag{5.113}$$

where

$$a' = b^2 a \tag{5.114}$$

is the relation that defines the renormalization of the coupling constant a.

The partition function can be finally rewritten as

$$Z = e^{-F_\Lambda} \prod_{|\mathbf{k}'|<\Lambda} \int dm'(\mathbf{k}')\, dm'^*(\mathbf{k}') e^{-\beta\mathcal{H}'[m'(\mathbf{k}')]}. \tag{5.115}$$

As we anticipated, the renormalization group method is definitely simpler in $\mathbf{k}$-space than in real space. The procedure that we have sketched here for the Landau theory at lowest order can be extended, with additional technical difficulties, to the case where we consider higher-order terms, for example, the quartic one. Treating in detail this problem in view of what we discuss later in this chapter is unnecessary. We just want to conclude this section by noting that, since in the Landau theory we deal with continuous variables, the rescaling parameter b can be taken as close as we want to 1, and since we are interested in determining just the rate of change of the coupling constants with b, the renormalization transformation can be finally cast in the form of a set of differential equations, as many as the number of coupling parameters. Determining their fixed points and the corresponding stable and unstable manifolds (as in a standard dynamical model) allows us to obtain a complete description of the renormalization group transformation in $\mathbf{k}$-space and to obtain a clear inference on its physical implications.

5.2.7 Equilibrium Phase Transitions Do Not Generally Exist in One Dimension

We conclude our short summary of equilibrium phase transitions by remembering that these important phenomena generally do not occur in one spatial dimension. Here below we are giving two proofs of this statement: one is based on an Ising-type model and is physically transparent; the other is based on the transfer matrix and it allows more easily to understand when the proof does not apply.

The core of the first proof is simple: In one spatial dimension, $d = 1$, the cost of the domain wall to vanish the magnetization of a macroscopic system is finite, while its entropy increases with system size L. Therefore, the free energy $F = U - TS$ is dominated by the second term and the system is disordered for any $T > 0$. In $d > 1$ we need a domain wall whose extension (and cost) is proportional to L^{d-1}, therefore also divergent with L: This explains why this argument does not apply for $d > 1$.

Earlier argument can be formalized considering the Ising Hamiltonian $H = -J \sum_{i=1}^{L} s_i s_{i+1}$ and evaluating the free energy difference ΔF between the ground state (all spins are equal to each other) and a general configuration with n domain walls (a domain wall is a point between two antiparallel, consecutive spins). Since the energy of a domain wall is $2J$, we can write

$$\Delta F = 2nJ - T \ln W, \qquad (5.116)$$

where $W = \binom{L}{n}$ is the number of distinct ways to locate n domain walls in L sites.

In the approximation $L \gg n$, we obtain

$$\Delta F \simeq 2nJ - nT \left(\ln(L/n) + 1 \right) \qquad (5.117)$$

so that

$$\frac{\partial \Delta F}{\partial n} = 2J - T \ln(L/n) \qquad (5.118)$$

and

$$n_{eq} \simeq L e^{-2J/T}. \qquad (5.119)$$

Therefore, as soon as $T > 0$, the entropy term creates a macroscopic number of domain walls that disorder the system.

A different, more formal way to obtain the same result is via the transfer matrix. Let us suppose that the Hamiltonian has the more general form

$$\mathcal{H} = \sum_{k=1}^{L} g(\sigma_k, \sigma_{k+1}) \qquad \sigma_{L+1} \equiv \sigma_1, \tag{5.120}$$

where σ is a local, discrete variable that can assume M different values. The partition function reads

$$Z(T, L) = \sum_{\sigma_1} \cdots \sum_{\sigma_L} e^{-\beta g(\sigma_1, \sigma_2)} \ldots e^{-\beta g(\sigma_L, \sigma_1)} \tag{5.121}$$

and its expression can be reworked defining the $(M \times M)$ matrix

$$\mathbb{P}_{\sigma, \sigma'} = e^{-\beta g(\sigma, \sigma')}, \tag{5.122}$$

which is real, symmetric, and strictly positive. In fact, we can write

$$Z(T, L) = \sum_{\sigma_1} \cdots \sum_{\sigma_L} \mathbb{P}_{\sigma_1, \sigma_2} \ldots \mathbb{P}_{\sigma_L, \sigma_1} \tag{5.123}$$

$$= \sum_{\sigma_1} (\mathbb{P}^L)_{\sigma_1, \sigma_1} = \mathrm{Tr}(\mathbb{P}^L) = \sum_{i=1}^{M} (\lambda_i)^L \tag{5.124}$$

where λ_i are the M (real) eigenvalues of the transfer matrix $\mathbb{P}$.

According to the Perron–Frobenius theorem, the maximal eigenvalue $\lambda_{max} \equiv \lambda_1$ is strictly positive, nondegenerate, all other eigenvalues are smaller in absolute value, and it is an analytic function of the transfer matrix elements. Therefore

$$\lim_{L \to \infty} Z(T, L) = (\lambda_{max})^L = e^{-\beta L f} \tag{5.125}$$

$$f = -T \ln \lambda_{max} \tag{5.126}$$

where f is an analytic function of all parameters and no phase transition may occur in the thermodynamic limit $L \to \infty$.

It should be stressed that the only hypothesis is that M is finite, which allows you to replace the whole summation $\sum_{i=1}^{M} (\lambda_i)^L$ with the dominant term $(\lambda_{max})^L$. The finiteness of M requires that interactions must be short-ranged, that the system must be of finite size in the spatial directions perpendicular to the one where matrix transfer applies, and that the site variable can only assume a finite number of different values.

5.3 The Model of Katz, Lebowitz, and Spohn

The Ising model, with binary variables and nearest-neighbor interactions, is the simplest model we can think of, and, in fact, it has a central role in equilibrium statistical physics. Therefore, it should not be surprising that an important role in nonequilibrium statistical physics is played by a model that is directly derived from it.

With a suitable change of variables, the Ising model is known to describe a lattice gas (Appendix H), which is a prototypical model for condensation: In dimension $d > 1$, there is a temperature T_c below which the gas condenses, forming an aggregate (a cluster) of particles. This model is described by the lattice-gas Hamiltonian $\mathcal{H}_{\mathrm{LG}}$, see Eq. (H.7), the equilibrium probability $p_{\mathrm{eq}}(s)$ of a microscopic state s is given by the Gibbs distribution, $p_{\mathrm{eq}}(s) = e^{-\beta \mathcal{H}_{\mathrm{LG}}(s)}/Z$, and detailed balance is valid for any (fictitious) dynamics $w^{\mathrm{eq}}_{s',s} = \omega\big(\beta(\mathcal{H}_{\mathrm{LG}}(s') - \mathcal{H}_{\mathrm{LG}}(s))\big)$ satisfying $\omega(x)/\omega(-x) = e^{-x}$, as discussed in Sections 2.3.6.

There are different ways to bring a Hamiltonian system out of equilibrium,[12] the simplest one being to attach the system to two different reservoirs, so as to induce macroscopic currents. The resulting NESS (nonequilibrium steady state) is clearly boundary driven, because it is generated by the imposed boundary conditions. In the case of short-range interactions (as it is the case for $\mathcal{H}_{\mathrm{LG}}$), local equilibrium is preserved and detailed balance is broken only due to interaction with the two reservoirs.

A different way to impose a macroscopic current (mass current, in this case) is through a bulk-driven mechanism, as done by the American Sheldon Katz and Joel Lebowitz and the German Herbert Spohn, who imagined driving the lattice gas in a given direction $\hat{x}$ (KLS model). In practice, this means favoring particles hopping in the $+\hat{x}$-direction and disadvantaging particles hopping in the opposite $(-\hat{x})$ direction, which can be obtained by modifying the original transition rates.

For simplicity we assume that $w^{\mathrm{eq}}_{s',s} \neq 0$ only if the final configuration s' differs from s just for the position of one particle, which has moved to a nearest-neighbor site. The drive must increase $w_{s',s}$ if the particle hops in the $+\hat{x}$-direction, and it must decrease it if the particle hops in the opposite $(-\hat{x})$ direction; Otherwise, $w_{s',s}$ remains unchanged. Since $\omega(x)$ is a decreasing function of x, a straightforward manner to implement such a rule is to write the transition rates of the driven lattice gas (DLG) as follows:

$$w^{\mathrm{DLG}}_{s',s} = \omega\left(\beta(\Delta\mathcal{H}_{\mathrm{LG}} - qE)\right), \tag{5.127}$$

where $E > 0$ measures the strength of the driving and $q = 0, \pm 1$, according to the move of the particle: $q = +1$ if the particle moves to the right, along the $\hat{x}$-axis; $q = -1$ if the particle moves to the left; $q = 0$ if the particle moves perpendicularly to the $\hat{x}$-axis.

If we define the potential energy associated with the field E,

$$\mathcal{H}_{\mathrm{E}} = -E \sum_i n_i x_i, \tag{5.128}$$

where x_i is the $\hat{x}$-coordinate of the ith site, it appears that

$$\Delta\mathcal{H}_{\mathrm{LG}} - qE = \Delta\mathcal{H}_{\mathrm{LG}} + \Delta\mathcal{H}_{\mathrm{E}}. \tag{5.129}$$

Therefore, it seems that driving the lattice gas has only the effect of modifying the total Hamiltonian,

$$\mathcal{H}_{\mathrm{DLG}} = \mathcal{H}_{\mathrm{LG}} + \mathcal{H}_{\mathrm{E}}, \tag{5.130}$$

which does not lead the system out of equilibrium, since $w^{\mathrm{eq}}_{s',s} = \omega\big(\beta(\mathcal{H}_{\mathrm{DLG}}(s') - \mathcal{H}_{\mathrm{DLG}}(s))\big)$.

---

[12] In Chapters 6 and 7 we are considering systems which are intrinsically out of equilibrium.

However, the earlier statement is true only if the system is virtually infinite, with no spatial bounds. For example, if we consider periodic boundary conditions, earlier argument no longer applies because a constant force acting on a particle moving on a ring cannot be derived by a potential.[13]

The previous considerations indicate that the driven lattice gas with periodic boundary conditions in the direction of the drive cannot have a steady state characterized by $p(s) = e^{-\beta \mathcal{H}_{\text{DLG}}(s)}/Z$, but they do not exclude that a different, non-Boltzmann-type, invariant measure might satisfy detailed balance. In order to exclude it, we need a different formulation of detailed balance that refers only to transition rates. Such a formulation and the proof of its equivalence are provided by Theorem 2.5.

In practice, to prove that the set of rates $w_{s',s}$ cannot satisfy detailed balance is sufficient to find a set $(s_1, \ldots, s_N)$ of microstates such that

$$\prod_{i=1}^{N} w_{s_{i+1},s_i} \neq \prod_{i=1}^{N} w_{s_{i-1},s_i}, \tag{5.131}$$

where $s_0 = s_N$ and $s_{N+1} = s_1$. We choose a particularly simple set of microstates: The system contains only one particle, hopping in the $\pm\hat{x}$-direction, along which periodic boundary conditions apply. Microstates are simply labeled by the particle position i along $\hat{x}$, $N = L$ is the linear size of the system in the x direction, and the two products of Eq. (5.131) correspond to the probabilities that the particle travels the whole ring of L sites returning to the starting point, either clockwise or anticlockwise.

If we write $w_{s_{i\pm1},s_i}^{\text{DLG}} = \omega_\pm$, where $\omega_+ = \omega(q = +1) \neq \omega_- = \omega(q = -1)$, we can finally evaluate the products

$$\prod_{i=1}^{L} w_{s_{i+1},s_i} = (\omega_+)^L \neq \prod_{i=1}^{L} w_{s_{i-1},s_i} = (\omega_-)^L, \tag{5.132}$$

proving that driving leads the system out of equilibrium, if periodic boundary conditions exist.

If instead the system is closed and finite along the driving force, the earlier argument does not apply, because the particle cannot return to its initial position always hopping in the same direction. In this case the system sets in an inhomogeneous, equilibrium state, and no macroscopic current appears. In Section 5.4, we will focus on nonequilibrium situations, either induced by periodic boundary conditions, see Section 5.4.1, or by open boundary conditions, see Section 5.4.2.

5.4 The (T)ASEP Models

The KLS model introduced in Section 5.3 is the simplest example of a system driven out of equilibrium. It is a lattice gas model where particles feel a force E favoring hops in

---

[13] A charged particle can rotate along a circle as an effect of a perpendicular magnetic field whose modulus increases linearly in time. Certainly, such circular motion cannot be obtained with an electrostatic potential nor with a gravitational potential (except in Escher's drawing *Ascending and descending*).

the $+x$-direction, while disfavoring hops in the opposite, $-x$-direction. This force, along with boundary conditions, breaks detailed balance, so that steady states are not equilibrium states.

In the following we will limit ourselves to one-dimensional systems so that the hops in directions perpendicular to x, not affected by the field, are absent. It is important to keep in mind that one-dimensional systems are very peculiar, because equilibrium phase transitions are not allowed for short-range interactions, as is the case here. However, we will show this is not true for nonequilibrium systems, so that our restriction to $d = 1$ makes the analysis simpler, but not too simple.

We will consider two different types of boundary conditions, both enforcing detailed balance breaking: periodic boundary conditions and open boundary conditions. In the former case we will consider a general force E, that is, a general degree of asymmetry between right and left hopping. In the latter case, particles are injected on the left boundary and extracted on the right boundary, so it is natural to consider a diverging force E: Particles can hop to the right only. The acronym for our models is ASEP (asymmetric simple exclusion process) or TASEP (totally ASEP), according to the degree of asymmetry.

5.4.1 ASEP with Periodic Boundary Conditions

We start with the simplest boundary conditions breaking detailed balance: periodic boundary conditions; see Fig. 5.5. The dynamical evolution works as follows. A particle in site k ($n_k = 1$) can move to the site on the right/left (if it is empty, $n_{k\pm1} = 0$) with rate $\nu_\pm$. After some transient, dependent on the initial conditions, we can imagine the system sets in a NESS, that is, a state whose statistical properties do not depend on time. Such a steady state must satisfy Eq. (2.131) for any s. The properties of this state for a single particle have been studied in Section 2.3.2 (random walk on a ring). Here we are interested in a system of N particles, interacting via the exclusion mechanism.

We can prove that for periodic boundary conditions, the steady state is characterized by a nonequilibrium stationary probability $p^{\text{ne}}(s)$, which does not depend on s: All microscopic states have the same probability. This proof makes reference to Fig. 5.5, where a general configuration s is charaterized by the number m_s of blocks of particles. A block is a sequence of adjoining particles. According to Eq. (2.131), $p^{\text{ne}}(s) = $ constant is a solution if

$$\sum_{s'} w_{s',s} = \sum_{s'} w_{s,s'}, \qquad (5.133)$$

where $w_{s',s} = \nu_+$ (ν_-) if s' differs from s for a single particle moved to the right (left). From any microstate s, we can attain $N_{\text{out}}^+(s)$ configurations by moving a particle on the right and $N_{\text{out}}^-(s)$ configurations by moving a particle on the left. Similarly, we can define the quantities $N_{\text{in}}^\pm(s)$ as the number of configurations from which it is possible to attain the state s. The stationarity condition (5.133) can therefore be rewritten as

$$N_{\text{out}}^+(s)\nu_+ + N_{\text{out}}^-(s)\nu_- = N_{\text{in}}^+(s)\nu_+ + N_{\text{in}}^-(s)\nu_-. \qquad (5.134)$$

It is easy to understand, making reference to Fig. 5.5, that

$$N_{\text{in}}^\pm(s) = N_{\text{out}}^\pm(s) = m_s, \qquad (5.135)$$

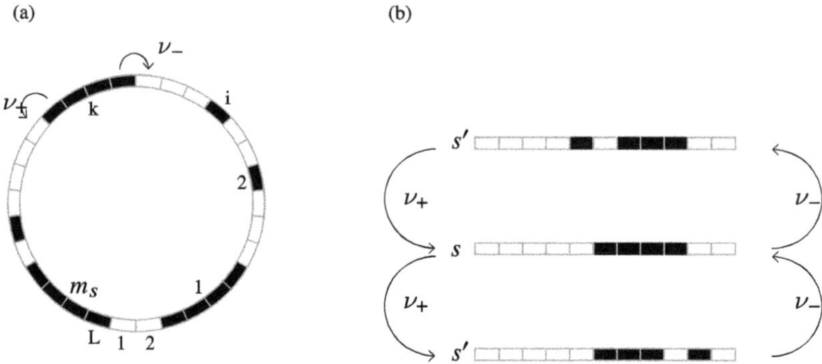

Fig. 5.5 The ASEP model with periodic boundary conditions. (a) Outer indexing refers to sites, $i = 1, 2, \ldots, L$. Inner indexing refers to blocks of particles, $k = 1, 2, \ldots, m_s$. Particle at site i can move to the site $i \pm 1$ with rate $\nu_\pm$, if the arrival site is empty. (b) Detail of a single block of particles (center), with the two configurations from which it originates and to which it evolves.

so that the stationarity condition is satisfied by the microcanonical ensemble (all microstates are equiprobable).

First, let us remark that $N^+_{\text{out}}(s) = N^-_{\text{in}}(s)$, because the reachable states s' with a single jump to the right are *exactly* the same configurations from which it is possible to arrive in s with a single jump to the left. For the same reason, $N^-_{\text{out}}(s) = N^+_{\text{in}}(s)$. Said this, Fig. 5.5(b) clarifies that any block of particles allows to have a single configuration to attain with a jump to the right and a single configuration to attain with a jump to the left, which proves Eq. (5.135).

It is noteworthy that earlier considerations apply whatever the rates $\nu_\pm$, therefore also if $\nu_+ = \nu_- = \nu$. However, in the symmetric case, detailed balance holds, because for any pair of configurations we have that $w_{s,s'} = w_{s',s} = w$, with $w = 0, \nu$. Therefore, the equilibrium distribution (for $\nu_+ = \nu_-$) and the stationary distribution (for $\nu_+ \neq \nu_-$) are the same: All microstates are equiprobable. This is due to the special periodic boundary conditions we have been considering.

The equiprobability of microscopic states implies the absence of correlations, for example, between the occupation probabilities of site i and site j, so that mean-field theory is exact. Let us check this point in more detail. Two quantities play a major role in ASEP models, the probability p_i that site i is occupied and the net current of particles,

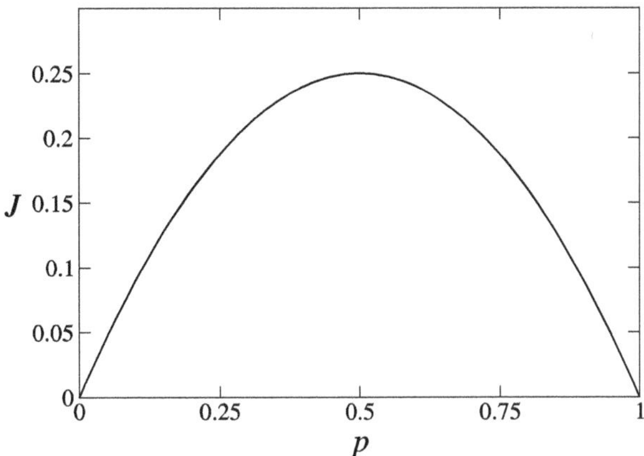

Fig. 5.6　The current $J(p) = p(1 - p)$ of the TASEP model.

$J_{i,i+1}$, between site i and site $i + 1$:

$$p_i = \langle n_i \rangle \tag{5.136}$$

$$J_{i,i+1} = \nu_+ \langle n_i (1 - n_{i+1}) \rangle - \nu_- \langle n_{i+1} (1 - n_i) \rangle. \tag{5.137}$$

The correlator $\langle n_i n_{i+1} \rangle$ has the general form

$$\langle n_i n_{i+1} \rangle = \frac{\sum_s p^{\mathrm{ne}}(s) n_i(s) n_{i+1}(s)}{\sum_s p^{\mathrm{ne}}(s)}. \tag{5.138}$$

Because of the equiprobability of microscopic states, $p^{\mathrm{ne}}(s)$ is constant and the earlier expression reduces to counting states,

$$\langle n_i n_{i+1} \rangle = \frac{\#\text{ states where both } i \text{ and } i + 1 \text{ are occupied}}{\text{total } \# \text{ of states}} = \frac{N(N - 1)}{L(L - 1)}. \tag{5.139}$$

The only correlation surviving in our model is due to finiteness: Once a site is occupied (with probability $p = N/L$), we are left with $N - 1$ particles and $L - 1$ sites. In the limit $N, L \gg 1$ we can forget these finite-size correlations and write $\langle n_i n_{i+1} \rangle = \langle n_1 \rangle \langle n_{i+1} \rangle$, so that $J_{i,i+1} = (\nu_+ - \nu_-)p(1 - p)$. This current is plotted in Fig. 5.6 for the TASEP case ($\nu_+ = 1, \nu_- = 0$), showing that the maximal current state is obtained for $p = \frac{1}{2}$, because this value maximizes the probability of finding a given site occupied and its right site empty.

5.4.2 TASEP with Open Boundary Conditions

In this section we study a system where particles are injected on one side and extracted on the other side. For this reason we consider the totally asymmetric case. The TASEP model on a ring is defined by only one parameter, the constant average density ρ, and no transition appears with varying it. A richer phenomenology appears as soon as we change boundary conditions, passing from periodic to open ones. This change is also relevant from an experimental point of view because the ring setup with a "force" always in the same clockwise direction is hardly feasible (see footnote 13).

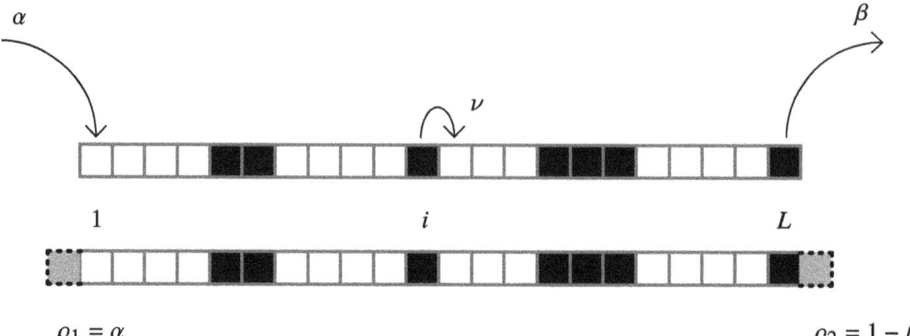

Fig. 5.7   The TASEP model with open boundary conditions, see also the main text. Black/white lattice sites correspond to occupied/empty sites. Particles are injected at rate α, hop with rate ν, and are removed at rate β. Boundary conditions (injection/extraction of particles) are equivalent to saying that the edge sites are in contact with reservoirs of particles (grey lattice sites) with densities $\rho_1 = \alpha$ and $\rho_2 = 1 - \beta$.

Therefore, we now consider a straight lattice of L sites ($i = 1, \ldots, L$), see Fig. 5.7: Particles are injected in site $i = 1$ (if empty) at rate α, they hop to the right with rate ν, and they are removed from site $i = L$ (if occupied) at rate β.[14] In practice, the site $i = 1$ is attached to a reservoir with particle density $\rho_1 = \alpha$ and site $i = L$ is attached to a reservoir with particle density $\rho_2 = 1 - \beta$. As sketched in Fig. 5.7, the two reservoirs correspond to two fictitious sites, located in $i = 0$ and in $i = L + 1$.

This model has been solved exactly, but results are in agreement with a much easier mean-field treatment in almost all aspects. Therefore, we are going to discuss the easiest one. Even restricting ourselves to the mean-field approximation, we can choose between a spatially discrete and a spatially continuous approach. The former can be faced with a method borrowed from the theory of dynamical systems and it can be implemented numerically very easily: It is treated in Appendix S.1. The latter allows a more physical analysis, through the space dependence of the density $\rho(x)$: It is covered later.

If $\rho_1 = \rho_2$, we expect a homogeneous density profile in a similar way to what happens on a ring where the particle density is exactly constant.[15] What happens if $\rho_1 \neq \rho_2$? In this case, boundaries impose a density gradient; therefore, $\rho = \rho(x)$, and the total current J is expected to be the sum of two terms,

$$J = j(\rho) - D\partial_x\rho, \tag{5.140}$$

where the first term, $j(\rho) = \rho(1 - \rho)$, is due to the (total) asymmetry of the hopping, see Section 5.4.1. The second term is the phenomenological term due to diffusion induced by density inhomogeneities, where D is a suitable diffusion constant, whose exact values do not affect the phase diagram.

---

[14] We assume $\nu = 1$, which corresponds to rescale time. If $\nu \neq 1$, in what follows α should be replaced by α/ν and β should be replaced by β/ν.

[15] The case $\rho_1 = \rho_2 = \rho$ is the canonical version of the model on a ring, where the density of particles, $\rho = N/L$, is exactly conserved, as in the microcanonical ensemble.

A steady state is characterized by a constant J and we expect that far from the boundaries the density profile is constant for large L, because in such a limit, the effect of the density gradient $(\rho_2 - \rho_1)/L$ must vanish. However, the effect of boundaries on the value of such constant density $\bar{\rho}$ and on the current J is felt also in the thermodynamic limit, $L \to \infty$.

The following line of reasoning is based on the hypothesis that $j(\rho)$ has a single maximum in the interval $\rho \in (0, 1)$. This is certainly true in the present case, where particles only interact via the exclusion principle, and the resulting current is $j(\rho) = \rho(1 - \rho)$,[16] which has a maximum, for $\rho = \rho_M = \frac{1}{2}$. The first remark is that $\rho(x)$ must be monotonous because in the presence of an extremum x^* in the interval $(0, L)$ there would be pairs of points, $x_1 < x^*$ and $x_2 > x^*$, where the density ρ is the same but where the gradient $\partial_x \rho$ has opposite signs; therefore, it would be impossible for the current J to be equal in such points. In conclusion, $\rho(x)$ is monotonously increasing (decreasing) if $\rho_2 > \rho_1$ $(\rho_2 < \rho_1)$.

The second observation leads to an "extremal current" principle. In fact, if $\bar{\rho}$ is the constant density in the flat region inside the system, we have

$$J = j(\bar{\rho}) = j(\rho) - D\partial_x \rho, \tag{5.141}$$

where ρ varies between ρ_1 and ρ_2 and the gradient term has a constant sign. Therefore, if $\rho_1 > \rho_2, \partial_x \rho \leq 0 \quad \forall x$ and $j(\bar{\rho})$ is the *maximal* current in the interval (ρ_1, ρ_2). Instead, if $\rho_1 < \rho_2, \partial_x \rho \geq 0 \quad \forall x$ and $j(\bar{\rho})$ is the *minimal* current in the interval (ρ_1, ρ_2). Using the "extremal current" principle for all possible combinations of ρ_1, ρ_2, and $\rho_M = \frac{1}{2}$, we obtain the phase diagram[17] of Fig. 5.8 and the properties of the different phases are summarized in Table 5.2.

It is customary to divide the phase diagram into three regions labeled by the acronyms HD, LD, and MC. The high-density (HD) phase is characterized by a density $p = \bar{\rho} > \frac{1}{2}$. The low-density (LD) phase is characterized by a density $p = \bar{\rho} < \frac{1}{2}$. The *absolutely* maximal current (MC) phase ($J = \frac{1}{4}$) is characterized by the density $p = \bar{\rho} = \frac{1}{2}$ and it is the easiest phase to identify through the "extremal current" principle. In fact, if $\alpha, \beta > \frac{1}{2}$, $\rho_2 < \rho_M < \rho_1$ and $\bar{\rho} = \frac{1}{2}$ is the density corresponding to the maximal current in the interval (ρ_1, ρ_2), which is also the maximal current in the whole domain $(0, 1)$.

In Fig. 5.9 we plot the stationary density profiles obtained numerically with the map method discussed in Appendix S.1. We now comment on some of their features. First, in the HD and LD regions, the spatial profile attains $\bar{\rho}$ exponentially fast, as attested by the inset in the HD panel. This is not the case for the MC region. This difference in behavior can be explained by inverting the function $\rho(x)$ and solving Eq. (5.140),

$$x = D \int_{\rho_1}^{\rho(x)} \frac{dc}{j(c) - J}. \tag{5.142}$$

---

[16] If we consider repulsive interactions between particles, $j(\rho)$ may display two maxima and a minimum in $\rho = \frac{1}{2}$.

[17] α and β are transition rates (or rather ratios between transition rates, see footnote 14), therefore they can be greater than one. On the other hand, the average occupancy of a site (within a simple exclusion model) cannot be greater than one. However, ρ_1 can be thought to be larger than one and ρ_2 can be thought to be even negative and the phase diagram is valid for arbitrarily large α and β. This is clearer in the discrete model, discussed in Appendix S.1.

Table 5.2 Different phases in the parameter space of TASEP model

Region	Name	Density p	Current J
$\alpha > \beta, \quad \beta < \dfrac{1}{2}$	HD	$1 - \beta$	$\beta(1 - \beta)$
$\alpha < \beta, \quad \alpha < \dfrac{1}{2}$	LD	α	$\alpha(1 - \alpha)$
$\alpha \geq \dfrac{1}{2}, \quad \beta \geq \dfrac{1}{2}$	MC	$\dfrac{1}{2}$	$\dfrac{1}{4}$

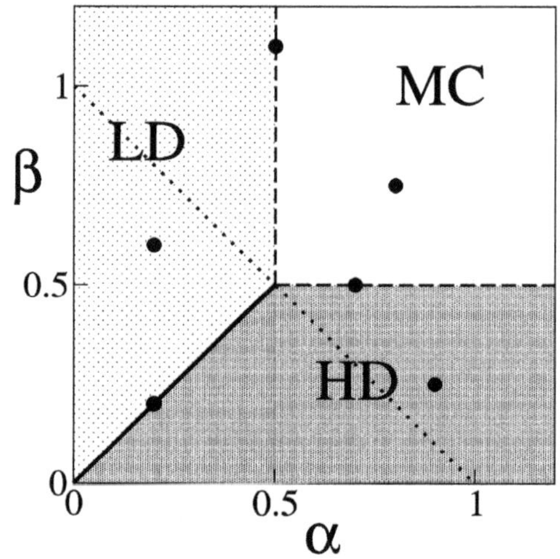

Fig. 5.8 Phase diagram of TASEP model. The dotted line separates the region where $\rho(x)$ is an increasing function ($\alpha + \beta < 1$) from the region where $\rho(x)$ is a decreasing function ($\alpha + \beta > 1$). The LD/HD phases can be characterized by an increasing as well as a decreasing profile. The MC phase can only emerge by the *maximal* current principle, therefore by a decreasing profile. The six full dots locate the points (α, β) that have been used in Fig. 5.9 to plot the mean-field density profiles, using the map method described in Appendix S.1.

In the HD phase, $\bar{\rho}$ is different from $\frac{1}{2}$ and it is attained for large x. In this limit ($x \gg 1, \rho \to \bar{\rho}$), we can therefore write $j(c) = j(\bar{\rho}) + j'(\bar{\rho})(c - \bar{\rho})$ and the integration of Eq. (5.142) gives an exponential tail,

$$\rho(x) = \bar{\rho} + (\rho_1 - \bar{\rho}) \exp\left(-\frac{|j'(\bar{\rho})|}{D}x\right). \tag{5.143}$$

In the MC current, $\bar{\rho} = \frac{1}{2}$ and $j'(\bar{\rho}) = 0$, therefore the expansion of the current in proximity of $\bar{\rho}$ is $j(c) = j(\bar{\rho}) + \frac{1}{2}j''(\bar{\rho})(c - \bar{\rho})^2$ and the integration of Eq. (5.142) now

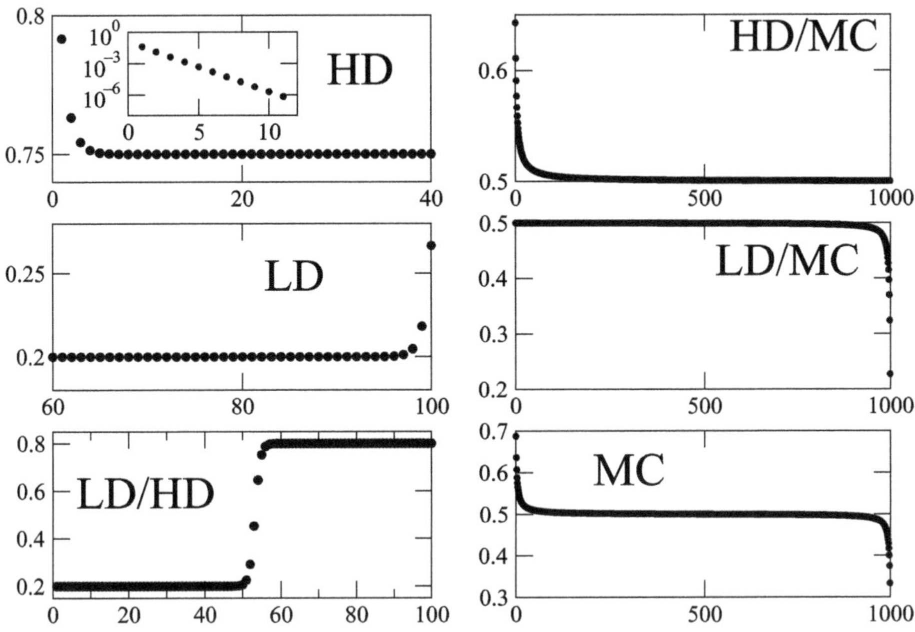

Fig. 5.9 Steady-state density profiles of the TASEP model in mean-field (discrete) approximation for the different phases (HD, LD, and MC) and in the transition lines between them.

gives a power-law tail,

$$\rho(x) = \bar{\rho} + \cfrac{1}{\cfrac{1}{\rho_1 - \bar{\rho}} + \cfrac{2|j''(\bar{\rho})|}{D}x}. \tag{5.144}$$

The exact treatment of the TASEP model confirms an exponential tail for the HD/LD phases and a power-law tail for the MC phase, but for the latter phase the exact result gives $\rho(x) - \bar{\rho} \sim 1/\sqrt{x}$ rather than $\rho(x) - \bar{\rho} \sim 1/x$.

Another point where mean-field theory is inaccurate concerns the transition line HD/LD. Along this line, $\rho_1 < \frac{1}{2} < \rho_2$ and $\rho_1 = 1 - \rho_2$, so there are two equivalent minima of the current in the interval (ρ_1, ρ_2), because $j(\rho_1) = j(\rho_2)$. The mean-field result gives a steady spatial profile interpolating between ρ_1 and ρ_2 with a narrow "domain wall" (see Fig. 5.9, bottom left panel). Numerical simulations of the *exact* model provide a linear profile $\rho(x)$, which can be understood in terms of a fluctuating domain wall. A few details about the implementation of a simple simulation are given in Appendix S.2.

5.5 Symmetry Breaking: The Bridge Model

We have discussed in detail the phase diagram of TASEP, because such a model is the building block of the bridge model, where spontaneous symmetry breaking appears. In order to study symmetry breaking, we need two classes of particles. We can imagine having

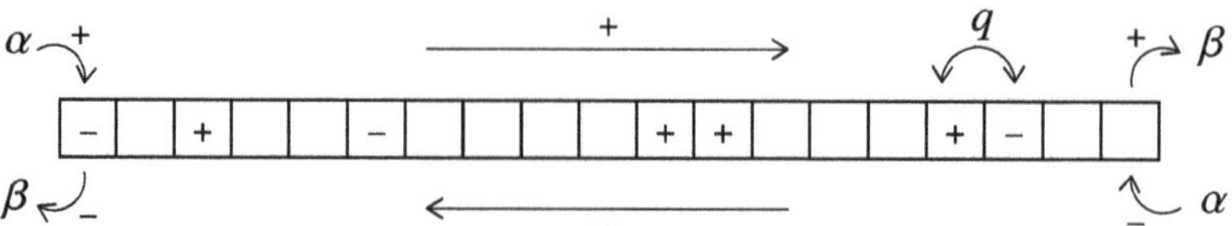

Fig. 5.10   Diagram of the bridge model; see Eqs. (5.145) and (5.146).

positively and negatively charged particles moving in opposite directions under the action of a field, but the name "bridge model" derives from a macroscopic model, where particles mimic vehicles traveling in opposite directions along a one-lane bridge. Traffic is not regulated by lights and vehicles may enter the bridge from opposite sides. You can imagine obtaining this model by combining two TASEP models with additional rules controlling the intersection of vehicles/particles.

In simple terms and making reference to Fig. 5.10, positive/negative particles are injected to the left/right with rate α and extracted from the right/left at rate β. Injection of a new particle is possible only if the site is empty. When different particles meet, they can exchange with probability q. We assume that the two classes of particles have the same injection/extraction rates, because we want to study the spontaneous symmetry breaking. As a matter of fact, if (for example) their injection rates were different, it would not be surprising that the resulting positive/negative currents and densities are also different. Instead, the problem of spontaneous symmetry breaking arises when kinetic rules for the two classes of particles are exactly the same.

More rigorously, the bridge model, is defined as follows. Each site $i = 1, \ldots, L$ is occupied by a positive particle $\oplus_i$, by a negative particle $\ominus_i$, or it is empty $\bigcirc_i$. We choose randomly an integer $k = 0, 1, \ldots, L$. If $k \neq 0, L$, the only microscopic processes giving rise to the evolution of the pair $k, k + 1$ are

$$
\begin{aligned}
\oplus_k \bigcirc_{k+1} &\;\rightarrow\; \bigcirc_k \oplus_{k+1} &\qquad& \text{with probability } 1 \\
\bigcirc_k \ominus_{k+1} &\;\rightarrow\; \ominus_k \bigcirc_{k+1} &\qquad& \text{with probability } 1 \\
\oplus_k \ominus_{k+1} &\;\rightarrow\; \ominus_k \oplus_{k+1} &\qquad& \text{with probability } q.
\end{aligned}
\tag{5.145}
$$

All other processes involving sites $k, k+1$ have zero probability. As for boundary conditions, they enter when $k = 0$ or $k = L$. The only permitted processes are

$$
\text{if } k = 0, \begin{cases} \ominus_1 \;\rightarrow\; \bigcirc_1 & \text{with probability } \beta \\ \bigcirc_1 \;\rightarrow\; \oplus_1 & \text{with probability } \alpha \end{cases}
$$

$$
\tag{5.146}
$$

$$
\text{if } k = L, \begin{cases} \oplus_L \;\rightarrow\; \bigcirc_L & \text{with probability } \beta \\ \bigcirc_L \;\rightarrow\; \ominus_L & \text{with probability } \alpha. \end{cases}
$$

Despite appearances, evolution rules are simple. Overall (see Eqs. (5.145)), particles move as in the TASEP with the additional condition that opposite particles can exchange their position with probability q. At boundaries (see Eqs. (5.146)), particles are injected with rate α and extracted with rate β (see Appendix T.2 for a short description of a rejection-free Kinetic Monte Carlo algorithm to simulate the Bridge model). One caveat should be stressed: The exchange of particles at boundaries is not allowed; therefore, a particle cannot get onto the bridge if the entry site is occupied by an opposite particle waiting to get off. We will come back to this point later.

In summary, we have two classes of particles traveling in opposite directions and having the same dynamical rules. Is it possible that a spontaneous symmetry breaking occurs, making currents and densities of positive and negative particles different? This is a natural question if we think of the regime $\beta \ll 1$ and $\alpha, q \approx 1$, because in this regime each flux of particles would tend to stabilize itself into the HD phase, which is manifestly impossible, because the system cannot sustain a density larger than $\frac{1}{2}$ for two different classes of particles at the same time. The solution might be an equal reduction of both densities. In fact, as we will see in MF, a symmetric solution exists for any point of the phase diagram (α, β), and in the region $\beta \ll \alpha, q$, it is a symmetric LD phase. However, this phase may be unstable in much the same way an equilibrium symmetric state with no magnetization is unstable below T_C in a ferromagnetic material.

We will see that this instability leads one type of particle to dominate, therefore producing a symmetry breaking. Which class of particles dominates depends on the initial conditions and on fluctuations: The same state may lead to a dominance of positive particles or to a dominance of negative particles. The key point, however, is that a true phase transition occurs only in the thermodynamic limit, so it will be of great interest to study the dynamics with varying the size L.

5.5.1 Mean-Field Solution

In the TASEP model we only need the variable $n_i = 1, 0$, which indicates whether site i is occupied or not by a positive particle, while in the bridge model we also need the variable $\tau_i = 1, 0$, which gives the same information for a negative particle. The currents of positive/negative particles between two adjacent sites $(i = 1, \ldots, L - 1)$ are

$$(J_+)_{i,i+1} = \langle n_i(1 - n_{i+1} - \tau_{i+1})\rangle + q\langle n_i \tau_{i+1}\rangle = \langle n_1(1 - n_{i+1} - (1 - q)\tau_{i+1})\rangle$$
$$(J_-)_{i+1,i} = \langle \tau_{i+1}(1 - \tau_i - n_i)\rangle + q\langle \tau_{i+1} n_i\rangle = \langle \tau_{i+1}(1 - \tau_i - (1 - q)n_i)\rangle,$$

where each current counts two distinct processes, the hopping toward an empty site and the exchange with an opposite particle.

These equations must be supplemented with boundary conditions, according to which

$$\begin{aligned} (J_+)_{01} &= \alpha\langle 1 - n_1 - \tau_1\rangle, & (J_-)_{10} &= \beta\langle \tau_1\rangle, & i &= 1 \\ (J_+)_{L,L+1} &= \beta\langle n_L\rangle, & (J_-)_{L+1,L} &= \alpha\langle 1 - n_L - \tau_L\rangle, & i &= L, \end{aligned} \tag{5.147}$$

where the notations for the current have been used in order to make valid the following evolution equations for any $i = 1, \ldots, L$,

$$\frac{d\langle n_i\rangle}{dt} = (J_+)_{i-1,i} - (J_+)_{i,i+1} \tag{5.148}$$

$$\frac{d\langle \tau_i\rangle}{dt} = (J_-)_{i+1,i} - (J_-)_{i,i-1}. \tag{5.149}$$

We are primarily interested in the stationary state, which is characterized by site-independent currents. Within a mean-field approach, variables on different sites are

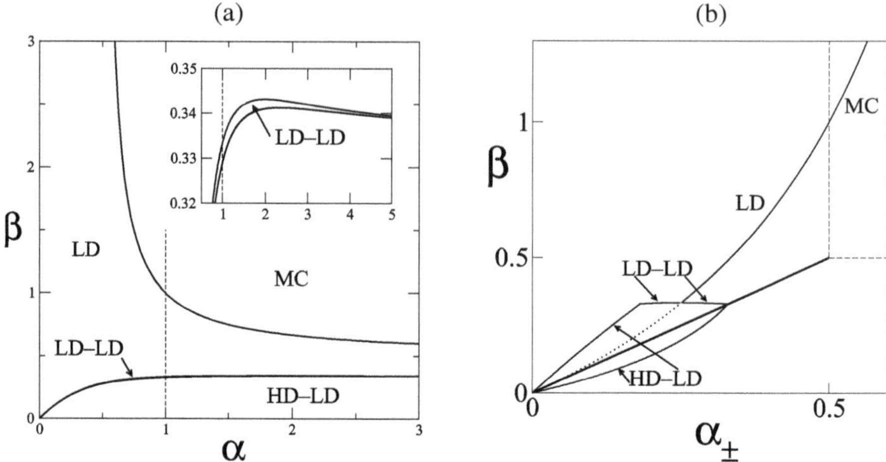

Fig. 5.11 (a) Mean-field phase diagram of the bridge model with a blow-up in the inset. (b) If we fix $\alpha = 1$ and let β increase from $\beta = 0$ to $\beta > 1$, we plot the corresponding $(\alpha_\pm, \beta)$ pairs as the thin straight lines. When the two curves join, we pass from an asymmetric phase to a symmetric one. The dotted line is the unstable, symmetric LD phase. The thick straight line and the dashed lines separate the different regions of the TASEP phase diagram; see Fig. 5.8.

independent and we can reduce to equations for the single-site average values

$$p_i = \langle n_i \rangle, \qquad m_i = \langle \tau_i \rangle. \tag{5.150}$$

In terms of p_i, m_i, Eqs. (5.148) and (5.149) can be rewritten as

$$
\begin{aligned}
J_+ &= \alpha(1 - p_1 - m_1) = p_i(1 - p_{i+1} - (1-q)m_{i+1}) = \beta p_L, \\
J_- &= \alpha(1 - p_L - m_L) = m_{i+1}(1 - m_i - (1-q)p_i) = \beta m_1,
\end{aligned}
\qquad i = 1, \ldots, L. \tag{5.151}
$$

The earlier equations show that the two fluxes of particles are coupled via boundary conditions and, if $q < 1$, because it is difficult to exchange their positions. If $q = 1$, their coupling in the bulk disappears, because a particle is no longer able to discriminate between an opposite particle and an empty site. In this case the above equations simplify to

$$
\begin{aligned}
J_+ &= \alpha(1 - p_1 - m_1) = p_i(1 - p_{i+1}) = \beta p_L, \\
J_- &= \alpha(1 - p_L - m_L) = m_{i+1}(1 - m_i) = \beta m_1,
\end{aligned}
\qquad i = 1, \ldots, L. \tag{5.152}
$$

We have already noticed how the model does not allow the exchange of particles at boundaries. We can now remark that allowing it would make the two fluxes completely independent for $q = 1$. In full general terms, the model might allow an exchange q_b at boundaries and an exchange q in the bulk. If $q_b = q = 1$, the model is trivially uncoupled and no phase transition may appear. The case we are going to study in a mean-field approximation corresponds to $q_b = 0, q = 1$: It allows an exact treatment and is not trivial.

We can now rephrase boundary conditions in a more tractable way. We start from boundary conditions in the present form

$$
\begin{aligned}
J_+ &= \alpha(1 - p_1 - m_1) = \beta p_L \\
J_- &= \alpha(1 - p_L - m_L) = \beta m_1,
\end{aligned}
\tag{5.153}
$$

to realize that coupling occurs through the entry site only, not through the exit site. Therefore, our goal is to encapsulate the coupling in effective injection rates $\alpha_\pm$, which means we want to write

$$J_+ = \alpha_+(1 - p_1)$$
$$J_- = \alpha_-(1 - m_L). \tag{5.154}$$

From Eqs. (5.153), we have $\frac{J_+}{\alpha} = (1 - p_1) - \frac{J_-}{\beta}$ and $\frac{J_-}{\alpha} = (1 - m_L) - \frac{J_+}{\beta}$. Using Eqs. (5.154) to express $(1 - p_1)$ and $(1 - m_L)$ in terms of J_+, α_+, and J_-, α_-, respectively, we can write

$$\alpha_+ = \frac{J_+}{\frac{J_+}{\alpha} + \frac{J_-}{\beta}} \qquad \text{and} \qquad \alpha_- = \frac{J_-}{\frac{J_-}{\alpha} + \frac{J_+}{\beta}}. \tag{5.155}$$

In conclusion, for $q = 1$ the bridge model can be reformulated in terms of two TASEP models with new injection rates $\alpha_\pm = g_\pm(\alpha, \beta, J_+, J_-)$. In Section 5.5, we have solved the TASEP model, so we know the functions $J_\pm = f(\alpha_\pm, \beta)$; see Table 5.2. We can therefore solve the bridge model self-consistently, assuming that the bridge model is in a given phase and using the above equations to find whether such a solution exists for some values of α and β.

The explicit calculations are made in Appendix T.1 for symmetric and asymmetric phases. A symmetric phase means that $\alpha_+ = \alpha_-$ and $J_+ = J_-$: The two types of particles are not distinguishable from their dynamics. In the appendix we find the LD and MC symmetric phases, while the HD symmetric phase cannot exist, as argued here earlier. An asymmetric phase means that $\alpha_+ \neq \alpha_-$ and $J_+ \neq J_-$: In the appendix we find two different types of asymmetric phases, the HD–LD and the LD–LD ones. In Fig. 5.11, we plot the different phases of the bridge model in the mean-field approximation. If we increase β, keeping $\alpha > \frac{1}{2}$, fixed, we first encounter the asymmetric HD–LD phase, then the asymmetric LD–LD phase, then the symmetric LD phase, and finally the symmetric MC phase. If $\alpha < \frac{1}{2}$, the earlier sequence of transitions stops with the symmetric LD phase.

The continuous or discontinuous character of the transitions can be determined by making reference to Fig. 5.11(b) while taking into account the phase diagram of the TASEP model; see Fig. 5.8. Using these figures, we can conclude that the first transition (HD–LD) $\rightarrow$ (LD–LD) is discontinuous, while the other ones are continuous.

5.5.2 Exact Solution for $\beta \ll 1$

The mean-field approximation has the advantage of providing a full phase diagram, but fluctuations can break the mean-field picture. In equilibrium systems, this is especially true in low dimensions, so an alternative proof that nonequilibrium symmetry breaking does occur in the bridge model would be welcome. Such evidence will be given in this section, where we focus on the limiting case $\beta \ll \alpha, q$.

In fact, in this limit there are two main time scales: a fast time scale, during which particles travel in the bulk and new particles possibly enter in the bridge, and a slow time scale, during which particles leave the bridge. This separation of time scales is better understood if we imagine setting $\beta = 0$ and monitoring the resulting dynamics, then switching on β. If $\beta = 0$, particles cannot leave the bridge, therefore as soon as a positive particle has attained the site

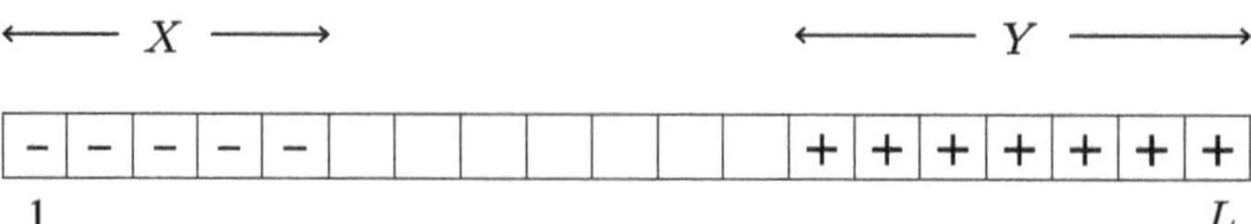

Fig. 5.12   Typical configuration of the bridge model for $\beta \to 0$.

$i = L$ and a negative particle has attained the site $i = 1$, the following dynamics conserves matter, because such particles act as a stopper, preventing the entry of new ones. From now on, dynamics leads to a final, frozen configuration where all (X) negative particles are grouped to the left, all (Y) positive particles are grouped to the right, and $(L - X - Y)$ empty sites are in the middle; see Fig. 5.12.

Now let us switch on β, therefore allowing particles to leave the bridge, but keeping $\beta \ll \alpha, q$. Now, after a time $\tau_{\text{ex}} \approx \beta^{-1}$, one of the two particles acting as a stopper may get out, therefore leaving an empty site in its place. This site can be taken up by the second particle of the same type in the queue for the exit or by a new particle of the opposite type entering into the bridge. In the former case, a new stopper again closes the exit. In the latter case, the new particle will travel freely toward its queue on the opposite side of the bridge while the entrance has been closed again. In both cases, it is necessary to wait a time τ_{ex}, because something new happens. The net result of having switched β has been to get the same type of configuration as before, with different lengths X, Y of the queues of positive and negative particles. In other words, the dynamics of the bridge model, if observed on a time $t > \tau_{\text{ex}}$, can be described in a simple, discrete two-dimensional phase space (X, Y), with suitable transition rates between a lattice point and its neighbors.

Before discussing dynamics, we need to clarify which aspects are relevant to settle the existence of a spontaneous symmetry breaking in the bridge model. For this reason we take inspiration from a system that is well known to have an ordered phase as a result of spontaneous symmetry breaking: the two-dimensional Ising model. The analogy with the bridge model is because this is only apparently a one-dimensional model, time playing the role of an effective additional dimension.

Let us consider the two-dimensional Ising model on a finite lattice of linear size L at temperature $T < T_c$. If $L \gg \xi(T)$, the correlation length, a typical snapshot of the system will provide a finite average magnetization m, very close to the equilibrium value $\overline{m}(T)$ for the infinite system. On a short time scale, fluctuations do not destroy order and as time goes by $m(t)$ fluctuates around $\overline{m}(T)$. However, since L is finite, the system has a nonvanishing probability of overcoming the barrier between the present state and the symmetric state with $m = -\overline{m}(T)$. The key point is the time $\tau(L)$ required for inverting the magnetization: It increases exponentially with L, so that finite but large systems practically stay in one minimum forever.

A second aspect of spontaneous symmetry breaking at equilibrium is worth discussing: the effect of a field. In fact, an arbitrarily small field destroys the phase transition. If we prepare the Ising model with negative magnetization and apply a small, positive field, the reversal of the magnetization occurs even in the thermodynamic limit, because it takes place via nucleation of the opposite phase. The nucleation process will be studied in detail in Section 8.8; here it suffices to say that at equilibrium the reversal time does not increase exponentially any longer with the size of the system if an arbitrarily small field is applied.[18]

Based on the earlier discussion, we are led to consider two problems: (1) we prepare the system filled with negative particles and evaluate the time $\tau(L)$ necessary to obtain the configuration where the system is filled with positive particles; (2) we do the same thing in the presence of a "field" favoring the HD phase of positive particles. Since the two problems can be treated in the same manner,[19] we discuss the probabilities for the transition $(X, Y) \to (X', Y')$ in the general case of a field, problem (2). If the field is zero, we recover problem (1). To begin with, what is a field in the context of the bridge model? It is something that favors a phase with respect to the other. The $+$ phase, for example, may be favored by an injection rate a_+ larger than the injection rate a_- for negative particles.[20]

Let us now describe the microscopic dynamics in the (X, Y) plane. If we start with X negative particles and Y positive particles, after a typical time τ_{ex}, one of the following four processes takes place:

1. *The particle $\ominus_1$ leaves the bridge and the particle $\ominus_2$ takes its place.* Then all the remaining negative particles automatically move to the left. The new state is $(X - 1, Y)$.
2. *The particle $\ominus_1$ leaves the bridge and a $\oplus$ particle is injected from the left.* This particle travels all along the bridge to join the other positive particles. The new state is $(X - 1, Y + 1)$.
3. *The particle $\oplus_L$ leaves the bridge and the particle $\oplus_{L-1}$ takes its place.* Then all other positive particles move to the right. The new state is $(X, Y - 1)$.
4. *The particle $\oplus_L$ leaves the bridge and a $\ominus$ particle is injected from the right.* This particle travels to the left until it joins the other negative particles. The new state is $(X + 1, Y - 1)$.

In order to determine the probabilities of different processes, we must not confuse competing dynamical processes with processes that automatically follow a given occurrence. The four competing dynamical processes are highlighted in italics earlier. Since the rate of hopping (processes 1 and 3) is equal to 1 and the rate of injecting a new particle is a_+ for process 2 and a_- for process 4, the probabilities p_i of the four processes are $p_1 = c, p_2 = ca_+, p_3 = c, p_4 = ca_-$, with c a suitable normalization constant. The existence of a "field" producing the asymmetry between a_+ and a_- can be made evident by writing $a_\pm = \alpha(1 \pm H)$. With this notation, the condition $\sum_{i=1}^{4} p_i = 1$ implies $c = 1/(2+2\alpha)$

---

[18] In the nonequilibrium context of the bridge model, there is another reason to consider the effect of a field, that is, an asymmetry between the two classes of particles. While in the Ising model there is a natural symmetry between up and down spins, deriving from the Hamiltonian describing the system, here there is no a priori reason to require that rules for positive and negative particles are *exactly* the same.

[19] This is another relevant difference with the equilibrium case, where the physics of magnetization reversal is different for $H = 0$ and $H \neq 0$.

[20] We use $a_\pm$ rather than $\alpha_\pm$ to avoid confusion with the Section 5.5.1, where such notation had a completely different meaning.

and the probabilities finally are as follows:

$$(X,Y) \rightarrow \begin{cases} (X-1,Y), & p_1 = \dfrac{1}{2(1+\alpha)} \\[2mm] (X-1,Y+1), & p_2 = \dfrac{\alpha(1+H)}{2(1+\alpha)} \\[2mm] (X,Y-1), & p_3 = \dfrac{1}{2(1+\alpha)} \\[2mm] (X+1,Y-1), & p_4 = \dfrac{\alpha(1-H)}{2(1+\alpha)}. \end{cases} \tag{5.156}$$

Possible moves are described in Fig. 5.13. Because of the constraints on allowed moves, the motion of the system is confined to a triangular region in the quadrant $X,Y \geq 0$ and it can cross its border only through the horizontal axis or through the vertical axis (not through the origin). In the former case ($Y = 0$), all $\oplus$ particles are out, no stopper exists on the right and the system fills with $\ominus$ particles; see the horizontal arrow crossing all sites $(X,0)$. A symmetric process happens for the $X = 0$ axis. In conclusion, all configurations $(X,0)$ immediately pass to $(L,0)$ and all configurations $(0,Y)$ immediately pass to $(0,L)$.

The system is initially in a full negative phase and we wonder how the probability of passing to a full positive phase scales with the system size L. This means we start with the system in $(L,0)$ (see the gray box in Fig. 5.13), and we wonder what the probability is of leaving the triangle crossing the Y-axis rather than the X-axis (these two possibilities are sketched in the same figure, on the right). The key point is that the system diffuses with a net drift in both directions, because

$$\langle \Delta X \rangle = \sum_{i=1}^{4} p_i (\Delta X)_i = -\frac{1+2\alpha H}{2(1+\alpha)} \tag{5.157a}$$

$$\langle \Delta Y \rangle = \sum_{i=1}^{4} p_i (\Delta Y)_i = -\frac{1-2\alpha H}{2(1+\alpha)}. \tag{5.157b}$$

Therefore, the system has a horizontal drift in the negative direction and a vertical drift that is negative for $H < 1/(2\alpha)$. If we monitor the motion in the (X,Y) directions, the problem of passing from the negative to the positive phase can be rephrased as follows. The X variable starts from value L and the Y value starts from zero. If Y is moved from zero to one, what is the probability that X attains zero before Y comes back to zero? We can get a simple, semi-quantitative answer if we treat the motions in the X and Y directions as independent and we approximate the motion along X with its average over noise.

In a continuum picture, with variables $x(t), y(t)$, dynamics is written in terms of Langevin equations,

$$\begin{aligned} \dot{x}(t) &= -v_x + \eta_x(t), & \langle \eta_x(t) \rangle &= 0, & \langle \eta_x(t)\eta_x(t') \rangle &= \Gamma_x \delta(t-t') \\ \dot{y}(t) &= -v_y + \eta_y(t), & \langle \eta_y(t) \rangle &= 0, & \langle \eta_y(t)\eta_y(t') \rangle &= \Gamma_y \delta(t-t'), \end{aligned} \tag{5.158}$$

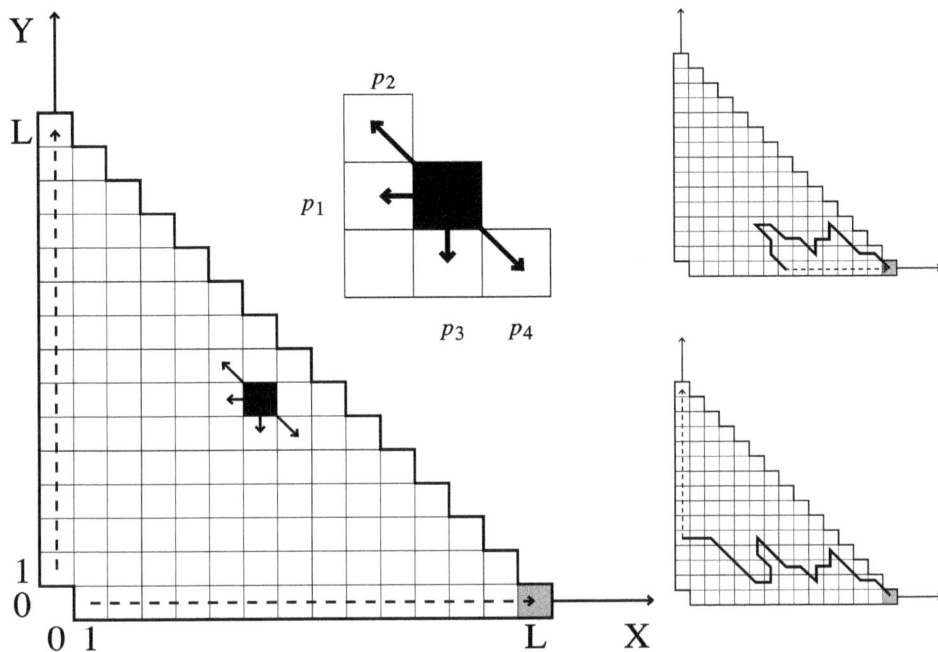

Fig. 5.13 Phase space of the bridge model for $\beta \ll 1$ and schematic of dynamics. The gray site of coordinates $(L, 0)$ is the initial configuration (bridge filled with negative particles) and the black site is a generic configuration whose allowed transitions are indicated by the four arrows. Each move has a probability p_i, given in Eq. (5.156). Dashed arrows represent the fast dynamics allowing to fill the system with negative or positive particles, when particles of the opposite sign are missing. On the right we sketch two possible trajectories: In the top panel, the representative point crosses the horizontal axis and the system fails to invert its population, while in the bottom panel the representative point crosses the vertical axis and the system has a population inversion, passing from a dominance of negative particles to a dominance of positive ones.

where

$$-v_x = \frac{\langle \Delta x \rangle}{\Delta t} = \langle \Delta X \rangle, \qquad \Gamma_x = \langle (\Delta X)^2 \rangle$$
$$-v_y = \frac{\langle \Delta y \rangle}{\Delta t} = \langle \Delta Y \rangle, \qquad \Gamma_y = \langle (\Delta Y)^2 \rangle. \qquad (5.159)$$

Variances are easily calculated as the average values above,

$$\langle (\Delta X)^2 \rangle = \langle (\Delta Y)^2 \rangle = \frac{1 + 2\alpha}{2(1 + \alpha)}. \qquad (5.160)$$

The motion along the x-direction starts at $x(0) = L$ and the average time to reach the origin is

$$t^* = \frac{L}{v_x}. \qquad (5.161)$$

We wonder what the probability $S(t^*)$ is that a particle moving in the y-direction according to Eq. (5.158) and such that $y(0) = y_0 > 0$ has not crossed the origin $y = 0$ within time t^*. Using Eq. (M.65), we get

$$S(t^*) \simeq \frac{y_0}{\sqrt{\pi}} \frac{\sqrt{2\Gamma_y}}{v_y^2} \frac{\exp\left(\frac{y_0 v_y}{\Gamma_y}\right)}{(t^*)^{3/2}} \exp\left(-\frac{v_y^2}{2\Gamma_y}t^*\right). \tag{5.162}$$

For $H = 0$, we obtain

$$S(L) \simeq y_0 \left(\frac{2}{\pi}(1 + 2\alpha)\right)^{1/2} \frac{\exp\left(\frac{y_0}{1 + 2\alpha}\right)}{L^{3/2}} \exp\left(-\frac{1}{2(1 + 2\alpha)}L\right) \equiv \frac{e^{-g(\alpha)L}}{f(\alpha)L^{3/2}}. \tag{5.163}$$

The average time to pass from one pure phase, for example, the negative one, to another pure phase, the positive one, is

$$\tau_{-\to+}(L) = \frac{1}{\beta}S(L)^{-1} = \frac{1}{\beta}f(\alpha)L^{3/2}e^{g(\alpha)L} \tag{5.164}$$

and it grows exponentially with the size of the system. This exponential growth is the key result, proving that a phase transition indeed exists in the limit of vanishing β. The exponential growth arises, because the motion along the y-direction has a negative drift v_y. Since

$$v_y = -\frac{1 - 2\alpha H}{2(1 + \alpha)}, \tag{5.165}$$

the earlier scenario is preserved as long as $v_y < 0$, that is, for

$$H < \frac{1}{2\alpha}. \tag{5.166}$$

Therefore, critical nonequilibrium systems display two important differences with respect to critical equilibrium systems: We may have a spontaneous symmetry breaking even in one dimension and the application of a small field does not destroy the transition.[21]

5.6 Bibliographic Notes

An extensive and basic illustration of critical phenomena, Landau theory, mean-field approximations, and the renormalization group can be found in K. Huang, *Statistical Mechanics*, 2nd ed. (Wiley, 1987). A deeper and more specialized insight about renormalization group methods is contained in D. Amit, *Field Theory, the Renormalization Group, and Critical Phenomena*, 2nd ed. (World Scientific Publishing, 1984), while a

---

[21] The reader may have observed that the symmetry-breaking field of the equilibrium Ising model acts on all spins, while here such field only acts on the boundaries. However, this is not really relevant because we obtain the same result by introducing an asimmetry between the hopping rates of positive and negative particles. What is decisive is that the application of an external field in the equilibrium Ising model triggers a new mechanism for reversing the magnetization.

historical introduction to critical phenomena is given in the book by C. Domb, *The Critical Point* (CRC Press, 1996).

A complete classification of phase transitions in two space dimensions relies on the solid theoretical ground of conformal field theory. Details can be found in the book by M. Henkel, *Conformal Invariance and Critical Phenomena* (Springer, 1999). See also the pioneering papers by A. A. Belavin, A. M. Polyakov, and A. B. Zamolodchikov, Infinite Conformal Symmetry in Two-dimensional Quantum Field-theory, *Nuclear Physics B*, **241** (1984) 333–380, and J. L. Cardy, Conformal Invariance and Universality in Finite-size Scaling, *Journal of Physics A: Mathematical and General*, **17** (1984) L385–L388.

A reference book about driven diffusive systems and their critical properties is B. Schmittmann and R. K. P. Zia, *Statistical Mechanics of Driven Diffusive Systems*, Phase Transitions and Critical Phenomena, vol. 17 (Academic Press, 1995).

An overview about the modern evolution of Markov theory of stochastic processes toward interacting particle models such as spin models, the voter model, and contact processes is contained in T. M. Liggett, *Interacting Particle Systems* (Springer, 1985).

The open TASEP solution using the continuum MF-approach can be found in J. Krug, Boundary-Induced Phase Transitions in Driven Diffusive Systems, *Physical Review Letters*, **67** (1991) 1882–1885.

The exact solution of the TASEP model in one dimension can be found in the papers by G. Schütz and E. Domany, Phase Transitions in an Exactly Soluble One-dimensional Exclusion Process, *Journal of Statistical Physics*, **72** (1993) 277–296, and by B. Derrida, M. R. Evans, V. Hakim, and V. Pasquier, Exact Solution of a 1D Asymmetric Exclusion Model Using a Matrix Formulation, *Journal of Physics A: Mathematical and General* **26** (1993) 1493–1517. It is also worth reading the short review by B. Derrida, An Exactly Soluble Nonequilibrium System: The Asymmetric Simple Exclusion Process, *Physics Reports* **301** (1998) 65–83. The exact solution of the (partially) ASEP model was given by J. de Gier and F.H.L. Essler, Bethe Ansatz Solution of the Asymmetric Exclusion Process with Open Boundaries, *Physical Review Letters*, **95** (2005) 240601.

The original paper where the bridge model was introduced is C. Godréche, J. M. Luck, M. R. Evans, D. Mukamel, S. Sandow, and E. R. Speer, Spontaneous Symmetry-breaking: Exact Results for a Biased Random Walk Model of an Exclusion Process, *Journal of Physics A: Mathematical and General* **28** (1995) 6039–6072.

6

Absorbing Phase Transitions

This chapter is entirely devoted to discuss a wide class of nonequilibrium phase transitions, characterized by the presence of absorbing states. This jargon is inspired by the analogy of the models under scrutiny with natural processes, where the evolution of an active pattern may be eventually absorbed or it may propagate indefinitely by tuning a suitable control parameter. In Section 6.1 we illustrate this scenario making reference to percolation processes. While isotropic percolation belongs to the realm of equilibrium phase transitions (it is in one of the universality classes of the Potts model), directed percolation (DP) is a classical example of a nonequilibrium absorbing phase transition. The Domany–Kinzel probabilistic cellular automaton (CA) described in Section 6.2 summarizes the different kinds of DP processes, including those known as bond and site percolation. This model amounts to a parallel (in time) stochastic evolution of active sites, ruled by suitable parameters, tuning the probability of generating an active offspring from one active ancestor or by a pair of active ancestors. The main feature of this model is that the critical exponents associated with the nonequilibrium phase transition from an active to an inactive phase are the same, independent of the details of the microscopic evolution rule. This enforces the conjecture that universality is at work also for nonequilibrium phase transitions. Other models that exhibit the same critical properties of DP are the so-called contact processes (see Section 6.2.1), which are characterized by a sequential, rather than parallel probabilistic evolution, mimicking the stochastic dynamics of epidemic spreading.

In Sections 6.3.1 and 6.3.2, the whole machinery of the phenomenological scaling theory is shown to be successfully applied to the study of the critical exponents characterizing DP-like models. It is also important to point out that a mean-field formulation of this class of models can be studied by a suitable Langevin-like equation for the density of active sites in the presence of multiplicative noise. More precisely, noise has to vanish with this density, so that no fluctuation can reactivate a fully inactive state, thus ensuring the existence of this absorbing state.

In Section 6.4, we survey scenarios of nonequilibrium phase transitions other than DP. They may be characterized by the existence of more than a single absorbing state, as for compact directed percolation (CDP) or for epidemic processes where the healing mechanism is introduced. New universality classes can also emerge by additional symmetries, as it happens in equilibrium critical phenomena. These examples do not cover all possible situations, but they aim at making the reader aware of the richness and peculiarities of nonequilibrium phase transitions.

We conclude this chapter by describing self-organized critical (SOC) models (see Section 6.5), which are introduced using a simple model of an absorbing phase transition, then adding driving and dissipation to it. SOC phenomena have no counterpart in equilibrium statistical mechanics. They cover a wide range of physical processes, such as avalanches and earthquakes as well as evolutionary competition among species, characterized by a scale-invariant dynamics, ruled by specific power-law behaviors and by peculiar statistical properties.

6.1 From Isotropic to Directed Percolation

The class of models we deal with is inspired by the physical mechanism of a fluid percolating through a porous medium, like ground coffee, paper, or cloth. In the first case, our goal might be to make a cup of coffee using a flip coffee pot; in the other cases, we might want to filter the fluid, a process that depends on the network of active pores, that is the channels letting the fluid pass through. A simple model of such a process is represented by a lattice where each site is either wet or dry and a wet site can wet a neighboring site if they are connected by an active bond (pore). This model, called bond percolation, is well studied in equilibrium statistical physics, where it is known to be equivalent to a Potts model, which is a q-state Ising-like model. In practice, each bond is active with probability p, and for p larger than some critical value, there is an infinite cluster of wet (active) sites spanning the entire system, which means that a fluid may percolate through it.[1] Alternatively, we may start from an active site and proceed to wet the system according to the earlier rule: Does the wetting process stop ($p < p_c$) or does it spread ($p > p_c$)? This question may correspond to a specific experimental setup, for example, to a horizontal sheet of paper. However, if we tilt the system (or, in general, if the fluid can flow in the vertical direction), it is clear that gravity plays a role, because the fluid will tend to flow in a certain direction. The two cases are called, respectively, isotropic (bond) percolation and directed (bond) percolation and are illustrated in Fig. 6.1.

The same questions can be posed for different experimental setups. For example, if the bond is metallic, the wetting process may correspond to the flow of an electric current induced by the application of an electric potential difference, ΔV. Isotropic and directed percolation, see Fig. 6.1, would correspond to apply ΔV between the central site and a circle centered on it (isotropic percolation) or between the central site and a row of sites (DP).

In DP, the flow direction can be interpreted as the arrow of time, so that DP in $d + 1$ dimensions corresponds to a nonequilibrium process in d spatial dimension ($d = 1$ in Fig. 6.1). The example of DP is therefore specially suited to illustrate a statement often made in the context of nonequilibrium phase transitions: Time takes the form of an additional dimension. The universality class of DP is generally characterized by a phase transition from a fluctuating active phase to an inactive phase, where the system is eventually trapped

---

[1] A similar model is called site percolation: In this case sites rather than bonds are made active with probability p. The critical values for the formation of an infinite cluster are different in the two models.

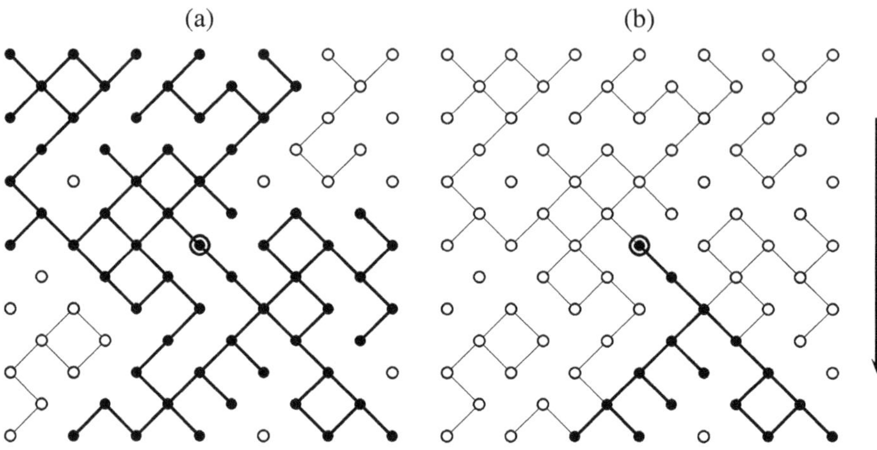

Fig. 6.1 Schematic of isotropic (a) and directed (b) bond percolation. The circled point in the center is the site where percolation originates; that is, it is the only wet site when dynamics starts. Active bonds between nearest neighbor sites are indicated by solid lines. A site becomes wet if it is linked to a wet site via an active bond. In isotropic percolation (a), the spread of wet sites proceeds isotropically, while in directed percolation (b), wetting proceeds only in the direction of the arrow. For more clarity, active sites are indicated as full dots and existing bonds between active sites are indicated as thick solid lines.

into an absorbing state. In the specific model of DP, the (only) absorbing state corresponds to all dry (inactive) sites.[2]

In general terms, we can introduce the variable $s_i(t) = 0, 1$ to specify if site i is active ($s_i = 1$) or inactive ($s_i = 0$) at time t. The system is usually initialized with a single active site, $s_i(0) = \delta_{i,i_0}$, or with a finite, uniform density of active sites, $s_i(0) = 1$ with probability $\rho_0 \leq 1$. One can easily realize that the total number of active sites at time t, $N(t) = \sum_{i=0}^{L} s_i(t)$, and the density of active sites, $\rho(t) = N(t)/L$ are important observables.

It is straightforward that the fully inactive (dry) state plays the role of the absorbing state: If the system eventually falls into such a state, it will be trapped there forever, since no active sites can be produced. It is just as simple to understand that the fully active (wet) state is not a stationary state. We stress that the very existence of an absorbing state in the dynamical evolution rule of DP processes tells us that detailed balance is violated. Actually, the absorbing state can be entered from any previously active configuration with some finite probability, while the time-reversed process has probability zero. As a consequence, phase transitions in such models cannot be described in terms of equilibrium ensemble theory.

---

[2] Which makes the terminology unfortunate.

Before passing to a more general and quantitative description of DP, it is instructive to argue that this class of nonequilibrium phase transitions goes beyond the description of percolative systems. If we look at percolation dynamics, see Fig. 6.1(b), we can remark that the evolution of active (inactive) sites A (0) can be described through basic processes of reaction and diffusion: $A \to 0$ (death), $A \to 2A$ (offspring), $2A \to A$ (coalescence), $2A \to 0$ (annihilation), and $A0 \to 0A$ (i.e., diffusion). The description in terms of reaction–diffusion dynamics is typical in the kinetics of chemical reactions. Instead, if we interpret active/inactive sites as infected/healthy individuals, percolation dynamics reads as an epidemic spreading. We will come back later to both of these descriptions.

6.2 The Domany–Kinzel Model of Cellular Automata

Many basic models of nonequilibrium phase transitions can be formulated as dynamical processes of interacting particles moving on a lattice. When a lattice site can be either occupied by a single particle or empty (exclusion process) and the evolution rule is local, synchronous, and Markovian, the model at hand is just a CA. "Local" means that the evolution rule of a site depends only on neighboring sites; "synchronous" means that sites are updated in parallel[3]; "Markovian" means that the configuration at time $t + 1$ depends on the configuration at time t only.

Deterministic CA were introduced in the last century by the great Hungarian mathematician John von Neumann to show that even a deterministic, discrete system evolving in discrete time with a local rule may display an unpredictable evolution.[4] Some decades later, the Israeli Eytan Domany and the German Wolfgang Kinzel introduced a very simple stochastic CA model that epitomizes several "contact processes," including the bond/site DP models described in Section 6.1. In this perspective, the Domany–Kinzel (DK) model has played a very important role in the scientific literature, as a general pedagogical example for different stochastic models of nonequilibrium phase transitions.

The DK model is defined on a tilted square lattice, as shown in Fig. 6.2, which evolves by parallel updates. Let us consider a given lattice configuration at time t, which can be coded as a sequence of binary symbols $s_i(t) \in \{0, 1\}$: If $s_i(t) = 0\,(1)$ at time t, we have an empty/inactive (occupied/active) lattice site at position i. The evolution rule is defined in terms of conditional probabilities involving the two neighbors of each lattice site,

---

[3] A synchronous evolution rule could be thought of as a time-rescaled sequential one, where the time unit has been replaced by the number of lattice sites.

[4] Since on a finite lattice of size L, the number of possible dynamical configurations in a deterministic Boolean CA is finite and certainly bounded from above by 2^L, its deterministic evolution must be periodic. On the other hand, the unpredictability of certain deterministic CA rules emerges when the thermodynamic limit is performed and the transient time needed to reach a periodic state, once averaged over random initial conditions, is typically found to grow exponentially with L. CA are a milestone of modern mathematics because of their conceptual consequences and implications for the theory of computation and, more generally, in many other fields of science including computer science, physics, biology, chemistry, and so on.

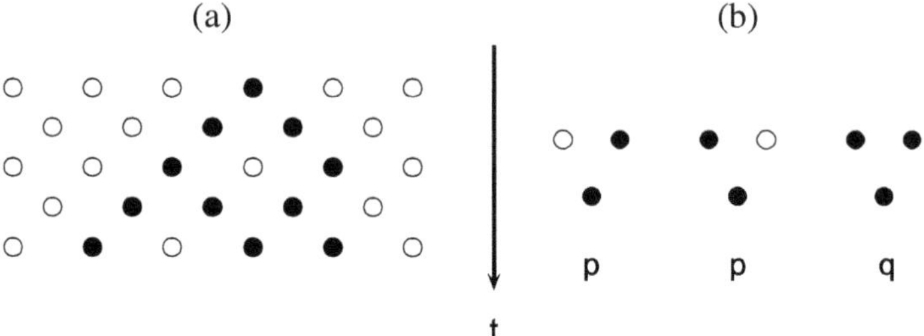

Fig. 6.2 The Domany–Kinzel CA model. (a) Possible evolution of a system where only one site is active at time $t = 0$. (b) Probabilistic evolution rules.

$$P(s_i(t + 1)|s_{i-1}(t), s_{i+1}(t)),$$

$$P(1|0,0) = 0$$
$$P(1|0,1) = P(1|1,0) = p \tag{6.1}$$
$$P(1|1,1) = q,$$

with $P(0|\cdot, \cdot) = 1 - P(1|\cdot, \cdot)$. One can easily realize that due to the first rule in Eq. (6.1), a fully empty configuration corresponds to the "absorbing state." Moreover, a site can be occupied at time $t + 1$ with probability p if one of its neighbor sites was occupied at time t or with probability q if both neighbor sites were occupied at time t.[5]

The phase diagram of the DK model can be represented in the plane of parameters p and q, as shown in Fig. 6.3. The transition line $p_c(q)$ separates the absorbing phase (on its left) from the active one (on its right). In the absorbing phase, clusters of active sites eventually vanish, while in the active phase one observes a fluctuating steady state with an average fraction of active sites depending on the choice of parameter values of the DK model.

There are some special cases in this phase diagram.

(i) The line $q = p(2-p)$ corresponds to the bond DP process. In fact, if one identifies with p the probability of an open channel, the coalescence process of two active sites occurs if at least one channel is open, namely, with probability $q = 1 - (1 - p)^2 = p(2 - p)$.

(ii) The line $q = p$ defines the site DP process, but we must be careful because a mechanistic reading of site percolation requires to distinguish between wet sites and

---

[5] We should make explicit a feature of CA models, which may be easily overlooked. The "tilted" graphical representation of the evolution actually means that even and odd sites are represented in the following rows, because they are updated in turn. So, the real time unit to update all sites is $dt = 2$, not $dt = 1$.

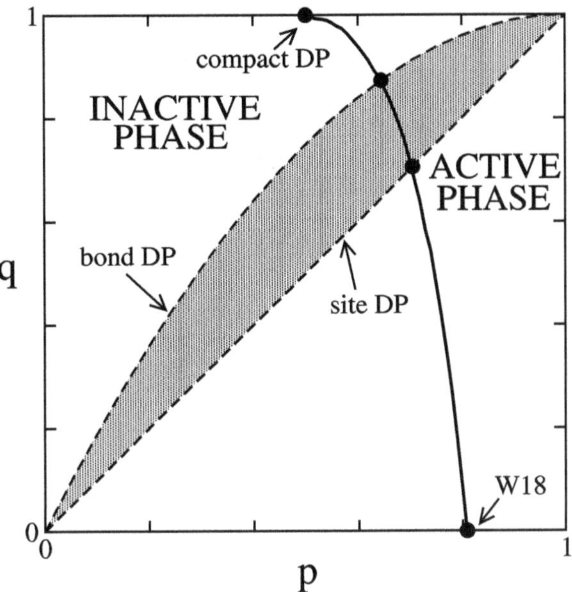

Fig. 6.3 Phase diagram of the DK model. The thick solid line is the curve $p_c(q)$, separating the absorbing (inactive) phase from the active one. The dashed lines correspond to the bond DP ($q = p(2 - p)$, upper line) and to the site DP percolation ($q = p$, lower line). In the region between them, the evolution can be understood in terms of a mechanistic mixed bond/site percolation and finite time reversal symmetry is valid (see the main text for details and caveats). Original data courtesy of Haye Hinrichsen. Haye Hinrichsen, Non-equilibrium Critical Phenomena and Phase Transitions into Absorbing States, *Advances in Physics*, **49** (2010) 815–958.

active (or existing) sites. Sites exist with probability p, and an existing (active) site may be wet or not. Making reference to Fig. 6.2, two neighbouring active and wet sites at time t produce a wet site at time $t + 1$ if such a site exists; therefore, $q = p$. It is not necessary to make this distinction for bond percolation, where the existence or otherwise of a bond between two sites is independent of whether sites are wet/active or not. We will return to this point when discussing time reversal symmetry.

(iii) The region between the two lines, $q = p$ and $q = p(2 - p)$, can be matched to a mixed bond/site percolation, where sites are permeable with probability p_s and bonds are open with probability p_b. In this case, $p = p_s p_b$ and $q = p_s p_b(2 - p_b)$. These equations can be reversed,

$$p_s = \frac{p^2}{2p - q}, \qquad p_b = \frac{2p - q}{p}, \tag{6.2}$$

and writing $q = p(2-p_b)$, it is easy to find that $q \geq p$ (because $p_b \leq 1$) and $q \leq p(2-p)$ (because $p_b \geq p$). The limiting cases $q = p$ and $q = p(2 - p)$ correspond to site and bond percolation, respectively. Therefore, the mixed site-bond percolation corresponds to

$$p \leq q \leq p(2 - p), \tag{6.3}$$

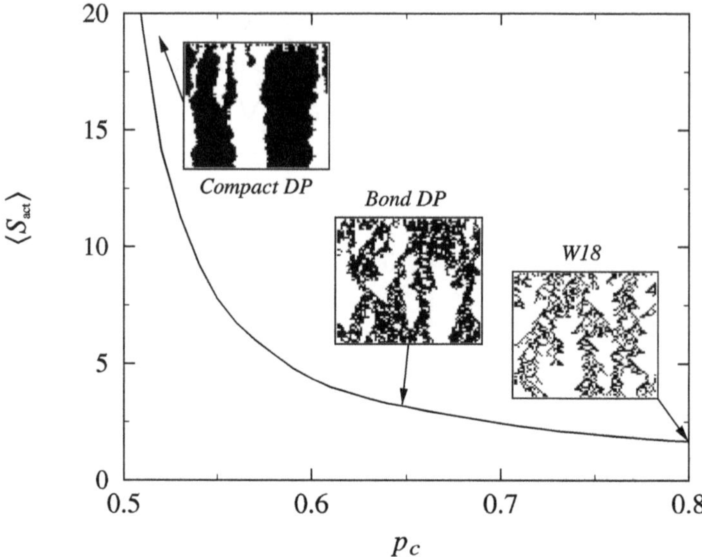

Fig. 6.4 Numerical estimates for the average size of active spots $\langle S_{act}\rangle$ in the DK model measured along the phase transition line. The insets show typical clusters for three special cases discussed in the text. Original data courtesy of Haye Hinrichsen. Haye Hinrichsen, Non-equilibrium Critical Phenomena and Phase Transitions into Absorbing States, *Advances in Physics*, **49** (2010) 815–958.

and this region is the shaded area shown in Fig. 6.3.

(iv) For $q = 0$, two active sites always annihilate. This situation is a stochastic variant of the rule called "W18" according to the Wolfram classification of CA.

(v) For $q = 1$, two active sites always coalesce, and the system has two absorbing states: In addition to the state where all sites are inactive, also the state where all sites are active is now an absorbing state. Dynamics is invariant under the transformation *active* $\leftrightarrow$ *inactive* site and $p \leftrightarrow (1 - p)$. Because of this symmetry, $p_c(q = 1) = \frac{1}{2}$.

The first qualitative feature of the phase diagram (6.3) is that the value of the critical point, $p_c(q)$, is a decreasing function of q. This is not surprising because a larger q means that coalescence prevails over annihilation, contributing to an increase in the number of active sites, therefore favoring the transition from the inactive to the active phase. The most important feature of the phase diagram concerns, however, the nature of the phase transition along the transition line, that is, for different values of q. An exact solution is lacking, but careful numerical studies provide strong evidence that all versions of the DK model belong to the same universality class of bond DP, except for $q = 1$, whose additional symmetry particle $\leftrightarrow$ hole changes the critical behavior of the system. In short, with the caveat that $q < 1$, q is an irrelevant parameter of the DK model.

Belonging to the same universality class means that the critical behavior along the transition line exhibits the same kind of long-range correlations; namely, it is characterized by the same critical exponents. More precisely, in the limits[6] $L \to +\infty$ and $t \to +\infty$,

---

[6] The order in which these limits are performed is important, because in general they do not commute with each other. In the absence of any prior knowledge about the dynamical features of the problem at hand, the

the percolation clusters at $p_c(q)$ can be mapped onto each other by a suitable scale transformation, which amounts to a renormalization procedure analogous to that employed for equilibrium critical phenomena, although in this case space and time coordinates play different roles and yield different scaling properties.

We should also stress that, at variance with long-range correlations, short-range correlations at $p_c(q)$ are found to depend on q. This can be easily checked by measuring the average size of active sites, $\langle S_{\text{act}} \rangle$, in a critical percolating cluster, for different values of q (see the insets in Fig. 6.4). This quantity is found to increase when decreasing $p_c(q)$, that is, when passing from the stochastic W18 rule ($q = 0$) to the special case $q = 1$ through the bond DP case. As shown in the insets of the figure, an increasing value of $\langle S_{\text{act}} \rangle$ means that critical percolating clusters become increasingly dense. In fact, the model with $q = 1$ is also called compact directed percolation, and it will be treated in more detail in Section 6.4.1, because it belongs to a different universality class.

The DK cellular automata can be generalized to $d \geq 2$ space dimensions making use of suitable conditional probabilities to define the evolution rule as follows:

$$P(1|n) = \begin{cases} 0, & n = 0 \\ p_n, & n \geq 1, \end{cases} \tag{6.4}$$

where $0 \leq n \leq 2d$ is the number of occupied sites in the neighborhood of the evolving site. The overall evolution rule depends on the $2d$ parameters p_n, while the absorbing state remains a fully unoccupied lattice. The generalization of bond DP in d dimensions is obtained by taking $p_n = 1 - (1 - p)^n$, which is the probability that at least one of the n channels, connecting the site being updated to the neighboring occupied site, is open. The case of site DP corresponds to a constant, $p_n = p$. In more than one space dimension, one has the freedom of choosing a great deal of evolution rules. On the other hand, one can conjecture that, apart from the special case $P(1|2d) = p_{2d} = 1$, where two absorbing states coexist, the critical properties associated with any evolution rule belong to the same universality class, namely, the one of DP in $d + 1$ dimensions.

6.2.1 Contact Processes

The relevance of the DP universality class goes beyond the DK CA, whose evolution is synchronous, meaning that all lattice sites are updated in one time step. In fact, there are models where DP critical properties emerge from sequential update rules, meaning that one (randomly chosen) lattice site at a time is updated. However, even if critical properties are the

physically suitable procedure amounts to first making the limit $L \to +\infty$ and then the limit $t \to +\infty$, in order to include any possible propagation process emerging from the adopted evolution rule. In many cases, however, we can use some preliminary information to define a proper way of performing these limits: For instance, if we are in the presence of ballistic propagation phenomena with a limit upper velocity $v_{\max}$, the two limits can be performed in such a way that $\lim_{t \to +\infty} L/t \geq v_{\max}$; in the case of a generalized diffusive behavior, we can impose the condition $\lim_{t \to +\infty} L/t^\alpha \geq D_{\max}$, where $\alpha < 1$ and $D_{\max}$ is an upper estimate of the corresponding diffusion parameter. Earlier considerations are particularly relevant for numerical simulations, where we generally proceed the other way round: We fix L and we go to large enough t so that we attain a steady state, but if L is not large enough, results may be affected by finite size effects.

same, short-range correlations are expected to be highly influenced by the adopted scheme, because the sequential update is certainly more effective in destroying local memory effects.

The simplest sequential evolution rule reproducing DP scaling properties at criticality is the contact process, which was introduced as a model of local epidemic spreading without partial or total immunization. The binary occupied/empty state at each lattice site adopted for describing percolating fluids in a porous medium can be turned to the binary infected/healthy state in the language of epidemic propagation in a population, whose individuals are in contact with a finite neighborhood of other individuals. Each infected individual has the possibility to heal itself or infect one of its neighbors according to some assigned probability rates. The possibility of the infection to propagate depends on the choice of these probability rates.

The contact process can be defined on a d-dimensional square lattice, whose sites are labeled by the integer index i. At any time t, on each lattice site i, we can find either an infected $(s_i(t) = 1)$ or a healthy individual $(s_i(t) = 0)$. The state of this population is updated at each time step by choosing at random one site and by assigning to it a new state $s_i(t+1) = 0, 1$. The outcome depends on $s_i(t)$, on the number of infected sites in its neighborhood, $n_i(t) = \sum_{j \in \{i\}} s_j(t)$,[7] and on certain transition rates $w(s_i(t) \rightarrow s_i(t+1), n_i(t))$.

A customary way to define w is to take an infection rate that is barely linear with the fraction of infected neighbors, while the recovery rate is some value r independent of the neighborhood. Thus,

$$\begin{aligned} w(1 \rightarrow 0, n) &= r \\ w(0 \rightarrow 1, n) &= \lambda \frac{n}{2d}. \end{aligned} \tag{6.5}$$

Numerical investigations based on Monte Carlo methods and series expansions indicate that for the contact process in $1+1$ dimensions, one finds a phase transition in the universality class of DP at a critical value of the ratio $(\lambda/r)_c \simeq 3.29785$.

6.3 The Phase Transition in DP-Like Systems

6.3.1 Control Parameters, Order Parameters, and Critical Exponents

The DP class of nonequilibrium phase transitions to one single absorbing state is characterized by typical scaling properties at the transition point that are widely reminiscent of critical phenomena in the equilibrium case (see Section 5.2). For instance, the ferromagnetic transition in the Ising model is found to occur at a critical temperature T_c, where the magnetization vanishes as $M \sim (T_c - T)^\beta$. The temperature T and the magnetization M are the control and the order parameter of this phase transition, respectively. The divergence of the correlation length as $\xi \sim |T - T_c|^{-\nu}$ implies that

---

[7] The symbol $\{i\}$ identifies the set of sites j that are neighbors of i, so that $0 \leq n_i(t) \leq 2d$.

very close to T_c there is no typical macroscopic length scale; that is, the physical system is invariant under scale transformations.

From the discussion in Section 6.2.1, one can easily infer that in the DP class, the natural control parameter, analogous to the temperature T in the equilibrium case, is some probability p, for example, the probability of an open channel in bond DP, the parameter $p(q)$ in the DK model, or the ratio (λ/r) in the contact process. As in equilibrium phenomena, the critical value of the control parameter, p_c, is a model-dependent quantity.

As for the order parameter, there are two possible ways to define it. (1) We can count the active sites and evaluate asymptotically in time their number (or their density), which must vanish in the inactive phase. (2) If we start from a single active site, we can also evaluate the probability of not yet having reached the absorbing phase at diverging time t. More precisely, the first choice depends on the initial conditions, which may be characterized by a vanishing or a finite density of active sites. In the former case (think of the limiting case of one single active site at $t = 0$), the morphology does not scale with the size of the system and we must simply count the number of active sites,

$$N(t) = \left\langle \sum_i s_i(t) \right\rangle, \tag{6.6}$$

where the ensemble average $\langle \bullet \rangle$ is performed over many realizations of the stochastic evolution. For homogeneous initial conditions, instead, we should use the density of active sites,[8]

$$\rho(t) = \frac{1}{L} \sum_i \langle s_i(t) \rangle = \frac{N(t)}{L}, \tag{6.7}$$

where L is the total number of sites.

The second choice for the order parameter is the survival probability for a trajectory whose initial state has one active site only, $s_i(0) = \delta_{i,k}$. It can be formally defined as

$$P(t) = \left\langle 1 - \prod_i (1 - s_i(t)) \right\rangle, \tag{6.8}$$

where $P(t)$ is the ensemble average of an observable that is equal to 1 until an active site is present, while it vanishes only when the system evolves into the fully inactive absorbing state. In other words, $P(t)$ is the fraction of the ensemble of stochastic evolutions that at time t have not yet reached the absorbing state.

In the limit $t \to \infty$, these order parameters can be associated with critical exponents as

$$\rho(\infty) \sim (p - p_c)^\beta, \tag{6.9}$$

$$P(\infty) \sim (p - p_c)^{\beta'}. \tag{6.10}$$

These relations indicate that when approaching the critical point p_c from the active phase, $p > p_c$, both order parameters vanish according to a power law.

---

[8] The density of active sites can actually be evaluated even for a single, initial active site if we normalize $N(t)$ with respect to t, which is the maximal possible number of active sites at time t, starting from a single active site.

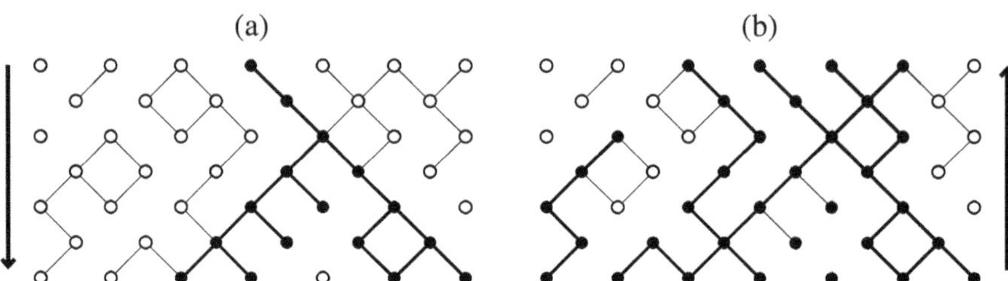

Fig. 6.5   (a) Directed (bond) percolation process starting from a single active site. (b) Time-reversed process of the left one, starting from a fully active state. If there is one directed path from top to bottom (a), there must be a directed path from bottom to top (b).

It is not obvious to say a priori if β and β' should be equal or not; in fact, this depends on the universality class, as revealed by the DK model: For $q < 1$, the DK model belongs to the DP universality class, has only one absorbing state, and, as discussed later, $\beta = \beta' \simeq 0.276$; for $q = 1$, there are two absorbing states, the model belongs to the compact DP universality class, and $\beta \neq \beta'$, as proved in Section 6.4.1.

The equality of the exponents β and β' in DP is a consequence of the fact that in the limit $t \to \infty$, $P(t)$ and $\rho(t)$ are proportional. For bond percolation, it is possible to show that $P(t) = \rho(t)$ at any time, as attested by the direct evaluation of $P(1)$ and $\rho(1)$ for the DK models. $P(1)$ is the probability that a single active site at $t = 0$ survives at $t = 1$, that is, $P(1) = 1 - (1 - p)^2$; $\rho(1)$ is the density of active sites at $t = 1$ starting from $\rho(0) = 1$, that is, $\rho(1) = q$. Therefore, $P(1) = \rho(1)$ if $q = 1 - (1 - p)^2$, that is, for bond percolation. The proof for any time t can be done using time reversal symmetry, using a mechanistic reading of bond percolation, see Fig. 6.5.

In this case, it is possible to establish a priori the evolution from time zero to time t setting existing bonds as a sort of quenched disorder, which has the same effect on a percolation process proceeding downward or upward: In fact, if a pair of sites A, B belongs to a percolating downward path (Fig. 6.5(a)), the same must be true for a percolating upward path (Fig. 6.5(b)), and vice versa. This means that for a given realization of disorder, the percolating downward paths are all and only the percolating upward paths. Said this, let us show how time-reversal symmetry implies $P(t) = \rho(t)$.

The quantity $P(t)$ is evaluated in direct time (Fig. 6.5(a)) wetting a single site and determining if there is a directed path through leading to the opposite (bottom) side. If we are in the thermodynamic limit, the average over disorder realizations can be replaced, using self-averaging, by an average over the starting site, so $P(t)$ is just the fraction of initial sites that are connected to the opposite side. Now we revert the time arrow, as shown in Fig. 6.5(b): We start from all wet sites ($\rho(0) = 1$) and evaluate the density of wet sites at time t, $\rho(t)$. It is clear that a "top" site is now wet if and only if there is a directed path connecting it to the bottom side, which is exactly the condition because, in direct time, it contributes to $P(t)$.[9]

Let us now pass to study correlations. The main feature that makes nonequilibrium critical phenomena different from equilibrium ones is the presence of independent spatial ($\xi_\perp$) and time ($\xi_\parallel$) correlation lengths, where the $\perp$ and $\parallel$ symbols refer to the time arrow. These quantities are associated with the asymptotic behavior of the space and time correlation functions, $c(r)$ and $c(t)$, respectively. Assuming to be in a stationary regime, they write

$$c(r) = \left\langle \overline{(s_i(t) - \bar{s})(s_{i+r}(t) - \bar{s})} \right\rangle \sim e^{-r/\xi_\perp} \tag{6.11a}$$

$$c(t) = \left\langle \overline{(s_i(t_0) - \bar{s})(s_i(t_0 + t) - \bar{s})} \right\rangle \sim e^{-t/\xi_\parallel}, \tag{6.11b}$$

where the brackets $\langle \ldots \rangle$ mean an ensemble average, a horizontal bar $\overline{(\ldots)}$ means a spatial and temporal average,

$$\overline{s_i(t)s_{i+r}(t)} = \lim_{T,L \to \infty} \frac{1}{TL} \sum_{t=t_0}^{t_0+T} \sum_{i=1}^{L} s_i(t)s_{i+r}(t), \tag{6.12}$$

and $\bar{s} = \overline{s_i(t)}$.

The physical interpretation of $\xi_\perp$ and $\xi_\parallel$ is illustrated in Fig. 6.6. In the inactive phase, $p < p_c$, the clusters of active sites generated by a single initially active site (panel (a)) typically extend in space up to a size $\xi_\perp$ and in time up to $\xi_\parallel$, before eventually being absorbed. In the active phase, $p > p_c$, and for the same kind of initial condition (panel (b)), the surviving clusters grow within a "cone" whose opening angle is characterized by the ratio $\xi_\perp/\xi_\parallel$. The correlation lengths can be identified also when using homogenous initial conditions. In fact, for $p < p_c$ (panel (c)), $\xi_\parallel$ amounts to the typical decay time of active clusters, while in the stationary state of the active phase (panel (d)), $\xi_\perp$ identifies the typical size of inactive islands and $\xi_\parallel$ identifies their duration.

---

[9] Earlier considerations about a path between sites A, B at different times apply for any combination of bond and site percolation, that is, a mechanism by which we can go from A to B (and vice versa!) if there is a path connecting them where all bonds and all sites are open / can be wetted. If also the sites A and B are open, this means that a seed at A will wet B, and after time reversal, a seed at B will wet A. That is the meaning of time reversal symmetry. The reason why DK model satisfies such symmetry for bond percolation only can be traced back to the difference between wet and open/active sites, which arises when site percolation is in play and only for the "initial" configuration (see the discussion at point (2), page 246). However, from a physical point of view, it is correct to say that finite time reversal symmetry is valid for any combination of site and bond percolation: In terms of the DK model, this means for any point (p, q) where $q_s(p) \le q \le q_b(p)$, where $q_s(p) = p$ corresponds to site percolation and $q_b(p) = p(2 - p)$ corresponds to bond percolation.

Very close to p_c both correlation lengths are found to diverge as

$$\xi_\perp \sim |p - p_c|^{-\nu_\perp}, \tag{6.13a}$$

$$\xi_\parallel \sim |p - p_c|^{-\nu_\parallel}. \tag{6.13b}$$

From a practical point of view, the most efficient order parameter for determining the critical point of DP processes by numerical simulation is the average cluster mass (6.6), which is found to grow algebraically at p_c as

$$N(t)|_{p=p_c} \sim t^\theta. \tag{6.14}$$

Instead, for $p < p_c$, $N(t)$ asymptotically decreases and vanishes after an initial increase, while for $p > p_c$, $N(t)$ grows exponentially. Therefore, plotting $N(t)$ versus t in a log-log scale, we obtain a linear behavior at criticality, a positive curvature above criticality, and a negative curvature below criticality.

All these critical exponents, characterizing nonequilibrium phase transitions in systems with absorbing states, are not independent of each other. In equilibrium critical phenomena, we could use well-known thermodynamic and statistical mechanical relations to derive the Josephson, Fisher, Widom, and Rushbrooke relations. In the present, nonequilibrium context there is only one hyperscaling relation, that is, a relation that involves the dimension d of the system (like Josephson's). The machinery required to get it is difficult and we just report it,

$$\theta = \frac{d\nu_\perp - \beta - \beta'}{\nu_\parallel}. \tag{6.15}$$

6.3.2 Phenomenological Scaling Theory

Similarly to the procedure adopted in equilibrium critical phenomena, we can work out a phenomenological scaling theory by making explicit the scale invariance engendered by the divergence of space and time correlation lengths close to p_c. In practice, we can assume

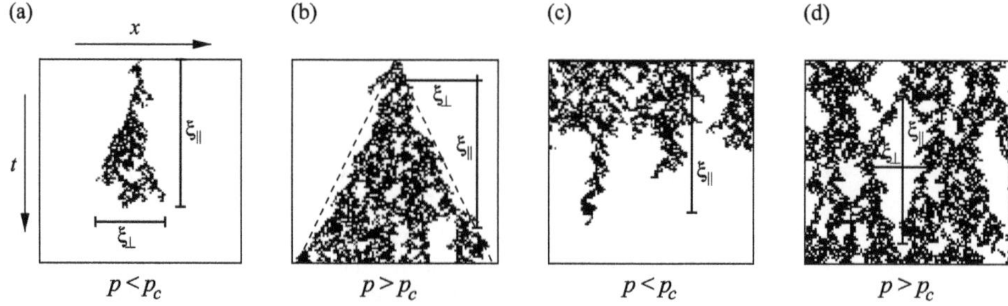

Fig. 6.6   Pictorial description of the correlation lengths $\xi_\parallel$ and $\xi_\perp$ in a DP process for different initial conditions, below and above criticality. The explanations for part labels (a)–(d) are discussed in the text. From Haye Hinrichsen, Non-equilibrium Critical Phenomena and Phase Transitions into Absorbing States, *Advances in Physics*, **49** (2010) 815–958.

that close to the critical point, the macroscopic properties of DP processes are invariant under scaling transformations of the following form

$$\Delta \to \Lambda \Delta, \quad x \to \Lambda^{-\nu_\perp} x, \quad t \to \Lambda^{-\nu_\parallel} t, \quad \rho \to \Lambda^{\beta} \rho, \quad P \to \Lambda^{\beta'} P, \tag{6.16}$$

where we have introduced the quantity $\Delta = |p - p_c|$, which plays the role of the reduced temperature in the equilibrium scaling theory. This means that if we rescale the distance from the critical point by an arbitrary scaling factor Λ, we can recover the same macroscopic properties of the original DP process by rescaling all other physical quantities (space x, time t, density of active sites ρ, and survival probability P) making use of the Eqs. (6.9), (6.10), (6.11), and (6.13).

Anyway, there is a degree of arbitrariness in choosing the quantity to be rescaled, similar to the equilibrium case, where we used either the rescaled temperature (Section 5.2.5) or the correlation length (Section 5.2.6) as building blocks of a scaling theory. For example, we could impose an equivalent assumption of scale invariance by first rescaling the space variable x ($x \to \Lambda x$), then rescaling all the other quantities accordingly. One can easily realize that this new set of scale transformations would be the same as (6.16), modulo a redefinition of the scale parameter, $\Lambda \to \Lambda^{-1/\nu_\perp}$.

Scaling relations allow us to determine the dependence on time of the order parameters close to the critical point. For instance, the average density of active sites $\rho(t)$ starting from a homogeneous initial condition has to be scale invariant at the critical point. Using Eq. (6.16), we obtain

$$\rho(\Lambda^{-\nu_\parallel} t) = \Lambda^{\beta} \rho(t). \tag{6.17}$$

We can choose Λ such that $\Lambda^{-\nu_\parallel} t = 1$ and we immediately obtain how this quantity decays in time at the critical point,[10]

$$\rho(t) = t^{-\beta/\nu_\parallel} \rho(1) \sim t^{-\delta}. \tag{6.18}$$

In a similar way, we can obtain the decay of the survival probability $P(t)$,

$$P(t) \sim t^{-\delta'}, \tag{6.19}$$

with $\delta' = \beta'/\nu_\parallel$. Since in DP processes $\beta = \beta'$, we have also $\delta = \delta'$ and the hyperscaling Eq. (6.15) simplifies to

$$\theta = \frac{d\nu_\perp - 2\beta}{\nu_\parallel}. \tag{6.20}$$

It is important to point out that the knowledge of the critical exponents provides relevant information about the behavior of the order parameters close to the critical point and in a finite-size system. Actually, in these conditions, ρ and P should depend on three parameters, namely, the time t, the distance from the critical point Δ, and the system size V (equal to L, in $d = 1$). On the other hand, the property of scale invariance implies that one of these

---

[10] Note that $\rho(t)$ in the absorbing phase typically vanishes exponentially in time, with a decay rate dependent on the value of the parameters of the model.

parameters can be expressed in terms of the others, thus yielding the expressions

$$\rho(t, \Delta, V) \sim t^{-\beta/\nu_{\parallel}} f(\Delta t^{1/\nu_{\parallel}}, t^{-d/z} V) \tag{6.21}$$

$$P(t, \Delta, V) \sim t^{-\beta'/\nu_{\parallel}} g(\Delta t^{1/\nu_{\parallel}}, t^{-d/z} V), \tag{6.22}$$

where f and g are suitable scaling functions whose explicit expression is unknown.

6.3.3 Mean-Field Theory

In analogy with equilibrium phase transitions, it has been conjectured that the notion of universality applies also to continuous nonequilibrium phase transitions. This means that scaling properties characterizing the behavior close to the critical point depend on basic properties only and are independent of the details of the model at hand. It is worth pointing out that such a conjecture has been successfully checked in many cases by careful numerical studies, but a rigorous mathematical proof is still unknown, because the dynamical renormalization group method encounters more serious technical difficulties to be worked out than the static one.

DP processes are characterized by a transition from a fluctuating active phase to a unique absorbing state and the order parameter is a nonnegative scalar quantity, while the evolution rule involves short-range interactions (like those between nearest-neighbor sites in the DK model). No other symmetries or conservation laws should be present in the model. In fact, models sharing such basic properties have been found to belong to the DP universality class.

By further extending the analogy with equilibrium phase transitions, we can wonder if there is a general mean-field-like approach to DP processes, analogous to the field-theoretic formulation of critical phenomena provided by the Landau theory. Since we want to deal with nonequilibrium processes, where time plays a crucial role, we certainly need a sort of mean-field dynamical equation. Earlier, we have widely discussed the Langevin equation as an effective continuous-time, coarse-grained formulation of microscopic stochastic processes (e.g., the random-walk model of diffusion). We can proceed in a similar way for DP or contact processes by introducing a phenomenological Langevin equation for the density of active sites at position $\mathbf{x}$ at time t, $\rho(\mathbf{x}, t)$. In full generality, we assume here that $\mathbf{x}$ is a vector in d space dimensions. Due to the hypothesis of scaling invariance at the critical point, we have to think about the quantity $\rho(\mathbf{x}, t)$ as a space–time coarse-grained average of the number of active sites in the microscopic configuration. Rather than following a rigorous mathematical procedure based on the master equation of the contact process, here we prefer to derive this equation by heuristic arguments. Let us start by making reference to the microscopic mechanism of contact processes. According to Eq. (6.5), r is the recovery rate of an infected site, while the infection rate of a site is equal to $\lambda(n/2d)$, therefore being proportional to the fraction of infected neighbor sites. In a mean-field formulation, the infection process occurs at a rate $\lambda\rho(1 - \rho)$, because it requires the site in question to be inactive (which occurs with probability $1 - \rho$) and the fraction of infected neighbors is the density ρ itself. Even more simply, the recovery process occurs at a rate $r\rho$. We can sum up the different contributions and obtain the equation

$$\frac{d\rho}{dt} = \lambda\rho(1-\rho) - r\rho \equiv a\rho - \lambda\rho^2, \tag{6.23}$$

where $a = \lambda - r$. If $\lambda < r$, $d\rho/dt < 0$, and the only steady solution is $\rho_1^* = 0$, corresponding to all inactive sites (the absorbing state). If $\lambda > r$, the absorbing state is unstable because $d\rho/dt > 0$ for a small density of active sites, and a new steady solution appears, $\rho_2^* = a/\lambda = 1 - r/\lambda$, which is stable. Therefore, this simple mean-field picture provides a transition between an absorbing and an active phase at the critical threshold of the control parameter, $(\lambda/r)_c = 1$, which should be compared (in $d = 1$) with the exact value $(\lambda/r)_c = 3.298$, determined numerically.

In the earlier formulation, we have put aside two ingredients that are important: the spatial dependence of the density ρ and the stochastic nature of these processes. The first ingredient is taken into account by adding a diffusive term, the second one adding a noise term, and both terms require a few words of explanation. In DK models, a single active site can diffuse, while in our basic contact model of epidemic spreading, diffusion seems absent. In fact, a more formal derivation of the time evolution of ρ should take into account that the sites neighboring $\mathbf{x}$ are one lattice constant far from $\mathbf{x}$, so the density evaluated there should be Taylor expanded, which would give a diffusion-like term on the right-hand side of Eq. (6.23).

The noise term is less trivial than we can expect on the basis of the Langevin equations we have studied until now. In fact, the dynamics of an absorbing state is a trivial, deterministic dynamics that simply reproduces the same state at all times: Therefore, the noise term must be switched off in proximity to an absorbing state. This is possible because for contact/DP/DK models, noise is not due to some "external" source as in the case of a Brownian particle; instead, it is due to the density itself of the active sites and must vanish when $\rho(\mathbf{x}, t) = 0$. We can now write a phenomenological Langevin equation for contact processes as

$$\frac{\partial\rho(\mathbf{x}, t)}{\partial t} = a\,\rho(\mathbf{x}, t) - \lambda\rho^2(\mathbf{x}, t) + D\nabla^2\rho(\mathbf{x}, t) + \eta(\mathbf{x}, t), \tag{6.24}$$

where D is a diffusion constant and the stochastic field $\eta(\mathbf{x}, t)$ amounts to a zero average process with a correlation function proportional to $\rho(\mathbf{x}, t)$,

$$\langle\eta(\mathbf{x}, t)\rangle = 0 \tag{6.25}$$

$$\langle\eta(\mathbf{x}, t)\eta(\mathbf{x}', t')\rangle = \Gamma\rho(\mathbf{x}, t)\delta(\mathbf{x} - \mathbf{x}')\delta(t - t'), \tag{6.26}$$

where Γ has the physical dimension of inverse time.

The last relation indicates that the stochastic field $\eta(\mathbf{x}, t)$ is proportional[11] to $\sqrt{\rho(\mathbf{x}, t)}$. This square-root dependence can be argued to be a consequence of the central limit theorem (see Section 3.2) . More precisely, since we are assuming that $\rho(\mathbf{x}, t)$ is a mesoscopic representation of the density of active sites, the large number of individual, independent noise contributions emerging from the microscopic stochastic process on a coarse-grained scale sum up to a Gaussian distribution, with a variance proportional to the number of active sites in this mesoscopic region.

---

[11] For this reason, the noise is sometimes written as $\eta(\mathbf{x}, t) = \sqrt{\rho(\mathbf{x}, t)}\,\bar{\eta}(\mathbf{x}, t)$, where $\bar{\eta}(\mathbf{x}, t)$ is delta-correlated.

The stochastic field $\eta(\mathbf{x}, t)$ is an example of multiplicative noise, where the effective amplitude of fluctuations is modulated by the density field $\rho(\mathbf{x}, t)$ itself: a peculiar situation where one of the competing "ground states" of the phase transition is set at zero temperature. This feature modifies significantly the nature of the Langevin equation with respect to the standard case of a purely additive noise (as in the model of generalized Brownian motion), and it has a crucial influence on the critical properties of contact processes.

It is now worth stressing that the same continuum description (6.24) can be easily obtained for reaction–diffusion dynamics, once we have the rates for the elementary processes described at the end of Section 6.1: death (ν_d), offspring (ν_o), coalescence (ν_c), and annihilation (ν_a). It is straightforward to write that the linear term in ρ depends on unary processes, $a = \nu_0 - \nu_d$, while the quadratic term in ρ depends on binary processes, $\lambda = \nu_c + 2\nu_a$.

The next step is to apply the scale transformations Eq. (6.16) to Eq. (6.24), taking into account that the distance Δ from the critical point is now equal to $a = \lambda - r$ because the critical point is defined precisely by the condition $a = 0$. With the relation $\Delta = a$ in mind, the rescaled Equation (6.24) is

$$\Lambda^{\beta+\nu_\parallel} \frac{\partial \rho(\mathbf{x}, t)}{\partial t} = \Lambda^{\beta+1} a\,\rho(\mathbf{x}, t) - \lambda\Lambda^{2\beta}\rho^2(\mathbf{x}, t) + D\Lambda^{\beta+2\nu_\perp}\nabla^2\rho(\mathbf{x}, t) + \Lambda^\gamma\eta(\mathbf{x}, t), \quad (6.27)$$

where the exponent $\gamma = (\beta + d\nu_\perp + \nu_\parallel)/2$ stems from Eq. (6.26), while taking into account the property of the Dirac delta distribution $\delta(cx) = \frac{1}{|c|}\delta(x)$.

Dividing all terms by $\Lambda^{\beta+\nu_\parallel}$, we obtain

$$\frac{\partial \rho(\mathbf{x}, t)}{\partial t} = \Lambda^{1-\nu_\parallel} a\,\rho(\mathbf{x}, t) - \lambda\Lambda^{\beta-\nu_\parallel}\rho^2(\mathbf{x}, t) + D\Lambda^{2\nu_\perp-\nu_\parallel}\nabla^2\rho(\mathbf{x}, t) + \Lambda^{\gamma-\beta-\nu_\parallel}\eta(\mathbf{x}, t). \quad (6.28)$$

If we impose that the deterministic part of the Langevin equation is invariant under scale transformations, we must impose the vanishing of the relative scaling exponents, which allows us to obtain

$$\beta = 1, \quad \nu_\perp = \frac{1}{2}, \quad \nu_\parallel = 1. \quad (6.29)$$

These values define the mean-field critical exponents for the DP universality class. Their validity depends on the possibility of disregarding fluctuations, which in turn depends on the exponent of the term renormalizing the noise. In fact, we expect noise is irrelevant if

$$\gamma - \beta - \nu_\parallel > 0, \quad (6.30)$$

that is,

$$d > \frac{\beta + \nu_\parallel}{\nu_\perp}. \quad (6.31)$$

Using the mean-field values (6.29), we obtain the condition $d > 4$.

In conclusion, dimensional analysis indicates that noise is irrelevant for $d > 4$ (i.e., approaching the critical point, it scales to zero as a positive power of Λ); it is marginal for $d = 4$; and it is relevant for $d < 4$. Since, as in equilibrium phase transitions, fluctuations determine the critical behavior, these considerations amount to obtaining the Ginzburg criterion for identifying the upper critical dimension, $d_c = 4$, for contact processes.

Table 6.1 Numerical and mean-field critical exponents for the DP universality class

Exponent	$d = 1$	$d = 2$	$d = 3$	Mean Field
β	0.276	0.583	0.813	1
$\nu_\perp$	1.097	0.733	0.584	$\frac{1}{2}$
$\nu_\parallel$	1.734	1.295	1.110	1
θ	0.314	0.229	0.114	0^a

[a] In the mean-field regime, the hyperscaling relation is valid only for $d = d_c^u = 4$.

Accordingly, we expect that the mean-field critical exponents become exact only for $d > 4$, while for lower dimension, they are expected to depart significantly from exact ones, as shown in Table 6.1. For $d = 4$, we expect subleading logarithmic corrections to mean-field predictions.

Earlier we have not mentioned the exponent θ because it should be determined by the hyperscaling relation (6.15), but the latter is not valid in the mean-field regime. In practice, such a relation can be used with mean-field exponents for $d = 4$ only. In this case, we obtain $\theta = (d - 4)/2 = 0$, which is in agreement with the observed decrease of $\theta(d)$ for $d < 4$.

6.4 Beyond the DP Universality Class

Earlier in this chapter, we focused on the model of DP, because it can be formulated in simple terms and because many other models share the same long-range properties, therefore the same critical exponents: The Domany–Kinzel cellular automata and the contact processes are two relevant examples. We now want to discuss other universality classes, but in order to go beyond DP, it is necessary to introduce some relevant changes. One possibility has already been mentioned in Section 6.2, where we have shown that for $q = 1$, the DK model passes from one to two absorbing states. A second possibility is to introduce a conservation law. These two new universality classes are discussed in Sections 6.4.1 and 6.4.2, respectively.

6.4.1 More Absorbing States

Compact Directed Percolation

The DK cellular automata trivially have one absorbing state, where all sites are inactive ($\circ$), because $\circ\circ \rightarrow \circ$. However, for $q = 1$, the same process with active sites occurs too, $\bullet\bullet \rightarrow \bullet$, implying that the fully active state is an absorbing state as well. Even more important, there is a symmetry between the fully inactive and the fully active state, because dynamics is invariant under the exchange of particles and holes if $p \rightarrow 1 - p$. This symmetry implies that it must be $p_c(q = 1) = 1 - p_c(q = 1)$, that is, $p_c(1) = \frac{1}{2}$.

The simplest way to show that the case $q = 1$ (called compact directed percolation, CDP) lies in a universality class other than DP is to prove that $\beta(q = 1) \neq \beta'(q = 1)$. The CDP model can be rephrased in terms of domain walls between active and inactive regions. In fact, new walls cannot be created, because this would require the formation of inactive (active) sites in the middle of an active (inactive) region, which is forbidden by the absorbing character of both states: This is the reason why this is called compact directed percolation. Since the creation of domain walls is prohibited, their number can only decrease through annihilation processes, which occur when an isolated active or inactive site disappears.

The exponent β' can be easily evaluated by considering the evolution of a single active site, which produces a compact cluster of width $L(t)$, and we wonder what the probability $P(p)$ is that it grows to infinity rather than dying out. In close proximity to $p_c = \frac{1}{2}$, such probability has the expression $P(p) \approx (p - p_c)^{\beta'}$, which defines the sought-after exponent β'. Using the evolution rules of DK cellular automata for $q = 1$ (see Fig. 6.2), we can say that a cluster of L active sites at time t: (1) increases its size at time $t + 1$ if both processes occurring at the bounds of the cluster produce active sites, $\circ\bullet \to \bullet$ (Fig. 6.2(a)) and $\bullet\circ \to \bullet$ (Fig. 6.2(b)); (2) decreases its size if both such processes produce inactive sites; (3) maintains its size if one process produces an active site and the other process produces an inactive site. Summarizing,

$$L(t + 1) = \begin{cases} L(t) + 1, & \text{with probability } p^2 \\ L(t) - 1, & \text{with probability } (1 - p)^2 \\ L(t), & \text{with probability } 2p(1 - p). \end{cases} \tag{6.32}$$

We therefore have that $L(t)$ performs an asymmetric random walk with a probability $r = p^2/(p^2 + (1 - p)^2)$ to move to the right and a probability $1 - r$ to move to the left.[12] We wonder what the probability is that $L(t = \infty) = \infty$ knowing that $L(0) = 1$ and that the random walk has an absorbing barrier for $L = 0$. The answer is given in Section 2.3.2, Eq. (2.92),

$$P(p) \equiv 1 - \mathfrak{p}_2 = 1 - \left(\frac{1 - r}{r}\right) = \frac{2}{p^2}\left(p - \frac{1}{2}\right). \tag{6.33}$$

Close to the critical value $p_c = \frac{1}{2}$, we have $P(p) \simeq \frac{2}{p_c^2}(p - p_c)$, so that $\beta' = 1$.

The exponent β instead refers to the asymptotic, time-independent fraction of active sites when we start from a homogeneous state with a finite density of active sites, $\rho(p) \simeq (p - p_c)^\beta$. In the language of domain walls, at $t = 0$ we have an ensemble of particles separating active and inactive regions, and for $p > p_c$, they diffuse so as to favor the active phase with respect to the inactive one: The final state is fully active,[13] so $\rho(p > p_c) = 1$ and a discontinuity appears, which means $\beta = 0$. In conclusion, $\beta = 0$ and $\beta' = 1$, proving that CDP does not belong to the DP universality class, whose critical exponents are given in Table 6.1 (in particular, $\beta(\text{DP}) = \beta'(\text{DP}) \simeq 0.276$ in $d = 1$).

---

[12] The probability to keep the same value of L does not affect the evaluation of $P(p)$, it simply modifies $\xi_\parallel$ by a constant prefactor.

[13] The dynamics of CDP in proximity to the critical point is equivalent to the zero-temperature, spin-flip dynamics of the kinetic Ising model (see Section 8.4.2): Active/inactive sites correspond to positive/negative spins in the presence of a magnetic field $H = p - p_c$. The result $\beta = 0$ is equivalent to say that a positive magnetice field always favors the phase with up spins.

Also, the other critical exponents can be found by using the properties of random walkers. For simplicity we are confined to an initial condition corresponding to a single active site, whose dynamics is summarized in Eq. (6.32). The exponent θ rules how the size L of an active region increases in time at criticality ($p = p_c$), $L(t) \sim t^{\theta}$. Taking advantage of the equivalence of the CDP model with a random walker on the line, starting at position $x_0 = 1$ (because $L(0) = 1$) and with a trap in $x = 0$ (because the cluster dies), we can use Appendix M.4 to determine $L(t)$ in a continuum picture. At criticality, where the walker diffuses symmetrically (see Eq. (6.32)), the result is that the average value of $L(t)$, corresponding to the average position of the walker, is constant in time. Therefore, $\theta = 0$.

The exponents $\nu_{\|}$ and $\nu_{\perp}$ describe the divergence of the correlation lengths. Below criticality they are defined as $\xi_{\|,\perp} \sim (p_c - p)^{-\nu_{\|,\perp}}$ and the quantity $\xi_{\|}$ is nothing but the average lifetime of the walker, while $\xi_{\perp}$ is the average distance attained before the process dies out; see Fig. 6.6 for a graphical representation. In Appendix M.5, Eq. (M.52), we find that in the limit of large times the trapping probability at time t decays as $e^{-v^2 t/4D} \equiv e^{-t/\xi_{\|}}$, which gives $\xi_{\|} \sim 1/v^2$. The drift v is proportional to the asymmetry δ,[14] defined as the normalized difference between the rate of hopping to the right and to the left (see Eq. (6.32)),

$$\delta = \frac{p^2 - (1-p)^2}{p^2 + (1-p)^2} = -\frac{1 - 2p}{1 + 2p(1-p)}. \tag{6.34}$$

In proximity to the critical value $p_c = \frac{1}{2}$, $\delta \simeq -(p_c - p)$. Summarizing, $\xi_{\|} \sim v^{-2} \sim |\delta|^{-2} \sim (p_c - p)^{-2}$, so $\nu_{\|} = 2$. As for the orthogonal correlation length, $\xi_{\perp}$, it is sufficient to note that in the limit $p \to p_c$ the size of the cluster performs a standard random walk, therefore attaining a maximal size of the order $\xi_{\perp} \sim \xi_{\|}^{1/2} \sim (p_c - p)^{-1}$, so $\nu_{\perp} = 1$.

In conclusion, for the CDP model we have the following critical exponents: $\beta = 0, \beta' = 1$, $\theta = 0, \nu_{\|} = 2$, and $\nu_{\perp} = 1$. These exponents satisfy the hyperscaling relation (6.15) among critical exponents.

The Z_2-symmetric Directed Percolation

The CDP model has the value of being exactly solvable, but the feature of compact clusters limits its generality. For this reason we mention another change of universality class induced by having more than one absorbing state. If we pass from DP to CDP making the fully active state an absorbing state, a different scenario is to have more than one inactive state (which is absorbing by definition). We may think to have two inactive states (I_1 and I_2) and one active state (A). As in DP, I_k sites can reproduce only themselves ($P(I_k|I_k, I_k) = 1$) and an (active, inactive) pair can produce an active site, $P(A|I_k, A) = p$, or the *same* inactive site, $P(I_k|I_k, A) = 1 - p$. Furthermore, two active sites can reproduce ($P(A|A, A) = q$) or give rise to an inactive one ($P(I_1|A, A) = P(I_2|A, A) = (1 - q)/2$). These rules are the natural generalization of DK evolution rules (see Eq. (6.1)), but we also need to specify what the result is of a pair of different inactive sites. The model DP2 (Z_2−symmetric directed percolation) corresponds to implementing a sort of interfacial noise, with $P(A|I_1, I_2) =$

---

[14] See the problem of random walk on a ring in Section 2.3.2.

$P(A|I_2, I_1) = 1$. In the absence of interfacial noise, the distinction between the two different types of inactive sites would be completely fictitious.

It is worth mentioning that the two inactive (absorbing) states are equivalent and this is what makes DP2 a universality class different from DP. If we create an asymmetry between I_1 and I_2, for example, by imposing $P(I_1|A, A) > P(I_2|A, A)$, this asymmetry would induce a preference for I_1 that would rule over I_2 and the model would fall in the DP universality class.

Dynamical Percolation

In Section 6.2.1, we have shown that stochastic processes like DP can also be interpreted as contact processes modeling, for example, an epidemic spreading: The activation of a site in a lattice by an active neighboring looks quite similar to the mechanism of infection transmission by contact from an infected individual to a healthy one. In fact, the probability per unit time of activating a neighboring site can be read as the infection rate of other individuals. Moreover, the probability per unit time that an active site turns into an inactive state is analogous to the rate of immunization of infected individuals. On the basis of these simple considerations, one can appreciate the strong similarity of DP to epidemic spreading processes. On the other hand, the mechanisms of real epidemics depend on additional ingredients. For instance, real epidemics spread in a highly disordered environment rather than in a regular lattice and individuals may have different responses to infection, according to their attitude of adopting immunization strategies, like vaccination. Moreover, infection can be transmitted by short-range as well as long-range mechanisms. It is well known that many epidemics may propagate through international travel via the transport of pathogens to geographical areas different from their origin.

In general, epidemic processes without immunization belong to the universality class of DP. As soon as immunization is introduced, the overall mechanism changes. In fact, we can assume that there are two different probabilities, p_1 and p_2 for the first and the second infection, respectively. According to common experience, the condition $p_1 > p_2$ is usually assumed. As soon as we introduce immunization, we obtain an infinite number of absorbing states, because any combination of healthy individuals (either immunized or not) cannot evolve.

Perfect immunization corresponds to the case $p_2 = 0$, where previously infected sites, once recovered, cannot be infected any longer. This means that the number of recovered sites can only increase and in a finite size system, all sites are first infected, then recovered. This process is called dynamical percolation (DyP) and we can exemplify this spreading process by considering an initial state where all sites are equally susceptible to infection and inserting a single infected site (the so-called patient zero) at the origin: How does the epidemic spread through the lattice? Since previously infected sites become immune, the infection front propagates and leaves behind a cluster of immune sites, but the precise form of this cluster depends on the infection probability p_1. In the supercritical phase ($p_1 > p_{1c}$), there is a finite probability that the infected front propagates to infinity, while in the subcritical phase the infection process eventually stops after a finite time. The DyP universality class is robust with respect to a small value of the reinfection rate $p_2 > 0$. If p_2

is finite, all sites are asymptotically infected or recovered and there is a finite probability to pass from one state to the other (and vice versa).

6.4.2 Conservation Laws

A different strategy to modify a DP model in order to enter a new universality class is to add a symmetry. The simplest example makes use of the representation of DP in terms of reaction–diffusion processes, as discussed at the end of Section 6.1 (the specific model is called branching annihilating random walk). In the DP case, there are processes (death, offspring, coalescence) where the number of active sites changes by one unit, therefore changing the even/odd parity of the number of active sites, $N(t)$. The simplest reaction–diffusion model giving rise to an absorbing phase transition is the branching annihilation random walk, summarized by the elementary processes: $A \rightarrow (n + 1)A$, $2A \rightarrow 0$, and $A0 \leftrightarrow 0A$. If n is odd, we fall in the DP universality class, but an even n gives rise to new universality classes, called parity conserving (PC). Let us describe operationally an algorithm for $n = 2$.

Particles are located on a regular one-dimensional lattice and one particle at a time is randomly chosen. The particle can either diffuse to a nearby site with probability p or generate two offspring at the neighboring sites with probability $1 - p$. In the diffusion case, the particle moves with equal probability to the left site or to the right one: If the selected site is occupied by another particle, they annihilate with probability r; otherwise, the particle remains in its site. In the branching case, the two offspring occupy the two nearest-neighbor sites. If one or both sites are occupied by particles, they annihilate with probability r; otherwise, no new birth occurs.

In the plane (r, p), there is an active phase for $p < p_c(r)$. Numerical estimates of the critical exponents in 1+1 dimensions indicate that our PC class is quite different from the DP class of nonequilibrium critical phenomena, as shown by the values

$$\beta = \beta' \approx 0.92, \quad \nu_\| \approx 3.22, \quad \nu_\perp \approx 1.83, \tag{6.35}$$

if they are compared with DP exponents, see Table 6.1.

The slip exponent θ and the survival probability exponent δ' are defined at the critical point and highlight some effects of the conservation constraint. In fact, because of the parity conservation, starting from an odd number of particles, we can never attain the complete absorbing phase ($N = 0$). If we initialize the system with a pair of particles, $N(0) = 2$, at criticality we find $\delta' = 0.285$ and $\theta = 0$.[15] In particular, we see that the scaling relation $\delta' = \beta'/\nu_\|$ is satisfied. Conversely, if we start with a single particle, $N(0) = 1$, the survival probability can not vanish, so $\delta' = 0$ and the above scaling relation is *not* satisfied. On the other hand, it is found $\theta = 0.285$, so the roles of the two exponents are surprisingly exchanged when we modify the parity of the initial state.

Another surprising relaxation property concerns the inactive phase, below the critical point, where the particle density decays as a power law, $\rho(t) \sim 1/\sqrt{t}$, rather than exponentially as it happens in the DP class, where $\rho(t) \sim e^{-t/\xi_\|}$. This occurs because in the PC case, the asymptotic absorption of active sites is governed by the annihilation

---

[15] The vanishing of θ means that $N(t)$ increases logarithmically.

process $2A \rightarrow 0$, which requires the diffusion of particles, from which the density decays is the square root of time.[16] Instead in the DP class, the unary process $A \rightarrow 0$ leads to a much faster decay.

6.5 Self-Organized Critical Models

The critical phenomena studied up to now, whether equilibrium or nonequilibrium, had the common feature that the fine-tuning of a suitable control parameter was necessary in order to get a critical behavior, characterized by scale invariance. There are, however, driven-dissipative systems that spontaneously evolve toward a critical dynamics, characterized by a power-law distribution of relaxation events, and this phenomenon does not require the tuning of any parameter, hence it is named self-organized criticality (SOC).

This phenomenon was discovered in some seminal papers by the Danish Per Bak and coworkers at the end of the 1980s. Since then, several models have been proposed to provide a mathematical theory for a class of phenomena that seem to be ubiquitous in nature. In fact, SOC phenomena have been identified in geology, paleontology, cosmology, evolutionary biology, plasma physics, neurobiology, and so on. For instance, already in the 1950s, the German Beno Gutenberg and the American Charles Francis Richter had found an empirical law according to which the probability to have (in any given region and time period) an earthquake of at least magnitude E obeys a universal power-law behavior,

$$P(E) \sim E^{-\gamma}, \tag{6.36}$$

over quite a large range of values of E, with $\gamma \approx 1$.

The bad news is that such a power-law distribution implies a non-negligible probability that an earthquake magnitude is sensibly large. More generally, the main consequence of this behavior is that the magnitude distribution of earthquakes is not characterized by a typical size and it does not obey the Gaussian statistics imposed by the central limit theorem (see Section 3.2). We must therefore conclude that earthquakes are not the outcome of random independent events and their occurrence is the result of well-correlated microevents.

On a physical ground, we can observe that the basic mechanism for producing SOC behavior is the presence of a slow driving process allowing for the accumulation of energy or matter, which is suddenly released through an avalanche-type process and which leads toward a quiescent, absorbing-like state.[17] Therefore, a typical feature of SOC is the wide separation between the driving and the relaxation time scales.[18] Once we inject new energy or matter, the whole process restarts and leads to a new absorbing state. For this reason it has been argued that SOC is closely related to nonequilibrium phase transitions into infinitely

---

[16] This decay corresponds to an average distance between particles growing as $t^{1/2}$. The same law, with the same "diffusional" explanation, will be found in Chapter 8.

[17] For instance, during storms a large amount of electrostatic energy is accumulated by cloud friction in the surrounding layers of the atmosphere and it is suddenly released when the voltage difference with the ground overtakes the ionization threshold of the air, thus producing lightnings.

[18] The request that the ratio between the dissipation time t_d and the time of energy injection t_e vanishes, $t_d/t_e \rightarrow 0$, somewhat weakens the statement that SOC does not require the tuning of any parameter.

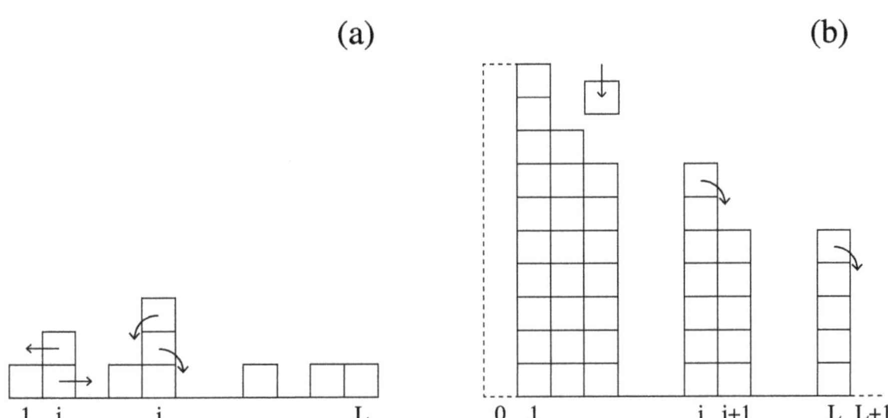

Fig. 6.7 (a) Sketch of the model, without driving or dissipation, introduced in Section 6.5 and preparatory to SOC models. We distribute N bricks on a lattice of fixed size and dimension d ($d = 1$ in the figure). Sites whose height h is larger than or equal to a given value h^* ($h^* = 2$ in the figure) are active sites. An active site is chosen randomly and its height is reduced by two bricks, which are distributed one per neighboring site. In the specific case, there are two active sites, i and j, and such redistributions make active the sites $i - 1$ and $j - 1$. (b) The Bak–Tang–Wiesenfield (BTW) model, with the three basic processes occurring during its dynamics: the toppling of a new sand grain (driving) on a column; the move of a sand grain from site i to site $i + 1$, if $z_i = h_i - h_{i+1} > Z$ ($Z = 2$ in the figure); the removal of a sand grain (dissipation) from site L, if $h_L > Z$. Dashed lines depict boundary conditions, introducing fictitious sites in $i = 0$ (with $h_0 = h_1$) and in $i = L + 1$ (with $h_{L+1} = 0$).

many absorbing states. In this sense, SOC can be viewed as an extreme case of the theory of nonequilibrium critical phenomena.

Before going on to describe a few systems displaying SOC, it is useful to start with a model without driving or dissipation, that is, a model of fixed energy or mass density ϵ, which acts as a control parameter. Let us consider a d-dimensional square lattice of linear size L embedded in a space of dimension $(d + 1)$, the extra dimension corresponding to a sort of deposition direction.[19] We distribute with some criterion N bricks on the L^d sites and at the end of the process, each site has a height h_i and is defined to be an active site if $h_i \geq h^*$ (see Fig. 6.7(a)). Dynamics proceeds as follows: An active site is randomly chosen and its height is reduced by the quantity h^*, which is distributed among neighboring sites (periodic boundary conditions are usually assumed). The simplest rule is to choose

---

[19] Chapter 7 will treat several examples of true deposition processes.

$h^* = 2d$ and distribute the $2d$ bricks, one for each neighboring site. The avalanches induced by active sites can be enforced either sequentially (as assumed here earlier) or parallely (as assumed in what follows), in which case the relaxation is fully deterministic.

For small $\epsilon = N/L^d$, we expect there are no active sites or that some transient exists before dynamics is completely frozen. For large ϵ instead, the cascade following the process $h_i \to h_i - h^*$ has a finite probability to activate new sites, producing a self-sustaining process with continuous cascades. Therefore, we have an usual nonequilibrium phase transition from an absorbing phase to an active phase, at some critical value ϵ_c.[20]

We can now switch on dissipation, allowing bricks to leave the system through its boundaries. If $\epsilon > \epsilon_c$, because of dissipation ϵ is going to decrease until $\epsilon < \epsilon_c$, when the system attains an absorbing state and dynamics stops. However, if we also switch on driving, that is, the inflow of new energy or matter, ϵ can increase, exceed the critical value, and be back in the active phase.

When both inflow and outflow are allowed, the system places itself at the critical point, regardless of what $\epsilon(t = 0)$ is, and dynamics can be interpreted as a continuous passage through the critical point, in both directions. In fact, if $\epsilon < \epsilon_c$, there is no dissipation, but there is injection, so $d\epsilon/dt > 0$. If $\epsilon > \epsilon_c$, on the other hand, there is no injection on the time scale of dissipation, so $d\epsilon/dt < 0$. Such ability to attain the critical point independently of any control parameter is the most relevant feature of SOC. However, it should be stressed again that this is due to the perfect separation of time scales, with dissipation much faster than driving. This property allows to draw an analogy between sandpiles and earthquakes. Sand grains are added to the system at a constant small rate, but they leave the system in a very irregular way, with long periods of apparent inactivity. A similar behavior can be seen in earthquakes, where the build-up of stress due to tectonic motion of the continental plates is a slow steady process, but the release of stress occurs sporadically in outbreaks of various sizes. In both cases, the length of rest periods and the size of avalanches are unpredictable quantities because of the complexity of the processes leading from the driving to the dissipation.

Before passing to discuss specific models, we introduce the basic observables to describe quantitatively the dynamics of systems displaying SOC. The stationary state is critical and it is composed of avalanches separated by seemingly rest periods, where driving is the only dynamical process taking place. An avalanche is composed of a series of toppling events, and it can be quantified in different ways. The two most studied observables are the size n and the duration τ of the avalanche,[21] whose probability distributions are $\mathcal{P}_s(n, L)$ and $\mathcal{P}_d(\tau, L)$, respectively. In these expressions, L is the linear size of the system and it affects the finite-size scaling through the explicit relations (valid for $n \gg n_0$ and $\tau \gg \tau_0$, where

---

[20] There are indications that this phase transition does not belong to the DP universality class.

[21] The size n is the total number of toppling. In parallel dynamics, active sites at time t topple together; therefore, the duration τ differs from n. There are two other observables that are much less studied: the area of the avalanche, corresponding to the number of *distinct* toppling sites, and the radius of gyration, which quantifies the distance between toppling sites during the same avalanche.

n_0, τ_0 are fixed lower cutoff)

$$\mathcal{P}_s(n, L) = \frac{1}{n^\theta} \, \mathcal{G}_s\left(\frac{n}{L^D}\right), \tag{6.37}$$

$$\mathcal{P}_d(\tau, L) = \frac{1}{\tau^\eta} \, \mathcal{G}_d\left(\frac{\tau}{L^z}\right). \tag{6.38}$$

Here above, the exponents θ and η rule the power-law decay of the distributions when L is in the thermodynamic limit, because $\mathcal{G}_{s,d}(0)$ are usually finite. Instead, the exponent D (the avalanche dimension) and the dynamical exponent z control the finite-size effects that come into play when $n \gtrsim L^D$ and $\tau \gtrsim L^z$. In these regimes, the functions $\mathcal{G}_{s,d}(x)$ ensure an effective cutoff of the distributions $\mathcal{P}_{s,d}$.

Avalanches can be labeled either with their size n or with their duration τ, which implies that n and τ are not independent, because we can count avalanches using both labels so that $\mathcal{P}_s(n, L)dn = \mathcal{P}_d(\tau, L)d\tau$. In the thermodynamic limit, $L \to \infty$, this means

$$\frac{dn}{n^\theta} \sim \frac{d\tau}{\tau^\eta} \tag{6.39}$$

and this determines a power law relation between n and τ, $n(\tau) \sim \tau^\mu$. In fact, using this relation in Eq. (6.39), we immediately find

$$\mu = \frac{\eta - 1}{\theta - 1}. \tag{6.40}$$

In the following we are not going to enter into a detailed account of SOC phenomena; we limit ourselves to give a few details for the Bak–Tang–Wiesenfeld (BTW) model of avalanches in sandpiles, introduced very schematically in Fig. 6.7(b) and which is strictly related to the model discussed in Fig. 6.7(a). In both systems the driving mechanism is performed by adding sand grains at random positions, but in Fig. 6.7(b), its competition with the loss of particles (forbidden in Fig. 6.7(a)) through the boundaries of the sandpile drives the system to its SOC behavior. However, in BTW, the criterion for producing an avalanche is described in terms of the local slope of the sandpile (as is the case in a real one) rather than in terms of the local height (as is in Fig. 6.7(a)). Some quantitative aspects concerning this model in dimensions $d > 1$ are still not completely understood and are debated.

6.5.1 The Bak–Tang–Wiesenfeld Model

The BTW model, introduced by Per Bak and by the Chinese Chao Tang and the American Kurt Wiesenfeld, originally aimed at describing the phenomenon of the occurrence of avalanches in a sandpile. Pictorially, we can think about a sandpile contained in a box, where sand grains (particles) topple to lower heights and may eventually escape from the box boundaries. If the toppling process stops, we proceed by adding single sand grains at random positions until a new avalanche forms.

The evolution rule of the avalanche amounts to a cellular automaton, whose bulk dynamics conserves the number of particles. For pedagogical reasons we describe the BTW rule in a $d = 1$ lattice made of L sites, see Fig. 6.7(b). We define the integer height h_i $(i = 1, \ldots, L)$ of the sandpile at site i, that is, the number of sand grains piled up at that position. The toppling process from site i to site $i + 1$ occurs if the height difference

$z_i = h_i - h_{i+1}$ exceeds a threshold value $Z > 0$.[22] In this case a sand grain moves from site i to site $i + 1$, thus yielding the bulk evolution rules ($i \neq 1, L$, see Fig. 6.7(b))

$$z_i \rightarrow z_i - 2 \tag{6.41}$$

$$z_{i\pm1} \rightarrow z_{i\pm1} + 1. \tag{6.42}$$

The boundary conditions are open on the right and closed on the left, because particles can leave the system from $i = L$ if $h_L > Z$, while they cannot roll to $i = 1$ from the left. These boundary conditions can be formally implemented by writing $z_0 = 0$ (i.e., $h_0 = h_1$) and $z_L = h_L$ (i.e., $h_{L+1} = 0$). Once an avalanche has terminated, in order to reactivate the toppling process, we add one sand grain at a randomly selected site i, so that

$$z_i \rightarrow z_i + 1 \quad (i = 1, \ldots, L) \tag{6.43}$$

$$z_{i-1} \rightarrow z_{i-1} - 1 \quad (i > 1). \tag{6.44}$$

We can notice that any configuration with $z_i \leq Z, \forall i$ is inactive and stops the avalanche. Accordingly, the total number of stable states is Z^L. In one dimension, the BTW model exhibits quite simple features. In fact, starting from any configuration and proceeding with toppling and deposition as described earlier, we always attain a steady state S_0 characterized by a constant slope, $z_i = Z, \forall i$. This is a stable configuration; no active site exists, but the addition of a grain at site $i = k$ has a unique evolution leading to S_0 again. In fact, the added grain simply goes down the steps until $i = L$, then it exits, restoring the configuration S_0. Any other inactive configuration S, characterized by $z_i \leq Z, \forall i$, has not such a property and dynamics leads S to S_0.[23]

The simple properties of the steady state S_0 allow to determine the probability distribution functions, Eqs. (6.37) to (6.38), in a straightforward manner, just considering that the size n and duration τ of the avalanche induced by deposition at site k are $n = \tau = L - k + 1$. Therefore, if deposition is random, we simply have $\mathcal{P}_s(n, L) = (1/L)\Theta(L - n)$ and $\mathcal{P}_d(\tau, L) = (1/L)\Theta(L - \tau)$, where Θ is the Heaviside function. Rewriting previous expressions as

$$\mathcal{P}_s(n, L) = \frac{1}{n}\frac{n}{L}\Theta\left(1 - \frac{n}{L}\right) \tag{6.45}$$

$$\mathcal{P}_d(\tau, L) = \frac{1}{\tau}\frac{\tau}{L}\Theta\left(1 - \frac{\tau}{L}\right) \tag{6.46}$$

we derive that the exponents are all equal to one, $\theta = \eta = D = z = 1$, and that

$$\mathcal{G}_{s,d}(x) = x\Theta(1 - x). \tag{6.47}$$

These conclusions are formally correct, but the $d = 1$ BTW, model is very peculiar, because in the steady state there are no spatiotemporal correlations between avalanches.

---

[22] The introduction of a threshold value for the height difference of neighboring sites is motivated by the existence of an angle of repose θ_c in the piles of granular materials: Slants smaller than or equal to θ_c correspond to stable piles. In the BTW model, $\theta_c = \tan^{-1} Z$.

[23] This statement can be supported by the evolution analysis of a configuration S_{i^*} close to S_0: $z_i = Z, \forall i \neq i^*$ and $z_{i^*} = Z - 1$. It is straightforward to see that a grain deposited in $k > i^*$ exits while a grain deposited in $k \leq i^*$ leads to S_{i^*-1}. Therefore, dynamics induces the following sequence of inactive configurations: $S_{i^*} \rightarrow S_{i^*-1} \rightarrow S_{i^*-2} \cdots \rightarrow S_1$. The final step $S_1 \rightarrow S_0$ is determined by the deposition of a grain in $k = 1$.

For $d > 1$, the BTW model (whose exact rules are not given here) exhibits completely different features, because the long-time robustness of S_0 is lost, and in the thermodynamic limit, the size n and duration τ of avalanches are now found to be distributed according to power-law distributions,

$$P(n) \sim \frac{1}{n^\theta}, \qquad P(\tau) \sim \frac{1}{\tau^\eta}, \tag{6.48}$$

with $\theta \simeq 0.98$ and $\eta \simeq 0.97$ in $d = 2$, while $\theta \simeq 1.35$ and $\eta \simeq 1.59$ in $d = 3$.

We want to point out that one of the main difficulties associated with the study of this model is the control of finite-size effects, which are particularly disturbing when performing numerical simulations. In particular, the determination of the values of the exponent η in $d = 2, 3$ and even the robustness of this SOC behavior is still the object of a long-standing debate. Here we want to mention that the BTW model has the merit of highlighting the existence of a new class of simple dynamical models, which exhibit interesting similarities with some natural phenomena. Moreover, despite their simplicity, this class of models seems to capture basic mechanisms that belong to a still widely unexplored realm of mathematical sciences.

In addition to the original BTW model, several other similar models have been introduced, showing that even apparently tiny differences may be relevant. Here we limit ourselves to cite the abelian sandpile models studied by the Indian Deepak Dhar. An example of such models has been depicted in Fig. 6.7(a). The abelian property means that microscopic dynamics is invariant under change of the deposition sequence. The original BTW model is not abelian as shown by a simple counterexample.[24] The exponents for the abelian model are different, for example, the exponent for the size distribution in $d = 2$ is $\theta \simeq 1.2$, which is definitely larger than the BTW value ($\theta \simeq 0.98$).

6.6 Bibliographic Notes

A recent reference book about the topic of this chapter is M. Henkel, H. Hinrichsen, and S. Lübeck, *Non-Equilibrium Phase Transitions,* volume 1: *Absorbing Phase Transitions* (Springer, 2008).

The so-called DP conjecture about the basic ingredients characterizing the universality class of DP is contained in the papers by H.-K. Janssen, On the Nonequilibrium Phase Transition in Reaction–diffusion Systems with an Absorbing Stationary State, *Zeitschrift für Physik B,* **42** (1981) 151–154, and by P. Grassberger, On Phase Transitions in Schlögl's Second Model, *Zeitschrift für Physik B,* **47** (1982) 365–374.

---

[24] Let us consider the one-dimensional BTW model with $L = 2$, $Z = 1$, initial configuration $h_1 = h_2 = 0$, and deposition of three grains (a grain is deposited only after conclusion of the possible avalanche-induced by the deposition of the previous grain). The deposition sequence $(2, 1, 1)$ induces no avalanche and the final configuration is $h_1 = 2$, $h_2 = 1$. The deposition sequence $(1, 1, 2)$ induces two avalanches of size one and the final configuration is $h_1 = 1$, $h_2 = 1$.

An account about the many facets concerning driven lattice gases, reaction–diffusion, catalysis, and contact processes is contained in the book by J. Marro and R. Dickman, *Nonequilibrium Phase Transitions in Lattice Models* (Cambridge University Press, 1999).

The computational problems concerning the study of the critical properties of DP are illustrated in the paper by I. Jensen, Low-density Series Expansions for Directed Percolation: I. A New Efficient Algorithm with Applications to the Square Lattice, *Journal of Physics A: Mathematical and General*, **32** (1999) 5233–5250.

A detailed overview of percolation processes in the framework of a field theory approach can be found in H. K. Janssen and U. C. Täuber, *Annals of Physics*, **315** (2005) 147–192.

Field theoretical arguments about nonequilibrium hyperscaling relations are discussed in the paper by J. F. F. Mendes, R. Dickman, M. Henkel, and M. C. Marques, Generalized Scaling for Models with Multiple Absorbing States, *Journal of Physics A: Mathematical and Theoretical*, **27** (1994) 3019–3028.

For what concerns the determination of the value of the critical point in the contact process, see the papers by R. Dickman and I. Jensen, Time-dependent Perturbation Theory for Nonequilibrium Lattice Models, *Physical Review Letters*, **67** (1991) 2391–2394, and R. Dickman and J. K. da Silva, Moment Ratios for Absorbing-State Phase Transitions, *Physical Review E*, **58** (1998) 4266–4270.

A review on deterministic cellular automata with selected papers can be found in S. Wolfram, *Theory and Applications of Cellular Automata* (World Scientific Publishing, 1986). A more recent and comprehensive illustration of this field of mathematics, including probabilistic models, can be found in A. Ilachinski, *Cellular Automata: A Discrete Universe* (World Scientific Publishing, 2001).

A book containing an overview about SOC phenomena is P. Bak, *How Nature Works: The Science of Self-Organized Criticality* (Copernicus Press, 1996). The most recent and complete account about SOS is G. Pruessner, *Self-Organized Criticality: Theory, Models and Characterization* (Cambridge University Press, 2012). Finally, the model to pass from absorbing phase transitions to SOC is A. Vespignani et al., Absorbing-State Phase Transitions in Fixed-energy Sand-Piles, *Physical Review E*, **62** (2000) 4564–4582.

7 Stochastic Dynamics of Surfaces and Interfaces

Throughout this book we have studied the stochastic dynamics of a Brownian particle (then generalized to the Brownian motion of a general physical quantity) and the stochastic dynamics of an ensemble of interacting particles. In this chapter, we address the stochastic motion of a line or a surface. More precisely, we intend to study the dynamics of a driven interface in the presence of noise. It is interesting that noise plus drive can induce a delocalization of the interface, as evidenced by the standard deviation of the height fluctuations of the interface. This quantity is called roughness (W): It depends on the time t and on the spatial scale L, over which such fluctuations are evaluated and it may occur that $W(L, t)$ diverges in the limit $L, t \to \infty$. When this happens, we have a kinetic roughening phenomenon.

In addition to the process of delocalization, the main consequence of kinetic roughening is to determine a self-affine interface, that is, an object that is statistically the same under a suitable anisotropic rescaling of space, time, and "growth direction." This property means that fluctuations have universal features and the function $W(L, t)$ is a homogeneous function, characterized in the limit $L, t \to \infty$ by suitable critical exponents, which define universality classes, much in the same way as in equilibrium and nonequilibrium phase transitions.

The phenomenon of kinetic roughening, giving several examples, is introduced in Section 7.1, while the main physical quantities that are pertinent to this topic are introduced in Section 7.2. A domain of physical processes that is both pedagogically useful and practically important is that of deposition models, because they allow to (easily) distinguish among universality classes and to emphasize the importance of symmetries and conservation laws. Section 7.3 is devoted to them. However, continuum approaches are unavoidable to define universality classes and to understand why certain equations play a dominant role. This analysis is done in Sections 7.4 and 7.5 and it sheds light on two universality classes: the Edwards–Wilkinson (EW) and the Kardar–Parisi–Zhang (KPZ). The former, studied in Section 7.6, describes systems that tend to grow in a specific direction and where dynamics conserves the volume of the aggregate. The latter, studied in Section 7.7, is relevant for systems tending to growth perpendicularly to the local orientation of the surface and the volume of the aggregate is not conserved by the dynamics, either because matter is not conserved or because voids or overhangs form. All physical processes entering in the previous universality classes are local, so the final Section 7.8 is devoted to an important nonlocal phenomenon: the diffusion limited aggregation (DLA).

7.1 The Physics of Growth Processes

A freshman student is likely to have a partial notion of what surfaces and interfaces are in science: A surface is a mathematical object, a two-dimensional manifold, while an interface might be what allows us to interact with an electronic device. A student in physics certainly meets these concepts when studying fluids (surface tension) and solid state ($p-n$ junctions). In this chapter we use the two terms, surfaces and interfaces, interchangeably, and we adopt a broad point of view, according to which an interface separates two regions with some different physical property. Restrictions come next: We are interested in the dynamics of these objects when they are set in motion in the presence of noise. What type of driving and what type of noise will be discussed later, but driving and noise are the key ingredients.

These few words should have clarified the title of this chapter. Now let's give a freewheeling list of examples of interfaces and of the driving forces setting them in motion. A class of examples, which will be pedagogically useful, concerns deposition and growth processes, where a solid phase is in contact with either a vapor phase or a liquid phase. In the former case, the solid grows, because particles attach to its surface and the flux of particles is the driving force. In the latter case, we can imagine a solid in contact with its melt, because a temperature gradient is applied. If the solid/melt interface is set in motion with respect to the gradient, the solid phase can grow and the liquid recedes. This method can be used for the production of monocrystalline silicon.

A domain wall in a ferromagnet, separating regions where the magnetization has different orientations, is another typical example of interface, whose motion can be easily tuned by applying a magnetic field (or an electric field in a ferroelectric material); see Fig. 7.1. A conceptually similar example is the interface between different turbulent states in a nematic liquid crystal undergoing a convection process. We will come back to this later.

There are other examples in different domains and at different scales: a flame or burning front, driven by a contact process and possibly by air currents; a wetting front in a porous medium, driven by capillarity, by gravity, or by imposing a pressure gradient to the flowing fluid; a liquid front on an inclined plane, driven by gravity; or the growing front of a bacterial colony in a Petri dish, driven by nutrients.

As for the noise, that is, fluctuations, it may have different origins. The most familiar one is certainly thermal noise: For instance, due to thermal fluctuations, the interface between magnetic domains in an Ising system cannot be a straight line. A second type of fluctuations, which has a relevant role in the statistical description of a disordered system, is the so-called *quenched* disorder, which exemplifies the role of randomly distributed structural defects in a material. A simple illustration of quenched disorder is given in Fig. 7.1, where the motion of the domain wall is hindered by defects. If the driving force (the magnetic field, in this case) is large enough, the interface can constantly grow. However, if the driving is small, there are parts of the interface that may feel a sufficiently strong "negative" noise hindering its motion, therefore pinning the interface. With increasing the driving, we may observe a depinning transition (see also the legend in Fig. 7.1).

Deposition and growth processes undergo kinetic fluctuations, whose nature is fairly similar to the fluctuating force acting on a Brownian particle, studied in Chapter 2. In

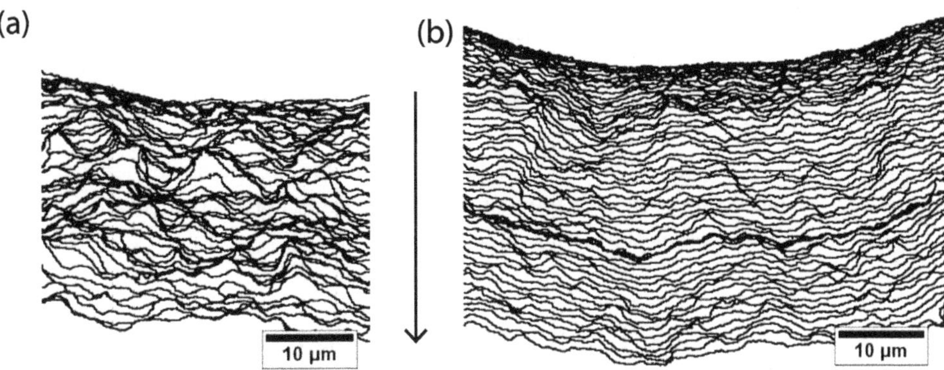

Fig. 7.1 Domain wall propagation in Pt/Co(0.65 nm)/Pt films. The Co ultrathin magnetic film is first magnetized in the down direction, then a magnetic field is applied in order to nucleate domains of positive magnetization. Once such a domain has been created, the domain wall is set in motion (from top to bottom) by subsequent applications of a field pulse $H_a = 77$ Oe (a) and $H_b = 260$ Oe (b). Pulse durations were between 2 and 8 s in (a) and between 20 and 100 µs in (b). The fields H_a and H_b are smaller than the depinning field, $H_{\text{dep}} \approx 750$ Oe, which plays the role of the static friction F_c when we pull a body in contact with a surface with a force F. In this case, if $F < F_c$, the (macroscopic) body remains at rest, but in the microscopic world of magnetic domains, the wall can move even if $H < H_{\text{dep}}$, provided $T > 0$, through a thermally activated dynamics. Courtesy of Peter Metaxas. P. J. Metaxas, Domain Wall Dynamics in Ultrathin Ferromagnetic Film Structures: Disorder, Coupling, and Periodic Pinning, PhD Thesis, The University of Western Australia and Université Paris-Sud 11, 2009.

practice, this noise is the result of more microscopic degrees of freedom. This is the main type of noise we focus on in this chapter.

Finally, strange as it may seem, noise can actually be present even if it should be absent, because the dynamics is deterministic. This is the case for deterministic chaos, and in some cases it is possible to make a rigorous connection between a chaotic, deterministic evolution and a stochastic evolution due to kinetic noise. The most relevant example (and here perfectly appropriate) is given by the Kuramoto–Sivashinsky equation,[1] which exhibits a deterministic, spatiotemporal chaotic dynamics: This is found to be statistically equivalent to the stochastic dynamics produced by the KPZ equation, which is one of the main subjects of this chapter.

The statement about the statistical equivalence of a deterministic, chaotic equation and a stochastic one should be understood as a sign of universality and a suggestion toward a treatment of stochastic dynamics, which is as general as possible. But what are the main features of dynamics and the main physical quantities we are interested in? The central physical concept appearing in this chapter is kinetic roughening, a dynamic phenomenon appearing when the surface is set in motion in the presence of noise; in the absence of noise, the interface would stay flat, but its presence tends to make the surface rough.

---

[1] In one spatial dimension, the Kuramoto–Sivashisnky equation has the form: $\partial_t v(x,t) = -\partial_{xx} v - \partial_{xxxx} v - v \partial_x v$.

An interface in d dimensions is defined by its local position,[2] $h(\mathbf{x}, t)$, and it is rough when mean square fluctuations of its position diverge on large space (L) and time (t) scales. Therefore, in the thermodynamic limit, a rough interface never attains a stationary state. However, for large L and t, dynamics has the important feature to be scale invariant, and this property allows us to define appropriate roughness exponents, much in the same way equilibrium critical exponents and universality classes can be defined in the proximity of a second-order phase transition.

7.2 Roughness: Definition, Scaling, and Exponents

Let us imagine having a d-dimensional interface moving in a $(d + 1)$-dimensional space. The interface position is defined by the function $h(\mathbf{x}, t)$, where $\mathbf{x} \in V = L^d$, L being the linear size of the system. For simplicity we can assume periodic boundary conditions in all spatial dimensions. Once we have defined the average height of the surface, $\overline{h}$, and the average square height, $\overline{h^2}$,

$$\overline{h}(t) = \frac{1}{V} \int_V d\mathbf{x}\, h(\mathbf{x}, t), \qquad \overline{h^2}(t) = \frac{1}{V} \int_V d\mathbf{x}\, h^2(\mathbf{x}, t), \tag{7.1}$$

the roughness $W(L, t)$ is given by

$$W^2(L, t) = \left\langle \frac{1}{V} \int_V d\mathbf{x} \left(h(\mathbf{x}, t) - \overline{h}(t) \right)^2 \right\rangle \tag{7.2}$$

$$= \left\langle \overline{\left(h(\mathbf{x}, t) - \overline{h}(t) \right)^2} \right\rangle \tag{7.3}$$

$$= \left\langle \overline{h^2}(t) - \overline{h}^2(t) \right\rangle. \tag{7.4}$$

Two different types of average appear in the definition of $W(L, t)$: the spatial average, $\overline{(\cdots)}$, and the average over noise, $\langle \cdots \rangle$. While the meaning of the former is evident from Eqs. (7.1), the meaning of the latter requires a few more words. The evolution of the dynamics is stochastic: Starting from the same initial conditions, we obtain different functions $h(\mathbf{x}, t)$. This is obvious in a simulation and the average over noise just means averaging over different simulation runs.[3] In analytical calculations, as shown in the following, we will average over a suitable function $\eta(\mathbf{x}, t)$, which represents noise. When trying to measure $W(L, t)$ in an experiment, the meaning of average over noise is not so clear, because the researcher may well not have the possibility of repeating the experiment many times. However, in an experiment, the interface may have a total linear size that may

---

[2] This possibility should not be taken for granted, because real or simulated systems may be characterized by overhangs, therefore resulting in a multivalued function. In these cases, the function $h(\mathbf{x}, t)$ should be understood as the local position of an effective interface.

[3] Noise might also affect the initial condition, chosen randomly within a given ensemble.

extend over a very large scale, so that we can evaluate $\left(\overline{h^2}(t) - \overline{h}^2(t)\right)$ for different regions of linear size L and average over them (self-averaging).[4]

We stress that the "average over noise" and the "space average" can be taken in any order, because they are independent of each other. If $A(\mathbf{x}, \eta)$ is a generic space-dependent and stochastic function, we have

$$\left\langle \overline{A(\mathbf{x}, \eta)} \right\rangle = \left\langle \frac{1}{V} \int_V d\mathbf{x} A(\mathbf{x}, \eta) \right\rangle = \frac{1}{V} \int_V d\mathbf{x} \left\langle A(\mathbf{x}, \eta) \right\rangle = \overline{\left\langle A(\mathbf{x}, \eta) \right\rangle}. \tag{7.5}$$

In order to discuss the evolution of $W(L, t)$, it is worth making reference to Fig. 7.2, where we plot the roughness for a specific (one-dimensional) deposition model with relaxation, to be introduced and discussed in Section 7.3. In the main panel we plot $W(L, t)$ versus t for different linear sizes L of the system. We remark on a few properties, which are common to different physical systems undergoing kinetic roughening. (1) If we fix L, W first increases in time following a power law, then it saturates to $W_{\text{sat}}(L)$. (2) If we use a larger L, the short time behavior of W is the same, but saturation occurs at a larger time. (3) The saturation value depends on L, following a power law (see the inset).

If t_c is the crossover time between the first regime (W increases in time) and the second regime (W saturates), we can summarize the results as

$$W(L, t) \approx \begin{cases} t^\beta, & t \ll t_c \\ L^\alpha, & t \gg t_c. \end{cases} \tag{7.6}$$

These expressions define the growth exponent β and the roughness exponent α. In Fig. 7.2, we have $\beta \simeq \frac{1}{4}$ and $\alpha \simeq \frac{1}{2}$. At the crossing time we must have $t_c^\beta \approx L^\alpha$, which implies

$$t_c \approx L^z, \qquad z = \frac{\alpha}{\beta}, \tag{7.7}$$

where z is called a dynamic exponent. In this case, $z \simeq 2$. We will see that earlier numerical exponents define the EW universality class in $d = 1$.

We can go further, because Fig. 7.2 suggests we might rescale $W(L, t)$ by $W_{\text{sat}}(L)$ so as to obtain a common asymptotic value and to rescale t by $t_c(L)$ so as to have a common crossover time. These rescalings do not affect the power-law behaviors (7.6) and should lead to a collapse of different curves, which is what we observe in Fig. 7.3, plotting $W(L, t)/L^\alpha$ versus t/L^z. Therefore, the roughness is expected to satisfy the scaling form, suggested by the Iranian Fereydoon Family and the Hungarian Tamás Vicsek,

$$W(L, t) = L^\alpha w \left(\frac{t}{L^z}\right), \tag{7.8}$$

with

$$w(u) \approx \begin{cases} u^\beta, & u \ll 1 \\ \text{const.}, & u \gg 1. \end{cases} \tag{7.9}$$

If instead we plot W versus L for different times t, we obtain curves very much similar to those shown in Fig. 7.2, with $W \approx L^\alpha$ for small L and $W \approx t^\beta$ for large L. The

---

[4] This self-averaging is effective only if different regions are essentially decoupled, which means that L is much larger than the in-plane correlation length, $L \gg \xi_\parallel$. See the continuation of this section for the definition of $\xi_\parallel$.

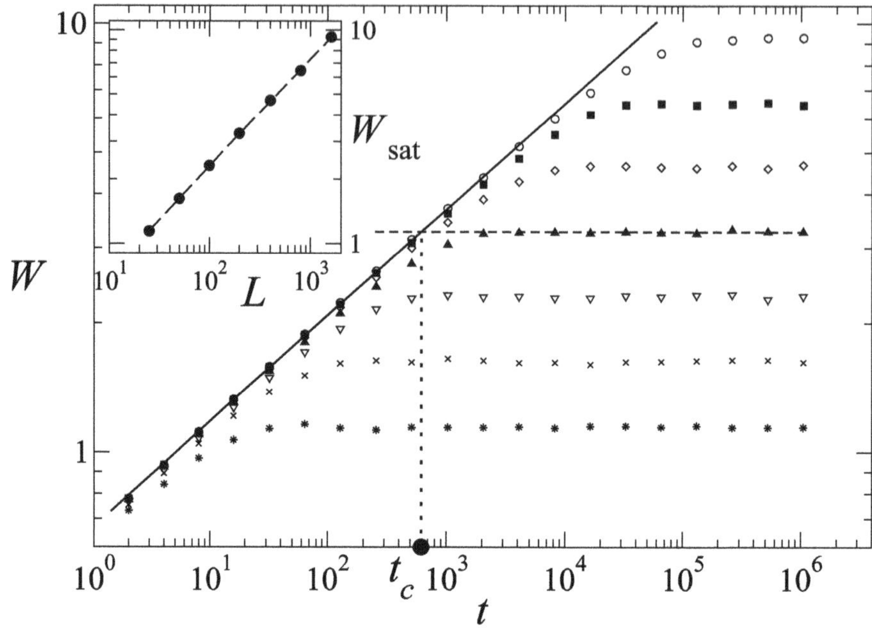

Fig. 7.2 Simulation of a deposition model with relaxation, to be introduced in Section 7.3. The main graph shows roughness as a function of time for $L = 25$ (asterisks), 50, 100, 200, 400, 800, and 1600 (open circles). The fit of the short time regime gives $\beta \simeq 0.247$. The crossing between the saturation value $W(L, t_\infty)$ and the short-time, power-law behavior defines the crossover time, $t_c(L)$. In the inset we plot $W(L, t_\infty)$ versus L, obtaining the roughness exponent $\alpha \simeq 0.502$.

crossover length L_c is defined by $L_c^\alpha \approx t^\beta$, so $L_c \approx t^{1/z}$. The correlation length is just this crossover length, $\xi_\parallel = L_c$. Therefore, we have efficient self-averaging (see footnote 4) if $L \gg \xi_\parallel \approx t^{1/z}$, that is, if $t \ll t_c(L)$. In other words, self-averaging is effective to determine β, not to determine α.

In equilibrium critical phenomena, we can define the lower (d_c^ℓ) and upper (d_c^u) critical dimensions for a given universality class. An equilibrium phase transition is the outcome of the destabilizing action of thermal fluctuations, which break long-range order. The effect of these fluctuations depends on the physical dimension d of the system and it increases with decreasing d. If d is small enough ($d < d_c^\ell$), fluctuations are so important that an ordered phase may survive at $T = 0$ K only. Conversely, if d is large enough ($d > d_c^u$), fluctuations are negligible and the exponents can be determined within a mean-field theory.

Lower and upper critical dimensions can be defined for kinetic roughening as well (see Fig. 7.4). If $d > d_c^u$, fluctuations do not roughen the surface on large time and length scales, because $\lim_{L,t \to \infty} W(L, t) = \text{const}$. The upper critical dimension is signaled by the vanishing of α and β, that is, $\alpha(d_c^u) = \beta(d_c^u) = 0$. Instead, if $d < d_c^\ell$, fluctuations are so important that the surface is *super rough*, because $\alpha > 1$. In this case, not only does the difference of height $(h(\mathbf{x}, t) - h(\mathbf{x}', t))$ diverge on large scales $|\mathbf{x} - \mathbf{x}'|$, but the slope $|h(\mathbf{x}, t) - h(\mathbf{x}', t)|/|\mathbf{x} - \mathbf{x}'|$ also diverges.

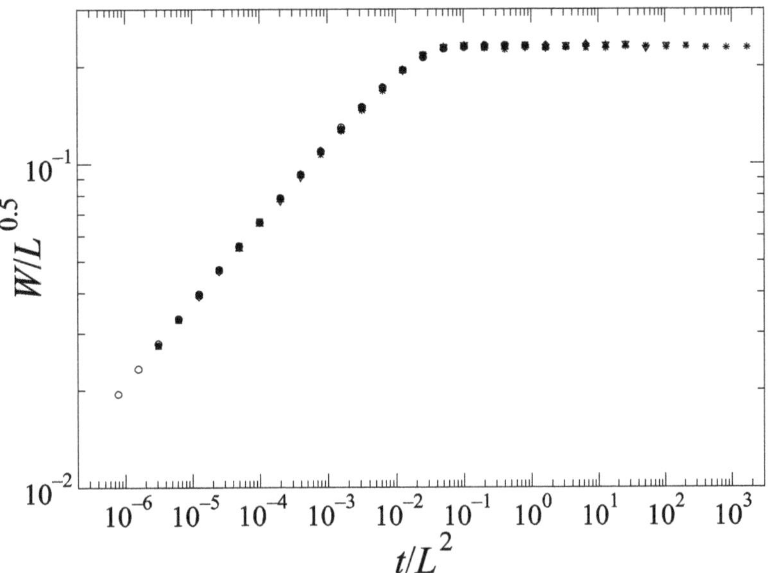

Fig. 7.3   Dynamic scaling through data collapse of curves shown in Fig. 7.2. We plot W/L^α versus t/L^z, using the values $\alpha = \frac{1}{2}$ and $z = 2$.

The use of the condition $\alpha(d_c^\ell) = 1$ to define the lower critical dimension should not come as a surprise, because this is also true for equilibrium, symmetry-breaking systems if we focus on the interface separating equivalent phases (e.g., two domains of an Ising model where the average magnetization has opposite directions). In fact, the interface separating two equivalent domains of size L has a certain roughness defined by the exponent α_E, where the subscript emphasizes that we are now at equilibrium. This means that the fluctuations of the domain wall in the direction perpendicular to it are of order L^{α_E}, but these fluctuations must be smaller than L; otherwise, a domain of size L cannot be defined, because the domain must have a size L in any direction. When $\alpha_E = 1$, phase coexistence breaks: We have attained the lower critical dimension.

We can comment about the choice of $W(L, t)$ to describe kinetic roughening phenomena. In fact, we could also use the correlation function,

$$G(\mathbf{r}, t) = \left\langle \overline{(h(\mathbf{x} + \mathbf{r}, t) - h(\mathbf{x}, t))^2} \right\rangle, \tag{7.10}$$

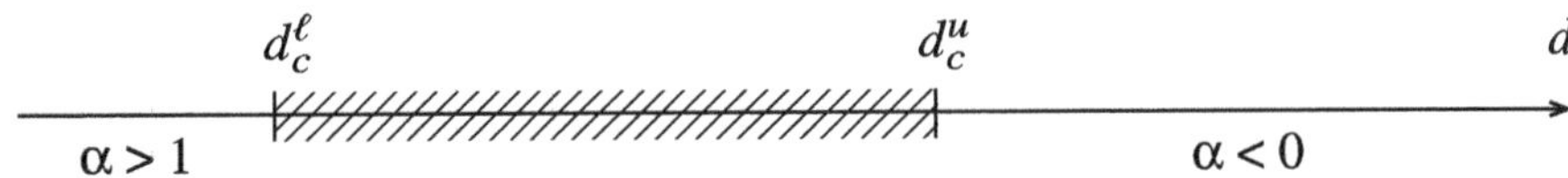

Fig. 7.4   Sketch for the lower and upper critical dimensions.

where, as before, $(\overline{\cdots})$ means spatial average over $\mathbf{x}$ and $\langle \cdots \rangle$ means average over the noise. In the limit of diverging $\mathbf{r}$ we have

$$G(\infty, t) = 2W^2(\infty, t), \tag{7.11}$$

while for general (large) $\mathbf{r}, t$, we have

$$G(\mathbf{r}, t) = r^{2\alpha} g\left(\frac{t}{r^z}\right), \tag{7.12}$$

with

$$g(u) \approx \begin{cases} u^{2\beta}, & u \ll 1 \\ \text{const} & u \gg 1. \end{cases} \tag{7.13}$$

It is interesting to observe (and, in Section 7.6.2 we will show) that the lower critical dimension has an additional signature in terms of $G(\mathbf{r}, t)$, because the constant appearing in Eq. (7.13) diverges when $d < d_c^{\ell}$. Thus,

$$\begin{aligned} \lim_{r,t \to \infty} G(\mathbf{r}, t) = \infty \qquad &\text{for} \qquad d \le d_c^u \\ \lim_{t \to \infty} G(\mathbf{r}, t) = \infty \qquad &\text{for} \qquad d \le d_c^{\ell}. \end{aligned} \tag{7.14}$$

A final remark about scaling and universality is in order. Similarly to equilibrium critical phenomena, a universality class of kinetic roughening defines the (d-dependent) critical exponents, the scaling functions (7.8) and (7.12), the lower and upper critical dimensions. However, the scaling functions $w(u)$ and $g(u)$ are defined up to two constants: a constant c_1 multiplying the function itself and a constant c_2 multiplying the argument of the function. Therefore, it is more correct to write the Family–Vicsek scaling as $W(L, t) = c_1 L^{\alpha} w(c_2 u)$. The constants $c_{1,2}$ take into account the irrelevant parameters distinguishing different models belonging to the same universality class. Later, we will have the opportunity to return to this aspect.

7.3 Deposition Models

Deposition models represent the simplest examples of a stochastically moving interface and allow to illustrate the concepts of noise, universality, and irrelevant parameters. We can imagine having a flat, d-dimensional substrate of size L^d and depositing particles in a sequential way, one particle after the other. First, we randomly choose a substrate site. Then, we assign some rule for attaching the incoming particle to the growing solid.

7.3.1 The Random Deposition Model

The simplest case (called random deposition, RD) is the one in which the new particle adds to the selected site, in the lowest allowed position, with no constraint and no postdeposition process. There are no correlations between different sites, each column of particles, h_i, growing independently. After deposition of N particles on a substrate of $V = L^d$ sites, the probability that $h_i = h$ is

$$P(h, N) = \binom{N}{h} p^h (1 - p)^{N-h}, \tag{7.15}$$

where $p = 1/V$ is the probability that $h_i \rightarrow h_i + 1$ after the deposition of a single particle.

It is easy to calculate the first two moments of the variable h, $\langle h \rangle = Np$ and $\langle h^2 \rangle = Np(1 - p) + N^2 p^2$. If the flux corresponds to a new layer per unit time, $\langle h \rangle = t$ and the (square) roughness is given by

$$W^2(t) = \langle h^2 \rangle - \langle h \rangle^2 = Np(1 - p) \simeq t, \tag{7.16}$$

so that we obtain the growth exponent $\beta = \frac{1}{2}$.

Due to the absence of correlations between different sites, the growth of the roughness is not bounded by the finite size of the system and the exponent α is not defined. In order to obtain a meaningful α (and a more realistic model), we should introduce some postdeposition rule.

7.3.2 Random Deposition with Relaxation

In this class of models, we allow deposited particles to choose a neighboring incorporation site, if this move allows the particle to decrease its height. This process, called relaxation, can be done in many different ways. We will mention two of them making reference to Fig. 7.5: a downward funnelling model and a minimum model. In the downward funnelling model (DF), the deposited particle may reduce its height by moving to a neighboring site for a maximum of δ hops. If the particle has several options to lower its height, a random choice is made. The data reported in Figs. 7.2 and 7.3 refer to the DF model with $\delta = 1$ (DF1). In the minimum model (MIN), the particle is incorporated at the site of minimum height within a "distance" δ.[5] For equivalent minima, one site is chosen at random. The data reported in Fig. 7.6 refer to four different models: DF1, DF10, MIN1, and MIN2.

All DF and MIN models belong to the same universality class, the EW one. This means that the critical exponents and the scaling function (7.8) are the same. However, as explained at the end of Section 7.2, a perfect data collapse occurs only for the same model: This explains why in Fig. 7.3 the data for four different models collapse on four different curves. In order to obtain a single curve, we should take into account the nonuniversal constants $c_{1,2}$ (see below Eqs. (7.14)), which depend on δ and are different for DF and MIN models. It is worthstressing that different models may also have different finite size effects, so that a model may be more suitable than another to determine the critical exponents (which, remember, characterize dynamics in the limit $L, t \rightarrow \infty$).

It is interesting to remark that relaxation might be governed not by the decrease of the height of the deposited particle, but by the increase of the number of its bonds, that is, the number of occupied neighbors. This procedure is implemented by the Wolf-Villain (WF) model and it leads to a different universality class, see below Eq. (7.76).

---

[5] In $d = 1$, if k is the deposition site the search for the minimum is done in the interval $[k - \delta, k + \delta]$. In $d > 1$, we must evaluate the height of all neighbours of order n up to $n = \delta$.

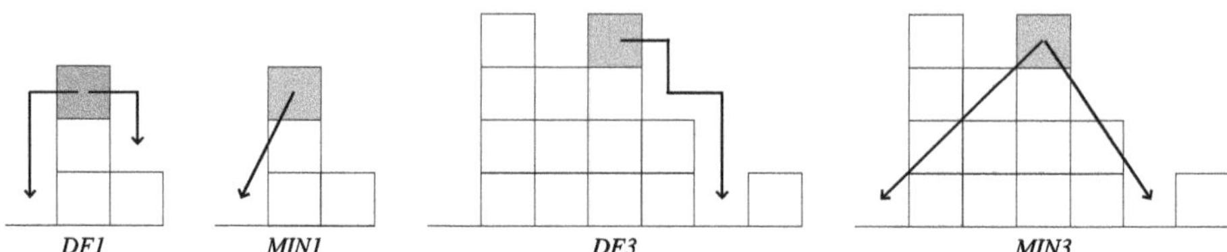

Fig. 7.5   Schematic diagrams for the particle movements in downward funneling (DF) model and for the minimum (MIN) model with $\delta = 1, 3$. Gray particles are newly deposited particles. When more than one arrow starts from the deposition particle, a random choice among different incorporation sites must be done. In this case, the DF model does not discriminate between lower sites of different heights.

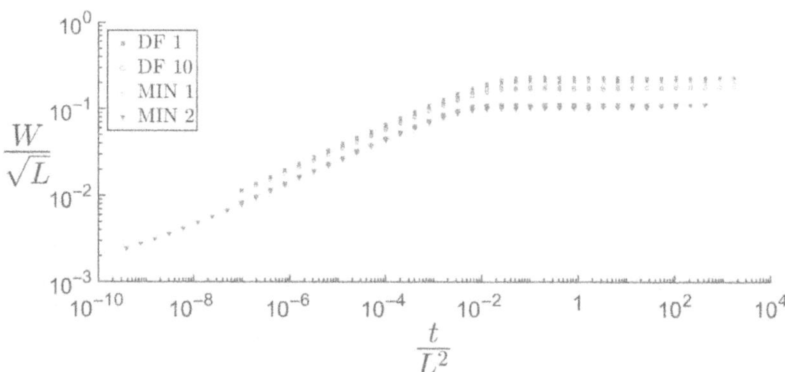

Fig. 7.6   Dynamic scaling for four different models of random deposition with relaxation.

A final comment concerns the types of noise that come into play in these models. Stochasticity plays a role during the deposition process (choice of the deposition site) and possibly during the relaxation process (choice among different, possible relaxation paths). It is the first type of noise that generally makes the surface rough, while the second type of noise (when present) is negligible, as it will be discussed in Sections 7.5 and 7.6.

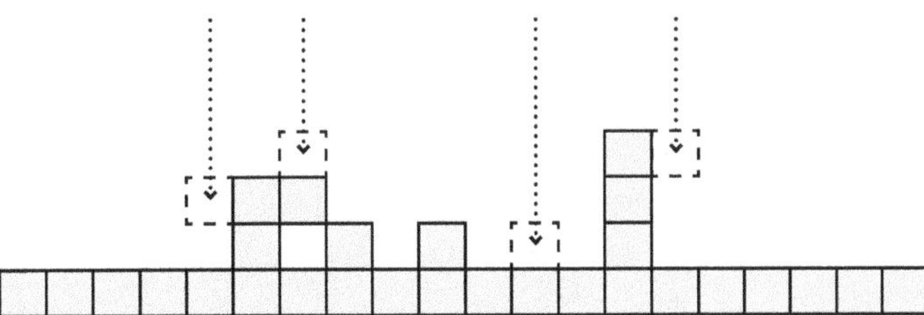

Ballistic deposition model. Gray bricks are deposited particles at time t. A void (white brick) is clearly visible. Dashed bricks represent different, possible deposition events at time $t + 1$. In two cases an overhang forms.

7.3.3 Deposition without Conservation

Deposition models discussed so far have the common feature to conserve matter, because all deposited particles are finally incorporated in the growing structure.

As we have already seen previously, conservation laws have a preeminent role in nonequilibrium critical phenomena and kinetic roughening is no exception. Strictly speaking, since the relevant quantity is the local height $h(x, t)$, we should say volume conservation rather than mass conservation. The two models we are going to introduce clarify the difference.

Let us consider the RD model, with the caveat that the arriving particle sticks to the closest, incorporated site found by the particle during its ballistic trajectory. The evolution rule of this model is easily implemented. If $\mathbf{x}$ is the deposition site,

$$h(\mathbf{x}) \to \max_{\mathrm{nn}}\{h(\mathbf{x}) + 1, h(\mathbf{x} + \boldsymbol{\delta}_{\mathrm{nn}})\}, \tag{7.17}$$

where $\boldsymbol{\delta}_{\mathrm{nn}}$ links a site to its nearest neighbors. In this model, matter is conserved (all deposited particles are incorporated), but volume is not, because voids and overhangs appear, as shown in Fig. 7.7. This model is called ballistic deposition (BD).

In Fig. 7.8 we plot $W(L, t)$ for the BD model, extracting the exponents $\beta \simeq 0.32$ and $\alpha \simeq 0.45$. If we compare them with the RDR model, β is approximately 30% larger and α is approximately 10% smaller: What should we conclude? The growth exponents β seem to be definitely different in the two models, with $\beta_{\mathrm{BD}} \simeq \frac{1}{3}$ against $\beta_{\mathrm{RDR}} \simeq \frac{1}{4}$. On the other hand, it would not be fair to draw a precise conclusion for α and we should rather perform larger-scale simulations. As a matter of fact, the BD model is known to suffer important finite-size effects[6] affecting α, which explain the discrepancy between the value found here above and the asymptotic value $\alpha_{\mathrm{BD}} = \frac{1}{2}$. Using the values $\beta_{\mathrm{BD}} = \frac{1}{3}$ and $\alpha_{\mathrm{BD}} = \frac{1}{2}$, in Fig. 7.9 we show the dynamic scaling.

---

[6] These finite-size effects can possibly be traced back to a relevant *intrinsic width* of the BD model, that is, to the residual roughness, when L is small and the scaling term $L^{\alpha} w(t/L^z)$ virtually vanishes. This intrinsic roughness is expected to be important for the BD model, since the difference in height of neighboring sites, $(\Delta h)_{\mathrm{nn}}$, can be quite large. For the same reason, the RSOS model (introduced a little later) is expected to have a small intrinsic width, because $(\Delta h)_{\mathrm{nn}} \leq 1$. In fact, the RSOS model does not suffer equally important finite-size effects. On the contrary, it is used for large-scale simulations.

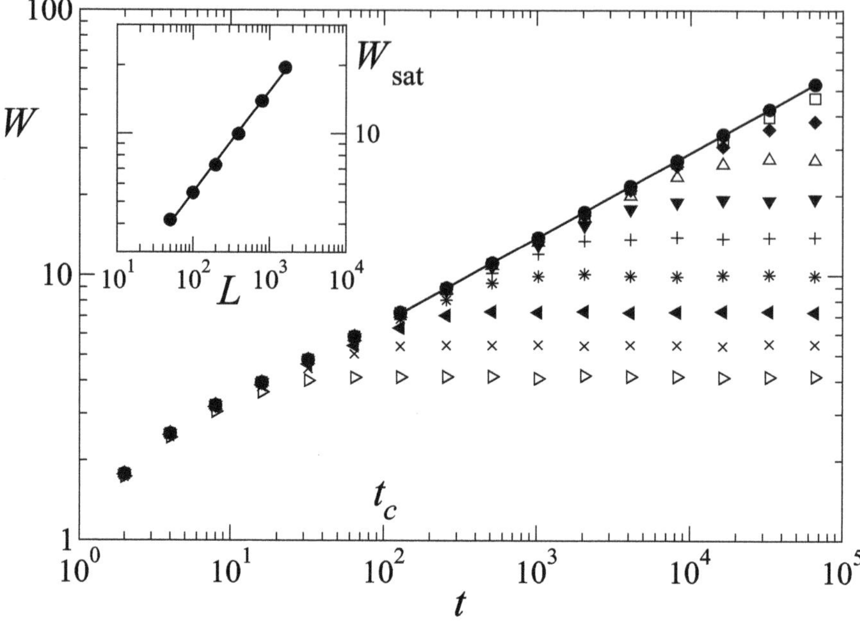

Fig. 7.8 Simulation of the BD model. Roughness as a function of time for $L = 50$ (bottom curve), 100, 200, 400, 800, 1 600, 3 200, 6 400, 12 800, and $L = 51 200$ (top curve). The fit of the short time regime gives $\beta \simeq 0.32$. In the inset we plot $W(L, t_\infty)$ versus L, providing the roughness exponent $\alpha \simeq 0.45$.

We conclude this section with the so-called restricted solid on solid model (RSOS), used for large-scale simulations, thanks to its weak finite-size effects. In this case, random deposition is not accompanied by a relaxation (as in the RDR model) and the new particle is incorporated on top of the particles deposited in the same site, in contrast to the BD model. However, it differs from the RD model because of a constraint on the height difference between neighboring sites, which must be preserved during growth: $|h(\mathbf{x} + \boldsymbol{\delta}_{nn}) - h(\mathbf{x})| \leq \Delta$, $\forall \boldsymbol{\delta}_{nn}$, where Δ is a fixed positive integer. If the deposition of a particle in the chosen site breaks this constraint, the particle is not deposited and a new lattice site is chosen. Because of this sticking rule, the attachment rate depends on the local environment, and the resulting growth process does not conserve matter. More precisely, the sticking rule clearly suggests a slope-dependent "evaporation" process, with a deposited particle that is more likely to be rejected if the surface is tilted.

A final, discrete growth model can be derived by the asymmetric simple exclusion process (ASEP) introduced in Chapter 5: The mapping is illustrated in Fig. 7.10. In the upper part of the figure, we sketch the topology and the dynamics of ASEP: (1) We have a one-dimensional lattice where sites can be occupied by a particle (gray sites) or they can be empty (white sites), and (2) each particle can move to its right with probability c_+ and to the left with probability c_-, if the arrival site is empty. Now, the mapping works as follows: An occupied (empty) site i corresponds to a piece of surface of negative (positive)

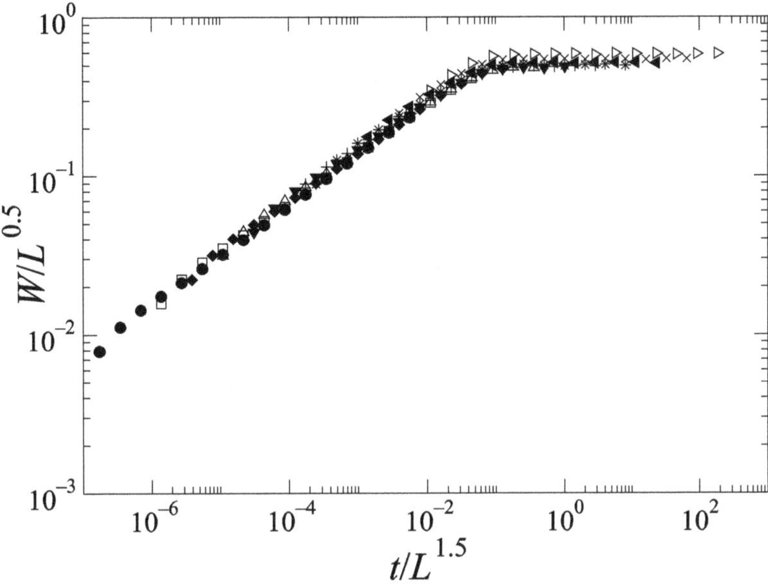

Fig. 7.9 Simulation of the BD model. We show dynamic scaling through data collapse plotting $W/\sqrt{L}$ versus $t/L^{3/2}$.

slope $m_i = -1$ ($m_i = +1$). As for dynamics, moving a particle to the right in the ASEP model corresponds to depositing a rhomboidal particle in a local minimum in the growth model and moving a particle to the left corresponds to evaporating a particle from a local maximum. In order to implement periodic boundary conditions in the growth model, we require an equal number of occupied and empty sites in the ASEP model; otherwise, helical boundary conditions should be used. In any case, the number of particles is a constant of motion and it determines the average slope of the surface. The resulting growth model is called single-step (SS) model.

The three models described here earlier, BD, RSOS, and SS, all belong to the same universality class, called Kardar–Parisi–Zhang (KPZ). In Section 7.7.2 we will explicitly tie the SS model to the KPZ equation, via the Burgers equation.

7.4 Self-Similarity and Self-Affinity

The reader should be familiar with self-similarity and scale invariance, because these concepts appear when studying equilibrium phase transitions. As discussed in Chapter 5, according to the scaling hypothesis, the correlation length ξ of a critical system is the only characteristic length. Since $\xi \to \infty$ when $T \to T_c$, at criticality the system is scale invariant. In the proximity of the critical point, that is, for vanishing reduced temperature[7] $t^* = (T - T_c)/T_c$ and field H, thermodynamic quantities are homogeneous functions of t^*

---

[7] We use the symbol t^* instead of t to avoid confusion with time.

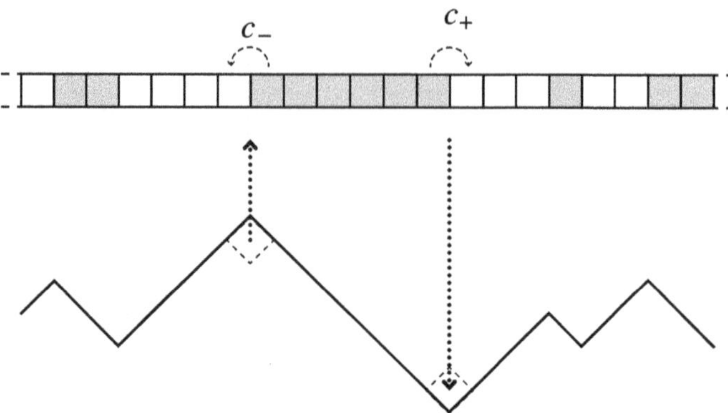

Fig. 7.10 Mapping of ASEP to the single-step model. A move of a particle to the right (left) corresponds to the deposition (evaporation) of a particle in a local minimum (from a local maximum). Driving the particle model out of equilibrium means creating an asymmetry between hops to the right and to the left: In the surface growth model, this asymmetry implies an unbalancing between deposition and evaporation.

and H. In the context of kinetic roughening, the critical point corresponds to large time t and large system size L. Furthermore, as the limits $H, t^* \to 0$ do not commute, neither do the limits $t, L \to \infty$.

Roughness $W(L, t)$ in some sense plays the role of the free energy density, $f(H, t^*)$. In order to determine the correct scaling of $\mathbf{x}, t$, and h to obtain self-similarity, let us consider the scaling form (7.8) for the roughness, which we rewrite here:

$$W(L, t) = L^\alpha w \left(\frac{t}{L^z} \right).$$

The question can be put in the following terms: If we rescale space by a factor[8] b, how should we rescale t and h in order to have a statistically similar profile? Since W depends on time through the function $w(t/L^z)$, if $\mathbf{x} \to b\mathbf{x}$, that is, $L \to bL$, we need $t \to b^z t$ so that the ratio (t/L^z) remains unchanged. With space rescaling, we also have $W \to b^\alpha W$. On the other hand, W is linear in the height h (see definition (7.4)); therefore, it must be $h \to b^\alpha h$. In conclusion, the scaling relations are

$$\mathbf{x} \to b\mathbf{x} \tag{7.18a}$$

$$t \to b^z t \tag{7.18b}$$

$$h \to b^\alpha h. \tag{7.18c}$$

Let us comment on transformations (7.18). If we are in the stationary regime ($t \gg L^z$), time transformation (7.18b) has no effect and invariance under Eqs. (7.18a) and (7.18c) just means that a snapshot of the interface profile is statistically invariant under a suitable, anisotropic magnification (the horizontal axis is scaled by a factor b, the vertical axis by a factor b^α). If we are not confined to the stationary regime, a given snapshot, after earlier

---

[8] Here, the factor $b \to \infty$ plays a similar role to $\Lambda \to 0$ in Eq. (5.68).

anisotropic rescaling, should be compared with a snapshot taken at a different time, and Eq. (7.18b) tells us what the appropriate time is.

The fact that the magnification of a snapshot of the profile is anisotropic, with $\mathbf{x}$ and h scaling differently, means that the stationary profile is not self-similar but is self-affine. The simplest example of a random, self-affine object is the trajectory of a one-dimensional random walk. If we plot its position as a function of time, we obtain the nonanalytic curve drawn in Fig. 2.3. It is well known that during the time interval Δt the position of the particle changes by a quantity Δx, with $\langle \Delta x \rangle = 0$ and $\langle (\Delta x)^2 \rangle \sim \Delta t$. This means that rescaling the horizontal axis by a factor b must be accompanied by a rescaling of the vertical axis by a factor $b^{1/2}$, in order to get a statistically equivalent profile.

In the previous paragraph, when discussing random walks, we have deliberately talked about rescaling of horizontal/vertical axes, rather than rescaling of time and position, because we will see that two different deposition models are characterized by the same value $\alpha = \frac{1}{2}$ in $d = 1$. This means that in the stationary regime, $t \gg t_c(L)$, the height profile $h(x, t)$ of the interface is self-affine under rescaling of the horizontal axis by a factor b and rescaling of the vertical axis by a factor $b^{1/2}$. In other words, if $\alpha = \frac{1}{2}$ the long-time height profile is statistically equivalent to the trajectory of a random walk.

7.5 Symmetries and Power Counting: Toward Langevin-Type Equations

Discrete models are successful for simulations, but continuum models are easier to be treated analytically. Here we plan to write a sufficiently general equation for a moving interface, that is, a partial differential equation for the interface position $h(\mathbf{x}, t)$,

$$\partial_t h(\mathbf{x}, t) = \mathcal{A}[h] + \eta(\mathbf{x}, t), \tag{7.19}$$

where $\mathcal{A}[h]$ is a sum of terms; see later. First, we use symmetries and invariances in order to reduce the number of terms, then scale transformation and power counting will provide us with the most relevant terms on large scales.

Eq. (7.19) is not the most general one, because higher-order time derivatives might appear on the left-hand side. However, such terms would be negligible with respect to $\partial_t h$, much in the same way that the term d^2x/dt^2 is asymptotically negligible with respect to dx/dt in the Langevin equation for a Brownian particle. As a matter of fact, Eq. (7.19) appears to be a generalized Langevin equation, and some known tools, like the Fokker–Planck equation, will be used (see Sections 2.4.3 and 2.4.5 for their introduction).

As for the noise, we limit ourselves here to additive, white, and uncorrelated noise, that is, to a function $\eta(\mathbf{x}, t)$ having the properties[9]

$$\langle \eta(\mathbf{x}, t) \rangle = 0$$
$$\langle \eta(\mathbf{x}, t)\eta(\mathbf{x}', t') \rangle = \Gamma \delta(\mathbf{x} - \mathbf{x}')\,\delta(t - t'). \tag{7.20}$$

The first, robust hypothesis is that the functional $\mathcal{A}$ does not depend explicitly on $\mathbf{x}$ and t. This means we assume invariance under time shift ($t \to t + t_0$) and spatial translation ($\mathbf{x} \to \mathbf{x} + \mathbf{x_0}$). A time-dependent incoming flux in a deposition process or a space-dependent wind in a flame front would break such invariances. A second, less robust hypothesis is the spatial invariance in the "growth" direction, that is, under the transformation $h(\mathbf{x}, t) \to h(\mathbf{x}, t) + h_0$. This assumption implies that $\mathcal{A}[h]$ cannot depend explicitly on h, but only on its spatial derivatives. This condition is not satisfied, for example, by deposition processes where incoming particles interact with the substrate. A typical example is the so-called heteroepitaxial growth, where a chemical species A is deposited on top of a different species B. The epitaxial constraint forces the growth process to occur under compression or tension conditions, inducing long-range, elastic effects. However, $\mathcal{A}$ does not depend on h in most of the examples cited in Section 7.1 and we will assume such independence also here. In conclusion, our starting point is

$$\partial_t h(\mathbf{x}, t) = \mathcal{A}[\partial_i h, \partial_{ij} h, \dots] + \eta(\mathbf{x}, t), \tag{7.21}$$

where $\partial_i \equiv \partial/\partial x_i$, with x_i denoting the ith component of $\mathbf{x}$. The noise term is now set aside, our focus being on $\mathcal{A}$, which can be expanded as follows, according to the order of the derivative and to its power,

$$\mathcal{A} = A_0 + \sum_{i=1}^{d} A_i \partial_i h + \sum_{i,\,j=1}^{d} A_{ij}\partial_i h \partial_j h + \sum_{i,\,j=1}^{d} B_{ij}\partial_{ij} h + \cdots. \tag{7.22}$$

Now we wonder if a scale transformation as Eq. (7.18) allows us to single out certain terms as the most relevant ones on large scales, that is, $b \to \infty$. First, applying Eq. (7.18b), we find that a time derivative of order n, $\partial_t^n h$, gains a factor b^{-nz}, which explains why we kept only the lowest order time derivative, $n = 1$, on the left-hand side. Before applying Eqs. (7.18a) and (7.18c), we can easily get rid of the first two terms on the right-hand side of Eq. (7.22) with a suitable redefinition of h. The constant term, A_0, disappears if we define $h'(\mathbf{x}, t) = h(\mathbf{x}, t) - A_0 t$, which corresponds to "measuring" the height of the surface with respect to a reference moving at speed A_0. Similarly, the linear term $\sum_i A_i \partial_i h$, disappears under a Galilean transformation,

$$h'(\mathbf{x}, t) = h(\mathbf{x} - \mathbf{v}t, t), \tag{7.23}$$

---

[9] For reasons similar to those for which $\partial_{tt} h$ is asymptotically negligible with respect to $\partial_t h$, here we can neglect conserved noise with respect to $\eta(\mathbf{x}, t)$. Conserved noise is responsible, for example, for the stochasticity of the relaxation process in the RDR model, and it will be explicitly evaluated in Section 7.6. We do not even consider the possibility of multiplicative noise.

which implies

$$\partial_t h' = \partial_t h - \sum_i v_i \partial_i h,$$
$$\partial_i h' = \partial_i h. \tag{7.24}$$

Therefore, if we choose $v_i = A_i$, the second term on the right-hand side of Eq. (7.22) disappears and we are finally left with the expression

$$\mathcal{A} = \sum_{i,\,j=1}^{d} A_{ij}(\partial_i h)(\partial_j h) + \sum_{i,\,j=1}^{d} B_{ij}\partial_{ij}h + \sum_{i,\,j,\,k=1}^{d} A_{ijk}(\partial_i h)(\partial_j h)(\partial_k h) + \cdots. \tag{7.25}$$

It is useful to label each term of Eq. (7.25) with a pair of integers $\{p, q\}$, where p is the number of factors h appearing in the term (regardless of the derivatives) and q is the total number of spatial derivatives. Using this notation, Eq. (7.25) reads

$$\mathcal{A} = \{2, 2\} + \{1, 2\} + \{3, 3\} + \cdots \tag{7.26}$$

It is noteworthy that it must be $p \le q$, otherwise an explicit dependence on h would appear. Think, for example, of the term $\{2, 1\} = \sum_i \tilde{A}_i h(\partial_i h)$.

Under the transformations $\mathbf{x} \to b\mathbf{x}$ and $h \to b^\alpha h$, we obtain

$$\{p, q\} \to b^{p\alpha - q}\{p, q\}. \tag{7.27}$$

Since $p \le q$, for any fixed p and large scale ($b \to \infty$), we have

$$\{p, p\} \gg \{p, p + 1\} \gg \{p, p + 2\} \ldots, \tag{7.28}$$

with the term $\{p, p\}$ transforming according to

$$\{p, p\} \to b^{p(\alpha - 1)}\{p, p\}. \tag{7.29}$$

If $\alpha > 1$, no specific term can be singled out and the expansion (7.22) is useless. If $\alpha < 1$, that is, for $d > d_c^\ell$, Eqs. (7.28) and (7.29) give a clear outcome of our expansion: The most relevant term is the term $\{p, p\}$ with the smallest, nontrivial p. The first two terms on the right-hand side of Eq. (7.22) correspond to $\{0, 0\}$ and $\{1, 1\}$, respectively. These terms are trivial, because we got rid of them with a simple redefinition of h. Therefore, we expect the most relevant term to be

$$\{2, 2\} \equiv \sum_{i,\,j=1}^{d} A_{ij}(\partial_i h)(\partial_j h). \tag{7.30}$$

Since the matrix A_{ij} is real and can be symmetrized,[10] it can be diagonalized, corresponding to a rotation of axes x_i. After this rotation, we have

$$\{2, 2\} \equiv \sum_{i=1}^{d} A_{ii}(\partial_i h)^2, \tag{7.31}$$

---

[10] If not symmetric, it suffices to rewrite $\sum_{ij} A_{ij}\partial_i h\partial_j h = \sum_{ij} \frac{1}{2}(A_{ij} + A_{ji})\partial_i h\partial_j h$ and the new matrix $\frac{1}{2}(A_{ij} + A_{ji})$ *is symmetric.*

and the resulting equation is

$$\partial_t h(\mathbf{x}, t) = \sum_{i=1}^{d} A_{ii} (\partial_i h)^2 + \eta(\mathbf{x}, t), \tag{7.32}$$

which reads, in one and two dimensions,

$$\begin{aligned}
\partial_t h(x, t) &= A_{11} (\partial_x h)^2 + \eta(x, t), & d &= 1 \\
\partial_t h(x, y, t) &= A_{11} (\partial_x h)^2 + A_{22} (\partial_y h)^2 + \eta(x, y, t), & d &= 2.
\end{aligned} \tag{7.33}$$

Changing the sign of the interface profile ($h \rightarrow -h$) changes the sign of all coefficients A_{ij} on the right-hand side (a change of sign of the noise is irrelevant), and rescaling x and y ($x \rightarrow a_x x$, $y \rightarrow a_y y$) allows us to rescale separately the coefficients to make $|A_{11}| = |A_{22}|$, without changing their signs. These facts allow us to rewrite Eqs. (7.33) as

$$\begin{aligned}
\partial_t h(x, t) &= |A_{11}| (\partial_x h)^2 + \eta(x, t), & d &= 1 \\
\partial_t h(x, y, t) &= |A_{11}| \left[(\partial_x h)^2 \pm (\partial_y h)^2 \right] + \eta(x, y, t), & d &= 2.
\end{aligned} \tag{7.34}$$

where the sign $+$ ($-$) applies if A_{11} and A_{22} have the same (opposite) sign.

As already noted (and exploited), Eq. (7.32) does not satisfy the reflection symmetry $h \rightarrow -h$, while it satisfies the symmetry $\mathbf{x} \rightarrow -\mathbf{x}$ (in fact, it satisfies the stronger symmetry $x_i \rightarrow -x_i$ for each direction i, separately). In case the $h \rightarrow -h$ symmetry does hold (and that depends on the specific physical process under investigation), the term $\{2, 2\}$ (see Eq. (7.30)) cannot survive, because p must be odd. If we also require the symmetry $\mathbf{x} \rightarrow -\mathbf{x}$, q must be even. Finally, the most relevant term would be

$$\{1, 2\} \equiv \sum_{i,j=1}^{d} B_{ij} \partial_{ij} h, \tag{7.35}$$

which simplifies to

$$\{1, 2\} \equiv \sum_{i=1}^{d} B_{ii} \partial_{ii} h \tag{7.36}$$

after a rotation of the axes or if the stronger symmetry $x_i \rightarrow -x_i$ applies.

In the following, we consider rotational invariance in the d-dimensional space, so that the term $\{2, 2\}$ is proportional to $(\nabla h)^2$ and the term $\{1, 2\}$ is proportional to $\nabla^2 h$. In conclusion, we are led to consider the following two equations,

$$\partial_t h(\mathbf{x}, t) = \nu \nabla^2 h + \eta(\mathbf{x}, t) \qquad\qquad \text{EW} \tag{7.37}$$

$$\partial_t h(\mathbf{x}, t) = \nu \nabla^2 h + \frac{\lambda}{2} (\nabla h)^2 + \eta(\mathbf{x}, t). \quad \text{KPZ} \tag{7.38}$$

Equation (7.37) is called Edwards–Wilkinson (EW) equation, after the British Samuel Frederick Edwards and David Robert Wilkinson. It looks like the diffusion equation with a noise term, which explains why we require $\nu > 0$. It is a linear equation that satisfies the reflection symmetry $h \rightarrow -h$ and has the conserved form $\partial_t h(\mathbf{x}, t) = -\nabla \cdot \mathbf{J} + \eta(\mathbf{x}, t)$.

Equation (7.38) is called KPZ equation, after the Iranian Mehran Kardar, the Italian Giorgio Parisi, and the Chinese Yi-Cheng Zhang. It does *not* satisfy the symmetry $h \rightarrow -h$ and it does not have a conserved form. KPZ is nonlinear and the sign of λ is irrelevant, because changing the sign of h also changes the sign of λ. It is usual to assume $\lambda > 0$. Despite its irrelevance at large scales, the linear term is present, because it regularizes the equation at small scales, as discussed in Appendix I.

In addition to the earlier two features (symmetry breaking and nonconservation), the KPZ equation is generally associated with a third feature, lateral growth, which is sketched and compared with the EW-type of growth in Fig. 7.11. The EW case (a) corresponds to attachment via deposition and the growth direction is $\hat{z}$, regardless of the slope of the interface. This picture is appropriate for processes where deposition and attachment have a specified direction. The KPZ case (b) corresponds to a surface that grows orthogonally to itself. This picture applies when growth can occur in any direction. We have, respectively,

$$
\begin{aligned}
&\text{(case a)} && \partial_t h = v_0 \\
&\text{(case b)} && \partial_t h = v_0\sqrt{1+(\nabla h)^2} \approx v_0 + \frac{v_0}{2}(\nabla h)^2 .
\end{aligned}
\tag{7.39}
$$

The equation for case (b), with the small slope expansion justified by power counting, tells us that the KPZ nonlinearity is the natural outcome (and signature) of a dynamical process where *lateral growth* occurs.

In Section 7.3 we mentioned two deposition models falling in the KPZ universality class: BD and RSOS. If for the first model (BD) lateral growth is evident, for the second model (RSOS), it is not obvious. However, in both models there is an attachment rule (not a relaxation rule) depending on the comparison between the height of the chosen site and the height of neighboring sites. In BD, deposition is favored by a local tilt, while in RSOS it is unfavored. This difference reflects in the (irrelevant) opposite sign of λ.

Equation (7.39) also suggests that λ, the coefficient of KPZ nonlinearity, is related to the interface velocity. In fact, for a surface of negligible curvature and average slope $\mathbf{m} = \overline{\nabla h}$,

$$
\overline{(\nabla h)^2} \simeq (\overline{\nabla h})^2 = m^2 .
\tag{7.40}
$$

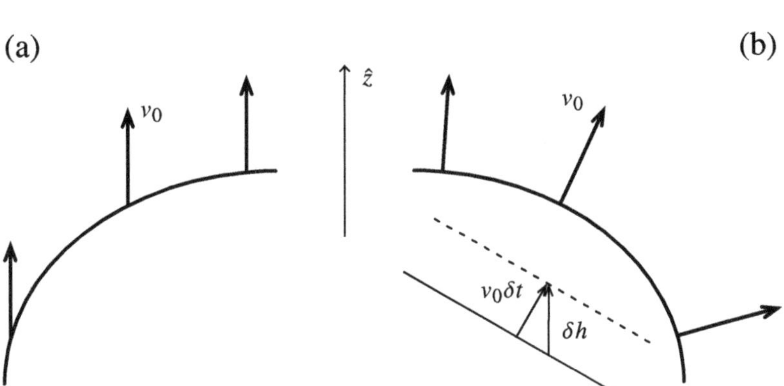

Fig. 7.11   Comparison of EW- and KPZ-type growth. In the absence of relaxation processes, EW-type growth (a) is characterized by a constant, vertical velocity, while KPZ-type growth (b) is characterized by a constant, lateral growth, that is, perpendicular to the local profile of the interface. We also show how a lateral velocity v_0 determines the growth rate, $\partial_t h = v_0\sqrt{1+(\nabla h)^2} \simeq v_0 + (v_0/2)(\nabla h)^2$.

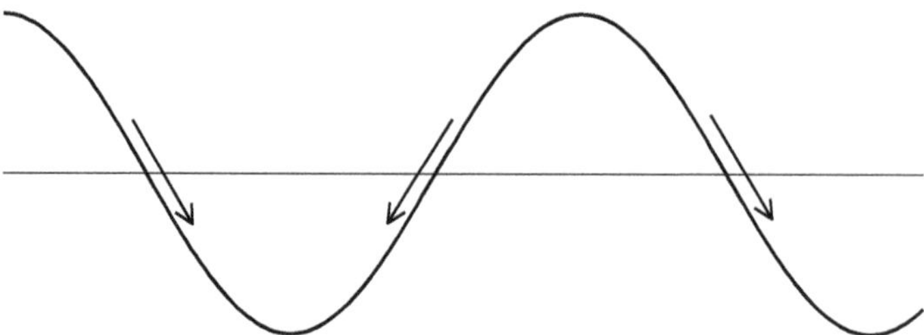

Fig. 7.12   Pictorial image of the current J of the EW model.

Therefore, the average interface velocity is proportional to the square of the slope,

$$\overline{\partial_t h} = \frac{\lambda}{2}\overline{(\nabla h)^2} \simeq \frac{\lambda}{2}m^2, \tag{7.41}$$

and the constant of proportionality is just $\lambda/2$. This relation allows us to test if a given deposition model has an orientation-dependent average velocity by tuning the orientation of the substrate.

7.6 The Edwards–Wilkinson Equation

The simplest, nontrivial continuum equation is the EW equation,

$$\partial_t h(\mathbf{x}, t) = \nu\nabla^2 h + \eta(\mathbf{x}, t), \tag{7.42}$$

which satisfies the symmetry $h \to -h$ and is rotationally invariant.

Its deterministic part is conserved,[11] because the equation has the form

$$\partial_t h = -\nabla \cdot \mathbf{J} + \eta, \tag{7.43}$$

with $\mathbf{J} = -\nu\nabla h$. This implies that the average height of the interface is constant,

$$\frac{d\overline{h}}{dt} = \frac{1}{V}\int_V d\mathbf{x}\,\eta(\mathbf{x}, t) = 0, \tag{7.44}$$

because the integral of the divergence vanishes for periodic boundary conditions (or in the limit of diverging V).

The deterministic part looks like the diffusion equation and its form means there is a net current of matter from high regions (large h) to low regions (small h), as exemplified in Fig. 7.12. This current produces a smoothing of the surface and competes with noise, which tends to roughen the interface, as clearly shown by the random deposition model. In a deposition process, it seems natural to relate $\mathbf{J}$ to the relaxation mechanisms studied in Fig. 7.5: After deposition a particle moves to a neighboring site if it can reduce its height.

---

[11] The noise itself might conserve matter. This is not the case in deposition processes, where noise represents fluctuations of an external source. However, in Section 7.6.1, we also consider the case of conserved noise to show that it does not make rough a surface with EW relaxation dynamics.

If this reasoning is correct, we expect that the RDR models belong to the EW universality class, therefore sharing the same exponents α, β: This is actually what happens. The hypothesis that RDR models are correctly described by the EW equation can be understood if we give h_i a different meaning: not the local height of site i, but the number of particles in "box" i. Then, RDR models correspond to a diffusion-like process in the presence of a uniform supply of particles from some external reservoir. Within this interpretation, the parameter ν is exactly the diffusion coefficient and it is also clear why the flux does not break the up/down symmetry.

Before determining the roughness exponents and the scaling function for the EW equation, we discuss a further property of this equation: It is derivable from a potential. In fact, the equation has the form (see Appendix I.1)

$$\partial_t h = -\frac{\delta \mathcal{F}}{\delta h} + \eta \tag{7.45}$$

with

$$\mathcal{F}[h] = \frac{\nu}{2} \int_V d\mathbf{x} (\nabla h)^2. \tag{7.46}$$

This form implies that the dynamics evolves in order to minimize the potential, because

$$\frac{d\mathcal{F}}{dt} = \int_V d\mathbf{x} \frac{\delta \mathcal{F}}{\delta h} \frac{\partial h}{\partial t} \tag{7.47}$$

$$= -\int_V d\mathbf{x} \left(\frac{\delta \mathcal{F}}{\delta h}\right)^2 + \int_V d\mathbf{x} \frac{\delta \mathcal{F}}{\delta h} \eta(\mathbf{x}, t), \tag{7.48}$$

so that[12]

$$\left\langle \frac{d\mathcal{F}}{dt} \right\rangle \leq 0. \tag{7.49}$$

The meaning of $\mathcal{F}$ is simple, because it represents the total area S of the surface $h(\mathbf{x})$, in the small slope approximation,

$$S = \int_V d\mathbf{x} \sqrt{1 + (\nabla h)^2} = V + \frac{1}{2} \int_V d\mathbf{x} (\nabla h)^2 + O\left((\nabla h)^4\right). \tag{7.50}$$

Therefore, neglecting terms that are irrelevant for power counting, we can write

$$\mathcal{F} = \nu S, \tag{7.51}$$

and the minimization of $\mathcal{F}$ acquires the trivial meaning of minimizing the total surface area.

The fact that dynamics is driven by the potential (7.46) has a simple consequence for the properties of the stationary state (which is attained at $t \gg t_c \simeq L^z$): It is statistically equivalent to an equilibrium state described by the Hamiltonian $\mathcal{H} = \mathcal{F}$.

---

[12] The second term in Eq. (7.48) may be positive and occasionally dominate the deterministic, negative term, but it averages to zero.

7.6.1 Dimensional Analysis

Let us rewrite the EW equation and the noise correlation function,

$$\partial_t h(\mathbf{x}, t) = \nu \nabla^2 h + \eta(\mathbf{x}, t) \tag{7.52}$$

$$\langle \eta(\mathbf{x}, t)\eta(\mathbf{x}', t') \rangle = \Gamma \delta(\mathbf{x} - \mathbf{x}')\delta(t - t'). \tag{7.53}$$

Two parameters enter in these expressions, the "diffusion coefficient" ν and the strength of noise $\sqrt{\Gamma}$, appearing in the noise correlation function. Indicating with $[A]$ the dimension of the physical quantity A, we can start writing $L = [\mathbf{x}]$, $T = [t]$, and $H = [h]$. The first two relations are trivial: We just state that $\mathbf{x}$ is a spatial variable and t is a time variable. Instead, the third one may appear arbitrarily redundant: Why should we introduce a *new* dimension H for a height variable? Shouldn't we simply write $H = L$? In fact, no: The EW equation describes a variety of different physical problems, not only deposition processes, and there is no reason to assume that h represents a length. There is also another motivation to keep H and L distinct, even in deposition processes, where h can be measured in centimeters as is x: The EW equation describes the dynamics of a self-affine interface, not a self-similar one; so, the space parallel to the substrate behaves differently from the space orthogonal to it.

We can now impose that the different terms appearing in Eqs. (7.52) and (7.53) have the same dimensions. We can write, respectively,

$$\frac{H}{T} = \frac{[\nu]\,H}{L^2} = [\eta] \tag{7.54a}$$

$$[\eta]^2 = \frac{[\Gamma]}{L^d T}, \tag{7.54b}$$

which give

$$[\nu] = \frac{L^2}{T} \tag{7.55a}$$

$$[\Gamma] = \frac{L^d H^2}{T}. \tag{7.55b}$$

We should now use the Family–Vicsek scaling for $W(L, t)$ including an explicit dependence on Γ and ν,

$$W(L, t) = \Gamma^{a_1} \nu^{b_1} L^\alpha w\left(\Gamma^{a_2} \nu^{b_2} \frac{t}{L^z}\right), \tag{7.56}$$

and hopefully determine all the exponents by imposing that the left-hand side and right-hand side have the same dimension. This amounts to saying that the argument of the function w is adimensional and the factor multiplying w has the dimension $[W] = H$,

$$[\Gamma]^{a_1} [\nu]^{b_1} L^\alpha = H, \tag{7.57a}$$

$$[\Gamma]^{a_2} [\nu]^{b_2} T = L^z. \tag{7.57b}$$

Using the earlier results for $[\nu]$ and $[\Gamma]$, we can finally write

$$\left(\frac{\mathrm{L}^d \mathrm{H}^2}{\mathrm{T}}\right)^{a_1} \left(\frac{\mathrm{L}^2}{\mathrm{T}}\right)^{b_1} \mathrm{L}^\alpha = \mathrm{H}, \tag{7.58a}$$

$$\left(\frac{\mathrm{L}^d \mathrm{H}^2}{\mathrm{T}}\right)^{a_2} \left(\frac{\mathrm{L}^2}{\mathrm{T}}\right)^{b_2} \mathrm{T} = \mathrm{L}^z. \tag{7.58b}$$

We can see that the consistency conditions on H impose $a_1 = \frac{1}{2}$ and $a_2 = 0$. The skeptical reader will be reassured by an alternative derivation of such results: In fact, $a_1 = \frac{1}{2}$ and $a_2 = 0$ mean that $W(L, t)$ is proportional to $\sqrt{\Gamma}$; see Eq. (7.56). The linearity between W and the strength of the noise simply comes from the linear character of the EW equation. Once we have determined a_1 and a_2, Eq. (7.58b) gives $b_2 = 1$ and $z = 2$, while Eq. (7.58a) gives $b_1 = -\frac{1}{2}$ and $\alpha = \frac{2-d}{2}$. Finally, using the scaling relation $z = \alpha/\beta$, we obtain β. In conclusion, a simple dimensional analysis has allowed us to find the roughness exponents

$$\alpha = \frac{2-d}{2}, \quad \beta = \frac{2-d}{4}, \quad z = 2. \qquad \text{(EW equation)} \tag{7.59}$$

From the conditions $\alpha(d_c^\ell) = 1$ and $\alpha(d_c^u) = \beta(d_c^u) = 0$, we find that the lower and upper critical dimensions are $d_c^\ell = 0$ and $d_c^u = 2$. For $d = 1$, we obtain $\alpha = \frac{1}{2}$ and $\beta = \frac{1}{4}$, which agree with the numerical values found for the RDR models, thus attesting that it belongs to the EW universality class.

We have also been able to write the scaling function in the universal form

$$W_{\mathrm{EW}}(L, t) = \sqrt{\frac{\Gamma L^{2-d}}{\nu}}\, w_{\mathrm{EW}}\left(\frac{\nu t}{L^2}\right). \tag{7.60}$$

The explicit expression of the function w_{EW} will be derived in Section 7.6.2.

The EW Equation with Conserved Noise

It is worth considering the (sole) effect of a conserved noise, which is unable to make rough a surface evolving accordingly to the same deterministic part of the EW equation. Let us therefore consider the equation $\partial_t h = -\nabla \cdot \mathbf{J}$, with $\mathbf{J} = -\nu \nabla h + \boldsymbol{\xi}$, where the noise term $\boldsymbol{\xi}$ is now included in the current, thus ensuring perfect conservation. If

$$\langle \xi_i(\mathbf{x}, t)\xi_j(\mathbf{x}', t')\rangle = \Gamma \delta_{ij}\delta(\mathbf{x} - \mathbf{x}')\delta(t - t'), \tag{7.61}$$

we can repeat the dimensional analysis. It is actually enough to realize that in Eq. (7.54a), we should replace $[\eta]$ with $[\xi]/\mathrm{L}$ and in Eq. (7.54b), we should replace $[\eta]$ with $[\xi]$. Therefore, Eq. (7.55a) is unchanged while the new Eq. (7.55b) corresponds to replace d with $d + 2$.

This allows us to go directly to Eq. (7.59) and obtain the following expressions for the roughness exponents of the conserved EW equation,

$$\alpha = -\frac{d}{2}, \quad \beta = -\frac{d}{4}, \quad z = 2. \qquad \text{(conserved EW equation)} \tag{7.62}$$

The exponents are always negative, reporting the inability of the conserved noise to rough the surface. If both types of noise are present (as in all RDR models), the conserved noise can be neglected.

7.6.2 The Scaling Functions

The conserved character of the EW equation implies $\overline{h(\mathbf{x}, t)} \equiv 0$, and the roughness has a simpler expression,

$$W^2(L, t) = \left\langle \overline{h^2(\mathbf{x}, t)} \right\rangle = \overline{\langle h^2(\mathbf{x}, t) \rangle}, \tag{7.63}$$

where we have used the possibility of exchanging the two types of average (see Section 7.2). In fact, if the system is translationally invariant, once we average $h^2(\mathbf{x}, t)$ over the noise realizations, the spatial dependence disappears, so we simply have

$$W^2(L, t) = \langle h^2(\mathbf{x}, t) \rangle. \tag{7.64}$$

The linear EW equation can be solved in Fourier space, so we write

$$h(\mathbf{x}, t) = \frac{1}{(2\pi)^d} \int d\mathbf{q}\, e^{i\mathbf{q}\cdot\mathbf{x}} h(\mathbf{q}, t), \tag{7.65}$$

and, from Eq. (7.64),

$$W_{\mathrm{EW}}^2(L, t) = \frac{1}{(2\pi)^{2d}} \int d\mathbf{q} \int d\mathbf{q}'\, e^{i(\mathbf{q}+\mathbf{q}')\cdot\mathbf{x}} \langle h(\mathbf{q}, t) h(\mathbf{q}', t) \rangle. \tag{7.66}$$

Replacing the Fourier expansion in the EW equation, we find that each Fourier component $h(\mathbf{q}, t)$ satisfies the equation

$$\partial_t h(\mathbf{q}, t) = -\nu q^2 h + \eta(\mathbf{q}, t), \tag{7.67}$$

where $q^2 = \mathbf{q} \cdot \mathbf{q}$. Its solution is[13]

$$h(\mathbf{q}, t) = h(\mathbf{q}, 0) e^{-\omega_{\mathbf{q}} t} + \int_0^t dt'\, \eta(\mathbf{q}, t') e^{-\omega_{\mathbf{q}}(t - t')}, \tag{7.68}$$

where we have used the simplified notation $\omega_{\mathbf{q}} = \nu q^2$. The first term on the right-hand side vanishes for an initially flat surface ($h(\mathbf{q}, 0) \equiv 0$), or it vanishes anyway after a transient, because of the exponential factor. Therefore, for long times the second term on the right-hand side always dominates, which also proves the linearity between h and the noise. So, we can write

$$h(\mathbf{q}, t) = \int_0^t dt'\, \eta(\mathbf{q}, t') e^{-\omega_{\mathbf{q}}(t - t')} \tag{7.69}$$

and the correlator appearing in Eq. (7.66) takes the form

$$\langle h(\mathbf{q}, t) h(\mathbf{q}', t) \rangle = \int_0^t dt'\, e^{-\omega_{\mathbf{q}}(t - t')} \int_0^t dt''\, e^{-\omega_{\mathbf{q}'}(t - t'')} \langle \eta(\mathbf{q}, t') \eta(\mathbf{q}', t'') \rangle. \tag{7.70}$$

---

[13] The solution of the differential equation $\dot{g}(t) = -\omega g + f(t)$, found by the method of variation of parameters, is $g(t) = g(0) e^{-\omega t} + \int_0^t dt'\, f(t') e^{-\omega(t - t')}$.

Therefore, we need to evaluate the correlator of noise in the Fourier space. Since $\eta(\mathbf{q}, t) = \int_V d\mathbf{x}\, e^{-i\mathbf{q}\cdot\mathbf{x}}\eta(\mathbf{x}, t)$, we find

$$
\begin{aligned}
\langle \eta(\mathbf{q}, t')\eta(\mathbf{q}', t'')\rangle &= \int d\mathbf{x}\int d\mathbf{x}'\, e^{-i\mathbf{q}\cdot\mathbf{x}}e^{-i\mathbf{q}'\cdot\mathbf{x}'}\langle \eta(\mathbf{x}, t')\eta(\mathbf{x}', t'')\rangle \\
&= \Gamma\delta(t' - t'')\int d\mathbf{x}\, e^{-i\mathbf{q}\cdot\mathbf{x}}e^{-i\mathbf{q}'\cdot\mathbf{x}} \\
&= (2\pi)^d\Gamma\delta(t' - t'')\delta(\mathbf{q} + \mathbf{q}'),
\end{aligned}
\tag{7.71}
$$

which can be replaced in Eq. (7.70), obtaining

$$
\begin{aligned}
\langle h(\mathbf{q}, t)h(\mathbf{q}', t)\rangle &= (2\pi)^d\Gamma\int_0^t dt'\, e^{-\omega_\mathbf{q}(t-t')}e^{-\omega_{-\mathbf{q}}(t-t')}\delta(\mathbf{q} + \mathbf{q}') \\
&= (2\pi)^d\Gamma\left(\frac{1 - e^{-2\omega_\mathbf{q}t}}{2\omega_\mathbf{q}}\right)\delta(\mathbf{q} + \mathbf{q}').
\end{aligned}
\tag{7.72}
$$

Finally, this result can be inserted in Eq. (7.66), which yields

$$
W_{\text{EW}}^2(L, t) = \frac{\Gamma}{(2\pi)^d}\int_{V_\mathbf{q}} d\mathbf{q}\,\frac{1 - e^{-2\omega_\mathbf{q}t}}{2\omega_\mathbf{q}},
\tag{7.73}
$$

where the domain of integration is $V_\mathbf{q} \equiv \{\frac{\pi}{L} \leq |q_i| \leq \frac{\pi}{a_0}\}$, a_0 being a cutoff for avoiding integration over arbitrarily small wavelengths, and it is justified by the ultimate discrete nature of matter. As expected, the right-hand side of Eq. (7.73) does not depend on $\mathbf{x}$ even if no spatial average has been performed: Averaging over noise restores translational invariance.

Result (7.73) has a much larger applicability than expected. In fact, it is valid for any equation of the form

$$
\partial_t h(\mathbf{x}, t) = \mathcal{L}[h] + \eta(\mathbf{x}, t)
\tag{7.74}
$$

where $\mathcal{L}[h]$ is a generic[14] linear operator. Fourier transforming the equation,

$$
\partial_t h(\mathbf{q}, t) = -\omega_\mathbf{q}h(\mathbf{q}, t) + \eta(\mathbf{q}, t),
\tag{7.75}
$$

we obtain the function $\omega_\mathbf{q}$, which encapsulates the properties of $\mathcal{L}$ and appears in Eq. (7.73). Let us consider a possible generalization of the EW equation, called the (stochastic) Herring–Mullins equation,

$$
\partial_t h(\mathbf{x}, t) = \nu\nabla^2 h - K(\nabla^2)^2 h + \eta(\mathbf{x}, t).
\tag{7.76}
$$

The minus sign in front of the quartic term is necessary to have a positive $\omega_\mathbf{q} = \nu q^2 + Kq^4$, so that the deterministic component of each Fourier mode $h(\mathbf{q}, t)$ decays in time as $\exp(-\omega_\mathbf{q}t)$ while the stochastic component fluctuates; see Eq. (7.68). From a physical point of view, the quadratic and quartic terms represent different atomistic processes. The quadratic term, as shown in Fig. 7.12, comes from a surface current proportional to the slope, while the quartic term, as discussed in more detail in Section 8.2.1, comes from a chemical potential related to the curvature of the surface. In terms of atomistic models, see Section 7.3, the quadratic

---

[14] In fact, in our derivation we have used $\omega_\mathbf{q} = \omega_{-\mathbf{q}}$, which means the operator $\mathcal{L}$ should be invariant under parity, $\mathbf{x} \to -\mathbf{x}$.

term is due to a relaxation driven by the minimization of the height of the deposited particle, while the quartic term reflects a relaxation driven by the maximization of its atomic bonds. Both cases have been mentioned in the presentation of deposition models with relaxation, see page 279.

According to power counting, the higher-order term $(\nabla^2)^2 h$ should be negligible with respect to the EW linearity $\nabla^2 h$, and this is reflected in the smallness of the q^4 term with respect to the q^2 term for large spatial scales, that is, for $q \to 0$. However, the microscopic processes determining the EW linearity might be absent, or, if present, ν might be small. In the latter case we have a crossover from a regime dominated by the quartic term (for $q \gg \sqrt{\nu/K}$) to the EW regime (for $q \ll \sqrt{\nu/K}$). Here we will limit ourselves to considering a single, dominant linear term with spatial derivatives of order δ, so that $\omega_{\mathbf{q}} = cq^\delta$.

If the system has to be rough, W must diverge with increasing L and t. This means we require (see Eq. (7.73)) that,

$$W^2(\infty, \infty) = \lim_{L \to \infty} \frac{\Gamma}{(2\pi)^d} \int_{V_{\mathbf{q}}} \frac{d\mathbf{q}}{2\omega_{\mathbf{q}}} = \infty, \tag{7.77}$$

and such divergence comes out from small-$\mathbf{q}$ integration. A simple evaluation of the small-$\mathbf{q}$ contribution allows us to determine the upper critical dimension. In fact,

$$W^2(\infty, \infty) \approx \int_0 \frac{d\mathbf{q}}{\omega_{\mathbf{q}}} \approx \int_0 dq \frac{q^{d-1}}{q^\delta}, \tag{7.78}$$

which diverges if $d < \delta$. So, $d_c^u = \delta$, extending the result $d_c^u = 2$, based on dimensional arguments we found for EW. In the following we assume $d < \delta$, in which case the integral has a divergence for small $\mathbf{q}$, while it converges for large $\mathbf{q}$. This allows us to replace the upper limit of the integral, of order π/a_0, with infinity. Let's now define $V_{\mathbf{q}}^\infty \equiv \left\{ |q_i| \geq \frac{\pi}{L} \right\}$ and write

$$W^2(L, t) = \frac{\Gamma}{(2\pi)^d} \int_{V_{\mathbf{q}}^\infty} d\mathbf{q} \frac{1 - e^{-2\omega_{\mathbf{q}} t}}{2\omega_{\mathbf{q}}}, \qquad d < \delta. \tag{7.79}$$

With a simple change of variable, $\mathbf{s} = L\mathbf{q}$, we can prove that W satisfies the Family–Vicsek scaling form:

$$\begin{aligned}
W^2(L, t) &= \frac{\Gamma}{(2\pi)^d} \int_{|q_i| > \pi/L} d\mathbf{q} \frac{1 - e^{-2cq^\delta t}}{2cq^\delta} \\
&= \frac{\Gamma L^{\delta-d}}{2c} \frac{1}{(2\pi)^d} \int_{|s_i| > \pi} d\mathbf{s} \frac{1 - e^{-2\frac{ct}{L^\delta} s^\delta}}{s^\delta} \\
&\equiv \left[\sqrt{\frac{\Gamma L^{\delta-d}}{2c}} w\left(\frac{ct}{L^\delta} \right) \right]^2,
\end{aligned} \tag{7.80}$$

with[15]

$$w^2(u) = \frac{1}{(2\pi)^d} \int_{|s_i| > \pi} d\mathbf{s} \frac{1 - e^{-2us^\delta}}{s^\delta}. \tag{7.81}$$

---

[15] We consistently denote the modulus of $\mathbf{s}$ by s.

Equation (7.80) suggests that $\alpha = (\delta - d)/2$ and $z = \delta$. However, in order to complete the proof, we have to show that $w(u) \approx u^\beta$ for $u \ll 1$ and $w(u)$ goes to a constant for large u; see Eq. (7.9). The limit $u \gg 1$ is trivial, because $w(\infty)$ is finite,

$$w^2(\infty) = \frac{1}{(2\pi)^d} \int_{|s_i|>\pi} \frac{d\mathbf{s}}{s^\delta} < \infty, \tag{7.82}$$

for $d < \delta$. The opposite limit $u \ll 1$ requires a new change of variable, $\mathbf{y} = u^{1/\delta}\mathbf{s}$, so

$$w^2(u) = \frac{1}{(2\pi)^d} u^{\frac{\delta-d}{\delta}} \int_{|y_i|>\pi u^{1/\delta}} d\mathbf{y} \frac{1 - e^{-2y^\delta}}{y^\delta}. \tag{7.83}$$

In the limit $u \ll 1$ the integral goes to a constant and

$$w(u) \approx u^{\frac{\delta-d}{2\delta}}, \tag{7.84}$$

therefore satisfying the relation $w(u) \approx u^\beta$, with $\beta = \alpha/z$.

The earlier calculations have allowed us to find the roughness exponents and the scaling function for a general linear theory whose dominant term has the form $(\nabla^2)^{\delta/2} h$. The result for the exponents is[16]

$$\alpha = \frac{\delta - d}{2}, \qquad \beta = \frac{\delta - d}{2\delta}, \qquad z = \delta. \tag{7.85}$$

The lower and upper critical dimensions are $d_c^\ell = \delta - 2$ and $d_c^u = \delta$. Therefore, passing from the EW equation ($\delta = 2$) to the quartic, stochastic Herring–Mullins equation ($\delta = 4$) shifts the physically relevant interval (d_c^ℓ, d_c^u) from $(0, 2)$ to $(2, 4)$, while the scaling function $w(u)$ depends on δ according to Eq. (7.81).

We conclude this part by evaluating the scaling form for the correlation function of the interface height,

$$G(\mathbf{r}, t) = \left\langle \overline{(h(\mathbf{x}, t) - h(\mathbf{x} + \mathbf{r}, t))^2} \right\rangle \tag{7.86}$$

where, as we have done for computing the roughness, we can exchange the two averages and remove the spatial average, because of translational invariance. Using Eq. (7.65), we can write

$$h(\mathbf{x}, t) - h(\mathbf{x} + \mathbf{r}, t) = \frac{1}{(2\pi)^d} \int d\mathbf{q} e^{i\mathbf{q}\cdot\mathbf{x}} \left(1 - e^{i\mathbf{q}\cdot\mathbf{r}}\right) h(\mathbf{q}, t), \tag{7.87}$$

so

$$G(\mathbf{r}, t) = \frac{1}{(2\pi)^{2d}} \int d\mathbf{q} e^{i\mathbf{q}\cdot\mathbf{x}} \left(1 - e^{i\mathbf{q}\cdot\mathbf{r}}\right) \int d\mathbf{q}' e^{i\mathbf{q}'\cdot\mathbf{x}} \left(1 - e^{i\mathbf{q}'\cdot\mathbf{r}}\right) \langle h(\mathbf{q}, t) h(\mathbf{q}', t) \rangle. \tag{7.88}$$

The correlation function appearing in Eq. (7.88) has been evaluated in Eq. (7.72), therefore yielding the result

$$G(\mathbf{r}, t) = \frac{\Gamma}{(2\pi)^d} \int d\mathbf{q} \left(1 - e^{i\mathbf{q}\cdot\mathbf{r}}\right) \left(1 - e^{-i\mathbf{q}\cdot\mathbf{r}}\right) \left(\frac{1 - e^{-2\omega_q t}}{2\omega_q}\right), \tag{7.89}$$

---

[16] The exponents could also be found by generalizing the dimensional analysis of Section 7.6.1, but the explicit expression of the scaling function can be found only with the present analysis.

Finally, since

$$\left(1 - e^{i\mathbf{q}\cdot\mathbf{r}}\right)\left(1 - e^{-i\mathbf{q}\cdot\mathbf{r}}\right) = 4\sin^2\left(\frac{\mathbf{q}\cdot\mathbf{r}}{2}\right), \tag{7.90}$$

we get

$$G(\mathbf{r}, t) = \frac{2\Gamma}{(2\pi)^d}\int d\mathbf{q}\,\sin^2\left(\frac{\mathbf{q}\cdot\mathbf{r}}{2}\right)\frac{\left(1 - e^{-2\omega_\mathbf{q} t}\right)}{\omega_\mathbf{q}}. \tag{7.91}$$

The $\mathbf{q}$ integration in the previous expression is not bounded from below, because the dependence on $\mathbf{r}$ is now explicit and the spatial integration in the definition of G (see Eq. (7.86)) is extended to the entire physical space, which is not restricted to a sample of linear size L. As before, we will assume the general form $\omega_\mathbf{q} = cq^\delta$.

The upper critical dimension can be recovered from Eq. (7.91), determining when such an integral diverges at large r and t. In this limit, $\sin^2\left(\frac{\mathbf{q}\cdot\mathbf{r}}{2}\right) \to \frac{1}{2}$ and

$$G(\infty, \infty) = \frac{\Gamma}{(2\pi)^d}\int\frac{d\mathbf{q}}{cq^\delta}. \tag{7.92}$$

The integral diverges in $\mathbf{q} = 0$ if $d < \delta$, confirming that $d_c^u = \delta$. With this constraint on the dimension, the upper bound of the integral, of order $1/a_0$, can be extended to infinity. With the change of variable $\mathbf{s} = r\mathbf{q}$, we obtain the scaling form

$$G(r, t) = \frac{2\Gamma r^{\delta-d}}{c}g\left(\frac{ct}{r^\delta}\right) \tag{7.93}$$

$$g(u) = \frac{1}{(2\pi)^d}\int d\mathbf{s}\,\sin^2\left(\frac{\mathbf{s}\cdot\hat{\mathbf{r}}}{2}\right)\frac{\left(1 - e^{-2us^\delta}\right)}{s^\delta}, \tag{7.94}$$

where the direction $\hat{\mathbf{r}} = \mathbf{r}/r$ is irrelevant.

The earlier scaling form for the correlation function should be compared with the scaling function for $W^2(L, t)$; see Eqs. (7.80) and (7.81). The main point we are making here is about the limit $t = \infty$ for finite r. Unlike the roughness, which is always finite for finite L, the correlation function may diverge for finite r. We must take the limit

$$\lim_{u\to\infty}g(u) = \frac{1}{(2\pi)^d}\int d\mathbf{s}\,\sin^2\left(\frac{\mathbf{s}\cdot\hat{\mathbf{r}}}{2}\right)\frac{1}{s^\delta} \tag{7.95}$$

and focus on a possible divergence in $\mathbf{s} = 0$. Evaluating the contribution of the vanishing s region, we can expand the sine for a small argument and obtain

$$g(\infty) = \frac{1}{4(2\pi)^d}\int_{S_{d-1}}d\Omega\,\cos^2\theta\int_0\frac{ds}{s^{\delta-d-1}} + \text{finite contribution}, \tag{7.96}$$

where S_d is the d-dimensional sphere and $d\Omega$ is the solid angle. It is straightforward to conclude that $g(\infty) = \infty$ for $d < \delta - 2$, that is, $d < d_c^\ell$. It is noteworthy that such divergence requires an infinite sample: If we evaluate $\left\langle(h(\mathbf{x}, t) - h(\mathbf{x}+\mathbf{r}, t))^2\right\rangle$ for a finite sample of size $L > r$, the divergence disappears.

We can sum up the difference between the scaling forms of $W^2(L, t)$ and $G(r, t)$ as follows. For $d < d_c^u$, the interface is rough and both scaling forms have the same asymptotic behaviors. In particular, for finite L, r and infinite t,

$$W^2(L, \infty) = \frac{\Gamma}{2c} w^2(\infty) L^{\delta-d} \tag{7.97a}$$

$$G(r, \infty) = \frac{2\Gamma}{c} g(\infty) r^{\delta-d}. \tag{7.97b}$$

While $w(\infty)$ is always finite, $g(\infty)$ diverges for $d < d_c^\ell$. As we already noticed, this divergence can be healed by taking a finite sample, that is, adding the length scale L. If we do that, for $d < d_c^\ell = \delta - 2$, we can write a more complicated scaling function involving r, t, and L. This is the so-called anomalous scaling.

7.7 The Kardar–Parisi–Zhang Equation

According to the results of power counting, Section 7.5, $(\nabla h)^2$ dominates over $\nabla^2 h$; therefore, we expect KPZ to be in a different universality class than EW, which means different critical exponents, different critical dimensions, and possibly different other relevant features. The first effect has already been noticed through the nonconserved deposition models studied in Section 7.3, where the simulation of the one-dimensional BD model clearly had different critical exponents than RDR models. In Section 7.7.1, we provide an overview of the KPZ universality class, based on scale transformation and perturbative renormalization group. This allows to obtain a few important results: a relation between critical exponents valid in any d, the roughness exponents in $d = 1$, and the existence of a nonperturbative critical point in $d \geq 2$. The first two results will be confirmed by exact procedures in Sections 7.7.2 and 7.7.3, respectively. The question of the upper critical dimension is briefly discussed at the end of Section 7.7.1. Finally, the existence of subclasses will be discussed in Section 7.7.4, which includes results that testify to a revival of interest toward KPZ in recent years.

7.7.1 Overview of KPZ versus Dimension d

It is useful to start the discussion by considering the effect of scale transformations (7.18) on (all terms of) the KPZ equation. In order to do that, we need to know how the noise scales, which can be deduced from its correlation function

$$\langle \eta(\mathbf{x}, t) \eta(\mathbf{x}', t') \rangle = \Gamma \delta(\mathbf{x} - \mathbf{x}') \delta(t - t'). \tag{7.98}$$

If $\eta \to b^\chi \eta$, using Eqs. (7.18), the condition that left-hand side and right-hand side scale in the same way implies

$$b^{2\chi} = \frac{1}{b^d} \frac{1}{b^z}, \tag{7.99}$$

so

$$\chi = -\frac{d+z}{2}. \tag{7.100}$$

Therefore, we can write

$$b^{\alpha-z}\partial_t h(\mathbf{x},t) = \nu b^{\alpha-2}\nabla^2 h + \frac{\lambda}{2}b^{2\alpha-2}(\nabla h)^2 + b^{-\frac{d+z}{2}}\eta(\mathbf{x},t), \tag{7.101}$$

and by dividing all terms by $b^{\alpha-z}$, we obtain

$$\partial_t h(\mathbf{x},t) = \nu b^{z-2}\nabla^2 h + \frac{\lambda}{2}b^{\alpha+z-2}(\nabla h)^2 + b^{(z-d-2\alpha)/2}\eta(\mathbf{x},t). \tag{7.102}$$

If we use the EW exponents, $\alpha_{\text{EW}} = (2-d)/2$ and $z_{\text{EW}} = 2$, we obtain

$$\partial_t h(\mathbf{x},t) = \nu\nabla^2 h + \frac{\lambda}{2}b^{(2-d)/2}(\nabla h)^2 + \eta(\mathbf{x},t), \tag{7.103}$$

which tells us two things. The first thing is that EW exponents might be determined by simply requiring that the equation is invariant under a scale transformation. This is due to the linearity of the equation itself. The second thing is that Eq. (7.103) seems to suggest that the KPZ nonlinearity is irrelevant for $d > 2$, that is, when $\alpha_{\text{EW}} < 0$. However, this statement is questionable, because we used the EW result, to evaluate the scale factors. As a matter of fact, this procedure has a perturbative spirit and it can be studied more rigorously within a perturbative renormalization group (pRG) approach, which confirms that a sufficiently small KPZ nonlinearity is irrelevant for $d > 2$. The detailed calculations are not reported: We limit here to report and discuss the results, while the procedure is sketched in Appendix R.

If we define $\ell = \ln b$, the ℓ-dependence of the coupling parameters ν, λ, and Γ is

$$\frac{d\nu}{d\ell} = \nu\left[z - 2 + \frac{K_d(2-d)}{4d}g^2\right] \equiv \nu f_\nu(g) \tag{7.104a}$$

$$\frac{d\lambda}{d\ell} = \lambda(\alpha + z - 2) \tag{7.104b}$$

$$\frac{d\Gamma}{d\ell} = \Gamma\left[z - d - 2\alpha + \frac{K_d}{4}g^2\right] \equiv \Gamma f_\Gamma(g), \tag{7.104c}$$

where K_d is a numerical factor and $g^2 = \Gamma\lambda^2/\nu^3$ measures the strength of the nonlinearity. In the linear case (EW equation, $\lambda = 0 = g^2$), we are left with two, uncoupled equations for ν and Γ whose solutions are equivalent to the result obtained in Eq. (7.102) via simple scale transformation.[17]

In the fully nonlinear case (KPZ equation, $\lambda, g^2 \neq 0$), we immediately remark that λ has the solution given by simple scale transformation, $\lambda(b) = \lambda(1)b^{\alpha+z-2}$: This means that λ does not renormalize and the fixed point of the KPZ equation is characterized by the relation

$$\alpha + z = 2, \tag{7.105}$$

which is an exact relation, as proved in Section 7.7.2.

In order to determine the perturbative KPZ fixed point (if any) and the critical exponents, it is necessary to analyze Eqs. (7.104): By combining (a) and (c) and using the relation

---

[17] More precisely, we obtain $\nu(b) = \nu(1)b^{\nu(z-2)}$ and $\Gamma(b) = \Gamma(1)b^{(z-d-2\alpha)/2}$. In Eq. (7.102), the scale factor of noise corresponds to $\sqrt{\Gamma(b)/\Gamma(1)}$, because the strength of the noise is equal to $\sqrt{\Gamma}$, not to Γ.

(7.105) we find

$$\frac{dg}{d\ell} = \frac{2-d}{2}g + \frac{(2d-3)K_d}{4d}g^3. \tag{7.106}$$

This equation always has the solution $g = 0$, corresponding to the EW fixed point, but it is stable only for $d > 2$. More precisely, we attain such a fixed point if we start with a sufficiently small $g(0)$, that is, if λ is sufficiently small: This is the formal proof of the statement made on the basis of Eq. (7.103).

For $d = 2$, Eq. (7.106) says nothing because $g = 0$ is unstable for arbitrarily small g, but no other stable fixed point is given by our perturbative approach. Instead, for $d = 1$, we are able to work out Eq. (7.106), find a nontrivial (stable) fixed point, $g^* = \sqrt{2/K_1} = 2\sqrt{\pi}$, and imposing $f_\nu(g^*) = 0$ (see Eq. (7.104a)), we find $z_{\text{KPZ}} = \frac{3}{2}$; then, using the Eq. (7.105), $\alpha_{\text{KPZ}} = \frac{1}{2}$. These two values attest that the BD model, discussed in Section 7.3, belongs to the KPZ universality class.

A variety of approaches has been used to gain information on KPZ behavior for $d > 1$ ($d = 1$ being the only case where a clear analytical picture is available): simulation of suitable deposition models, direct numerical integration of the KPZ equation, nonperturbative RG and real space RG, and mapping to the directed polymer problem. All these tools agree about the phase diagram for $d < 4$: They confirm the pRG picture and a strong coupling, rough phase. The behavior of the KPZ equation for $d \geq 4$ is still controversial, however, and the main bone of contention is the value of d_c^u: While field-theoretic approaches seem to suggest[18] $d_c^u = 4$, numerics (direct integration or simulation of deposition models) support $d_c^u > 4$ and the real-space RG method gives $d_c^u = \infty$.

Computation simplicity has given a special role to the simulation of the RSOS model, introduced in Section 7.3. Figure 7.13(a) plots $W(L, t)$ of the RSOS model in $d = 4$, giving a firm support to the claim that $d_c^u > 4$ in this model. Figure 7.13(b) also shows data collapse using $\alpha \simeq 0.25$ and $z = 2 - \alpha$. More precise values for all exponents are given in Table 7.1. These values also suggest that a simple, rational conjecture made by the South Korean Jin Min Kim and the British John Kosterlitz, $\alpha = \frac{2}{d+3}$, is not correct.

7.7.2 The Galilean (or Tilt) Transformation

Before proving the invariance of the KPZ equation under this transformation and discussing its consequence in terms of the scaling relation among critical exponents, let's take a look back to the mapping between the ASEP and the SS model. This allows us to derive the one-dimensional KPZ equation from the continuum description of ASEP dynamics, but most of all it relates KPZ to the Burgers equation, which is (physically) at the origin of the invariance under Galilean transformation.[19]

In the ASEP model, the probability that site i is occupied, p_i, depends on time through the unbalance of the net currents $J_{i-1,i}$ (between sites $i - 1$ and i) and the net current $J_{i,i+1}$.

---

[18] Or at least that something physically relevant occurs at $d_c = 4$.
[19] This explains the double name of the transformation: Galilean refers to the Burgers equation, Tilt refers to the KPZ equation.

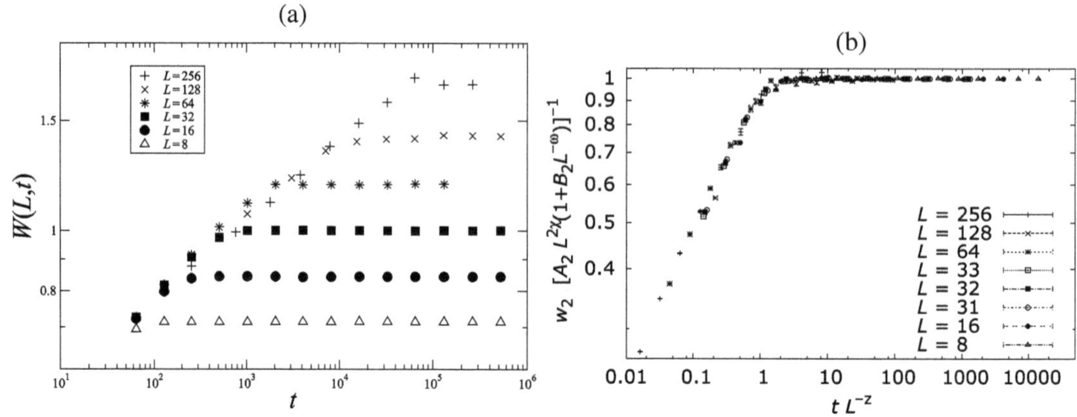

Fig. 7.13 (a) $W(L, t)$ for the RSOS model in $d = 4$. The asymptotic value increases with the size L, clearly showing that $\alpha > 0$ and that $d = 4$ is not the upper critical dimension. Data courtesy of Andrea Pagnani. (b) The data collapse of the square roughness $W^2(L, t)$. It is necessary to introduce the subleading term, as explained in the original paper, A. Pagnani and G. Parisi, Multisurface Coding Simulations of the Restricted Solid-on-solid Model in Four Dimensions, *Physical Review E*, **87** (2013) 010102.

Table 7.1 Numerical values of the roughness exponents for the RSOS model

d^a	α^b	z^c	β^d
1	$\dfrac{1}{2}$	$\dfrac{3}{2}$	$\dfrac{1}{3}$
2	0.3869(4)	1.6131	0.2398
3	0.3135(15)	1.6865	0.186
4	0.2537(8)	1.7463	0.1453

[a]For $d = 1$ we report the exact values.

[b]The numbers within parentheses are the standard deviation. Numerical data from A. Pagnani and G. Parisi, Numerical estimate of the Kardar–Parisi–Zhang universality class in (2+1) dimensions, *Physical Review E*, **92** (2015) 010101, for $d = 2$; from E. Marinari, A. Pagnani, and G. Parisi, Critical Exponents of the KPZ Equation via Multi-surface Coding Numerical Simulations, *Journal of Physics A: Mathematical and General*, **33** (2000) 8181–8192, for $d = 3$; from A. Pagnani and G. Parisi, Multisurface Coding Simulations of the Restricted Solid-on-solid Model in Four Dimensions, *Physical Review E*, **87** (2013) 010102, for $d = 4$.

[c]Determined by the relation $z = 2 - \alpha$.

[d]Determined by the relation $\beta = \alpha/z$.

More precisely,

$$\tau_0 \frac{dp_i}{dt} = c_+[p_{i-1}(1 - p_i) - p_i(1 - p_{i+1})] + c_-[p_{i+1}(1 - p_i) - p_i(1 - p_{i-1})], \quad (7.107)$$

where the time constant τ_0 has been introduced to make the relation dimensionally correct. To pass to the continuum, we write $p_i(t) = p(x,t)$ and $p_{i\pm1} = p(x,t) \pm a_0\, p'(x,t) + \frac{1}{2}a_0^2\, p''(x,t)$, where a_0 is the lattice constant and primes indicate spatial derivatives. If we write $c_\pm = \frac{1}{2} \pm \delta$, we obtain

$$\tau_0 \frac{\partial p}{\partial t} = -2\delta a_0 p' + \frac{1}{2}a_0^2 p'' + 4\delta a_0 pp'. \tag{7.108}$$

We can now pass from the variable density, p, to the variable slope, $m = 1 - 2p$, arriving at the so-called Burgers equation,

$$\partial_t m(x,t) = \nu m_{xx} + \lambda m m_x, \tag{7.109}$$

where[20] $\nu = a_0^2/(2\tau_0)$ and $\lambda = -(2\delta a_0)/\tau_0$.

Earlier mapping indicates that m is a slope rather than a height h. Therefore, we write $m(x,t) = \partial_x h(x,t)$ and integrate both sides of Eq. (7.109), obtaining

$$\partial_t h(x,t) = \nu h_{xx} + \frac{\lambda}{2}h_x^2, \tag{7.110}$$

which is recognized as the (deterministic) KPZ equation (see Appendix I).

In general terms (and general dimension d), the Burgers equation is the differential equation describing the dynamics of an incompressible fluid when pressure effects and external forces, for example, gravity, can be neglected,[21]

$$\frac{d\mathbf{u}}{dt} \equiv \left(\frac{\partial}{\partial t} + \mathbf{u} \cdot \nabla\right)\mathbf{u} = \nu\nabla^2\mathbf{u}. \tag{7.111}$$

The operator d/dt is the material time derivative, typical of the mechanics of continuous media, while the linear term on the right-hand side accounts for viscous friction. The presence of the material derivative ensures the invariance under the Galilean transformation

$$\mathbf{u}'(\mathbf{x}, t) = \mathbf{u}(\mathbf{x} - \mathbf{u}_0 t, t) + \mathbf{u}_0. \tag{7.112}$$

In fact, since

$$\partial_t \mathbf{u}' = \partial_t \mathbf{u} - (\mathbf{u}_0 \cdot \nabla)\mathbf{u} \tag{7.113a}$$

$$\partial_i \mathbf{u}' = \partial_i \mathbf{u}, \tag{7.113b}$$

we can write

$$\begin{aligned}
(\partial_t + \mathbf{u}' \cdot \nabla)\mathbf{u}' &= (\partial_t - \mathbf{u}_0 \cdot \nabla)\mathbf{u} + \big((\mathbf{u} + \mathbf{u}_0) \cdot \nabla\big)\mathbf{u} \\
&= (\partial_t + \mathbf{u} \cdot \nabla)\mathbf{u} \\
&= \nu\nabla^2\mathbf{u} \\
&= \nu\nabla^2\mathbf{u}',
\end{aligned} \tag{7.114}$$

---

[20] The continuum limit must be handled with care, because if we take $a_0, \tau_0 \to 0$, it is manifest that the nonlinear term dominates over the linear one. This is a well-known fact when we consider anisotropic diffusion: In fact, along with $a_0 \to 0$, we should also assume a vanishing drift, $\delta \to 0$. The physical reason is that the combination of a finite drift with a vanishing lattice constant a_0 would lead to a divergent driving force in the continuum limit.

[21] The reader can start from the Navier–Stokes equation (W.1) and remove the pressure term and the gravity term.

therefore confirming the invariance of the Burgers equation under Eq. (7.112).

We now want to show that the Galilean invariance for the Burgers equation implies the invariance of the KPZ equation under a more complicated transformation, called tilt transformation. The first step is to establish the relation between the two equations in a generic dimension d, because here above we were confined to $d = 1$. In order to make the relation more transparent, let us define $\mathbf{m} = \nabla h$ and take the (partial) derivative ∂_i of both sides of KPZ equation,

$$\partial_i\,(\partial_t h) = \partial_i\left(\nu\nabla^2 h + \frac{\lambda}{2}\sum_j (\partial_j h)(\partial_j h) + \eta\right) \tag{7.115a}$$

$$\partial_t m_i = \nu\nabla^2 m_i + \lambda\sum_j m_j\partial_j m_i + \partial_i\eta \tag{7.115b}$$

$$\partial_t\mathbf{m} = \nu\nabla^2\mathbf{m} + \lambda(\mathbf{m}\cdot\nabla)\mathbf{m} + \nabla\eta, \tag{7.115c}$$

and finally,

$$(\partial_t - \lambda(\mathbf{m}\cdot\nabla))\,\mathbf{m} = \nu\nabla^2\mathbf{m} + \nabla\eta. \tag{7.116}$$

The earlier equation is manifestly related to the Burgers equation. In fact, if we define $\mathbf{u} = -\lambda\mathbf{m} \equiv -\lambda\nabla h$, we obtain

$$(\partial_t + (\mathbf{u}\cdot\nabla))\,\mathbf{u} = \nu\nabla^2\mathbf{u} - \lambda\nabla\eta. \tag{7.117}$$

In conclusion, if h satisfies the KPZ equation, $\mathbf{u} = -\lambda\nabla h$ satisfies the Burgers equation with conserved noise. What does Galilean invariance for the Burgers equation imply for KPZ? Let us rewrite Eq. (7.112) in terms of h as follows,

$$-\lambda\nabla h'(\mathbf{x}, t) = -\lambda\nabla h(\mathbf{x} - \mathbf{u}_0 t, t) + \mathbf{u}_0 \tag{7.118}$$

or, after integration over the space variable $\mathbf{x}$, as

$$h'(\mathbf{x}, t) = h(\mathbf{x} - \mathbf{u}_0 t, t) - \frac{1}{\lambda}\mathbf{u}_0\cdot\mathbf{x} + a(t), \tag{7.119}$$

where the integration constant a may depend on time. The function $a(t)$ can be determined by replacing the earlier expression in the KPZ equation, obtaining $\dot{a}(t) = u_0^2/(2\lambda)$. Finally, the KPZ equation is invariant under the transformation[22]

$$h'(\mathbf{x}, t) = h(\mathbf{x} - \mathbf{u}_0 t, t) - \frac{1}{\lambda}\mathbf{u}_0\cdot\mathbf{x} + \frac{u_0^2}{2\lambda}t. \tag{7.120}$$

We should not forget noise, which transforms according to

$$\eta'(\mathbf{x}, t) = \eta(\mathbf{x} - \mathbf{u}_0 t, t). \tag{7.121}$$

If η is δ-correlated in time, η' has the same properties,[23] but if η has finite time correlations, η' has different spectral properties. Therefore, the tilt transformation is actually valid only for δ-correlated noise, which is what we are interested in here.

---

[22] A direct proof of this statement is greatly simplified for an infinitesimal transformation, $u_0 \to 0$, because quadratic terms in u_0 can be neglected.

[23] We have $\langle\eta'(\mathbf{x}, t)\eta'(\mathbf{x}', t')\rangle = \langle\eta(\mathbf{x} - \mathbf{u}_0 t, t)\eta(\mathbf{x}' - \mathbf{u}_0 t', t')\rangle = \Gamma\delta(\mathbf{x} - \mathbf{x}' - \mathbf{u}_0(t - t'))\delta(t - t') = \Gamma\delta(\mathbf{x} - \mathbf{x}')\delta(t - t')$. The last passage requires δ-correlation in time.

If we have devoted so much space to prove the invariance of the KPZ equation with respect to the tilt transformation, it is because it has a consequence of primary importance: the possibility of deriving a relation between roughness exponents, which is exact in any dimension. In fact, since the tilt transformation explicitly depends on λ and the invariance must be preserved at all length scales, this combination of facts implies that λ cannot be renormalized by a change of scale.[24] Going back to Eq. (7.102), this requires that $b^{\alpha+z-2} = 1$, that is,

$$\alpha + z = 2. \tag{7.122}$$

In conclusion, this relation implies that only one exponent is actually undetermined in the KPZ model, the other being fixed by Eq. (7.122). Then, Section 7.7.3 shows that in $d = 1$ it is possible to determine α, therefore solving the problem, while for generic d the problem is still open.

7.7.3 Exact Exponents in $d = 1$

In one dimension it is possible to determine α using the Fokker–Planck formalism. The generalized Langevin equation

$$\partial_t h(\mathbf{x}, t) = \mathcal{N}[h] + \eta(\mathbf{x}, t) \tag{7.123a}$$

$$\langle \eta(\mathbf{x}, t) \rangle = 0 \tag{7.123b}$$

$$\langle \eta(\mathbf{x}, t)\eta(\mathbf{x}', t') \rangle = \Gamma\delta(\mathbf{x} - \mathbf{x}')\delta(t - t') \tag{7.123c}$$

is associated with the Fokker–Planck equation,

$$\frac{\partial P}{\partial t} = \int d\mathbf{x}\, \frac{\delta}{\delta h}\left[-\mathcal{N}P + \frac{\Gamma}{2}\frac{\delta P}{\delta h} \right], \tag{7.124}$$

where $P(h, t)$ is the probability that a given profile $h(\mathbf{x})$ occurs at time t. In principle, the knowledge of $P(h, t)$ allows us to determine any average over noise,

$$\langle F(h) \rangle = \frac{\int \mathcal{D}h F(h) P(h, t)}{\int \mathcal{D}h P(h, t)}, \tag{7.125}$$

therefore, in particular, $F(h) = \overline{h^2(\mathbf{x}, t)}$ and $F(h) = \left(\overline{h(\mathbf{x}, t)}\right)^2$, which are the two building blocks to evaluate the roughness.

Because of Eq. (7.122), it is sufficient to find α, which characterizes the asymptotic steady state, corresponding to the time-independent solution of the Fokker–Planck equation. This is actually possible for the EW equation, because, as shown later, its derivability from a Lyapunov functional automatically provides the stationary solution of Eq. (7.124). The

---

[24] If $h(\mathbf{x}, t)$ is a solution of the KPZ equation, then its tilt-transformed $h^*_\lambda(\mathbf{x}, t)$ is also a solution. Let us now perform a coarse graining procedure of $h(\mathbf{x}, t)$, obtaining a function $H(\mathbf{x}, t)$, which is solution of the KPZ equation, but with coupling parameters λ', ν', γ'. If we want that the tilt-transformed of H, $H^*_{\lambda'}$ is equal to h^*_λ after coarse graining, we must require that $\lambda' = \lambda$, that is, that λ does not change under the flow of the renormalization group.

important point we want to discuss here is that, in $d = 1$ (and only in $d = 1$!), KPZ has the same stationary solution as EW. Therefore, it must be

$$\alpha_{\mathrm{KPZ}} = \alpha_{\mathrm{EW}} = \frac{1}{2}, \qquad d = 1. \tag{7.126}$$

The rest of this section is devoted to proving that in $d = 1$, the Fokker–Planck equations for EW and KPZ have the same stationary solution.

A stationary solution $P_s(h)$ of (7.124) is surely found if we are able to solve the equation

$$-\mathcal{N} P_s + \frac{\Gamma}{2} \frac{\delta P_s}{\delta h} = 0 \tag{7.127}$$

or

$$\frac{\delta P_s}{\delta h} = \frac{2}{\Gamma} \mathcal{N} P_s. \tag{7.128}$$

For a problem with one degree of freedom and order parameter $h(t)$, $\mathcal{N}$ is a function of h and so is P, and the functional derivative is replaced by the usual derivative. Every function $\mathcal{N}(h)$ would be integrable,

$$\mathcal{U}(h) = - \int dh \mathcal{N}(h), \tag{7.129}$$

and the solution of Eq. (7.128) would simply be $P_s(h) = e^{-2\mathcal{U}(h)/\Gamma}$. For our generalized Langevin equation, the "integrability" of $\mathcal{N}[h]$ is equivalent to saying that Eq. (7.123a) is derivable from a Lyapunov functional $\mathcal{F}$. If this happens,

$$\mathcal{N}[h] = -\frac{\delta \mathcal{F}}{\delta h} \tag{7.130}$$

and

$$P_s = e^{-\dfrac{2\mathcal{F}}{\Gamma}} \tag{7.131}$$

is the stationary solution of the Fokker–Planck equation.

For the EW equation,

$$\partial_t h = \mathcal{N}_{\mathrm{EW}} + \eta, \tag{7.132}$$

and we know that $\mathcal{N}_{\mathrm{EW}} = -\delta \mathcal{F}_{\mathrm{EW}}/\delta h$, where

$$\mathcal{F}_{\mathrm{EW}} = \frac{\nu}{2} \int d\mathbf{x} (\nabla h)^2, \tag{7.133}$$

so that we also know the stationary solution

$$P_s^{\mathrm{EW}} = e^{-\dfrac{2\mathcal{F}_{\mathrm{EW}}}{\Gamma}}. \tag{7.134}$$

The KPZ equation,

$$\partial_t h = \mathcal{N}_{\mathrm{KPZ}} + \eta, \tag{7.135}$$

admits the same Fokker–Planck stationary solution if (see the right-hand side of Eq. (7.124))

$$\int d\mathbf{x}\,\frac{\delta}{\delta h(\mathbf{x})}\left[-\mathcal{N}_{\text{KPZ}}P_s^{\text{EW}} + \frac{\Gamma}{2}\frac{\delta P_s^{\text{EW}}}{\delta h(\mathbf{x})}\right] = 0. \tag{7.136}$$

Since $\mathcal{N}_{\text{KPZ}} = \mathcal{N}_{\text{EW}} + \frac{\lambda}{2}(\nabla h)^2$, the earlier equation corresponds to

$$-\frac{\lambda}{2}\int d\mathbf{x}\,\frac{\delta}{\delta h(\mathbf{x})}\left[(\nabla h(\mathbf{x}))^2 P_s^{\text{EW}}\right] + \int d\mathbf{x}\,\frac{\delta}{\delta h(\mathbf{x})}\left[-\mathcal{N}_{\text{EW}}P_s^{\text{EW}} + \frac{\Gamma}{2}\frac{\delta P_s^{\text{EW}}}{\delta h(\mathbf{x})}\right] = 0. \tag{7.137}$$

The second integral vanishes automatically, because P_s^{EW} is the stationary solution of the Fokker–Planck equation for EW. Finally, P_s^{EW} is also a stationary solution of the Fokker–Planck equation for KPZ if (and only if)

$$\int d\mathbf{x}\,\frac{\delta}{\delta h(\mathbf{x})}\left[(\nabla h(\mathbf{x}))^2 P_s^{\text{EW}}\right] = 0. \tag{7.138}$$

Taking the functional derivative, we obtain[25]

$$\int d\mathbf{x}\,\frac{\delta}{\delta h(\mathbf{x})}\left[(\nabla h(\mathbf{x}))^2 P_s^{\text{EW}}\right] = -2P_s^{\text{EW}}\delta(0)\int d\mathbf{x}\nabla^2 h(\mathbf{x}) + \frac{2\nu}{\Gamma}P_s^{\text{EW}}\int d\mathbf{x}(\nabla h(\mathbf{x}))^2\nabla^2 h(\mathbf{x})$$

$$= \frac{2\nu}{\Gamma}P_s^{\text{EW}}\int d\mathbf{x}(\nabla h(\mathbf{x}))^2\nabla^2 h(\mathbf{x}), \tag{7.139}$$

where the integral $\int d\mathbf{x}\nabla^2 h(\mathbf{x})$ vanishes because of periodic boundary conditions and the origin of the term $\delta(0)$ is discussed in Appendix I.1. Let us now evaluate the second integral in $d = 1$ and $d = 2$:

$$\int d\mathbf{x}(\nabla h)^2\nabla^2 h = \begin{cases} \int dx\, h_x^2 h_{xx} = \int dx\,\frac{1}{3}(h_x^3)_x = 0, & d = 1 \\ \int d\mathbf{x}(h_x^2 + h_y^2)(h_{xx} + h_{yy}) \neq 0, & d = 2. \end{cases} \tag{7.140}$$

Therefore, in $d = 1$, the KPZ and the EW equations share the same long-time statistics of stationary fluctuations. This property implies that $\alpha_{\text{KPZ}} = \alpha_{\text{EW}} = \frac{1}{2}$. Since $\alpha_{\text{KPZ}} + z_{\text{KPZ}} = 2$, we can derive $z_{\text{KPZ}} = \frac{3}{2}$ and consequently $\beta_{\text{KPZ}} = \frac{1}{3}$.

In conclusion, in $d = 1$, $\alpha_{\text{KPZ}} = \alpha_{\text{EW}}$ and $\beta_{\text{KPZ}} > \beta_{\text{EW}}$. This means that in one dimension a KPZ surface roughens faster than an EW surface, but the final, stationary value scales with L in the same way. In higher dimensions, both KPZ exponents are larger than EW exponents (which actually vanish for $d \geq 2$). The fact that a KPZ surface is generally rougher than an EW surface can be traced to the weaker smoothing mechanism of its deterministic part. The EW smoothing process is the typical diffusional smoothing, which induces an exponentially fast relaxation, while the parabolas appearing in Fig. I.1 flatten following a power law.[26] It is therefore reasonable to speculate that in KPZ, noise should lead to a rougher surface than in EW.

---

[25] The treatment of the first term of the right-hand side, proportional to $\delta(0)$, can be made more rigorous within a discrete notation in the Fourier space, but the second term (see Eq. (7.140)) is more manageable in the continuum, real space.

[26] Only parabolas of negative curvature flatten; we refer the reader to Appendix I for a detailed discussion.

7.7.4 Beyond the Exponents

The determination of the roughness exponents is an important step, but it does not complete the solution and the comprehension of a given model. The exponents α, β represent average properties: How does the standard deviation of the height fluctuations depend on the time t and on the size L of the region under observation? A complete study should also provide the distribution of fluctuations, $\rho(\chi)$, where $\chi = (h(\mathbf{x}, t) - \overline{h})/W(L, t)$. The factor W in the denominator allows to have $\langle \chi^2 \rangle = 1$.

Brownian motion is a useful example of what we mean. Its properties are not limited to stating that the average distance Δx traveled by a Brownian particle in time t is of order $\sqrt{t}$; we also know the probability distribution of Δx, which follows a Gaussian distribution. In fact, we can even say where its universal properties come from: the central limit theorem (see Appendix 3.2), according to which the sum of *independent* variables follows a normal distribution. We expect the same to hold for the Edwards–Wilkinson equation. In fact, if we combine Eqs. (7.65) and (7.69), we obtain (for $h(\mathbf{x}, 0) \equiv 0$)

$$h(\mathbf{x}, t) = \frac{1}{(2\pi)^d} \int d\mathbf{q} e^{i\mathbf{q}\cdot\mathbf{x}} \int_0^t dt' \eta(\mathbf{q}, t') e^{-\omega_q(t-t')}, \tag{7.141}$$

which shows that the height, at given point $\mathbf{x}$ and time t, is a linear combination of independent random variables $\eta(\mathbf{q}, t')$. Therefore, the random variable $h(\mathbf{x}, t)$ follows a Gaussian distribution with zero mean and standard deviation σ_h^{EW} equal to the roughness,

$$\sigma_h^{\text{EW}} = \sqrt{\langle h^2(\mathbf{x}, t) \rangle} \equiv W(L, t). \tag{7.142}$$

If we now come back to the (one-dimensional) KPZ equation, we immediately acknowledge that $h(x, t)$ is no longer a linear function of noise and there is no reason to expect the random variable $h(x, t)$ to follow a Gaussian distribution. Nonetheless, if the system is finite, in the asymptotic steady state the distribution of h must be Gaussian, because we have shown that EW and KPZ fluctuations are described by the same stationary, Fokker–Planck solution; see Eq. (7.131). The open question is: What is the distribution of h in the thermodynamic limit (or for $t \ll L^z$, if L is finite)? This problem has given new life to "KPZ and all that" for a couple of reasons. First, because such a distribution, called Tracy–Widom (TW) distribution (see Section 3.4), can be found in completely different domains. Second, because the concept of universality has been somewhat updated.

In fact, a one-dimensional KPZ interface is characterized by different distributions of fluctuations according to its growth geometry, which may be either flat or curved, see Fig. 7.15. The two distributions make reference to the so-called Gaussian orthogonal ensemble (GOE) and Gaussian unitary ensemble (GUE), respectively (see Section 3.4). The role of the growth geometry also appears in the mapping between the ASEP and the SS model, see the end of Section 7.3. Flat interface corresponds, for example, to a periodic initial ASEP condition, where an occupied site alternates with an empty site (the SS initial condition is a perfect sawtooth of period two). Instead, a curved interface corresponds to a perfectly separated initial state: all particles on the left, all holes on the right (the SS initial state is perfectly V-shaped).

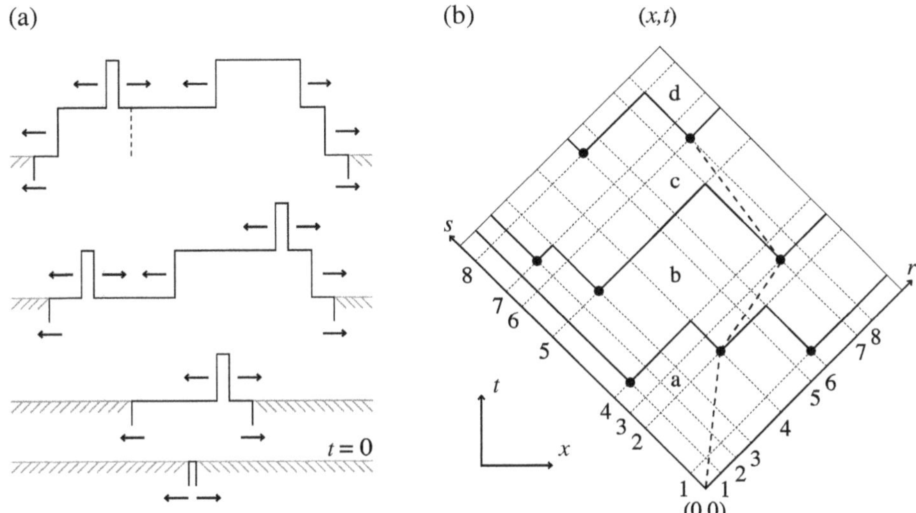

Fig. 7.14 (a) Sketch of the PNG model for a circular interface. (b) Graphical representation of the height at point x at time t through the occurrence of nucleation events. $h(x, t)$ is influenced by nucleations occurring at (x', t'), with $|x'| \leq t'$ and $|x - x'| \leq (t - t')$. Using lightlike coordinates, $r = (t' + x')/\sqrt{2}$ and $s = (t' - x')/\sqrt{2}$, we obtain the external rectangle of the figure. Thick bold lines separate regions of different heights. The dashed line connects nucleations of increasing i_r and i_s and it is an example of the mapping between PNG and directed polymers.

Let us now introduce the polynuclear growth (PNG) model, which allows to gain some insights into the relation among KPZ and other, apparently far away, problems. The one-dimensional PNG model can be easily understood using Fig. 7.14(a), where it is described in the case of a circular interface, whose initial condition corresponds to a layer of zero height ($h = 0$) and vanishing spatial extension. At $t = 0^+$, its extension increases linearly in time and there is a finite probability per unit time and space to nucleate a new layer ($h = 1$). When it happens, the new layer is limited by two steps moving away at unitary speed. We therefore have a layer-by-layer growth because multiple nucleation events on the same layer lead to the "completion" of the upper layer and nucleations on the top layer lead to the overall growth of the surface. This model does not conserve mass/volume because the growth velocity, that is, the speed of a step, is independent of its distance with other steps: PNG falls in the KPZ universality class.[27]

The height of the interface at time t at point x can be graphically represented as in Fig. 7.14(b), where we plot nucleation events using lightlike coordinates, as explained in

---

[27] The flat surface geometry corresponds to an initial condition where the layer of zero height has infinite spatial extension and as soon as $t = 0^+$ nucleation events may occur everywhere.

the caption. Thick straight lines separate the different layers and the integer numbers i_r (i_s) label nucleation events through the ascending r (s) coordinate. The thick dashed line (which is not unique) moves between nucleations characterized by increasing i_r and i_s and it looks like a directed polymer, where nucleations play the role of random defects. This mapping can be made rigorous by assigning a fixed negative energy to every nucleation event, thus establishing a connection between a nonequilibrium growth problem (in dimension d) and the equilibrium problem of directed polymers moving in a disordered $(d+1)$–dimensional space.[28] However, a more astonishing connection can be found by the following observation: If nucleations are ordered by increasing r, the integer string $[i_s] = (4, 7, 5, 2, 8, 1, 3, 6)$ is a permutation of the string $[i_r] = (1, 2, 3, 4, 5, 6, 7, 8)$ and the height $h(x, t) = 3$ is equal to the length ℓ of the longest increasing subsequence in $[i_s]$, corresponding to $(1, 3, 6)$.[29] Since nucleations are uniformly distributed in the rectangle, their number has a Poisson distribution and all permutations are equally likely. This allows to relate the distribution of the height h to the distribution of the length ℓ, which is a well-studied problem (and apparently very far from a growth problem). The solution of the combinatorial problem brings up certain probability distributions, known in random matrix theory, see Section 3.4.

7.7.5 Experimental Results

KPZ $d = 1$

Convincing experimental evidence of KPZ behavior for a one-dimensional interface has been found in the study of the electroconvection of nematic liquid crystals. Here, the convection is induced by an ac electric field applied to a thin container of liquid crystal, and turbulence sets in. Two different, spatiotemporal chaotic states may appear, one is named DSM1 and is metastable, and the other is named DSM2 and is stable. The stable state can be nucleated from the metastable one by inducing a defect with a laser pulse. Once a DSM2 cluster is created, it grows and the interface DSM1/DSM2 gets rough. The laser pulse can be shot so as to produce a circular geometry or a flat geometry, which are visualized in the two snapshots on top of Fig. 7.15, the dark region corresponding to DSM2.

The Family–Vicsek scaling is shown in Fig. 7.16, both for the roughness (a, c) and the correlation function (b, d). The exponents are the same, independently of the circular (a, b) or flat (c, d) geometry. The geometry is relevant instead for the distribution of height fluctuations, as shown in Fig. 7.15. In both cases the curves follow Tracy–Widom distributions: The circular geometry is related to the GUE and the flat geometry is related to the GOE.

---

[28] This mapping highlights two important points at the same time, both of which have already been found: the role of time as additional dimension and the possibility to have the same scenario in the presence of noise sources of a completely different nature.

[29] The same result is obtained with the inverse procedure, of course: label nucleations by increasing s and look for the longest increasing subsequence in $[i_r]$.

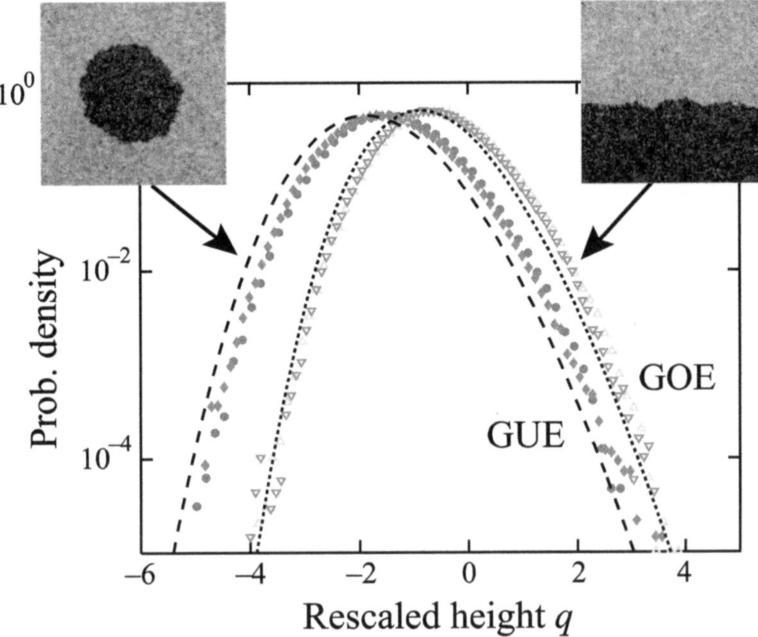

Fig. 7.15 Histogram of the rescaled local height $q \equiv (h - v_\infty t)/(\Gamma t)^{1/3}$ for the circular (solid symbols) and flat (open symbols) interfaces. The dashed and dotted curves show the GUE and GOE TW distributions, respectively. The snapshots show the connection between the geometry of the interface and the pertinent ensemble to evaluate the TW distribution. From K. A. Takeuchi and M. Sano, Evidence for Geometry-dependent Universal Fluctuations of the Kardar–Parisi–Zhang Interfaces in Liquid-crystal Turbulence, *Journal of Statistical Physics*, **147** (2012) 853–890.

KPZ $d = 2$

A deposition process on a real, two-dimensional substrate seems to be the best candidate to look for KPZ behavior, but this has remained elusive for a long time despite the expected universal character of this equation. In addition, it is not possible to determine the universality class of a given roughening process by simply looking at approximate roughness exponents; checking some distribution or correlation function is essential. Figure 7.17 plots height fluctuations and roughness distribution for the growth process of vapor-deposited organic thin films. The extracted roughness exponents, $\alpha = 0.45 \pm 0.04$ and $\beta = 0.28 \pm 0.05$, are in marginal agreement with numerical KPZ values (see Table 7.1), but the two distributions shown in the figure are more convincing. The left panel shows the distribution of normalized height fluctuations in the regime $t \ll t_c = L^z$, L being the sample size, and it is the equivalent of Fig. 7.15. The right panel refers to the regime $t \gg t_c = L^z$, that is, $L \ll \xi_\parallel$, where in this case L is the sampling scale. In this regime, fluctuations are stationary and the distribution of the normalized fluctuations of the roughness does not depend on L. Experimental results are compared with simulation and analytical results.

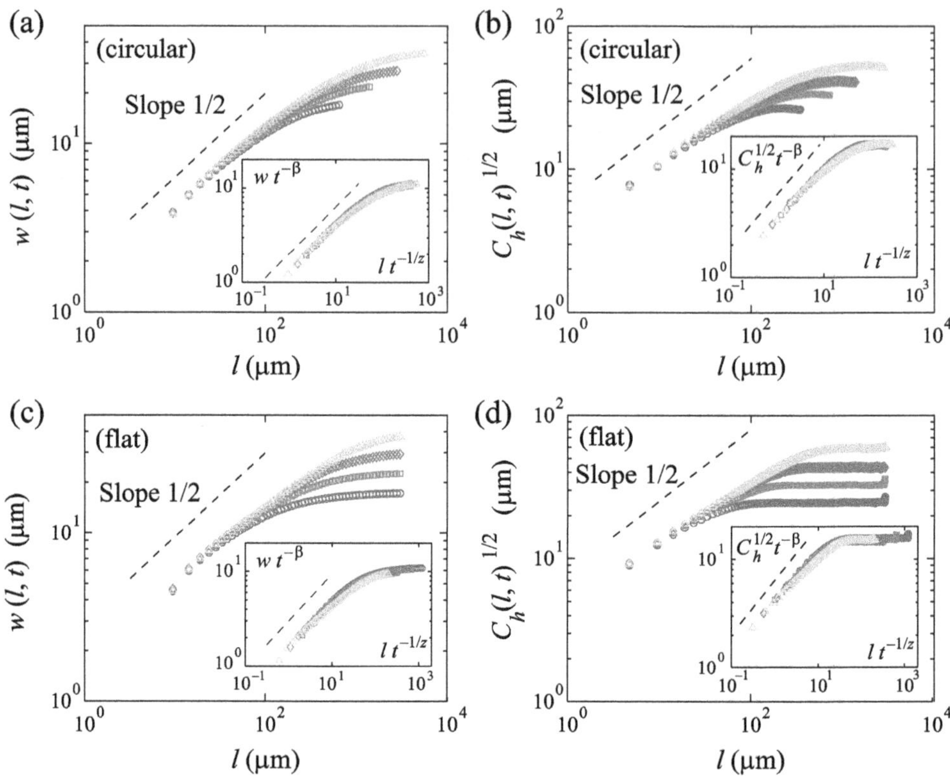

Fig. 7.16 The interface width $w(l, t)$ (a, c) and the square root of the correlation function $C_h(l, t)^{1/2}$ (b, d) are shown for different times t for the circular (a, b) and the flat (c, d) interfaces. Time varies from $t = 2$ s to $t = 30$ s for the circular interface and from $t = 4$ s to $t = 60$ s for the flat interface, time increasing from bottom to top. The insets show the same data with the rescaled axes, with the KPZ exponents $\beta = \frac{1}{3}$ and $z = \frac{3}{2}$. The dashed lines are guides for the eyes indicating the slope for the KPZ exponent $\alpha = \frac{1}{2}$. From K. A. Takeuchi and M. Sano, Evidence for Geometry-dependent Universal Fluctuations of the Kardar–Parisi–Zhang Interfaces in Liquid-crystal Turbulence, *Journal of Statistical Physics*, **147** (2012) 853–890.

7.8 Nonlocal Models

The deposition models and the equations studied earlier in this chapter all have a common feature: They are local models. For a discrete model, this means that the evolution rule of a given site depends on the height of the site itself and the height of neighboring sites, within a finite distance. For example, in the RDR model, we may choose a site and the deposited particle will be incorporated in the lowest height site within a certain distance. In a continuum description, this means that the equation has a finite number of relevant terms. This is why EW and KPZ models have the form $\partial_t h = \mathcal{N}[h] + \eta$, where $\mathcal{N}[h]$ is a single (linear) term for EW and is the sum of two terms for KPZ. We can add other terms, but they are irrelevant in the sense of the dynamical renormalization group.

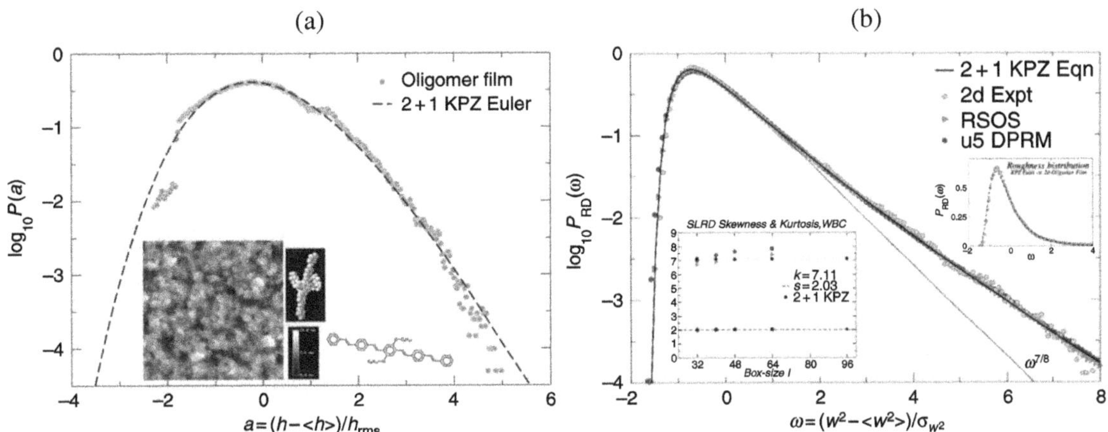

Fig. 7.17 (a) Height fluctuation probability distribution function in the short time regime: 2 + 1 KPZ equation versus kinetic roughening experiment. Inset: Atomic force microscopy image, 50-nm-thick oligomer film; lateral scan direction 2 μm. (b) Roughness distribution in the long time regime. For more details, see T. Halpin-Healy and G. Palasantzas, Universal Correlators and Distributions as Experimental Signatures of (2 + 1)-dimensional Kardar–Parisi–Zhang Growth, *Europhysics Letters*, **105** (2014) 50001.

However, there are *nonlocal* processes where the growth velocity at a given point depends on the profile of the entire system. A simple way to obtain one such process is to modify the growth rule of the BD model, where particles are deposited ballistically along the vertical direction and stick to the aggregate as soon as the particle occupies a site close to the aggregate itself. There are two main modifications we can do that both lead to a nonlocal model: either change the flux orientation of incoming particles or change from ballistic to diffusional deposition. The former case is illustrated in Fig. 7.18(a) with the so-called grass model, where each blade grows in proportion to the light it receives (in a particle model this is equivalent to saying that incoming particles arrive uniformly from all possible directions). Here nonlocality is due to shadowing effects: A high part of the profile may capture a larger fraction of an oblique flux than a low part of the profile, much in the same way a tall building shades a short one if the sun is out of the zenith. The latter case corresponds to particles that arrive not ballistically but diffusionally, as illustrated in the right panel of the same figure[30]: As in the BD model, particles stop and are incorporated in the growing aggregate as soon as they occupy a nearest-neighbor site of the substrate or of the aggregate. Here nonlocality is a natural outcome of diffusion, and in the rest of this section we will focus on this case.

The resulting morphology of diffusional growth is fairly different from those discussed previously in this chapter. Two striking differences are immediately visible: Diffusional growth produces a fractal, rather than a compact morphology, and the growing aggregate is

---

[30] Here we are limited to considering the case where the total dimension of space (where diffusion takes place) is equal to $d + 1$, but we might have a total space of dimension $d + n$. For example, the case $d = 1$, $n = 2$, would correspond to three-dimensional diffusion and attachment to a one-dimensional wire.

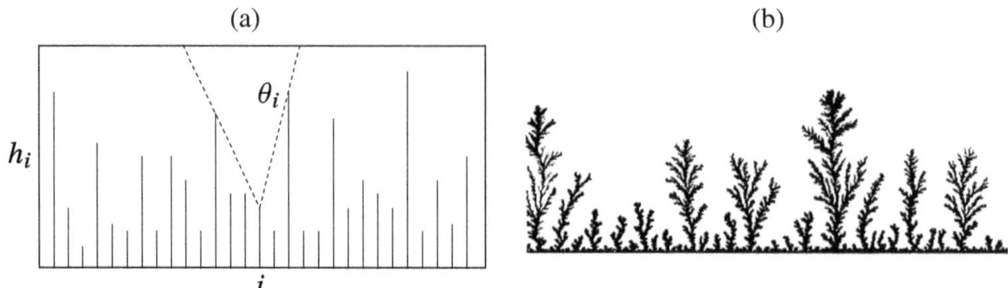

Fig. 7.18 (a) The grass model for a one-dimensional lattice: Each blade of grass has a height h_i and grows as a function of the light it receives, which is proportional to the angle θ_i. (b) A forest of zinc metal trees obtained by electrodeposition and representative of the diffusional growth for a one-dimensional substrate. A similar morphology can be obtained with a simulation where particles are released, one by one, at a large distance above the substrate. Each particle performs a standard random walk until it sticks to the aggregate or to the substrate. At that point the particle is incorporated and a new particle is released. Reprinted from M. Matsushita, Y. Hayakawa, and Y. Sawada, Fractal Structure and Cluster Statistics of Zinc-metal Trees Deposited on a Line Electrode, *Physical Review A*, **32** (1985) 3814–3816.

an ensemble of separated (but strongly interdependent) tree-shaped clusters.[31] The different morphology implies that we also need different quantities to characterize the aggregate as the fluctuations of the local height are no longer appropriate. The main relevant quantity is now the cluster-size distribution function, $n_s(N)$: If V is the total number of substrate sites and we have deposited N particles per substrate site, we have $N_s(N)$ clusters of size s, whence the function

$$n_s(N) = \frac{N_s(N)}{V}. \tag{7.143}$$

In practice, N plays the role of time. In fact, in a simulation, N is exactly the number of Monte Carlo steps.

The earlier quantity is found to exhibit the scaling form

$$n_s(N) = \frac{1}{s^\tau} f\left(\frac{s}{s^*(N)}\right), \tag{7.144}$$

where $s^* \sim N^\sigma$ is the mean cluster size and $f(x)$ is a function that is a constant for small arguments ($x \ll 1$) and it vanishes exponentially fast for large arguments ($x \gg 1$), because it is extremely unlikely to have clusters much larger than s^*. In practice we may think that $n_s(N)$ is a power-law function with an upper cut-off that scales with s^*. The normalization condition imposes a relation between σ and τ, because

$$N = \sum_s s n_s(N) = \sum_s s \frac{1}{s^\tau} f\left(\frac{s}{s^*(N)}\right) \simeq \int^{s^*} ds \frac{1}{s^{\tau-1}} \simeq (s^*)^{2-\tau}, \tag{7.145}$$

so $s^* \simeq N^{1/(2-\tau)}$ and $\sigma = (2-\tau)^{-1}$. This relation implies $\tau < 2$ because s^* must increase with N.

---

[31] In this context, a cluster can be defined as a collection of "deposited" particles connected to the same substrate site through nearest neighbors.

If a cluster of size s has a typical linear size (therefore also a typical height) $h(s) \simeq s^\theta$, the average height of the whole system, which is a function of N, will be given by

$$\bar{h}(N) = \frac{1}{N} \sum_s sh(s)n_s(N) \simeq \frac{1}{N} \int^{s^*} ds \frac{ss^\theta}{s^\tau} \simeq N^{\frac{\theta}{2-\tau}}. \tag{7.146}$$

Therefore, we have introduced two exponents, θ and τ, to characterize the statistics of the morphology resulting from diffusional growth. However, up to now we have not yet used the fractal character of such morphology. We do it now, finding that both exponents can be expressed in terms of the fractal dimension D of the aggregate and of the Euclidean dimension d of the substrate.

Since each cluster is a fractal of dimension D, the quantity of matter contained in a volume of linear size $h(s)$ scales according to $s \sim h^D$, so $\theta = 1/D$. The whole aggregate has a fractal dimension D as well, so the condition that in a volume of linear size $\bar{h}(N)$ there must be N particles per substrate site is

$$\frac{\bar{h}^D}{\bar{h}^d} \simeq N, \tag{7.147}$$

which means $\bar{h}(N) \simeq N^{1/(D-d)}$. The relations $\theta = 1/D$ and $\theta/(2-\tau) = 1/(D-d)$ imply

$$\tau = 1 + \frac{d}{D}. \tag{7.148}$$

In order to focus on the essential features of the "diffusional" morphology, it is easier to consider a simpler geometry, where the aggregate grows from a single seed, rather than a whole substrate. In fact, this is the geometry originally used by the American Thomas A. Witten and Leonard M. Sander in 1981 to define the celebrated DLA model. Here the aggregate is originally composed of a single seed placed at the origin of a regular lattice and new particles are released homogeneously, far from the aggregate. Once released a particle diffuses until it sticks to the aggregate (or is lost into outer space). Rather than plotting a simulated DLA geometry (whose branches look very similar to the trees shown in Fig. 7.18 (b)), we refer the reader to the experimental morphologies shown in Fig. 7.19. The panel on the left is a good experimental realization of the DLA model: Gold atoms are deposited on a rubidium substrate and diffuse at its surface, and when two gold atoms meet, they stop diffusing and form the seed for a growing layer, which will collect other diffusing Au atoms.[32] The reason why such an intricate structure emerges is apparently simple: Once a fjord-like recess is formed, it will be harder to be filled by diffusing particles.

Figure 7.19(b) displays a similar morphology, for a similar reason: a growing process related to a diffusive field. The experiment of bacterial colony growth consists of preparing an agar-based growth medium in a Petri dish at whose center a droplet of bacteria is inoculated. Then, the presence of nutrients allows the bacterial colony to grow and multiply. With an appropriate tuning of parameters, first of all the food concentration, it is possible to obtain the morphology shown in the figure. Without pretending to oversimplify a process

---

[32] The experiment is not a perfect realization of DLA for two main reasons: First, Au atoms may be deposited close to the growing cluster (if not on the top of it); second, once Au atoms stick the aggregate they may have a residual diffusion motion along the cluster edge. Both effects contribute to smooth the morphology of the growing cluster, making it "less fractal."

(a)   (b)

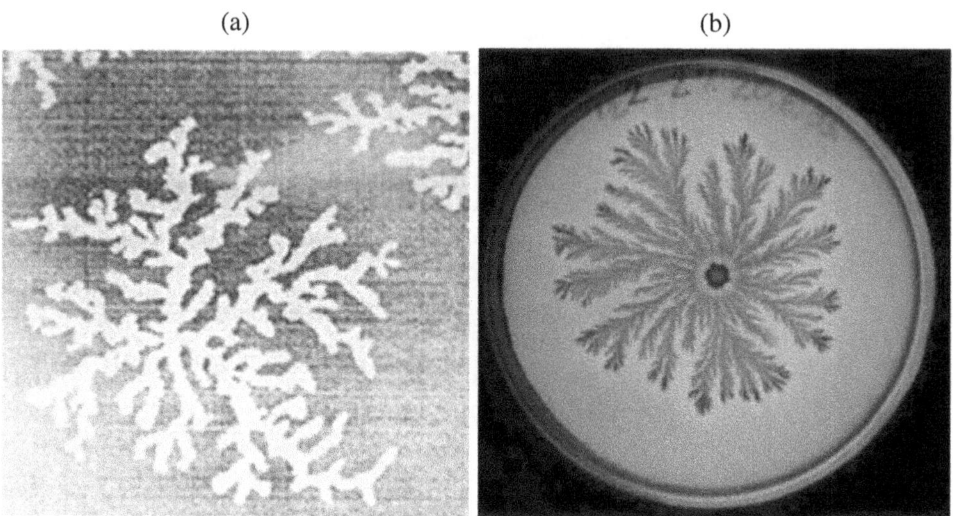

Fig. 7.19 (a) Gold atoms deposited on top of a rubidium substrate form DLA aggregates. Reprinted from R. Q. Hwang et al., Fractal Growth of Two-dimensional Islands: Au on Ru(0001), *Physical Review Letters*, **67** (1991) 3279–3282. (b) Similar morphology in a bacterial colony growing in a Petri dish. Reprinted from E. Ben-Jacob et al., Generic Modelling of Cooperative Growth Patterns in Bacterial Colonies, *Nature*, **368** (1994) 46–49.

that is more complex than DLA, it is nevertheless correct to say that both experiments of Fig. 7.19 display a branching instability of diffusional character. Diffusion is what makes a straight profile unstable and is also what makes the problem nonlocal. The instability mechanism is simple: If a fluctuation creates a small amplitude wave-like profile, the bumps collect more diffusing material and growth is faster, therefore reinforcing the wavy structure of the profile (positive feedback). This mechanism would not be active if the quantity attaching to the front moved ballistically rather than diffusionally.[33] Nonlocality comes from the screening effect that the growing front exerts on itself.

Let us now give a more formal description of a diffusional growth process, because it allows us to understand its generality, making reference to the DLA model. In a continuum picture, the concentration $c(\mathbf{x}, t)$ of diffusing particles satisfies the equation

$$\frac{\partial c}{\partial t} = D\nabla^2 c, \tag{7.149}$$

with the boundary condition, $c(\mathbf{x}_F, t) = 0$ at the points $\mathbf{x}_F$ of the growing front, denoting that a sticking particle is definitely incorporated in the aggregate. The local growth velocity of the front, v, is perpendicular to the front itself and is equal to the current of atoms attaching to it,

$$v = D\frac{\partial c}{\partial n}, \tag{7.150}$$

---

[33] In that case, depending on the orientation of the flux, we may have shadowing effects, which are what characterize the grass model.

where ∂_n means the derivative along the outward normal. In DLA and also in some experiments, the growth velocity v is weak enough and we can replace the diffusion equation (7.149) with the Laplace equation, $\nabla^2 c(\mathbf{x}, t) = 0$. This approximation is justified if the typical time τ_f to relax the field $c(\mathbf{x}, t)$ to the stationary solution, corresponding to fixed boundary conditions, is much smaller than the time τ_b required to move the boundary. In DLA, τ_f is the time interval between two consecutive releases of a particle. In order to move the boundary, on average, by a lattice constant, we need a number of releases equal to the number of boundary sites, N_b, which increases and diverges in time. Therefore, $\tau_f/\tau_b \approx 1/N_b \to 0$. In conclusion, the approximation becomes asymptotically exact.

It is therefore clear why DLA and similar models are examples of "Laplacian growth," whose continuum formulation reduces to studying an electrostatic problem with a growth velocity proportional to the local electric field. This electrostatic equivalence explains why electrodeposition patterns may give rise to DLA-like morphologies. In fact, we might go even further and find similar morphologies as the result of the injection of an inviscid fluid in a viscous one. The resulting interface is unstable and viscous fingering forms. However, these phenomena fall in the domain of pattern formation, studied in Chapter 9, rather than in the domain of kinetic roughening, because they originate from a deterministic instability.

7.9 Bibliographic Notes

A general introduction to the topics discussed in this chapter is A.-L. Barabási and H. E. Stanley, *Fractal Concepts in Surface Growth* (Cambridge University Press, 1995). Here the reader can find references to experiments and many simulations.

The book by Paul Meakin, *Fractals, Scaling and Growth Far from Equilibrium* (Cambridge University Press, 1998), has a special emphasis on models and an extensive list of references.

The book by A. Pimpinelli and J. Villain, *Physics of Crystal Growth* (Cambridge University Press, 1998), is not formal and contains many remarks and physical considerations. However, it is oriented to growth processes, rather than to generic processes of kinetic roughening.

A detailed discussion of EW and KPZ equations can be found in the review paper by J. Krug, Origins of Scale Invariance in Growth Processes, *Advances in Physics*, **46** (1997) 139–282, which is also a good source for the topic of scale invariance in general.

The details of the perturbative RG approach to KPZ can be found in the books by Barabasi and Stanley and in Pimpinelli and Villain.

Recent results and useful physical discussions about KPZ in $d = 1$ can be found in K. A. Takeuchi, M. Sano, T. Sasamoto, and H. Spohn, Growing Interfaces Uncover Universal Fluctuations behind Scale Invariance, *Scientific Reports*, **1** (2011) 34. A review paper on KPZ that connects this chapter to the previous one is T. Kriecherbauer and J. Krug, A Pedestrian's View on Interacting Particle Systems, KPZ Universality and Random Matrices,

Journal of Physics A: Mathematical and Theoretical, **43** (2010) 1–41. A short history of KPZ in the context of kinetic roughening is T. Halpin-Healy and K. A. Takeuchi, A KPZ Cocktail–Shaken, Not Stirred . . . , *Journal of Statistical Physics*, **160** (2015) 794–814. A recent review is K. A. Tacheuchi, An Appetizer to Modern Developments on the KPZ Universality Class, *Physica A: Statistical Mechanics and its Applications*, **504** (2018) 77-105.

A simple account of DLA is in L. M. Sander, Diffusion-Limited Aggregation: A Kinetic Critical Phenomenon?, *Contemporary Physics*, **41** (2000) 203–218.

8

Phase-Ordering Kinetics

This chapter addresses the basic question of relaxation to equilibrium when the sudden change of some parameter brings a physical system into the region of a phase diagram that is inaccessible at equilibrium. The simplest example is a ferromagnet in the absence of a magnetic field and whose temperature is quenched from the paramagnetic, disordered phase to the ferromagnetic, ordered phase: The system will tend to order, but thermal fluctuations are unable to globally drive it to one of the equivalent equilibrium states because (the final) T is lower than the critical temperature T_c, so that an infinite energy barrier separates them. The system will therefore tend to order locally through the formation of domains whose size increases in time.

If we consider the condensation transition of a gas or the order-disorder transition of a binary alloy, we have an order parameter that must be conserved during the relaxation process: the density of the fluid or the concentration of each component of the alloy. This allows to perform the quenching at different values of such density/concentration and this translates in a postquenching state that may be either unstable (as in the case of the ferromagnet) or metastable. In the former case, we have a homogeneous process called spinodal decomposition. In the latter case, relaxation must be triggered by an activated process, the nucleation of one of the two phases. At long times the two mechanisms lead to the same process, coarsening, where the size of ordered regions increases in time, typically following a power law. The exponent will be seen to depend on the conservation or not of the order parameter, but not on the spatial dimension of the physical system.

Section 8.1 is a short primer on equilibrium phase diagrams, introducing and making examples of the region that is inaccessible at equilibrium. It is followed by Section 8.2, where we discuss the process of quenching and its possible outcomes. We also introduce the continuum equations, which are used to study the phenomenon of phase ordering. The correlation function and its Fourier transform, the structure factor, are tools that play an important role also in this context: They are studied in Section 8.3. The following two sections are the core of this chapter: They discuss the coarsening process and determine the coarsening exponents for scalar systems, when the order parameter is not conserved, Section 8.4, and when it is conserved, Section 8.5. The size of domains is important, but the distribution of domain sizes gives a more complete description of the coarsening process: It is studied, with much less detail, in Section 8.6. Nonscalar systems are treated in Section 8.7. Finally, the nucleation process appearing when after quenching the state is metastable rather than unstable is treated in Section 8.8, where we limit to discuss the classical Becker–Döring theory.

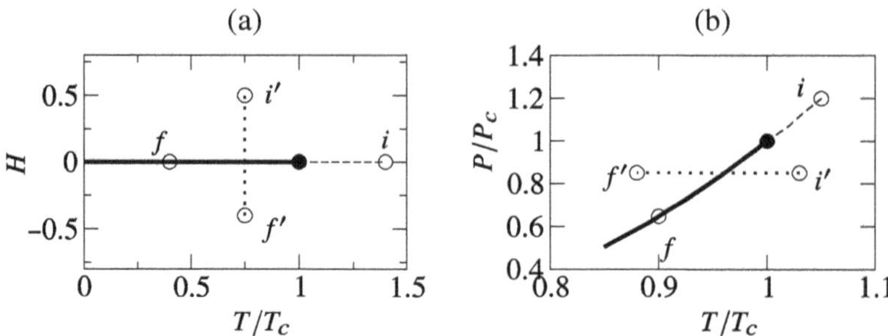

Fig. 8.1 Critical lines for the para-ferromagnetic transition (a) and for the gas–fluid transition (b). In the latter case, the thick line defines the critical isochore, $P_{\text{iso}}(T)$. The dashed lines indicate a transformation between state i and state f. At equilibrium (quasi-static transformation), it is a second-order transition; at nonequilibrium, it indicates a critical quench. The dotted lines indicate a first-order transition between states i' and f.

8.1 The Equilibrium Phase Diagrams

Phase transitions are among the most relevant equilibrium phenomena in a many-body system. In Fig. 8.1 we plot (a portion of) the phase diagram of a magnetic system (left panel) and of a simple one-component fluid (right panel). In the former case, the system undergoes a para-ferromagnetic transition and the thermodynamic state is described by the magnetic field H and the temperature T. In the latter case, the system has a gas–liquid transition and it is described by pressure P and T, for a given volume V. The two phases are separated by a critical line that terminates in a critical point. A magnetic system typically satisfies the symmetry $M(T, -H) = -M(T, H)$ where M is the magnetization; therefore, the critical line corresponds to $H = 0$, while the fluid does not have a similar property. This fact has an often overlooked consequence. Since an experiment or a simulation is generally performed at constant H or constant P, in the magnetic case we can follow the critical line with a constant, vanishing magnetic field. This is not possible for a fluid, where we should perform a peculiar thermodynamic path to *enter* in the critical line, see the dashed lines connecting i and f in the two panels.

The critical lines are characterized by the coexistence of different phases: magnetization *up* and *down* for a (uniaxial) magnetic system; gas and liquid for a fluid. However, in the magnetic case it is more correct to speak of equivalence rather than of coexistence, because magnetization is generally not conserved. To be more precise, let us see what happens when we perform a quasi-static transformation between state i and state f. In the magnetic case, if we cross the critical point and keep $H = 0$, for $T < T_c$ we have two equivalent states and the energy barrier between them vanishes at T_c: The choice between the two ordered states is a random choice made by fluctuations, which are symmetric in the absence of stray fields. In the fluid case, the conservation of matter requires the density to keep constant if we work at constant volume. Therefore, when passing the critical point, the fluid separates in a liquid part and in a gas part. We stress that magnetization may be forced to be constant, at least in a simulation as we will see, but in this case the magnetic field is irrelevant.

The earlier scenarios correspond to a continuous (second-order) phase transition, because at the critical point the equilibrium magnetization vanishes and the liquid and gas phases have the same density. The dotted paths going from i' to f', see again Fig. 8.1, do not cross the critical point and correspond to a discontinuous (first order) phase transition. In the fluid system, this is the typical scenario of everyday life because it occurs at constant pressure. For the magnetic system, it may occur if we keep constant the temperature $T < T_c$ and we invert the magnetic field. When we cross the critical line (not at the critical point), the change of state is accompanied by metastability (further details will be given when discussing Figs. 8.3 and 8.4).

It may be of some help to refer to a real phase diagram. In Fig. 8.2 we plot the magnetization curves for an ultrathin magnetic film Fe/W, a good experimental realization of a two-dimensional Ising model. Because of the symmetry $M(T, -H) = -M(T, H)$, we limit ourselves to $M > 0, H > 0$. The thick line, called the curve of spontaneous magnetization, is the coexistence curve, because if $H = 0$ and $T < T_c$, the equilibrium state corresponds to two distinct values of the magnetization, $|M(T, 0)|$ and $-|M(T, 0)|$. The coexistence curve separates two different regions of the phase diagram: The states for $T < T_c$ and $|M| < M(T, H = 0^+)$ are nonequilibrium states, which cannot be attained with a reversible transformation. Instead, it is possible to tune temperature and field in order

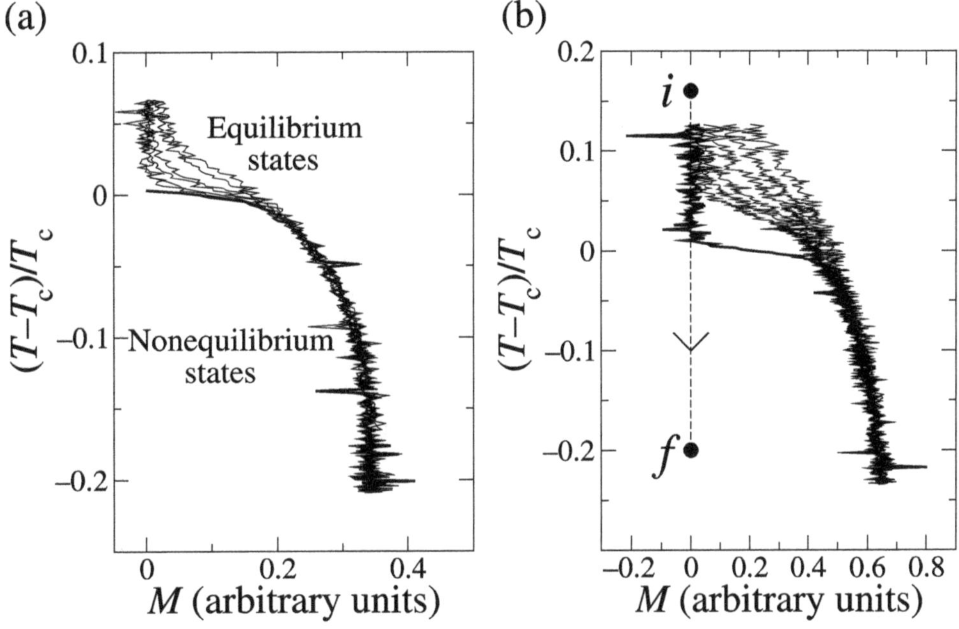

Fig. 8.2   Magnetization of equilibrium of a Fe/W(110) magnetic film (1.8 ± 0.1 atomic layers) as a function of temperature T and field H for two different samples. (a) Sample 1. $H = 0, 0.05, 0.1, 0.5, 1, 3$, and 5 Oe. (b) Sample 2. $H = 0, 5, 10, 20, 50, 100$, and 200 Oe. Thick lines correspond to $H = 0$, thin lines depart increasingly more with increasing H. The dashed line in (b) represents a variation of the temperature from $T_i > T_c$ to $T_f < T_c$. Raw experimental data courtesy of Danilo Pescia. C. H. Back et al., *Experimental Confirmation of Universality for a Phase Transition in Two Dimensions*, *Nature*, **378** (1995) 597–600.

to attain, at equilibrium, all other states.[1] It is possible to attain the inner, nonequilibrium region with a sudden change of control parameters, a procedure called quenching. Indeed, the goal of this chapter is to study the dynamics after a similar quenching, but in this section we focus on equilibrium phase diagrams.

In Appendix Q, we derive the Helmholtz free energy density for a magnet, starting from the microscopic Hamiltonian of a ferromagnet and performing a coarse graining. The homogeneous part of the free energy, up to prefactor f_0, which can be taken as an energy unit, is

$$f(m) = -\frac{m^2}{2} + \frac{T}{T_c}\left[-\ln 2 + \frac{1}{2}(1+m)\ln(1+m) + \frac{1}{2}(1-m)\ln(1-m) \right]. \tag{8.1}$$

Using the thermodynamic relation $H = \partial f/\partial m$, where H is the magnetic field, we can derive the mean-field equation of state,

$$H = -m + \frac{T}{T_c}\left[\frac{1}{2}\ln(1+m) - \frac{1}{2}\ln(1-m) \right], \tag{8.2}$$

which can be rewritten in the more familiar form,

$$m = \tanh\left(\beta T_c(H+m)\right). \tag{8.3}$$

In Fig. 8.3 we plot the free energy (top) and the equation of state (center) for $T > T_c$ (dotted lines) and for $T < T_c$ (solid lines). For $H = 0$ and $T < T_c$, the two minima of the free energy are equivalent and we have phase coexistence. Passing from $H = 0^-$ to $H = 0^+$, the order parameter m displays a discontinuity (first-order transition) unless $T = T_c$ (second-order transition). The central panel shows the typical Maxwell construction, which is fairly trivial in this case because of the symmetry of the problem, according to which $m(-H, T) = -m(H, T)$. Therefore, we have phase coexistence for $H = 0$, regardless of the temperature T.

The top and central figures allow us to define, for any $T < T_c$, special values of the order parameter: the equilibrium values, which coexist for $H = 0$ and which are defined by $f'(m) = 0, f''(m) > 0$, and the values for which $f''(m) = 0$, which define the spinodal curve. The former (equilibrium) values are indicated by black circles and by the thick solid line in the bottom figure. The spinodal values, which limit the metastability region, are indicated by gray circles and by the dot-dashed line in the bottom panel.

In order to draw the panels corresponding to Fig. 8.3 for a gas–fluid transition, it is simpler to start from the equation of state of a nonideal gas (see Section 5.2.1 and Appendix P), rather than from its free energy, which is poorly known. So, the starting point is the van der Waals equation,

$$P = \frac{T}{v-b} - \frac{a}{v^2}. \tag{8.4}$$

It is worth recalling that in the earlier equation, b represents the excluded volume effects (which reduce the volume per particle v, accessible to other particles) and a represents the attractive force between molecules, which reduces the pressure.

---

[1] With the obvious constraint that M is smaller than the maximum physically acceptable value, corresponding to the value attained at the lowest T.

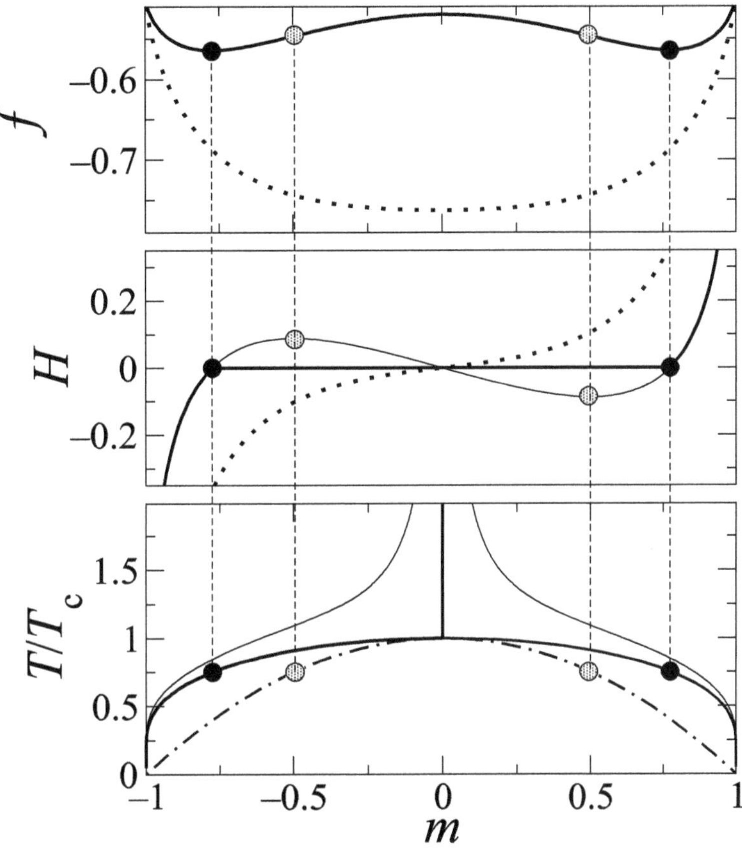

Fig. 8.3 In the top panel we plot the free energy density of a magnet, Eq. (8.1), for $T < T_c$ (solid line) and for $T > T_c$ (dotted line). In the center panel we plot the equation of state, Eq. (8.2), for $T < T_c$ (solid line) and for $T > T_c$ (dotted line). The thick, horizontal line corresponds to the Maxwell construction, which is fairly trivial in the case of a magnet, because of the symmetry $m(T, -H) = -m(T, H)$. The thin line corresponds to metastable states (segments between black and gray circles) or to unstable states (segment between the two gray circles). In both panels, black circles indicate the free energy minima, which coexist for $H = 0$ and $T < T_c$; gray circles indicate the change of curvature of f and the stability limit of metastable states when dynamics is conserved. In the bottom panel, for each temperature T, we plot the values of the magnetization corresponding to coexistence (black circles) and metastability limit (gray circles), obtaining the coexistence curve (thick, solid line) and the spinodal line (dot-dashed line). The thin solid lines give the equilibrium curves $m(T, H)$ for a positive and a negative field H.

The function $P(v)$, much in the same way as the function $H(m)$, is a monotonic function for high T and displays the typical van der Waals loop for low T; see the central panel in Fig. 8.4. As shown in Appendix P, the free parameter form of the equation of state and the

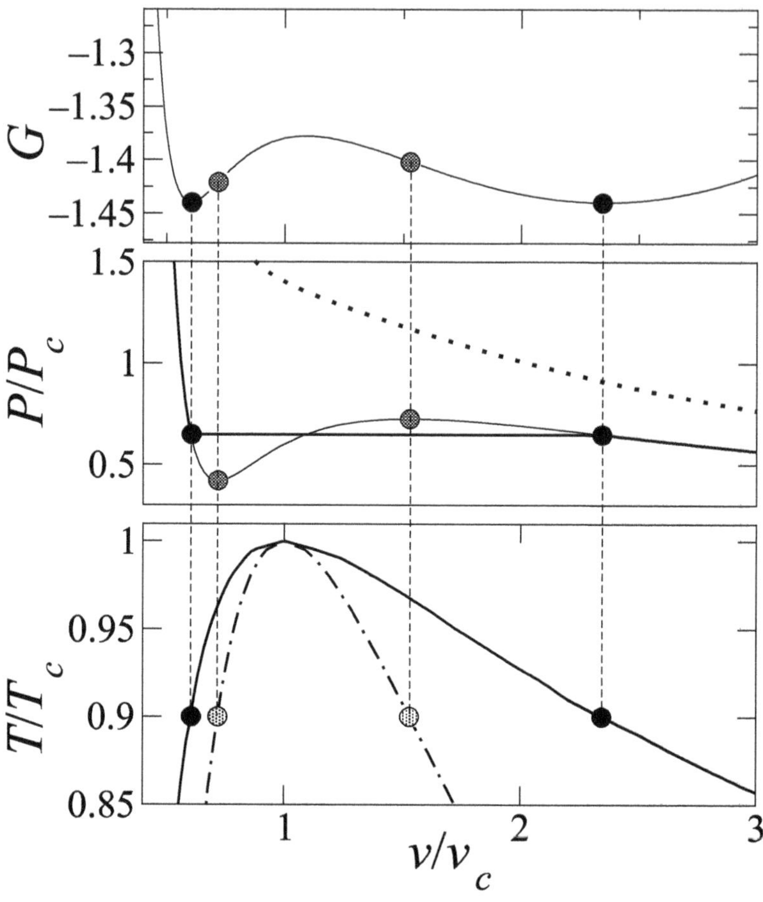

Fig. 8.4   This set of three panels is the equivalent of Fig. 8.3 for a van der Waals fluid. Free energy (top), equation of state (center), and coexistence/spinodal curves (bottom) for a fluid described by the van der Waals equation (8.5). Black circles indicate the free energy minima and define the coexistence curves (solid line, bottom panel). Gray circles indicate the change of curvature of G and the stability limit of metastable states; they define the spinodal line (dot-dashed line, bottom panel). In the central panel, the thick, solid line refers to $T < T_c$, the dotted line to $T > T_c$.

Gibbs free energy are

$$P^* = \frac{8}{3} \frac{T^*}{v^* - \frac{1}{3}} - \frac{3}{(v^*)^2} \tag{8.5}$$

$$G = P^* v^* - \frac{8}{3} T^* \ln\left(v^* - \frac{1}{3}\right) - \frac{3}{v^*}. \tag{8.6}$$

In the top panel of Fig. 8.4, we plot G for the point $(T^*, P^*) = (0.9, 0.647)$ of the critical line. In the middle panel, we plot the equation of state (8.5) for $T^* = 0.9$ (solid line) and for $T^* > 1$ (dotted line). In the bottom panel, we plot the coexistence and metastability curves.

The thick, solid lines in the bottom panels of Figs. 8.3 and 8.4 are the equilibrium coexistence lines. The region inside such curves corresponds to nonequilibrium states, which can be attained by a quench and which decay toward equilibrium via a phase-ordering process including spinodal decomposition or nucleation. The region outside such curves corresponds to equilibrium states.

8.2 Quenching and Relaxation to Equilibrium

In Section 8.1, we have shown that a critical line in the plane (T, H) (or in the plane (T, P)) determines a region that is inaccessible at equilibrium. The main goal of this chapter is to understand what happens if we prepare the system to be in such a region. This can be done through an abrupt change of some parameters, a procedure called quenching and an example of which is depicted by the dashed line in Fig. 8.2(b): We initialize the system in the paramagnetic phase $(T_i > T_c, H = 0)$, then suddenly reduce the temperature in the ferromagnetic phase, $T_f < T_c$.

Phase ordering is not limited to magnetic systems, of course, because it occurs whenever there is phase coexistence. A binary alloy, which undergoes an order–disorder transition, is a nice and useful additional example. In brief, an AB solid alloy can be described by a lattice model, where each site is occupied by an A or B atom and a pair of nearest-neighbor atoms has an energy ϵ_{AA}, ϵ_{BB}, or ϵ_{AB}, depending on the type of atoms. This model is equivalent to an Ising model, where an up (down) spin corresponds to an A (B) atom; see Appendix H. If $\epsilon_{AA} + \epsilon_{BB} < 2\epsilon_{AB}$, the system exhibits a demixing transition with decreasing temperature, equivalent to the ferromagnetic transition. The total numbers of A and B atoms, N_A and N_B, are constant, which means, in the magnetic language, that the total magnetization is constant.

The main question is: How does the system relax to equilibrium after a quenching in the region inaccessible at equilibrium? The answer depends on at least two factors: (1) if the order parameter is conserved or not, and (2) if after the quenching, the system is inside or outside the spinodal line.[2] As discussed in Section 8.1, after a quenching, the arrival state is unstable if it is inside the spinodal line and it is metastable if it is outside such line (i.e., if the quenching is strongly asymmetric). The ensuing dynamics is different in the two cases, because the destabilization of the unstable state is a uniform, not activated process that tends to decrease its free energy by creating ordered regions of increasing size $L(t)$. This process, called spinodal decomposition (or coarsening), is the main feature of phase-ordering kinetics if we perform a symmetric quenching (or in any case within the spinodal line) and it is the main focus of this chapter. An example of the coarsening process is visible in Fig. 8.5 for a nonconserved order parameter and in Fig. 8.16 for a conserved order parameter; both figures have been obtained by kinetic Monte Carlo (KMC) simulations.

In the case of strongly asymmetric quenching to a state outside the spinodal line, the arrival state is metastable and its destabilization is an activated process, which in order to start requires the formation of nuclei of one of the two stable phases, a process that is spatially localized. In Fig. 8.6 we show some well-known snapshots of the phase separation process of a $Fe_x Al_{1-x}$ alloy, for different concentrations x (as visible from the white to

---

[2] For nonconserved order parameter, we only consider the possibility of a symmetric quenching. In fact, the concept of spinodal line does not apply to nonconserved systems.

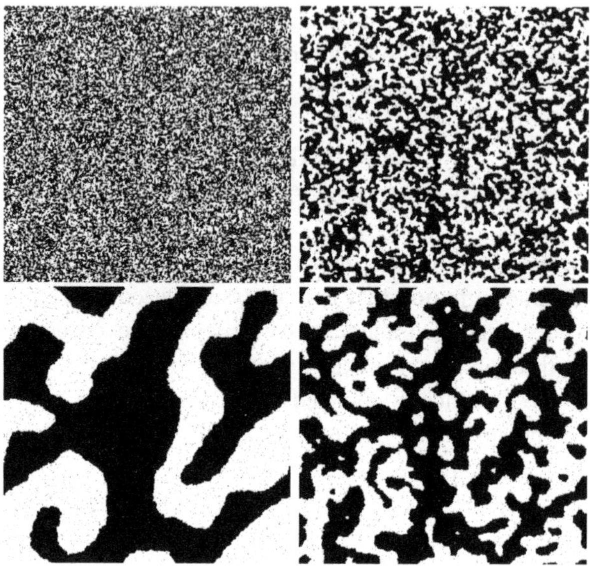

KMC simulations of a $1\,000 \times 1\,000$ Ising model with nonconserved, spin-flip dynamics, after quenching from $T_i = \infty$ to $T_f = 0.661T_c$. Clockwise, from top left: Snapshots of the system at times $t = 10, 10^2, 10^3, 10^4$. Simulation data are a courtesy of Federico Corberi.

black regions ratio). The middle column and the two side columns have a clearly different morphology. The middle sequence of images from top to bottom is qualitatively similar to the coarsening process of a magnet, described here earlier and visualized in Figs. 8.5 and 8.16. The two side columns correspond to qualitatively different initial conditions, where the two phases are strongly unbalanced. This case will be considered in Section 8.8.

Although it is pedagogically useful to distinguish between the two mechanisms of phase ordering (spinodal decomposition for unstable states, nucleation for metastable states), it is useful to make some comments that make this distinction less strong than it might seem. First, the spinodal line is a mean-field concept. Second, if the two mechanisms are clearly different in the early stages of phase separation, the long-time dynamics is not: In both cases there is a coarsening process. Finally, metastable states may also appear starting from an initially linearly unstable state: Think of Fig. 8.5 and imagine that a straight domain wall forms parallel to a boundary: This structure is metastable and if all domain walls are similarly straight lines, the whole system is in a metastable state.

Until now we have only mentioned Ising-type systems. Even if these models are the focus of this chapter, we should not forget there are many systems of interest that require a more complicated modeling. The simplest generalization, which will be studied in Section 8.7, is a nonscalar system with $O(n)$ symmetry, that is, $n > 1$. A magnetic system described by the Heisenberg Hamiltonian is an example for the case $n = 3$. A more complex example is represented by a binary liquid mixture. In this case the order parameter, that is, the density, must be coupled to the fluid velocity, which satisfies the Navier–Stokes equation. Convective motion allows for a faster transport of the order parameter, which asymptotically speeds up the growth of $L(t)$. However, at short times, convective effects can be neglected and the

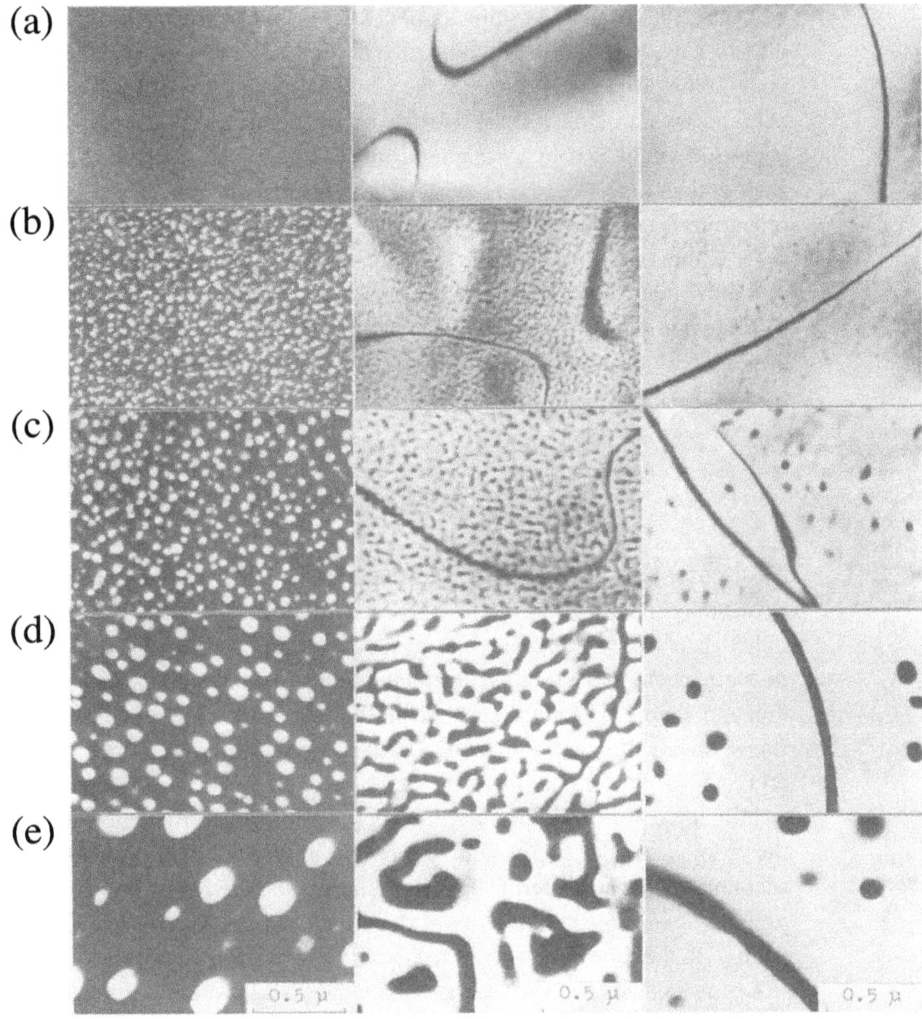

Fig. 8.6 Electron microscopy study of the phase separation in a Fe–Al alloy. The temperature is quenched from $T_i = 630°C$ to $T_f = 570°C$ (left and center images) and to $T_f = 568°C$ (right images). The fraction of Al is 23% (left panel), 24.7% (center panel), and 24.9% (right panel). The center panels correspond to a near-critical quench, the case discussed in these pages. The left and right panels correspond to off-critical quenches, discussed in Section 8.3: (a) as quenched; (b) annealed for 15 min in left and right columns and for 10 min in center column; (c) 100 min; (d) 1 000 min; (e) 10 000 min. From K. Oki, H. Sagane, and T. Eguchi, Separation and Domain Structure of $\alpha + B_2$ Phase in Fe–Al Alloys, *Journal de Physique Colloques*, **38** (1977) C7-414–C7-417.

results discussed in this chapter can be applied. This is especially true for high-viscosity fluids, where convection is delayed. An example where convective effects are relevant is given in Fig. 8.7, where we show the phase separation of a barium-borosilicate glass.

The main feature of the coarsening process is self-similarity (in time), which has been introduced in Chapter 5. Self-similarity and the derived property of dynamic scaling will be seen to hold at large times, when $L(t)$ is much greater than any other scale. Which other scales? To begin with, the lattice constant of a discrete system (which is the smallest scale of the problem). Another scale is the width of a domain wall, the region between two neighboring domains. Its size depends on temperature, but not on time. Finally, the domain wall itself might be rough, which introduces a new scale depending on time, but this scale is smaller than $L(t)$, because above the lower critical dimension, see Chapter 7, the roughness of an interface of linear size L cannot be larger than L^α, with $\alpha < 1$.

In simple terms, self-similarity means that two pictures taken at different times, t_1 and t_2, look statistically the same if we rescale them so as $L(t_1) = L(t_2)$, as we do in Fig. 8.8. This means that $L(t)$ is the only relevant scale in the problem and time enters in a statistical average only through $L(t)$. We will discuss more widely and formalize self-similarity in Section 8.3.

8.2.1 The Langevin Approach

As discussed in Chapter 4, the Langevin approach is a generalization of the description of a Brownian particle to a problem where the unknown is a field depending on space and time, rather than a set of discrete variables depending on time (the three-dimensional position of the particle and possibly its momentum). We have already seen several examples of this approach: In Section 6.3.3 we studied the contact process with a phenomenological Langevin equation and in Chapter 7 the continuum study of kinetic roughening was based on Langevin-type equations.

In many cases, it is possible to write an appropriate free energy $\mathcal{F}$, which depends on the order parameter and possibly on remaining slow variables, which cannot be taken into account by the noise term. The free energy should reflect the basic symmetries of the problem, while information about dynamics and conservation laws enters when we pass from $\mathcal{F}$ to the time evolution of the order parameter(s) and of the slow variables.

Let us focus on the simple case of a single, scalar order parameter $\phi(\mathbf{x}, t)$ without slow fields and with fully dissipative dynamics, which is equivalent to the overdamped motion of a Brownian particle: This is the case representing the physical systems discussed throughout

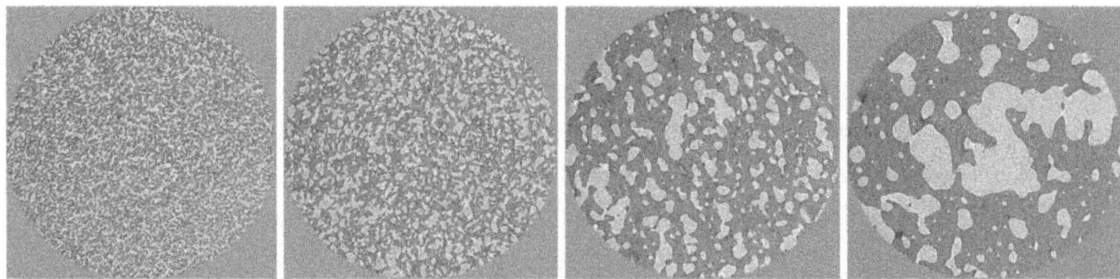

Fig. 8.7　Phase separation of a barium-borosilicate glass under heat treatment at 1 030°C. Interrupted in-situ X-ray tomographic measurements have been performed after 2 min, 8 min, 23 min, and 45 min. Courtesy of David Bouttes. From David Bouttes, Micro-tomographie d'un borosilicate de baryum démixé: du mûrissement à la fragmentation, Ph. D. Thesis (Université Pierre et Marie Curie, 2014).

Fig. 8.8 A simple visual example of self-similarity. We compare the configurations of the nonconserved model at $t = 10^3$ and $t = 10^4$ (see Fig. 8.5) by enlarging a piece of the $t = 10^3$ configuration by a factor $L(t = 10^4)/L(t = 10^3)$. The two images are statistically indistinguishable.

this chapter. If the order parameter is not conserved, that is,

$$\frac{d}{dt} \int d\mathbf{x}\phi(\mathbf{x}, t) \neq 0, \tag{8.7}$$

we can write

$$\frac{\partial \phi}{\partial t} = -M \frac{\delta \mathcal{F}}{\delta \phi} + \eta(\mathbf{x}, t), \tag{8.8}$$

which is the simplest equation representing a dynamical, dissipative process that decreases the free energy. Here,

$$\mathcal{F} = \int d\mathbf{x} f(\phi, \nabla\phi, \nabla^2\phi, \dots), \tag{8.9}$$

so the functional derivative means (see Appendix I.1)

$$\frac{\delta \mathcal{F}}{\delta \phi} = \frac{\partial f}{\partial \phi} - \sum_i \partial_i \frac{\partial f}{\partial(\partial_i \phi)} + \sum_{ij} \partial_{ij} \frac{\partial^2 f}{\partial(\partial_{ij}\phi)} + \cdots, \tag{8.10}$$

where M is the mobility and η is the noise, taking into account the fast degrees of freedom. As usual, the noise has zero average and its correlation function has the form[3]

$$\langle \eta(\mathbf{x}, t)\eta(\mathbf{x}', t')\rangle = \Gamma\delta(\mathbf{x} - \mathbf{x}')\delta(t - t'). \tag{8.11}$$

Since the physical system is evolving toward equilibrium, the mobility M and the strength of the noise Γ must be related by the fluctuation–dissipation condition, similar to what we have seen in Chapter 2 for the Brownian particle,[4]

$$\Gamma = 2MT. \tag{8.12}$$

---

[3] Without noise, we are back to the mean-field approach sketched in Section 8.1.

[4] In Section 2.2.2 (see Eqs. (2.22) and (2.28)), we have $\dot{v} = -\gamma v + \eta$ and $\Gamma = 2\gamma T/m$. However, if we want to write the Langevin equation in the form (8.8), we must introduce the energy $E = mv^2/2$, so $\dot{v} = -(\gamma/m)E'(v) + \eta$. Introducing the quantity $M = \gamma/m$, we have the fluctuation–dissipation relation, $\Gamma = 2MT$. It is also instructive to check that Eq. (8.12) is dimensionally correct.

Dynamics decreases the free energy, because we have

$$\frac{d\mathcal{F}}{dt} = \int d\mathbf{x}\,\frac{\delta\mathcal{F}}{\delta\phi}\frac{\partial\phi}{\partial t} = -\int d\mathbf{x}\left(\frac{\delta\mathcal{F}}{\delta\phi}\right)^2 + \int d\mathbf{x}\,\frac{\delta\mathcal{F}}{\delta\phi}\eta(\mathbf{x},t). \tag{8.13}$$

The meaning of $d\mathcal{F}/dt$ is that we evaluate the free energy on the trajectory of the system, that is, on a specific function $\phi(\mathbf{x},t)$, solution of Eq. (8.8). The deterministic part on the right-hand side of Eq. (8.13) leads to a decreasing $\mathcal{F}$. The noise term may occasionally lead to its momentary increase, but on average the noise term vanishes and

$$\left\langle\frac{d\mathcal{F}}{dt}\right\rangle = -\int d\mathbf{x}\left(\frac{\delta\mathcal{F}}{\delta\phi}\right)^2 \leq 0. \tag{8.14}$$

If instead the order parameter is conserved, we must start by enforcing its conservation via a continuity equation,[5]

$$\frac{\partial\phi(\mathbf{x},t)}{\partial t} = -\nabla\cdot\mathbf{j}, \tag{8.15}$$

where the current $\mathbf{j}$ transports the order parameter ϕ. The current, in turn, is proportional to the gradient of the chemical potential, which is the thermodynamic force,

$$\mathbf{j} = -M\nabla\mu + \boldsymbol{\eta}(\mathbf{x},t), \tag{8.16}$$

with the noise components η_i satisfying the usual relations,

$$\langle\eta_i(\mathbf{x},t)\eta_j(\mathbf{x}',t')\rangle = \Gamma\delta_{ij}\delta(\mathbf{x}-\mathbf{x}')\delta(t-t'). \tag{8.17}$$

A couple of remarks are in order here. First, the mobility M may depend on ϕ, but we neglect this possibility. Second, inserting the noise term in Eq. (8.16) rather than in Eq. (8.15) means that also noise is conserved, which is certainly correct if the only source of noise is thermal fluctuations in the transport process of the order parameter. Finally, the chemical potential μ can be expressed as the functional derivative of the free energy,

$$\mu = \frac{\delta\mathcal{F}}{\delta\phi}. \tag{8.18}$$

Summing up, we find

$$\frac{\partial\phi(\mathbf{x},t)}{\partial t} = M\nabla^2\frac{\delta\mathcal{F}}{\delta\phi} + \zeta(\mathbf{x},t), \tag{8.19}$$

where $\zeta(\mathbf{x},t) = -\nabla\cdot\boldsymbol{\eta}(\mathbf{x},t)$ and its correlation function is[6]

$$\langle\zeta(\mathbf{x},t)\zeta(\mathbf{x}',t')\rangle = \langle\nabla_{\mathbf{x}}\cdot\boldsymbol{\eta}(\mathbf{x},t)\nabla_{\mathbf{x}'}\cdot\boldsymbol{\eta}(\mathbf{x}',t')\rangle \tag{8.20}$$

$$= \partial_{x_i}\partial_{x_j'}\langle\eta_i(\mathbf{x},t)\eta_j(\mathbf{x}',t')\rangle \tag{8.21}$$

$$= -\Gamma\delta_{ij}\partial_{x_i}\partial_{x_j}\delta(\mathbf{x}-\mathbf{x}')\delta(t-t') \tag{8.22}$$

$$= -\Gamma\nabla_{\mathbf{x}}^2\delta(\mathbf{x}-\mathbf{x}')\delta(t-t'). \tag{8.23}$$

---

[5] If phase separation occurs in a fluid where convective motion cannot be neglected, as in Fig. 8.7, we should rather write: $\partial_t\phi(\mathbf{x},t) + \mathbf{v}\cdot\nabla\phi = -\nabla\cdot\mathbf{j}$, where the fluid velocity $\mathbf{v}$ is ruled by the Navier–Stokes equations.

[6] In order to distinguish between the derivatives with respect to $\mathbf{x}$ and with respect to $\mathbf{x}'$, we use the notations $\nabla_{\mathbf{x}}, \nabla_{\mathbf{x}'}, \partial_{x_i}, \partial_{x_i'}$. Once the function $\delta(\mathbf{x}-\mathbf{x}')$ appears, $\partial_{x_i'} = -\partial_{x_i}$.

Again, the fluctuation–dissipation relation requires that $\Gamma = 2MT$.

The order parameter is conserved, because of the continuity equation, and dynamics decreases the free energy:

$$\left\langle \frac{d\mathcal{F}}{dt} \right\rangle = \left\langle \int d\mathbf{x} \left(\frac{\delta \mathcal{F}}{\delta \phi} \right) \frac{\partial \phi}{\partial t} \right\rangle \tag{8.24}$$

$$= M \int d\mathbf{x} \left(\frac{\delta \mathcal{F}}{\delta \phi} \right) \nabla^2 \left(\frac{\delta \mathcal{F}}{\delta \phi} \right) \tag{8.25}$$

$$= -M \int d\mathbf{x} \left(\nabla \frac{\delta \mathcal{F}}{\delta \phi} \right)^2 \le 0. \tag{8.26}$$

In the earlier considerations, we have not made explicit the free energy in order to be as general as possible in the derivation of Eqs. (8.8) and (8.19). We focus on the Ginzburg–Landau free energy, which can be written as

$$\mathcal{F} = \mathcal{F}_{\mathrm{GL}} \equiv \int d\mathbf{x} \left[\frac{1}{2} (\nabla \phi)^2 - \epsilon \frac{\phi^2}{2} + \frac{\phi^4}{4} \right], \tag{8.27}$$

where $\epsilon \equiv (T_{\mathrm{c}} - T)$.

Equations (8.8) and (8.19) read, respectively,

$$\frac{\partial \phi}{\partial t} = M(\nabla^2 \phi + \epsilon \phi - \phi^3) + \eta \qquad \text{model A (TDGL)} \tag{8.28}$$

$$\frac{\partial \phi}{\partial t} = -M\nabla^2(\nabla^2 \phi + \epsilon \phi - \phi^3) + \zeta \qquad \text{model B (CH)} \tag{8.29}$$

The nonconserved Eq. (8.28) is called[7] model A of the dynamics and its deterministic version ($\eta \equiv 0$) is called time-dependent Ginzburg–Landau (TDGL) equation. The conserved Eq. (8.29) is called model B of the dynamics and its deterministic version ($\zeta \equiv 0$) is called Cahn–Hilliard (CH) equation. In the final section of this chapter, it will be useful to study the effect of a magnetic field H on a spherical domain. Because of it, the free energy $\mathcal{F}$ acquires the term $-\int d\mathbf{x} H\phi(\mathbf{x})$ whose functional derivative $\delta/\delta\phi$ simply gives $-H$, so that models A and B gain a term H between parentheses in their right-hand side, see Eqs. (8.28) to (8.29). A constant magnetic field, as expected, does not affect the conserved model while it modifies Eq. (8.28) as follows,

$$\frac{\partial \phi}{\partial t} = M(\nabla^2 \phi + \epsilon \phi - \phi^3 + H) + \eta \quad \text{model A with field.} \tag{8.30}$$

The evolution equations for conserved and nonconserved dynamics are nonlinear and their general, exact solutions are not known, so we must proceed through approximations. In Section 8.3, for example, we will solve Eq. (8.28) in the spherical approximation, appropriate for the dynamics of a single spherical domain. Here we are going to solve both of them in an appropriate linear approximation, which is valid at short times. This approximation amounts to perturb a constant solution $\phi(x, t) \equiv \phi_0$. For the TDGL equation, only $\phi_0 = 0$

---

[7] According to the classification of Pierre Hohenberg and Bertrand Halperin. In the case of a fluid, see footnote 5, we might obtain model H.

is solution, while for the CH equation, any constant function is solution: This is due to the conserved character of the Cahn–Hilliard equation. Let us therefore start with Eq. (8.28), assuming that $\phi(\mathbf{x}, t)$ is small so that the cubic term can be neglected and we obtain

$$\frac{\partial \phi}{\partial t} = \nabla^2 \phi + \epsilon \phi + \eta. \tag{8.31}$$

It is useful to rewrite it in the Fourier space,

$$\frac{\partial \phi(\mathbf{q}, t)}{\partial t} = \sigma(q)\phi + \eta(\mathbf{q}, t), \tag{8.32}$$

where $\sigma(q) = \epsilon - q^2$. The full solution of this equation can be derived from Eq. (2.23), because Eq. (8.32) is equivalent to Eq. (2.22):

$$\phi(\mathbf{q}, t) = e^{\sigma(q)t}\phi(\mathbf{q}, 0) + \int_0^t d\tau e^{\sigma(q)(t-\tau)}\eta(\mathbf{q}, t). \tag{8.33}$$

This expression will be useful in Section. 8.3, where we derive the structure factor. Here we limit to comment its deterministic part, that is, the first term on the right-hand side, which determines the linear stability of the constant solution $\phi \equiv 0$ through the sign of $\sigma(q)$. This quantity is negative for $T > T_c$, meaning stability because any initial condition decays to zero (apart from stochastic fluctuations). For $T < T_c$, $\epsilon > 0$ and initial perturbations of sufficiently small wavevector, $q < \sqrt{\epsilon}$, grow exponentially over time and maximal growth rate is attained for $q = q^* = 0$. This result signals a delocalized instability that leads to the formation of domains and to coarsening.

The same analysis can be carried out for model-B. In this case any constant function ϕ_0 [8] is a solution of Eq. (8.29), so linear analysis proceeds by writing $\phi(\mathbf{x}, t) = \phi_0 + \psi(\mathbf{x}, t)$ and assuming ψ a small quantity. Neglecting higher order terms, we get the equation

$$\frac{\partial \psi}{\partial t} = -(\nabla^2)^2 \psi + (3\phi_0^2 - \epsilon)\nabla^2 \psi + \zeta, \tag{8.34}$$

which rewrites in the Fourier space as

$$\frac{\partial \psi(\mathbf{q}, t)}{\partial t} = \tilde{\sigma}(q)\psi + \zeta(\mathbf{q}, t), \tag{8.35}$$

with $\tilde{\sigma}(q) = (\epsilon - 3\phi_0^2)q^2 - q^4$. Again, the stability of the uniform solution is set by the sign of $\tilde{\sigma}(q)$, which is positive (meaning instability) if $3\phi_0^2 < \epsilon$. If $T < T_c$, this occurs for "central quenching" (small ϕ_0). For strongly asymmetric quenching ($\phi_0^2 > \epsilon/3$), the initial state is linearly stable and its destabilization passes through the nucleation process. The line $3\phi_0^2 = \epsilon$ is the spinodal line and it corresponds to the vanishing of the second derivatives of the GL potential; see Section 8.1. We remark that the most unstable mode q^*, that is, the q-value corresponding to the positive maximum of $\tilde{\sigma}(q)$, is not $q^* = 0$ as in the conserved case, it is a finite one. Since Eq. (8.35) is equivalent to Eq. (8.32), its solution is also given by Eq. (8.33) with the appropriate change of notations. The explicit expression will be used in Section 8.3.

---

[8] The minima of the GL potential are obtained for $\phi \equiv \phi_M = \pm\sqrt{\epsilon}$, therefore a quenching can be done in a region where $|\phi_0| \leq |\phi_M| = \sqrt{\epsilon}$.

8.3 The Correlation Function and the Structure Factor

A relevant function that allows a more quantitative analysis of phase-ordering is the equal time correlation function,

$$C(r, t) = \langle \phi(\mathbf{x}, t)\phi(\mathbf{x} + \mathbf{r}, t) \rangle - \langle \phi(\mathbf{x}, t) \rangle^2, \tag{8.36}$$

which depends on $r = |\mathbf{r}|$ only, because of translational and rotational invariances in space. It is useful to write $C(r, t)$ as the sum of the equilibrium contribution plus a nonequilibrium contribution,

$$C(r, t) = C_{eq}(r) + C_{aging}(r, t). \tag{8.37}$$

Before discussing the two contributions, it is worth noting that $\langle \phi(\mathbf{x}, t) \rangle = \phi_0$, the average value of the order parameter, vanishing for a critical quenching and nonvanishing for an off-critical one.[9] Instead, $\langle \phi(\mathbf{x}, t) \rangle_{eq} = m(T)$, the equilibrium, temperature-dependent order parameter. Therefore, $C(r, t)$ varies between $C(0, t) = 1 - \phi_0^2$ and $C(\infty, t) = 0$ and this decay occurs on a scale of order of the domain size $L(t)$; instead, $C_{eq}(r)$ varies between $C_{eq}(0) = 1 - m^2(T)$ and $C_{eq}(\infty) = 0$ and this decay occurs on a scale of order of the equilibrium correlation length, $\xi(T)$.

If quenching is well inside the ordered phase, $\xi(T)$ is of the order of the lattice constant and $m(T) \simeq m(0) = 1$. In this case $C_{eq}(r)$ can be neglected and $C_{aging}(r, t) \simeq C(r, t)$. However, we should keep in mind that the scaling behavior of the correlation functions discussed here later is entirely related to the nonequilibrium contribution, $C_{aging}(r, t)$. Therefore, if we want to test scaling in a simulation or in an experiment, it is necessary either to verify that $C_{eq}(r)$ is actually negligible or to take it into account, if not negligible.[10] In Figs. 8.9 and 8.10, $\phi_0 = 0$ and $m(T) \simeq 0.99$, therefore both $C(r, t)$ and $C_{aging}(r, t)$ practically vary between 1 and 0.[11]

We can now use scaling to further simplify the expression of $C(r, t)$, thanks to self-similarity: Two snapshots of the system at different times $t, t' = bt$ are statistically equivalent if we rescale space appropriately, $\mathbf{x}' = b^n \mathbf{x}$. Therefore,[12]

$$C(r', t') = C(b^n r, bt) = C(r, t). \tag{8.38}$$

This relation must be valid for any b, in particular for $b = 1/t$, which implies

$$C\left(\frac{r}{t^n}, 1\right) = C(r, t). \tag{8.39}$$

---

[9] In the conserved case, $\langle \phi(\mathbf{x}, t) \rangle$ is (by definition) constant in time. In the nonconserved case, it must be $\langle \phi(\mathbf{x}, t) \rangle = 0$, otherwise ϕ_0 acts as a time-dependent effective magnetic field.

[10] With this caveat, the temperature after quenching as well as the initial temperature (before quenching) are irrelevant, as proved by studies of the dynamical renormalization group applied to continuum models.

[11] This is due to the fact that in the two-dimensional Ising model, the equilibrium magnetization is practically saturated even at two third of the critical temperature.

[12] In the most general case, scaling would imply a rescaling of the function itself, $C(r', t') = b^z C(b^n r, bt)$. This is the case, for example, for the roughness function studied in Chapter 7 (see Eq. (7.8)), or for the scaling form of the Gibbs free energy in equilibrium critical phenomena. In general terms, such rescaling is related to the fractal morphology of the structure under study, but in the present context domains have a fractal structure only if the temperature after quenching is the critical temperature, $T_f = T_c$. We are not considering this case.

In conclusion, if we define the length $L(t) = t^n$, we have

$$C(r,t) = f\left(\frac{r}{L(t)}\right), \tag{8.40}$$

a result that is also valid for logarithmic coarsening. Similar scaling arguments can be applied to the two-time correlation function,

$$C(t,t_w) = \langle \phi(\mathbf{x},t)\phi(\mathbf{x},t_w)\rangle = f^*\left(\frac{L(t)}{L(t_w)}\right), \tag{8.41}$$

where, by convention, time t is larger than the waiting time t_w.

In Fig. 8.9 we plot $C(r,t)$ and $C(t,t_w)$ for the simulations of the two-dimensional Ising model with nonconserved order parameter. The inset of Fig. 8.9(a) shows the data collapse when $C(r,t)$ is plotted as $C(r/L(t))$, where the length $L(t)$ has been evaluated through the condition[13]

$$C(L,t) = \frac{C(0,t)}{2}. \tag{8.42}$$

Fig. 8.9(b) shows directly data collapse when $C(t,t_w)$ is plotted as a function of $L(t)/L(t_w)$. We do not comment here about the asymptotic, power law behavior, $f^*(x) \sim x^{-5/4}$, but when discussing the relaxation dynamics of the one-dimensional (nonconserved) Ising model, we will find the exact expressions of the correlation functions.

In Fig. 8.10, we plot the spatial correlation function of the conserved model, which appears to be oscillating, in contrast to the nonconserved model, Fig. 8.9, where $C(r)$ is monotonously decreasing. These different behaviors are related to the different distributions of domain size, $n(A)$, where A is the area of the domain. This connection is discussed in Section 8.4.

It is now useful to rewrite Eq. (8.40) in the $\mathbf{q}$-space, because the structure factor $S(\mathbf{q},t)$, defined as the Fourier transformation of the correlation function, is of interest for scattering experiments, being proportional to the scattering intensity. In $\mathbf{q}$-space the dynamic scaling has the form

$$\begin{aligned}
S(\mathbf{q},t) &= \int d\mathbf{r}\, e^{-i\mathbf{q}\cdot\mathbf{r}} C(\mathbf{r},t) = \int d\mathbf{r}\, e^{-i\mathbf{q}\cdot\mathbf{r}} f\left(\frac{r}{L(t)}\right) \\
&= L^d \int d\mathbf{s}\, e^{-iL\mathbf{q}\cdot\mathbf{s}} f(s) \\
&= L^d f(qL),
\end{aligned} \tag{8.43}$$

where $q = |\mathbf{q}|$, $s = |\mathbf{s}|$, and $f(q)$ is the d-dimensional Fourier transform of $f(s)$. We can also relate the structure factor to the correlation function of the order parameter in the $\mathbf{q}$-space. If

$$\phi(\mathbf{q},t) = \int d\mathbf{x}\, e^{-i\mathbf{q}\cdot\mathbf{x}} \phi(\mathbf{x},t), \tag{8.44}$$

---

[13] If scaling were perfect, the choice of the factor 2 in the denominator would be as good as any other. In practice, finite size and finite statistics effects spoil the simulation data for $C(r,t)$ at large r. It is common practice to determine $L(t)$ through Eq. (8.42) or, if $C(r,t)$ oscillates, through its first zero, $C(L,t) = 0$.

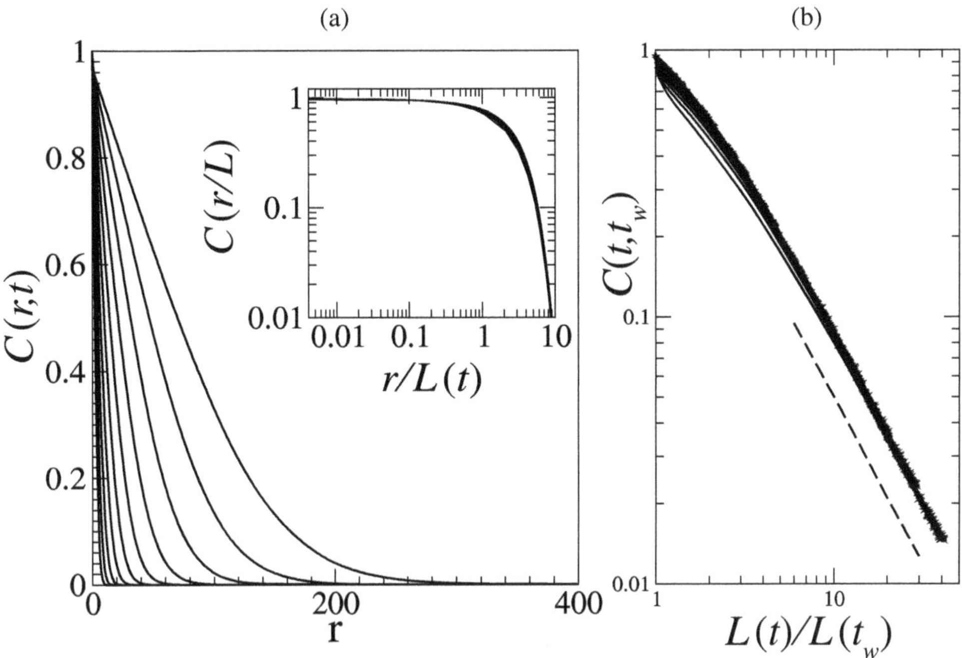

Fig. 8.9 (a) The correlation function $C(r, t)$ for the two-dimensional Ising model with nonconserved (Glauber) dynamics, after quenching from $T_i = \infty$ to $T_f = 0.661 T_c$, for different times $10 \le t \le 14\,000$ (with increasing time the curve moves to the right). In the inset, C is plotted as a function of the reduced variable $r/L(t)$ to illustrate scaling. (b) The two-time correlation function, $C(t, t_w)$, with $t > t_w$, plotted as a function of $L(t)/L(t_w)$. Different curves correspond to different values of the waiting time, $10 \le t_w \le 14338$. The dashed line corresponds to a power-law decay with exponent equal to $-\frac{5}{4}$. Simulation data courtesy of Federico Corberi.

because of translational invariance we get

$$\langle \phi(\mathbf{q}, t)\phi(\mathbf{q}', t)\rangle = \int d\mathbf{x}\, e^{-i\mathbf{q}\cdot\mathbf{x}} \int d\mathbf{r}\, e^{-i\mathbf{q}'\cdot(\mathbf{x}+\mathbf{r})} \langle \phi(\mathbf{x}, t)\phi(\mathbf{x} + \mathbf{r}, t)\rangle$$

$$= \int d\mathbf{x}\, e^{-i(\mathbf{q}+\mathbf{q}')\cdot\mathbf{x}} \int d\mathbf{r}\, e^{-i\mathbf{q}'\cdot\mathbf{r}} (C(r, t) + \phi_0^2)$$

$$= (2\pi)^d \delta(\mathbf{q} + \mathbf{q}') \left[S(\mathbf{q}, t) + (2\pi)^d \delta(\mathbf{q})\phi_0^2 \right]. \tag{8.45}$$

In Fig. 8.11, we offer experimental evidence of the dynamic scaling of the structure factor for the phase separation process of a Mn–Cu binary alloy. The quantity Q_{max} is proportional to $1/L(t)$, so the data collapse shown is equivalent to Eq. (8.43). We stress that dynamic scaling is a long-time, asymptotic property, as attested by the bottom panel and discussed in the figure legend.

The short-time behavior of the structure factor can be determined using the solution of the linear approximation, see Eq. (8.33). Using this equation for the nonconserved dynamics,

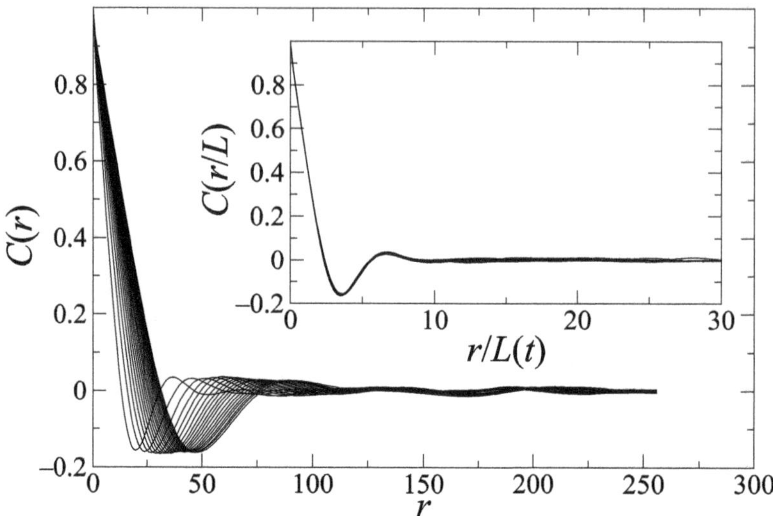

Fig. 8.10 The correlation function $C(r,t)$ for the two-dimensional Ising model with conserved (Kawasaki) dynamics, after quenching from $T_i = \infty$ to $T_f = 0.661 T_c$, for different times $5 \times 10^4 \leq t \leq 10^6$ (with increasing time the curve moves to the right). In the inset, C is plotted as a function of the reduced variable $r/L(t)$ to illustrate scaling. Simulation data courtesy of Federico Corberi.

it is easily found that

$$\langle \phi(\mathbf{q},t)\phi(\mathbf{q}',t)\rangle = \langle \phi(\mathbf{q},0)\phi(\mathbf{q}',0)\rangle e^{2\sigma(q)t} + (2\pi)^d \Gamma \delta(\mathbf{q}+\mathbf{q}')\frac{1}{2\sigma(q)}\left(e^{2\sigma(q)t}-1\right),$$

$$(8.46)$$

where Γ is the noise correlator, see Eq. (8.11). Using Eq. (8.45) with $\phi_0 = 0$, we finally obtain

$$S(\mathbf{q},t) = S(\mathbf{q},0)e^{2\sigma(q)t} + \frac{\Gamma}{2\sigma(q)}\left(e^{2\sigma(q)t}-1\right). \qquad (8.47)$$

For conserved dynamics we must start again from Eq. (8.33) where now $\sigma(q) = \epsilon - q^2$ is replaced by $\tilde{\sigma}(q) = (\epsilon - 3\phi_0^2)q^2 - q^4$ and where the noise η with correlator (8.11) is replaced by the noise ζ with correlator (8.23). Furthermore, with conserved dynamics ϕ_0 may also be different from zero. Now we obtain

$$\langle \phi(\mathbf{q},t)\phi(\mathbf{q}',t)\rangle = \langle \phi(\mathbf{q},0)\phi(\mathbf{q}',0)\rangle e^{2\tilde{\sigma}(q)t} + (2\pi)^d \Gamma \delta(\mathbf{q}+\mathbf{q}')\frac{q^2}{2\tilde{\sigma}(q)}\left(e^{2\tilde{\sigma}(q)t}-1\right)$$

$$(8.48)$$

and comparing with Eq. (8.45) we finally get

$$S(\mathbf{q},t) = S(\mathbf{q},0)e^{2\tilde{\sigma}(q)t} + \left(\frac{\Gamma q^2}{2\tilde{\sigma}(q)} + (2\pi)^d \phi_0^2 \delta(\mathbf{q})\right)\left(e^{2\tilde{\sigma}(q)t}-1\right). \qquad (8.49)$$

Finally, let us discuss the short distance (small r, large q) behavior of the correlation function $C(r,t)$. Since this regime may be affected by $C_{eq}(r,t)$, we explicitly suppose to be at $T = 0$ and to consider a symmetric quenching, $\phi_0 = 0$, so that $C(0,t) = 1$. However, our result also applies to Figs. 8.9 and 8.10, where $T/T_c = \frac{2}{3}$; in fact, such figures

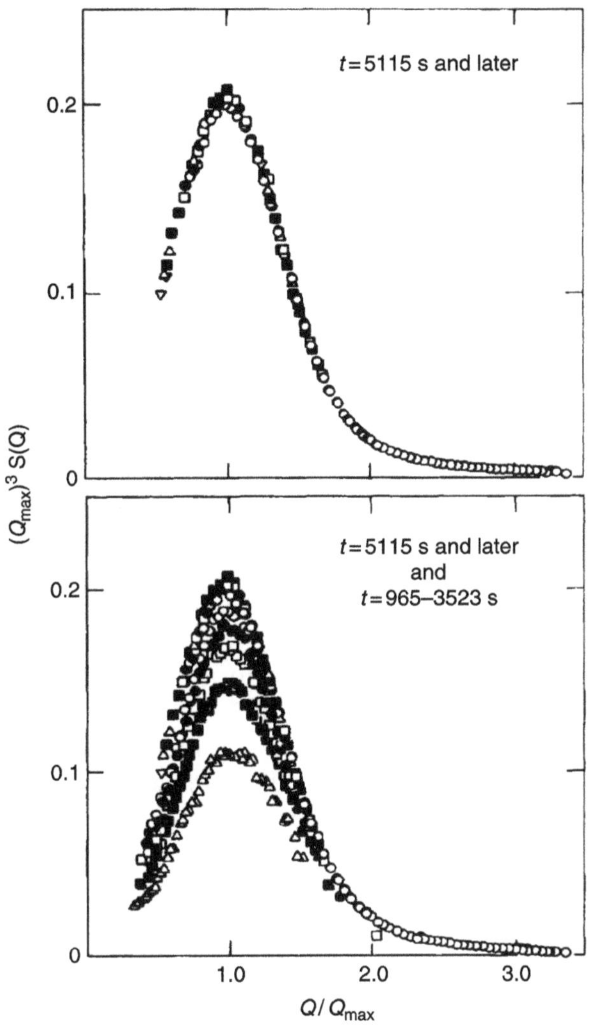

Fig. 8.11 Dynamic scaling of the coarsening process in a binary alloy ($Mn_{0.67}Cu_{0.33}$). The structure factor is proportional to the intensity of the neutron scattering experiment. The top panel shows clear dynamic scaling at late times (5115 s $< t < 8429$ s). The bottom panel shows results at five different earlier times. There is no collapse and the departure from the scaling expression grows with decreasing time. Reprinted from B. D. Gaulin, S. Spooner, and Y. Morii, Kinetics of Phase Separation in $Mn_{0.67}Cu_{0.33}$, *Physical Review Letters*, **59** (1987) 668–671.

clearly display a small r linear behavior, more precisely a linearity in r/L. The behavior $C(0,t) - C(r,t) \simeq r/L$ is easily explained by the observation that for two points $\mathbf{x}, \mathbf{x} + \mathbf{r}$ at small distance r, the product $\langle \phi(\mathbf{x}, t)\phi(\mathbf{x} + \mathbf{r}, t)\rangle$ is equal to -1 if a (single) domain wall crosses the segment joining them; otherwise, the product is equal to $+1$ (for $r \ll L$, we can disregard the possibility that such a segment is crossed by more than one domain wall).

Therefore, we can write

$$C(r,t) \simeq -\frac{r}{L} + \left(1 - \frac{r}{L}\right) = 1 - \frac{2r}{L}. \tag{8.50}$$

This (nonanalytic) short r behavior of $C(r,t)$ reflects in a power-law tail in the structure factor, which can be determined by Eq. (8.43) by simple power counting, obtaining the so-called Porod law,

$$S(\mathbf{q},t) \simeq \frac{1}{Lq^{d+1}}. \tag{8.51}$$

8.4 The Coarsening Law in Ising-Like Systems: Nonconserved Dynamics

8.4.1 The Continuum Picture in Any Dimension

Following the adage "A picture is worth a thousand words," in Fig. 8.5 we plot some snapshots at different times for the relaxation process of an Ising model with nonconserved dynamics. They refer to a quenching from $T_i = \infty$ to $T_f = 0.661T_c$. A quantitative analysis of such images via the correlation function was made in Section 8.3. Here we want to focus on the coarsening process, which is clearly visible and is made quantitative in Fig. 8.12. The numerical data suggest a coarsening exponent $n = 1/2$, a value that, with some caveats, is independent on the dimension d.

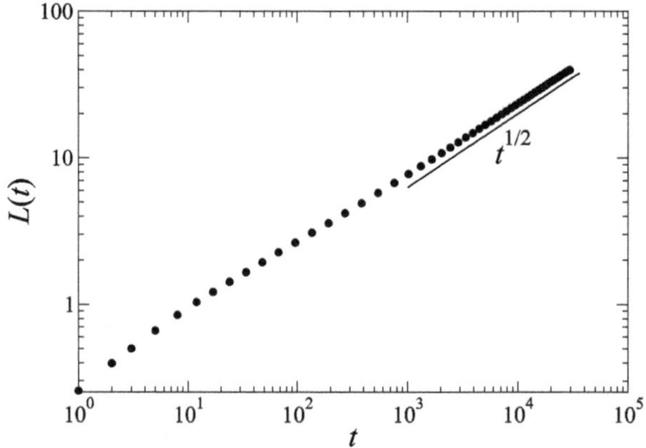

Fig. 8.12   KMC simulation of a $2\,000 \times 2\,000$ Ising model with nonconserved, spin-flip dynamics, after quenching from $T_i = \infty$ to $T_f = 0.661T_c$. The length scale L is determined by the width at half height of the correlation function $C(r,t)$, $C(L,t) = C(0,t)/2$. Simulation data courtesy of Federico Corberi.

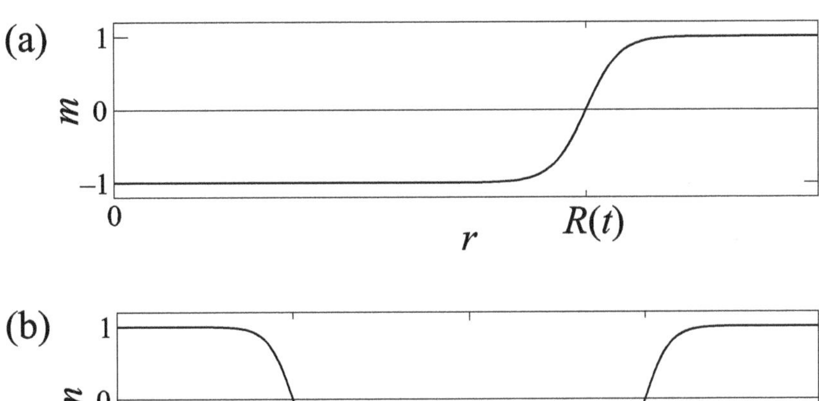

Fig. 8.13　Profiles of the order parameter for a single negative domain in a positive sea. (a) In $d > 1$, the domain is spherical and the profile of $m(r)$ is approximated by a kink centered in $r = R(t)$. (b) In $d = 1$, the domain is a segment and can be obtained by superposing two kinks: a negative kink centered in $x = -R(t)$ and a positive kink centered in $x = +R(t)$. Note that in $d > 1$, $r > 0$ because it is the radius in spherical coordinates, while in $d = 1$ x is the Cartesian coordinate, so it may be negative.

Let us start with Eq. (8.28). Rescaling variables, we obtain

$$\frac{\partial m}{\partial t} = \nabla^2 m + m - m^3 + \eta \tag{8.52}$$

$$\equiv \nabla^2 m - U'(m) + \eta \quad \text{with} \quad U(m) = -\frac{m^2}{2} + \frac{m^4}{4}. \tag{8.53}$$

The general goal would be to solve it with an initial condition corresponding to a disordered state, but we will not try to face such a general problem here. We would rather focus on a simpler one: determining the time scale $t(L)$ necessary for a domain of negative magnetization in a sea of positive magnetization to disappear. Because of scaling, there is only one relevant length scale at time t, the average domain size L. The relation between L and t, the coarsening law we are looking for, is the same as we can find from such a simplified problem.

Let us therefore consider the problem of a single spherical domain of radius R. For spherical symmetry, $m(\mathbf{r}, t) = m(r, t)$, but what about the noise? Noise breaks such symmetry, but on average it preserves it. This amounts to replace $\eta(\mathbf{x}, t)$ with its average over angular coordinates,

$$\bar{\eta}(r, t) = \frac{1}{S_d} \int d\Omega\, \eta(\mathbf{x}, t), \tag{8.54}$$

where S_d is the surface of the d-dimensional hypersphere of radius 1. Its first two moments are

$$\langle \tilde{\eta}(r,t) \rangle = 0, \qquad \langle \tilde{\eta}(r,t)\tilde{\eta}(r',t') \rangle = \frac{\Gamma}{S_d}\delta(r-r')\delta(t-t'). \tag{8.55}$$

In spherical coordinates, Eq. (8.53) is rewritten as

$$\partial_t m(r,t) = \partial_{rr} m + \frac{d-1}{r}\partial_r m - U'(m) + \tilde{\eta}(r,t). \tag{8.56}$$

The potential $U(m)$ has minima in $m = \pm 1$, so a domain of radius $R(t)$ corresponds to a profile $m(r,t)$, which varies with continuity between $m = -1$ for $r \ll R$ to $m = +1$ for $r \gg R$; see Fig. 8.13. More precisely, what is the form $m(r,t)$? We can answer in $d = 1$ and assume the answer is qualitatively correct for any d. This approach is somewhat confirmed in Appendix U.

In $d = 1$, neglecting noise, we have

$$\partial_t m(x,t) = \partial_{xx} m + m - m^3, \tag{8.57}$$

which has the exact stationary solution

$$m(x) = \tanh\left(\frac{x-R}{\sqrt{2}}\right). \tag{8.58}$$

The domain wall has a finite size, the profile $m(x)$ converging exponentially toward ± 1. Let us now come back to Eq. (8.56), with that in mind. We assume that while the domain shrinks, that is, $R(t)$ decreases, the value of $m(r,t)$ just depends on the distance between r and the center of the domain wall,

$$m(r,t) = m(r - R(t)), \tag{8.59}$$

so that $\partial_t m = -\dot{R}(t)\partial_r m$, and Eq. (8.56) can be rewritten as

$$-\dot{R}m' = m'' + \frac{d-1}{r}m' - U'(m) + \tilde{\eta}, \tag{8.60}$$

where the easier notation $m' = \partial_r m(r - R)$ has been used.

A differential equation for $R(t)$ can be obtained by multiplying both terms of Eq. (8.60) by m' and integrating between zero and infinity:

$$-\dot{R}\int_0^\infty dr(m')^2 = \int_0^\infty drm'\left[m'' + \frac{d-1}{r}m' - U'(m) + \tilde{\eta}\right] \tag{8.61}$$

$$= \left[\frac{1}{2}(m')^2 - U(m)\right]_{r=0}^{r=\infty} + (d-1)\int_0^\infty dr\frac{(m')^2}{r} + \int_0^\infty drm'\tilde{\eta}. \tag{8.62}$$

The integral proportional to $(d-1)$ can be simplified, because $(m')^2$ is nonvanishing only in a small, finite region around $r = R$, so it can be evaluated by replacing r with R in the denominator. Defining $I = \int_0^\infty dr(m')^2$ and using that $m'(+\infty) = 0$, we obtain

$$-\dot{R}I = -\frac{1}{2}(m'(r=0))^2 + U(m(r=0)) - U(m(r=\infty)) + \frac{d-1}{R}I + \int_0^\infty drm'\tilde{\eta}. \tag{8.63}$$

Since the domain wall profile converges exponentially to the minima of U (see Eq. (8.58)), both $m'(r = 0)$ and $[U(m(r = 0)) - U(m(r = \infty))]$ are exponentially small with respect to $1/R$, so it is necessary to distinguish between $d > 1$ and $d = 1$.

If $d > 1$, such terms are negligible and

$$\dot{R} = -\frac{d-1}{R} + \xi(t), \tag{8.64}$$

where $\xi(t) = -I^{-1}\int_0^\infty dr\, m'(r - R(t))\tilde{\eta}(r, t)$, with $\langle \xi(t) \rangle = 0$ and

$$\langle \xi(t)\xi(t') \rangle = \delta(t - t')\frac{\Gamma}{S_d I^2}\int_0^\infty dr\,(m'(r - R(t)))^2 \equiv \frac{\Gamma}{S_d I}\delta(t - t'). \tag{8.65}$$

It is noteworthy that each term on the right-hand side of Eq. (8.64) gives a closure time $t(L)$ that scales quadratically with the initial size L of the domain. In fact, if we neglect noise, the exact solution is

$$R^2(t) = R^2(0) - 2(d-1)t \tag{8.66}$$

and $R(0) = L$ gives $t(L) = L^2/(2(d-1))$. If instead we neglect the deterministic force, we obtain the Langevin equation for a random walker and, again, the typical time to travel a distance L scales as L^2. Inverting this relation, we conclude that the typical size of domains grows as $L(t) \simeq t^{1/2}$, giving a coarsening exponent $n = 1/2$, as expected by numerical simulations for $d = 2$. Earlier treatment for the dynamics of a spherical domain is generalized in Appendix U to a domain of arbitrary shape, confirming the exponent n.

Before passing to $d = 1$, we should remember that the critical temperature of the one-dimensional Ising model vanishes, $T_c(d = 1) = 0$; therefore, it is not possible to make a quenching of the temperature *below* T_c. We will discuss this point in more detail later; now let us consider Eq. (8.63) for $d = 1$. In one dimension the variable r is replaced by the spatial coordinate x and the profile $m(x, t)$ of a domain located between $x = -R$ and $x = R$ is symmetric so $m'(x = 0)$ is exactly zero, leading to the equation

$$\dot{R} = -\frac{1}{I}\left[U(m(x = 0)) - U(m(x = \infty))\right] + \xi(t). \tag{8.67}$$

In order to obtain a negative domain between $x = -R$ and $x = R$, we have to suitably superpose a kink centered in $x_0 = R$ and a kink centered in $x_0 = -R$ (see Fig. 8.13(b)),

$$m(x) = \tanh\left(\frac{x - R}{\sqrt{2}}\right) - \tanh\left(\frac{x + R}{\sqrt{2}}\right) + 1, \tag{8.68}$$

where the constant term 1 is required to have $m(x) \to 1$ for $x \to \pm\infty$. Using Eq. (8.68), for large R, we obtain

$$m(0) = 1 - 2\tanh(R/\sqrt{2}) \approx -1 + 4\exp(-\sqrt{2}R) \tag{8.69}$$

and

$$U(m(x = 0)) - U(m(x = \infty)) = 16e^{-2\sqrt{2}R}, \tag{8.70}$$

so that, using Eq. (8.67), we obtain

$$\dot{R} = -\frac{16}{I}e^{-2\sqrt{2}R} + \xi(t). \tag{8.71}$$

In $d = 1$ we therefore have an exponentially small (negative) drift that tends to close the domain and a standard noise term. In the absence of noise, the equation is easily integrated by separation of variables, giving

$$\frac{1}{2\sqrt{2}}\left[e^{2\sqrt{2}R(t)} - e^{2\sqrt{2}R(0)}\right] = -\frac{16}{I}t. \tag{8.72}$$

If $R(0) = L/2$, the domain disappears in a time $t(L)$ that increases exponentially,

$$t(L) = \frac{I}{32\sqrt{2}}\left(e^{\sqrt{2}L} - 1\right) \simeq \frac{I}{32\sqrt{2}}e^{\sqrt{2}L} \tag{8.73}$$

and the coarsening is logarithmically slow,

$$L(t) \approx \ln t. \tag{8.74}$$

It is obvious that the noise term dominates, allowing a closure time growing quadratically and giving again a coarsening exponent $n = 1/2$. We will comment on the logarithmic coarsening in Section 8.4.2.

In view of the last section on nucleation, it is worth considering the effect of a magnetic field on the evolution of a single domain of radius R, which means to start from Eq. (8.30) rather than from Eq. (8.28). It is straightforward that Eq. (8.56) acquires the factor H on the right-hand side and Eq. (8.63) gains the quantity

$$\int_0^{\infty} drm'H \simeq \int_{-\infty}^{\infty} drm'H = 2H. \tag{8.75}$$

Such a constant term dominates all other terms, which decrease with R. This means that in the presence of a magnetic field, a spherical domain with magnetization opposite to the field closes with constant velocity,

$$v_0 = -\frac{2H}{I}. \tag{8.76}$$

8.4.2 The Discrete Picture in $d = 1$

The one-dimensional Ising model has peculiar features, because the equilibrium ordering temperature vanishes, $T_c(d = 1) = 0$. Despite this, it is possible to study phase-ordering at zero or vanishingly small T, gaining useful insights on the effect of a conservation law and on the comparison between deterministic and stochastic dynamics.

Since $T_c = 0$, the model is ordered at $T = 0$ only. At any finite temperature, the equilibrium state is disordered and has zero magnetization, but it is composed of large domains of up and down spins, whose typical size, equal to the correlation length ξ, diverges exponentially when $T \to 0$,

$$\xi(T) \simeq \exp(2J/T). \tag{8.77}$$

This means that the coarsening process in a one-dimensional Ising model cannot last forever, because it will stop when $L(t) \approx \xi(T)$. Zero temperature coarsening might be perpetual because $\xi(T = 0) = \infty$, but $T = 0$ dynamics is restricted to processes that conserve or lower the energy, and, as we will see, conservation of the order parameter forbids coarsening at all. Here we focus on nonconserved dynamics, whose simplest possible microscopic move is a single-spin flip, see Appendix H.

The low-T dynamical picture is made easier if we represent a given spin configuration using domain walls (DWs), that is, fictitious points located between spins of opposite orientation (see Fig. 8.14). There are three types of elementary processes for DWs: (a) The hopping of a DW to its right or to its left. This corresponds to flipping a spin i (the thick spin in Fig. 8.14) between two spins of opposite orientations ($s_{i+1} = -s_{i-1}$). This process does not change the energy, $\Delta E = 0$. (b) The annihilation of two neighboring DWs. This corresponds to flipping a spin i located between two spins, both antiparallel to it ($s_{i+1} = s_{i-1} = -s_i$). This process decreases the energy, $\Delta E < 0$. (c) The creation of two neighboring DWs. This corresponds to flipping a spin i that is parallel to its neighbors ($s_{i-1} = s_i = s_{i+1}$). This process increases the energy, $\Delta E > 0$.

At $T = 0$, process (c) is forbidden, but processes (a) and (b) are sufficient to drive the system toward the ordered state. Therefore, perpetual coarsening occurs at $T = 0$ with nonconserved dynamics. Zero temperature dynamics in terms of DWs is easy: These (pseudo)particles diffuse; when two of them occupy the same site, they annihilate, leading to a decreasing number of DWs, that is, to an increase in their average distance $L(t)$, which is nothing but the average size of ordered regions (magnetic domains). At $T > 0$, process (c) would also allow the creation of new DWs: This is the mechanism by which a finite correlation length $\xi(T)$ eventually arises, since large magnetic domains are broken into smaller parts by this process.

Following the same line of reasoning used in Section 8.4.1 and based on self-similarity, we wonder what is the typical closure time of a domain of size L, see Fig. 8.15. In the DW language, we have two particles at distance L that can diffuse, and we wonder what is the time t required to ensure they meet and annihilate, that is, such that $L(t) = 0$? Since the quantity L itself performs a random walk, the time necessary to pass from $L(0) = L$ to $L(t) = 0$ scales as L^2, $t \sim L^2$. Assuming that there is just one length scale at time t, the

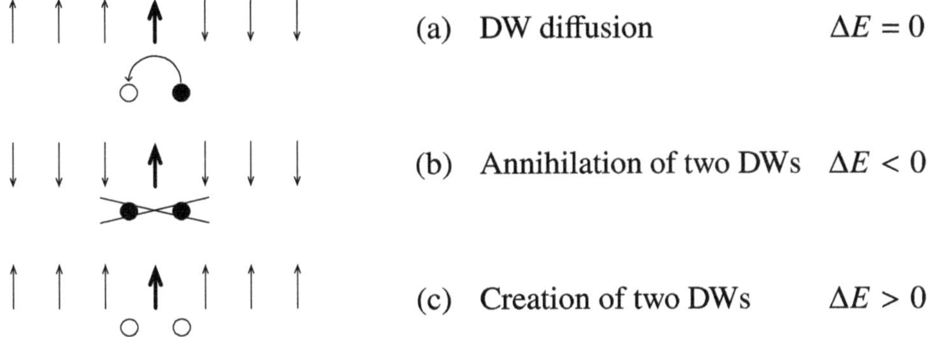

Fig. 8.14   Nonconserved, spin-flip (Glauber) dynamics. We sketch the possible transitions using spin language and domain wall (DW) language. The thick spin is reversed and circles represent the DWs before (solid circles) and after (open circles) the flipping process.

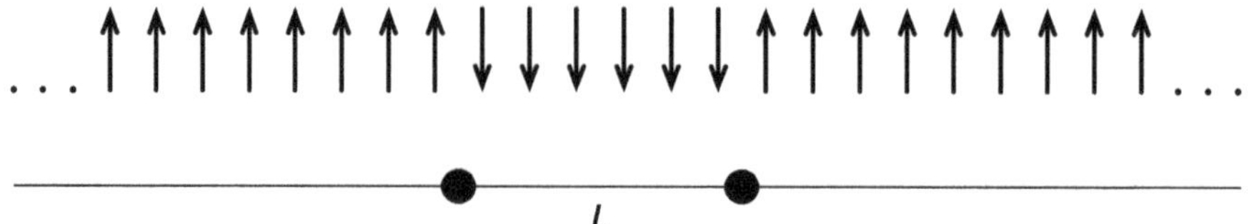

Fig. 8.15　A domain of down spins of size L is represented through two pseudo-particles, located at DW positions.

coarsening law can simply be found by reversing the above relation, giving

$$L(t) \approx t^n \qquad \left(n = \frac{1}{2}, \quad \text{nonconserved dynamics}\right). \tag{8.78}$$

In conclusion, the (nonconserved) coarsening exponent $n = \frac{1}{2}$ simply comes from the diffusive motion of DWs.[14]

Let us now compare the results for the discrete and continuous pictures in $d = 1$. The discrete picture corresponds to Eq. (8.71) without the deterministic driving, $\dot{R} = \xi(t)$. Such exponentially small driving is formally due to the kink profile that has exponential tails, while a discrete picture of the Ising model gives a kink that is infinitely narrow. To get a similar term in the discrete model, we should replace the nearest neighbor interaction of the Ising model with a coupling exponentially decreasing with the distance. In this case a logarithmic coarsening can appear as a transient regime,[15] because the noise term always dominates asymptotically.

The Glauber Model

The Glauber model corresponds to the dynamics of the Ising model, with spin-flip dynamics and an evolution rule satisfying detailed balance, see Eqs. (2.115) and (2.116), with $\omega(x)$ given by Eq. (2.119). The evolution of the probability $P(S, t)$ to find the system at time t in the configuration $S \equiv (s_1, \dots, s_N)$ is governed by a master equation of the form

$$\frac{dP(S,t)}{dt} = \sum_{S'} \left[P(S',t)W_{S,S'} - P(S,t)W_{S',S} \right] \tag{8.79}$$

where $W_{S',S}$ is the transition probability from S to S'. Using Eq. (2.119) and that each move involves only a single spin-flip, such probability has the form $W(s_i \to -s_i) = \frac{1}{2}(1 - s_i \tanh(\beta h_i))$, where $h_i = J(s_{i-1} + s_{i+1})$ is the field acting on the flipping spin i and where we have made explicit that in a move only one spin may flip. Using this result we can simplify Eq. (8.79) and obtain

$$\frac{dP(S,t)}{dt} = \frac{1}{2} \sum_i \left[P(S^*,t)(1 + s_i \tanh \beta h_i) - P(S,t)(1 - s_i \tanh \beta h_i) \right], \tag{8.80}$$

---

[14] The reader familiar with random walks can object that the chosen configuration, one single minority domain in an infinite system, is *too* simple. In fact, if a random walker is in $x = L$ at $t = 0$, the mean time to cross the origin is actually infinite, while the median time scales as L^2, as shown in Appendix M.4. In order to have a finite mean time, we should consider a finite sample of size $2L$ with periodic boundary conditions.

[15] The coupling between spins results in a similar coupling between domain walls, which determines an L-dependent drift favouring the closure of the domain. If L is not too large, the drift dominates over diffusion.

where $S \equiv (s_1, \ldots, s_i, \ldots s_N)$ and $S^* \equiv (s_1, \ldots, -s_i, \ldots s_N)$.

The knowledge of $P(S, t)$ allows to find, for example, the spatial correlation function,

$$C_{i,j}(t) = \langle s_i(t)s_j(t) \rangle = \mathrm{Tr}\left\{ P(S, t)s_i s_j \right\}. \tag{8.81}$$

In fact, multiplying both sides of Eq. (8.80) by $s_i s_j$ and performing the trace, we obtain

$$\frac{dC_{i,j}}{dt} = -2C_{i,j} + \langle s_i \tanh \beta h_j \rangle + \langle s_j \tanh \beta h_i \rangle. \tag{8.82}$$

This expression is valid in any dimension and for any $T = 1/\beta$,[16] but in $d = 1$, the earlier equations are closed and linear. In fact, in $d = 1$ we have $\tanh \beta h_i = \frac{1}{2}(\tanh \beta J)(s_{i-1} + s_{i+1})$ and for vanishing T, we obtain

$$\frac{dC_{i,j}}{dt} = -2C_{i,j} + \frac{1}{2}(C_{i,j+1} + C_{i,j-1} + C_{i-1,j} + C_{i+1,j}). \tag{8.83}$$

If quenching has been done from a completely disorder system, $P(S, 0) = 1/2^N$, $C_{i,j}(t) = C(r, t)$ with $r = |i - j|$, and the Eq. (8.83) writes

$$\frac{\partial C(r, t)}{\partial t} = -2C(r, t) + C(r + 1, t) + C(r - 1, t), \tag{8.84}$$

or (in the continuum limit), $\partial_t C = \partial_{rr} C$, with $C(0, t) = 1$. This equation allows for a solution having the scaled form $C(r, t) = f(r/\sqrt{t})$, because the diffusion equation rewrites $f''(x) = -\frac{x}{2}f'(x)$. Using $f(0) = 1$ and $f(\infty) = 0$, we finally obtain

$$C(r, t) = \mathrm{erfc}\left(\frac{|r|}{2\sqrt{t}} \right), \tag{8.85}$$

which is the sought-after scaling form for the spatial correlation function, see Eq. (8.40), with $L(t) \simeq \sqrt{t}$.

Using a similar approach and similar calculations, we can also determine the two-time correlation function, see Eq. (8.41),

$$C(t, t_w) = \langle s_i(t)s_i(t_w) \rangle \xrightarrow[t \gg t_w]{} f^*(L(t)/L(t_w)), \tag{8.86}$$

with $f^*(x) = \frac{2\sqrt{2}}{\pi}x^{-1}$.

In general, $f^*(x) \sim x^{-\lambda}$ for large x, that is, for $t \gg t_w$. Comparison of Eq. (8.86) (valid for Ising $1d$) with Fig. 8.9 on the right (valid for Ising $2d$) shows that the exponent λ is different in different spatial dimensions, in opposition to the coarsening exponent n.

---

[16] For $d > 1$, the local field has a clear different form of that given above.

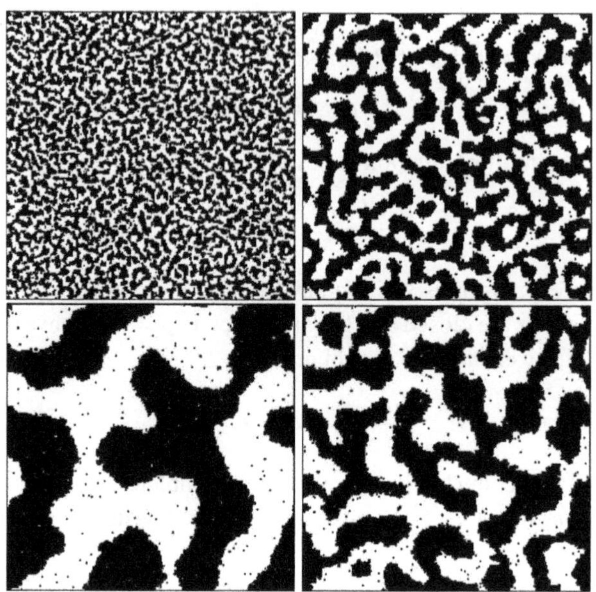

KMC simulations of a 256×256 Ising model with conserved, spin-exchange dynamics, after a quenching from $T_i = \infty$ to $T_f = 0.661 T_c$. Clockwise, from top left: Snapshots of the system at times $t = 10^2, 10^4, 10^5, 10^6$. Simulation data courtesy of Federico Corberi.

8.5 The Coarsening Law in Ising-Like Systems: Conserved Dynamics

8.5.1 The Continuum Picture in $d > 1$

We begin this section with some snapshots of a two-dimensional Ising model, obtained with a KMC simulation of conserved dynamics, after a quench into the ordered phase; see Fig. 8.16. The numerical analysis of the growth of $L(t)$ (see Fig. 8.17) suggests $n = 1/3$, therefore a coarsening exponent that is different from the nonconserved case.

While a quantitative comparison of the domain morphology with the nonconserved case is deferred to the Section 8.5.2, we want first to offer an analytical derivation of the coarsening exponent. The continuum analysis of the nonconserved case was based on the study of model A, Eq. (8.28), in spherical symmetry. A similar analysis of model B, Eq. (8.29), would be more elaborate; therefore, here we prefer to focus on diffusion, the physical process leading to coarsening in the conserved system, trying to perform a calculation similar in spirit to what we did for the nonconserved model. In that case it was sufficient to consider a single domain of the minority phase and determine the typical time because it disappears in favor of the majority phase. When the order parameter is conserved, we need to consider (at least) two domains of the same phase and ask what is the typical time because they merge in a single domain.

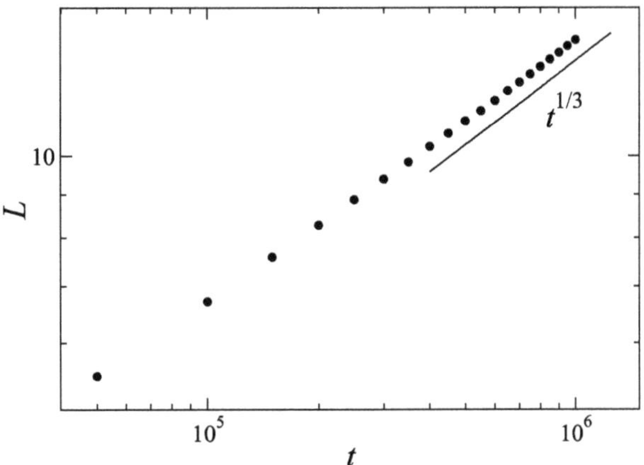

Fig. 8.17   KMC simulation of a 512 × 512 Ising model with conserved, spin-exchange dynamics, after quenching from $T_i = \infty$ to $T_t = 0.661T_c$. The length scale L is determined by the width at half height of the correlation function $C(r, t)$, $C(L, t) = C(0, t)/2$. Simulation data courtesy of Federico Corberi.

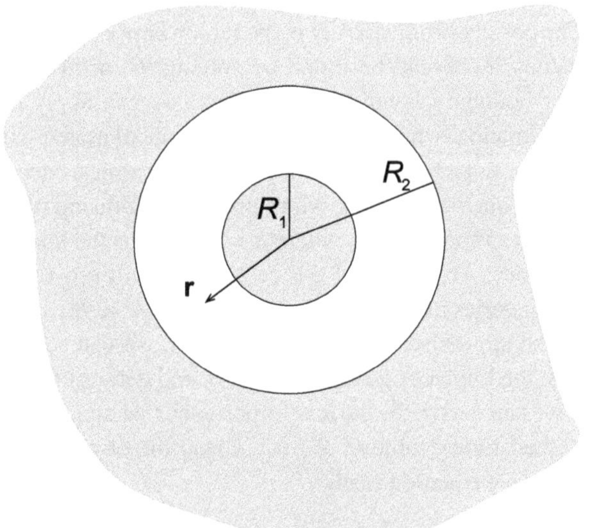

Fig. 8.18   Two-sphere geometry to study the conserved coarsening in a continuous model, in dimension $d > 1$.

Thanks to the mapping between Ising and lattice gas models, a domain of, for example, up spins is equivalent to a cluster of particles and the conserved spin-exchange dynamics (see Appendix H) amounts to particle diffusion. Therefore, the earlier problem can be reformulated as follows: Given two clusters of initial size L, what is the time $t(L)$ because all mass is in a single cluster? In this section we are going to face this problem in $d > 1$ and within a continuum approach, while in Section 8.5.2, we face the same problem in $d = 1$ and within a discrete approach.

In $d > 1$, the appropriate geometry to make the problem solvable is a spherical geometry where the external domain is infinite and is called a reservoir; see Fig. 8.18. The idea is to model the process of particle detachment from the small domain of radius R_1, the particle diffusion in the empty region $R_1 < r < R_2$, and the particles' final attachment to the reservoir, of inner radius R_2. The main question is: Why is there a net flux of particles between domain and reservoir? The answer is qualitatively simple and we discuss it here. It is less trivial to make it quantitative, because we must refer to the so-called Gibbs–Thomson effect, discussed in Appendix V.

Particles diffuse from high-density to low-density regions, and the density ρ of diffusing particles close to a domain depends on the curvature of the domain itself (Gibbs–Thomson effect). If a domain wall is straight, the equilibrium density of particles close to it has some value $\rho = \rho_0$, depending on temperature. If we now bend the domain, we can expect an effect on ρ, which can be understood from a microscopic point of view. If we impose a positive curvature, as for the domain of radius R_1, particles at the surface of the inner domain feel a decreasing average number of nearest neighbors; therefore, it is easier to detach and this leads to a larger ρ. If we impose a negative curvature, as for the reservoir, the reverse is expected. For this reason, $\rho(R_1) > \rho(R_2)$, leading to a current of order $J \approx -\nabla \rho \approx -(\rho(R_2) - \rho(R_1))/(R_2 - R_1)$. This current shrinks the domain in a time $t(R_1(0))$, whose determination is the final goal of this section.

The density profile $\rho(r, t)$ in the region between the droplet of radius R_1 and the reservoir of radius R_2 should be found by solving the diffusion equation, $\partial_t \rho = D\nabla^2 \rho$, with the earlier boundary conditions. However, we can apply here a sort of Born–Oppenheimer approximation, where domains play the role of atomic nuclei and diffusing particles that of electrons. In fact, we can decouple their dynamics, because diffusive motion is much faster than domain wall motion. More precisely, diffusing particles relax to a stationary density profile in a time $\tau_0 \approx R^2$, where $R = R_2 - R_1$ is the linear size of the region where diffusion takes place. At the end of this section, we will prove self-consistently that during time τ_0 the boundaries move by a quantity of order $1 \ll R$.

In the approximation discussed earlier, we can solve the stationary diffusion equation, that is, the Laplace equation, $\nabla^2 \rho = 0$, and determine the steady profile $\rho(r, R_1, R_2)$. From that we can derive the current of particles that attach to/detach from boundaries and, at the end, their time evolution $R_{1,2}(t)$. Using the obvious spherical symmetry of the problem, the Laplace equation reads

$$\rho'' + \frac{d-1}{r}\rho' = 0, \tag{8.87}$$

where r is the radius in spherical coordinates. Boundary conditions, qualitatively anticipated here above, are derived in Appendix V. The result is

$$\begin{aligned} \rho(R_1) &= \rho_0 + \frac{c}{R_1} \\ \rho(R_2) &= \rho_0 - \frac{c}{R_2}, \end{aligned} \tag{8.88}$$

with $c > 0$.

Once we know $\rho(r)$, we can determine the current in the vicinity of the inner domain edge,

$$j = -D\rho'(r)|_{R_1} \tag{8.89}$$

and finally get the evolution equation for R_1 via the relation $dV/dt = -jS$, where V and S are, respectively, the volume and the surface of the domain of radius R_1. Since

$$\frac{dV}{dt} = \frac{dV}{dR_1}\frac{dR_1}{dt} = S\dot{R}_1,\tag{8.90}$$

we find $-jS = S\dot{R}_1$, that is,

$$\frac{dR_1}{dt} = D\rho'(R_1).\tag{8.91}$$

Equation (8.91), supplemented by the initial condition $R_1(0) = L$, allows us to determine the time $t(L)$ required to close the inner domain. As a matter of fact, we do not really need to determine the exact expression of $t(L)$; it is enough to obtain the functional dependence of t on L. Since we expect a power-law dependence, our task is equivalent to determining how $t(L)$ rescales when $R_1(0)$ rescales: If $t(L) \simeq L^{1/n}$, where n is the sought-after coarsening exponent, then

$$t(R_1(0) = L) = t(R_1(0) = 1)L^{1/n}.\tag{8.92}$$

The determination of the prefactor $t(R_1(0) = 1)$ requires us to solve the full problem, but the determination of $1/n$ can be done with a simple rescaling of variables. If $r = Lx$ and $R_{1,2} = Lx_{1,2}$, the boundary conditions (8.88) suggest the rescaling of the density[17]

$$\rho(r,t;L) = \rho_0 + \frac{1}{L}p(x,t).\tag{8.93}$$

Using the new variables p and x, the problem is rewritten as

$$p_{xx} + \frac{d-1}{x}p_x = 0$$
$$p(x_1) = +\frac{c}{x_1}\tag{8.94}$$
$$p(x_2) = -\frac{c}{x_2}.$$

The central point is that Eq. (8.94) do not depend on L anymore. Therefore, Eq. (8.91) can be rewritten with the new variables as

$$L\frac{dx_1}{dt} = \frac{D}{L^2}p_x(x_1).\tag{8.95}$$

The last step is to rescale time so as to get rid of L in this equation as well. It is straightforward to show that the correct rescaling is $t = L^3\tau$, from which we get

$$\frac{dx_1}{d\tau} = Dp_x(x_1).\tag{8.96}$$

Equations (8.94) and (8.96) allow us to determine $t(R_1(0) = 1)$, while the earlier scaling tells us that when $R_1(0) \to LR_1(0)$, $t(R_1(0)) \to L^3 t(R_1(0))$. In conclusion,

$$t(L) = t(1)L^3,\tag{8.97}$$

---

[17] Here (and only here) we make explicit that the density profile of the gas phase also depends on the initial condition, $R_1(t = 0) = L$.

which means that the closure time of a domain of initial radius L is L^3 times larger than the closure time of a domain of unit initial radius. Therefore, inverting earlier relation, the coarsening exponent is found to be $n = \frac{1}{3}$, regardless of the dimension d.

Since a time of order L^3 is required to close an inner domain, whose size is of order L, on a time scale of order L^2, the radius of the inner domain varies by a quantity of order one. This is the result we have used at the beginning of this section to justify the approximation of replacing the diffusion equation with the Laplace equation.

In one dimension, the curvature effect on the particle density in proximity to a cluster is not effective; the earlier treatment would give a constant density profile in the region $R_1 < r < R_2$ and no net current between the cluster and the reservoir. For this reason the one-dimensional case requires a different treatment, as it will be seen in Section 8.5.2.

The specific problem we have studied in this section, the exchange of particles between a central cluster and the reservoir surrounding it, was justified by its spherical symmetry and by the scaling hypothesis. In general, we should have a distribution of (finite) clusters that exchange particles, with a net current directed from clusters smaller than average toward clusters larger than average. This process, called Ostwald ripening, leads to the disappearance of the smallest clusters and to the increase in time of the average size $\overline{R}(t)$ of clusters, according to the same law we have just found, $\overline{R}(t) \simeq t^{1/3}$. It is also possible to find the limit of the self-similar distribution of cluster sizes, but we do not discuss it here.

8.5.2 The Discrete Picture in $d = 1$

We now have "real" particles that can diffuse, attach, and detach. The hopping of a particle in an empty site corresponds, in spin language, to exchange an up spin with a neighboring down spin. In one spatial dimension, different types of processes (see Fig. 8.19) can be listed according to the variation $\Delta\bar{z}$ in the number of nearest neighbors of the hopping particle, that is, to the change of the energy. Since hopping is restricted to nearest neighbors, a hopping particle cannot be enclosed by two particles, neither before nor after hopping. So, the possible processes are: (a) particle/hole diffusion ($\Delta\bar{z} = 0$). If $\bar{z} = 0$ we speak of particle diffusion, if $\bar{z} = 1$ we speak of hole diffusion. This process does not change energy; $\Delta E = 0$. (b) Particle attachment ($\Delta\bar{z} = 1$): The particle gains one neighbor after hopping, lowering the energy; $\Delta E < 0$. (c) Particle detachment ($\Delta\bar{z} = -1$): The particle loses one neighbor after hopping, raising the energy; $\Delta E > 0$.

Therefore, there are similarities and differences between lattice gas dynamics and domain wall dynamics. Similarities refer to single particle dynamics, which is free diffusion in both cases. Differences refer to the "coupling" between particles: In the lattice gas we have real particles with an attractive nearest-neighbor interaction; in the domain wall case we have fictitious particles, which annihilate when they meet and which are created in pairs via a thermally activated process.[18]

Let us now review the one-dimensional lattice gas dynamics, making reference to Fig. 8.19. At $T = 0$, only diffusion, process (a), and attachment, process (b), are permitted, but

---

[18] There is also another difference: The lattice gas model is valid in any dimension d, while domain walls can be represented as fictitious particles only in $d = 1$.

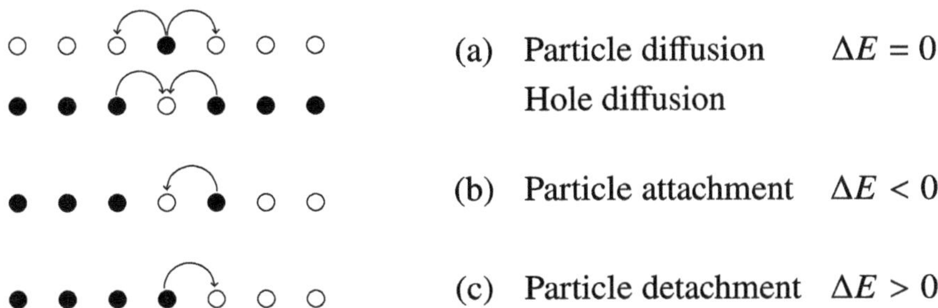

Fig. 8.19 Conserved, spin-exchange (Kawasaki) dynamics. We plot the possible transitions using the lattice gas picture. Solid circles are particles (up spins), open circles are holes (down spins). Arrows depict possible moves of a single particle.

they are not sufficient to provide a meaningful dynamics. In fact, zero temperature dynamics stops as soon as all particles are attached to other particles and all holes are attached to other holes; in other words, when domains have a minimal length of two. Therefore, if we want to study coarsening in this conserved model, we have to switch the temperature on, being aware that phase separation will not last forever: It will stop at the equilibrium correlation length.

Because of conservation, we need at least two domains of particles that can exchange matter. The simplest configuration giving rise to coarsening is plotted in Fig. 8.20, where two particle domains of size L are separated by two hole domains of the same size, with periodic boundary conditions. It is apparent that the system cannot evolve at $T = 0$, because dynamics requires the detachment of particles from domain edges, a process that costs energy. The probability of such a process, $\exp(-2J/T)$, sets a time scale $\tau_0 \approx \exp(2J/T)$, which is extremely long at low temperature. For a fixed size L of the domains, we can choose T such that τ_0 is much longer than any other, nonthermally activated process, as particle diffusion (which scales as L^2) or particle attachment (which is of order 1). Once a particle has detached by one of the four edges (A, B, C, D), there are two possible outcomes: Either the particle attaches (with probability p) to the next domain or it goes back (with probability $1 - p$) to the starting site.

Dynamics can be summarized as follows. Every time τ_0 a particle detaches from an edge, with probability p, it attaches to the other domain. Therefore, every time $\Delta t = \tau_0/p$, $L_1 \rightarrow L_1 - 1$ and $L_2 \rightarrow L_2 + 1$ (if the particle passes from the first to the second domain) or $L_1 \rightarrow L_1 + 1$ and $L_2 \rightarrow L_2 - 1$ (if the particle passes from the second to the first domain). Therefore, the length of each block of particles performs a random walk, and after a time $t(L) \approx L^2 \Delta t$, one of the two intervals has disappeared, leading to coarsening. We still have to calculate Δt, that is, the probability $p(L)$ that a detached particle attaches to the neighboring interval, rather than going back to the same interval. This is equivalent to saying that the detached particle attains a distance L before coming back to the starting site.

The probability $p(L)$ can be easily evaluated as follows (see Fig. 8.21). Suppose that particle in site 1 detaches from the left domain (top). In order to join the other domain, namely, to reach site L, it must first reach site $L/2$ (bottom). Once the particle has attained the

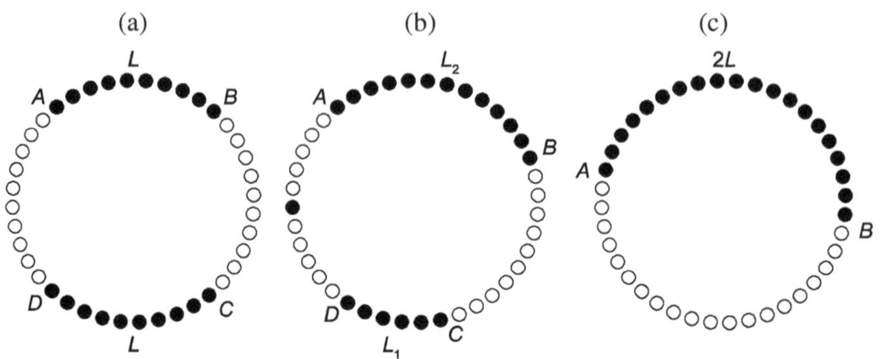

Fig. 8.20   (a) Two domains of particles of initial size L are separated by a distance L, with periodic boundary conditions. (b) Particles can detach at domain edges A, B, C, D, diffuse and either attach to the other domain or reattach to the same domain. (c) After a time $t(L)$, one domain has evaporated and all particles are gathered in a domain of size $2L$.

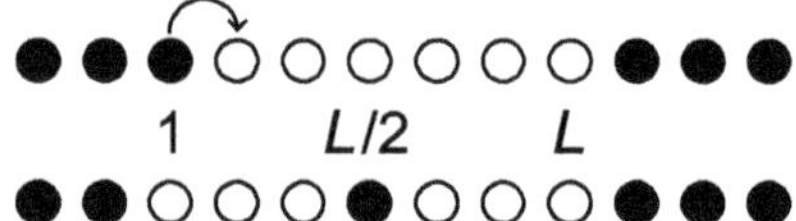

Fig. 8.21   Two domains are separated by holes. Domains can exchange matter if a particle detaches, for example, by the left domain and attains the right domain (site L) before coming back to the original site 1. If the particle is in site $L/2$, the probabilities to attain the starting domain or the arrival domain are the same.

distance $L/2$ from the starting point, which occurs with probability $p(L/2)$, for symmetry reasons, it has the same probability of attaining the starting domain or the other domain. Therefore,

$$p(L) = \frac{p(L/2)}{2}, \tag{8.98}$$

whose solution is $p(L) = c/L$. Since $p(2) = \frac{1}{2}$, $p(L) = 1/L$. In conclusion, $\Delta t = \tau_0/p = \tau_0 L$ and $t(L) \approx L^2 \Delta t \approx \tau_0 L^3$, with τ_0 depending on temperature, not on L. This relation leads to

$$L(t) \approx t^n \qquad \left(n = \tfrac{1}{3}, \quad \text{conserved dynamics}\right), \tag{8.99}$$

as in $d > 1$.

8.6 Domain Size Distribution

The average size of domains, $L(t)$, is a relevant physical quantity, but the full distribution of domain sizes, $n_d(A,t)$, certainly contains much more information. In Fig. 8.22 we reproduce the size distribution of domains in the two-dimensional Ising model, both in the nonconserved (left) and in the conserved (right) case. The main difference between the two distributions is the presence of a maximum at finite size in the conserved case, while in

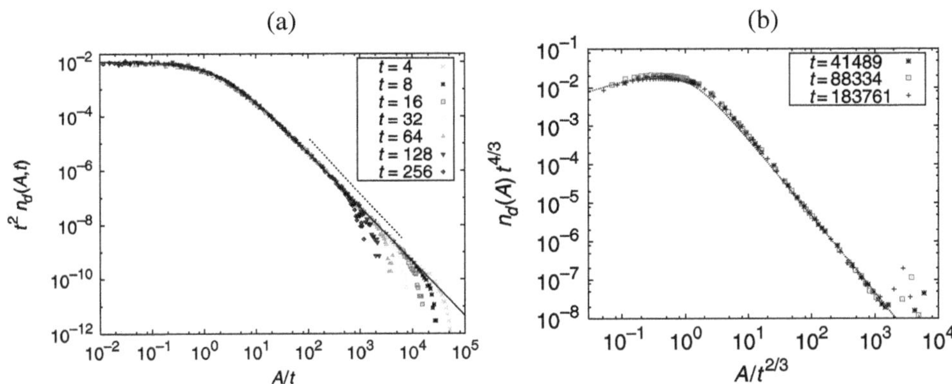

Fig. 8.22 The size distribution of domains in the two-dimensional Ising model, in the nonconserved (a) and in the conserved (b) case. Panel on left reprinted from J. J. Arenzon et al., Exact results for curvature-driven coarsening in two dimensions, *Physical Review Letters*, **98** (2007) 145701. Panel on right reprinted from A. Sicilia et al., Geometry of phase separation, *Physical Review E*, **80** (2009) 031121.

the nonconserved case the function $n_d(A, t)$ decreases monotonically. In other words, the domain pattern is more regular for conserved dynamics. Why is it so?

The explanation must lie in the conservation law, which imposes a constraint on the size of neighboring domains of the same type: Since they exchange particles, the shortening of one domain implies the lengthening of the other. In addition, the conservation law surely has an effect on the size distribution at short times, as shown by the linear stability analysis reported in Section 8.2.1: The comparison of $\sigma(q)$, Eq. (8.32), and $\tilde{\sigma}(q)$, Eq. (8.35), indicates that the latter (conserved case) has a peak for finite q, that is, for finite size.

It is also interesting to note a possible connection between $n_d(A, t)$ and $C(r, t)$, exemplified in $d = 1$ by two limiting cases. In the first case, we consider a perfectly periodic configuration: Positive domains of length L alternate with negative domains of the same length. The distribution $n_d(\ell)$ is a simple Dirac delta, and the correlation function is an oscillating triangle wave. We can expect that destroying a bit of the periodic domain configuration will preserve the oscillating character of the correlation function, but it will make $C(r, t)$ vanish at large distances.

The second, limiting case is opposed to the first one, because it corresponds to a fully disordered state: Each spin is equally likely to be positive or negative. Now a domain has the probability $n_d(\ell) = 1/2^\ell$ (a decreasing function) to have length ℓ, and the correlation function is trivially equal to $C(r) = 1$ if $r = 0$ and $C(r) = 0$ if $r \neq 0$; that is, it does not oscillate. We guess that the different shapes of the correlation function for the nonconserved (Fig. 8.9) and for the conserved (Fig. 8.10) models are related to the different behaviors of $n_d(A, t)$ (see Fig. 8.22): A peaked distribution $n_d(A, t)$ in the conserved case produces an oscillating $C(r, t)$.

More rigorous considerations on domain size can be made for the two-dimensional nonconserved model, using the Allen–Cahn equation (see Appendix U). According to Eq. (U.6), the local velocity of a domain wall is proportional to its curvature, $v = -\frac{\lambda}{2\pi} K$. We have introduced the constant $\frac{\lambda}{2\pi}$, because Eq. (U.6) has been derived from the parameter-free TDGL equation, while the constant is necessary if we want to compare with numerics or experiments. It is now simple to derive the time variation of the area A surrounded by a domain wall, because

$$\frac{dA}{dt} = \oint v\,dl = -\frac{\lambda}{2\pi} \oint K\,dl = -\lambda, \tag{8.100}$$

where we have used the Gauss–Bonnet theorem to evaluate the line integral of the curvature. Therefore, the area A decreases linearly in time and the distribution of areas is a function of $(A + \lambda t)$, because

$$n_a(A, t) = n_a(A + \lambda t, 0). \tag{8.101}$$

It is important to stress that the area A is defined as the region surrounded by any closed domain wall, which in turn may contain other domain walls. Therefore, the distribution n_a is not equal to the distribution of domains, $n_d(A, t)$, which only counts areas of equal order parameter. However, n_d and n_a are clearly related, and there is numerical evidence they are almost identical. The reason to focus on n_a is that we can derive analytically Eq. (8.101), while this is not possible for $n_d(A, t)$. The final, missing piece is the initial condition, $n_a(A, 0)$. We state without further detail that for a quenching from $T_i \to \infty$, the area distribution can be approximated by the expression $n_a(A, 0) = c/A^2$, so that

$$n_a(A, t) = \frac{c}{(A + \lambda t)^2}, \tag{8.102}$$

where c is a known constant, $c = (4\pi\sqrt{3})^{-1} \simeq 0.046$. Figure 8.22 (left) is in very good agreement with this function.

Analytical treatments of the size distribution of domains for conserved dynamics (not reported here) support the presence of a maximum and show that the distribution depends on the value of the order parameter: In the case of a strongly asymmetric quench, the minority phase is diluted in the majority phase with a droplet morphology and the size distribution has a cut-off.

8.7 The Coarsening Law in Nonscalar Systems

So far we have discussed only Ising-type models, that is, models with a scalar order parameter, while in this section we go beyond this limitation and consider $O(n)$ models. These models are characterized by an n-component, vector order parameter ϕ, which can freely rotate in an n-dimensional space and which lives in a d-dimensional space. We adopt a continuum description, making use of the Ginzburg–Landau free energy, suitably generalized to be applied to nonscalar models. In practice, since dynamics is purely dissipative, the continuum description of the ordering process for $O(n)$ models is a generalization of the TDGL (8.28) and CH (8.29) equations.

The starting point is the Ginzburg–Landau free energy for the vectorial order parameter, $\phi(\mathbf{x}, t)$,

$$\mathcal{F}[\phi] = \int \left[\frac{1}{2}(\nabla\phi)^2 + V(\phi) \right] d\mathbf{x} \equiv \int f\,d\mathbf{x}, \tag{8.103}$$

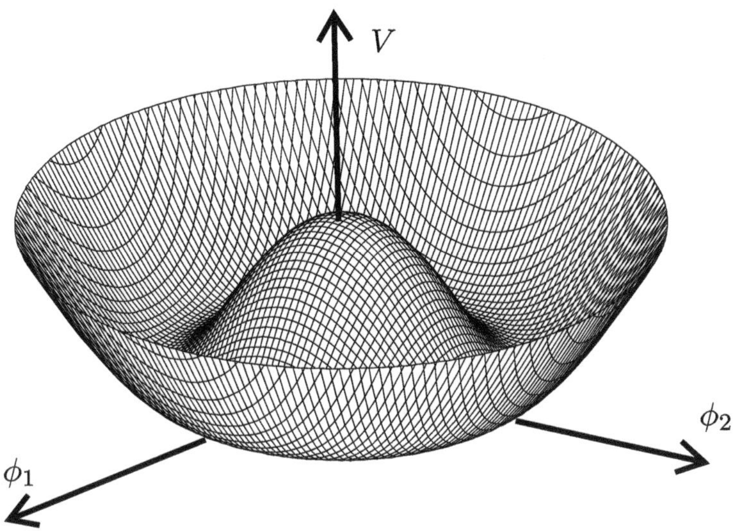

Fig. 8.23   Plot of the "Mexican hat" potential for $n = 2$, $V(\boldsymbol{\phi}) = V_0(\boldsymbol{\phi}^2 - 1)^2$.

where

$$(\nabla\boldsymbol{\phi})^2 = \sum_{i=1}^{d}\sum_{j=1}^{n}(\partial_i\phi_j)^2 \tag{8.104}$$

with $\partial_i \equiv \partial/\partial x_i$ and $V(\boldsymbol{\phi}) = V_0(\boldsymbol{\phi}^2 - 1)^2$ is the usual, rotationally invariant, "Mexican hat" potential[19]; see Fig. 8.23.

As for the scalar models, dynamics is obtained by the equation[20] $\partial_t\boldsymbol{\phi} = -\delta\mathcal{F}/\delta\boldsymbol{\phi}$ for a nonconserved order parameter and by the equation $\partial_t\boldsymbol{\phi} = \nabla^2\delta\mathcal{F}/\delta\boldsymbol{\phi}$ for a conserved one. Here we focus on the nonconserved case, whose coarsening exponent can be derived without too much difficulty. A few words on the conserved models will be given at the end. The notation of vectorial functional derivative, $\delta\mathcal{F}/\delta\boldsymbol{\phi}$, means a vector, whose components are

$$\frac{\delta\mathcal{F}}{\delta\phi_i} = \frac{\partial f}{\partial\phi_i} - \sum_j \partial_j \frac{\partial f}{\partial(\partial_j\phi_i)}. \tag{8.105}$$

With this expression in mind, we find

$$\frac{\delta\mathcal{F}}{\delta\phi_i} = -\left(\nabla^2\phi_i - \frac{\partial V}{\partial\phi_i}\right), \tag{8.106}$$

so the equation for the $O(n)$ nonconserved model is

$$\frac{\partial\boldsymbol{\phi}}{\partial t} = \nabla^2\boldsymbol{\phi} - \frac{\partial V}{\partial\boldsymbol{\phi}}. \tag{8.107}$$

The corresponding conserved model is obtained by applying $-\nabla^2$ to the right-hand side.

---

[19] We write it in this form so that the free energy of the ground state ($\boldsymbol{\phi}^2 = 1$, $\nabla\boldsymbol{\phi} = 0$) vanishes.

[20] The mobility M (see Section 8.2.1) has been absorbed in the time t, to make these equations parameter-free.

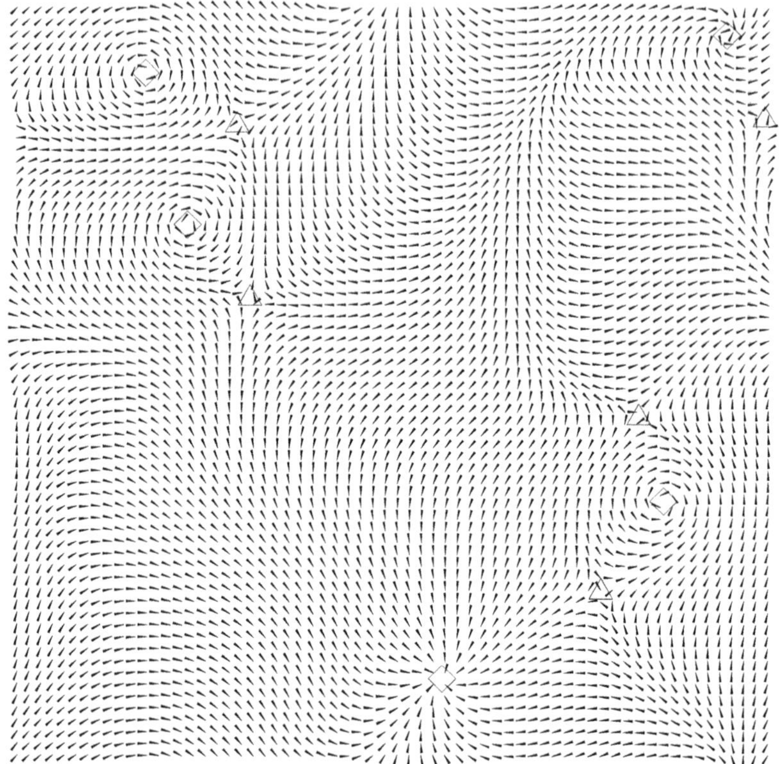

Fig. 8.24 Snapshot of the ordering dynamics of the XY model. The squares and the triangles are the core regions of vortices and anti-vortices, respectively, where the magnitude of the order parameter is close to zero. For graphical reasons, not all the lattice sites are shown. From H. Qian and G. F. Mazenko, Vortex Dynamics in a Coarsening Two-dimensional XY Model, *Physical Review E*, **68** (2003) 021109.

Now that we have written the equations defining phase-ordering dynamics of $O(n)$ models, what are the relevant features making it different from that of the scalar model? In Fig. 8.24 we show a snapshot of the two-dimensional XY-model (i.e., $d = n = 2$) after a quench from the disordered to the ordered phase. This figure reveals ordered regions with $(\phi)^2 \simeq 1$, but with different orientations, separated by zones where the order parameter connects different minima of $V(\phi)$. We might think that the rotational symmetry should allow passage from one minimum to the other with continuity, but this is not the case if topological defects are stable, which is the case in which we are interested.

In Fig. 8.25 we plot a sketch of topological defects for $n = 1$ (the domain wall, or kink, which should be familiar), for $n = 2$ (the vortex), and for $n = 3$ (the monopole). It is worth stressing that each defect is strictly linked to a given value of n: The vortex cannot exist for $n = 1$ (which is fairly obvious), but if spins were allowed to have a third component ($n \geq 3$), they could rotate continuously and become parallel, so as to make the defect disappear. The same would happen with a kink for $n > 1$. Since the dimension of a topological defect

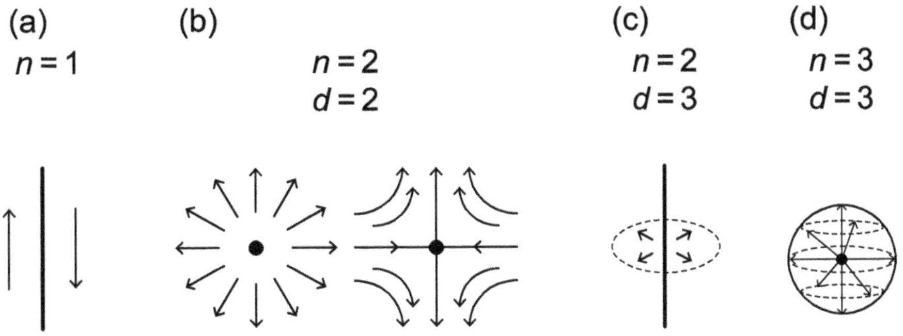

Fig. 8.25 Types of topological defects in the $O(n)$ model: (a) domain wall ($n = 1$), (b) vortex and antivortex ($n = 2$), (c) vortex ($n = 2$) in $d = 3$, (d) monopole or "hedgehog" ($n = 3$).

is $(d - n)$, a defect can only exist if $d \geq n$. If d is strictly larger than n, the defect is translationally invariant in the remaining $(d - n)$ dimensions: See the case of the vortex in three dimensions in Fig. 8.25 or think of a domain wall in two or three dimensions. The evaluation of the energy and of the dynamical properties, for example, the friction, of a defect are the required steps to determine the coarsening laws. We are going to accomplish this task.

From a formal point of view, a defect is a time-independent, radially symmetric function,

$$\boldsymbol{\phi}_{\text{def}}(\mathbf{r}) = \hat{\mathbf{r}} f(r), \tag{8.108}$$

with $r = |\mathbf{r}|$ and $\hat{\mathbf{r}} = \mathbf{r}/r$, which is the solution of Eq. (8.107),

$$\nabla^2 \boldsymbol{\phi}_{\text{def}}(\mathbf{r}) - \frac{\partial V}{\partial \boldsymbol{\phi}_{\text{def}}(\mathbf{r})} = 0, \tag{8.109}$$

with the boundary conditions $f(0) = 0$ and $f(r = \infty) = 1$. These conditions correspond to saying that the order parameter $\boldsymbol{\phi}$ must vanish in the center of the defect and must tend to a minimum of the potential ($|\boldsymbol{\phi}| = 1$) for large r.

Using Eq. (8.108), we can write

$$\partial_i \left(\hat{r}_j f(r) \right) = \frac{\delta_{ij}}{r} f + \frac{r_i r_j}{r} \left(\frac{f}{r} \right)', \tag{8.110}$$

$$\partial_{ii} \left(\hat{r}_j f(r) \right) = \delta_{ij} \left(\frac{f}{r} \right)' \frac{r_i}{r} + \frac{r_j}{r} \left(\frac{f}{r} \right)' + \frac{r_i \delta_{ij}}{r} \left(\frac{f}{r} \right)' + \frac{r_i^2 r_j}{r} \left(\frac{1}{r} \left(\frac{f}{r} \right)' \right)' \tag{8.111}$$

where the prime means the derivative with respect to r. From Eq. (8.111), summing over i, we find the expression

$$\nabla^2 \boldsymbol{\phi}_{\text{def}}(\mathbf{r}) = \hat{\mathbf{r}} \left[f''(r) + \frac{n-1}{r} f'(r) - \frac{n-1}{r^2} f(r) \right], \tag{8.112}$$

while it is easier to write

$$\frac{\partial V}{\partial \boldsymbol{\phi}_{\text{def}}(\mathbf{r})} = \hat{\mathbf{r}} V'(f). \tag{8.113}$$

Combining Eqs. (8.112) and (8.113), we conclude that the profile $f(r)$ must satisfy the ordinary differential equation,

$$f''(r) + \frac{n-1}{r} f'(r) - \frac{n-1}{r^2} f(r) - V'(f) = 0. \tag{8.114}$$

The equation determining the kink shape for $n = 1$ is $f''(r) - V'(f) = 0$, whose solution (see Eq. (8.58)) is equal to the hyperbolic tangent: It tends exponentially to the limiting value $f = 1$ for $r \to \infty$. This means that $1 - f(r)$ is exponentially small for large r and the defect is localized, while this is not true for $n > 1$, as we are going to prove. If $f(r) = 1 - \epsilon(r)$, at the lowest order in ϵ we find

$$-\epsilon''(r) - \frac{n-1}{r} \epsilon'(r) - \frac{n-1}{r^2} + V''(1)\epsilon(r) = 0, \tag{8.115}$$

where we have used that $V'(1) = 0$. At the leading order, we obtain

$$\epsilon(r) = \frac{n-1}{V''(1)} \frac{1}{r^2}, \tag{8.116}$$

proving that the profile $f(r)$ attains the asymptotic value with a power law of exponent two.

The next step is to determine the energy of the defect, because dynamics is driven by the minimization of the energy of the system, which is concentrated in the defects. Since an isolated, single defect is translationally invariant in the $(d - n)$ directions orthogonal to the n-dimensional plane of the defect itself, let us determine its energy per unit length in such orthogonal directions (or, equivalently, let us assume that $d = n$). First (see Eq. (8.103)), we should evaluate $(\nabla \phi_{\text{def}})^2$. Taking the square of Eq. (8.110), we find

$$(\nabla \phi_{\text{def}})^2 = \sum_{ij} (\partial_i \phi_j)^2 = \frac{n-1}{r^2} f^2(r) + |\nabla f|^2, \tag{8.117}$$

so that, integrating Eq. (8.103) over angular variables, we obtain

$$E^n_{\text{def}} = S_n \int r^{n-1} \left[\frac{n-1}{2r^2} f^2(r) + \frac{1}{2} |\nabla f|^2 + V(f) \right] dr, \tag{8.118}$$

where S_n is the surface of an n-dimensional sphere of unit radius.

Using the asymptotic form, $f(r) = 1 - \epsilon(r)$, with $\epsilon \approx r^{-2}$, we understand that the dominant term in square brackets here above is the first one, f^2/r^2, which decays as $1/r^2$ and makes the integral divergent for $n > 1$. So, what is the correct length scale we should use to evaluate the energy of the defect? Because of the scaling hypothesis, there is just one length scale, the size $L(t)$ of ordered regions, which means

$$E^n_{\text{def}} = \begin{cases} \text{const}, & n = 1 \\ \ln L, & n = 2 \\ L^{n-2}, & n > 2, \end{cases} \tag{8.119}$$

where the superscript n indicates it is the defect energy in the n-dimensional plane of the defect. If $d > n$, the defect is translationally invariant in the remaining $(d - n)$ spatial directions and the total energy of the defect is obtained multiplying E^n_{def} by its linear size in such orthogonal directions elevated to the power $(d - n)$. Invoking scaling again, such linear size must be L and we finally obtain

$$E_{\text{def}} = E_{\text{def}}^n \times L^{d-n} = \begin{cases} L^{d-1}, & n = 1 \\ L^{d-2} \ln L, & n = 2 \\ L^{d-2}, & n > 2, \end{cases} \tag{8.120}$$

where we always assume that $d \geq n$.

Dynamics is driven by the fact that E_{def} depends on L and the driving force is $-dE_{\text{def}}/dL$. Since dynamics is dissipative, we may think that the closing speed of a domain, $v = dL/dt$, and the driving force F are related by

$$F(L) = -\eta(L)\frac{dL}{dt}, \tag{8.121}$$

where we have indicated that the friction η (yet to be determined) may itself be dependent on the size L of the domain. A warning is in order here: The force $F(L)$ appearing in Eq. (8.121) is the force per unit length/surface of the $(d - n)$ directions, that is,

$$F(L) = \frac{1}{L^{d-n}}\left(-\frac{dE_{\text{def}}}{dL}\right) = \begin{cases} 0, & d = n = 1 \\ L^{-1}, & d > n = 1 \\ L^{-1}, & d = n = 2 \\ L^{-1} \ln L, & d > n = 2 \\ L^{n-3}, & d \geq n > 2. \end{cases} \tag{8.122}$$

Before evaluating $\eta(L)$, we may focus on the case $n = 1$. For $d = 1$, $F(L) = 0$ and we obtain $dL/dt = 0$, which implies no coarsening. In fact, we know that for $n = d = 1$ it is necessary to take into account the exact profile of the kink, which has an exponential tail, while we have approximated the kink to a region of finite size, neglecting such exponentially small corrections. If they are taken into account, as we did in Section 8.4, we would get a logarithmically slow coarsening, $L(t) \sim \ln t$, rather than no coarsening at all. For $d > 1$, if the friction is assumed to be constant, we find $dL/dt = -1/L$, which gives $L(t) \sim t^{1/2}$, that is, the result we already found in Section 8.4. This suggests that the friction should be constant for $n = 1$. As we are now going to show, this is no longer true for $n > 1$.

The friction $\eta(L)$ is now evaluated with regard to the scheme where a defect moves with constant velocity v_0 in a given direction (e.g., x_1), and we ask how its energy varies with time. As the force $F(L)$ appearing in Eq. (8.121) is the force per unit length/surface perpendicular to the n-dimensional space, the same "geometry" should be used to evaluate the friction $\eta(L)$. For this reason, the integrals below are restricted to such a space, $\mathbf{r} = (x_1, \ldots, x_n)$.

We can now evaluate the desired quantity,

$$\frac{dE_{\text{def}}}{dt} = \int d\mathbf{r} \left.\frac{\delta \mathcal{F}}{\delta \phi}\right|_{\text{def}} \cdot \frac{\partial \phi_{\text{def}}}{\partial t} \tag{8.123}$$

$$= -\int d\mathbf{r} \left(\frac{\partial \phi_{\text{def}}}{\partial t}\right)^2 \tag{8.124}$$

$$= -v_0^2 \int d\mathbf{r} \left(\frac{\partial \phi_{\text{def}}}{\partial x_1}\right)^2 \tag{8.125}$$

$$\equiv -\eta(L)v_0^2, \tag{8.126}$$

where we have assumed that the defect moves rigidly, $\phi_{\text{def}}(\mathbf{r}, t) = \phi_{\text{def}}(x_1 - v_0 t, x_2, \dots)$. The result $d\mathcal{F}_{\text{def}}/dt = -\eta(L)v_0^2$, which defines the friction $\eta(L)$, should not be a surprise, because it is the standard expression for the dissipated power by a particle moving at velocity v_0 and subject to the friction force $-\eta v_0$.

The last step is therefore to evaluate $\eta(L)$ for different values of n. Since

$$\eta(L) \equiv \int d\mathbf{r} \left(\frac{\partial \phi_{\text{def}}}{\partial x_1}\right)^2 = \frac{1}{n} \int d\mathbf{r}(\nabla \phi_{\text{def}})^2, \tag{8.127}$$

using Eq. (8.118), we observe that $\eta(L)$ has the same L-dependence as E_{def}^n, given by Eq. (8.119). Finally, using Eqs. (8.121) and (8.122), we obtain

$$\frac{dL}{dt} = -\frac{1}{\eta(L)}F(L) = \begin{cases} 0, & d = n = 1 \\ -L^{-1}, & d > n = 1 \\ -(L \ln L)^{-1}, & d = n = 2 \\ -L^{-1}, & d > n = 2 \\ -L^{-1}, & d \geq n > 2. \end{cases} \tag{8.128}$$

Therefore, apart from the (already commented on) special case $n = d = 1$ and a logarithmic correction for $n = d = 2$, we always have $dL/dt = -1/L$, whence $L(t) \sim t^{1/2}$. This result can be understood as follows. If we rewrite Eq. (8.128) as

$$\frac{dL}{dt} = -\frac{1}{\eta(L)} \frac{1}{L^{d-n}} \frac{d}{dL}(L^{d-n} E_{\text{def}}^n), \tag{8.129}$$

the fact that $\eta(L)$ and $E_{\text{def}}^n(L)$ scale alike implies $dL/dt \sim -1/L$, where the right-hand term $1/L$ comes from the derivative, because the involved quantities are powers of L. We can summarize the coarsening laws for the nonconserved case as

$$L(t) \sim \begin{cases} \ln t \quad \text{versus} \quad t^{1/2}, & n = d = 1 \\ \left(\dfrac{t}{\ln t}\right)^{1/2}, & n = d = 2 \\ t^{1/2}, & \text{otherwise.} \end{cases} \tag{8.130}$$

For $n = d = 1$ we have taken into account the correct results discussed in Section 8.4, distinguishing between deterministic and noisy coarsening. As for the logarithmic

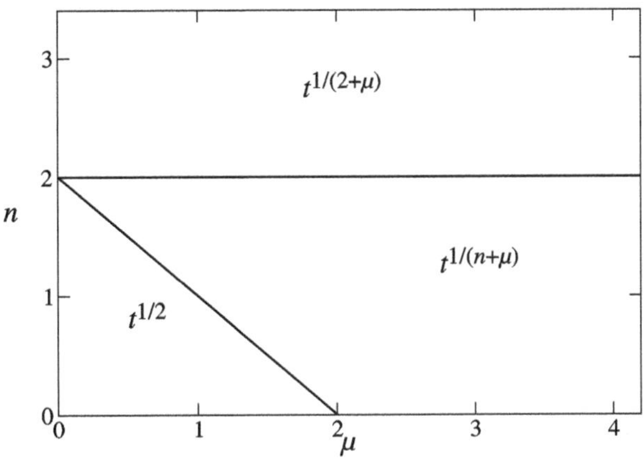

Fig. 8.26 Coarsening laws for purely dissipative dynamics, as a function of the dimension n of the order parameter and of the conservation law, represented by μ; see Eq. (8.131). Along the solid lines there are logarithmic corrections, which are related to scaling violations.

correction for $n = d = 2$, it is due to a violation of the scaling, reminiscent of the ordered Kosterlitz–Thouless phase occurring in two dimensions in the XY- model.

For completeness, in Fig. 8.26 we graphically report the results, due to the British Alan Bray, for the coarsening laws for nonconserved and conserved models. The conservation law is represented by the parameter μ, which appears in the general equation[21]

$$\frac{\partial \phi}{\partial t} = (-\nabla^2)^{\mu/2} \left[\nabla^2 \phi - \frac{\partial V}{\partial \phi} \right]. \tag{8.131}$$

The (nonconserved) Ginzburg–Landau equation and the (conserved) Cahn–Hilliard equation correspond to $\mu = 0$ and $\mu = 2$, respectively. Larger values of μ correspond to higher-order conservation rules, which we do not discuss here. However, there are two major remarks to be made with respect to the results shown in Fig. 8.26. The first, qualitative observation is that as soon as $\mu \neq 0$, the coarsening law depends on n. The second, more quantitative observation is that, except for the triangle covering the region of small μ and n, the coarsening law has the form $L(t) \sim t^{1/(p+\mu)}$, where p is some integer. Such μ-dependence simply corresponds to replacing the operator $(-\nabla^2)^{\mu/2}$ with the dimensional factor $1/L^\mu$, which would simply rescale t. Nonetheless, this simple dimensional analysis does not work for $n = 1, \mu = 0$; this is the reason why for Ising-like systems, when passing from the nonconserved ($\mu = 0$) to the conserved ($\mu = 2$) case, we pass from $t^{1/2}$ to $t^{1/3}$ and not to $t^{1/4}$.

We conclude this part on nonscalar systems by remarking that a value $n > 1$ also modifies the Porod law, because the argument leading to Eq. (8.50) no longer applies. We limit ourselves to observe that the dimension of a topological defect is $d - n$ and Eq. (8.51) now

---

[21] If μ is not an even integer, we have a fractional partial differential equation, which we have already encountered in Section 3.5.1.

writes

$$S(\mathbf{q}, t) \simeq \frac{1}{L^n q^{d+n}}. \tag{8.132}$$

8.8 The Classical Nucleation Theory

We have seen how a strongly asymmetric quench, beyond the spinodal decomposition line, leads to a state that is metastable, rather than linearly unstable. Therefore, the mechanism that triggers phase separation is different from what we have discussed until now, because an infinitesimal perturbation is no longer effective to destabilize the uniform state. What we need is a localized perturbation of finite amplitude and a thermally activated process, which requires an energy barrier to be overcome. This process is called *nucleation*, because it goes through the formation of nuclei of the new phase.

We can provide a qualitative picture of the nucleation process by making reference to a system described by a free energy, which has two equivalent minima if parameters are suitably tuned. For instance, in a ferromagnet, such a condition corresponds to $T < T_c$ and $H = 0$, while in a fluid it corresponds to the critical isochore $P_{iso}(T)$; see Fig. 8.1. If $H \neq 0$, the two minima are not equivalent, and at equilibrium the system is magnetized in the direction of the field. Analogously, if $T < T_c$ and $P < P_{iso}(T)$ ($P > P_{iso}(T)$), the fluid is entirely in the gas (liquid) phase.

Let us now suppose the system is in phase 1 (negative magnetization or gas phase). If we reverse the magnetic field or increase the pressure at constant T, the equilibrium state is characterized by a different phase 2 (positive magnetization or liquid phase), while phase 1 is now metastable. How does the transition between the metastable phase 1 and the stable phase 2 occur? Since phase 1 is metastable, such a transition cannot occur via a perturbation of vanishing amplitude; we need a perturbation of finite amplitude, which must necessarily be localized in space, otherwise its energetic cost would be infinite in a macroscopic sample.

The mechanism is easily exemplified by the Ising model, where, according to the above picture, we should imagine having a negatively magnetized state in the presence of a positive field. If we reverse an ensemble of spins, we gain in terms of bulk free energy (reversed spins are now aligned with the field), but we lose in terms of surface free energy, because interface spins are antiparallel. Overall, we lose energy for a small cluster of reversed spins and we gain energy if the cluster is large enough. Let's see it in detail.

If $\delta g = g_1 - g_2 > 0$ is the difference in the Gibbs free energy density between the two phases and σ is the surface tension, the energy of a three-dimensional cluster of phase 2 and radius r within a sea of phase 1 is[22]

$$\Delta G(r) = -\frac{4}{3}\pi r^3 \delta g + 4\pi r^2 \sigma. \tag{8.133}$$

This function initially grows with r, it has a maximum for $r = r^* = 2\sigma/\delta g$ and it is negative for $r > \frac{3}{2}r^*$. The size r^* defines the *critical* nucleus, because for $r > r^*$ the growth of the nucleus implies a lowering of its energy, so that it can grow spontaneously.

The quantity

$$\Delta G(r^*) = \frac{16\pi}{3}\frac{\sigma^3}{(\delta g)^2} \tag{8.134}$$

represents the energy barrier to be overcome to pass from phase 1 to phase 2, and the probability p to form a critical droplet is given by the Arrhenius formula (2.257),

$$p \sim \exp(-\Delta G(r^*)/T). \tag{8.135}$$

If we make reference to an Ising model with ferromagnetic exchange coupling J in the presence of a field H, the loss (per unit surface) in interface free energy is proportional to the former, $\sigma \sim J$, while the gain (per unit volume) in bulk free energy is proportional to the latter, $\delta g \sim H$. Therefore, the radius of the critical nucleus is $r^* \sim (J/H)$, the energy barrier is $\Delta G(r^*) \sim J^3/H^2$, and $p \sim \exp(-aJ^3/H^2)$, where a is a numerical prefactor.

The nucleation rate $\tilde{I}$, that is, the number of critical nuclei formed per unit time and volume in the metastable phase, is proportional to p. In Section 8.8.1, we describe the classical Becker–Döring theory to determine $\tilde{I}$ in the conserved case, using a kinetic description. Here below we answer a question concerning the late-time dynamics of the nonconserved case: If $\tilde{I}$ is the nucleation rate of the stable phase, what is the fraction of the system in the stable phase at time t? In fact it is easier to evaluate what is the probability P that a given point has not reversed its magnetization. In poor terms if the magnetization in $\mathbf{x}_0$ is still opposite to the field, it means that no nucleation has occurred until time t, *sufficiently close* to $\mathbf{x}_0$. Since the radius of a magnetic domain parallel to the field grows linearly with velocity v_0 given by Eq. (8.76), only nucleation events occurring at distance smaller than $v_0 t$ might reverse the magnetization in $\mathbf{x}_0$. More precisely, a nucleation event at distance r must have occurred no later than time $t' = t - r/v_0$. We can therefore write

$$P = \prod_{\substack{r < v_0 t \\ 0 < t' < t - r/v_0}} (1 - \tilde{I}\,d\mathbf{r}dt'). \tag{8.136}$$

---

[22] A couple of comments regarding the conserved case are appropriate. First, compact nuclei are reasonable only far from the spinodal line, where surface tension vanishes. Second, the nucleation process has some peculiar features. Using the binary mixture language (appropriate for the experimental results shown in Fig. 8.6), the initial, metastable state is a disordered, homogeneous state made up, for example, of 80% of atoms A and 20% of atoms B. This ratio is conserved and dynamics occurs through diffusion: The formation of a nucleus of the minority phase is accompanied by a local increase of the majority phase in its surrounding. Regarding the difference of the free energy densities, $\delta g = g_1 - g_2$, in the present context g_1 should be understood as the free energy of the homogeneous mixture 80/20 and g_2 as the free energy of the nucleated new phase, which is 100% of atoms B. Similar remarks can be made for the interface energy σ between the two phases.

Taking the log of both terms, we obtain

$$\ln P = -\tilde{I} \int_0^{v_0 t} 4\pi r^3 dr \int_0^{t-r/v_0} dt' \tag{8.137}$$

$$= -\frac{\pi}{3} v_0^3 \tilde{I} t^4, \tag{8.138}$$

and the fraction of the system *not* having reversed its magnetization is given by the Kolmogorov–Avrami formula,

$$P = \exp\left(-\frac{\pi}{3} v_0^3 \tilde{I} t^4\right). \tag{8.139}$$

8.8.1 The Becker–Döring Theory

In this section we pass from purely thermodynamic considerations to a kinetic description, studying the formation of nuclei of the condensed, liquid phase. For simplicity and by tradition, we use a language where "molecules" can attach or detach from clusters. Therefore, rather than using a continuous variable for the radius of the cluster, it is more useful to consider the number ℓ of condensed molecules. Therefore, Eq. (8.133) is replaced by the energy of a cluster of size ℓ, given by[23]

$$\epsilon_\ell = -\delta\mu(\ell - 1) + \gamma(\ell - 1)^{2/3}, \tag{8.140}$$

where we have written $(\ell - 1)$ rather than ℓ in the right-hand side, because we are interested in the excess energy with respect to the gas phase, so it should be $\epsilon_1 = 0$. For $\delta\mu > 0$, ϵ_ℓ has a maximum for

$$\ell^* - 1 = \left(\frac{2\gamma}{3\,\delta\mu}\right)^3; \tag{8.141}$$

see Fig. 8.27.

The number of clusters composed of ℓ particles (from now on, ℓ-clusters) is indicated by $n_\ell(t)$; it can increase because of the condensation process of an atom on an $(\ell - 1)$-cluster and because of the evaporation of an atom from an $(\ell + 1)$-cluster, while condensation/evaporation on/from the ℓ-cluster itself leads to a decrease of n_ℓ. The processes of condensation and evaporation are described by certain kinetic coefficients, respectively, R_ℓ and R'_ℓ, which may depend on the size of the cluster, and the net balance between an ℓ- and an $(\ell + 1)$-cluster defines the current J_ℓ. All processes are graphically depicted in Fig. 8.28 and formally summarized in the following equations:

$$\frac{\partial n_\ell(t)}{\partial t} = J_{\ell-1}(t) - J_\ell(t) \tag{8.142}$$

$$J_\ell(t) = R_\ell n_\ell(t) - R'_{\ell+1} n_{\ell+1}(t). \tag{8.143}$$

Equations such as those above are called rate equations and are applicable to a wide variety of physical processes, whose details are contained in the explicit expressions of the

---

[23] If v_0 is the volume per particle/spin, $\delta\mu = v_0 \delta g$ and $\gamma = (36\pi)^{1/3} v_0^{2/3} \sigma$.

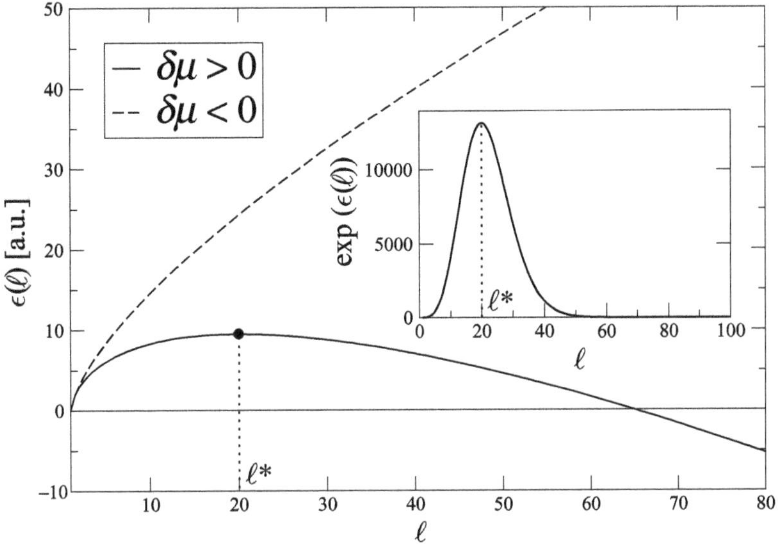

Fig. 8.27   Schematic plot of $\epsilon(\ell)$, Eq. (8.140). If $\delta\mu > 0$, it has a maximum for $\ell = \ell^*$. Inset: The function $\exp(\epsilon(\ell))$. Since $\epsilon(\ell)$ is in arbitrary units, the function is also representative of the inverse of the Boltzmann factor, $\exp(\epsilon(\ell)/T)$.

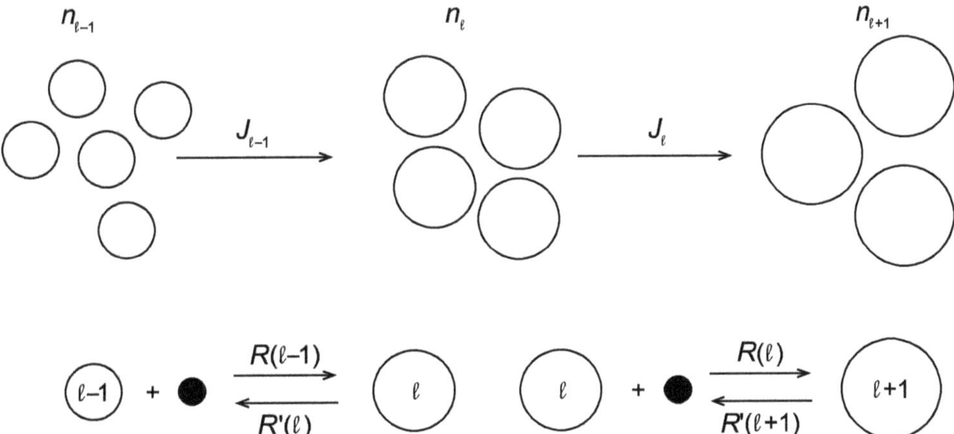

Fig. 8.28   Graphical description of rate equations governing the transitions between cluster (ℓ) and clusters $(\ell \pm 1)$: $J(\ell)$ is the net flux between (ℓ) and $(\ell + 1)$ and includes the condensation process of a molecule on (ℓ) (at rate $R(\ell)$) and the evaporation process of a molecule from $(\ell + 1)$ (at rate $R'(\ell + 1)$).

kinetic coefficients R_ℓ, R'_ℓ. We are not going to make a specific ansatz on them; rather, we assume their values are the same as in equilibrium. This assumption allows us to find a relation between R and R', because at equilibrium a detailed balance holds, that is, $J_\ell = 0$ for any ℓ. This condition is written as

$$R_\ell \langle n_\ell \rangle - R'_{\ell+1} \langle n_{\ell+1} \rangle = 0, \tag{8.144}$$

where $\langle n_\ell \rangle$ is the equilibrium value determined by the relation

$$\langle n_\ell \rangle = \langle n_1 \rangle e^{-\epsilon(\ell)/T}. \tag{8.145}$$

Therefore, kinetic coefficients must satisfy the relation

$$\frac{R_\ell}{R'_{\ell+1}} = \exp\left(\frac{\epsilon_{\ell+1} - \epsilon_\ell}{T}\right), \tag{8.146}$$

with ϵ_ℓ given in Eq. (8.140).

Then, next step is to assume that, after a transient, the time evolution of the densities n_ℓ (see Eqs. (8.142) and (8.143)) yields a nonequilibrium steady state, therefore characterized by an ℓ-independent current, $J_{\ell-1} = J_\ell$, and by a steady population of clusters, $\bar{n}_\ell$. It is important to distinguish between the equilibrium steady state, defined by $J_\ell = 0$ and characterized by the population $\langle n_\ell \rangle$ given by the Gibbs distribution (8.145), and the nonequilibrium steady state, defined by $J_\ell = \tilde{I} \; \forall \ell$ and characterized by the population $\bar{n}_\ell$. Finding the explicit expression of $\bar{n}_\ell$ is now our task.

Using Eqs. (8.143) and (8.146), we get a set of coupled, discrete equations

$$R_\ell \left[\bar{n}_\ell - \bar{n}_{\ell+1} \exp\left(\frac{\epsilon_{\ell+1} - \epsilon_\ell}{T}\right)\right] = \tilde{I}. \tag{8.147}$$

If we pass to a continuum variable ℓ, we can use standard methods for solving ordinary differential equations, so we expand at first order,

$$\bar{n}_{\ell+1} = \bar{n}_\ell + \frac{d\bar{n}}{d\ell}, \qquad \epsilon_{\ell+1} = \epsilon_\ell + \frac{d\epsilon}{d\ell}, \tag{8.148}$$

and we finally get

$$\frac{d\bar{n}(\ell)}{d\ell} = -\left(\frac{1}{T}\frac{d\epsilon}{d\ell}\right)\bar{n}(\ell) - \frac{\tilde{I}}{R(\ell)}, \tag{8.149}$$

where we have assumed that both $d\bar{n}/d\ell$ and $d\epsilon/d\ell$ are small quantities. Let us note that, in the earlier equation, $\epsilon(\ell)$ is the free energy of a cluster composed of ℓ atoms and its explicit expression is given in Eq. (8.140). The quantity $R(\ell)$ is the rate of the transition $(\ell) \to (\ell+1)$ through adsorption of a molecule; see Fig. 8.28.

Equation (8.149) has the form

$$\frac{d\bar{n}}{d\ell} = A'(\ell)\bar{n}(\ell) + B(\ell), \tag{8.150}$$

with $A(\ell) = -\epsilon(\ell)/T$ and $B(\ell) = -\tilde{I}/R(\ell)$. The general solution of the homogeneous equation is $\bar{n}(\ell) = ce^{A(\ell)}$ and a particular solution of the nonhomogeneous equation is found via variation of constants. Combining the two, we have

$$\bar{n}(\ell) = ce^{A(\ell)} - \int_\ell^\infty dx\, B(x) e^{(A(\ell)-A(x))}. \tag{8.151}$$

Since $A(\ell) \to +\infty$ for diverging ℓ, the condition $\bar{n}(\infty) = 0$ requires us to impose $c = 0$, while the finiteness of the integral requires that $B(x)$ vanishes for large x, that is, that $R(x)$ diverges in the same limit. We will come back to this condition later. Finally, we find

$$\bar{n}(\ell) = \tilde{I} \int_\ell^\infty \frac{dx}{R(x)} \exp\left(\frac{\epsilon(x) - \epsilon(\ell)}{T}\right). \tag{8.152}$$

The earlier expression for the steady-state distribution is exact, but the integral cannot be calculated exactly. It will be evaluated in the limits $\ell \ll \ell^*$ and $\ell \gg \ell^*$ assuming that $R(\ell)$ has a smooth behavior, while the function $\exp(\epsilon(x)/T)$ is strongly peaked in $x = \ell^*$ and rapidly vanishes for $x > \ell^*$; see Fig. 8.27. If $\ell < \ell^*$, the point $x = \ell^*$ is in the integration domain and we can evaluate the integral with the saddle point method. Therefore, we expand $\epsilon(x)$ around $x = \ell^*$,

$$\epsilon(x) = \epsilon(\ell^*) + \frac{1}{2}\epsilon''(\ell^*)(x - \ell^*)^2, \tag{8.153}$$

we evaluate $R(x)$ in $x = \ell^*$, and we change the integration limits to $\pm\infty$, getting a Gaussian integral,

$$\bar{n}(\ell) \simeq \frac{\tilde{I}}{R(\ell^*)} e^{(\epsilon(\ell^*) - \epsilon(\ell))/T} \int_{-\infty}^{\infty} dx \exp\left(\frac{\epsilon''(\ell^*)}{2T}(x - \ell^*)^2\right)$$

$$\simeq \boxed{\frac{\tilde{I}}{R(\ell^*)} \sqrt{\frac{2\pi T}{-\epsilon''(\ell^*)}} \exp\left[\frac{1}{T}(\epsilon(\ell^*) - \epsilon(\ell))\right], \quad \ell < \ell^*} \tag{8.154}$$

If $\ell \gg \ell^*$, the dominant contribution to the integral (8.152) comes from the region x close to the lower limit, so we can expand $\epsilon(x)$ to the linear term,

$$\epsilon(x) = \epsilon(\ell) + \epsilon'(\ell)(x - \ell), \tag{8.155}$$

and obtain

$$\bar{n}(\ell) \simeq \frac{\tilde{I}}{R(\ell)} \int_{\ell}^{\infty} dx \exp\left(-\frac{1}{T}|\epsilon'(\ell)|(x - \ell)\right)$$

$$\simeq \boxed{\frac{\tilde{I}}{R(\ell)} \frac{T}{|\epsilon'(\ell)|}, \quad \ell \gg \ell^*} \tag{8.156}$$

It is now interesting to compare the equilibrium distribution $\langle n_\ell \rangle$ (see Eq. (8.145)) with the newfound nonequilibrium stationary distribution. For $\ell \gg \ell^*$, $\epsilon(\ell)$ is large and negative, so $\langle n_\ell \rangle$ has a divergence, which has been healed by the nonequilibrium treatment: $\bar{n}(\ell)$ does not diverge. From Eq. (8.156)), we also observe that since $|\epsilon'(\ell)|$ is a constant for large ℓ, $\bar{n}(\ell)$ vanishes in the same limit only if $R(\ell)$ diverges. In Eq. (8.160) we provide the explicit expression of $R(\ell)$, valid for gas–liquid transition, which clearly shows such a divergence. As for the opposite limit of small ℓ, Eq. (8.154) shows that $\bar{n}(\ell)$ and $\langle n_\ell \rangle$ have the same ℓ-dependence, both being proportional to the Boltzmann factor,

$$\bar{n}(\ell) \approx \langle n_\ell \rangle \approx \exp\left[-\frac{\epsilon(\ell)}{T}\right], \quad \ell < \ell^*. \tag{8.157}$$

This result is certainly not accidental, because in the limit $\ell^* \gg 1$ (a limit that is tacitly assumed when we pass from discrete to continuum), the nucleation barrier $\epsilon(\ell^*)$ is high and the nucleation rate $\tilde{I}$ is low. If we recall that for $\tilde{I} = 0$ the equilibrium steady state distribution must be recovered, for $\ell \ll \ell^*$ the two quantities $\bar{n}(\ell)$ and $\langle n_\ell \rangle$ must coincide.

Therefore, in this limit $\bar{n}(\ell)$ and $\langle n_\ell \rangle$ must be equal,

$$\frac{\tilde{I}}{R(\ell^*)} \sqrt{\frac{2\pi T}{-\epsilon''(\ell^*)}} \exp\left[\frac{1}{T}\left(\epsilon(\ell^*) - \epsilon(\ell)\right)\right] = \langle n_1 \rangle \exp\left[-\frac{\epsilon(\ell)}{T}\right]. \tag{8.158}$$

From this equation, we obtain the nucleation rate,[24]

$$\tilde{I} = \langle n_1 \rangle \sqrt{\frac{-\epsilon''(\ell^*)}{2\pi T}} R(\ell^*) \exp\left[-\frac{\epsilon(\ell^*)}{T}\right], \tag{8.159}$$

with $\epsilon(\ell^*) = \dfrac{4}{27}\dfrac{\gamma^3}{(\delta\mu)^2}$ and $\epsilon''(\ell^*) = -\dfrac{9}{8}\dfrac{(\delta\mu)^4}{\gamma^3}$. The quantity $\langle n_1 \rangle \exp\left[-\frac{\epsilon(\ell^*)}{T}\right]$ is the average number of nuclei attaining the top of the energy barrier by thermal fluctuations; not all of them continue to grow, but only the fraction given by the square root, $\sqrt{-\epsilon''(\ell^*)/(2\pi T)}$, called Zeldovich factor.

In the case of the gas–liquid transition, the expression for $R(\ell)$ can be found within the kinetic theory, assuming that it is given by the number of gas atoms colliding with the ℓ-cluster per unit time, that is,

$$R(\ell) \simeq \frac{1}{6}\langle n_1 \rangle v (4\pi r^2), \tag{8.160}$$

where $v \simeq \sqrt{3T/m}$ is the thermal velocity[25] and r is the radius of the ℓ-cluster.

The most notable feature of the nucleation rate, Eq. (8.159), is its extreme sensitivity to the difference of chemical potential $\delta\mu$, which is in first approximation linear with $T_c - T$, the temperature difference between the liquid–vapor transition temperature T_c and the actual temperature T. The strong dependence on $\delta\mu$ is due to the exponential[26] $\exp(-\epsilon(\ell^*)/T)$, so it is useful to write

$$\tilde{I} = I_0 e^{-\frac{\epsilon(\ell^*)}{T}} = I_0 e^{-\frac{4\gamma^3}{27(\delta\mu)^2 T}}. \tag{8.161}$$

Since $\delta\mu$ vanishes when $T \to T_c$, $\tilde{I}$ is extremely sensitive to $\delta\mu$. This fact also points out a weak point of the above treatment, where the quantity $\delta\mu$ is assumed to be constant during the nucleation process. In reality it depends on cluster density, because of the depletion effect.

Finally, we should consider that our treatment considers homogeneous nucleation, that is, the formation of a nucleus of the new phase in an ideally perfect system. In a real system, the escape from the metastable phase occurs via heterogeneous nucleation, induced by the walls of the container (which may reduce the surface tension) or by impurities (which may expedite the formation of critical nuclei).

---

[24] Note the analogy with the Kramers–Arrhenius formula, discussed in Section 2.5.1.

[25] With the notation used in Chapter 1, the thermal velocity is the square root of the average square velocity, $\langle v^2 \rangle^{1/2}$; see Eq. (1.6).

[26] For the gas–liquid transition, if we use Eq. (8.160), this is the only dependence on $\delta\mu$. In fact, we obtain $R(\ell^*) \sim (\delta\mu)^{-2}$, which cancels the $\delta\mu$-dependence of the Zeldovich factor.

8.9 Bibliographic Notes

The most classical reference on continuum theories of phase-ordering is the review paper by A. J. Bray, Theory of Phase-ordering Kinetics, *Advances in Physics*, **51** (2002) 481–587. Here one can find a lot on continuum models and different theoretical approaches to treat them, including the dynamical renormalization group. The main limitation is that nucleation and off-critical quenches are not considered.

Another classical reference is the review paper by P. C. Hohenberg and B. I. Halperin, Theory of Dynamical Critical Phenomena, *Reviews of Modern Physics*, **49** (1977) 435–479.

The original paper on the kinetic Ising model is R. J. Glauber, Time-dependent Statistics of the Ising Model, *Journal of Mathematical Physics*, **4** (1963) 294–307.

The argument for finding the coarsening exponent $n = 1/3$ for the conserved kinetic Ising model has been derived from S. J. Cornell, K. Kaski, and R. B. Stinchcombe, Domain Scaling and Glassy Dynamics in a One-dimensional Kawasaki Ising Model, *Physical Review B*, **44** (1991) 12263–12274.

A more recent collection of different contributions covering the topics of the present chapter is the book by S. Puri and V. Wadhawan, eds., *Kinetics of Phase Transitions* (Taylor & Francis, 2009).

An article with extensive simulations on the nonconserved Ising model and an analysis of the role of initial state and final quench temperature is F. Corberi and R. Villavicencio-Sanchez, Role of Initial State and Final Quench Temperature on Aging Properties in Phase-ordering Kinetics, *Physical Review E*, **93** (2016) 052105.

A book that is a useful bridge between this chapter and the following one is R. C. Desai and R. Kapral, *Dynamics of Self-Organized and Self-Assembled Structures* (Cambridge University Press, 2009). In particular, the authors discuss two topics that have been hardly touched on here: the droplet-like morphology appearing in a slightly off-critical quenching of a conserved model and the self-similar distribution of clusters that results from Ostwald ripening.

For the nucleation process the reader is referred to J. S. Langer, An Introduction to the Kinetics of First-order Phase Transitions, in Claude Godrèche, ed., *Solids Far from Equilibrium* (Cambridge University Press, 1991), pp. 297–363.

A rapid overview of coarsening phenomena is given in L. F. Cugliandolo, Coarsening Phenomena, *Comptes Rendus Physique*, **16** (2015) 257–266, where the author also discusses the relation between the morphology resulting from a quenching of a $d = 2$ Ising model and the percolation morphology.

More generally, the reader can browse all contributions to the special issue on coarsening dynamics, by Federico Corberi and Paolo Politi, eds., *Comptes Rendus Physique*, **16** (2015) 255–342.

9

Highlights on Pattern Formation

This last chapter deals with a topic, pattern formation, which, by itself, is covered by entire books. In the present and more general context of nonequilibrium phenomena, rather than providing an overview of the whole subject, we have preferred to focus on a few relevant concepts. The first ones are the control parameter and the order parameter, which should be familiar to the reader after Chapters 5 and 6 dealing with equilibrium and nonequilibrium phase transitions. However, these concepts are introduced here again in reference to specific pattern-forming systems.

In brief, pattern formation appears when an out-of-equilibrium system passes from a homogeneous to a nonhomogeneous state while tuning a suitable external parameter. This transition (or, more precisely, this bifurcation) typically occurs because the homogeneous solution loses its stability, so that an infinitesimal perturbation is enough to drive the system far from there. We therefore need to perform a linear stability analysis of the homogeneous state. Usually, this is not a hard task, but in some cases, depending on the physics and the geometry of the problem, it certainly is. Two nontrivial examples are provided: the Turing patterns, Section 9.3, where difficulties arise from the vectorial character of the order parameter, and the Rayleigh–Bénard instability, Appendix W, whose description requires starting from the Navier–Stokes equations. The linear stability analysis not only determines the critical value of the control parameter beyond which a given solution loses stability; it also allows us to gain information on the new state and, from a general point of view, allows classification of the bifurcations.

But what is the fate of the instability? The answer concerns the nonlinear dynamics of the model and depends on the type of bifurcation, in particular, if the instability is stationary or oscillatory, if the unstable range of wave vectors includes $q = 0$ or not, and if there are conservation laws. For the sake of simplicity and space, here we limit ourselves to stationary instabilities, and a special role is played by periodic steady states, which arise when the homogeneous state is unstable. Therefore, the first nonlinear step is to determine such periodic patterns and, afterward, to study their stability using suitable perturbative techniques. If periodic patterns are unstable, a secondary instability appears, the primary instability being the loss of stability of the homogeneous solution.

9.1 Pattern Formation in the Laboratory and the Real World

Previous chapters show that driven, nonequilibrium systems may display a rich and varied phenomenology, ranging from phase transitions with symmetry breaking to kinetic roughening. In the present, final chapter we consider pattern formation, that is, the appearance of a nontrivial spatial pattern when tuning a suitable control parameter, related to the driving force.

Spatial patterns are widespread in nature. A well-known example in an extended system are sand dunes (including patterns underwater), whose appearance can be easily connected to wind or currents, which are the driving forces destabilizing the otherwise flat surface of the sand. Another example is the formation of snowflakes, with the growing ice nucleus that does not remain spherical as a water droplet but forms arms with the sixfold symmetry typical of the most common ice structure. In this case the driving force, supersaturation, is less evident to the unskilled observer. Patterns are also visible in living organisms; think of the formation of regular spots and stripes on animal skin or fur or of the regular structure of some leaves. Even the kitchen can be a place to observe patterns: If we cook a thin layer of oil in a pan over low heat, we can notice the formation of hexagonal convection cells, which can be made more visible by adding, for example, some cinnamon to the oil.

From now on, in order to be more specific and to introduce the most relevant concepts, we make reference to a few experimental examples, shown in Fig. 9.1. The first example, Fig. 9.1(a), is present in any discussion of pattern formation: A liquid is confined in a closed container, whose height is much smaller than its horizontal sizes, and it is brought out of equilibrium by heating the bottom. The main difference with a pan on the fire is that the upper surface of the liquid is not free but confined. If $\Delta T = T_{\mathrm{inf}} - T_{\mathrm{sup}}$ is the temperature difference between the lower and upper surfaces, the system is out of equilibrium as soon as $\Delta T > 0$, but for small ΔT, the fluid remains at rest and heat is transported by conduction. The experiment shows the existence of a critical value $(\Delta T)_{\mathrm{c}}$ above which the fluid is set in motion and convection cells form.

Convection cells are known to exist in the atmosphere and in the Earth's mantle, but their size is orders of magnitude bigger than cells described here. In all cases a convection cell is characterized by a hot current of fluid going from a warm to a cold region and by a balancing cold current in the opposite direction. In the experiment discussed here, the process of cell formation is the result of an instability, called the Rayleigh–Bénard instability.[1] If we tune ΔT from $\Delta T < (\Delta T)_{\mathrm{c}}$ to $\Delta T > (\Delta T)_{\mathrm{c}}$, the resulting instability has two main features: At short times, the initially homogeneous (conductive) state is made unstable and convection cells appear; at longer times, there is a readjustment of the size of convection cells through coalescence or splitting. Finally, the system settles in a nonequilibrium stationary state.

More formally, some key points can be stressed:

- We can define a control parameter, ΔT, and an order parameter, the field velocity $\mathbf{u}$ of the fluid.
- There exists a critical value of the control parameter, $(\Delta T)_{\mathrm{c}}$, at which the order parameter passes from a homogeneous state (in this case, $\mathbf{u}(\mathbf{r}, t) \equiv 0$) to a nonhomogeneous state showing a spatial pattern (convection cells).
- The ensuing dynamics, if ΔT is not much higher than $(\Delta T)_{\mathrm{c}}$, terminates at a spatially periodic, stationary state.

---

[1] After the British Lord Rayleigh and the French Henri Bénard.

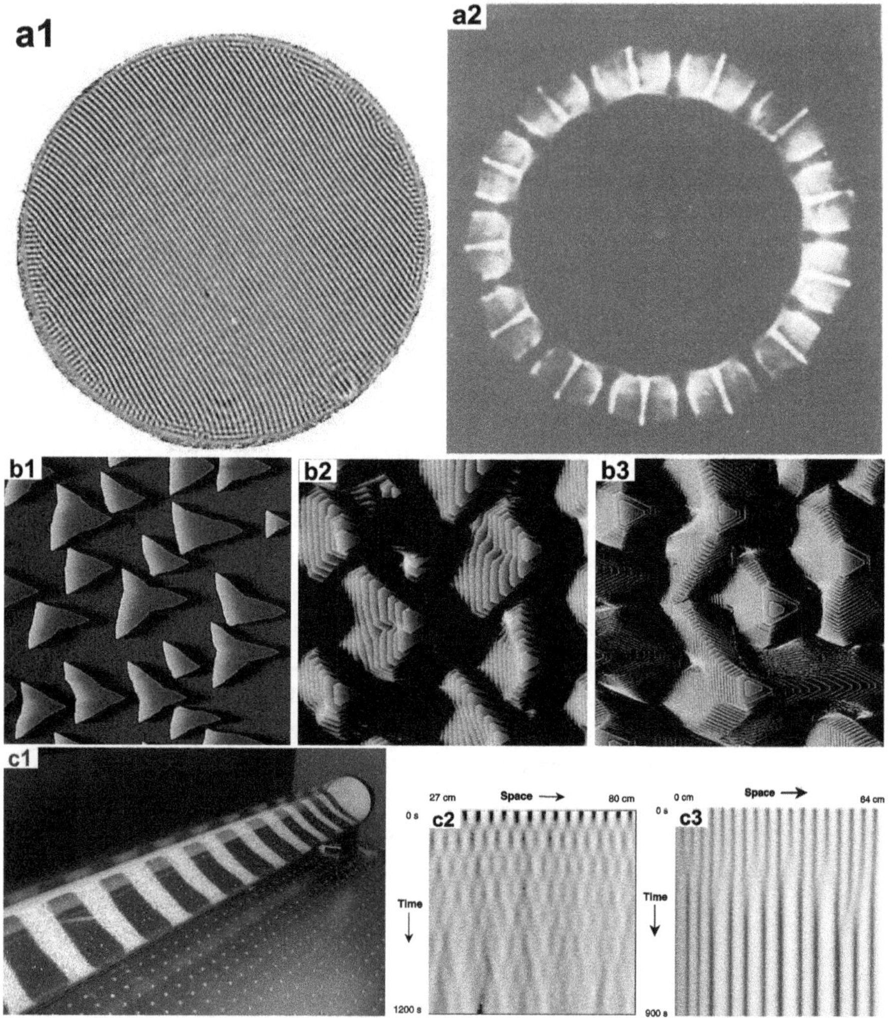

Fig. 9.1 (a1) Nearly straight rolls appear just above the threshold of the Rayleigh–Bénard instability ($r = 0.043$; see Eq. (9.3)) in a two-dimensional, circular cell. Reprinted from S. W. Morris et al., The Spatio-temporal Structure of Spiral-defect Chaos, *Physica D*, **97** (1996) 164–179. (a2) Convective rolls in a one-dimensional, annular geometry. Reprinted from S. Ciliberto, F. Bagnoli, and M. Caponeri, Pattern Selection in Thermal Convection in an Annulus, *Il Nuovo Cimento D*, **12** (1990) 781–792. (b1–b3) Scanning tunneling microscopy images of Pt/Pt(111) at (b1) 0.35 monoLayer (ML), (b2) 12 ML, and (b3) 90 ML. The area is 2900×2900 Å. Reprinted from M. Kalff et al., No Coarsening in Pt(111) Homoepitaxy, *Surface Science*, **426** (1999) L447–L453. (c1) Granular demixing. Segregation patterns of a mixture of different types of sands. In this system we may have (c2) traveling waves in the opposite direction so as to create standing waves or (c3) band merging. Panel (c1) courtesy of C. R. J. Charles and S. Morris. Panels (c2) and (c3) are reprinted from K. Choo et al., Dynamics of Granular Segregation Patterns in a Long Drum Mixer, *Physical Review E*, **58** (1998) 6115–6123.

The first two points are fully general and represent the distinctive features of pattern formation. The third point is peculiar and different pattern-forming systems may display different nonlinear dynamics, that is, different outcomes of the instability.

The second example, Fig. 9.1(b), shows the result of a growth process of a crystal surface of platinum. Starting from a platinum substrate, whose high-symmetry orientation is characterized by the Miller indices (111), platinum atoms are deposited onto it in ultra-high vacuum conditions. Tuning the intensity Φ of the flux of deposited particles, it is possible to make the flat surface of the growing crystal unstable, leading to the formation of mounds/pyramids. In this case, Φ is the control parameter and the local height $z(\mathbf{r}, t)$ plays the role of the order parameter. The uniform state corresponds to $z(\mathbf{r}, t) = z_0 + \Phi t$. As for the dynamics following the instability, in the specific case discussed here we get mounds of constant wavelength and increasing height. With other materials or in different experimental conditions, the same instability may give rise to pyramids of increasing wavelength and constant slope, therefore leading to a coarsening process that strongly resembles a phase separation process for the order parameter $\mathbf{m}(\mathbf{r}, t) = \nabla z(\mathbf{r}, t)$.

In the third example, Fig. 9.1(c), we deal with a granular system, partly filling a rotating drum. It is a mixture of large (black) and small (transparent) glass spheres. In this case it is not possible to identify a general control parameter, but to stay at a simple level we can imagine that the role of the control parameter is played by the rotational speed Ω: If it is large enough, an initially homogeneous mixture will tend to segregate, producing a pattern of alternating black and white bands. Identifying the order parameter, which should quantify the asymmetry between the two types of spheres, is easier. If $N_{1,2}$ is their total number, we can define the quantity

$$u(\mathbf{r}, t) = (n_1 c_2 - n_2 c_1)/(n_1 c_2 + n_2 c_1), \tag{9.1}$$

where $n_{1,2}(\mathbf{r}, t)$ are the local number densities of large and small spheres and $c_{1,2} = N_{1,2}/(N_1 + N_2)$ are the global relative fractions, respectively. For a symmetric mixture, $c_1 = c_2$, we simply have $u(\mathbf{r}, t) = (n_1 - n_2)/(n_1 + n_2)$. In general, u varies from $u = -1$, if only type-2 particles are locally present at site $\mathbf{r}$ ($n_1 = 0$), to $u = +1$, if only type-1 particles are at site $\mathbf{r}$ ($n_2 = 0$). If $u = 0$, the local concentrations are equal to the average concentrations, $n_i = c_i$. Once those bands have appeared (see Fig. 9.1(c1)), their size can remain constant or increase very slowly. It is also possible to have more complex dynamical behaviors, with the birth of oscillations or traveling waves.

A fourth and final example we want to mention here are the so-called Turing patterns, named after the British Alan Turing, who wrote a seminal paper in 1952 with the title "The Chemical Basis of Morphogenesis," proposing a basic model to account for the formation of natural patterns. According to Turing, such patterns can be the result of an instability of a homogeneous state, through a process of reaction–diffusion, where different ingredients may diffuse and react among them. A suitable combination of different diffusivity and different reaction properties can give rise to the emergence of a steady pattern of well-defined wavelength. Reaction–diffusion models will be studied with some detail in Section 9.3, but we can anticipate that they require a description based on more than one scalar order parameter.

In fact, throughout the rest of this chapter, the theoretical study of pattern formation in spatially extended systems will be based on a continuum approach, where the evolution of a scalar order parameter is described by a suitable partial differential equation, such as

$$\partial_t u(\mathbf{x}, t) = \mathcal{A}[u, r], \tag{9.2}$$

where $\mathcal{A}$ is a generic functional of the field u (which is assumed here to be a scalar, just for the sake of simplicity), and it also depends on the (reduced) control parameter r. The latter is usually defined so that its critical value is zero. For instance, in the case of the Rayleigh–Bénard instability, we can define it as

$$r = \frac{\Delta T - (\Delta T)_c}{(\Delta T)_c}. \tag{9.3}$$

Making reference to the examples shown in Fig. 9.1 and discussed earlier, the homogeneous solution corresponds to heat conduction in (a), to a growing, flat surface in (b), and to a homogeneous mixture in (c). If $u(\mathbf{x}, t)$ represents, respectively, the velocity field in (a), the local height with respect to the average height ($u(\mathbf{x}, t) = z(\mathbf{x}, t) - \langle z(\mathbf{x}, t) \rangle$) in (b), and the quantity (9.1) in (c), then the homogeneous solution corresponds to $u(\mathbf{x}, t) \equiv 0$ in all cases.

We should stress that the homogeneous solution is expected to be a solution of the problem, $\mathcal{A}[0, r] \equiv 0$, for any r. What changes between negative and positive r (i.e., below and above the threshold) is the stability character of the homogeneous solution: Conventionally, it should be stable for $r < 0$ and unstable for $r > 0$. Unstable means that some weak perturbation leads the system far from it. Since fluctuations are unavoidable in any real or simulated system, an unstable solution is physically unobservable; this is why the homogeneous solution disappears for $r > 0$ and a new, spatially modulated solution arises.

The earlier, simplified picture allows us to clarify why our study of pattern formation is basically made of two parts. The first part analyzes the stability character of the homogeneous solution with varying the control parameter and classifies the type of instability. This linear analysis also allows us to predict what the relevant space and time scales of the emerging pattern are. The second part goes beyond the linear stability analysis of the homogeneous solution and addresses the so-called nonlinear regime, trying to show the ensuing dynamics. The linear analysis is valid at short times, the nonlinear analysis at longer times.

9.2 Linear Stability Analysis and Bifurcation Scenarios

Two of the three experimental examples shown in Fig. 9.1 and discussed in the previous section display a phenomenology that resembles a phase separation process, discussed at length in Chapter 8. In phase-ordering, the homogeneous state becomes unstable after quenching the system from the disordered to the ordered phase. In a pattern-forming system, a homogeneous state is unstable when the system is driven out of equilibrium and

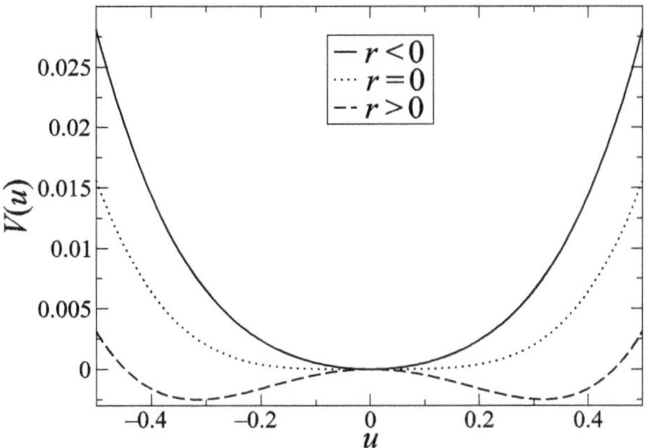

Fig. 9.2   The potential $V(u)$ for negative, vanishing, and positive r; see Eq. (9.6).

a suitable control parameter exceeds a critical value. In phase separation, the instability produces ordered regions, whose size increases in time through a coarsening process. In cases shown in Fig. 9.1(b,c), we may have a spatial structure characterized by regions of constant order parameter, and it may happen that their size increases in time, therefore producing a dynamics similar to a phase separation process.

At first sight this statement is confusing, because phase-ordering dynamics in a system relaxing toward equilibrium is controlled by its free energy, but a driven, out-of-equilibrium system has no free energy. In spite of this, it appears that some out-of-equilibrium dynamics may be characterized by the decreasing of a suitable functional, called the Lyapunov functional or potential functional, and in some cases this functional looks like the Ginzburg–Landau free energy. In fact, this result is not so strange if we think of the GL free energy as a power expansion in the order parameter. However, there is an even simpler reason to reintroduce here the time-dependent Ginzburg–Landau (TDGL) and the CH models: These equations allow us to present the concepts of linear stability analysis and bifurcation scenarios.

Let us start by rewriting these two equations in the simple one-dimensional case, making explicit the dependence on the control parameter,

$$\frac{\partial u}{\partial t} = \frac{\partial^2 u}{\partial x^2} + ru - u^3 \equiv -\frac{\delta \mathcal{F}_{\text{GL}}}{\delta u} \qquad \text{(TDGL)} \tag{9.4}$$

$$\frac{\partial u}{\partial t} = -\frac{\partial^2}{\partial x^2}\left(\frac{\partial^2 u}{\partial x^2} + ru - u^3\right) \equiv \frac{\partial^2}{\partial x^2}\frac{\delta \mathcal{F}_{\text{GL}}}{\delta u} \qquad \text{(CH)} \tag{9.5}$$

$$\mathcal{F}_{\text{GL}}[u] = \int dx \left[\frac{1}{2}\left(\frac{\partial u}{\partial x}\right)^2 - r\frac{u^2}{2} + \frac{u^4}{4}\right] \equiv \int dx \left[\frac{1}{2}\left(\frac{\partial u}{\partial x}\right)^2 + V(u)\right], \tag{9.6}$$

and recalling that dynamics minimizes the functional $\mathcal{F}_{\text{GL}}[u]$, $d\mathcal{F}_{\text{GL}}[u]/dt \leq 0$; see Section 8.2.1. The quantity r, the control parameter, allows passage from the single-well potential for $r < 0$, where $V(u)$ has a single minimum in $u = 0$, to a double-well potential for $r > 0$, where $V(u)$ has two equivalent minima in $u = \pm\sqrt{r}$; see Fig. 9.2.

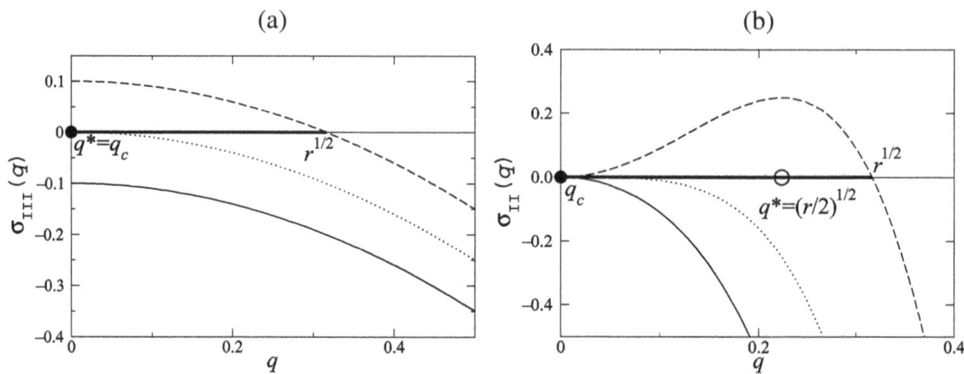

Fig. 9.3   The stability spectra for the TDGL equation (a) and the CH equation (b), with increasing the control parameter: $r < 0$ (solid line), $r = 0$ (dotted line), and $r > 0$ (dashed line). The thick line segments represent the instability region of the spectrum ($\sigma(q) > 0$), the solid circle indicates the critical wavevector, q_c, and the open circle (when different from the solid circle) indicates the most unstable wavevector, q^*. These two scenarios are representative of type-II (b) and type-III (a) bifurcations; see also Table 9.1.

Let's now focus on the partial differential equations, observing that $u = 0$ is always a solution of both equations, for any value of r. Its relevance in the dynamics depends on its stability character, that is, on the evolution of a profile that is initially close to it. This analysis is called linear stability analysis, because it amounts to writing $u(x,t) = \epsilon \tilde{u}(x,t)$ and keeping only terms that are linear in the small parameter ϵ. The physical grounds of this approximation will be clear later. We can therefore write

$$\epsilon \partial_t \tilde{u}(x,t) = \epsilon \left(\tilde{u}_{xx} + r\tilde{u} \right) + O(\epsilon^3) \tag{9.7}$$

$$\epsilon \partial_t \tilde{u}(x,t) = \epsilon \left(-\tilde{u}_{xxxx} - r\tilde{u}_{xx} \right) + O(\epsilon^3), \tag{9.8}$$

where the terms we are going to disregard are of order ϵ^3. The resulting equations are, of course, linear and can be easily solved with the help of Fourier analysis. The evolution of a single Fourier component is determined by assuming $\tilde{u}(x,t) = u_0 e^{\sigma t} e^{iqx}$. It is straightforward to see that each spatial derivative corresponds to multiplying by (iq) and each time derivative corresponds to multiplying by σ,

$$\partial_t \leftrightarrow \sigma \tag{9.9}$$

$$\partial_x \leftrightarrow iq. \tag{9.10}$$

Therefore, for Eqs. (9.7) and (9.8), we obtain the stability spectra

$$\sigma = r - q^2 \qquad \text{(TDGL)} \tag{9.11}$$

$$\sigma = rq^2 - q^4. \qquad \text{(CH)} \tag{9.12}$$

If the TDGL/CH models are written in generic dimension d, it is sufficient to replace ∂_{xx} with ∇^2 in Eqs. (9.4) and (9.5), and the linear stability analysis remains unchanged. For general d, the perturbation has the form $\tilde{u}(\mathbf{x}, t) = u_0 e^{\sigma t} e^{i\mathbf{q} \cdot \mathbf{x}}$; Eq. (9.10) is replaced by $\nabla \leftrightarrow i\mathbf{q}$; and spectra (9.11) and (9.12) are still valid, with $q = |\mathbf{q}|$. We can now comment on such spectra, making reference to Fig. 9.3.

The first, basic remark is that in both the nonconserved and conserved cases, $\sigma < 0$ for all $q \neq 0$ when $r \leq 0$ and $\sigma > 0$ for some q when $r > 0$. A positive (negative) $\sigma(q)$ means that the amplitude of the solution exponentially grows (decreases). A given stationary solution is stable if the amplitude of any perturbation goes to zero with time; otherwise, it is unstable. The earlier results mean that the homogeneous solution $u = 0$ is stable for negative r and unstable for positive r, and this statement is valid for both the TDGL and CH equations. In other words, if we integrate one of such equations starting from a profile $u(x, 0)$ that is "small" everywhere, that is, $|u(x, 0)| \ll 1 \ \forall x$, we expect the evolution to converge to $u = 0$ if $r \leq 0$, while it increasingly moves away from $u = 0$ for $r > 0$. What the ensuing dynamics is for $r > 0$ depends on the nonlinear terms in the equation, which have been disregarded in the earlier analysis.

A second, basic remark is that $\sigma(q)$ is real. An imaginary part would correspond to an oscillating component of the amplitude. If $\sigma = \sigma_R + i\sigma_I$, we would have $\tilde{u}(x, t) = u_0 \cos(\sigma_I t) \exp(\sigma_R t)$. Therefore, in general terms, the real and imaginary parts of σ are different pieces of information about the stability of the solution. The earlier equations provide a purely real spectrum, because the linear part is just a sum of even spatial derivatives; therefore, only terms of the form $(iq)^{2n} = (-1)^n q^{2n}$ appear in $\sigma(q)$. More precisely, in TDGL we have $n = 0, 1$ and in CH we have $n = 1, 2$.

We can generalize these considerations to any equation having the form

$$u_t = \sum_k c_k \partial_x^k u + N[u] \equiv L[u] + N[u], \tag{9.13}$$

where $L[u]$ is a generic, local, and linear operator and $N[u]$ is the nonlinear part. We should stress that L and N might not be naturally separated, as in the following example:

$$u_t = u_{xx} + \frac{u}{1 + u^2}. \tag{9.14}$$

Here,

$$L[u] = u + u_{xx} \tag{9.15}$$

$$N[u] = \frac{u}{1 + u^2} - u = -\frac{u^3}{1 + u^2}. \tag{9.16}$$

The Fourier analysis of Eq. (9.13) gives the spectrum

$$\sigma(q) = \sum_{n \geq 0} c_{2n}(-1)^n q^{2n} + i \sum_{n \geq 0} (-1)^n c_{2n+1} q^{2n+1} \equiv \sigma_R + i\sigma_I, \tag{9.17}$$

where we have explicitly separated the real and the imaginary parts. From the previous expression, we can draw a couple of conclusions: (i) $\sigma_R(q)$ is an even function and $\sigma_I(q)$ is an odd function, so a real growth rate σ is necessarily an even function of q. (ii) Since $\partial_x \leftrightarrow iq$, if $\sigma(q)$ is a real function, only even-order derivatives appear in $L[u]$, which implies (at least for the linear part) that the equation is invariant under the symmetry $x \to -x$. These conclusions are strictly correct for Eq. (9.13), that is, a local equation for a scalar order parameter in $d = 1$. In Section 9.3, we consider the Turing instability, which is described by a pair of coupled order parameters. The linear analysis will show that an oscillatory behavior ($\sigma_I \neq 0$) may occur even if the model has the symmetry $x \to -x$.

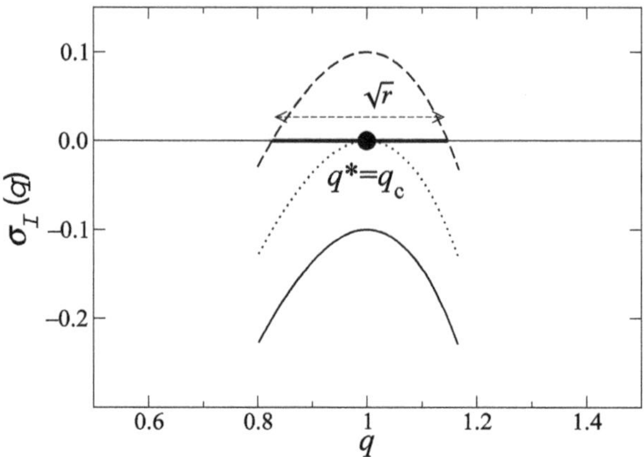

Fig. 9.4 Type I bifurcation scenario, valid for a static instability, with $q_c \neq 0$. The full, nonlinear equation we are referring to is the Swift–Hohenberg (SH) equation, introduced in Section 9.4. With increasing the control parameter, we pass from $r < 0$ (solid line), $r = 0$ (dotted line), and $r > 0$ (dashed line). The thick line segment represents the instability region of the spectrum ($\sigma(q) > 0$), which is of order $\sqrt{r}$, and the solid circle represents the critical wavevector, q_c, and the most unstable wavevector, q^*, which coincide in this model.

We have discussed the stability spectrum in some detail because, on the one hand, it plays a major role in determining the type of instability and, on the other hand, it determines the linear part of the partial differential equation. For this reason we proceed discussing further features of $\sigma(q)$ for TDGL and CH. Let us suppose that the control parameter is positive and very small, $0 < r \ll 1$.[2] The unstable part of the spectrum corresponds to $|q| < \sqrt{r}$, therefore a small interval around $q_c = 0$. In other words, Fourier modes of arbitrarily large wavelength are unstable. There is, however, a crucial difference between TDGL and CH, because the existence of a conservation law, $\partial_t \int dx u(x,t) = 0$, implies that a spatially constant profile cannot move, as explicitly seen from Eqs. (9.4) and (9.5) by assuming $u(x,t) \equiv u(t)$. TDGL gives $\dot{u} = ru - u^3$ and CH gives $\dot{u} = 0$. For $q = 0$, the linear stability analysis reduces to studying the stability of the solution $u = 0$ of these ordinary differential equations. For the CH equation, any constant value is a solution, which corresponds to saying that by perturbing $u = 0$, the amplitude neither grows nor decreases: $\sigma = 0$.

We can conclude that the form of $\sigma(q)$ depends on two main factors: (i) the value q_c of the most unstable mode at the instability threshold ($r \to 0^+$) and (ii) the presence or absence of the conservation law of the order parameter. TDGL and CH correspond to $q_c = 0$, the former without and the latter with the conservation law. It is now natural to wonder what an equation must look like if $q_c \neq 0$. Understanding this point is our next goal.

As an example, in Fig. 9.4 we schematically plot the expected behavior of $\sigma(q)$ when $q_c \neq 0$. We focus on the unstable region close to q_c, assuming that for q far from q_c, $\sigma(q) < 0$. More precisely, we assume that there is no conservation law ($\sigma(0) \neq 0$), that we are dealing with a static instability, and that $\sigma = \sigma(q^2)$. These conditions are sufficient to

---

[2] If r is not adimensional, we may always write $r = r_0\delta$, with $\delta \ll 1$.

write a function $\sigma(q)$ having a maximum in $q = q_c$ and such that $\sigma(q = q_c, r = 0) = 0$. In fact, since $q^2 - q_c^2 \simeq 2q_c(q - q_c)$ close to q_c, we can guess that

$$\sigma(q) = r - c_1(q^2 - q_c^2)^2, \tag{9.18}$$

with $c_1 > 0$. Now, we can perform a linear stability analysis in reverse order, asking what the linear operator providing such a spectrum is. Since we know that $\partial_x \leftrightarrow iq$, we easily obtain the equation

$$\partial_t u(x,t) = \left[r - c_1(\partial_{xx} + q_c^2)^2\right] u(x,t). \tag{9.19}$$

If we now rescale x and t,

$$x \to \alpha x \qquad \text{and} \qquad t \to \beta t, \tag{9.20}$$

we obtain

$$\partial_t u(x,t) = \left[\beta r - \frac{\beta c_1}{\alpha^4}(\partial_{xx} + \alpha^2 q_c^2)^2\right] u(x,t). \tag{9.21}$$

Therefore, we can choose α and β so that

$$\frac{\beta c_1}{\alpha^4} = 1 \qquad \text{and} \qquad \alpha^2 q_c^2 = 1. \tag{9.22}$$

However, for ease of presentation, we prefer to keep q_c explicit. This corresponds to not rescaling x, $\alpha = 1$, while $\beta = 1/c_1$. There is no loss of generality to redefine $\beta r = r/c_1$ as r, so we finally obtain

$$\partial_t u(x,t) = \left[r - (\partial_{xx} + q_c^2)^2\right] u(x,t). \tag{9.23}$$

Our results are summarized in Table 9.1. They do not cover all possible scenarios of the rising of an instability, of course. They just represent the simplest cases of stationary instabilities, in one spatial dimension and for a single, scalar order parameter. The central idea is to classify them according to two criteria: (i) the value of the critical wavevector q_c, which can be zero (types II and III) or not zero (type I), and (ii) the conservation (type II) or nonconservation (types I and III) of the order parameter. At a linear level, conservation implies that $\sigma(q = 0) = 0\ \forall r$. We do not consider here the conserved case with $q_c \neq 0$. The

Table 9.1 Different types of static bifurcations

Type[a]	Critical q [b]	σ	$\mathcal{L}$	q^* [c]
I	q_c	$r - (q^2 - q_c^2)^2$	$r - (\partial_{xx} + q_c^2)^2$	q_c
II	0	$rq^2 - q^4$	$-r\partial_{xx} - \partial_{xxxx}$	$\sqrt{\frac{r}{2}}$
III	0	$r - q^2$	$r + \partial_{xx}$	0

[a] We are using the classification introduced in M. C. Cross and P. C. Hohenberg, Pattern Formation Outside of Equilibrium, *Reviews of Modern Physics*, **65** (1993) 851–1112.

[b] The critical wavevector is defined as the q value corresponding to the maximum of $\sigma(q)$ at the threshold, $r \to 0^+$.

[c] It is the possibly r-dependent wavevector maximizing $\sigma(q)$. $q^*(r = 0)$ is the critical wavevector.

main message of the linear analysis is that it is possible to classify the linear spectrum $\sigma(q)$ according to the earlier criteria and that the knowledge of $\sigma(q)$ determines unambiguously the linear part of the partial differential equation. Therefore, the three classes of spectra, given in the third column of the table, determine the linear operators $\mathcal{L}$, explicitly written in the fourth column.

The linear analysis provides other relevant information about the instability. First, there is a clear difference between $q_c = 0$ and $q_c \neq 0$. In the former case, all modes with a wavelength larger than $2\pi/\sqrt{r}$ up to the size L of the physical system are unstable. That means very different unstable scales, which all contribute to the nonlinear behavior of the system. In the latter case, $q_c \neq 0$, unstable modes are in the approximate interval $(q_c - \delta, q_c + \delta)$, with $\delta \simeq \sqrt{r}/(2q_c)$. Therefore, even if the size of the unstable q-interval is approximately the same as for types II and III (i.e., of order $\sqrt{r}$), it is evident that for type I, all length scales are approximately equal to $2\pi/q_c$. We will see that this fact has relevant consequences for the nonlinear dynamics.

If an unstable Fourier mode q has a time dependence $e^{\sigma t}$, the quantity $\tau_q = 1/\sigma(q)$ is the time scale to initiate the instability on that length scale. Therefore, we can expect that the mode q^*, corresponding to the shortest instability time, $\sigma'(q^*) = 0$, will dominate the spatial pattern in the linear regime. Again, we should distinguish between $q^* = 0$ (type III) and $q^* \neq 0$ (types I and II); see column five of Table 9.1. If $q^* = 0$, there is no well-defined length scale dominating over the others, so we expect that many different scales may appear, producing an initial "disordered" pattern. If $q^* \neq 0$, there is a clear length scale $\lambda^* = 2\pi/q^*$ that dominates. This scale is simply set by q_c and does not depend on r for type I instabilities, while it scales as $\lambda^* \simeq 1/\sqrt{r}$ for type II instabilities.

In all cases, regardless of q^*, the time $\tau^* = 1/\sigma(q^*)$ gives the typical time, because the instability develops. When $t \gtrsim \tau^*$, the amplitude of the modes is of order one and disregarding nonlinear terms is no longer justified. Therefore, our linear analysis is valid for times smaller than τ^*.

9.3 The Turing Instability

In this section we deal with an important class of pattern-forming systems, the so-called *reaction–diffusion* models, which are primarily used to study the reaction dynamics of diffusing chemicals (hence their name). The main goal here is to investigate the conditions giving rise to a type I spectrum (see Table 9.1 and Fig. 9.4), because this is a necessary condition to get a pattern of defined size: In this section we use the wording "pattern formation" as a shorthand for systems displaying a type I spectrum.

Each chemical species is characterized by its concentration $u_i(\mathbf{x}, t)$, whose time variation has a diffusive term plus a local term due to the interaction (reaction) among species,

$$\frac{\partial u_i}{\partial t} = f_i(u_1, u_2, \dots) + D_i \nabla^2 u_i. \tag{9.24}$$

In the limiting case of just one species, we get

$$\frac{\partial u}{\partial t} = f(u) + \nabla^2 u, \tag{9.25}$$

which is the generalized TDGL equation, whose one-dimensional version is studied in Section 9.8; see Eq. (9.164). This equation cannot give rise to a type I spectrum in any dimension, as easily seen by its linearization around a constant solution $\bar{u}$, that is, such that $f(\bar{u}) = 0$. If $u(\mathbf{x}, t) = \bar{u} + \epsilon(\mathbf{x}, t)$, we get

$$\partial_t \epsilon = f'(\bar{u})\epsilon + D\nabla^2 \epsilon, \tag{9.26}$$

whose solution $\epsilon(\mathbf{x}, t) = \epsilon_0 \exp(\sigma t + i\mathbf{q} \cdot \mathbf{x})$ gives the spectrum

$$\sigma(\mathbf{q}) = f'(\bar{u}) - D\mathbf{q}^2, \tag{9.27}$$

which is a type III spectrum, always centered in $\mathbf{q} = 0$.

This is the reason why we need to introduce more than one species. We also remark that if diffusion is isotropic, spatial dimension does not affect the linear stability analysis. So, in the following, q will be the modulus of a d-dimensional wavevector, $q^2 = \mathbf{q} \cdot \mathbf{q}$.

The analysis of this section is focused on the linear stability of a general homogeneous solution, $u_i(\mathbf{x}, t) = \bar{u}_i$. We are searching the conditions for the homogeneous solution being unstable in a finite range of wavelengths, that is, in a range (q_1, q_2) with $q_1 \neq 0$. Since it must be $\sigma(0) < 0$ and since diffusion certainly stabilizes the homogeneous solution for large enough q, we require some destabilizing mechanism induced by diffusion, but mediated by reactions. We are now going to study a two-species model with the goal of determining the physical ingredients to obtain pattern formation. The general conditions will be derived in Appendix X.

9.3.1 Linear Stability Analysis

The most general reaction–diffusion equations for two species are

$$\begin{aligned}
\frac{\partial u_1}{\partial t} &= f_1(u_1, u_2) + D_1 \nabla^2 u_1, \\
\frac{\partial u_2}{\partial t} &= f_2(u_1, u_2) + D_2 \nabla^2 u_2.
\end{aligned} \tag{9.28}$$

We assume the existence of some homogeneous steady state $(\bar{u}_1, \bar{u}_2)$, which must therefore satisfy the conditions

$$\begin{aligned}
f_1(\bar{u}_1, \bar{u}_2) &= 0 \\
f_2(\bar{u}_1, \bar{u}_2) &= 0.
\end{aligned} \tag{9.29}$$

We are now going to perturb this state, writing

$$\begin{aligned}
u_1(\mathbf{x}, t) &= \bar{u}_1 + \epsilon_1(\mathbf{x}, t) \\
u_2(\mathbf{x}, t) &= \bar{u}_2 + \epsilon_2(\mathbf{x}, t),
\end{aligned} \tag{9.30}$$

then linearizing with respect to $\epsilon_{1,2}$,

$$\frac{\partial \epsilon_1}{\partial t} = a_{11}\epsilon_1 + a_{12}\epsilon_2 + D_1 \nabla^2 \epsilon_1$$
$$\frac{\partial \epsilon_2}{\partial t} = a_{21}\epsilon_1 + a_{22}\epsilon_2 + D_2 \nabla^2 \epsilon_2, \tag{9.31}$$

where

$$a_{ij} = \left. \frac{\partial f_i}{\partial u_j} \right|_{\bar{u}_1,\bar{u}_2}. \tag{9.32}$$

Equation (9.31) can be solved by assuming an exponential dependence on space and time,

$$\epsilon_1(\mathbf{x},t) = \epsilon_1^0 e^{\sigma t} e^{i\mathbf{q}\cdot\mathbf{x}}$$
$$\epsilon_2(\mathbf{x},t) = \epsilon_2^0 e^{\sigma t} e^{i\mathbf{q}\cdot\mathbf{x}}, \tag{9.33}$$

whose replacement in Eq. (9.31) gives

$$\begin{pmatrix} a_{11} - D_1 q^2 & a_{12} \\ a_{21} & a_{22} - D_2 q^2 \end{pmatrix} \begin{pmatrix} \epsilon_1^0 \\ \epsilon_2^0 \end{pmatrix} \equiv \mathbf{A_q} \begin{pmatrix} \epsilon_1^0 \\ \epsilon_2^0 \end{pmatrix} = \sigma \begin{pmatrix} \epsilon_1^0 \\ \epsilon_2^0 \end{pmatrix}. \tag{9.34}$$

The solvability condition is

$$\det(\mathbf{A_q} - \sigma \mathbf{I}) = 0, \tag{9.35}$$

which can be developed as

$$\sigma^2 - \mathrm{Tr}(\mathbf{A_q})\sigma + \det(\mathbf{A_q}) = 0. \tag{9.36}$$

This equation has two solutions, $\sigma_1(q)$ and $\sigma_2(q)$. The homogeneous steady state $(\bar{u}_1, \bar{u}_2)$ is linearly stable if both solutions have a negative real part for any q. Conversely, it is linearly unstable if at least one solution has a positive real part for some q. The next step is to discuss conditions under which pattern formation occurs.

Coupling between Chemicals

The first condition is somewhat, but not completely, trivial: We require that each species acts on the dynamics of the other species. In fact, if a_{12} (or a_{21}) vanishes, the product of the off-diagonal terms of $\mathbf{A_q}$ vanishes and Eq. (9.35) trivially gives ($i = 1, 2$)

$$\sigma_i(q) = a_{ii} - D_i q^2. \tag{9.37}$$

These are the type III spectra (see Table 9.1) of the single species problem, which are not what we are looking for. Therefore, both a_{12} and a_{21} must be nonvanishing in order to get pattern formation.

Different Diffusion Rates

The stability condition says that the real part of both solutions $\sigma_{1,2}(q)$ is negative. Since $\sigma_{1,2}$ are either real or complex conjugates, both of the following quantities,

$$\sigma_1 + \sigma_2 = \mathrm{Tr}(\mathbf{A_q}) \equiv \mathcal{T}_q$$
$$\sigma_1 \sigma_2 = \det(\mathbf{A_q}) \equiv \mathcal{D}_q, \tag{9.38}$$

are real. Therefore, the stability condition is equivalent to imposing

$$\mathcal{T}_q = a_{11} + a_{22} - (D_1 + D_2)q^2 < 0 \tag{9.39a}$$

$$\mathcal{D}_q = (a_{11} - D_1 q^2)(a_{22} - D_2 q^2) - a_{12}a_{21} > 0. \tag{9.39b}$$

We are interested in understanding the conditions for which these inequalities are satisfied for $q = 0$ and are *not* satisfied for finite q, having in mind that they are surely satisfied for large q, since in this limit diffusion is dominant, so that

$$\mathcal{T}_q \approx -(D_1 + D_2)q^2 \tag{9.40a}$$

$$\mathcal{D}_q \approx D_1 D_2 q^4. \tag{9.40b}$$

We are now going to show that for equal diffusion rates, stability at $q = 0$ implies stability at any q, which excludes pattern formation. This is trivial and even more general for the trace, because

$$\mathcal{T}_q = \mathcal{T}_0 - (D_1 + D_2)q^2, \tag{9.41}$$

so diffusion always has a stabilizing effect on the trace. This means that a change of stability with increasing q must be determined by a change of sign of $\mathcal{D}_q$.

As for the determinant, if $D_1 = D_2 = D$, we get

$$\mathcal{D}_q = \mathcal{D}_0 - \mathcal{T}_0(Dq^2) + (Dq^2)^2, \tag{9.42}$$

so that stability at $q = 0$ ($\mathcal{T}_0 < 0$, $\mathcal{D}_0 > 0$) also implies $\mathcal{D}_q > 0$ for any q.

In conclusion, we must consider different diffusion rates if we want to observe pattern formation. However, this is not enough, as we are going to describe hereafter.

Local Activation, Long-Range Inhibition

We have clarified that coupling and different diffusion coefficients are necessary conditions to have stability at vanishing q and instability for intermediate q. Let's take a further step. Since it must be $D_1 \neq D_2$, there is no loss of generality if we assume $D_1 > D_2$. While the general case will be considered in Appendix X, here it is easier to focus on the limit $D_1 \gg D_2$. The stability condition (9.39b) on the determinant can be written

$$\mathcal{D}_q = \mathcal{D}_0 - (a_{22}D_1 + a_{11}D_2)q^2 + D_1 D_2 q^4 > 0. \tag{9.43}$$

If $\mathcal{D}_0 > 0$ and $D_1 \gg D_2$, we may have $\mathcal{D}_q < 0$, at increasing q if and only if[3] $a_{22} > 0$, which in turn imposes $a_{11} < 0$, because it must be that $\mathcal{T}_0 = a_{11} + a_{22} < 0$.

Since the sign of a_{ii} indicates whether the presence of the ith species reinforces ($a_{ii} > 0$, positive feedback) or suppresses ($a_{ii} < 0$, negative feedback) itself, we can summarize earlier results as follows: *We must have reaction between an activator that spreads slowly and an inhibitor that spreads quickly.* As for the cross-coupling terms, a_{12} and a_{21}, since

$$\mathcal{D}_0 = a_{11}a_{22} - a_{12}a_{21} > 0, \tag{9.44}$$

---

[3] For fixed a_{11} and a_{22}, if D_1 is sufficiently large, we have $|a_{22}|D_1 \gg |a_{11}|D_2$.

the negative sign of the product $a_{11}a_{22}$ implies that two cross terms must have opposite signs. In conclusion, if $D_1 \gg D_2$, we have two possible cases for the emergence of the Turing instability,

$$a_{11} < 0, \quad a_{22} > 0, \quad \begin{cases} a_{12} > 0 \quad a_{21} < 0 \\ a_{12} < 0 \quad a_{21} > 0. \end{cases} \tag{9.45}$$

The two cases differ in the expression of the eigenvalues. In fact, at criticality, $\sigma_1(q_c) = 0$ and Eq. (9.34) is

$$(a_{11} - D_1 q_c^2)\epsilon_1^0 + a_{12}\epsilon_2^0 = 0. \tag{9.46}$$

For large D_1, it is manifest that

$$\frac{\epsilon_1^0}{\epsilon_2^0} = (\text{positive factor}) \times a_{12}, \tag{9.47}$$

so the ratio $\epsilon_1^0/\epsilon_2^0$ has the same sign as a_{12}. In Appendix X, we prove the validity of Eqs. (9.45) and (9.47) for the general case $D_1 > D_2$.

We can now describe in words the instability mechanism making reference to the upper case of (9.45): $a_{12} > 0$ and $a_{21} < 0$. A positive fluctuation of the activator (species 2) tends to grow and also to reinforce a positive fluctuation of the inhibitor (species 1), while the inhibitor tends to suppress both species, and the combined balance at $q = 0$ must result in a stable situation (because this is what we require). However, if the inhibitor has a large diffusion coefficient, it is no longer effective to suppress the growth of the activator, and an instability may appear. If the signs of the cross-coupling terms, a_{12} and a_{21}, are reversed, it is enough to realize (see the linear Eqs. (9.31)) that it is equivalent to change the sign to one of the two perturbations, ϵ_1^0 or ϵ_2^0, which also explains why the ratio $\epsilon_1^0/\epsilon_2^0$ is related to the sign of a_{12}.

In Fig. 9.5 we give a pictorial summary of the bifurcation scenarios of the reaction–diffusion equations (9.28) for two species, plotting the space $(\mathcal{T}_q, \mathcal{D}_q)$, which define unambiguously $\sigma_{1,2}(q)$, that is, the linear stability properties of Eq. (9.28). For complex eigenvalues, $\sigma_{1,2} = R \pm iI$, we have $\mathcal{D}_q = \frac{1}{4}\mathcal{T}_q^2 + I^2$, while for real eigenvalues, $\sigma_{1,2} = R_{1,2}$, we have $\mathcal{D}_q = \frac{1}{4}\mathcal{T}_q^2 - \frac{1}{4}(R_1 - R_2)^2$. Therefore, the parabola $\mathcal{D}_q = \frac{1}{4}\mathcal{T}_q^2$ (dashed line) separates an upper region with complex eigenvalues and a lower region with real eigenvalues. Since the stable region corresponds to $\mathcal{T}_q < 0$, $\mathcal{D}_q > 0$ (upper left quadrant with dots), by tuning a suitable control parameter we may leave the stability region either (i) crossing the axis $\mathcal{D}_q = 0$ (vertical arrow) or (ii) crossing the axis $\mathcal{T}_q = 0$ (horizontal axis). In the former case, the imaginary part of the eigenvalues is zero and we have a stationary bifurcation; in the latter case, the imaginary part is nonzero and the bifurcation is oscillating. As shown earlier, a change of stability at finite q_c is determined by the change of sign of $\mathcal{D}_q$, while the change of sign of $\mathcal{T}_q = \mathcal{T}_0 - (D_1 + D_2)q^2$ implies $q_c = 0$. Therefore scenario (i) corresponds to a type I instability and scenario (ii) corresponds to a type III-o instability, that is, an oscillating type III bifurcation. In order to obtain an oscillating Turing instability (type I-o), characterized by a finite q_c and a nonvanishing imaginary part $I(q_c)$, we should consider the dynamics of three chemical species.

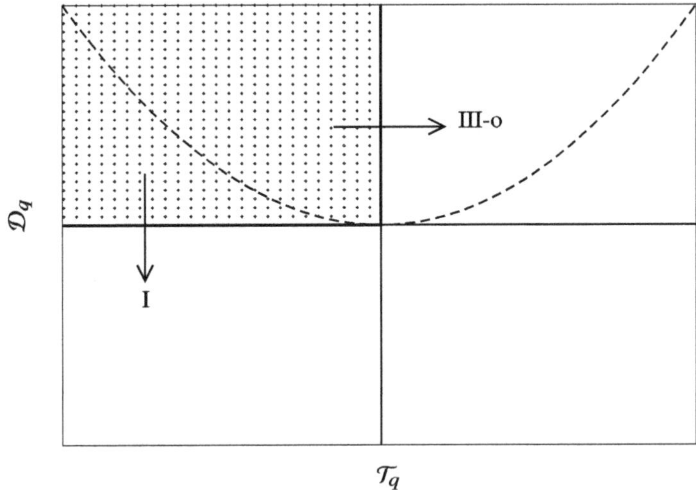

Fig. 9.5 The bifurcation scenarios in the plane $\mathcal{T}_q, \mathcal{D}_q$. The stability (dotted) region corresponds to $\mathcal{T}_q < 0, \mathcal{D}_q > 0$. We can pass from stability to instability following one of the two arrows, which correspond to a type I instability (vertical arrow crossing the axis $\mathcal{D}_q = 0$) or to a type III-o instability (horizontal arrow crossing the axis $\mathcal{T}_q = 0$).

9.3.2 The Brusselator Model

In this section we treat a specific example of the reaction–diffusion process of two species, the Brusselator model, presenting a detailed linear analysis of its unique homogeneous steady state.

The Brusselator is a minimal model for an autocatalytic reaction that involves two intermediates. Using the typical notation of chemical reactions, we write

$$A \rightarrow X \tag{9.48}$$

$$B + X \rightarrow Y + C \tag{9.49}$$

$$2X + Y \rightarrow 3X \tag{9.50}$$

$$X \rightarrow D. \tag{9.51}$$

Earlier, A and B are the initial reactants, C and D are the final products, and X and Y are the intermediates, which are the variables to be described by the coupled system (9.28). If u_1, u_2 represent the spatial densities of X, Y, respectively, and we include diffusion, we obtain the following pair of coupled equations,

$$\partial_t u_1 = \tilde{a} - (\tilde{b} + \tilde{d})u_1 + \tilde{c}u_1^2 u_2 + \tilde{D}_1 \nabla^2 u_1 \tag{9.52}$$

$$\partial_t u_2 = \tilde{b}u_1 - \tilde{c}u_1^2 u_2 + \tilde{D}_2 \nabla^2 u_2, \tag{9.53}$$

where the nonlinear terms originate from Eq. (9.50) and the (positive) quantities $\tilde{a}, \tilde{b}, \tilde{c}$, and $\tilde{d}$ denote the chemical reaction rates. It is possible to reduce the number of parameters by a suitable rescaling of u_1, u_2, t. More precisely, if

$$u_{1,2} \to \sqrt{\frac{\tilde{d}}{\tilde{c}}}\, u_{1,2} \qquad \text{and} \qquad t \to \frac{1}{\tilde{d}}\, t, \tag{9.54}$$

we can rid the system of two of the four reaction rates,

$$\partial_t u_1 = a - (b+1)u_1 + u_1^2 u_2 + D_1 \nabla^2 u_1 \tag{9.55}$$
$$\partial_t u_2 = b u_1 - u_1^2 u_2 + D_2 \nabla^2 u_2, \tag{9.56}$$

where $a = \tilde{a}\sqrt{\tilde{c}}/\tilde{d}^{3/2}$, $b = \tilde{b}/\tilde{d}$, $D_1 = \tilde{D}_1/\tilde{d}$, and $D_2 = \tilde{D}_2/\tilde{d}$. As for the diffusion constants, a rescaling of the space variable $\mathbf{x}$ would allow us to put either D_1 or D_2 equal to one. This is not common in the literature, so we maintain both coefficients, keeping in mind that only their ratio will be relevant to determining the stability of a given solution. In conclusion, the previous equations have the form (9.28), with

$$f_1 = a - (b+1)u_1 + u_1^2 u_2 \tag{9.57}$$
$$f_2 = b u_1 - u_1^2 u_2. \tag{9.58}$$

The only homogeneous fixed point $(\bar{u}_1, \bar{u}_2)$ is obtained when both f_1 and f_2 vanish, that is,

$$\bar{u}_1 = a, \qquad \bar{u}_2 = \frac{b}{a}, \tag{9.59}$$

and the elements $a_{ij} = (\partial f_i/\partial u_j)_{\bar{u}_1,\bar{u}_2}$ of the linearization matrix are

$$\mathbf{A}_0 = \begin{pmatrix} b-1 & a^2 \\ -b & -a^2 \end{pmatrix}. \tag{9.60}$$

We observe that the second species (Y) is the inhibitor $(a_{22} < 0)$, while the first species (X) is the activator $(a_{11} > 0)$ if $b > 1$. We therefore expect that $D_2 > D_1$. As for the cross terms, a_{12} and a_{21}, they have opposite signs, as they should be.

We can now be more quantitative and impose the three conditions (X.6) to determine the parameter region where a Turing pattern can appear. The condition $\mathcal{T}_0 = a_{11} + a_{22} < 0$ yields $b - 1 - a^2 < 0$, that is,

$$b < 1 + a^2, \tag{9.61}$$

while the condition $\mathcal{D}_0 = a_{11}a_{22} - a_{12}a_{21} > 0$ yields $-(b-1)a^2 + ba^2 = a^2 > 0$ and is therefore trivially satisfied. Finally, we should impose that the determinant $\mathcal{D}_q$, evaluated at its minimum $q = q_{\mathrm{m}}$, is negative, that is, $(b-1)D_2 - a^2 D_1 > 2a\sqrt{D_1 D_2}$. This condition can be rewritten as

$$b > \left(1 + a\sqrt{\frac{D_1}{D_2}}\right)^2. \tag{9.62}$$

The two conditions (9.61) and (9.62) are consistent if

$$1 + a^2 > \left(1 + a\sqrt{\frac{D_1}{D_2}}\right)^2, \tag{9.63}$$

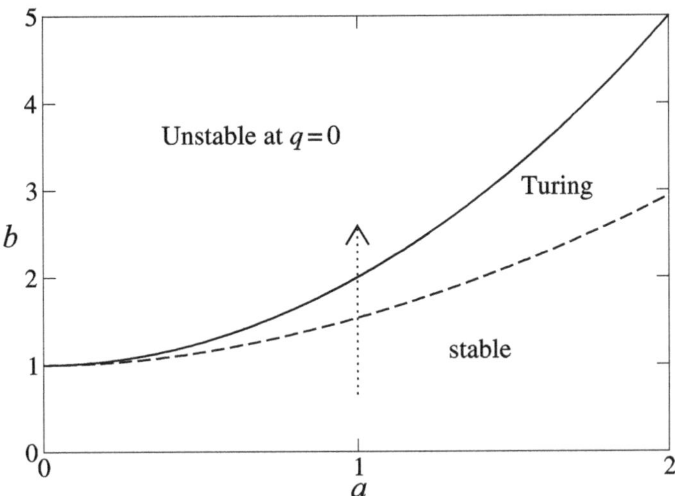

Fig. 9.6 The three regions of the parameter space (a, b). The solid line denotes the change of sign of $\mathcal{T}_0$, Eq. (9.61), which does not depend on the diffusion coefficients. The dashed line denotes the change of sign of $\mathcal{D}_{q_m}$, Eq. (9.62), for $D_1/D_2 = \frac{1}{3}(D_1/D_2)_c$. For $D_1/D_2 \to (D_1/D_2)_c$ the two lines coincide and the Turing region vanishes; for $D_1/D_2 \to 0$ the dashed line becomes the horizontal line $b = 1$.

and recalling that a is positive, we obtain

$$\sqrt{\frac{D_1}{D_2}} < \sqrt{\left(\frac{D_1}{D_2}\right)_c} \equiv \frac{1}{a}\left(\sqrt{1 + a^2} - 1\right) - 1 < 1. \tag{9.64}$$

This inequality confirms that the diffusion coefficient D_1 of the activator must be smaller than that of the inhibitor, D_2. How much smaller it has to be, that depends on the parameter a. If the ratio D_1/D_2 is just slightly lower than the critical value $(D_1/D_2)_c$, the Turing region is small, while for a vanishingly small ratio D_1/D_2, the Turing region is given by $1 < b < 1 + a^2$. In Fig. 9.6, we plot the different stability regions in the plane (a, b). Following the dotted arrow, we first meet the stable region, where $\sigma_{1,2}(q) < 0$ for all q; then, we have a type I instability and enter the Turing region; finally, the unstable q-region extends up to $q = 0$.

9.4 Periodic Steady States

In Section 9.2 we have provided a sort of classification of stationary linear instabilities for a scalar order parameter $u(\mathbf{x}, t)$ and we explained that the validity of such linear analysis is limited to times shorter than some scale τ^*, because at longer times nonlinear terms are no longer negligible. Therefore, in this section we need to consider the full partial differential equation describing the dynamics of $u(\mathbf{x}, t)$, having in mind that in only a few cases can we derive the evolution equation from first principles. In most cases the evolution equation is the result of a phenomenological approach, where considerations based on symmetries and conservation laws play a major role.

We may wonder, for example, what nonlinear term we can add to the linear operator appearing in type I instabilities. The equation

$$\partial_t u = [r - (\partial_{xx} + q_c^2)^2]u \qquad (9.65)$$

satisfies two symmetries, $x \to -x$ and $u \to -u$. The simplest possible nonlinear term still satisfying both symmetries is a cubic term, proportional to u^3, giving

$$\partial_t u = [r - (\partial_{xx} + q_c^2)^2]u + c_3 u^3. \qquad (9.66)$$

If we require that the diverging amplitude of unstable Fourier modes, $u_0 e^{\sigma(q)t}$, is counterbalanced by the nonlinear term, it is necessary that its contribution to $\partial_t u$ is negative for $u > 0$ and positive for $u < 0$. This requires $c_3 < 0$, which can be set to $c_3 = -1$ by rescaling u. Therefore we get the equation

$$\partial_t u = [r - (\partial_{xx} + q_c^2)^2]u - u^3, \qquad \text{(SH)} \qquad (9.67)$$

which is well known in the literature as the Swift–Hohenberg (SH) equation.

We can observe that such a stabilizing term is necessary even if the symmetry $u \to -u$ is broken. In that case a quadratic term, $c_2 u^2$, should be added to the linear SH operator, but it cannot guarantee the saturation of the unstable amplitude, because it contributes to $\partial_t u$ with a constant sign, independent of u. Therefore, the SH equation (9.67) is the simplest nonlinear equation of type I satisfying the symmetry $u \to -u$. If this symmetry is broken, the term $c_2 u^2$ should be added to its right-hand side.

Once we have turned our attention to the nonlinear term u^3 for the type I instability, it should not be surprising that the very same nonlinearity appears in Eqs. (9.4) and (9.5). In conclusion, simple considerations of possible instability scenarios, accompanied by the effect of symmetries and conservation laws, allow us to focus on three partial differential equations: the SH, the Cahn–Hilliard (CH), and the TDGL equations, corresponding, respectively, to type I (SH), type II (CH), and type III (TDGL) instabilities.

The study of the dynamics close to the homogenous, steady solution $u = 0$ has revealed a rich source of information even for $r > 0$, that is, when such a solution is linearly unstable: It has allowed us to classify instabilities and to extract relevant space and time scales. It is therefore not surprising that locating other steady states is often the first step beyond the linear study faced in Section 9.2, and it is also not surprising that spatially periodic steady states play a special role in pattern-forming systems. The rest of this section will be devoted to discussing the emergence of such periodic states.

Finding the steady states for TDGL is particularly simple, and in certain, physically relevant conditions they are also the steady states of the CH equation. In fact, TDGL and CH steady states are determined by the equations

$$u_{xx} + ru - u^3 = 0 \qquad \text{(TDGL)} \qquad (9.68)$$

$$\partial_{xx}[u_{xx} + ru - u^3] = 0. \qquad \text{(CH)} \qquad (9.69)$$

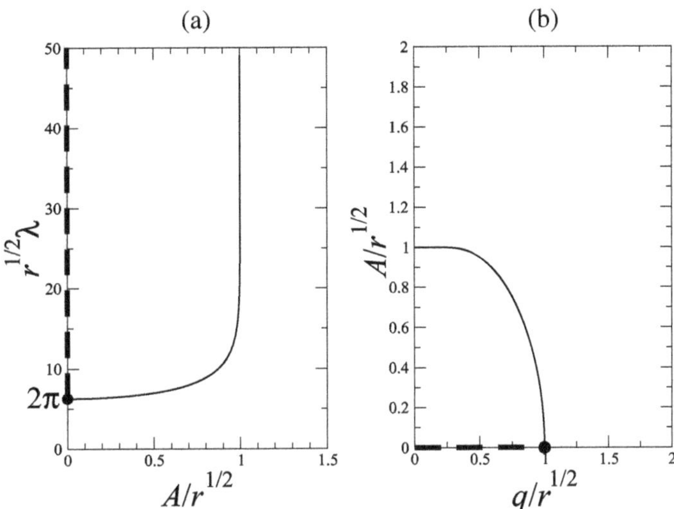

Fig. 9.7   The branch of steady states for TDGL, in the plane (λ, A) (a) and in the plane (A, q) (b). The dashed line intervals identify the unstable regions for the homogeneous solution, $u \equiv 0$.

Equation (9.68) is easily acknowledged as Newton's equation for a (fictitious) particle of position u at "time" x, oscillating in the potential

$$\tilde{V}(u) = -V(u) = r\frac{u^2}{2} - \frac{u^4}{4}. \tag{9.70}$$

In this case periodic solutions are the only bounded steady states, and if we denote by A the amplitude of such oscillations, we can say that: (i) There is a periodic steady state for any amplitude $0 < A < \sqrt{r}$. (ii) The wavelength is an increasing function of A (because the potential is less and less steep). (iii) $\lambda(A \to 0) = 2\pi/\sqrt{r}$, which is the smallest unstable wavelength in the linear spectrum of the homogeneous solution, that is, $\sigma_{\mathrm{III}}(q) > 0$ $\forall q < 2\pi/\sqrt{r}$. (iv) $\lambda(A \to \sqrt{r}) = \infty$, because (in the mechanical analogy) the solution would correspond to the particle leaving the left maximum of the potential and reaching the right maximum of the potential after an infinite time (or the other way around). These limiting solutions correspond to the positive and negative kinks discussed in Chapter 8. For large, but finite λ, the periodic solution corresponds to a sequence of positive and negative kinks at distance $\lambda/2$.

The exact expression of $\lambda(A)$, derived in Appendix Y, is

$$\lambda(A) = 4\sqrt{\frac{2}{2r - A^2}} K\left(\frac{A}{\sqrt{2r - A^2}}\right), \tag{9.71}$$

where $K(\cdots)$ is the complete elliptic integral of the first kind. In Fig. 9.7 we plot $\lambda(A)$ and $A(q)$ using rescaled variables so as to stress the intrinsic independence of these functions on r. In particular, the function $A(q)$ shows its similarity with the linear spectrum for the nonconserved model, $\omega(q) = r - q^2$, making evident that there is a periodic steady state for any q such that $\sigma_{\mathrm{III}}(q) > 0$ and that the amplitude $A(q)$ vanishes when $\sigma(q)$ vanishes.

Let us now pass to the steady states of the CH model, Eq. (9.69), which satisfy the equation

$$u_{xx} + ru - u^3 = c_1 x + c_2.\tag{9.72}$$

The constant c_1 must vanish in order to get confined solutions. In this case the mechanical analogy is introduced by the asymmetric potential $\tilde{V}_{\mathrm{CH}}(u) = \tilde{V}(u) - c_2 u$. Since CH dynamics conserves the order parameter and it is physically relevant to focus on symmetric steady states, that is, such that their spatial average $\langle u(x)\rangle = 0$, we can choose $c_2 = 0$. Therefore, symmetric steady states for the CH model are the same steady states as for the TDGL model.

If we finally turn to the SH model, we realize that in this case it is not possible to find steady states analytically for general r. However, it is possible for vanishing r, because we can use a perturbative approach. Since the approach is fairly general, we discuss it in some detail using a formalism that is appropriate for a large class of equations having the form

$$\partial_t u = \mathcal{L}[u] - u^3.\tag{9.73}$$

In the absence of the nonlinear term, the basic solution would be $u(x, t) = a(t)\cos(qx)$, with $\dot{a} = \sigma(q)a$. If q belongs to the unstable spectrum, that is, $\sigma(q) > 0$, we need the nonlinear term to compensate for the linear, exponential growth. For a single harmonic we would have

$$u^3 = a^3(t)\cos^3(qx) = \frac{a^3(t)}{4}\left(3\cos(qx) + \cos(3qx)\right).\tag{9.74}$$

Therefore, the nonlinear term has a double effect: It contributes to the dynamics of the base harmonic through the term $\cos(qx)$ and it also induces the rising of higher order harmonics through the term $\cos(3qx)$. The first effect goes in the expected direction, because it counterbalances the unstable growth of $a(t)$. However, the appearance of the term $\cos(3qx)$ tells us we should modify the ansatz by writing

$$u(x, t) = a(t)\cos(qx) + a_3(t)\cos(3qx),\tag{9.75}$$

whose cube is now more involved,

$$u^3 = \frac{3}{4}(a^3 + a^2 a_3 + 2aa_3^2)\cos(qx) + \frac{1}{4}(a^3 + 6a^2 a_3 + 3a_3^3)\cos(3qx)$$
$$+ \frac{3}{4}(a^2 a_3 + aa_3^2)\cos(5qx) + \frac{3}{4}aa_3^2\cos(7qx) + \frac{1}{4}a_3^3\cos(9qx).\tag{9.76}$$

As expected, even higher-order harmonics now appear, which implies we should write the whole series:

$$u(x, t) = a(t)\cos(qx) + \sum_{n\geq 1} a_{2n+1}(t)\cos((2n+1)qx).\tag{9.77}$$

The idea is that we can truncate the expansion (9.77) if $a \gg a_3 \gg a_5 \gg a_7 \cdots$. Let's now forget a_5 and higher-order harmonics and show that $a_3 \ll a$. So, let us approximate Eq. (9.76) with the first two terms on the right-hand side and replace them into Eq. (9.73).

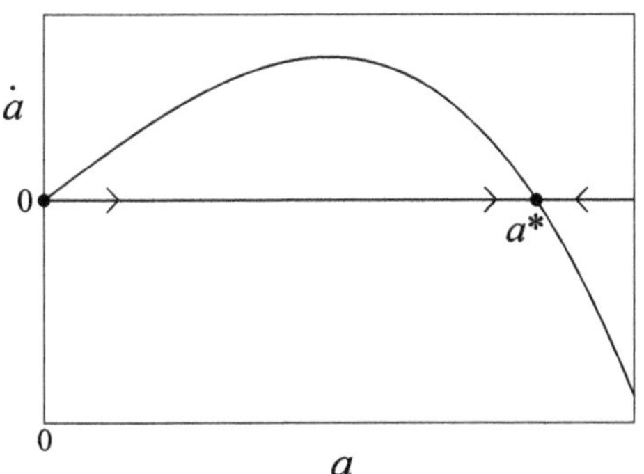 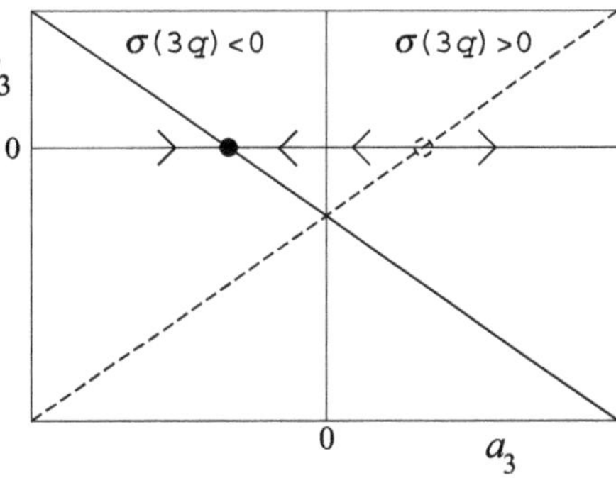

Fig. 9.8   Flows for $a(t)$ (see Eq. (9.80)) and $a_3(t)$ (see Eq. (9.81)).

We get two coupled equations for $a(t)$ and $a_3(t)$,

$$\dot{a} = \sigma(q)a - \frac{3}{4}(a^3 + a^2 a_3 + 2aa_3^2), \tag{9.78}$$

$$\dot{a}_3 = \sigma(3q)a_3 - \frac{1}{4}(a^3 + 6a^2 a_3 + 3a_3^3). \tag{9.79}$$

As part of the hypothesis (to be confirmed) that $a_3 \ll a$, we also have $a_3^2 a \ll a^2 a_3 \ll a^3$, so that we can keep only nonlinear terms proportional to a^3,

$$\dot{a} = \sigma(q)a - \frac{3}{4}a^3, \tag{9.80}$$

$$\dot{a}_3 = \sigma(3q)a_3 - \frac{1}{4}a^3. \tag{9.81}$$

The equation for $a(t)$ has two fixed points (stationary solutions), $a = 0$ and $a = a^* = \sqrt{4\sigma(q)/3}$. The trivial one, $a = 0$, is unstable, because $\sigma(q) > 0$. The nontrivial solution, $a = a^*$ (which exists only if $\sigma(q) > 0$), is stable, as easily shown graphically in Fig. 9.8, left panel.

Let us now focus on the equation for the amplitude a_3. The key point is the sign of $\sigma(3q)$. In the SH equation close enough to the threshold, $3q$ is certainly outside the unstable interval, regardless of the value of q belonging to the unstable interval. Therefore, $\sigma(3q) < 0$ and the resulting (negative) fixed point $a_3 = a_3^* = (a^*)^3/(4\sigma(3q))$ is stable; see Fig. 9.8, right panel. If it was $\sigma(3q) > 0$, the now positive fixed point a_3^* would be unstable and we should take into account higher-order harmonics. The proof is completed once we remark that

$$\frac{a_3^*}{a^*} \simeq \frac{(a^*)^2}{\sigma(3q)} \simeq \frac{\sigma(q)}{\sigma(3q)}. \tag{9.82}$$

In conclusion, if $\sigma(q)$ is small and $\sigma(3q)$ is negative and not small, $|a_3^*/a^*| \ll 1$ and the earlier expansion is consistent.

We stress that the condition for the validity of the calculation (which goes beyond the specific SH equation) is twofold: $\sigma(q)$ must be small and $\sigma(3q)$ must be negative (and not small). These conditions are surely satisfied by the SH equation close to the threshold, that

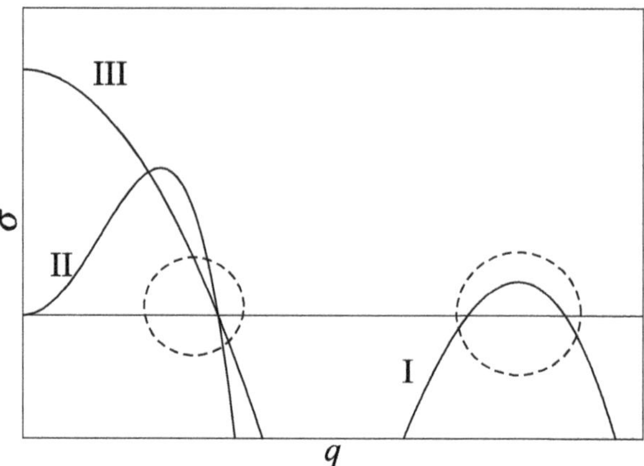

Fig. 9.9 Graphical sketch of the stability curves for the three models I, II, and III: The dashed areas highlight the q-regions, where the weakly nonlinear analysis can be applied.

is, for $r \to 0$, but they are also satisfied, for example, by the TDGL equation for any r, if $q \to \sqrt{r}$; see Fig. 9.9.

A fairly similar perturbative approach can be performed for the CH equation, if $q \to \sqrt{r}$. In that case, Eq. (9.73) is replaced by

$$\partial_t u = \mathcal{L}[u] + \partial_{xx}(u^3). \tag{9.83}$$

If we start again from Eq. (9.75), Eq. (9.76) should be replaced by

$$-\partial_{xx}(u^3) = \frac{3q^2}{4}(a^3 + a^2 a_3 + 2a a_3^2)\cos(qx) + \frac{9q^2}{4}(a^3 + 6a^2 a_3 + 3a_3^3)\cos(3qx), \tag{9.84}$$

where higher-order harmonics have been neglected.

We can now go directly to Eqs. (9.80) and (9.81) and rewrite them as

$$\dot{a} = \sigma(q)a - \frac{3q^2}{4}a^3 \tag{9.85}$$

$$\dot{a}_3 = \sigma(3q)a_3 - \frac{9q^2}{4}a^3. \tag{9.86}$$

Finally, we have

$$a^* = \sqrt{\frac{4\sigma(q)}{3q^2}} \tag{9.87}$$

$$a_3^* = \frac{9q^2(a^*)^3}{4\sigma(3q)} \tag{9.88}$$

$$\left|\frac{a_3^*}{a^*}\right| \simeq \frac{(a^*)^2}{|\sigma(3q)|} \simeq \frac{\sigma(q)}{|\sigma(3q)|} \ll 1. \tag{9.89}$$

We conclude that the same method can be consistently applied to the CH equation. In Fig. 9.9 we plot $\sigma(q)$ for the different models and highlight when the weakly nonlinear analysis can be applied: For model I, if $r \ll 1$, it can be applied to the whole unstable spectrum; for models II and III, it can be applied close to the upper limit of the unstable spectrum, for any r.

We should not be misled by this result and conclude that the above solutions are stable, because our analysis was limited to amplitude fluctuations. However, the earlier analysis is useful, because it proves the existence of periodic steady states in suitable limits and gives the explicit expression of these steady states.

9.5 Energetics

Dynamics can be accompanied by the minimization of a suitable Lyapunov functional. If so, we might think to circumvent the difficult task of determining the nonlinear evolution by minimizing such a functional, imagining that the "ground state" is the final outcome of dynamics. In fact, investigating the Lyapunov functional surely allows us to gain insights in the model, but it is not a shortcut to the dynamical problem for several reasons. First, as discussed in Chapter 8, for discrete models, metastable states may trap the system and block dynamics either forever (at $T = 0$ K) or for a long time. Second, the final ground state may be attained in a time that diverges with the system size; in this case, the way to it is more interesting than the final destination itself (think about the coarsening process occurring for TDGL and CH models). And finally, even transient dynamics may have its own interest.

For those who have not read Chapter 8 on phase-ordering, we start with the Lyapunov functional for the TDGL and CH equations; see Eqs. (9.4)–(9.6). It is straightforward to evaluate $\mathcal{F}_{GL}$ for a periodic configuration of large wavelength, because both the "domain wall" term, $\frac{1}{2}(\partial_x u)^2$, and the excess potential energy with respect to the ground state, $V(u) - V(u \equiv \pm\sqrt{r})$, are concentrated in the finite region where the order parameter passes from a minimum of the potential to the other minimum. Therefore, we can write the average value of the free energy as

$$\langle \mathcal{F}_{GL} \rangle = \frac{1}{\lambda} \int_0^\lambda dx \left[\frac{1}{2} u_x^2 + V(u) - V(\sqrt{r}) \right] + V(\sqrt{r}) \tag{9.90}$$

$$= \frac{\text{const}}{\lambda} + O\left(\frac{1}{\lambda^2}\right) + V(\sqrt{r}), \tag{9.91}$$

where we have used the fact that for large λ the integral $\int_0^\lambda [\cdots]$ of Eq. (9.90) is independent of λ. In conclusion, if we evaluate the average value of the Ginzburg–Landau free energy along the curve of the periodic steady state (see Fig. 9.7), energetic considerations based on the dynamical minimization of the Lyapunov functional would immediately lead us to establish a coarsening process where λ increases in time. Although this logical step is essentially true for TDGL/CH models, it is not true in general, as exemplified by the SH equation. So, we are now going to study the energetics for this equation, while the next sections are devoted to studying its dynamics.

The SH model has the variational structure

$$\frac{\partial u}{\partial t} = \left[r - (\partial_{xx} + q_c^2)^2 \right] u - u^3$$

$$\equiv -\frac{\delta \mathcal{F}_{\text{SH}}}{\delta u},$$

(9.92)

where

$$\mathcal{F}_{\text{SH}} = \int dx \left[\left(q_c^4 - r \right) \frac{u^2}{2} + \frac{u^4}{4} - q_c^2 u_x^2 + \frac{1}{2} u_{xx}^2 \right].$$

(9.93)

The form of the pseudo free-energy $\mathcal{F}_{\text{SH}}$ is not easily understandable, in particular the negative sign of the "tension" term u_x^2. Another peculiar feature is related to the u-dependent, potential part, which has a positive quadratic term for small r; that is, it is a single-well potential, not the classical double-well potential we are more used to. This is not surprising if we realize that the unstable q-interval of the homogeneous solution is centered in $q = q_c$ rather than in $q = 0$. In fact, for $r > q_c^4$ (strongly unstable regime, not considered here), the unstable interval extends up to $q = 0$ and the potential term gains the standard double-well form.

Let us now evaluate $\mathcal{F}_{\text{SH}}$ for a periodic steady state, in the approximation of small r, because in this limit we know the explicit expression of the steady state (see Section 9.4):

$$u(x) = a(q) \cos(qx) = \sqrt{\frac{4\sigma(q)}{3}} \cos(qx), \qquad \text{for } r \ll 1.$$

(9.94)

Since $\mathcal{F}_{\text{SH}}$ diverges in the thermodynamic limit, it is customary to evaluate its average value, as we did in Eq. (9.90),

$$\langle \mathcal{F}_{\text{SH}} \rangle = \frac{1}{L} \int_0^L dx [\cdots] = \frac{1}{\lambda} \int_0^\lambda dx [\cdots],$$

(9.95)

where the last equality is valid for periodic configurations of period λ. Using Eq. (9.93), we find

$$\langle \mathcal{F}_{\text{SH}} \rangle = \left(q_c^4 - r \right) \frac{\langle u^2 \rangle}{2} + \frac{\langle u^4 \rangle}{4} - q_c^2 \langle u_x^2 \rangle + \frac{1}{2} \langle u_{xx}^2 \rangle,$$

(9.96)

where the average values can be evaluated using (9.94),

$$\langle u^2 \rangle = a^2 \langle \cos^2(qx) \rangle = \frac{1}{2} a^2$$

$$\langle u^4 \rangle = a^4 \langle \cos^4(qx) \rangle = \frac{3}{8} a^4$$

$$\langle u_x^2 \rangle = a^2 q^2 \langle \sin^2(qx) \rangle = \frac{q^2}{2} a^2$$

$$\langle u_{xx}^2 \rangle = a^2 q^4 \langle \cos^2(qx) \rangle = \frac{q^4}{2} a^2.$$

(9.97)

Since $a^2(q) = \frac{4}{3}\sigma(q)$, we finally get

$$
\begin{aligned}
\langle \mathcal{F}_{\mathrm{SH}} \rangle &= -\frac{r}{4}a^2 + \frac{1}{4}a^2(q^2 - q_c^2)^2 + \frac{3}{32}a^4 \\
&= -\frac{1}{3}\sigma(q)\left[r - (q^2 - q_c^2)^2\right] + \frac{1}{6}\sigma^2(q) \\
&= -\frac{1}{6}\sigma^2(q),
\end{aligned}
\tag{9.98}
$$

where we have used the explicit expression for $\sigma(q)$, given in Table 9.1. Therefore, $\langle \mathcal{F}_{\mathrm{SH}} \rangle$ is minimal where $\sigma(q)$ is maximal,[4] that is, for $q = q_c$.

This result does not mean that the ground state, $u_{\mathrm{GS}}(x) = a(q_c)\cos(q_c x)$ is attained, because the deterministic dynamics may stop at a stationary state, if it is locally stable. This is why it is important to perform a dynamical investigation, which is the goal of Sections 9.6 and 9.7. At the end we will resume this discussion on energetics versus dynamics.

9.6 Nonlinear Dynamics for Pattern-Forming Systems: The Envelope Equation

The instability scenario I, qualitatively depicted in Fig. 9.4, is the prototype of bifurcation giving rise to a pattern of well-defined wavelength, and the simplest equation characterized by such a scenario is the SH equation. However, what is most relevant here is the universality of the dynamics close to the threshold. We are going to derive the relevant equation, called the amplitude or envelope equation, starting from the SH model, and then we discuss it in terms of symmetries, which is the only way to bring universality out.

Type I bifurcation is characterized by an interval of unstable q-vectors of order $\sqrt{r}$ and a growth rate of order $\sigma(q_c) = r$. In simple words, we expect to have a slow dynamics involving a weak modulation of an oscillation of wavelength $2\pi/q_c$. Dynamics is slow, because the time scale is set by the inverse of the growth rate, $\sigma(q_c)$, which is vanishingly small for $r \to 0$. The modulation is weak, because the active wave vectors, that is, such that $\sigma(q) > 0$, have a small variability for $r \to 0$. The last point can be corroborated by a simple calculation. If we sum two oscillating terms,

$$
u(x) = a_1 \cos(q_1 x) + a_2 \cos(q_2 x),
\tag{9.99}
$$

and define the mean wavevector $q_0 = (q_1 + q_2)/2$ and the half-difference $\delta = (q_1 - q_2)/2$, simple trigonometry gives

$$
u(x) = (a_1 + a_2)\cos(\delta x)\cos(q_0 x) + (a_2 - a_1)\sin(\delta x)\sin(q_0 x).
\tag{9.100}
$$

In our problem $q_0 \simeq q_c$ and $\delta \simeq \sqrt{r} \ll q_c$, which means we have oscillations of wavevector q_0 with a weakly varying amplitude. With that in mind we can anticipate the form of the solution,

$$
u(x, t) = A(x, t)e^{iq_c x} + \text{c.c.},
\tag{9.101}
$$

---

[4] Let us remember that steady states exist only where $\sigma(q) > 0$.

where the amplitude $A(x, t)$ depends weakly on x and t. The correct formalism to treat this type of problem is multiscale analysis, which is briefly introduced in Appendix Z.

The underlying idea (see Eq. (9.100)) is that the spatial variable x appears in the carrier wave as $q_c x$ and in the modulating amplitude as $\delta x \approx \sqrt{r} x$. It is therefore natural to introduce a fast scale $x_0 = x$ (because q_c is of order one) and a slow spatial scale $X = \sqrt{r} x$. As for time, the carrier wave is stationary, so no fast time scale appears, while the slow temporal scale is set by the product $\sigma(q_c)t \approx rt$, so we can define $T = rt$. We should finally remark that the amplitude A itself is small, close to the threshold: $A \approx u^* \approx \sqrt{\sigma(q_c)} \approx \sqrt{r}$.

To avoid handling square roots when performing the perturbative expansion, it is customary to put $r = \epsilon^2$. Under this notation, we can define the variables

$$
\begin{aligned}
x_0 &= x, && \text{fast scales} \\
X &= \epsilon x \quad T = \epsilon^2 t, && \text{slow scales}
\end{aligned}
\tag{9.102}
$$

and perform a standard expansion of the solution u,

$$
u = \epsilon u_1 + \epsilon^2 u_2 + \epsilon^3 u_3.
\tag{9.103}
$$

In fact, since the leading nonlinear term is of order ϵ^3, we expect the ϵ-expansion of u should arrive at such order.

We can now evaluate the time and space derivatives,

$$
\partial_t = \epsilon^2 \partial_T \quad \text{and} \quad \partial_x = \partial_{x_0} + \epsilon \partial_X,
\tag{9.104}
$$

and apply them to the SH equation,

$$
\partial_t u = [r - (\partial_{xx} + q_c^2)^2]u - u^3.
\tag{9.105}
$$

Finally, we can write

$$
\epsilon^2 \partial_T u = \epsilon^2 u - ((\partial_{x_0} + \epsilon \partial_X)^2 + q_c^2)^2 u - u^3
\tag{9.106}
$$

$$
= \epsilon^2 u - \left\{ (\partial_{x_0^2}^2 + q_c^2)^2 + 4\epsilon \partial_{x_0 X}^2 (\partial_{x_0^2}^2 + q_c^2) \right.
$$

$$
\left. + \epsilon^2 \left[2\partial_{XX}^2 (\partial_{x_0^2}^2 + q_c^2) + (2\partial_{x_0 X}^2)^2 \right] \right\} u - u^3
\tag{9.107}
$$

and expanding u we get

$$
\epsilon^3 \partial_T u_1 = \epsilon^3 u_1 - (\partial_{x_0^2}^2 + q_c^2)^2 (\epsilon u_1 + \epsilon^2 u_2 + \epsilon^3 u_3)
$$

$$
- 4\epsilon \partial_{x_0 X}^2 (\partial_{x_0^2}^2 + q_c^2)(\epsilon u_1 + \epsilon^2 u_2)
$$

$$
- \epsilon^3 \left[2\partial_{XX}^2 (\partial_{x_0^2}^2 + q_c^2) + (2\partial_{x_0 X}^2)^2 \right] u_1 - \epsilon^3 u_1^3 + O(\epsilon^4)
\tag{9.108}
$$

We should now proceed order by order.

Order ϵ

$$
(\partial_{x_0^2}^2 + q_c^2)^2 u_1 \equiv \mathcal{L}_0[u_1] = 0,
\tag{9.109}
$$

where we denote with $\mathcal{L}_0$ the unperturbed part of the linear evolution operator defined in Eq. (9.13). The general solution of (9.109) is

$$u_1 = A(X,T)e^{iq_c x_0} + \text{c.c.} \tag{9.110}$$

It is worth noting that this result, in terms of standard variables u, x, t means that

$$u(x,t) = \epsilon A(\epsilon x, \epsilon^2 t)e^{iq_c x} + \text{c.c.} + O(\epsilon^2). \tag{9.111}$$

This expression is clearly in agreement with the result for the steady state found in the Section 9.5, as we are going to argue at the end of this analysis.

Order ϵ^2

$$-\mathcal{L}_0[u_2] - 4\partial^2_{x_0 X}\left(\partial^2_{x_0^2} + q_c^2\right)u_1 = 0, \tag{9.112}$$

which implies

$$\mathcal{L}_0[u_2] = 0, \tag{9.113}$$

because the physically relevant solutions of the equation $(\partial^2_{x_0^2} + q_c^2)^2 u_1 = 0$ actually satisfy the stronger condition $(\partial^2_{x_0^2} + q_c^2)u = 0$. Finally, using that $\mathcal{L}_0[u_{1,2}] = 0$, at the next order, we obtain

Order ϵ^3

$$\partial_T u_1 = u_1 - \mathcal{L}_0[u_3] - \left(2\partial^2_{x_0 X}\right)^2 u_1 - u_1^3 \tag{9.114}$$

or, equivalently

$$\mathcal{L}_0[u_3] = -\partial_T u_1 + u_1 - \left(2\partial^2_{x_0 X}\right)^2 u_1 - u_1^3 \equiv f_3(x_0, X, T). \tag{9.115}$$

Since f_3 contains terms that are proportional to $e^{\pm iq_c x_0}$, which are solutions of the homogeneous equation $\mathcal{L}_0[u_3] = 0$, we must require that such terms vanish. Therefore, let us evaluate f_3 with $u_1 = Ae^{iq_c x_0} + A^*e^{-iq_c x_0}$,

$$f_3 = -\left(\partial_T Ae^{iq_c x_0} + \partial_T A^*e^{-iq_c x_0}\right) + \left(Ae^{iq_c x_0} + A^*e^{-iq_c x_0}\right)$$
$$+ 4q_c^2\left(\partial^2_{XX}Ae^{iq_c x_0} + \partial^2_{XX}A^*e^{-iq_c x_0}\right) - \left(Ae^{iq_c x_0} + A^*e^{-iq_c x_0}\right)^3 \tag{9.116}$$
$$= \left(-\partial_T A + A + 4q_c^2\partial^2_{XX}A - 3|A|^2 A\right)e^{iq_c x_0} - A^3 e^{i3q_c x_0} + \text{c.c.} \tag{9.117}$$

The vanishing of the term between brackets in the last line implies the so-called envelope (or amplitude) equation,

$$\frac{\partial A}{\partial T} = A(1 - 3|A|^2) + 4q_c^2\frac{\partial^2 A}{\partial X^2}. \tag{9.118}$$

It is worth stressing the generality of this equation, which depends not on the explicit form of the SH equation (9.67) but on its symmetries and, of course, on the type I linear spectrum. Let us show how the amplitude equation may be determined using the two main symmetries of the SH equation: the spatial inversion symmetry, $x \to -x$, and the translational invariance, $x \to x + \bar{x}$. Eq. (9.118) has been derived from the starting ansatz

$$u(x,t) = A(X,T)e^{iq_c x_0} + \text{c.c.}, \tag{9.119}$$

where $x_0 = x$, $X = \epsilon x$, and $T = \epsilon^2 t$. The symmetry $x \to -x$ implies that the envelope equation for $A(X,T)$ must satisfy the same symmetry, $X \to -X$. Furthermore, since $x \to -x$ implies $x_0 \to -x_0$, A^* must satisfy the same equation. Finally, the translational invariance means that $A \exp(iq_c \bar{x})$ must be a solution, if A is. We can now write the simplest equation for $A(X,T)$ satisfying these constraints,

$$\partial_T A = c_1 A + c_2 \partial_{XX} A + c_3 |A|^2 A, \tag{9.120}$$

where all coefficients c_i must be real. The signs of the coefficients of the linear terms are fixed by the condition that $A = 0$ is linearly unstable for large wavelength, so both c_1 and c_2 must be positive, while the coefficient of the nonlinear term must be negative in order to provide nonlinear saturation to the instability. The actual value of the coefficients is not relevant, because they can all be set equal to unity ($c_1 = c_2 = -c_3 = 1$) by rescaling X, T, and A. Finally, we remark that we have not considered the possible symmetry $u \to -u$. This means that Eq. (9.118) is still valid if we add a quadratic term to the SH equation (9.67).

Let us conclude this section by verifying that the amplitude equation (9.118) allows for simple steady-state solutions, already found in the Section 9.5. If we assume that $A(X) = A_0 e^{iQX}$ and replace it in Eq. (9.118), we find

$$|A_0|^2 = \frac{1 - 4q_c^2 Q^2}{3}, \tag{9.121}$$

so that the full solution, restoring old variables, is

$$u(x) = \epsilon A_0 e^{i(q_c + \epsilon Q)x} + \text{c.c.} \tag{9.122}$$

If we recognize that $q = q_c + \epsilon Q$, we can also write

$$\sigma(q) = r - (q^2 - q_c^2)^2 \tag{9.123}$$

$$\simeq \epsilon^2 - 4q_c^2 (q - q_c)^2 \tag{9.124}$$

$$= \epsilon^2 \left(1 - 4q_c^2 Q^2\right) \tag{9.125}$$

so that (apart from a constant, irrelevant phase factor)

$$\epsilon A_0 = \sqrt{\frac{\sigma(q)}{3}} \tag{9.126}$$

and

$$u(x) = \sqrt{\frac{\sigma(q)}{3}} e^{iqx} + \text{c.c.} = \sqrt{\frac{4\sigma(q)}{3}} \cos(qx), \tag{9.127}$$

which is the result found with a different approach in the Section 9.5 (see Eq. (9.94)). The key point here is that Eq. (9.118) gives more than the steady state; it gives the dynamics around the steady states, which allows us to determine their stability.

9.7 The Eckhaus Instability

We can now address the stability problem of the steady states of the amplitude equation. Let us rewrite the envelope equation,

$$\frac{\partial A}{\partial T} = A(1 - 3|A|^2) + 4q_c^2 \frac{\partial^2 A}{\partial X^2}, \tag{9.128}$$

whose time-independent solutions are

$$A_s(X) = \sqrt{\frac{1 - 4q_c^2 Q^2}{3}} e^{iQX} \equiv A_0 e^{iQX}. \tag{9.129}$$

We are going to perturb these solutions and study the resulting dynamics, retaining only the terms that are linear in the perturbations. However, rather than writing $A(X,T) = A_s(X) + \Delta(X,T)$, we perturb separately the amplitude and the phase of $A_s(X)$,

$$A(X,T) = A_0(1 + \delta)e^{i(QX+\phi)} \tag{9.130}$$

where both $\delta = \delta(X,T)$ and $\phi = \phi(X,T)$ are real.

Keeping only terms that are linear in δ, ϕ, we can write

$$|A|^2 = A_0^2(1 + 2\delta) \tag{9.131}$$

$$\partial_T A = A_0 e^{i(qX+\phi)} \left[\frac{\partial \delta}{\partial T} + i \frac{\partial \phi}{\partial T} \right] \tag{9.132}$$

$$\partial_{XX} A = A_0 e^{i(qX+\phi)} \left[\frac{\partial^2 \delta}{\partial X^2} - Q^2(1 + \delta) - 2Q \frac{\partial \phi}{\partial X} + i \frac{\partial^2 \phi}{\partial X^2} + 2iQ \frac{\partial \delta}{\partial X} \right]. \tag{9.133}$$

Assembling them in the envelope equation (9.128) and imposing the vanishing of the real and imaginary parts, we obtain two coupled, linear differential equations,

$$\frac{\partial \delta}{\partial T} = -6A_0^2 \delta + 4q_c^2 \frac{\partial^2 \delta}{\partial X^2} - 8q_c^2 Q \frac{\partial \phi}{\partial X} \tag{9.134a}$$

$$\frac{\partial \phi}{\partial T} = 4q_c^2 \frac{\partial^2 \phi}{\partial X^2} + 8q_c^2 Q \frac{\partial \delta}{\partial X}. \tag{9.134b}$$

The stability analysis performed in Section 9.4 corresponds to perturb only the amplitude, making it dependent on time, which means that $\phi = $ constant and $\delta = \delta(T)$. In this case, Eq. (9.134) reduce to $\partial_T \delta = -6A_0^2 \delta$; that is, the solution $A_s(X)$ is stable for pure amplitude fluctuations. An equally simple analysis is not possible for the phase, because $\delta = 0$ implies that ϕ is a constant. In simple terms, the dynamics of the phase must take into account variations of the amplitude. It is possible to simplify the above equations to obtain an effective one for the phase only, but we will do that at the end of a more rigorous analysis.

The coupled linear equations (9.134) are studied by standard Fourier analysis, that is, assuming

$$\delta = \bar{\delta} e^{\alpha T} e^{iKX} \tag{9.135}$$

$$\phi = \bar{\phi} e^{\alpha T} e^{iKX}, \tag{9.136}$$

which give a (2×2) linear system,

$$(\alpha + 6A_0^2 + 4q_c^2 K^2)\bar{\delta} + 8iq_c^2 QK\bar{\phi} = 0$$
$$-8iq_c^2 QK\bar{\delta} + (\alpha + 4q_c^2 K^2)\bar{\phi} = 0. \tag{9.137}$$

For $K = 0$, we get a more formal version of the discussion of the previous paragraph, because the linear system simplifies to

$$(\alpha + 6A_0^2)\bar{\delta} = 0 \quad \text{and} \quad \alpha\bar{\phi} = 0, \tag{9.138}$$

whose two solutions are

$$\alpha = \alpha_1 = -6A_0^2 \qquad \text{with} \quad (\bar{\delta}, \bar{\phi}) = (1, 0) \tag{9.139}$$
$$\alpha = \alpha_2 = 0 \qquad \text{with} \quad (\bar{\delta}, \bar{\phi}) = (0, 1). \tag{9.140}$$

The first solution corresponds to stable amplitude fluctuations and the second solution to neutral phase fluctuations. When $K \neq 0$, we have two branches, $\alpha_{1,2}(K)$. Our goal is now to study $\alpha_2(K)$ to evaluate its sign at finite K.

The system of two equations (9.137) has solutions only if the determinant vanishes, that is, if

$$(\alpha + 6A_0^2 + 4q_c^2 K^2)(\alpha + 4q_c^2 K^2) - (8q_c^2 QK)^2 = 0 \tag{9.141}$$

or

$$\alpha^2 + 2\alpha(4q_c^2 K^2 + 3A_0^2) + 4q_c^2 K^2(6A_0^2 + 4q_c^2 K^2 - 16q_c^2 Q^2) = 0, \tag{9.142}$$

which gives the two branches

$$\alpha_{1,2} = -(4q_c^2 K^2 + 3A_0^2) \mp \left[(4q_c^2 K^2 + 3A_0^2)^2 - 8q_c^2 K^2(1 + 2q_c^2 K^2 - 12q_c^2 Q^2)\right]^{1/2}. \tag{9.143}$$

The branch $\alpha_1(K)$, corresponding to the minus sign, is negative for any K, therefore giving stable dynamics. Instead, we are interested in the branch $\alpha_2(K)$, corresponding to the plus sign and such that $\alpha_2(0) = 0$. For large K, it is straightforward to check that Eq. (9.141) simplifies to $(\alpha + 4q_c^2 K^2)^2 = 0$, giving $\alpha_{1,2} = -4q_c^2 K^2$. Therefore, a possible instability may arise only from a positive α_2 at small K. In this limit we can neglect K^4 terms, so that the quantity in square brackets of Eq. (9.143), $[\dots]$, and its square root are written as

$$[\cdots] = (1 - 4q_c^2 Q^2)^2 + (8q_c^2 QK)^2 \tag{9.144}$$
$$[\cdots]^{1/2} = (1 - 4q_c^2 Q^2) + \frac{1}{2}\frac{(8q_c^2 Q)^2}{1 - 4q_c^2 Q^2}K^2. \tag{9.145}$$

Finally, we obtain

$$\alpha_2(K) = -4q_c^2\left(\frac{1 - 12q_c^2 Q^2}{1 - 4q_c^2 Q^2}\right)K^2 \equiv -4q_c^2\frac{\mathcal{N}(Q)}{\mathcal{D}(Q)}K^2. \tag{9.146}$$

Let us start by discussing the role of the different wavevectors appearing here: q_c, Q, and K.

- q_c appears in the SH equation and it is the critical wavevector: The dispersion relation of the linear stability analysis of the solution $u = 0$ gives $\sigma(q) = \epsilon^2 - (q^2 - q_c^2)^2$, therefore $\sigma > 0$ and the solution is unstable in the vicinity of $q = q_c$.

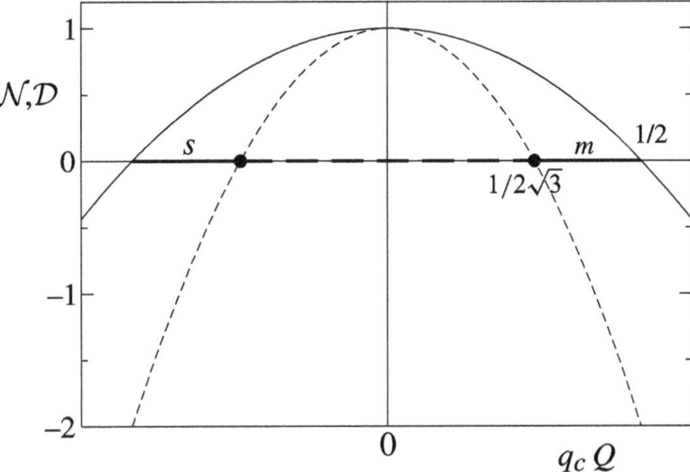

Fig. 9.10 $\mathcal{D}(q)$, solid line, and $\mathcal{N}(q)$, dashed line (see Eq. (9.146)). Thick dashed segment indicates Eckhaus stable steady states. Thick solid segments (regions s and m) indicate Eckhaus unstable steady states.

- Q labels the steady states appearing in the range where $\sigma(q) > 0$ and is related to q via the relation $q = q_c + \epsilon Q$. The explicit expression of the steady state is given in Eq. (9.129). The quantity at the denominator of Eq. (9.146) is proportional to $\sigma(q)$, therefore it is positive in the range of application of such an equation.
- The stability of steady solutions labeled by Q has been studied in a linear approximation, where the perturbation has spatial and time dependences of the form e^{iKX} and $e^{\alpha T}$, where $\alpha = \alpha(K)$ much in the same way that $\sigma = \sigma(q)$ for the stability analysis of the homogeneous solution $u = 0$. The main difference is that we have two branches, $\alpha_{1,2}(K)$, representing amplitude fluctuations (α_1) and phase fluctuations (α_2), with $\alpha_1(K) < 0$. Eq. (9.146) gives the spectrum of phase fluctuations at small but finite K.

We can now come back to $\alpha_2(K)$ (see Eq. (9.146)) and draw the relevant conclusions. Its sign depends on the numerator $\mathcal{N}(q)$. If Q is small enough, that is, q is very close to the critical value q_c, $\alpha_2 < 0$ and the steady state is stable. On the contrary, if Q is "large" (but such that the denominator $\mathcal{D}(q)$ is positive, of course), then $\alpha_2(K)$ is positive and the steady state is unstable. In Fig. 9.10 we plot $\mathcal{D}(q)$ and $\mathcal{N}(q)$. $\mathcal{D}(q)$ is essentially equal to $\sigma(q)$ in the limit $\epsilon \to 0$; see Eq. (9.125). Steady-state solutions exist if and only if $\mathcal{D}(q) > 0$ and they are stable if $\mathcal{N}(q) > 0$, that is, for $q_c Q < 1/2\sqrt{3}$ (thick, dashed line segment); otherwise they are unstable (thick, solid line segments).

Since the stability depends on the branch $\alpha_2(K)$, which refers to phase fluctuations, the Eckhaus instability is a *phase* instability, leading to changes in the local wavevector of the structure. If the wavector is too small (region s), it will tend to increase by the *splitting* of a cell in two (a cell is the spatial region corresponding to a single period of oscillation). Conversely, if the wavevector is too large (region m), it will tend to decrease by the *merging* of two neighboring cells. We should not think these processes (splitting and merging) involve all cells: If so, the wavevector of the structure would double (splitting) or it would halve (merging), leading the system to a region where the homogeneous solution is stable, which is not possible. We should rather think of local processes occurring, where cells, because of fluctuations, are smaller or larger than the optimal value, corresponding to q_c.

The earlier qualitative discussion of the Eckhaus instability can be accompanied by energetic considerations, according to which the minimum of the Lyapunov functional is attained for $q = q_c$, that is, for $Q = 0$. Stability analysis tells us that a finite region around $Q = 0$ is linearly stable, which may prevent the system from attaining the "ground state." In fact, we should distinguish between an adiabatic change of the control parameter and a sudden increase to a small but finite value of r. In the former case, cells of the optimal value $q = q_c$ appear as soon as $r = 0^+$ and their size is maintained afterward. In the latter case, the system passes abruptly from a stable to an unstable state, characterized by a finite unstable interval. Because of fluctuations, the instability will lead the system to display cells of various size, and the splitting/merging process described earlier is visible. This readjustment of the local size is called Eckhaus instability and is a secondary instability, because it occurs after the Rayleigh–Bénard instability (the primary one) has led to passing from the homogeneous (conductive) state to a nonhomogeneous (convective) state. In the absence of thermal noise, the system may be frozen in the Eckhaus stable region without attaining the ground state, that is, $q = q_c$.

We can now go back to coupled Eq. (9.134). As anticipated, we cannot study phase dynamics assuming a constant amplitude, $\delta = 0$. However, we know that for any given, constant value of the phase, $\phi = \phi_0$, the amplitude is stable, while phase dynamics has a diffusive character. This means that amplitude dynamics is *enslaved* to phase dynamics, and on the time scale relevant for phase dynamics, the amplitude can be considered as constant, so $\partial_T \delta = 0$ in Eq. (9.134a). Furthermore, we can focus on small K, so the term proportional to $\partial_{XX}\delta$ can be neglected with respect to the term proportional to δ. We are left with the simplified Eq. (9.134a),

$$-6A_0^2\delta - 8q_c^2 Q \frac{\partial \phi}{\partial X} = 0, \tag{9.147}$$

which gives the "slavery relationship" between δ and ϕ,

$$\delta = -\frac{4q_c^2 Q}{3A_0^2} \frac{\partial \phi}{\partial X}. \tag{9.148}$$

Finally, if we replace it in Eq. (9.134b), we get the phase diffusion equation

$$\frac{\partial \phi}{\partial T} = 4q_c^2 \frac{\partial^2 \phi}{\partial X^2} + 8q_c^2 Q \frac{\partial \delta}{\partial X} \tag{9.149}$$

$$= \left(4q_c^2 - 8q_c^2 Q \frac{4q_c^2 Q}{3A_0^2} \right) \frac{\partial^2 \phi}{\partial X^2} \tag{9.150}$$

$$= D(Q) \frac{\partial^2 \phi}{\partial X^2}, \tag{9.151}$$

where the phase diffusion coefficient, $D(Q)$, using Eq. (9.129), is given by

$$D(Q) = 4q_c^2 \frac{1 - 12q_c^2 Q^2}{1 - 4q_c^2 Q^2} \equiv 4q_c^2 \frac{\mathcal{N}(Q)}{\mathcal{D}(Q)}. \tag{9.152}$$

Therefore, Eq. (9.146) can be rewritten as

$$\alpha_2(K) = -D(Q)K^2,\tag{9.153}$$

so that the condition for phase instability, $\alpha_2(K) > 0$, is equivalent to having a negative phase diffusion coefficient, $D(Q) < 0$.

9.8 Phase Dynamics

All models we have considered so far, TDGL, CH, and SH, are characterized by a branch of periodic steady states, $u_s(x, q) = u_s(x+\frac{2\pi}{q}, q)$, for all q that make the homogeneous solution $u = 0$ unstable, that is, $\sigma(q) > 0$. These solutions have been shown in suitable limits to be stable against amplitude fluctuations, and all relevant dynamics is phase dynamics, with local changes of the wavelength. This process may produce an endless coarsening – this is the case for TDGL and CH – or a transient rearrangement until the wavelength is in the stable domain – this is the case for SH. The stability of the amplitude implies its dynamics is adiabatically enslaved to phase dynamics. The key point is that phase dynamics is slow, which allows a multiscale perturbative treatment. Let us investigate why.

Translational invariance means that $u_s(x + c, q)$ is still a solution for any constant c. Therefore, if c acquires a weak spatial dependence such that c is almost constant on a scale $\lambda = 2\pi/q$, we expect the resulting dynamics to be slow. In a slightly more formal way, we can assume that dynamics occurs in the functional space

$$u(x, t) = u_s(x, q(X, T)),\tag{9.154}$$

where, as before, $X = \epsilon x$ and $T = \epsilon^2 t$. Two main features of this approach to phase dynamics should be stressed: First, it is applicable to patterns arbitrarily far from the threshold, and second, the meaning of ϵ is not obvious. In fact, we are used to perturbative treatments, where a small parameter appears in the differential equation or in the Hamiltonian, while here it does not appear explicitly. Its meaning is, rather, related to the weakness of the perturbation, or better still, to the weakness of the variability of q. We are now going to be more precise.

The steady state $u_s(x, q)$ is defined by a constant q, and the phase of the pattern is simply $\phi = qx$. In the general case of a (slowly) varying q, we can define the phase

$$\phi = \int dx\, q\tag{9.155}$$

so that $q = \partial_x \phi$. We also define the slow phase

$$\psi = \epsilon\phi,\tag{9.156}$$

so $q = \partial_X \psi$ (actually, both ψ and X are of order ϵ and their ratio is of order one). The expansion of the order parameter is written as

$$u(x, X, T) = u_0(\phi, q(X, T)) + \epsilon u_1(\phi, q(X, T))\tag{9.157}$$

and derivatives are written as

$$\partial_x = q\partial_\phi + \epsilon\partial_X \tag{9.158}$$

$$\partial_{xx} = q^2\partial_{\phi\phi} + \epsilon(q^2\partial_\phi\partial_X + \partial_X q\partial_\phi) + O(\epsilon^2) \tag{9.159}$$

$$= q^2\partial_\phi\partial_X + \epsilon(2q\partial_X + (\partial_X q))\partial_\phi \tag{9.160}$$

$$= q^2\partial_\phi\partial_X + \epsilon\psi_{XX}(1 + 2q\partial_q)\partial_\phi \tag{9.161}$$

$$\partial_t = \epsilon^2\partial_T\phi\partial_\phi + \epsilon^2\partial_T \tag{9.162}$$

$$= \epsilon\partial_T\psi + O(\epsilon^2). \tag{9.163}$$

We can now apply this method to any model displaying a branch of steady states. We could do that for the SH equation and we would get a result similar to, but more general than, Eq. (9.151), because it would be valid for any value of the control parameter. However, we will illustrate the method and its power with a simple generalization of the TDGL equation,

$$\partial_t u = u_{xx} + F(u), \tag{9.164}$$

the standard model corresponding to $F(u) = u - u^3$.

If we now use the earlier expressions for derivatives and the u-expansion, we obtain

$$\epsilon\frac{\partial\phi}{\partial T}\partial_\phi u_0 = \left[q^2\partial_{\phi\phi} + \epsilon\psi_{XX}(1 + 2q\partial_q)\partial_\phi\right](u_0 + \epsilon u_1) + F(u_0 + \epsilon u_1) \tag{9.165}$$

$$= \left[q^2\partial_{\phi\phi}u_0 + F(u_0)\right]$$
$$+ \epsilon\left\{\left[q^2\partial_{\phi\phi} + F'(u_0)\right]u_1 + \psi_{XX}(1 + 2q\partial_q)\partial_\phi u_0\right\} + O(\epsilon^2) \tag{9.166}$$

The vanishing of the zero-order term,

$$\mathcal{N}[u_0] \equiv q^2\partial_{\phi\phi}u_0 + F(u_0) = 0, \tag{9.167}$$

trivially corresponds to the condition defining steady states, so $u_0(\phi, q) = u_s(\phi = qx, q)$. The vanishing of the first-order term gives instead

$$\frac{\partial\phi}{\partial T}\partial_\phi u_0 = \left[q^2\partial_{\phi\phi} + F'(u_0)\right]u_1 + \psi_{XX}(1 + 2q\partial_q)\partial_\phi u_0, \tag{9.168}$$

which is more transparent if rewritten as

$$\left[q^2\partial_{\phi\phi} + F'(u_0)\right]u_1 = \frac{\partial\phi}{\partial T}\partial_\phi u_0 - \psi_{XX}(1 + 2q\partial_q)\partial_\phi u_0. \tag{9.169}$$

It therefore has the form

$$\mathcal{L}[u_1] = g(u_0, \psi_T, \psi_{XX}), \tag{9.170}$$

where the linear operator

$$\mathcal{L} \equiv q^2\partial_{\phi\phi} + F'(u_0) \tag{9.171}$$

is the so-called Fréchet derivative of the operator $\mathcal{N}[u_0]$ defined in Eq. (9.167). In fact, if we take the derivative of such an equation with respect to ϕ, we get

$$q^2\partial_{\phi\phi\phi}u_0 + F'(u_0)\partial_\phi u_0 = \mathcal{L}[\partial_\phi u_0] = 0. \tag{9.172}$$

According to the Fredholm alternative theorem (see Appendix Z), Eq. (9.170) has a solution only if g, defined as the right-hand side of Eq. (9.169), is orthogonal to $\partial_\phi u_0$,

$$\langle(\partial_\phi u_0)g\rangle \equiv \frac{1}{2\pi}\int_0^{2\pi} d\phi(\partial_\phi u_0)g = 0. \tag{9.173}$$

We must therefore require that

$$\langle(\partial_\phi u_0)^2\rangle \partial_T\psi - \langle\partial_\phi u_0(1+2q\partial_q)\partial_\phi u_0\rangle \partial_{XX}\psi = 0, \tag{9.174}$$

which implies

$$\langle(\partial_\phi u_0)^2\rangle \partial_T\psi = \partial_q \langle q(\partial_\phi u_0)^2\rangle \partial_T\psi. \tag{9.175}$$

We finally get a diffusion equation for the slow phase ψ,

$$\frac{\partial\psi}{\partial T} = \frac{\partial_q \langle q(\partial_\phi u_0)^2\rangle}{\langle(\partial_\phi u_0)^2\rangle} \frac{\partial^2\psi}{\partial X^2} \equiv D(q)\frac{\partial^2\psi}{\partial X^2}, \tag{9.176}$$

where $D(q)$ is the phase diffusion coefficient and its sign is related to the stability ($D(q) > 0$) or instability ($D(q) < 0$) of the pattern of period $\lambda = 2\pi/q$ with respect to large wavelength modulations. Since it has the form $D(q) = (\partial_q D_1(q))/D_2(q)$, with $D_1(q) = \langle q(\partial_\phi u_0)^2\rangle$ and $D_2(q) = \langle(\partial_\phi u_0)^2\rangle > 0$, the sign of D depends on the increasing or decreasing behavior of $D_1(q)$. Before analyzing it, let us remark that the above procedure allows us to gain dynamical information from the analysis of the branch of steady states.

We have applied this method to Eq. (9.164), because its stationary solutions are particularly simple. In fact, the condition of time independence writes $u_{xx} + F(u) = 0$, which is Newton's equation of motion for a fictitious particle with position u at time x and subject to the force $-F(u)$, that is, moving in the potential $V(u) = \int du F(u)$. For the standard TDGL equation, $F(u) = u - u^3$ and $V(u) = u^2/2 - u^4/4$, but here the only hypothesis is that the trivial solution $u \equiv 0$ is linearly unstable, which means that $F(u) \simeq c_1 u$ for small u, with $c_1 > 0$. In other words, $V(u)$ is a harmonic potential for small oscillations, but we make no assumptions about the nonlinear part of $F(u)$.

We are now going to prove that the phase diffusion coefficient $D(q)$ (see Eq. (9.176)) is related to the amplitude dependence of the wavelength of steady states.[5] If we revert to the old variable x, we find

$$D_1 = \langle q(\partial_\phi u_0)^2\rangle = \frac{1}{2\pi}\int_0^{2\pi} d\phi q(\partial_\phi u_0)^2 = \frac{1}{2\pi}\int_0^\lambda dx(\partial_x u_0)^2 \equiv \frac{J}{2\pi}, \tag{9.177}$$

where J is the well-known action variable. In a similar way, we find $D_2 = (2\pi)^{-2}\lambda J$. Following classical mechanics results, the derivative of J with respect to the energy of the particle gives the period of oscillation, λ, so the following expression is finally established

$$D(q) = -\frac{\lambda^2 F(A)}{J(\partial_A\lambda)}, \tag{9.178}$$

where A is the amplitude of the oscillation, that is, the (positive) maximal value of $u_0(x)$.

---

[5] In the mechanical analogy, this means the amplitude dependence of the period of oscillation of the particle.

In conclusion, the stability is ruled by $\partial_A \lambda$. If the wavelength of the steady states is an increasing function of the amplitude, the steady states are unstable and a coarsening process takes place; otherwise, no phase instability occurs, the wavelength remains constant, and the amplitude diverges. These two scenarios are easily understood if $F(u) = c_1 u + c_3 u^3$, because when $c_3 < 0$ it is characteristic of the TDGL case, while for $c_3 > 0$ there is no nonlinear saturation of the linear stability. However, the approach is valid for *any* function $F(u)$, including the cases where $V(u)$ is such that $\partial_A \lambda$ changes sign. Furthermore, the above approach allows us to find the asymptotic coarsening law, $L \simeq t^n$, through a simple dimensional ansatz,

$$|D(q)| \sim \frac{L^2}{t}, \tag{9.179}$$

where $q = 2\pi/L$. If we apply this ansatz to the TDGL equation, $F(u) = u - u^3$, we recover the logarithmic coarsening found in Chapter 8.

9.9 Back to Experiments

This chapter, more than the others, has referred to a few experimental examples in order to make explicit the meaning of phrases such as pattern formation and (in this context) control parameter and order parameter. We did not want to limit ourselves to a purely formal enunciation of such concepts. Although it is not our intention to provide a theory for specific pattern-forming systems, we want to close the chapter by going back to some classes of experiments.

Experiments on the convective instability in a fluid heated from below are fairly old and we would like to stress the difference between the setup where the fluid is confined and where its upper surface is free. In the confined geometry case we speak of Rayleigh–Bénard instability, which is due to the temperature dependence of the density of the fluid, which favors the exchange of the lower/warmer/lighter parcels with the upper/colder/heavier parcels. In the nonconfined geometry, the upper surface of the fluid is free and we speak of Bénard–Marangoni instability. Now the convection is triggered by the temperature dependence of the surface tension at the free surface, which is in contact with the air. According to the Marangoni effect, the gradient of surface tension induces a mass transfer at the interface, but mass conservation requires a simultaneous transfer of mass in the vertical direction, which causes the rising of convection cells.

Two main types of cells appear in convection experiments, rolls and hexagons, which differ in up/down symmetry (it is present in rolls and absent in hexagons). In fact, in the latter case, the fluid rises (falls) in the center of the hexagons and falls (rises) at the edges. In standard conditions, Rayleigh–Bénard cells are roll-like, while hexagons appear when the up/down symmetry is broken by boundary conditions, either directly (the case of Bénard–Marangoni instability) or indirectly, because, for example, the viscosity depends on temperature. Although convection cells have a constant wavelength, their orientation may vary from one area to another, and, for example, domains with rolls of different orientation may appear. In this case, the size of these domains increases in time through a coarsening process, similar to what we have discussed in the previous chapter on phase separation.

Finally, if we further increase the control parameter, for example, the temperature difference ΔT between the lower and upper surfaces, new instabilities appear, until a turbulent state develops.

Experiments on patterns at the surface of a crystal surface brought out of equilibrium, either by a growth or by an erosion process, are more recent, because they need experimental conditions and probes that have been developed only since the 1980s. Furthermore, the popularity of such experiments has also been due to the possibility of testing the kinetic roughening models discussed in Chapter 7. In this chapter we have been interested in *deterministic* pattern-forming systems, whose phenomenology at a crystal surface is very broad. In fact, the mechanisms leading to an instability, then to pattern formation, range from kinetic to energetic mechanisms, passing through athermal ones. Unlike the hydrodynamic instabilities, which can be faced, in principle, by making use of the Navier–Stokes equations, there is no general, microscopic theory describing, for example, a growing surface far from equilibrium. We must therefore resort to using phenomenological equations or, in some cases, mesoscopic models based on a clear separation of time scales between the fast dynamics of surface atoms and the slow dynamics of steps. We may cite two different cases, the homoepitaxial growth[6] of a metal and the heteroepitaxial growth of a semiconductor. In the former case, a kinetic instability may be at the origin of a growth dynamics with some resemblance to a phase separation process (see Chapter 8), where the local surface slope $\mathbf{m} = \nabla z(\mathbf{x}, t)$ plays the role of a (vectorial) order parameter, with the additional constraint to be an irrotational field, $\nabla \times \mathbf{m} \equiv 0$. The linear instability falls into category II, but only if nonconservative processes, such as evaporation, are negligible. Nonlinear dynamics is a phase dynamics like that discussed in Section 9.8, with the additional complication that $\mathbf{m}$ is an irrotational vector field. In the latter case, heteroepitaxial growth of a semiconductor, we have an energetic instability driven by elasticity, whose final product is potentially of great interest, because it might be a self-organized way to produce quantum dots. Elasticity makes the problem fairly complicated even at a linear level, and the classes I–III introduced in Section 9.2 are not sufficient to cover the present problem. In fact, we need a type I scenario for a conserved order parameter.

The third example of pattern formation shown in panels (c1–c3) of Fig. 9.1 is a case of athermal dynamics in a granular system. Here the attempts to give a continuum description of dynamics, even at a linear level, are not satisfactory, and the system should rather be studied with simulations such as molecular dynamics. Despite the inherent difficulties due to the lack of common conceptual tools (free energy, temperature, etc.), in the dynamics of granular media, it is possible to recognize phenomena that are familiar after reading Chapters 8 and 9.

We conclude this section with the celebrated Turing patterns. Together with those emerging from hydrodynamic Bénard instabilities, they are the most famous stationary patterns. Although the original article by Turing dates from 1952, it has been necessary to

---

[6] Homoepitaxy means an ordered growth of a material fitting to the lattice structure of a substrate of the same type. Heteroepitaxy means the substrate is of a different type.

Fig. 9.11 Stationary chemical patterns. (a, b) Hexagons; (c) stripes (subject to a transverse instability); (d) mixed state (between spots and stripes). The bar beside each picture represents 1 mm. Reprinted from Q. Ouyang and H. L. Swinney, Transition from a uniform state to hexagonal and striped Turing patterns, *Nature*, **352** (1991) 610–612.

wait 40 years to have clear experimental evidence of Turing patterns, the main difficulty being the necessity to have species with significantly different diffusion coefficients. The experimental evidence of Turing patterns was obtained as a result of the chlorite–iodide–malonic acid (CIMA) reaction, where reagents are continuously fed on the top and the bottom of a thin hydrogel, where chemical species diffuse, while products are continuously removed. The continuous feeding and removal of chemicals is a necessary condition to have a stationary setup. In Fig. 9.11 we plot some stationary patterns in CIMA reactions, corresponding to different experimental conditions. The control parameter is the temperature, which allows us to pass from a homogeneous to a pattern configuration, when it is lowered below $T_c \simeq 18°C$.

9.10 Bibliographic Notes

The most important bibliographic resource to which we direct the reader is the book by M. Cross and H. Greenside, *Pattern Formation and Dynamics in Nonequilibrium Systems* (Cambridge University Press, 2009). Here you can find many physical considerations, discussions of experiments, and theoretical calculations. The historical review paper on the topic of this chapter is M. C. Cross and P. C. Hohenberg, Pattern Formation Outside of Equilibrium, *Reviews of Modern Physics*, **65** (1993) 851–1112.

For a more formal approach, see R. Hoyle, *Pattern Formation: An Introduction to Methods* (Cambridge University Press, 2006).

The book by C. Misbah, *Complex Dynamics and Morphogenesis* (Springer, 2017), devotes much attention to bifurcation theory, which is the formally correct way to approach pattern formation. This book is a useful introduction to nonlinear physics and is suited to students.

Another book devoted to various topics of nonlinear physics is P. Manneville, *Instabilities, Chaos and Turbulence* (Imperial College Press, 2004).

The reader interested in pattern formation in the real world should consult Philip Ball, *Nature's Patterns: A Tapestry in Three Parts* (Oxford University Press, 2009), where the three parts are branches, flow, and shapes.

For more details on Turing patterns, the first reference is the original paper by A. M. Turing, The Chemical Basis of Morphogenesis, *Philosophical Transactions of the Royal Society of London B: Biological Sciences*, **237** (1952), 37–72. Details about recent experiments can be found in the book by Cross and Greenside.

A detailed discussion of convective and other hydrodynamic instabilities is given in S. Chandrasekhar, *Hydrodynamic and Hydromagnetic Stability* (Dover Publications, 1981).

A formulation of the Fredholm alternative theorem is given in D. Zwillinger, *Handbook of Differential Equations*, 3rd ed. (Academic Press, 1997).

Binary Elastic Collisions in the Hard Sphere Gas

We describe the elastic collision between particles of an ideal gas by a simple classical model. In practice we approximate the short-range repulsive potential acting between molecules in a real gas by a hardcore interaction between massive particles. This is quite a reliable approximation in view of the small dimensions of the molecules. The hypothesis of elastic collisions implies that the total momentum and the total (kinetic) energy are conserved quantities. We assume that all particles have the same mass m, so that the conservation laws simplify to:

$$\mathbf{v} + \mathbf{u} = \mathbf{v}' + \mathbf{u}' \tag{A.1}$$

$$\mathbf{v}^2 + \mathbf{u}^2 = \mathbf{v}'^2 + \mathbf{u}'^2, \tag{A.2}$$

where unprimed and primed variables indicate the particle velocities before and after the collision, respectively.

Taking the square of Eq. (A.1) and using Eq. (A.2), we obtain

$$\mathbf{v} \cdot \mathbf{u} = \mathbf{v}' \cdot \mathbf{u}'. \tag{A.3}$$

As a consequence of this equation and of Eq. (A.2), also the following equalities hold

$$(\mathbf{v} - \mathbf{u})^2 = (\mathbf{v}' - \mathbf{u}')^2 \quad \rightarrow \quad |\mathbf{v} - \mathbf{u}| = |\mathbf{v}' - \mathbf{u}'|. \tag{A.4}$$

It is instructive introducing the center of mass velocity and the relative velocity of the colliding particles before the collision,

$$\mathbf{W} = \frac{1}{2}(\mathbf{v} + \mathbf{u}), \qquad \mathbf{w} = \mathbf{v} - \mathbf{u}, \tag{A.5}$$

and after the collision,

$$\mathbf{W}' = \frac{1}{2}(\mathbf{v}' + \mathbf{u}'), \qquad \mathbf{w}' = \mathbf{v}' - \mathbf{u}'. \tag{A.6}$$

The conservation condition (A.1) can be rewritten as follows:

$$\mathbf{W} = \mathbf{W}' \tag{A.7}$$

and Eq. (A.4), which is equivalent to Eq. (A.2), takes the form

$$\mathbf{w}^2 = \mathbf{w}'^2 \quad \rightarrow \quad |\mathbf{w}| = |\mathbf{w}'|. \tag{A.8}$$

These equations tell us that a conservative binary collision can be equivalently expressed by the conservation of the velocity of the center of mass and of the modulus of the relative velocity. Moreover, the symmetry of transformations (A.5) and (A.6) indicate that their Jacobian determinants are equal

$$\frac{\partial(\mathbf{v}, \mathbf{u})}{\partial(\mathbf{W}, \mathbf{w})} = \frac{\partial(\mathbf{v}', \mathbf{u}')}{\partial(\mathbf{W}', \mathbf{w}')}, \tag{A.9}$$

while for the Jacobian determinant associated with transformations (A.7) and (A.8), we obtain

$$\frac{\partial(\mathbf{W}', \mathbf{w}')}{\partial(\mathbf{W}, \mathbf{w})} = 1. \tag{A.10}$$

By formally inverting Eq. (A.9) and using Eq. (A.10), we can write

$$\frac{\partial(\mathbf{v}', \mathbf{u}')}{\partial(\mathbf{v}, \mathbf{u})} = \frac{\partial(\mathbf{W}', \mathbf{w}')}{\partial(\mathbf{W}, \mathbf{w})} = 1. \tag{A.11}$$

This result shows that the elastic collision process satisfies Liouville's principle of volume conservation in the velocity space, namely

$$d\mathbf{v}\, d\mathbf{u} = d\mathbf{v}'\, d\mathbf{u}'. \tag{A.12}$$

We make use of this result in Section 1.4.

We can observe that the physically relevant quantity in a conservative binary collision is the relative velocity $\mathbf{w}$, because, without prejudice to generality, we can describe the collision in the reference frame of the center of mass velocity $\mathbf{W}$, which is conserved by the collision (see Eq. (A.7)). Since $\mathbf{w}$ determines just the modulus of $\mathbf{w}'$ (see Eq. (A.8)), two additional parameters should be specified for the complete knowledge of $\mathbf{w}'$. In order to accomplish this task, we should take into account explicitly the nature of the interaction forces acting during the collision. These details go beyond our interest, and in Fig. A.1 we just sketch the geometry of binary elastic collisions, described in terms of the relative velocities. We want to point out that in general the scattering angle θ is a function of the collision parameter b and on the azimuthal angle ϕ.

We can further simplify the classical model of binary conservative collisions by assuming that the interaction potential is a central one, that is, it depends only on the relative distance between the center of mass of the colliding particles. In this case, θ depends on b only, $b \equiv b(\theta)$. In particular, the overall information about binary conservative collisions is usually summarized by the differential cross section $\sigma(\theta)$, which can be formally expressed as follows in terms of the collision parameter b

$$\sigma(\theta) \sin(\theta)\, d\theta = b(\theta)\, db. \tag{A.13}$$

For instance, if we assimilate the gas particles to classical hard spheres of radius R, interacting by central forces (which amounts to a reasonable approximation for a real gas), the differential cross section simplifies to $\sigma(\theta) = \frac{R^2}{4}$ so that the total cross section results to be $\sigma_{tot} = \int \sigma(\theta)\, d\Omega = \frac{R^2}{4} \int d\Omega = \pi R^2$, where $d\Omega = \sin(\theta)\, d\theta\, d\phi$ is the differential of the solid angle Ω.

Anyway, for what is discussed in this book (see Section 1.4), we do not need to specify the nature of the interaction forces acting in the binary elastic collision: It is enough to point out that the only possible dependence of the differential cross section on the velocity variables goes through the modulus of the relative velocity, namely $\sigma \equiv \sigma(|\mathbf{w}|)$.

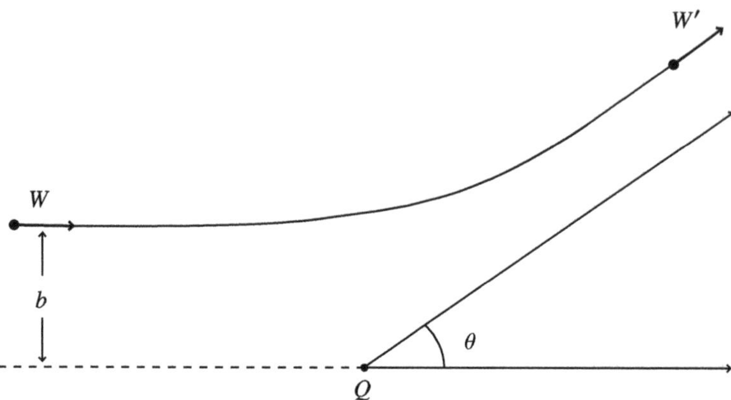

Fig. A.1 Geometrical representation of a binary elastic collision in the reference frame of the center of mass velocity. The relative velocities of the colliding particles before (**w**) and after (**w′**) the collision are co-planar vectors, b is a collision parameter, which depends on the forces determining the collision, whose center of interaction is the point Q, θ is the scattering angle. Assuming the simplifying hypothesis of central, that is, isotropic, elastic collisions, b and θ are the only relevant parameters, while no dependence on the azimuthal angle is present.

We conclude this appendix by showing that, if we assume that the binary conservative collision can be represented in terms of the instantaneous central interaction between hard spheres, the particle velocities before and after the collision are related to each other by simple mathematical expressions. In fact, in such a case, the momentum transferred to any sphere is directed along the unit vector **n**, joining the center of the two spheres at the instant of the impact. In particular, this implies that the difference between **v** and **v′** is proportional to **n**,

$$\mathbf{v} - \mathbf{v}' = -k\,\mathbf{n}, \tag{A.14}$$

so that also

$$\mathbf{u} - \mathbf{u}' = k\,\mathbf{n} \tag{A.15}$$

must hold (see Eq. (A.1)). The value of the parameter k can be determined by substituting these relations into Eq. (A.3), thus yielding

$$k^2 + k\mathbf{n} \cdot (\mathbf{v} - \mathbf{u}) = 0. \tag{A.16}$$

The nontrivial solution of this equation gives $k = \mathbf{n} \cdot (\mathbf{u} - \mathbf{v})$, and Eqs. (A.14) and (A.15) can be rewritten as follows:

$$\mathbf{v}' = \mathbf{v} + \mathbf{n}\left[\mathbf{n} \cdot (\mathbf{u} - \mathbf{v})\right] \tag{A.17}$$
$$\mathbf{u}' = \mathbf{u} - \mathbf{n}\left[\mathbf{n} \cdot (\mathbf{u} - \mathbf{v})\right]. \tag{A.18}$$

As a straightforward consequence of the time reversibility, associated with this conservative collision process, these equations can be also rewritten as follows:

$$\mathbf{v} = \mathbf{v}' + \mathbf{n}\left[\mathbf{n} \cdot (\mathbf{u}' - \mathbf{v}')\right] \tag{A.19}$$

$$\mathbf{u} = \mathbf{u}' - \mathbf{n}\left[\mathbf{n} \cdot (\mathbf{u}' - \mathbf{v}')\right]. \tag{A.20}$$

Maxwell–Boltzmann Distribution in the Uniform Case

In Section 1.5 we have found the explicit expression of the equilibrium distribution function for an ideal gas made of particles with equal mass m in the uniform case, which is known as the Maxwell–Boltzmann distribution

$$\bar{f}_{\mathbf{v}} = A \exp\left[-\frac{1}{2}m\beta(\mathbf{v} - \mathbf{v}_0)^2\right]. \tag{B.1}$$

We can attribute a physical interpretation to the set of parameters A, $\mathbf{v}_0$ and β making use of the conservation laws

$$n = \int d\mathbf{v}\, f(\mathbf{v}, t), \tag{B.2}$$

$$e = \int d\mathbf{v}\, \frac{1}{2}mv^2 f(\mathbf{v}, t), \tag{B.3}$$

$$\mathbf{p} = \int d\mathbf{v}\, m\mathbf{v} f(\mathbf{v}, t), \tag{B.4}$$

where n, e, and $\mathbf{p}$ are the number of particles, the energy, and the momentum per unit volume of the ideal gas, respectively. By substituting Eq. (B.1) into Eq. (B.2), one obtains

$$A = n\left(\frac{m\beta}{2\pi}\right)^{\frac{3}{2}}. \tag{B.5}$$

The velocity $\mathbf{v}_0$ coincides with the average velocity of the particles in the gas, namely

$$\langle \mathbf{v} \rangle = \frac{\int d\mathbf{v}\, \mathbf{v}\, f(\mathbf{v}, t)}{\int d\mathbf{v}\, f(\mathbf{v}, t)} = \left(\frac{m\beta}{2\pi}\right)^{\frac{3}{2}} \int d\mathbf{w}\, (\mathbf{w} + \mathbf{v}_0) \exp\left(-\frac{1}{2}m\beta w^2\right) = \mathbf{v}_0, \tag{B.6}$$

where we have introduced the change of variable $\mathbf{w} = \mathbf{v} - \mathbf{v}_0$ and we have used that $\int d\mathbf{w}\, \mathbf{w} \exp\left[-\frac{1}{2}m\beta w^2\right] = 0$ for symmetry reason. We can conclude that $\mathbf{p} = nm\langle \mathbf{v} \rangle = \rho \mathbf{v}_0$ and that $\mathbf{v}_0$ coincides with the velocity of the center of mass of the gas. Without prejudice to generality, we can assume to be in a reference frame where the center of mass of the gas is at rest, that is, $\mathbf{v}_0 = 0$, which amounts to make the average momentum vanish, that is, $\mathbf{p} = 0$.

It remains to associate the parameter β to a physical quantity. For this purpose we can compute the average kinetic energy of any particle in the ideal gas with $\mathbf{v}_0 = 0$, thus obtaining

$$\bar{\varepsilon} = \frac{\int d\mathbf{v}\, \frac{1}{2}m|\mathbf{v}|^2 f(\mathbf{v}, t)}{\int d\mathbf{v}\, f(\mathbf{v}, t)} = \frac{3}{2}\frac{1}{\beta}. \tag{B.7}$$

According to kinetic theory, this is a fundamental relation that implies equipartition of energy among particles of the ideal gas at thermodynamic equilibrium, namely

$$\frac{1}{2} m \langle v_x^2 \rangle = \frac{1}{2} m \langle v_y^2 \rangle = \frac{1}{2} m \langle v_z^2 \rangle = \frac{1}{2} T \tag{B.8}$$

where

$$\beta = \frac{1}{T} \tag{B.9}$$

is the so-called *inverse temperature* and the parameter A defined in Eq. (B.5) turns out to be expressed in terms of physical quantities as follows:

$$A = n \left(\frac{m}{2\pi T} \right)^{\frac{3}{2}}. \tag{B.10}$$

The equilibrium thermodynamics of the ideal gas can be completely derived from distribution (B.1). In fact, one can compute the pressure exerted by colliding particles on a unit surface, whose normal is directed along the x-axis. Despite the arbitrariness of this choice, there is no prejudice of generality, because at thermodynamic equilibrium the physical properties of the ideal gas have to be the same at any point and with respect to any direction in space. The number of particles with velocity component v_x along the x-axis colliding in the unit time on the unit surface is given by the expression $v_x \bar{f}(\mathbf{v})$, while the force exerted by each one of these particles in the collision on the unit surface is given by the variation of its momentum, that is $2mv_x$. Accordingly, the pressure force exerted by all particles colliding in the unit time on the unit surface is given by the formula

$$P = \int_{v_x > 0} d\mathbf{v} \, 2mv_x^2 \, \bar{f}(\mathbf{v}) = \frac{2}{3} n \bar{\varepsilon} = n T, \tag{B.11}$$

thus recovering the equation of state of an ideal gas at thermodynamic equilibrium.

Physical Quantities from the Boltzmann Equation in the Nonuniform Case

C.1 Pressure Tensor and Thermodynamics of the Ideal Gas

In Appendix B we have computed the pressure of the ideal gas in the uniform case (see Eq. (B.11)), relying upon the remarkable fact that the equilibrium Maxwell–Boltzmann distribution (B.1), as well as any equilibrium observable, must be independent of the space coordinates $\mathbf{r}$ and of time.

In the nonuniform case, when local equilibrium conditions described by the Boltzmann distribution function (1.86) apply, we need a more general approach for computing the pressure tensor.

The number of particles with velocity in the range $(\mathbf{v}, \mathbf{v} + d\mathbf{v})$, which collide in the time interval dt over the surface dS is given by $f(\mathbf{r}, \mathbf{v}, t)\,\mathbf{v} \cdot \mathbf{n}\,dt\,dS\,d\mathbf{v}$, where $\mathbf{n}$ is the unit vector normal to dS and pointing to the outside of dS. The total momentum of the particles, which in the time interval dt collide over dS, is then given by

$$dS\,dt \int_{+} m\,\mathbf{v}\,f(\mathbf{r}, \mathbf{v}, t)\,\mathbf{v} \cdot \mathbf{n}\,d\mathbf{v} \tag{C.1}$$

where the label "+" indicates that the integration is extended to the region where $\mathbf{v} \cdot \mathbf{n} > 0$. Analogously, the total momentum of the particles that in the time interval dt are reflected by the surface dS after the collision is given by

$$dS\,dt \int_{-} m\,\mathbf{v}\,f(\mathbf{r}, \mathbf{v}, t)|\mathbf{v} \cdot \mathbf{n}|d\mathbf{v} \tag{C.2}$$

where the label "-" indicates that the integration is extended to the region where $\mathbf{v} \cdot \mathbf{n} < 0$.

The difference between Eqs. (C.1) and (C.2) gives the variation in the time interval dt of the total momentum of the particles colliding on the surface dS, which is equivalent to the total momentum $d\mathbf{F}\,dt$ transferred by the particles to dS, where $d\mathbf{F}$ denotes the infinitesimal force vector acting on dS. Thus, we can write

$$d\mathbf{F} = dS \int_{+} m\,\mathbf{v}\,f(\mathbf{r}, \mathbf{v}, t)\,\mathbf{v} \cdot \mathbf{n}\,d\mathbf{v} - dS \int_{-} m\,\mathbf{v}\,f(\mathbf{r}, \mathbf{v}, t)\,|\mathbf{v} \cdot \mathbf{n}|\,d\mathbf{v}$$

$$= dS \int m\,\mathbf{v}\,f(\mathbf{r}, \mathbf{v}, t)\,\mathbf{v} \cdot \mathbf{n}\,d\mathbf{v} \tag{C.3}$$

where the last equality is obtained by considering that $f(\mathbf{r}, \mathbf{v}, t)$ is an even function of $\mathbf{v} \cdot \mathbf{n}$ (see Eq. (1.74)). It is useful rewriting Eq. (C.3) for any component of the vector $d\mathbf{F}$, namely

$$dF_i = m\,dS \int v_i\,\mathbf{v} \cdot \mathbf{n}\,f(\mathbf{r}, \mathbf{v}, t)\,d\mathbf{v} \quad , \quad i = x, y, z. \tag{C.4}$$

Let us denote with $\langle v_i \rangle$ the average value of the ith component of the velocity vector at point $\mathbf{r}$:

$$\langle v_i \rangle = \frac{1}{n(\mathbf{r})} \int v_i \, f(\mathbf{r}, \mathbf{v}, t) \, d\mathbf{v} \quad , \quad i = x, y, z \tag{C.5}$$

where the normalization factor $n(\mathbf{r})$ amounts to the density of particles at position $\mathbf{r}$ (see Eq. (1.103)). Since $f(\mathbf{r}, \mathbf{v}, t)$ is invariant under the transformation $\mathbf{v} \cdot \mathbf{n} \to -\mathbf{v} \cdot \mathbf{n}$ (see Eq. (1.74)), from the earlier definition it follows that $\langle \mathbf{v} \cdot \mathbf{n} \rangle = \langle \mathbf{v} \rangle \cdot \mathbf{n} = 0$. Making use of this result, we can write

$$\frac{dF_i}{dS} = m \int v_i \, \mathbf{v} \cdot \mathbf{n} f(\mathbf{r}, \mathbf{v}, t) \, d\mathbf{v} \tag{C.6}$$

$$= m \int v_i \, (\mathbf{v} - \langle \mathbf{v} \rangle) \cdot \mathbf{n} \, f(\mathbf{r}, \mathbf{v}, t) \, d\mathbf{v}$$

$$= m \int (v_i - \langle v_i \rangle) \, (\mathbf{v} - \langle \mathbf{v} \rangle) \cdot \mathbf{n} \, f(\mathbf{r}, \mathbf{v}, t) \, d\mathbf{v} + m \langle v_i \rangle \int (\mathbf{v} - \langle \mathbf{v} \rangle) \cdot \mathbf{n} \, f(\mathbf{r}, \mathbf{v}, t) \, d\mathbf{v}$$

$$= m \int (v_i - \langle v_i \rangle)(\mathbf{v} - \langle \mathbf{v} \rangle) \cdot \mathbf{n} \, f(\mathbf{r}, \mathbf{v}, t) \, d\mathbf{v} \tag{C.7}$$

where in the last passage we have used the identity $\int (\mathbf{v} - \langle \mathbf{v} \rangle) \cdot \mathbf{n} \, f(\mathbf{r}, \mathbf{v}, t) \, d\mathbf{v} \equiv 0$. Then we can rewrite

$$dF_i = (\hat{P} \cdot \mathbf{n})_i \, dS \quad , \quad i = x, y, z \tag{C.8}$$

where we have defined the components of the symmetric pressure tensor $\hat{P}$ by the expression

$$P_{ij} = m \int (v_i - \langle v_i \rangle)(v_j - \langle v_j \rangle) f(\mathbf{r}, \mathbf{v}, t) \, d\mathbf{v} \tag{C.9}$$

so that $(\hat{P} \cdot \mathbf{n})_i = m \sum_{j=x,y,z} \int (v_i - \langle v_i \rangle)(v_j - \langle v_j \rangle) n_j \, f(\mathbf{r}, \mathbf{v}, t) \, d\mathbf{v}$.

If the gas is at equilibrium, that is, if the distribution function is given by Eq. (1.107), one easily obtains

$$P_{ij} = P \, \delta_{ij} \quad , \quad dF_i = P \, n_i \, dS \tag{C.10}$$

where

$$P = m \int (v_i - v_{0i})^2 \, f_0(\mathbf{v}, \mathbf{r}) \, d\mathbf{v}$$

$$= m \, n(\mathbf{r}) \left(\frac{m\beta}{2\pi} \right)^{\frac{3}{2}} \int (v_i - v_{0i})^2 \, \exp\left[-\frac{1}{2} \beta m \, (\mathbf{v} - \mathbf{v}_0(\mathbf{r}))^2 \right] d\mathbf{v}$$

$$= \frac{n(\mathbf{r})}{\beta} \tag{C.11}$$

where in the last passage we have used Eq. (1.103) and well-known properties of the Gaussian integrals.

If external forces are absent (i.e., $n(\mathbf{r}) = n$ is independent of $\mathbf{r}$) and the gas is at rest (i.e., $\mathbf{v}_0(\mathbf{r}) = 0$), we finally obtain the equation of state of the ideal gas

$$P = \frac{N}{V} T, \tag{C.12}$$

with $n = \frac{N}{V}$ and $T = \frac{1}{\beta}$.

C.2 Mean Free Path from Boltzmann Equation

The number R of collisions per unit volume and per unit time at position $\mathbf{r}$ of particles with velocity $\mathbf{v}$ is given by Eq. (1.37), namely

$$R = f(\mathbf{r}, \mathbf{v}, t) \int d\Omega \int d\mathbf{u}\, \sigma(|\mathbf{v} - \mathbf{u}|, \Omega)\, f(\mathbf{r}, \mathbf{u}, t)\, |\mathbf{v} - \mathbf{u}|. \qquad (C.13)$$

In order to obtain the total number $\bar{\nu}$ of collisions per unit volume and per unit time at position $\mathbf{r}$, we have to sum this expression over all possible values of the velocity $\mathbf{v}$:

$$\bar{\nu} = \int d\mathbf{v} \int d\mathbf{u} \int d\Omega\, \sigma(|\mathbf{v} - \mathbf{u}|, \Omega)\, |\mathbf{v} - \mathbf{u}|\, f(\mathbf{r}, \mathbf{v}, t)\, f(\mathbf{r}, \mathbf{u}, t). \qquad (C.14)$$

Since we have n particles per unit volume, the average number of collisions experienced by one particle per unit time is given by $\bar{\nu}/n$, so that the average time between two subsequent collisions can be expressed as follows:

$$\tau = \frac{n}{\bar{\nu}}. \qquad (C.15)$$

Explicit calculations can be performed by assuming that we are in equilibrium conditions for the uniform case, that is, $f(\mathbf{r}, \mathbf{v}, t) \rightarrow \bar{f}(\mathbf{v})$ (see Eq. (1.66)), that the center of mass velocity of the gas is null, that is, $\mathbf{v}_0 = 0$, and that the cross-section of the collision process reduces to a pure geometrical factor, for example, $\sigma(|\mathbf{v} - \mathbf{u}|, \Omega) \rightarrow \sigma_{tot}$, as in the case of hard spheres:

$$
\begin{aligned}
\bar{\nu} &= \sigma_{tot} \int d\mathbf{v} \int d\mathbf{u}\, |\mathbf{v} - \mathbf{u}|\, \bar{f}(\mathbf{v})\, \bar{f}(\mathbf{u}) \\
&= \sigma_{tot}\, n^2 \left(\frac{m}{2\pi T}\right)^3 \int d\mathbf{v} \int d\mathbf{u}\, |\mathbf{v} - \mathbf{u}|\, \exp\left[-\frac{m}{2T}(v^2 + u^2)\right] \\
&= \sigma_{tot}\, n^2 \left(\frac{m}{2\pi T}\right)^3 \int d\mathbf{W} \int d\mathbf{w}\, |\mathbf{w}|\, \exp\left[-\frac{m}{2T}\left(2\,|\mathbf{W}|^2 + \frac{1}{2}\,|\mathbf{w}|^2\right)\right],
\end{aligned}
$$

$$\qquad (C.16)$$

where $\mathbf{W} = \frac{1}{2}(\mathbf{v} + \mathbf{u})$ and $\mathbf{w} = \mathbf{v} - \mathbf{u}$. By integrating this expression, we obtain

$$\bar{\nu} = 4\, n^2\, \sigma_{tot} \sqrt{\frac{T}{\pi m}} \qquad (C.17)$$

and

$$\tau = \frac{1}{4n\sigma_{tot}} \sqrt{\frac{\pi m}{T}}. \qquad (C.18)$$

Similarly, we can compute the expectation value of the modulus of the particles velocity at equilibrium

$$\langle |\mathbf{v}| \rangle = \int d\mathbf{v}\, |\mathbf{v}|\, \bar{f}(\mathbf{v}) = 4\pi \left(\frac{m}{2\pi T}\right)^{3/2} \int d|\mathbf{v}|\, |\mathbf{v}|^3\, \exp\left(-\frac{m}{2T} v^2\right) = \sqrt{\frac{8T}{\pi m}}, \qquad (C.19)$$

where the second equality is obtained by passing from Cartesian to polar coordinates. Thus, we finally obtain an explicit formula for the mean free path λ at equilibrium

$$\lambda = \langle |\mathbf{v}| \rangle \, \tau = \frac{1}{\sqrt{2}\, n\, \sigma_{tot}} \, , \tag{C.20}$$

which results to be independent of the temperature and inversely proportional to the density and to the total cross-section of collisions between particles. This equation reproduces the same result of elementary kinetic theory, see Eq. (1.13).

For the hard sphere gas, one has $\sigma_{tot} = 4\pi a^2$, where a is the sphere radius. Typical values for a diluted gas in equilibrium conditions at room temperature and pressure are $a = 10^{-8}$cm, $\lambda = 10^{-5}$cm, $\nu = 2\times10^{29}$cm^{-3}s^{-1}, $\tau = 7\times10^{-11}$s and $\langle |\mathbf{v}| \rangle = 1.7\times10^5$cm s^{-1}.

As discussed at the beginning of this section, the time scale associated with the collision term τ (see Eq. (1.108)) and the one associated with the stream term t_0 (see Eq. (1.113)) of the Boltzmann equation (1.45) are quite different from each other, in such a way that $\tau \ll t_0$. This allows us to recognize the ratio $\varepsilon = \frac{\tau}{t_0}$ as a natural perturbation parameter of Boltzmann theory. Moreover, we have also argued that, almost independently of the initial condition $f(\mathbf{r}(0), \mathbf{v}(0), 0)$, the distribution function evolves on the shorter (collisional) time scale τ to a form very close to $\bar{f}(\mathbf{r}, \mathbf{v}, t) = \bar{f}(n(\mathbf{r}, \mathbf{v}, t); \langle v \rangle (\mathbf{r}, \mathbf{v}, t); T(\mathbf{r}, \mathbf{v}, t))$ where in the last expression we have pointed out that, as one can conclude by direct inspection of (1.86), the dependence of $\bar{f}(\mathbf{r}, \mathbf{v}, t)$ on its arguments actually goes through the physical quantities associated to the first three of its moments with respect to the variable $\mathbf{v}$, namely the constant, that is, zeroth moment (associated with the mass density), the linear moment (associated with the momentum), and the second moment (associated with kinetic energy). Even if not based on mathematical rigor, it seems physically plausible to assume that for times larger than τ, the distribution function should remain dependent on these three moments, that is,

$$f(\mathbf{r}, \mathbf{v}, t) = f(n(\mathbf{r}, \mathbf{v}, t); \langle v \rangle (\mathbf{r}, \mathbf{v}, t); T(\mathbf{r}, \mathbf{v}, t)) \quad t > \tau. \tag{D.1}$$

As we are going to show in what follows, this assumption is necessary for the correct definition of the perturbative iterative Chapman–Enskog method, whose first two steps are the zero- and the first-order approximations described in Sections 1.7.2 and 1.7.3. Before passing to the outline of this method, it is worth introducing a more compact notation that can help the reader to grasp its basic structure, beyond details. First, we can rewrite the Boltzmann transport equation as follows:

$$\mathcal{D} f = \frac{1}{\varepsilon} J(f, f) \tag{D.2}$$

$$\mathcal{D} = \left(\frac{\partial}{\partial t} + \mathbf{v} \cdot \nabla_{\mathbf{r}} + \frac{\mathbf{F}(\mathbf{r})}{m} \cdot \nabla_{\mathbf{v}} \right) \tag{D.3}$$

$$J(f, f) \propto \left(\frac{\partial f}{\partial t} \right)_c \tag{D.4}$$

where, for the sake of simplicity, we have omitted the arguments of $f(\mathbf{r}, \mathbf{v}, t)$, while in (D.2) we have made explicit the presence of the dependence on the perturbation parameter ε. Note that, by making use of this compact notation, the theorem of additive invariants is summarized by the relation

$$\int d\mathbf{v} (\mathcal{D} f) \psi^{(i)}(\mathbf{v}) = 0, \qquad i = 0, 1, 2 \tag{D.5}$$

where $\psi^{(0)}(\mathbf{v}) = m$, $\psi^{(1)}(\mathbf{v}) = m\mathbf{v}$ and $\psi^{(2)}(\mathbf{v}) = \frac{1}{2}m|\mathbf{v}|^2$.

The Chapman—Enskog method is based on the basic idea that we can represent $f(\mathbf{r}, \mathbf{v}, t)$, defined in (D.1), as a perturbative expansion in powers of ε:

$$f(\mathbf{r}, \mathbf{v}, t) = \sum_{s=0}^{+\infty} \varepsilon^s f^{(s)}(\mathbf{r}, \mathbf{v}, t) \tag{D.6}$$

where $f^{(0)} = \bar{f}$ and $f^{(1)} = \bar{f} + g$, where g is defined in Eq. (1.158). As a consequence of assumption (D.1), the expectation values of the additive invariants $\psi^{(i)}(\mathbf{v})$ must be the same for any solution of the Boltzmann transport equation, namely

$$\int d\mathbf{v}\,(f - \bar{f})\psi^{(i)}(\mathbf{v}) = \int d\mathbf{v}\,\left(\sum_{s=1}^{+\infty} \varepsilon^s f^{(s)}\right)\psi^{(i)}(\mathbf{v}) = 0, \qquad i = 0, 1, 2 \tag{D.7}$$

where in the second equality, we have made use of Eq. (D.6). Since by definition both ε and $f^{(s)}$ are non-negative quantities, the last equation implies

$$\int d\mathbf{v}\, f^{(s)}\, \psi^{(i)}(\mathbf{v}) = 0, \qquad i = 0, 1, 2 \tag{D.8}$$

which expresses the identity $f^{(s)}(\mathbf{r}, \mathbf{v}, t) \equiv f^{(s)}(n(\mathbf{r}, \mathbf{v}, t); \langle v \rangle(\mathbf{r}, \mathbf{v}, t); T(\mathbf{r}, \mathbf{v}, t))$. If we substitute Eq. (D.6) into the definitions of the pressure tensor (1.130) and of the heat flux (1.134), we can formally write

$$P_{ij} = \sum_{s=0}^{+\infty} \varepsilon^s P_{ij}^{(s)}, \tag{D.9}$$

$$\mathbf{q} = \sum_{s=0}^{+\infty} \varepsilon^s \mathbf{q}^{(s)}. \tag{D.10}$$

We can also take into account the formal relations

$$\frac{\partial n}{\partial t} = \sum_{s=0}^{+\infty} \varepsilon^s \partial_t n^{(s)} \tag{D.11}$$

$$\frac{\partial \langle \mathbf{v} \rangle}{\partial t} = \sum_{s=0}^{+\infty} \varepsilon^s \partial_t \langle \mathbf{v} \rangle^{(s)} \tag{D.12}$$

$$\frac{\partial T}{\partial t} = \sum_{s=0}^{+\infty} \varepsilon^s \partial_t T^{(s)}. \tag{D.13}$$

The following step of the Chapman–Enskog method amounts to substitute all of these relations into the hydrodynamic equations Eqs. (1.127), (1.131) and (1.136): By equating the same powers of the perturbative parameter ε, we obtain the following equations:

$$\partial_t n^{(0)} = -\nabla_{\mathbf{r}} \cdot (n \langle \mathbf{v} \rangle) \tag{D.14}$$

$$\partial_t n^{(s)} = 0 \tag{D.15}$$

$$\partial_t \langle v_i \rangle^{(0)} = -\langle \mathbf{v} \rangle \cdot \nabla_{\mathbf{r}} \langle v_i \rangle - \frac{1}{\rho} \sum_{j=x,y,z} \frac{\partial P_{ij}^{(0)}}{\partial j} - \frac{1}{m} F_i \tag{D.16}$$

$$\partial_t \langle v_i \rangle^{(s)} = -\frac{1}{\rho} \sum_{j=x,y,z} \frac{\partial P_{ij}^{(s)}}{\partial j} \tag{D.17}$$

$$\partial_t T^{(0)} = -(\nabla_{\mathbf{r}} T) \cdot \langle \mathbf{v} \rangle - \frac{2}{3\kappa n} \sum_{i,j=x,y,z} P_{ij}^{(0)} \frac{\partial \langle v_i \rangle}{\partial j} \tag{D.18}$$

$$\partial_t T^{(s)} = -\frac{2}{3\kappa n} \left(\nabla_{\mathbf{r}} \cdot \mathbf{q}^{(s)} + \sum_{i,j=x,y,z} P_{ij}^{(s)} \frac{\partial \langle v_i \rangle}{\partial j} \right) \tag{D.19}$$

where in Eqs. (D.15), (D.17), and (D.19), $s = 1, 2, \ldots, \infty$ and in Eqs. (D.16), (D.17), $i = x, y, z$. This complete set of equations allows us to compute explicitly $\partial_t n^{(s)}$, $\partial_t \langle \mathbf{v} \rangle^{(s)}$, and $\partial_t T^{(s)}$, once we have obtained an expression for $f^{(s)}$ as a function of $\bar{f}$, which generalizes the procedure adopted to obtain $f^{(1)}$ from $f^{(0)} \equiv \bar{f}$, described in Section 1.7.3.

By substituting these expressions into Eqs. (D.9) and (D.10) and taking into account the following identity (see Eq. (D.1))

$$\frac{\partial f}{\partial t} \equiv \frac{\partial f}{\partial n} \frac{\partial n}{\partial t} + \sum_{i=x,y,z} \frac{\partial f}{\partial \langle v_i \rangle} \frac{\partial \langle v_i \rangle}{\partial t} + \frac{\partial f}{\partial T} \frac{\partial T}{\partial t} \tag{D.20}$$

we can write the formal relation

$$\frac{\partial f}{\partial t} = \sum_{s=0}^{+\infty} \varepsilon^s \partial_t f^{(s)} \tag{D.21}$$

where

$$\partial_t f^{(s)} = \sum_{r=0}^{s} \left[\frac{\partial f^{(r)}}{\partial n} \partial_t n^{(s-r)} + \sum_{i=x,y,z} \frac{\partial f^{(r)}}{\partial \langle v_i \rangle} \partial_t \langle v_i \rangle^{(s-r)} + \frac{\partial f^{(r)}}{\partial T} \partial_t T^{(s-r)} \right]. \tag{D.22}$$

Accordingly, we can also determine $\partial_t f^{(s)}$, once $f^{(1)}, f^{(2)}, \ldots, f^{(s)}$ are known, so that Eq. (D.21) provides us the power series expansion of $\frac{\partial f}{\partial t}$ in the perturbation parameter ε. Also the collision term can be written in the form

$$J(f, f) = \sum_{s=0}^{+\infty} \varepsilon^s \sum_{r=0}^{s} J(f^{(r)}, f^{(s-r)}). \tag{D.23}$$

Then, if we denote

$$\mathcal{D} f^{(s)} = \partial_t f^{(s)} + \mathbf{v} \cdot \nabla_{\mathbf{r}} f^{(s)} + \frac{\mathbf{F}(\mathbf{r})}{m} \cdot \nabla_{\mathbf{v}} f^{(s)} \tag{D.24}$$

we can rewrite Eq. (D.2) as follows

$$\mathcal{D}f^{(s-1)} + \sum_{r=0}^{s} J(f^{(r)}, f^{(s-r)}) = 0 \qquad \text{(D.25)}$$

so that

$$J(f^{(0)}, f^{(0)}) = 0$$
$$J(f^{(1)}, f^{(0)}) + J(f^{(0)}, f^{(1)}) = -\mathcal{D}f^{(0)}$$

$$\cdots\cdots\cdots\cdots$$

$$\sum_{r=0}^{s} J(f^{(s-r)}, f^{(r)}) = -\mathcal{D}f^{(s-1)}$$

$$\cdots\cdots\cdots\cdots \qquad \text{(D.26)}$$

Note that $\mathcal{D}f^{(s-1)}$ is a function of $f^{(0)}, f^{(1)}, \ldots, f^{(s-1)}$ only. In fact, the s-th equation in the earlier set is a linear equation in $f^{(s)}$ and by exploiting the condition (D.8), one can show that this equation admits a unique solution for $f^{(s)}$ as a function of $f^{(0)}, f^{(1)}, \ldots, f^{(s-1)}$. In practice, once n, $\langle \mathbf{v} \rangle$ and T are fixed from the solution of the first equation in Equation (D.26), one can iteratively obtain $f^{(s)}$ from the following equations at any order in s. Manifestly, the overall procedure is far from straightforward and it is usually implemented into quite elaborate numerical algorithms, in particular when the hydrodynamic equations are used for applications to meteorology or to real viscous fluids.

First-order Approximation to Hydrodynamics

In this appendix, we report the main details of the calculations yielding the results of the first-order approximation to the hydrodynamic equations discussed in Section 1.7.3. The starting point is Eq. (1.157) that we treat here as a true equality

$$g_{\mathbf{v}} = -\tau \left(\frac{\partial}{\partial t} + \mathbf{v} \cdot \nabla_{\mathbf{r}} + \frac{\mathbf{F}(\mathbf{r})}{m} \cdot \nabla_{\mathbf{v}} \right) \bar{f}_{\mathbf{v}} \tag{E.1}$$

where

$$\bar{f}_{\mathbf{v}} = \frac{n}{(2\pi m T)^{\frac{3}{2}}} \exp\left[-\frac{m}{2T}(\mathbf{v} - \langle \mathbf{v} \rangle)^2 \right] \tag{E.2}$$

is the local Maxwell–Boltzmann equilibrium distribution, depending on the mass density $\rho(\mathbf{r}, t) = m\, n(\mathbf{r}, t)$, on the average velocity $\langle \mathbf{v} \rangle(\mathbf{r}, t)$ and on the temperature $T(\mathbf{r}, t)$ (see Eq. (1.86)). In order to simplify the notation, we introduce the new velocity vector

$$\mathbf{w} = \mathbf{v} - \langle \mathbf{v} \rangle \tag{E.3}$$

which represents the particle velocity relative to its average local value, and we rewrite Eq. (E.2) as follows:

$$\bar{f}_{\mathbf{w}} = \frac{n}{(2\pi m T)^{\frac{3}{2}}} \exp\left(-\frac{m}{2T} \mathbf{w}^2 \right). \tag{E.4}$$

As a first step for computing explicitly the r.h.s. of Eq. (E.1), we compute the derivatives of $\bar{f}_{\mathbf{w}}$ w.r.t. its arguments, namely

$$\frac{\partial \bar{f}_{\mathbf{w}}}{\partial \rho} = \frac{\bar{f}_{\mathbf{w}}}{\rho} \tag{E.5}$$

$$\frac{\partial \bar{f}_{\mathbf{w}}}{\partial \langle v_i \rangle} = -\frac{\partial \bar{f}_{\mathbf{w}}}{\partial v_i} = \frac{m}{T} w_i \bar{f}_{\mathbf{w}} \quad , \tag{E.6}$$

$$\frac{\partial \bar{f}_{\mathbf{w}}}{\partial T} = \frac{1}{T} \left(\frac{m}{2T} \mathbf{w}^2 - \frac{3}{2} \right) \bar{f}_{\mathbf{w}} \tag{E.7}$$

Making use of these expressions, we can also write

$$\frac{\partial \bar{f}_{\mathbf{w}}}{\partial t} = \frac{\bar{f}_{\mathbf{w}}}{\rho} \frac{\partial \rho}{\partial t} + \frac{m}{T} \bar{f}_{\mathbf{w}} \mathbf{w} \cdot \frac{\partial \langle \mathbf{v} \rangle}{\partial t} + \frac{1}{T} \left(\frac{m}{2T} \mathbf{w}^2 - \frac{3}{2} \right) \bar{f}_{\mathbf{w}} \frac{\partial T}{\partial t} \tag{E.8}$$

$$\mathbf{v} \cdot \nabla_{\mathbf{r}} \bar{f}_{\mathbf{w}} = \frac{\bar{f}_{\mathbf{w}}}{\rho} \mathbf{v} \cdot \nabla_{\mathbf{r}} \rho + \frac{m}{T} \bar{f}_{\mathbf{w}} \sum_{i,j=x,y,z} w_i v_j \frac{\partial \langle v_i \rangle}{\partial j} + \frac{1}{T} \left(\frac{m}{2T} \mathbf{w}^2 - \frac{3}{2} \right) \bar{f}_{\mathbf{w}} \mathbf{v} \cdot \nabla_{\mathbf{r}} T$$

$$\nabla_{\mathbf{v}} \bar{f}_{\mathbf{w}} = -\frac{m}{T} \mathbf{w}\, \bar{f}_{\mathbf{w}} \tag{E.9}$$

We can further simplify the notation by introducing the material-derivative operator (see also Eq. (1.145)), acting on the generic function $A(\mathbf{r}, t)$

$$\mathbb{D}A = \left(\frac{\partial}{\partial t} + \mathbf{v} \cdot \nabla_{\mathbf{r}}\right) A(\mathbf{r}, t) \tag{E.10}$$

and by substituting the former equations, we can rewrite (E.1) as follows:

$$g_{\mathbf{w}} = -\tau \bar{f}_{\mathbf{w}} \left[\frac{1}{\rho}\mathbb{D}\rho + \frac{m}{T}\mathbf{w} \cdot \mathbb{D}\langle\mathbf{v}\rangle + \frac{1}{T}\left(\frac{m}{2T}\mathbf{w}^2 - \frac{3}{2}\right)\mathbb{D}T - \frac{1}{T}\mathbf{w} \cdot \mathbf{F}\right]. \tag{E.11}$$

Making use of the results obtained for the zero-order approximation presented in Section 1.7.2, it remains to obtain an explicit expression for

$$\mathbb{D}\rho = \mathbf{w} \cdot \nabla_{\mathbf{r}}\rho - \rho\nabla_{\mathbf{r}} \cdot \langle\mathbf{v}\rangle \tag{E.12}$$

$$\mathbb{D}\langle\mathbf{v}\rangle = -\frac{1}{\rho}\nabla_{\mathbf{r}} \cdot \hat{P}^{(0)} + \frac{\mathbf{F}}{m} + \mathbf{w} \cdot \nabla_{\mathbf{r}}\langle\mathbf{v}\rangle \tag{E.13}$$

$$\mathbb{D}T = \mathbf{w} \cdot \nabla_{\mathbf{r}}T - \frac{2}{3}T\nabla_{\mathbf{r}} \cdot \langle\mathbf{v}\rangle \tag{E.14}$$

where the first of these equations has been derived from the *continuity equation* (1.127), the second one from Euler's equation (1.142), where $P_{ij}^{(0)} = \delta_{ij} P = \delta_{ij} n T$ (see Eq. (1.141)), and the third one from (1.144). By substituting these results into Eq. (E.11) and by performing some algebraic simplifications, one eventually obtains an explicit expression for $g_{\mathbf{w}}$, in formulae

$$g_{\mathbf{w}} = -\tau \bar{f}_{\mathbf{w}} \left[\frac{1}{T}\left(\frac{m}{2T}|\mathbf{w}|^2 - \frac{5}{2}\right)\mathbf{w} \cdot \nabla_{\mathbf{r}}T + \frac{m}{T}\sum_{ij}\Lambda_{ij}\left(w_i w_j - \frac{1}{3}|\mathbf{w}|^2\delta_{ij}\right)\right] \tag{E.15}$$

(see also (1.158)).

In this appendix, we also illustrate some calculations yielding the hydrodynamic equations in the first-order approximation (1.173) to (1.175). Once the explicit expressions of the heat flux (1.161) and of the pressure tensor (1.171) are obtained in this approximation, they have to be substituted into the conservation equations. In particular, the equation of momentum conservation (1.129) contains an addendum of the form $(\nabla_{\mathbf{r}} \cdot \hat{P})_i$, whose components in the first-order approximation have the explicit expression

$$\frac{\partial P_{ij}^{(1)}}{\partial j} = \frac{\partial P}{\partial j} - \frac{2\alpha}{m}\left[\frac{\partial \Lambda_{ij}}{\partial j} - \frac{m}{3}\frac{\partial}{\partial j}(\nabla_{\mathbf{r}} \cdot \langle\mathbf{v}\rangle)\right] - \frac{2}{m}\frac{\partial \alpha}{\partial j}\left[\Lambda_{ij} - \frac{m}{3}\delta_{ij}\nabla_{\mathbf{r}} \cdot \langle\mathbf{v}\rangle\right]$$

$$= \frac{\partial P}{\partial j} - \alpha\left[\nabla^2\langle v_i\rangle + \frac{1}{3}\frac{\partial}{\partial i}\frac{\partial \langle v_j\rangle}{\partial j}\right] - \frac{2}{m}\frac{\partial \alpha}{\partial j}\left[\Lambda_{ij} - \frac{m}{3}\delta_{ij}\nabla_{\mathbf{r}} \cdot \langle\mathbf{v}\rangle\right] \tag{E.16}$$

where $i, j = x, y, z$, $\alpha = \tau n T$ is the viscosity (see Eq. (1.169)), and the last expression has been obtained by taking into account the explicit expression of the symmetric tensor (see Eq. (1.135))

$$\Lambda_{ij} = \frac{1}{2}m\left(\frac{\partial \langle v_i\rangle}{\partial j} + \frac{\partial \langle v_j\rangle}{\partial i}\right). \tag{E.17}$$

The equation of conservation of energy (1.136) contains two terms depending on the divergence of the heat flux and on the tensor product $\hat{P} \cdot \hat{\Lambda}$, which in the first-order approximation reads

$$\nabla_{\mathbf{r}} \cdot \mathbf{q}^{(1)} = -\nabla_{\mathbf{r}} \cdot (\kappa \nabla_{\mathbf{r}} T) = -\kappa \nabla^2 T - (\nabla_{\mathbf{r}} \kappa) \cdot (\nabla_{\mathbf{r}} T) \tag{E.18}$$

where $\kappa = \frac{5}{2} \tau n T$ is the thermal conductivity and

$$\sum_{ij} (P_{ij}^{(1)} \Lambda_{ij}) = m P \nabla_{\mathbf{r}} \cdot \langle \mathbf{v} \rangle - \frac{2\alpha}{m} \sum_{ij} (\Lambda_{ij} \Lambda_{ij}) + \frac{2\alpha}{3m} (\nabla_{\mathbf{r}} \cdot \langle \mathbf{v} \rangle)^2. \tag{E.19}$$

In Eq. (E.19), we have used Eq. (1.171) and the definition of Λ_{ij}, given in Eq. (1.135), which yields also the following result

$$\sum_{ij} (\Lambda_{ij} \Lambda_{ij}) = \frac{m^2}{2} \left[\nabla_{\mathbf{r}}^2 (\langle \mathbf{v} \rangle^2) - 2 \langle \mathbf{v} \rangle \cdot (\nabla_{\mathbf{r}}^2 \langle \mathbf{v} \rangle) - (\nabla_{\mathbf{r}} \wedge \langle \mathbf{v} \rangle)^2 \right]. \tag{E.20}$$

On the other hand, all the quantities that we have reported here contain many terms, among which the leading ones in the first-order approximation depend on ρ, $\langle \mathbf{v} \rangle$, T and their first derivatives w.r.t. space coordinates and time, together with α and κ. Conversely, all terms depending on higher powers of ρ, $\langle \mathbf{v} \rangle$, T and their derivatives can be neglected, together with those depending on the derivatives of α and κ. In fact, one can realize that the latter contributions to the hydrodynamic equations are negligible, because the transport coefficients α and κ are both proportional to the perturbative parameter τ (see Eqs. (1.169) and (1.162)), that is, they are already leading terms of the first-order approximation. Accordingly, maintaining only the leading contributions consistent with the first-order approximation, the hydrodynamic equations become

$$\frac{\partial \rho}{\partial t} + \nabla_{\mathbf{r}} \cdot (\rho \langle \mathbf{v} \rangle) = 0 \tag{E.21}$$

$$\left(\frac{\partial}{\partial t} + \langle \mathbf{v} \rangle \cdot \nabla_{\mathbf{r}} \right) \langle \mathbf{v} \rangle = \frac{\mathbf{F}}{m} - \frac{1}{\rho} \nabla_{\mathbf{r}} \left(P - \frac{\alpha}{3} \nabla_{\mathbf{r}} \cdot \langle \mathbf{v} \rangle \right) + \frac{\alpha}{\rho} \nabla_{\mathbf{r}}^2 \langle \mathbf{v} \rangle \tag{E.22}$$

$$\left(\frac{\partial}{\partial t} + \langle \mathbf{v} \rangle \cdot \nabla_{\mathbf{r}} \right) T = -\frac{1}{C_V} (\nabla_{\mathbf{r}} \cdot \langle \mathbf{v} \rangle) T + \frac{\kappa}{\rho C_V} \nabla_{\mathbf{r}}^2 T. \tag{E.23}$$

We consider a non-negative matrix W, $W_{ij} \geq 0 \ \forall i, j$, such that

$$\sum_i W_{ij} = 1 \quad \forall j. \tag{F.1}$$

We say that $\mathbf{w}^{(\lambda)} = \left(w_1^{(\lambda)} w_2^{(\lambda)} \cdots w_j^{(\lambda)} \cdots \right)$ is the right eigenvector of W with eigenvalue λ if the following relation holds:

$$\sum_j W_{ij} w_j^{(\lambda)} = \lambda w_i^{(\lambda)} \quad \forall i. \tag{F.2}$$

We want to prove the following properties:

(a) $|\lambda| \leq 1$.

Consider the (right) eigenvector $\mathbf{w}^{(\lambda)}$ of W; then, we can write

$$\sum_j W_{ij} |w_j^{(\lambda)}| \geq \left| \sum_j W_{ij} w_j^{(\lambda)} \right| = |\lambda| |w_i^{(\lambda)}|, \tag{F.3}$$

where the first inequality holds because $W_{ij} \geq 0$ and the last equality stems from the definition of eigenvector. By summing over the index i both sides of this inequality and considering (F.1), we finally obtain:

$$\sum_j |w_j^{(\lambda)}| \geq |\lambda| \sum_i |w_i^{(\lambda)}|, \tag{F.4}$$

that is, $|\lambda| \leq 1$.

(b) There is at least one eigenvalue $\lambda = 1$.

Let us consider the N-component vector $\mathbf{v} = (1, 1, \ldots, 1, \ldots, 1)$. Taking into account (F.1), we can write

$$\sum_i v_i W_{ij} = \sum_i W_{ij} = 1 = v_j, \tag{F.5}$$

which implies that $\mathbf{v}$ is a left eigenvector of W with eigenvalue 1. Since left and right eigenvalues coincide, there must be a (right) eigenvector with eigenvalue equal to 1.

(c) $\mathbf{w}^{(\lambda)}$ is either an eigenvector with eigenvalue 1, or it fulfills the condition $\sum_j w_j^{(\lambda)} = 0$.

The components of the eigenvector $\mathbf{w}^{(\lambda)}$ obey the relation

$$\sum_j W_{ij} w_j^{(\lambda)} = \lambda w_i^{(\lambda)}. \tag{F.6}$$

By summing both sides of this relation over i and considering (F.1), we obtain

$$\sum_j w_j^{(\lambda)} = \lambda \sum_i w_i^{(\lambda)}. \tag{F.7}$$

Accordingly, either $\lambda = 1$ or $\sum_j w_j^{(\lambda)} = 0$.

Let us consider a system with a discrete configuration space, where states are labeled by an index i. Let $p_i(t)$ be the probability of state i at time t and define the quantity

$$\mathcal{D}[t] = \sum_i \frac{1}{p_i^*}(p_i(t) - p_i^*)^2 = \sum_i \frac{p_i^2(t)}{p_i^*} - 1, \tag{G.1}$$

where the sum runs over all possible states of the system, p_i^* is the equilibrium distribution, and the last equality stems from the normalization condition

$$\sum_i p_i(t) = 1 \quad \forall t. \tag{G.2}$$

One can easily realize that $\mathcal{D} \geq 0$ and $\mathcal{D} = 0$ if and only if $p_i(t) = p_i^* \ \forall i$. Let us analyze how $\mathcal{D}$ changes in time,

$$\Delta \mathcal{D} \equiv \mathcal{D}[t+1] - \mathcal{D}[t] = \sum_i \frac{p_i^2(t+1)}{p_i^*} - \sum_i \frac{p_i^2(t)}{p_i^*}. \tag{G.3}$$

Since the evolution equation of a Markov chain reads (see Eq. (2.59))

$$p_i(t+1) = \sum_j W_{ij} p_j(t), \tag{G.4}$$

we can write Eq. (G.3) as

$$\begin{aligned}
\Delta \mathcal{D} &= \sum_i \frac{1}{p_i^*}\left(\sum_j W_{ij} p_j(t)\right)\left(\sum_k W_{ik} p_k(t)\right) - \sum_i \frac{p_i^2(t)}{p_i^*} \\
&= \sum_{ijk} W_{ij} W_{ik} \frac{p_j(t) p_k(t)}{p_i^*} - \sum_i \frac{p_i^2(t)}{p_i^*}.
\end{aligned} \tag{G.5}$$

Since p_i^* is the equilibrium distribution, the detailed balance condition (2.109), $p_a^* W_{ba} = p_b^* W_{ab}$, holds; a Markov chain whose stochastic matrix W obeys this condition is said to be reversible. We can use the detailed balance condition to write $W_{ij} W_{ik} = W_{ji} W_{ki} (p_i^*)^2/(p_j^* p_k^*)$, while, in the second term, we insert the normalization condition $\sum_j W_{ji} = 1$ and we exchange the indices $i \leftrightarrow j$, thus obtaining

$$\Delta \mathcal{D} = \sum_{ijk} W_{ji} W_{ki} p_i^* \frac{p_j p_k}{p_j^* p_k^*} - \sum_{ij} W_{ij} \frac{p_j^2}{p_j^*}. \tag{G.6}$$

The shortened notation $p_i \equiv p_i(t)$ has also been used.

In the second addendum on the right-hand side, we can use again the detailed balance condition $W_{ij} = (W_{ji}\, p_i^*)/p_j^*$ and introduce one more normalization condition $\sum_k W_{ki} = 1$, yielding

$$\Delta \mathcal{D} = \sum_{ijk} W_{ji} W_{ki} p_i^* \frac{p_j p_k}{p_j^* p_k^*} - \sum_{ijk} W_{ji} W_{ki} p_i^* \left(\frac{p_j}{p_j^*}\right)^2. \tag{G.7}$$

Finally, the last addendum on the right-hand side of Eq. (G.7) can be rewritten as the sum of 1/2 of itself and 1/2 of the same expression with $j \leftrightarrow k$, and we finally obtain

$$\Delta \mathcal{D} = -\frac{1}{2} \sum_{ijk} W_{ji} W_{ki} p_i^* \left(\frac{p_j}{p_j^*} - \frac{p_k}{p_k^*}\right)^2. \tag{G.8}$$

We can conclude that $\Delta \mathcal{D} \leq 0$; that is, during the evolution, the non-negative quantity $\mathcal{D}$ either decreases or remains constant. In particular, if we want that $\Delta \mathcal{D} = 0$, each term must vanish separately. If we assume that the matrix W_{ij} is ergodic, for any initial condition after a time t_0, we have $(W^{t_0})_{ij} > 0 \; \forall i, j$. Accordingly, from time t_0, the condition $\Delta \mathcal{D} = 0$ can be fulfilled if and only if $p_i = p_i^* \; \forall i$.

The Ising model on a regular lattice is defined by the Hamiltonian

$$\mathcal{H}_{\mathrm{I}} = -J \sum_{\langle ij \rangle} \sigma_i \sigma_j - H \sum_i \sigma_i, \tag{H.1}$$

where the variable $\sigma_i = \pm 1$ is a binary variable and the first sum runs over all unordered pairs of nearest-neighbor sites of the lattice. For establishing the equivalence with different models, it is useful to write

$$\mathcal{H}_{\mathrm{I}} = \sum_{\langle ij \rangle} \epsilon_{ij}, \tag{H.2}$$

with

$$\epsilon_{ij} = -J \sigma_i \sigma_j - \frac{H}{\gamma}(\sigma_i + \sigma_j) + K. \tag{H.3}$$

Here above γ is the coordination number of the lattice (i.e., the number of nearest neighbors per lattice site) and K is a constant for future use, which vanishes for the Ising model.

The Ising model is usually discussed using a magnetic language where $\sigma_i \pm 1$ is a spin variable and the ground state is a completely ferromagnetic state with all spins parallel. If $H = 0$, the two ferromagnetic states are degenerate, if $H \neq 0$, there is one single ground state with all spins parallel to H, that is $\sigma_i = \mathrm{sign}\,(H)$. The model with $H = 0$, in dimension $d > 1$, undergoes a second-order phase transition between a disordered, paramagnetic phase at high T and an ordered, ferromagnetic phase at low T.

If the order parameter is conserved, the total magnetization is a constant of motion and the energy term proportional to H is constant as well. In this case the ground state corresponds to a complete phase separation between two oppositely magnetized, ferromagnetic regions. There are two models, the binary alloy and the lattice gas, where the order parameter is naturally conserved, because magnetization is replaced by a matter density. Let us define these two models, which can be reformulated using the Ising language.

In the model of binary mixture, each lattice site may be occupied by a type 1 atom (e.g., Zn) or by a type 2 atom (e.g., Cu). Each pair of nearest-neighbor atoms has a different energy, according to the type: e_1, e_2, e_{12} for pairs (11), (22), and (12), respectively. If we associate a type 1 atom with an up spin and a type 2 atom with a down spin, we can make the identification

$$e_1 = \epsilon_{++} = -J - \frac{2H}{\gamma} + K \tag{H.4}$$

$$e_2 = \epsilon_{--} = -J + \frac{2H}{\gamma} + K \tag{H.5}$$

$$e_{12} = \epsilon_{+-} = J + K \tag{H.6}$$

and obtain $J = -\frac{1}{4}(e_1 + e_2) + \frac{1}{2}e_{12}$, $H = \frac{\gamma}{4}(e_2 - e_1)$, and $K = \frac{1}{4}(e_1 + e_2) + \frac{1}{2}e_{12}$. Therefore, we have a ferromagnetic interaction ($J > 0$) if the mean energy coupling between two particles of the same type is lower than the energy coupling between two different particles. The phase transition corresponds to a passage from a disordered mixture at high T to a separated phase at low T.

The lattice gas model describes a fluid, and the spin variable $s_i = \pm 1$ is replaced by the variable $n_i = 0, 1$, indicating whether the site is occupied by a particle or is empty. The energy is

$$\mathcal{H}_{\mathrm{LG}} = -\epsilon_0 \sum_{\langle ij \rangle} n_i n_j. \tag{H.7}$$

If we identify the particles with type 1 particles (or spin up) and the holes with type 2 particles (or down spins), that is, $n_i = (s_i + 1)/2$, the model can be mapped to the binary mixture case with $e_1 = -\epsilon_0$ and $e_2 = e_{12} = 0$, so that the corresponding Ising model has the parameters $J = \epsilon_0/4$, $H = (\gamma/4)\epsilon_0$, and $K = -\epsilon_0/4$. The phase transition corresponds to the passage from a gas phase at high T to a condensed phase at low T.

We conclude this general part on the Ising model commenting about its simulation with a Kinetic Monte Carlo algorithm, which is clearly different for the spin model (H.1) and for the lattice gas model (H.7): In the latter model, dynamics must conserve the number of particles, while magnetization is not conserved in the former model. Referring in both cases to a spin language, the basic dynamics which are relevant in the two cases are called spin-flip (or Glauber) dynamics in the nonconserved case, and spin-exchange (or Kawasaki) dynamics in the conserved case. Spin-flip dynamics means that we choose a random spin and we attempt its flipping; spin-exchange dynamics means that we choose a random pair of neighbouring spins and if they are antiparallel we attempt to exchange their orientation. In both cases, we must calculate the energy difference $(E_j - E_i)$ between (possible) final state j and initial state i, and evaluate the probability of the move, Eq. (2.115), which depends on the specific function $\omega(x)$ we have adopted.

H.1 The Ising Model in One Dimension

As mentioned in Section 5.2.2, the Ising model (5.10) in $d = 1$ for $H = 0$ exhibits a phase transition at $T_\mathrm{c} = 0$. This can be easily shown by computing explicitly the partition function

(5.12), while assuming periodic boundary conditions ($\sigma_{N+1} = \sigma_1$)

$$Z(N,T,J,H = 0) \equiv \sum_{\{\sigma_i\}} \exp(-\beta \mathcal{H}_i) = \sum_{\{\sigma_i\}} \exp\left(\beta J \sum_{i=1}^{N} \sigma_i \sigma_{i+1}\right), \tag{H.8}$$

where $\beta = 1/T$ and the volume V has been replaced by the number of spins in the one-dimensional lattice, whose spacing a can be set equal to 1 without prejudice of generality, so that $V \equiv N$. We can rewrite (H.8) as

$$Z(N,T,J,H = 0) = \sum_{\sigma_1}\sum_{\sigma_2}\cdots\sum_{\sigma_N} \prod_{i=1}^{N} \exp\left(\beta J \sigma_i \sigma_{i+1}\right)$$

$$= \sum_{\sigma_1}\sum_{\sigma_2}\cdots\sum_{\sigma_N} c^N \prod_{i=1}^{N} (1 + t\,\sigma_i\sigma_{i+1}), \tag{H.9}$$

where $c = \cosh(\beta J)$ and $t = \tanh(\beta J)$. Once we have noticed that $\sigma_i\sigma_{i+1} = \pm 1$, the last expression stems from the identity $\exp(\pm K) = \cosh(K)[1 \pm \tanh(K)]$ and from pulling c out of the product.

There are only two factors from the product of the N binomials that survive: those independent of any σ_i. In fact, any factor depending (linearly) on σ_i cancels out when performing the sums over the $\sigma_i = \pm 1$. So, the two surviving terms are the product of all "1"s and the product $\prod_{i=1}^{N} \sigma_i\sigma_{i+1} = \prod_{i=1}^{N} \sigma_i^2 = 1$, due to periodic boundary conditions. Then, we can write

$$Z(N,T,J,H = 0) = 2^N c^N (1 + t^N), \tag{H.10}$$

where the factor 2^N derives from the summations over σ_i. According to Eqs. (5.13) and (5.23), the free energy density has the expression

$$f(T,J,H = 0) = T \lim_{N\to\infty} \frac{1}{N} \ln Z = T \ln[2\cosh(\beta J)], \tag{H.11}$$

where we have used the result, valid for $|t| < 1$,

$$\lim_{N\to\infty} \frac{1}{N} \ln(1 + t^N) = 0. \tag{H.12}$$

The free energy density does not exhibit any singularity at finite temperature, because it is an analytic function of its variables for $0 < T \leq \infty$, so the Ising model in $d = 1$ has no phase transition for $T > 0$. On the other hand, for $T = 0$, the free energy reduces just to the internal energy and the equilibrium state coincides with the minimum of $\mathcal{H}_i$ for $H = 0$. In fact, there are two minima of $\mathcal{H}_i$ for $T = 0$, obtained for $\sigma_i = +1$ and $\sigma_i = -1$ for all i, that correspond to a magnetization density $m(T = 0, H = 0) = \pm 1$. We can conclude that the Ising model in $d = 1$ exhibits a discontinuous phase transition at $T = 0$, because it preserves $m(T, H = 0) = 0$ for any $T > 0$, while at $T = 0$, it exhibits a completely positive or negative magnetization.

This singular behaviour is a peculiar feature of thermodynamic fluctuations in $d = 1$, which can be illustrated by a simple argument. Let us suppose a fully magnetized state with $m = +1$ on a chain of length N with periodic boundary conditions and we want to evaluate the change of the free energy due to a fluctuation that makes n consecutive spins flip to

$\sigma_i = -1$. The internal energy increases by an amount $4J$, due to the opposite orientation of spins at the boundaries of the flipped domain, while the entropy of the new state, that is, the logarithm of the number of equivalent states, is of the order $\ln N$, because the equivalent states can be obtained by moving the reversed domain along the chain. Accordingly, for any finite T and in the thermodynamic limit the flipped state has a lower free energy than the fully magnetized one, because, despite the internal energy increasing by a finite amount, the entropic factor (i.e., the one that estimates fluctuations) increases by a macroscopic amount, proportional to $\ln N$. By iterating this argument, that is, by further flipping new domains, we can conclude that for any finite T the equilibrium state corresponds to $m(T, H = 0) = 0$, because the entropic contribution to minimizing the free energy is always favored with respect to the increase of the internal energy. This argument does not apply to the Ising model in $d > 1$, because in order to reverse a macroscopic fraction of spins, we need a domain wall, whose energy cost increases with the number of spins. Therefore, for $d > 1$, the internal energy and the entropy compete on an equal ground.

Since we know the exact solution of the Ising model in $d = 1$, studying its renormalization by the procedure discussed in Section 5.2.6 has only a pedagogical interest. Let us consider the partition function of the Ising model in $d = 1$ with an external magnetic field H,

$$Z(N, J, H) = \sum_{\{\sigma_i\}} \exp\left[\sum_{i=1}^{N}\left(J\sigma_i\sigma_{i+i} + \frac{H}{2}(\sigma_i + \sigma_{i+i})\right)\right], \tag{H.13}$$

where we have adopted the simplified notation $\beta J \rightarrow J$ and $\beta H \rightarrow H$. We can now construct a block transformation, in the spirit of the renormalization procedure described in Section 5.2.6. In particular, we can find an explicit expression of the renormalization transformation,

$$\mathcal{R}(J, H) = (J', H') \tag{H.14}$$

for a known value of the scale factor ℓ. This task can be accomplished by integrating the partition function (H.13) over all spins with even index. For this purpose, it is useful to rewrite (H.13) as

$$Z(N, J, H) = \sum_{\{\sigma_{2i+1}\}} \prod_{i=1}^{N/2} \sum_{\sigma_{2i}} \exp\left(J\sigma_{2i}(\sigma_{2i-1} + \sigma_{2i+1}) + \frac{H}{2}(\sigma_{2i-1} + 2\sigma_{2i} + \sigma_{2i+1})\right). \tag{H.15}$$

Each factor of the product has the form

$$\sum_{\sigma_{2i}=\pm 1} e^{A\sigma_{2i}+B} = e^{B}\sum_{\sigma_{2i}=\pm 1} e^{A\sigma_{2i}} = 2e^{B}\cosh(A) \tag{H.16}$$

with

$$A = J(\sigma_{2i-1} + \sigma_{2i+1}) + H, \qquad B = \frac{H}{2}(\sigma_{2i-1} + \sigma_{2i+1}). \tag{H.17}$$

We can also use the equation

$$2\cosh(A) = 2\cosh\left(J(\sigma_{2i-1} + \sigma_{2i+1}) + H\right) \tag{H.18}$$

$$= A_0 \exp\left(J'\sigma_{2i-1}\sigma_{2i+1} + \frac{\eta}{2}(\sigma_{2i-1} + \sigma_{2i+1})\right), \tag{H.19}$$

where

$$J' = \frac{1}{4} \ln \left(\frac{\cosh(2J + H)\,\cosh(2J - H)}{\cosh^2(H)} \right) \tag{H.20}$$

$$\eta = \frac{1}{2} \ln \left(\frac{\cosh(2J + H)}{\cosh(2J - H)} \right) \tag{H.21}$$

$$A_0 = 2\cosh(H)\,\exp(J'), \tag{H.22}$$

to recover the same form of the Ising Hamiltonian after having traced over all the spins of even index $2i$, with J' given by (H.20) and $H' = H + \eta$.

We want to point out that all of these equations provide an explicit definition of the renormalization transformation $\mathcal{R}$ with $\ell = 2$ (see Section 5.2.6). By this decimation procedure, the original Ising Hamiltonian defined on a lattice of N sites has been transformed into the new Ising Hamiltonian defined on a lattice of $N/2$ sites, whose spacing has been increased by a factor $\ell = 2$ and that has acquired an additional constant, thus,

$$\mathcal{H}(J, H) \to \mathcal{H}(J', H') + \ln A_0. \tag{H.23}$$

As a consequence, the free-energy densities of the original Hamiltonian and of the *decimated* Hamiltonian satisfy the relation

$$f(J, H) = \frac{1}{2} \ln A_0 + \frac{1}{2} f(J', H'), \tag{H.24}$$

where the factor 2 comes from $\ell = 2$ (see Eq. (5.97)), while the first term on the right-hand side is the additional constant mentioned in note 9 of Chapter 5.

Let us analyze the result of the renormalization procedure in the absence of an external field, that is, $H = 0$. The earlier equations simplify to

$$J' = \frac{1}{2} \ln \cosh(2J) \tag{H.25}$$

$$H' = 0 \tag{H.26}$$

$$A_0 = 2\sqrt{\cosh(2J)}. \tag{H.27}$$

The first equation can be interpreted as a mapping $J' = F(J)$ that has only two fixed points, that is, two solutions of the equation $J^* = F(J^*)$, $J_1^* = 0$ (which corresponds to $T = \infty$), and $J_2^* = +\infty$ (which corresponds to $T = 0^+$). On the other hand, starting from any finite value of J, the iteration of the renormalization transformation maps J to the stable fixed point $J_1^* = 0$, because, for any finite J, $J' < J$. The physical interpretation is that for any finite initial coupling J, the Ising Hamiltonian in $d = 1$ is renormalized to an infinite temperature state, where $m = 0$, with the exception of the case $T = 0$, where the model is fully magnetized, because its ground state energy is equivalent to the minimum of the free-energy density. This result, as expected, is consistent with the exact solution of the model.

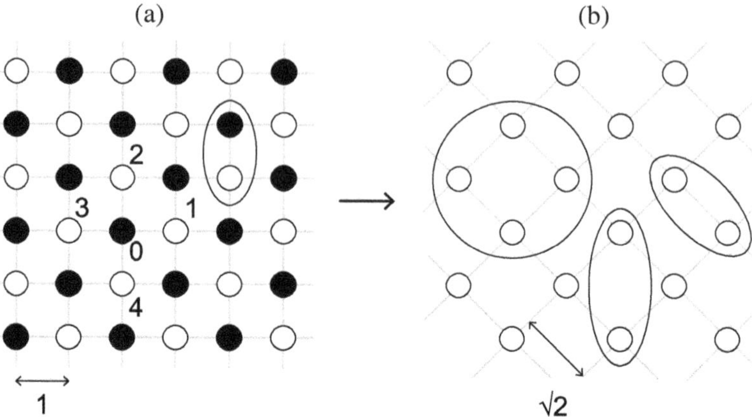

Fig. H.1 Illustration of one step of the renormalization group procedure for the Ising model on a square lattice, which allows passage from the (a) (lattice spacing equal to 1) to (b) (lattice spacing equal to $\sqrt{2}$). The decimation procedure is obtained by summing over the configurations of the spins located at the solid circles. Starting from a Hamiltonian involving two-spin interactions between nearest-neighbor spins, we obtain two-spin interaction between nearest and next-to-nearest neighbors and four-spin interactions.

H.2 The Renormalization of the Two-Dimensional Ising Model

The decimation procedure sketched in Section H.1 amounts to computing explicitly the contribution to the partition function of the spins located at even lattice sites in a spin chain with periodic boundary conditions. Apart from the additional constant, this procedure exactly transforms the original Hamiltonian into a renormalized one, which has the same form with new couplings that are functions of the original ones. This is a peculiar feature of the case $d = 1$ for a model with nearest-neighbor and local interactions, as in the Ising case with an external magnetic field. Here, we proceed in a pedagogical illustration of the technical problems inherent in the construction of an explicit renormalization procedure in $d > 1$ by considering the Ising model in $d = 2$ in the absence of an external magnetic field. In particular, we want to provide an example of how a decimation procedure requires us to include additional interactions and coupling constants in the renormalized Hamiltonian of the Ising model, thus yielding an effective description of the renormalization procedure that has to operate on an abstract space of renormalized Hamiltonians.

We consider a square lattice made of $N \times N$ sites and we impose periodic boundary conditions; that is, the lattice has the topology of a toroidal surface. In analogy with the decimation procedure adopted for the case $d = 1$ (see Appendix H.1), we can decimate the partition function by summing over all of the spins whose coordinates (i, j) are such that $|i + j|$ is an even integer. These sites form the sublattice $\mathcal{S}_e$ of solid circles in the left panel of Fig. H.1, and the scale factor associated with the renormalization transformation is $\ell = \sqrt{2}$; see Fig. H.1(b).

For the sake of simplicity, we can fix our attention on one of this sites, identified by the symbolic index 0, and its four nearest-neighbor sites, identified by the symbolic indices from 1 to 4, which, by construction, do not belong to $\mathcal{S}_e$. We can write the partition function in the following form:

$$Z(N^2, J) = \sum_{\sigma_1,\ldots,\sigma_4} \prod_{0 \in S_e} \sum_{\sigma_0 = \pm 1} \exp\left(J\sigma_0(\sigma_1 + \sigma_2 + \sigma_3 + \sigma_4)\right), \tag{H.28}$$

where the index "0" means a generic site belonging to S_e and "1, 2, 3, 4" are the indices of the neighboring sites of "0." Performing the sum over σ_0, we obtain

$$\sum_{\sigma_0 = \pm 1} \exp\left(J\sigma_0(\sigma_1 + \sigma_2 + \sigma_3 + \sigma_4)\right) = 2 \cosh\left(J(\sigma_1 + \sigma_2 + \sigma_3 + \sigma_4)\right). \tag{H.29}$$

We could recover the original Hamiltonian if it was possible to rewrite the right-hand side in the form

$$A_0 \exp\left(\frac{J'}{2}(\sigma_1\sigma_2 + \sigma_2\sigma_3 + \sigma_3\sigma_4 + \sigma_4\sigma_1)\right), \tag{H.30}$$

where the factor $1/2$ is due to the fact that decimated squares provide two contributions to each one of the new links between nearest-neighbor sites.

We can easily realize that this is not the case, because the function on the right-hand side of Eq. (H.29) can take three different values for $\sigma_i = \pm 1$ and $i = 1, \ldots, 4$,[1] so that we need at least three free parameters to be adjusted, not just two. Moreover, Eq. (H.29) is invariant under permutations of the coordinate indices $1, \ldots, 4$, and if we want to rewrite it in exponential form we have to include all interaction terms among the undecimated spins that keep this property, while maintaining the Z_2 symmetry ($\sigma \to -\sigma$) of the original Hamiltonian. Therefore, we should rather think to rewrite the right-hand side of Eq. (H.29) as

$$2 \cosh\left(J(\sigma_1 + \sigma_2 + \sigma_3 + \sigma_4)\right) \tag{H.31}$$

$$= A_0 \exp\left(\frac{J'}{2}(\sigma_1\sigma_2 + \sigma_2\sigma_3 + \sigma_3\sigma_4 + \sigma_4\sigma_1 + \sigma_1\sigma_3 + \sigma_2\sigma_4) + K\sigma_1\sigma_2\sigma_3\sigma_4\right).$$

Let us introduce the short-hand notations $\Sigma = \sigma_1 + \sigma_2 + \sigma_3 + \sigma_4$, $Q_2 = \sigma_1\sigma_2 + \sigma_2\sigma_3 + \sigma_3\sigma_4 + \sigma_4\sigma_1 + \sigma_1\sigma_3 + \sigma_2\sigma_4$ and $Q_4 = \sigma_1\sigma_2\sigma_3\sigma_4$. It is straightforward to realize that $|\Sigma|$ can take only three different values, namely $0, 2$, and 4, and that Σ determines univocally the values of Q_2 and Q_4. More precisely, the triples of allowed values are ($\Sigma = 0, Q_2 = -2, Q_4 = 1$), ($\Sigma = \pm 2, Q_2 = 0, Q_4 = -1$), and ($\Sigma = \pm 4, Q_2 = 6, Q_4 = 1$). By substitution in earlier equation, one obtains the relations defining the renormalization transformation:

$$J' = \frac{1}{4} \ln \cosh(4J) \tag{H.32}$$

$$K = \frac{1}{8} \ln \cosh(4J) - \frac{1}{2} \ln \cosh(2J) \tag{H.33}$$

$$A_0 = 2\left(\cosh(2J)\right)^{\frac{1}{2}} \left(\cosh(4J)\right)^{\frac{1}{8}}. \tag{H.34}$$

We want to point out that the new Hamiltonian of the renormalized partition function contains not only interactions between nearest-neighbor spins on the rescaled square lattice with spacing $\ell = \sqrt{2}$ but also those between next-to-nearest neighbors (i.e., spins on the diagonals of the square cell) and the interaction among the four spins together (see Fig. H.1). If we iterate the decimation procedure, the new renormalized Hamiltonian is expected

---

[1] In fact, $\sum_{i=1}^{4} \sigma_i = \pm 4, \pm 2, 0$ and $\cosh(x)$ is an even function of its argument.

to contain further and more involved interaction terms, thus yielding a larger and larger set of equations necessary to define the renormalization transformation. It is clear that we have to devise a consistent strategy to reduce this avalanche of equations to an effective equation for the basic coupling J in the form[2]

$$J' = R(J), \tag{H.35}$$

where R should epitomize the contribution of the new coupling constants generated by the renormalization procedure. The simplest way of projecting the renormalization transformation onto the subspace of the parameter J is to consider just Eq. (H.32). As in the case $d = 1$, this relation yields only two trivial fixed points, the stable one $J_1^* = 0$ (i.e., $T = +\infty$) and the unstable one $J_2^* = \infty$ (i.e., $T = 0$), because $J' < J$ for any finite J. Once again, we obtain no phase transition at finite temperature, thus failing our expectations.

A heuristic recipe to obtain a nontrivial result amounts to defining an effective renormalized nearest-neighbor coupling on the square lattice, $\bar{J}'$, such that its contribution to the ground-state energy (all aligned spins) is the same as the one given, in the decimated Hamiltonian, by the nearest- and next-to-nearest neighbor terms,[3]

$$\bar{J}'(\sigma_1\sigma_2 + \sigma_2\sigma_3 + \sigma_3\sigma_4 + \sigma_4\sigma_1) = J'(\sigma_1\sigma_2 + \sigma_2\sigma_3 + \sigma_3\sigma_4 + \sigma_4\sigma_1 + \sigma_1\sigma_3 + \sigma_2\sigma_4), \tag{H.36}$$

thus yielding the relation

$$\bar{J}' = \frac{3}{2}J' = \frac{3}{8}\ln\cosh(4J). \tag{H.37}$$

Now, we identify $\bar{J}'$ with J' and finally obtain the effective renormalization transformation,

$$J' = \frac{3}{8}\ln\cosh(4J). \tag{H.38}$$

As sketched in Fig. H.2, in addition to the usual fixed points of infinite temperature ($J_1^* = 0$) and zero temperature ($J_2^* = \infty$), this map admits a nontrivial fixed point, located at $J_c^* = 0.50689$: It corresponds to the expected finite critical temperature $T_c = (J_c^*)^{-1}$. It is an unstable fixed point, because $J' < J$ for $J < J_c^*$ and $J' > J$ for $J > J_c^*$. Despite the adopted heuristic recipe necessarily introducing some level of approximation, we want to point out that the estimated value of J_c^* is not far from the exact one of $J_c^* = 0.44068$ obtained by the Onsager solution of the Ising model in two dimensions.

Making use of Eqs. (5.100) and (5.102), we can obtain an estimate of the critical exponent ν. We can first estimate τ by expressing Eq. (5.100) in the form

$$J' - J_c^* = \left.\frac{dJ'}{dJ}\right|_{J_c^*}\left(J - J_c^*\right) + O\left((J - J_c^*)^2\right), \tag{H.39}$$

---

[2] In principle J should be a vector of different coupling constants, including the original nearest-neighbor coupling. We make things simple just considering one single coupling, the basic one.

[3] This might appear to be a reasonable physical assumption, because it is based on the idea of keeping constant the minimum value of the internal energy of the original and of the decimated Hamiltonian, while ignoring the four-body interaction term, usually called plaquette. On the other hand, the degree of arbitrariness of this heuristic recipe becomes evident as soon as we observe that also including the plaquette contribution to the ground state energy, we recover only the two trivial fixed points $J_1^* = 0$ (attractive) and $J_1^* = +\infty$ (repulsive).

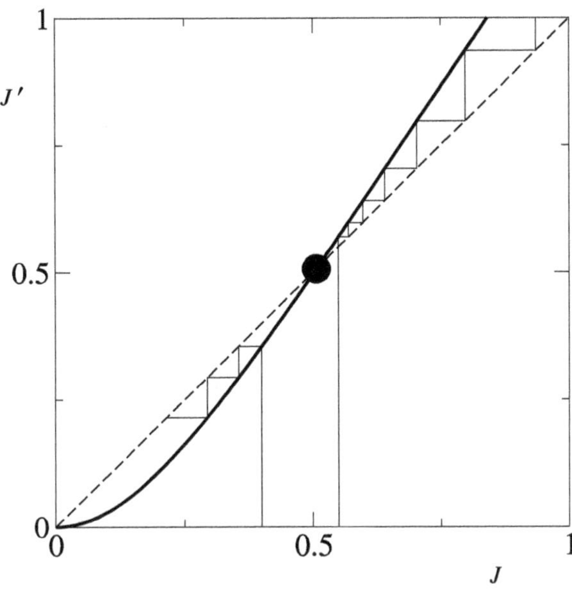

Fig. H.2 Plot of Eq. (H.38) (thick line), showing the existence of a nontrivial fixed point $J'(J_c^*) = J_c^*$, located at the intersection with the diagonal (dashed line). The unstable character of the fixed point is shown through the evolution of two values, one smaller than J_c^*, one larger than J_c^* (see thin lines).

where we have assumed $|J - J_c^*| \ll 1$, so that

$$\tau = \frac{dJ'}{dJ}\bigg|_{J_c^*} = 1.44892. \tag{H.40}$$

It follows that

$$\frac{1}{\nu} = \frac{\ln \tau}{\ln \ell} = \frac{\ln 1.44892}{\ln \sqrt{2}} \approx 1.06996, \tag{H.41}$$

to be compared with the exact result $\nu = 1$ and with the classical exponent of the Landau mean-field theory, $\nu = 1/2$.

This result allows us to appreciate the effectiveness of the renormalization group approach to the study of critical phenomena, even if affected by some tricky approximation.

The Deterministic KPZ Equation and the Burgers Equation

In this appendix, we provide some insights into the one-dimensional deterministic KPZ (dKPZ) equation,

$$\partial_t h(x,t) = \nu h_{xx} + \frac{\lambda}{2} h_x^2, \tag{I.1}$$

and into the Burgers equation, obtained from it by taking the spatial derivative of both sides,

$$\partial_t m(x,t) = \nu m_{xx} + \lambda m m_x. \tag{I.2}$$

We are going to argue that dynamics leads to the formation of negative curvature parabolas, which are connected by small regions of high positive curvature, corresponding to moving fronts in the Burgers equation. Later we prove that a parabola is a self-similar solution of the dKPZ equation and we derive the expression of a front solution for the Burgers equation. Finally, we use the Cole–Hopf transformation to provide a formally exact solution of the dKPZ equation for any initial condition and in any dimension d.

Qualitative considerations of the evolution of a parabola can be easily drawn by considering the different contributions of the linear and nonlinear terms. The linear term is constant, either positive or negative depending on the sign of the curvature. The nonlinear term is always positive and increases with the distance from the vertex of the parabola. The combination of the two terms is graphically depicted in Fig. I.1: A parabola of negative curvature moves in the negative direction and flattens, while a parabola of positive curvature moves in the positive direction and steepens. Let us be more quantitative, assuming the functional form

$$h(x,t) = A(t) + C(t)(x - x_0)^2. \tag{I.3}$$

Evaluating its temporal and spatial derivatives, we find

$$\dot{A} + \dot{C}(x - x_0)^2 = 2\nu C + 2\lambda C^2 (x - x_0)^2, \tag{I.4}$$

whose solution is

$$A(t) = A(0) - \frac{\nu}{\lambda} \ln \left| \frac{1}{C(0)} - 2\lambda t \right| \tag{I.5}$$

and

$$C(t) = \frac{1}{\dfrac{1}{C(0)} - 2\lambda t}. \tag{I.6}$$

Therefore, as anticipated, a parabola with negative curvature moves downward and it flattens with time, $|C(t)| \approx 1/t$, while a parabola with positive curvature moves upward and

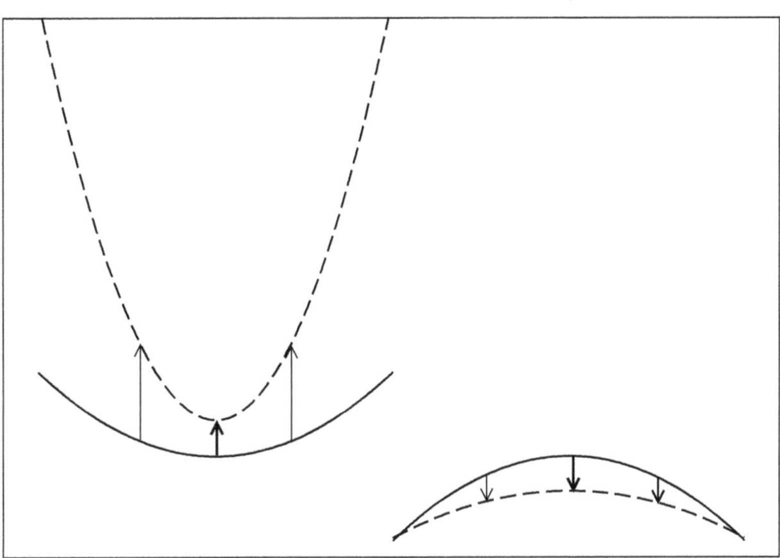

Fig. I.1 Evolution of two parabolas, with negative and positive curvature. The thick arrow at the vertex of the paraboloid is the constant contribution of the linear term, the nonlinear term vanishing there. Moving away from the center we must add the increasingly important contribution of the nonlinear term, which is always positive (for $\lambda > 0$).

its curvature increases in time, diverging at a finite time $\tau_c = 1/(2\lambda C(0))$. In a continuum profile, where regions with positive and negative curvature alternate, this divergence disappears and is replaced by a finite curvature dip, easily derivable in the context of the Burgers equation where it corresponds to a kink or a front. So, let us turn to Eq. (I.2) and look for a front-type solution, that is, a function $m(x, t) = u(x - vt)$ with $u(x) \to \pm u_\pm$ for $x \to \pm\infty$. Since $\partial_t u(x - vt) = -v\partial_x u$, the sought-after function $u(x)$ satisfies the equation

$$-vu_x = \nu u_{xx} + \lambda u u_x. \tag{I.7}$$

Integrating between $-\infty$ and $+\infty$, we find

$$v = -\frac{\lambda}{2}(u_+ - u_-). \tag{I.8}$$

So, perfectly antisymmetric boundary conditions ($u_+ = u_- = u_0$) induce a static kink, while an asymmetry sets the front in motion.

The form of the kink can be found by integrating Eq. (I.7). We do it explicitly for the static kink only, that is, $v = 0$. In this case, after integration, we obtain $\nu u_x = \frac{\lambda}{2}(u_0^2 - u^2)$, whose solution is

$$u(x) = u_0 \tanh\left(\frac{\lambda u_o}{2\nu}x\right), \tag{I.9}$$

which can also be integrated to provide the dip-like profile for the dKPZ equation,

$$h(x, t) = \frac{2\nu}{\lambda} \ln \cosh\left(\frac{\lambda u_o}{2\nu}x\right) + \frac{\lambda u_0^2}{2}t. \tag{I.10}$$

The first term on the right-hand side represents a function that interpolates between the straight lines $\mp u_0 x$, and the second term is a drift, quite similar to the drifts felt by the self-similar parabolas. In a "real" profile, the continuity between the pieces of parabola and the pieces of dip-like profile generally imposes asymmetric boundary conditions on the latter, therefore determining their drift and a sort of process where bigger parabolas "eat" smaller ones.

We conclude this appendix using the Cole–Hopf transformation to solve the dKPZ equation exactly in any dimension d,

$$\partial_t h(\mathbf{x}, t) = \nu \nabla^2 h + \frac{\lambda}{2} (\nabla h)^2. \tag{I.11}$$

The Cole–Hopf transformation is

$$H(\mathbf{x}, t) = e^{\frac{\lambda}{2\nu} h(\mathbf{x}, t)}, \tag{I.12}$$

which transforms the nonlinear Eq. (I.11) into a much simpler diffusion equation,

$$\partial_t H = \nu \nabla^2 H. \tag{I.13}$$

If H evolves according to the diffusion equation, it is obvious to conclude that a paraboloid is a self-similar solution of the dKPZ: In fact a Gaussian is a self-similar solution of the diffusion equation, and the logarithm of a Gaussian, $h(\mathbf{x}, t) = (2\nu/\lambda) \ln H(\mathbf{x}, t)$, is a paraboloid.

The general solution of H is straightforward if we recall that a Gaussian is the solution of the diffusion equation corresponding to a Dirac delta as the initial condition. Since any initial condition $H(\mathbf{x}, 0)$ is a linear combination of Dirac delta functions centered at all points $\mathbf{x}'$, $H(\mathbf{x}, 0) = \int d\mathbf{x}' \delta(\mathbf{x} - \mathbf{x}') H(\mathbf{x}', 0)$, the general solution of H is the same linear combination of the evolution of such a Dirac delta, that is, Gaussian functions:

$$H(\mathbf{x}, t) = \int \frac{d\mathbf{x}'}{(4\pi \nu t)^{d/2}} \exp\left[-\frac{(\mathbf{x} - \mathbf{x}')^2}{4\nu t} \right] H(\mathbf{x}', 0). \tag{I.14}$$

If we now invert the Cole–Hopf transformation, Eq. (I.12), we get the general solution of the dKPZ equation:

$$h(\mathbf{x}, t) = \frac{2\nu}{\lambda} \ln \left\{ \int \frac{d\mathbf{x}'}{(4\pi \nu t)^{d/2}} \exp\left[-\frac{(\mathbf{x} - \mathbf{x}')^2}{4\nu t} + \frac{\lambda}{2\nu} h(\mathbf{x}', 0) \right] \right\}. \tag{I.15}$$

This result might suggest a strong simplification of the stochastic equation (7.38) as well. This does not happen, because of noise transformation. If we start from

$$\partial_t h = \nu \nabla^2 h + \frac{\lambda}{2} (\nabla h)^2 + \eta, \tag{I.16}$$

we obtain

$$\partial_t H = \frac{\lambda}{2\nu} H \partial_t h \tag{I.17}$$

$$= \frac{\lambda}{2\nu} H \left(\nu \nabla^2 h + \frac{\lambda}{2} (\nabla h)^2 + \eta \right) \tag{I.18}$$

$$= \nu \nabla^2 H + \frac{\lambda}{2\nu} H \eta(\mathbf{x}, t), \tag{I.19}$$

which is known as the stochastic heat equation. Therefore, the stochastic term now depends on H itself: It is called multiplicative noise and can also be found in the mean-field description of directed percolation; see Section 6.3.3. Multiplicative noise, which is more difficult to treat than standard additive noise, also poses the problem of choosing the Itô or the Stratonovich representation of stochastic noise.

I.1 The Functional Derivative

Throughout the book, we meet quantities such as the following,

$$\mathcal{F}[h] = \int d\mathbf{x} f(\mathbf{x}, h(\mathbf{x}), \nabla h(\mathbf{x}), \dots), \tag{I.20}$$

and we may need to evaluate the variation of $\mathcal{F}[h]$ induced by an infinitesimal variation of $h(\mathbf{x})$,

$$\delta \mathcal{F}[h] = \mathcal{F}[h + \delta h] - \mathcal{F}[h] \equiv \int d\mathbf{x} \frac{\delta \mathcal{F}}{\delta h(\mathbf{x})} \delta h(\mathbf{x}). \tag{I.21}$$

The equality earlier is the definition of the functional derivative $\delta \mathcal{F} / \delta h(\mathbf{x})$.

In order to evaluate it, we introduce a small parameter ϵ that allows us to tune the strength of the variation of h, $\delta h(\mathbf{x}) = \epsilon \varphi(\mathbf{x})$, where $\varphi(\mathbf{x})$ is any smooth function (to be further commented on later). Therefore, we obtain

$$\int d\mathbf{x} \frac{\delta \mathcal{F}}{\delta h(\mathbf{x})} \varphi(\mathbf{x}) = \lim_{\epsilon \to 0} \frac{\mathcal{F}[h + \epsilon \varphi] - \mathcal{F}[h]}{\epsilon} = \frac{d}{d\epsilon} \mathcal{F}[h + \epsilon \varphi] \Big|_{\epsilon = 0}. \tag{I.22}$$

In order to be more specific, let us assume that f depends at most on the first spatial derivatives of h, so

$$\frac{d}{d\epsilon} \mathcal{F}[h + \epsilon \varphi] \Big|_{\epsilon = 0} = \frac{d}{d\epsilon} \int d\mathbf{x} f(\mathbf{x}, h + \epsilon \varphi, \nabla h + \epsilon \nabla \varphi) \Big|_{\epsilon = 0} \tag{I.23}$$

$$= \int d\mathbf{x} \left(\frac{\partial f}{\partial h} \varphi + \frac{\partial f}{\partial (\partial_i h)} \partial_i \varphi \right) \tag{I.24}$$

$$= \int d\mathbf{x} \left(\frac{\partial f}{\partial h} \varphi + \partial_i \left(\frac{\partial f}{\partial (\partial_i h)} \varphi \right) - \left(\partial_i \frac{\partial f}{\partial (\partial_i h)} \right) \varphi \right) \tag{I.25}$$

$$= \int d\mathbf{x} \left(\frac{\partial f}{\partial h} - \partial_i \frac{\partial f}{\partial (\partial_i h)} \right) \varphi. \tag{I.26}$$

Earlier we use the standard notation to sum over repeated indices, and from the third to the fourth line, we remove the total derivative, $\int d\mathbf{x} \partial_i (\cdots)$, that vanishes either for periodic boundary conditions or for perturbations $\varphi(\mathbf{x})$, which vanish at infinity.

Since earlier result must hold for any function $\varphi(\mathbf{x})$, we can finally write

$$\frac{\delta \mathcal{F}}{\delta h(\mathbf{x})} = \frac{\partial f}{\partial h} - \partial_i \frac{\partial f}{\partial (\partial_i h)} \equiv \frac{\partial f}{\partial h} - \nabla \cdot \frac{\partial f}{\partial (\nabla h)}. \tag{I.27}$$

To avoid cumbersome notation, we extend this result to a dependence of f on derivatives of generic order for one spatial dimension only. If

$$\mathcal{F}[h] = \int dx\, f(x, h, \partial_x h, \partial_x^2 h, \partial_x^3 h, \dots), \tag{I.28}$$

we have

$$\frac{\delta \mathcal{F}}{\delta h(x)} = \frac{\partial f}{\partial h} + \sum_{n=1}^{\infty} (-)^n \frac{\partial^n}{\partial x^n} \frac{\partial f}{\partial (\partial_x^n h)}. \tag{I.29}$$

Should the above derivation be too formal, we provide a simpler one that uses the discrete representation, the drawback being we focus on a specific function $\mathcal{F}$ in one spatial dimension,

$$\mathcal{F}[h] = \int dx \left[U(h) + \frac{K}{2} (\partial_x h)^2 \right], \tag{I.30}$$

which is rewritten as a discrete sum,

$$\mathcal{F}[h] = \sum_i \Delta \left[U(h_i) + \frac{K}{2} \left(\frac{h_{i+1} - h_{i-1}}{2\Delta} \right)^2 \right], \tag{I.31}$$

where Δ is the discrete lattice constant.

Let us now evaluate the quantity $\delta \mathcal{F}[h]$, Eq. (I.21), using δ_i as the discrete version of $\delta h(x)$,

$$\begin{aligned}
\delta \mathcal{F}[h] &= \sum_i \Delta \Bigg[U(h_i + \delta_i) - U(h_i) \\
&\quad + \frac{K}{2} \left(\left(\frac{h_{i+1} + \delta i + 1 - h_{i-1} - \delta_{i-1}}{2\Delta} \right)^2 - \left(\frac{h_{i+1} - h_{i-1}}{2\Delta} \right)^2 \right) \Bigg] \\
&= \sum_i \Delta \left[\frac{\partial U}{\partial h_i} \delta_i + \frac{K}{2} \frac{(h_{i+1} - h_{i-1})(\delta_{i+1} - \delta_{i-1})}{2\Delta^2} \right] \tag{I.32} \\
&= \sum_i \Delta \left[\frac{\partial U}{\partial h_i} - K \frac{(h_{i+2} + h_{i-2} - 2h_i)}{(2\Delta)^2} \right] \delta_i. \tag{I.33}
\end{aligned}$$

We can finally go back to continuum and write

$$\delta \mathcal{F}[h] = \int dx \left[\frac{\partial U}{\partial h} - K \frac{\partial^2 h}{\partial x^2} \right] \delta h(x), \tag{I.34}$$

that is to say,

$$\frac{\delta \mathcal{F}}{\delta h(x)} = \frac{\partial U}{\partial h} - K \frac{\partial^2 h}{\partial x^2}, \tag{I.35}$$

in agreement with the general formula (I.27).

In reference to Section 7.7.3, it is useful to define the functional derivative of f (see Eq. (I.20)) through the relation

$$\frac{\delta \mathcal{F}}{\delta h(\mathbf{x})} \equiv \int d\mathbf{x}'\, \frac{\delta f(h(\mathbf{x}'))}{\delta h(\mathbf{x})}. \tag{I.36}$$

For example, if $f = U(h(\mathbf{x}')) + \frac{K}{2}(\nabla h(\mathbf{x}'))^2$, we have

$$\frac{\delta f(h(\mathbf{x}'))}{\delta h(\mathbf{x})} = [U'(h(\mathbf{x}') + K\nabla^2 h(\mathbf{x}')]\delta(\mathbf{x} - \mathbf{x}'). \tag{I.37}$$

Stochastic Differential Equation for the Energy of the Brownian Particle

The Langevin equation obeyed by the kinetic energy $E(t) = \frac{1}{2}mv^2(t)$ of a Brownian particle can be derived from Eq. (2.21) and it takes the form

$$\frac{dE(t)}{dt} = -2\gamma E(t) + \sqrt{2mE(t)}\eta(t). \tag{J.1}$$

It is a simple example of a Langevin-like equation with multiplicative noise, because the noise term η is multiplied by a factor proportional to $\sqrt{E(t)}$. This problem can be reformulated in terms of the general formalism of stochastic differential equations described in Section 2.4.3 as follows

$$dE(t) = -2\gamma E(t)dt + \sqrt{2mE(t)}bdW, \tag{J.2}$$

where $b = \gamma\sqrt{2D}$, see Eq. (2.150). Dividing both sides of this stochastic equation by $\sqrt{E(t)}$, we obtain

$$\frac{1}{\sqrt{E(t)}}dE(t) = -2\gamma\sqrt{E(t)}dt + \sqrt{2m}bdW. \tag{J.3}$$

Adopting the Stratonovich formulation amounts to assume that the l.h.s. of this equation is a standard differential, that is, $\frac{1}{\sqrt{E}}dE(t) \equiv 2d\sqrt{E}$ and we can rewrite the previous equation as follows

$$d\sqrt{E(t)} = -\gamma\sqrt{E(t)}dt + \sqrt{\frac{m}{2}}bdW. \tag{J.4}$$

By recalling that for the Brownian particle $E(t) = \frac{1}{2}mv^2(t)$, we find that Eq. (J.4) reproduces the Langevin equation for the velocity of the Brownian particle Eq. (2.22), which is consistent with Einstein relation Eq. (2.29).

Conversely, if we adopt the Itô formulation, we have to apply the Itô formula (also called Itô Lemma) Eq. (2.174) with $X(t) \equiv E(t)$ and $f(X(t)) \equiv \sqrt{E(t)}$, thus yielding the stochastic differential equation

$$d\sqrt{E(t)} = -\gamma\sqrt{E(t)}dt - \frac{1}{\sqrt{E(t)}}\frac{b^2m}{4}dt + \sqrt{\frac{m}{2}}bdW. \tag{J.5}$$

This equation is obtained by considering that for solving a stochastic differential equation, we have to integrate it and average the result over the Wiener process. Accordingly, we can assume that in the Itô formula (2.174), the addendum proportional to $(dW)^2$ yields a contribution of order $dt = \langle(dW)^2\rangle$.

Note that the second addendum on the r.h.s of Eq. (J.5) makes the corresponding Langevin equation for the velocity of the Brownian particle acquire an additional contribution, which cannot be compatible with Einstein relation (2.29). This is why in this case, the Stratonovich formulation is the one to be adopted for reproducing this basic physical relation.

The Kramers–Moyal Expansion

We illustrate the method known as Kramer–Moyal expansion to obtain the backward Kolmogorov equation (2.229). We start from the Chapman–Kolmogorov equation (2.66) in the form

$$W(X_0, t_0 | X, t) = \int_{\mathbb{R}} dY \, W(X_0, t_0 | Y, t_0 + \Delta t_0) \, W(Y, t_0 + \Delta t_0 | X, t). \tag{K.1}$$

The Kramers–Moyal expansion makes use of the combination of the following two identities:

$$W(X_0, t_0 | Y, t_0 + \Delta t_0) = \int_{\mathbb{R}} dZ \, \delta(Z - Y) W(X_0, t_0 | Z, t_0 + \Delta t_0) \tag{K.2}$$

$$\delta(Z - Y) = \sum_{n=0}^{+\infty} \frac{(Z - X_0)^n}{n!} \left(\frac{\partial}{\partial X_0} \right)^n \delta(X_0 - Y), \tag{K.3}$$

where $\delta(x)$ denotes the Dirac-delta distribution of argument x and $\left(\frac{\partial}{\partial X_0} \right)^n$ is a shortened notation for the nth-order partial derivative with respect to X_0. While Eq. (K.2) is self-evident, Eq. (K.3) can be proved by considering a test function in the Schwartz space $\mathcal{S}$, $\phi(Y)$, for which the following relation holds:

$$\int dY \, \delta(Z - Y) \, \phi(Y) = \phi(Z). \tag{K.4}$$

In fact, the right-hand side of Eq. (K.3) yields

$$\int dY \sum_{n=0}^{+\infty} \frac{(Z - X_0)^n}{n!} \left(\frac{\partial}{\partial X_0} \right)^n \delta(X_0 - Y) \phi(Y)$$

$$= \sum_{n=0}^{+\infty} \frac{(Z - X_0)^n}{n!} \left(\frac{\partial}{\partial X_0} \right)^n \int dY \, \delta(X_0 - Y) \phi(Y)$$

$$= \sum_{n=0}^{+\infty} \frac{(Z - X_0)^n}{n!} \left(\frac{\partial}{\partial X_0} \right)^n \phi(X_0) \equiv \phi(Z),$$

where the left-hand side of last equality is the Taylor series expansion of $\phi(Z)$, thus proving the identity.

By substituting Eq. (K.3) into Eq. (K.2), we obtain

$$W(X_0, t_0 | Y, t_0 + \Delta t_0)$$

$$= \sum_{n=0}^{+\infty} \frac{1}{n!} \int_{\mathbb{R}} (Z - X_0)^n W(X_0, t_0 | Z, t_0 + \Delta t_0) dZ \left(\frac{\partial}{\partial X_0}\right)^n \delta(X_0 - Y)$$

$$= \left[1 + \sum_{n=1}^{+\infty} \frac{1}{n!} \int_{\mathbb{R}} (Z - X_0)^n W(X_0, t_0 | Z, t_0 + \Delta t_0) dZ \left(\frac{\partial}{\partial X_0}\right)^n\right] \delta(X_0 - Y) \quad \text{(K.5)}$$

where the last equality stems from the normalization condition

$$\int_{\mathbb{R}} dZ \, W(X_0, t_0 | Z, t_0 + \Delta t_0) = 1, \tag{K.6}$$

meaning that, starting from any initial condition (X_0, t_0), at any following time $t_0 + \Delta t_0$, the process will certainly be at some position Z among all the accessible ones.

We can insert Eq. (K.5) into the Chapman–Kolmogorov equation (K.1), thus obtaining

$$W(X_0, t_0 | X, t) - W(X_0, t_0 + \Delta t_0 | X, t)$$

$$= \sum_{n=1}^{+\infty} \frac{1}{n!} \int_{\mathbb{R}} dZ \, (Z - X_0)^n W(X_0, t_0 | Z, t_0 + \Delta t_0) \left(\frac{\partial}{\partial X_0}\right)^n$$

$$\times \int_{\mathbb{R}} dY \, \delta(X_0 - Y) W(Y, t_0 + \Delta t_0 | X, t)$$

$$= \sum_{n=1}^{+\infty} \frac{1}{n!} \int_{\mathbb{R}} dZ \, (Z - X_0)^n \, W(X_0, t_0 | Z, t_0 + \Delta t_0) \left(\frac{\partial}{\partial X_0}\right)^n W(X_0, t_0 + \Delta t_0 | X, t),$$

$$\tag{K.7}$$

where we have assumed that $W(Y, t_0 + \Delta t_0 | X, t) \in \mathcal{S}$. We define the nth-order momentum

$$\mu_n(X_0, t_0) = \lim_{\Delta t_0 \to 0} \frac{1}{\Delta t_0} \int_{\mathbb{R}} dZ (Z - X_0)^n \, W(X_0, t_0 | Z, t_0 + \Delta t_0), \tag{K.8}$$

and using Eq. (K.7), we can write

$$\frac{\partial W(X_0, t_0 | X, t)}{\partial t_0} = - \lim_{\Delta t_0 \to 0} \frac{W(X_0, t_0 | X, t) - W(X_0, t_0 + \Delta t_0 | X, t)}{\Delta t_0}$$

$$= - \sum_{n=1}^{+\infty} \frac{\mu_n(X_0, t_0)}{n!} \left(\frac{\partial}{\partial X_0}\right)^n W(X_0, t_0 | X, t). \tag{K.9}$$

If we assume that only the first two moments of the Kramers–Moyal expansion, $\mu_1(X_0, t_0) \equiv a(X_0, t_0)$ and $\mu_2(X_0, t_0) \equiv b^2(X_0, t_0)$, are the leading contributions (i.e., $\mu_n(X_0, t_0) \approx 0$ for $n \geq 3$), we finally obtain the backward Kolmogorov equation (2.229).

Let us consider a random variable x, distributed accordingly to the normalized distribution $p(x)$. The nth moment of the distribution is defined as the average value of x^n,

$$\langle x^n \rangle = \int dx\, x^n p(x) \tag{L.1}$$

and all moments can be derived from the moment-generating function $\phi(t)$,

$$\phi(t) = \langle e^{tx} \rangle = \int dx\, e^{tx} p(x) = \sum_{n=0}^{\infty} \frac{t^n}{n!} \langle x^n \rangle \tag{L.2}$$

$$\langle x^n \rangle = \frac{d^n \phi(t)}{dt^n}\bigg|_{t=0}. \tag{L.3}$$

The function $\phi(t)$ is strictly related to the characteristic function $\varphi(q)$, which is basically the Fourier transform of $p(x)$,

$$\varphi(q) = \langle e^{iqx} \rangle = \int dx\, e^{iqx} p(x) = \phi(iq), \tag{L.4}$$

and Eq. (L.3) can be rewritten in terms of the characteristic function,

$$\langle x^n \rangle = (-i)^n \frac{d^n \varphi(q)}{dq^n}\bigg|_{q=0}. \tag{L.5}$$

If x and y are independent random variables with moment-generating functions $\phi_x(t), \phi_y(t)$, we immediately find that

$$\phi_{x+y}(t) = \langle e^{(x+y)t} \rangle = \langle e^{xt} e^{yt} \rangle = \langle e^{xt} \rangle \langle e^{yt} \rangle = \phi_x(t)\phi_y(t) \tag{L.6}$$

and this property suggests to define the logarithm of the moment-generating function,

$$K(t) = \log \phi(t) = \log \langle e^{tx} \rangle. \tag{L.7}$$

The function $K(t)$ is called cumulant-generating function and has the trivial property

$$K_{x+y}(t) = K_x(t) + K_y(t), \tag{L.8}$$

where x, y are independent random variables, as earlier. Its name (*cumulant*-generating function) derives from its Taylor expansion,

$$K(t) = \sum_{n=0}^{\infty} \frac{t^n}{n!} M_n, \tag{L.9}$$

which defines the cumulants M_n of the distribution $p(x)$. The first four cumulants are[1]

$$M_1 = \langle x \rangle, \tag{L.10}$$

$$M_2 = \langle x^2 \rangle - \langle x \rangle^2, \tag{L.11}$$

$$M_3 = \langle (x - \langle x \rangle)^3 \rangle, \tag{L.12}$$

$$M_4 = \langle (x - \langle x \rangle)^4 \rangle - 3\langle (x - \langle x \rangle)^2 \rangle^2. \tag{L.13}$$

For the normal (Gaussian) distribution,

$$p_G(x) = \frac{1}{\sqrt{2\pi\sigma^2}} \exp\left(-\frac{(x - \mu)^2}{2\sigma^2} \right), \tag{L.14}$$

we can easily find the moment-generating function,

$$\phi_G(t) = \langle e^{tx} \rangle_G = \exp\left(\mu t + \frac{\sigma^2 t^2}{2} \right), \tag{L.15}$$

whose logarithm gives the cumulant-generating function,

$$K_G(t) = \mu t + \frac{\sigma^2 t^2}{2}. \tag{L.16}$$

Therefore, all Gaussian cumulants of order three or higher vanish.[2]

---

[1] The skewness of a distribution is equal to M_3/σ^3, where $\sigma^2 = M_2$ is the variance. The kurtosis of a distribution is equal to $\langle (x - \langle x \rangle)^4 \rangle / \sigma^4$.

[2] The skewness of $p_G(x)$ vanish and its kurtosis is equal to 3.

M.1 The Diffusion Equation with Drift: General Solution

In this appendix we offer the general solution of the diffusion equation with drift (see Eq. (2.48)):

$$\frac{\partial \rho(\mathbf{x}, t)}{\partial t} = -\mathbf{v}_0 \cdot \nabla \rho(\mathbf{x}, t) + D\nabla^2 \rho(\mathbf{x}, t). \tag{M.1}$$

As we will see, the solution is the combination of a standard diffusion process in the plane orthogonal to $\mathbf{v}_0$ and a diffusion with drift along $\mathbf{v}_0$. Because of isotropy we can assume the drift $\mathbf{v}_0$ is oriented along the $\hat{x}$-axis and we limit ourselves to two spatial dimensions, the generalization to $d > 2$ being straightforward. The earlier equation therefore reads

$$\frac{\partial \rho(x, y, t)}{\partial t} = -v_0 \partial_x \rho(x, y, t) + D(\partial_{xx} + \partial_{yy})\rho(x, y, t), \tag{M.2}$$

whose solution can be found by separation of variables, $\rho(x, y, t) = p(x, t)q(y, t)$, obtaining

$$\partial_t p(x, t) = -v_0 \partial_x p(x, t) + D\partial_{xx} p(x, t) \tag{M.3}$$

$$\partial_t q(y, t) = D\partial_{yy} q(y, t), \tag{M.4}$$

which we are going to solve with initial conditions $p(x, 0) = \delta(x - x_0)$ and $q(y, 0) = \delta(y - y_0)$. In fact, because of the linear character of the equations, any different initial profile can be written as a superposition of the Dirac delta's evolving, each separately from the other.

The equation for $q(y, t)$ is called diffusion equation or heat equation,

$$\frac{\partial q(y, t)}{\partial t} = D\frac{\partial^2}{\partial y^2} q(y, t), \tag{M.5}$$

which can be easily solved by introducing the spatial Fourier transform of $q(y, t)$,

$$q(k, t) = \int_{-\infty}^{+\infty} dy \, e^{-iky} q(y, t), \tag{M.6}$$

where i is the standard notation for the imaginary constant, $i^2 = -1$. By applying the Fourier transform to both sides of Eq. (M.5), one obtains

$$\frac{\partial q(k, t)}{\partial t} = -D k^2 q(k, t), \tag{M.7}$$

which yields the solution

$$q(k, t) = q(k, 0)e^{-D k^2 t}. \tag{M.8}$$

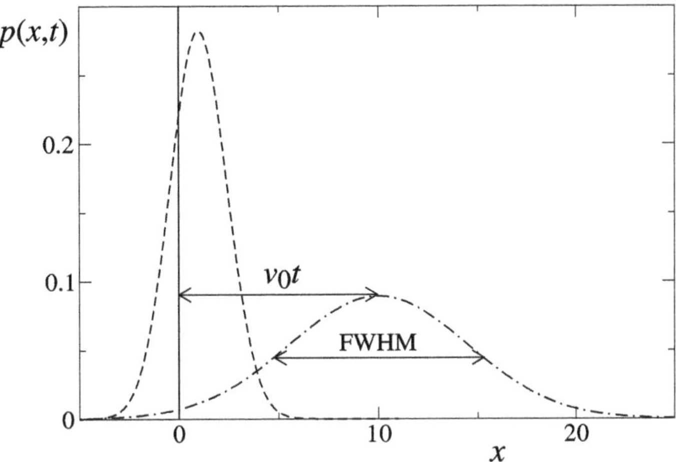

Fig. M.1 Plot of $p(x, t)$ (see Eq. (M.10)) at different times, for $v_0/D = 1$ and $x_0 = 0$: $Dt = 1$ (dashed curve) and $Dt = 10$ (dot-dashed curve). At $t = 0$, we have a Dirac delta, while at $t > 0$, we have a Gaussian centered in $x = v_0 t$, with a full width at half maximum (FWHM) equal to $4\sqrt{\ln 2}\sqrt{Dt}$.

If we assume the initial condition $q(y, 0) = \delta(y - y_0)$, we have $q(k, 0) = \exp(-i k y_0)$. The solution for $q(y, t)$ can be obtained by antitransforming expression (M.8):

$$q(y, t) = \frac{1}{2\pi} \int_{-\infty}^{+\infty} dk\, e^{i k(y-y_0) - D k^2 t} = \frac{1}{\sqrt{4\pi D t}} \exp\left(-\frac{(y - y_0)^2}{4 D t}\right). \tag{M.9}$$

We can now turn to the equation for $p(x, t)$ and remark that for a constant drift v_0, it can be solved by the simple relation $p(x, t) = q(x - v_0 t, t)$. In conclusion, we have

$$p(x, t) = \frac{1}{\sqrt{4\pi D t}} \exp\left(-\frac{(x - v_0 t - x_0)^2}{4 D t}\right) \tag{M.10}$$

$$q(y, t) = \frac{1}{\sqrt{4\pi D t}} \exp\left(-\frac{(y - y_0)^2}{4 D t}\right). \tag{M.11}$$

Solution (M.10) (see Fig. M.1) corresponds to a Gaussian distribution, which broadens and flattens as time increases, while its center translates at constant speed v_0.

M.2 Diffusion Equation: Fourier–Laplace Transform

The solution of the diffusion equation

$$\frac{\partial \rho(x, t)}{\partial t} = D \frac{\partial^2 \rho(x, t)}{\partial x^2} \tag{M.12}$$

has been found in Appendix M.1,

$$\rho(x, t) = \frac{1}{\sqrt{4\pi Dt}} \exp - \left[\frac{(x - x(0))^2}{4Dt}\right]. \tag{M.13}$$

Since $\rho(x, t)$ is actually a function of $(x - x(0), t)$, its Fourier–Laplace transform has the property

$$\rho(k, s) = e^{-ikx(0)}\rho_0(k, s), \tag{M.14}$$

where $\rho_0(x, t)$ corresponds to $x(0) = 0$. Here later we prove that the Laplace–Fourier transform of $\rho_0(x, t)$ is

$$\rho_0(k, s) = \frac{1}{s + Dk^2}, \tag{M.15}$$

without evaluating any integral, just using the fact that $\rho_0(x, t)$ satisfies Eq. (M.12). In fact, we can show that Eq. (M.13) is the antitransform of $\rho(k, s)$. As a first step we can rewrite Eq. (M.15) as

$$s\,\rho_0(k, s) = -D\,k^2\,\rho_0(k, s), \tag{M.16}$$

then we can antitransform with respect to Laplace both sides of this equation as

$$\frac{1}{2\pi i}\int_{a-i\infty}^{a+i\infty} ds\, e^{st}\, s\,\rho_0(k, s) = -Dk^2\,\frac{1}{2\pi i}\int_{a-i\infty}^{a+i\infty} ds\, e^{st}\,\rho_0(k, s), \tag{M.17}$$

where $a > a_c$, a_c is the abscissa of convergence of the inverse Laplace transform, that is, the highest value of the real part among the singularities of $\rho_0(k, s)$ in the complex plane of the variable s. We want to point out that the result of the two integrals in Eq. (M.17) is independent of the actual value of a and that a_c can be set to zero if $\rho_0(k, s)$ is a regular function of s. We can now observe that the left-hand side is a time derivative and rewrite this equation as

$$\frac{\partial}{\partial t}\rho_0(k, t) = -Dk^2\,\rho_0(k, t), \tag{M.18}$$

where

$$\rho_0(k, t) = \frac{1}{2\pi i}\int_{a-i\infty}^{a+i\infty} e^{st}\rho_0(k, s). \tag{M.19}$$

Now, we can antitransform with respect to Fourier both sides of Eq. (M.18),

$$\frac{\partial}{\partial t}\left[\frac{1}{2\pi}\int_{-\infty}^{+\infty} dk\, e^{ikx}\rho_0(k, t)\right] = -D\,\frac{1}{2\pi}\int_{-\infty}^{+\infty} dk\, k^2\, e^{ikx}\rho_0(k, t), \tag{M.20}$$

and observe that the right-hand side of this equation is a second derivative,

$$D\frac{\partial^2}{\partial x}\left[\frac{1}{2\pi}\int_{-\infty}^{+\infty} dk\, e^{ikx}\rho_0(k, t)\right], \tag{M.21}$$

so we finally obtain Eq. (M.12), where

$$\rho_0(x, t) = \frac{1}{2\pi}\int_{-\infty}^{+\infty} dk\, e^{ikx}\rho_0(k, t). \tag{M.22}$$

The reader who aims to prove Eq. (M.15) "more traditionally" should first Fourier-transform $\rho(x, t)$, which is a tabulated integral, so as to obtain $\rho_0(k, t) = e^{-k^2 Dt}$, then

Laplace transform,

$$p_0(k, s) = \int_0^\infty e^{-st} e^{-k^2 Dt} dt \tag{M.23}$$

$$= \frac{1}{s + Dk^2} . \tag{M.24}$$

M.3 Random Walk and Its Momenta

We describe here a generalization of the random walk problem discussed in Section 2.3.2, where the random walker at each unit time step walks a distance y, distributed according to a normalized distribution $\eta(y)$, that is,

$$\int_{-\infty}^{+\infty} dy\, \eta(y) = 1. \tag{M.25}$$

The probability that the walker is at position x at (discrete) time n is indicated by $p_n(x)$. Without prejudice to generality, we can assume that the walk starts at the origin, and we can write $p_0(x) = \delta(x)$. At time $t = 1$, the distribution of the new position of the walker is provided by $\eta(x)$, that is, $p_1(x) = \eta(x)$. More generally, we can conclude that $p_n(x)$ has to satisfy the following recursion relation:

$$p_n(x) = \int_{-\infty}^{+\infty} dy\, p_{n-1}(x - y)\, \eta(y) = \int_{-\infty}^{+\infty} dy\, p_{n-1}(y)\, \eta(x - y). \tag{M.26}$$

Fourier transforming both sides of this equation, we can write

$$p_n(k) = p_{n-1}(k)\, \eta(k), \tag{M.27}$$

because the integral on the right-hand side is a convolution product that transforms into a standard product of the Fourier transformed functions, where

$$p_n(k) = \int_{-\infty}^{+\infty} dx\, e^{-ikx}\, p_n(x) \tag{M.28}$$

and

$$\eta(k) = \int_{-\infty}^{+\infty} dx\, e^{-ikx}\, \eta(x), \tag{M.29}$$

which is known in the literature as the characteristic function of the probability density distribution $\eta(x)$ and corresponds to $\langle \exp(ikx) \rangle$.[1]

The solution of Eq. (M.27) is

$$p_n(k) = \eta^n(k), \tag{M.30}$$

---

[1] The sign in the exponential is irrelevant if the distribution $\eta(x)$ is an even function.

because $p_0(k) = 1$. It is useful to rewrite Eq. (M.29) by expanding the imaginary exponential inside the integral as

$$
\begin{aligned}
\eta(k) &= \int_{-\infty}^{+\infty} dx \left(1 - ik x - \frac{1}{2} k^2 x^2 + \frac{i}{6} k^3 x^3 + \cdots \right) \eta(x) \\
&= \sum_{n=0}^{+\infty} \frac{(-i)^n}{n!} \langle x^n \rangle k^n,
\end{aligned}
\tag{M.31}
$$

where the expression

$$
\langle x^m \rangle = \int_{-\infty}^{+\infty} dx\, x^m\, \eta(x)
\tag{M.32}
$$

defines the mth momentum of the probability density function $\eta(x)$.[2] According to Eq. (M.31), we can also write

$$
\langle x^m \rangle = i^m \left. \frac{d^m \eta(k)}{d k^m} \right|_{k=0}.
\tag{M.33}
$$

This is quite a useful formula, because it allows us to express the moments of the probability density function $\eta(x)$ in terms of the derivatives of its Fourier transform $\eta(k)$.

M.4 Isotropic Random Walk with a Trap

The equation describing the distribution probability $p(x, t)$ of a random walker moving freely in one dimension, with diffusion coefficient D, is

$$
\frac{\partial p}{\partial t} = D \frac{\partial^2 p}{\partial x^2}.
\tag{M.34}
$$

Its solution, if the particle is initially located in $x = x_0$, is given in Appendix M.1,

$$
p_{x_0}(x, t) = \frac{1}{\sqrt{4\pi Dt}} \exp\left(-\frac{(x - x_0)^2}{4Dt} \right).
\tag{M.35}
$$

If there is a trap in $x = 0$, the probability $p(x, t)$ must vanish in $x = 0$ at any t. The correct solution is easily obtained by taking the linear combination of $p_{x_0}(x, t)$ and $p_{-x_0}(x, t)$, because it automatically satisfies the boundary condition $p(0, t) = 0$ at all times. This combination is simply

$$
p(x, t) = \frac{1}{\sqrt{4\pi Dt}} \left[\exp\left(-\frac{(x - x_0)^2}{4Dt} \right) - \exp\left(-\frac{(x + x_0)^2}{4Dt} \right) \right].
\tag{M.36}
$$

The probability that the particle is trapped in $x = 0$ in the time interval $(t, t + dt)$ is called first-passage probability, $f(t)$, and it is equal to the current $|J|$ flowing in the origin, where $J = -D p'(x = 0)$, that is,

$$
f(t) = |J(t)| = \frac{x_0}{\sqrt{4\pi Dt^3}} \exp\left(-\frac{x_0^2}{4Dt} \right).
\tag{M.37}
$$

---

[2] We want to point out that Eq. (M.31) is a meaningful expression, provided all moments $\langle x^m \rangle$ are finite.

This probability is normalized to 1,

$$\int_0^\infty f(t)\,dt = \int_0^\infty \frac{x_0}{\sqrt{4\pi D t^3}}\,\exp\left(-\frac{x_0^2}{4Dt}\right)dt \tag{M.38}$$

$$= \frac{2}{\sqrt{\pi}}\int_0^\infty e^{-\tau^2}\,d\tau \tag{M.39}$$

$$= 1, \tag{M.40}$$

and the average trapping time diverges,

$$\langle t_{\rm tr}\rangle = \int_0^\infty t f(t)\,dt = +\infty, \tag{M.41}$$

because the integrand $t f(t) \sim 1/\sqrt{t}$ for large t. The median trapping time $t_{\rm med}$, defined through the relation

$$\int_0^{t_{\rm med}} f(t)\,dt = \int_{t_{\rm med}}^\infty f(t)\,dt, \tag{M.42}$$

scales as x_0^2/D.

Finally, the time dependence of the average position of the particle is easily evaluated from the definition:

$$\langle x(t)\rangle = \int_0^\infty dx\,\frac{x}{\sqrt{4\pi D t}}\left[\exp\left(-\frac{(x-x_0)^2}{4Dt}\right) - \exp\left(-\frac{(x+x_0)^2}{4Dt}\right)\right] \tag{M.43}$$

and performing the change of variable $x \to -x$ in the integral of the right exponential,

$$\langle x(t)\rangle = \int_0^\infty dx\,\frac{x}{\sqrt{4\pi D t}}\,\exp\left(-\frac{(x-x_0)^2}{4Dt}\right) + \int_{-\infty}^0 dx\,\frac{x}{\sqrt{4\pi D t}}\,\exp\left(-\frac{(x-x_0)^2}{4Dt}\right) \tag{M.44}$$

$$= \int_{-\infty}^\infty dx\,\frac{x}{\sqrt{4\pi D t}}\,\exp\left(-\frac{(x-x_0)^2}{4Dt}\right) \tag{M.45}$$

$$= x_0. \tag{M.46}$$

In conclusion, the presence of a trap does not modify the average position of the walker.

M.5 Anisotropic Random Walk with a Trap

We have a random walker moving in one dimension, with diffusion coefficient D and a negative drift $-v$ (with $v > 0$). Its distribution probability $p(x,t)$ is described by the convection–diffusion equation,

$$\frac{\partial p}{\partial t} = v\frac{\partial p}{\partial x} + D\frac{\partial^2 p}{\partial x^2}. \tag{M.47}$$

If the initial condition is $p(x,0) = \delta(x - x_0)$, its solution, in the absence of traps, is given in Appendix M.1,

$$p_{x_0}(x,t) = \frac{1}{\sqrt{4\pi D t}}\,\exp\left(-\frac{(x - x_0 + vt)^2}{4Dt}\right). \tag{M.48}$$

If the particle has a trap in $x = 0$, this means the probability $p(x, t)$ must vanish in $x = 0$ at any t. This problem can be solved with the image method, which consists of finding a suitable linear combination of $p_{x_0}(x, t)$ and $p_{-x_0}(x, t)$ that satisfies the boundary condition $p(0, t) = 0$ at all times. This combination is

$$p(x, t) = \frac{1}{\sqrt{4\pi Dt}} \left[\exp\left(-\frac{(x - x_0 + vt)^2}{4Dt}\right) - \exp\left(\frac{vx_0}{D}\right) \exp\left(-\frac{(x + x_0 + vt)^2}{4Dt}\right) \right].$$

(M.49)

The probability the particle is trapped in $x = 0$ in the time interval $(t, t + dt)$ is equal to the current $|J|$ flowing in the origin, where now $J = (-vp - Dp')_{x=0}$, that is,

$$|J| = D \left.\frac{\partial p}{\partial x}\right|_0$$

(M.50)

$$= \frac{D}{\sqrt{4\pi Dt}} \left[\exp\left(-\frac{(x_0 - vt)^2}{4Dt}\right) \frac{(x_0 - vt)}{2Dt} + \exp\left(\frac{vx_0}{D}\right) \exp\left(-\frac{(x_0 + vt)^2}{4Dt}\right) \frac{(x_0 + vt)}{2Dt} \right]$$

(M.51)

$$= \frac{x_0}{\sqrt{4\pi Dt^3}} \exp\left(-\frac{(x_0 - vt)^2}{4Dt}\right).$$

(M.52)

The probability the particle is trapped within time t is

$$\int_0^t dt' |J(t')| = \frac{x_0}{\sqrt{4\pi D}} \exp\left(\frac{x_0 v}{2D}\right) \int_0^t \frac{dt'}{(t')^{3/2}} \exp\left(-\left(\frac{x_0^2}{4D}\frac{1}{t'} + \frac{v^2}{4D}t'\right)\right)$$

(M.53)

$$= \frac{2}{\sqrt{\pi}} e^c \int_{\bar{u}}^{\infty} du \, \exp\left(-\left(u^2 + \frac{c^2}{4u^2}\right)\right),$$

(M.54)

where $\bar{u} = x_0/\sqrt{4Dt}$, $c = vx_0/(2D)$, and the passage between Eqs. (M.53) and (M.54) has been obtained with the change of variable $u^2 = x_0^2/(4Dt')$.

Since, for $c > 0$,

$$\int du \, \exp\left(-\left(u^2 + \frac{c^2}{4u^2}\right)\right) = \frac{\sqrt{\pi}}{4} \left[e^c \operatorname{erf}\left(\frac{c}{2u} + u\right) - e^{-c} \operatorname{erf}\left(\frac{c}{2u} - u\right) \right],$$

(M.55)

we can evaluate above integral, getting the result

$$\int_0^t dt' |J(t')| = \frac{e^c}{2} \left\{ e^c \left[1 - \operatorname{erf}\left(\frac{c}{2\bar{u}} + \bar{u}\right)\right] + e^{-c} \left[1 + \operatorname{erf}\left(\frac{c}{2\bar{u}} - \bar{u}\right)\right] \right\}.$$

(M.56)

Earlier, $\operatorname{erf}(x) = (2/\sqrt{\pi}) \int_0^x ds e^{-s^2}$ is the error function.

The total probability to be trapped is one, regardless of v (i.e., regardless of c),

$$\int_0^{\infty} dt' |J(t')| = 1 \qquad \forall v,$$

(M.57)

because $\mathrm{erf}(x \to +\infty) \to 1$. It is interesting to evaluate the average time to be trapped,

$$\langle t_{\mathrm{tr}} \rangle = \int_0^\infty dt\, t |J(t)| = \frac{x_0}{\sqrt{4\pi D}} \int_0^\infty \frac{dt}{\sqrt{t}} \exp\left(-\frac{(x_0 - vt)^2}{4Dt}\right) \tag{M.58}$$

$$= \frac{x_0}{\sqrt{4\pi D}} \exp\left(\frac{x_0 v}{2D}\right) \int_0^\infty \frac{dt}{\sqrt{t}} \exp\left(-\left(\frac{x_0^2}{4D}\frac{1}{t} + \frac{v^2}{4D}t\right)\right) \tag{M.59}$$

$$= \sqrt{\frac{x_0^3}{\pi D v}} \exp\left(\frac{x_0 v}{2D}\right) K_{1/2}\left(\frac{x_0 v}{2D}\right) \tag{M.60}$$

$$= \frac{x_0}{v}, \tag{M.61}$$

where $K_{1/2}(x)$ is the modified Bessel function of the second kind.

We are now interested in evaluating the trapping probability Eq. (M.56) in the long t limit, that is, in the limit of vanishing $\bar{u}$, in which case the arguments of the error functions diverge, so we can use the asymptotic expansion

$$\mathrm{erf}(x) \simeq 1 - \frac{e^{-x^2}}{\sqrt{\pi} x}, \qquad \text{for} \quad x \to +\infty \tag{M.62}$$

and get, for $\bar{u} \to 0$,

$$\int_0^t dt' |J(t')| \simeq 1 + \frac{e^c}{2\sqrt{\pi}} \exp\left(-\frac{c^2}{4\bar{u}^2}\right)\left[\frac{1}{\frac{c}{2\bar{u}} + \bar{u}} - \frac{1}{\frac{c}{2\bar{u}} - \bar{u}}\right] \tag{M.63}$$

$$\simeq 1 - \frac{4}{\sqrt{\pi}}\frac{e^c}{c^2}\bar{u}^3 \exp\left(-\frac{c^2}{4\bar{u}^2}\right). \tag{M.64}$$

We now pass to the survival probability $S(t) = 1 - \int_0^t dt' |J(t')|$, that is, the probability that the particle has not been trapped within time t. Making explicit the expressions of c and $\bar{u}$, we finally obtain

$$S(t) \simeq \frac{2}{\sqrt{\pi}}\frac{x_0 \sqrt{D}}{v^2}\frac{e^{\frac{v x_0}{2D}}}{t^{3/2}} \exp\left(-\frac{v^2}{4D}t\right). \tag{M.65}$$

The linear response formalism that we have described for classical systems can be extended to quantum systems by suitable adjustments. First, it is useful to work in the Heisenberg picture, where quantum operators are time dependent. The reason is that we want to take into account the response of the quantum system to an external time-dependent source of perturbation. If the unperturbed quantum system is described by the Hamiltonian operator $\hat{H}$, we can introduce a perturbation by considering the perturbed Hamiltonian operator

$$\hat{H}'(t) = \hat{H} - \hat{H}_s \tag{N.1}$$

$$\hat{H}_s = \sum_k \eta_k(t)\, \hat{X}_k(t), \tag{N.2}$$

where $\hat{X}_k(t)$ are Hermitian operators corresponding to physical observables of the quantum system and $\eta_k(t)$ are the associated source terms. This Hamiltonian is the quantum counterpart of Eq. (4.134), where the classical perturbation fields $h_k(t)$, which represent generalized thermodynamic forces, and the classical observables X_k have been replaced by the source terms $\eta_k(t)$ and by the quantum observables $\hat{X}_k(t)$, respectively.

In analogy with the classical case, we assume that fluctuations of any observable depend linearly on the sources of perturbation through a response function, thus

$$\delta\langle \hat{X}_k(t)\rangle = \int dt' \sum_l \Xi_{kl}(t,t')\, \eta_l(t'), \tag{N.3}$$

where $\delta\langle \hat{X}_k(t)\rangle = \langle \hat{X}_k(t)\rangle - \langle \hat{X}_k\rangle_0$ is the difference between the expectation values of $\hat{X}_k(t)$ in the perturbed and in the unperturbed systems, while $\Xi_{kl}(t,t')$ is the response function of the observable $\hat{X}_k(t)$ to the perturbation source $\eta_l(t)$. Also in the quantum case, we can assume the relation $\Xi_{kl}(t,t') \equiv \Xi_{kl}(t-t')$, as a consequence of the time-translation invariance of the quantum system. Therefore, we can rewrite Eq. (N.3) as

$$\delta\langle \hat{X}_k(t)\rangle = \int dt' \sum_l \Xi_{kl}(t-t')\, \eta_l(t'), \tag{N.4}$$

and by transforming in Fourier space, we obtain

$$\delta\langle \hat{X}_k(\omega)\rangle = \int dt' \int dt\, e^{-i\omega t} \sum_l \Xi_{kl}(t-t')\, \eta_l(t')$$

$$= \sum_l \int dt' \int dt\, e^{-i\omega(t-t')}\, \Xi_{kl}(t-t')\, e^{-i\omega t'}\, \eta_l(t')$$

$$= \sum_l \Xi_{kl}(\omega)\, \eta_l(\omega), \tag{N.5}$$

which is the analog of Eq. (4.140) for classical observables. In fact, as in classical systems, linear response theory implies that the response to a perturbation source of frequency ω is local in the frequency space; that is, the fluctuations of the corresponding quantum observable exhibit the same frequency of the perturbation source. Moreover, if we assume that the perturbation sources $\eta_l(t)$ are real quantities and the operator $\hat{X}_k(\omega)$ is Hermitian, that is, $\langle \hat{X}_k(t) \rangle$ is real, then also the response function $\Xi_{kl}(t - t')$ has to be real, and its Fourier transform $\Xi_{kl}(\omega)$ is such that $\Xi_{kl}^*(\omega) = \Xi_{kl}(-\omega)$. All considerations concerning causality and the analytic properties of $\Xi_{kl}(\omega)$, including the Kramers–Krönig relations (see Sections 4.4.1, 4.6, and Appendix O), also apply to the quantum case. In particular, the susceptibility can be defined again (see Eq. (4.67)) as the zero-frequency limit of the response function, namely,

$$\chi_{kl} = \lim_{\omega \to 0} \Xi_{kl}(\omega) = \lim_{\epsilon \to 0^+} \int_{-\infty}^{+\infty} \frac{d\omega'}{\pi} \frac{\Xi_{kl}^I(\omega')}{\omega' - i\epsilon}, \tag{N.6}$$

where $\Xi_{kl}^I(\omega)$ is the imaginary part of the Fourier transform of the response function.

Now, we want to obtain an explicit expression of the response function, in terms of sources and observables. For this purpose, we have to introduce the evolution operator associated with the source Hamiltonian $\hat{H}_s$ in Eq. (N.2),

$$\hat{U}(t, t_0) = \exp\left(-\frac{i}{\hbar} \int_{t_0}^{t} dt' \, \hat{H}_s(t') \right). \tag{N.7}$$

The evolution operator is defined to obey the differential operator equation

$$\frac{d\hat{U}}{dt} = -\frac{i}{\hbar} \hat{H}_s \, \hat{U}. \tag{N.8}$$

In the interaction representation, specified by the subscript I, time evolution is determined by the unperturbed Hamiltonian $\hat{H}$ and $\hat{U}(t, t_0)$ maps the quantum state of the system at time t_0 to the quantum state at time t,

$$|\psi(t)\rangle_I = \hat{U}(t, t_0) \, |\psi(t_0)\rangle_I. \tag{N.9}$$

In general, we can assume that the ensemble of states of the quantum system is described by a density matrix $\hat{\rho}(t)$. We can think that far in the past, $t_0 \to -\infty$, before the perturbation is switched on, the density matrix is given by $\hat{\rho}_0$. Accordingly, at any time t, we can write

$$\hat{\rho}(t) = \hat{U}(t) \, \hat{\rho}_0 \, \hat{U}^{-1}(t), \tag{N.10}$$

where $\hat{U}(t) = \hat{U}(t, t_0 \to -\infty)$. Having at our disposal an expression of the density matrix at time t, we can compute the expectation value of a perturbed observable as

$$\langle \hat{X}_k(t) \rangle = \mathrm{Tr} \, \hat{\rho}(t) \, \hat{X}_k(t) = \mathrm{Tr} \, \hat{\rho}_0 \hat{U}^{-1}(t) \, \hat{X}_k(t) \, \hat{U}(t), \tag{N.11}$$

where we have used the cyclic property of the trace. If we assume that the sources $\eta_k(t)$ are small quantities, we can take advantage of quantum perturbation theory and write the approximate formula

$$\langle \hat{X}_k(t) \rangle \approx \mathrm{Tr}\, \hat{\rho}_0 \left(\hat{X}_k(t) + \frac{i}{\hbar} \int_{-\infty}^{t} dt'\, [\hat{H}_s(t'), \hat{X}_k(t)] + \cdots \right)$$

$$= \langle \hat{X}_k(t) \rangle_0 + \frac{i}{\hbar} \int_{-\infty}^{t} dt'\, \langle [\hat{H}_s(t'), \hat{X}_k(t)] \rangle_0 + \cdots , \qquad (\mathrm{N}.12)$$

where $\langle \hat{X}_k \rangle_0 = \mathrm{Tr}\, \hat{\rho}_0 \hat{X}_k$ is the expectation value of the operator $\hat{X}_k$ in the unperturbed system, while $\langle [\hat{A}(t'), \hat{B}(t)] \rangle_0$ is the same kind of expectation value for the commutator between the unspecified operators $\hat{A}(t')$ and $\hat{B}(t)$. The dots in the previous equation stand for higher-order terms of the perturbative expansion.

Using the definition (N.2) and the Heaviside distribution $\Theta(t - t')$, which allows us to extend the upper limit of the integral to $+\infty$, we can rewrite Eq. (N.12) as the quantum version of the fluctuation–dissipation relation:

$$\delta \langle \hat{X}_k(t) \rangle = \frac{i}{\hbar} \int_{-\infty}^{+\infty} dt'\, \Theta(t - t') \sum_l \langle [\hat{X}_l(t'), \hat{X}_k(t)] \rangle_0\, \eta_l(t'). \qquad (\mathrm{N}.13)$$

Finally, by comparing Eq. (N.13) with Eq. (N.4), we have

$$\Xi_{kl}(t - t') = -\frac{i}{\hbar}\, \Theta(t - t') \langle [\hat{X}_k(t), \hat{X}_l(t')] \rangle_0, \qquad (\mathrm{N}.14)$$

which is the quantum version of Eq. (4.138): The time derivative of the correlation function of classical observables is replaced by the commutator of the quantum observables at different times. The inverse temperature β does not appear explicitly in this formula, because it has been derived making use of a density matrix for zero temperature quantum states. In fact, up to here we have derived information concerning the dynamics of linear response theory in the quantum case. As we discuss hereafter, in order to make β come into play, we have to take into account averages performed by the density matrix of a finite temperature thermodynamic state, $\hat{\rho} = \exp(-\beta \hat{H})$, which corresponds to the Boltzmann–Gibbs measure in the canonical ensemble.

As we have seen in Section 4.6, the simplest formulation of the fluctuation–dissipation theorem in the classical case is Eq. (4.142): The dissipative component of the response function, more precisely the imaginary part of its Fourier transform, is related to the fluctuations of the observables of the system through the Fourier transform of their correlation function. In what follows we want to find a quantum version of the fluctuation–dissipation theorem. Since $\Xi_{kl}(t - t')$ is invariant under time translation, we can set $t' = 0$ in Eq. (N.14) and we define the quantity

$$\Xi''_{kl}(t) = -\frac{i}{2}[\Xi_{kl}(t) - \Xi_{lk}(-t)]. \qquad (\mathrm{N}.15)$$

As we are going to show, the Fourier transform of this quantity is proportional to the anti-Hermitian part of the Fourier transform of the response function: The prefactor $-\frac{i}{2}$ makes the Fourier transform a real quantity. In fact,

$$\Xi''_{kl}(\omega) = -\frac{i}{2}[\Xi_{kl}(\omega) - \Xi^*_{lk}(\omega)]$$

$$= -\frac{i}{2}\int_{-\infty}^{+\infty} dt \, [\Xi_{kl}(t)e^{-i\omega t} - \Xi_{lk}(t)e^{i\omega t}]$$

$$= -\frac{i}{2}\int_{-\infty}^{+\infty} dt \, e^{-i\omega t}[\Xi_{kl}(t) - \Xi_{lk}(-t)] \tag{N.16}$$

$$\equiv \int_{-\infty}^{+\infty} dt e^{-i\omega t}\Xi''_{kl}(t), \tag{N.17}$$

where we have used the fact that $\Xi_{kl}(t)$ is a real function.

By substituting Eq. (N.14) into Eq. (N.15), we obtain

$$\Xi''_{kl}(t) = -\frac{1}{2\hbar}\Theta(t)\big[\langle \hat{X}_k(t)\,\hat{X}_l(0)\rangle_0 - \langle \hat{X}_l(0)\,\hat{X}_k(t)\rangle_0\big]$$

$$+\frac{1}{2\hbar}\Theta(-t)\big[\langle \hat{X}_l(-t)\,\hat{X}_k(0)\rangle_0 - \langle \hat{X}_k(0)\,\hat{X}_l(-t)\rangle_0\big]$$

$$= -\frac{1}{2\hbar}\langle \hat{X}_k(t)\,\hat{X}_l(0)\rangle_0 + \frac{1}{2\hbar}\langle \hat{X}_l(-t)\,\hat{X}_k(0)\rangle_0, \tag{N.18}$$

where we have used translation invariance, that is, $\langle \hat{X}_l(0)\,\hat{X}_k(t)\rangle_0 = \langle \hat{X}_l(-t)\,\hat{X}_k(0)\rangle_0$, and the identity $\Theta(t) + \Theta(-t) \equiv 1$. We can reorder the operators in the last term by considering that the average operation is performed in the canonical ensemble with a density matrix given by the Boltzmann weight $\hat{\rho} = \exp(-\beta\hat{H})$:

$$\langle \hat{X}_l(-t)\,\hat{X}_k(0)\rangle_0 = Z^{-1}\mathrm{Tr}\,e^{-\beta\hat{H}}\,\hat{X}_l(-t)\,\hat{X}_k(0)$$

$$= Z^{-1}\mathrm{Tr}\,e^{-\beta\hat{H}}\,\hat{X}_l(-t)\,e^{\beta\hat{H}}\,e^{-\beta\hat{H}}\,\hat{X}_k(0)$$

$$= Z^{-1}\mathrm{Tr}\,e^{-\beta\hat{H}}\,\hat{X}_k(0)\,\hat{X}_l(-t - i\hbar\beta)$$

$$= \langle \hat{X}_k(t + i\hbar\beta)\,\hat{X}_l(0)\rangle_0, \tag{N.19}$$

where $Z = \mathrm{Tr}\hat{\rho}$. The third equality earlier is obtained in a formal way, by interpreting the Boltzmann weight as a sort of evolution operator in the imaginary time $i\hbar\beta$, while the last one stems again from time-translation invariance in the complex plane. The formal trick of introducing an imaginary time is far from a rigorous procedure, but it provides us a shortcut through more complex calculations (i.e., solving the complete spectral problem) to obtain the quantum formulation of the fluctuation–dissipation theorem. In fact, coming back to Eq. (N.18), we can now write

$$\Xi''_{kl}(t)) = -\frac{1}{2\hbar}\big[\langle \hat{X}_k(t)\,\hat{X}_l(0)\rangle_0 - \langle \hat{X}_k(t + i\hbar\beta)\,\hat{X}_l(0)\rangle_0\big]. \tag{N.20}$$

We can finally compute the Fourier transform of both sides and obtain

$$\Xi''_{kl}(\omega) = -\frac{1}{2\hbar}\big[1 - e^{-\beta\hbar\omega}\big]C_{kl}(\omega), \tag{N.21}$$

where

$$C_{kl}(\omega) = \int_{-\infty}^{+\infty} dt e^{-i\omega t}C_{kl}(t), \tag{N.22}$$

with

$$C_{kl}(t) = \langle \hat{X}_k(t)\, \hat{X}_l(0)\rangle_0. \tag{N.23}$$

Equation (N.21) represents the quantum version of the fluctuation–dissipation theorem (4.142). A physical interpretation of Eq. (N.21) emerges if we rewrite this equation in the form

$$C_{kl}(\omega) = -2\hbar\big[n_B(\omega) + 1\big]\, \Xi''_{kl}(\omega), \tag{N.24}$$

where $n_B(\omega) = 1/(\exp(\beta\hbar\omega) - 1)$ is the Bose–Einstein distribution function. We can conclude that, in the quantum case, the contributions to fluctuations come not only from thermal effects, that is, from $n_B(\omega)$, but also from intrinsic quantum fluctuations, that is, from the "+1" term in the square brackets. In the limit $\hbar \to 0$, or (more physically) in the limit $\beta \to 0$, $[n_B(\omega) + 1] \approx 1/(\beta\hbar\omega)$ and we recover the classical equation Eq. (4.142)

$$C_{kl}(\omega) = -\frac{2}{\beta\omega}\Xi^I_{kl}(\omega), \tag{N.25}$$

provided we consider that the classical counterpart of $\Xi''_{kl}(\omega)$ is $\Xi^I_{kl}(\omega)$.

N.1 Examples of Linear Response in Quantum Systems

The previous discussion indicates a clear analogy between the classical and the quantum formulation of linear response theory. Due to space limitations in what follows we cannot provide an exhaustive discussion of the many applications of linear response theory to quantum systems. We sketch just a few selected examples that summarize conceptual similarities as well as technical peculiarities of the quantum case with respect to the classical one.

N.1.1 Power Dissipated by a Perturbation Field

In Section 4.4.2 we have discussed how the response function is related to the power dissipated by the external work made on a classical system by a perturbation field (see Eqs. (4.59) and (4.60)), and we have computed this quantity for the damped harmonic oscillator in the presence of a periodic driving force (see Eq. (4.93)).

Now we can study the problem of dissipated power for a general quantum system, represented by the Hamiltonian operator $\hat{H}'$. Its general expression amounts to the rate of change in time of the energy in the quantum system, namely,

$$W(t) = \frac{d}{dt}\,\mathrm{Tr}(\hat{\rho}\hat{H}') = \mathrm{Tr}\Big(\frac{d\hat{\rho}}{dt}\,\hat{H}' + \hat{\rho}\,\frac{d\hat{H}'}{dt}\Big), \tag{N.26}$$

where both operators $\hat{\rho}$ and $\hat{H}'$ are functions of time t. Evaluating a physical observable, such as $W(t)$ is independent of the chosen representation. In this case it is more convenient working in the Schrödinger one, where the density matrix evolves according to the equation

$$i\frac{d\hat{\rho}}{dt} = \left[\hat{H}',\hat{\rho}\right],\tag{N.27}$$

so that the first term on the right-hand side of Eq. (N.26) vanishes because of the cyclic property of the trace. On the other hand, we assume that $\hat{H}'$ varies in time due to the presence of a perturbation introduced by the source term (N.2), where, for the sake of simplicity, we assume that only one source term is active, namely,

$$\hat{H}_s = \eta(t)\,\hat{X}.\tag{N.28}$$

In fact, in the Schrödinger representation, the dependence on time is limited to the perturbation field η, while the observable $\hat{X}$ is a time-independent quantity, so that

$$\frac{d\hat{H}'}{dt} = \hat{X}\,\frac{d\eta(t)}{dt}.\tag{N.29}$$

Substituting into Eq. (N.26), we obtain

$$W(t) = \mathrm{Tr}\!\left(\hat{\rho}\,\frac{d\hat{H}'}{dt}\right) = \langle\hat{X}\rangle\frac{d\eta}{dt} = \left(\langle\hat{X}\rangle_0 + \delta\langle\hat{X}\rangle\right)\frac{d\eta}{dt}.\tag{N.30}$$

If we assume that we are dealing with a periodic forcing of period $2\pi/\Omega$, the source term has the form

$$\eta(t) = \mathrm{Re}\left(\eta_0\,e^{i\Omega t}\right) = \frac{1}{2}\left(\eta_0\,e^{i\Omega t} + \eta_0^*\,e^{-i\Omega t}\right)\tag{N.31}$$

and we can express the average dissipated power as

$$\overline{W} = \frac{\Omega}{2\pi}\int_0^{2\pi/\Omega} dt\, W(t) = \frac{\Omega}{2\pi}\int_0^{2\pi/\Omega} dt\left(\langle\hat{X}\rangle_0 + \delta\langle\hat{X}\rangle\right)\frac{d\eta}{dt}.\tag{N.32}$$

The contribution from the term $\langle\hat{X}\rangle_{\eta=0}$ cancels when integrated over a full period and we have

$$
\begin{aligned}
\overline{W} &= \frac{\Omega}{2\pi}\int_0^{2\pi/\Omega} dt \int_{-\infty}^{+\infty} dt'\, \Xi(t-t')\eta(t')\frac{d\eta}{dt}\\[4pt]
&= \frac{\Omega}{2\pi}\int_0^{2\pi/\Omega} dt \int_{-\infty}^{+\infty} dt'\,\frac{1}{2\pi}\int_{-\infty}^{+\infty} d\omega\,\Xi(\omega)e^{i\omega(t-t')}\\[4pt]
&\quad \times\frac{1}{4}\left[\eta_0\,e^{i\Omega t'} + \eta_0^*\,e^{-i\Omega t'}\right]\left[i\Omega\eta_0\,e^{i\Omega t} - i\Omega\eta_0^*\,e^{-i\Omega t}\right]\\[4pt]
&= \frac{i\Omega}{4}\frac{\Omega}{2\pi}\int_0^{2\pi/\Omega} dt \int_{-\infty}^{+\infty} d\tau\,\frac{1}{2\pi}\int_{-\infty}^{+\infty} d\omega\,\Xi(\omega)e^{-i\omega\tau}\\[4pt]
&\quad \times\left[\eta_0^2 e^{i\Omega(2t+\tau)} - \eta_0^{*2}e^{-i\Omega(2t+\tau)} + |\eta_0|^2 e^{-i\Omega\tau} - |\eta_0|^2 e^{i\Omega\tau}\right],
\end{aligned}\tag{N.33}
$$

where we have made a change of variable from t' to $\tau = t' - t$. We can now observe that the contributions from terms proportional to η_0^2 and η_0^{*2} vanish when integrated over t, while the terms proportional to $|\eta_0|^2$ are independent of t, and, when integrated over τ, they yield Dirac-delta distributions $\delta(\omega-\Omega)$ and $\delta(\omega+\Omega)$. We therefore obtain

$$\overline{W} = \frac{i\Omega}{4}\left[\Xi(-\Omega) - \Xi(\Omega)\right]|\eta_0|^2,\tag{N.34}$$

which can be further simplified observing that $\Xi^{\mathrm{I}}(\omega)$ is the odd component of $\Xi(\omega)$ (see Appendix O),

$$\overline{W} = \frac{\Omega}{2}\Xi^{\mathrm{I}}(\Omega)\,|\eta_0|^2. \tag{N.35}$$

If we perform work on a system, its energy has to increase and, accordingly, the quantity $\Omega\,\Xi^{\mathrm{I}}(\Omega)$ has to be positive. These results can be extended to the general case described by the perturbed Hamiltonian operator (N.1).

N.1.2 Linear Response in Quantum Field Theory

In order to describe transport problems in a quantum framework, we need to consider a quantum field theory (QFT) formulation, where we assume that the physical observables are quantum Hermitian operators that depend on the space coordinates $\mathbf{r}$ and on time t, that is, $\hat{X}(\mathbf{r}, t)$. In this case, the perturbative component of the Hamiltonian (see Eq. (N.2)) is

$$\hat{H}_s = \int d\mathbf{r} \sum_k \eta_k(\mathbf{r}, t)\,\hat{X}_k(\mathbf{r}, t), \tag{N.36}$$

while the response function $\Xi_{kl}(\mathbf{r}, t; \mathbf{r}', t')$ (see Eq. (N.3)) is defined by the relation

$$\delta\langle\hat{X}_k(\mathbf{r}, t)\rangle = \int d\mathbf{r}'dt' \sum_l \Xi_{kl}(\mathbf{r}, t; \mathbf{r}', t')\,\eta_l(\mathbf{r}', t'). \tag{N.37}$$

Again, all the considerations concerning causality and the analytic properties of the response function that hold in the classical as well in the quantum case straightforwardly extend to the QFT formulation. In particular, we can write the analog of Eq. (N.14) in the form

$$\Xi_{kl}(\mathbf{r}, \mathbf{r}'; t - t') = -\frac{i}{\hbar}\,\Theta(t - t')\langle[\hat{X}_k(\mathbf{r}, t), \hat{X}_l(\mathbf{r}', t')]\rangle_0, \tag{N.38}$$

where we have introduced explicitly the condition that the response function is invariant under time translations. For systems that are also translationally invariant in space, the response function depends on $\mathbf{r} - \mathbf{r}'$, rather than separately on both coordinate vectors, and, passing to Fourier-transformed quantities in space and time, Eq. (N.37) becomes

$$\delta\langle\hat{X}_k(\mathbf{k}, \omega)\rangle = \sum_l \Xi_{kl}(\mathbf{k}, \omega)\,\eta_l(\mathbf{k}, \omega). \tag{N.39}$$

Electric Conductivity

In order to describe electric conductivity, we have to consider a QFT with global $U(1)$ symmetry, associated with a gauge field, which is the electromagnetic four-potential. According to Noether's theorem (after the German mathematician Emmy Noether), there is an associated conserved four-current $j^\mu = (j^0, j^i)$ (with $i = x, y, z$), which is a conserved quantity, that is, $\partial_\mu j^\mu = 0$. We assume that the current j^μ is coupled to the background electromagnetic gauge field $A_\mu(\mathbf{r})$, yielding the perturbation Hamiltonian operator

$$H_s = \int d\mathbf{r} A_\mu\, j^\mu. \tag{N.40}$$

For the sake of simplicity, here we assume that A_μ is a fixed perturbation source under our control. On the other hand, the conserved current in the presence of a background field is given by

$$j^\mu = ie[\phi^\dagger \partial^\mu \phi - (\partial^\mu \phi^\dagger)\phi] - e^2 A^\mu \phi^\dagger \phi, \tag{N.41}$$

where e is the electric charge and ϕ is a free, relativistic, complex scalar field.

As a first step we can apply Eq. (N.37) to the case where the perturbation is Eq. (N.40), thus obtaining

$$\delta\langle j_\mu\rangle = \langle j_\mu\rangle - \langle j_\mu\rangle_0$$
$$= -\frac{i}{\hbar}\sum_\nu \int d\mathbf{r}' \int_{-\infty}^{t} dt' \,\langle[j_\mu(\mathbf{r},t), j_\nu(\mathbf{r}',t')]\rangle_0 \, A_\nu(\mathbf{r}',t'), \tag{N.42}$$

where the limits of the second integral on the right-hand side are imposed by the causality condition; see Eq. (N.38).

Let us first consider the contribution coming from the term $\langle j_\mu\rangle_0$. Even if the unperturbed situation has no currents, the second term in the right-hand side of Eq. (N.41) provides a contribution

$$\langle j_\mu\rangle_0 = -e^2 A_\mu \langle\phi^\dagger\phi\rangle_0 = -e\, A_\mu\, \rho, \tag{N.43}$$

where ρ is the background charge density that is evaluated in the unperturbed quantum state.

As a physical example, we want to describe the problem of electric conductivity, that is, Ohm's law, in the language of QFT. This amounts to studying linear response in the presence of a background electric field $\mathbf{E} = (E_x, E_y, E_z)$. In order to evaluate the right-hand side of Eq. (N.42), we can decide to fix the gauge $A_0 = 0$, so that the space components of the electric field are given by the expression

$$E_l = -\frac{dA_l}{dt} \;\rightarrow\; A_l(\omega) = -\frac{E_l(\omega)}{i\omega}. \tag{N.44}$$

In the absence of an external magnetic field, the system is invariant under rotation and parity transformation, so that the current has to be parallel to the applied electric field and the phenomenological equation in Fourier space describing Ohm's law reads

$$\langle j(\mathbf{k},\omega)\rangle = \sigma(\mathbf{k},\omega)\, E(\mathbf{k},\omega), \tag{N.45}$$

where $E(\mathbf{k},\omega)$ is the background electric field in Fourier space and σ is the conductivity. Note that in this simple setup, the dependence on the subscript indices can be omitted and Eq. (N.42) simplifies to

$$\delta\langle j\rangle = \langle j\rangle - \langle j\rangle_0$$
$$= -\frac{i}{\hbar}\int d\mathbf{r}' \int_{-\infty}^{t} dt' \,\langle[j(\mathbf{r},t), j(\mathbf{r}',t')]\rangle_0 \, A(\mathbf{r}',t')$$
$$= \int d\mathbf{r}' \int_{-\infty}^{+\infty} dt' \,\Xi(\mathbf{r}-\mathbf{r}',t-t')A(\mathbf{r}',t'). \tag{N.46}$$

Now we can Fourier-transform this equation to cast it in the form of Eq. (N.45) and we obtain that the conductivity has two contributions,

$$\sigma(\mathbf{k}, \omega) = -i\frac{e\rho}{\omega} + i\frac{\Xi(\mathbf{k}, \omega)}{\omega}, \tag{N.47}$$

where the first one stems from the background charge density, while the second one is associated with the Fourier transform of the response function,

$$\Xi(\mathbf{k}, \omega) = -\frac{i}{\hbar} \int_{-\infty}^{+\infty} dt\, d\mathbf{r}\, \Theta(t)\, e^{-i(\omega t + \mathbf{k}\cdot\mathbf{r})}\, \langle [j(\mathbf{r}, t), j(\mathbf{0}, 0)] \rangle_0 . \tag{N.48}$$

This equation can be interpreted as the Kubo formula for electric conductivity in QFT.

Viscosity

The procedure that we have just described to obtain an expression for the conductivity is quite instructive, because it can be followed to obtain, by suitable adjustments, other expressions for transport coefficients from linear response theory applied to QFT. As an example, here we want to sketch how to obtain an expression for the viscosity η in QFT. From the classical case (see Section 1.2.2), we know that viscosity is the phenomenological coefficient associated with the transport of momentum, which, in the hydrodynamic representation, is a conserved quantity (as electric charge is a conserved quantity in the previous example). Moreover, we have to consider a QFT that is invariant under space and time translations: In this case, Noether's theorem yields four currents that are associated with energy and momentum conservation. They are represented by the stress-energy tensor $T^{\mu\nu}$ and the conservation condition reads $\partial_\mu T^{\mu\nu} = 0$. We can proceed as for electric conductivity by replacing the electric current with the momentum current. In order to simplify the problem, we can assume a setup where momentum in the x-direction is transported along the z-direction, thus making immaterial any dependence on the y-direction (this setup is the same adopted in the classical case discussed in Section 1.2.2). In this case, we have to deal with a single, relevant component of the stress-energy tensor, namely, T^{xz}, which, in this case, plays the same role as the electric current in the previous example. Accordingly, the expression that we have obtained for the electric conductivity tensor can be formally turned into an expression for the viscosity η. Anyway, in this case, there are two main differences with respect to electric conductivity. First, there is no background charge density to be taken into account, because momentum current is a mechanical observable. The second difference is that viscosity, as in the classical setup, has to be computed in the presence of a constant driving force. This implies that we have to consider the limits $\omega \to 0$ and $\mathbf{k} \to 0$ in the expression analogous to Eq. (N.47). Thus, we can write

$$\eta = \lim_{\omega \to 0} \lim_{\mathbf{k} \to 0} \frac{\Xi_{xz,xz}(\mathbf{k}, \omega)}{i\omega}, \tag{N.49}$$

where

$$\Xi_{xz,xz}(\mathbf{k}, \omega) = -\frac{i}{\hbar} \int_{-\infty}^{+\infty} dt\, d\mathbf{r}\, \Theta(t)\, e^{-i(\omega t + \mathbf{k}\cdot\mathbf{r})}\, \langle [T_{xz}(\mathbf{r}, t), T_{xz}(\mathbf{0}, 0)] \rangle_0. \tag{N.50}$$

Equation (N.49) can be interpreted as the Kubo formula for viscosity in QFT.

Mathematical Properties of Response Functions

Taking into account the property of time translation invariance of the response function $\Xi(t, t')$, the right-hand side of Eq. (4.44) has the form of a convolution product

$$\langle X(t) \rangle = \int dt' \, \Xi(t - t') \, h(t'), \quad t > t', \tag{O.1}$$

which becomes a standard product for the Fourier-transformed functions

$$\langle X(\omega) \rangle = \Xi(\omega) \, h(\omega), \tag{O.2}$$

where the frequency ω is the dual variable of time t. This relation indicates that, if we perturb a thermodynamic observable with a field at frequency ω, it responds linearly, at the same frequency ω.[1]

$\Xi(t)$ is a real quantity, while $\Xi(\omega)$ is a complex one. Let us introduce the following notation:

$$\begin{aligned}
\Xi(\omega) &= \operatorname{Re} \Xi(\omega) + i \operatorname{Im} \Xi(\omega) \\
&\equiv \Xi^R(\omega) + i \Xi^I(\omega).
\end{aligned} \tag{O.3}$$

Both $\Xi^R(\omega)$ and $\Xi^I(\omega)$ have a physical interpretation. In fact, we can write

$$\begin{aligned}
\Xi^I(\omega) &= -\frac{i}{2} \left[\Xi(\omega) - \Xi^*(\omega) \right] \\
&= -\frac{i}{2} \int_{-\infty}^{+\infty} dt \, \Xi(t) \left[e^{-i\omega t} - e^{i\omega t} \right] \\
&= -\frac{i}{2} \int_{-\infty}^{+\infty} dt \, e^{-i\omega t} \left[\Xi(t) - \Xi(-t) \right],
\end{aligned} \tag{O.4}$$

where the superscript $*$ is the complex conjugate symbol. We see that the imaginary part of the Fourier transform of the response function depends on the part of the response function that is antisymmetric under the *time-reversal*, therefore $\Xi^I(\omega)$ is an odd function of its argument,

$$\Xi^I(\omega) = -\Xi^I(-\omega). \tag{O.5}$$

In the literature it is called the dissipative component of the response function, and it contains information about the dissipative processes associated with the perturbation field.

---

[1] This is not the case when nonlinear effects enter the game, making the overall treatment of the response problem much more complex.

Conversely, the real part of the response function is insensitive to the direction of time,

$$\Xi^{R}(\omega) = \frac{1}{2} \int_{-\infty}^{+\infty} dt\, e^{-i\,\omega\,t} \left[\Xi(t) + \Xi(-t) \right], \qquad (O.6)$$

namely, it is an even function of its argument:

$$\Xi^{R}(\omega) = \Xi^{R}(-\omega). \qquad (O.7)$$

Causality imposes that

$$\Xi(t) = 0 \qquad \forall t < 0. \qquad (O.8)$$

In the mathematical literature, such a function is called backward Green's function. Condition (O.8) attributes specific properties to $\Xi(\omega)$. They can be inferred from the expression of the inverse Fourier transform:

$$\Xi(t) = \frac{1}{2\pi} \int_{-\infty}^{+\infty} d\omega\, e^{i\,\omega\,t}\, \Xi(\omega). \qquad (O.9)$$

Without entering into mathematical detail, we can assume ω is a complex variable,[2] $\omega \to z = \omega - i\epsilon$ with $\epsilon \in \mathbb{R}^{+}$, by analytic extension of $\Xi(\omega)$ to the complex plane, that is, $\Xi(\omega) \to \Xi(z)$. As a consequence of the Jordan lemma, for $t < 0$, the integral on the right-hand side of Eq. (O.9) can be computed by considering a closed integration path Γ as shown in Fig. O.1 (the half-circle of radius R lies in the lower-half complex plane):

$$\Xi(t) = \frac{1}{2\pi} \lim_{R \to +\infty} \int_{\Gamma(R)} dz\, e^{i\,\omega\,t}\, e^{\epsilon t}\, \tilde{\Xi}(\omega - i\epsilon). \qquad (O.10)$$

The theorem of residues states that the right-hand side of Eq. (O.10) is equal to minus the sum of the residues of the poles of $\Xi(\omega - i\epsilon)$ in the lower-half complex plane. Accordingly, causality imposes that $\Xi(\omega - i\epsilon)$ is analytic in all of the lower-half plane of the complex variable $\omega - i\epsilon$ (i.e., no poles are present and the integral on the right-hand side of Eq. (O.10) is null for $\epsilon > 0$).

Now, we are going to discuss how analyticity of $\Xi(\omega)$ in the lower-half complex plane engenders the so-called Kramers–Krönig relations between $\Xi^{R}(\omega)$ and $\Xi^{I}(\omega)$. The starting

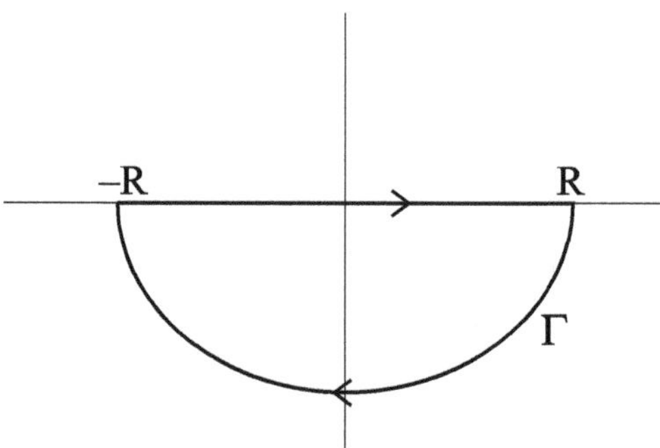

Fig. O.1 Integration path for Eq. (O.10).

2 With our notation for the Fourier transform, the response function is analytic for negative imaginary argument, so we prefer to write $z = \omega - i\epsilon$, with positive ϵ.

point is Cauchy's theorem, which provides us with the Cauchy integral representation of a complex function around any point z belonging to its analyticity domain,

$$\Xi(z) = \oint \frac{d\,z'}{2\pi i}\, \frac{\Xi(z')}{z' - z}, \tag{O.11}$$

where the contour integral is performed counter clockwise, that is, in the opposite orientation of the path $\Gamma(R)$ in Fig. O.1.

Due to causality, the analyticity domain of the response function is the lower-half complex plane, so that Eq. (O.11) holds for any closed contour in the lower-half complex plane. On the other hand, we want to consider the case where $z = \omega - i\epsilon$ approaches the real axis, namely, the limit $\epsilon \to 0^+$. In such a case, it is convenient to compute the right-hand side of Eq. (O.11) over a contour that is made by a segment of length $2R$, symmetric to the origin and parallel to the real axis at a distance $\delta < \epsilon$, closed by the semi-circle of radius R centered in the origin and lying in the lower-half of the complex plane (see Fig. O.1). If $|\Xi(z)| < 1/|z|$ for very large values of R, the integral on the semi-circle vanishes in the limit $R \to \infty$ and we can write

$$\Xi(\omega - i\epsilon) = -\frac{1}{2\pi i}\int_{-\infty}^{+\infty} d\omega'\, \frac{\Xi(\omega')}{\omega' - \omega + i\epsilon}, \tag{O.12}$$

where the minus sign in front of the integral appears, because the contour Γ is oriented clockwise. We can separate the real and imaginary part of the integrand by the relation

$$\frac{1}{\omega' - \omega + i\epsilon} = \frac{\omega' - \omega}{(\omega' - \omega)^2 + \epsilon^2} - i\frac{\epsilon}{(\omega' - \omega)^2 + \epsilon^2}. \tag{O.13}$$

Now we can make use of the following result,

$$\lim_{\epsilon \to 0^+} \frac{\epsilon}{(\omega' - \omega)^2 + \epsilon^2} = \pi\,\delta(\omega' - \omega), \tag{O.14}$$

where $\delta(\omega' - \omega)$ is the Dirac-delta distribution. For small enough ϵ, we can write

$$\Xi(\omega) = \frac{1}{2\pi i}\lim_{\epsilon \to 0^+}\int_{-\infty}^{+\infty} d\,\omega'\,\Xi(\omega')\,\frac{\omega - \omega'}{(\omega - \omega')^2 + \epsilon^2} + \frac{1}{2}\Xi(\omega). \tag{O.15}$$

We can separate the response function in its real and imaginary parts (see Eq. (O.3)), and we obtain the Kramers–Krönig relations

$$\Xi^{\mathrm{R}}(\omega) = \frac{1}{\pi}\lim_{\epsilon \to 0}\int_{-\infty}^{+\infty} d\omega'\Xi^{\mathrm{I}}(\omega')\,\frac{\omega - \omega'}{(\omega - \omega')^2 + \epsilon^2} \tag{O.16}$$

$$\Xi^{\mathrm{I}}(\omega) = -\frac{1}{\pi}\lim_{\epsilon \to 0}\int_{-\infty}^{+\infty} d\omega'\,\Xi^{\mathrm{R}}(\omega')\,\frac{\omega - \omega'}{(\omega - \omega')^2 + \epsilon^2}. \tag{O.17}$$

They are a direct consequence of causality and show that the dissipative (imaginary) and the reactive (real) components of the response function are related to each other by a nonlocal dependence in frequency space.

There is an alternative formulation of the Kramers–Krönig relations, namely,

$$\Xi(\omega) = \lim_{\epsilon \to 0^+}\int_{-\infty}^{+\infty} \frac{d\omega'}{\pi}\, \frac{\Xi^{\mathrm{I}}(\omega')}{\omega - \omega' - i\epsilon}, \tag{O.18}$$

where the response function is expressed just in terms of its dissipative (imaginary) component. Eq. (O.18) can be obtained by considering the relation

$$\lim_{\epsilon \to 0^+} \frac{1}{\omega - i\epsilon} = \lim_{\epsilon \to 0^+} \left[i\frac{\epsilon}{\omega^2 + \epsilon^2} + \frac{\omega}{\omega^2 + \epsilon^2} \right] = i\pi\delta(\omega) + \mathrm{PV}\left(\frac{1}{\omega}\right), \tag{O.19}$$

which holds in the sense of distributions. PV (principal value) is defined as

$$\mathrm{PV} \int_{-\infty}^{+\infty} \frac{d\omega}{\omega} \phi(\omega) = \lim_{\epsilon \to 0^+} \int_{-\infty}^{-\epsilon} \frac{d\omega}{\omega} \phi(\omega) + \int_{\epsilon}^{+\infty} \frac{d\omega}{\omega} \phi(\omega), \tag{O.20}$$

where $\phi(\omega)$ is a test function in Schwarz space, that is, a regular rapidly decreasing function of its argument.

In fact, if we specialize Eq. (O.19) to the case $\omega \to \omega - \omega'$ and assume that $\Xi^{\mathrm{I}}(\omega')$ has the properties of a test function in Schwarz space, we can write

$$\lim_{\epsilon \to 0^+} \int_{-\infty}^{+\infty} \frac{d\omega'}{\pi} \frac{\Xi^{\mathrm{I}}(\omega')}{\omega - \omega' - i\epsilon} = i\,\Xi^{\mathrm{I}}(\omega) + \mathrm{PV} \int_{-\infty}^{+\infty} \frac{d\omega'}{\pi} \frac{\Xi^{\mathrm{I}}(\omega')}{\omega - \omega'}$$

$$= i\,\Xi^{\mathrm{I}}(\omega) + \lim_{\epsilon \to 0^+} \int_{-\infty}^{+\infty} \frac{d\omega'}{\pi} \Xi^{\mathrm{I}} \frac{\omega - \omega'}{(\omega - \omega')^2 + \epsilon^2}. \tag{O.21}$$

Making use of Eq. (O.16), we finally find

$$\lim_{\epsilon \to 0^+} \int_{-\infty}^{+\infty} \frac{d\omega'}{\pi} \frac{\Xi^{\mathrm{I}}(\omega')}{\omega - \omega' - i\epsilon} = i\,\Xi^{\mathrm{I}}(\omega) + \Xi^{\mathrm{R}}(\omega') \equiv \Xi(\omega). \tag{O.22}$$

The Van der Waals Equation

The van der Waals equation for real gases can be obtained by considering that particles of finite size are subject to a reciprocal attracting interaction, which can be approximated by a square-well potential $U(r)$, as shown in Fig. P.1: E_0 is the well depth and d is the range of the interaction. At equilibrium, all particles have the same average energy, so the kinetic energy of a particle outside the interaction range (E_{out}) is related to the same quantity for a particle inside the interaction range (E_{in}) by the relation

$$E_{\text{out}} = E_{\text{in}} - E_0. \tag{P.1}$$

Accordingly, the average kinetic energy of a particle $\langle E \rangle$ is given by

$$\langle E \rangle = (1 - p)E_{\text{out}} + pE_{\text{in}} = E_{\text{out}} + pE_0, \tag{P.2}$$

where p is the fraction of particles inside the interaction range. This quantity can be estimated by considering that the volume of the interaction sphere associated with the square-well potential is given by $V_I = 4/3\pi d^3$. By assuming space isotropy and homogeneity, one can write

$$p = \frac{NV_I}{V}, \tag{P.3}$$

where N is the number of molecules in the gas and V is the volume of the container.

For an ideal gas, the pressure exerted by the particles on the walls of the container is given by

$$P = \frac{2}{3}\langle E \rangle \frac{N}{V}, \tag{P.4}$$

and it can be derived in kinetic theory, assuming $N/6$ particles travel toward a given wall and the force exerted on it is determined by the variation of momentum of each particle bouncing on it. This formula can now be modified for a real gas, assuming that when particles collide with the walls of the container, they are typically far from the other particles. This means that $\langle E \rangle \rightarrow E_{\text{out}}$ and V should be replaced by the effective volume available to particles, $V - Nb$, where b is the volume occupied by a particle. In conclusion,

$$P = \frac{2}{3}E_{\text{out}}\frac{N}{V - Nb} \tag{P.5}$$

$$= \frac{2}{3}\left(\langle E \rangle \frac{N}{V - Nb} - E_0 \frac{N^2 V_I}{V(V - Nb)}\right). \tag{P.6}$$

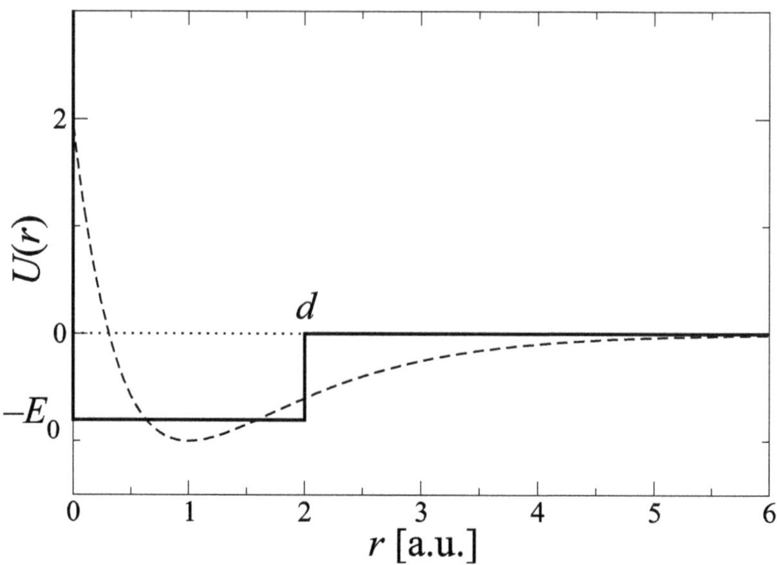

Fig. P.1 The phenomenological interaction potential between particles in a real gas (dashed line) and its square-well approximation (solid line).

According to the equipartition theorem,

$$\frac{2}{3}\langle E \rangle = T,\tag{P.7}$$

where T is the temperature of the gas. We can finally write the van der Waals equation for a real gas as

$$P = \frac{T}{v-b} - \frac{a}{v^2},\tag{P.8}$$

where $a = (2/3)E_0 V_I$, $v = V/N$ is the volume per particle, and in the second term of the right-hand side, we have neglected b (the "physical" volume of a particle) with respect to v, because, considering a and b as small parameters, it would be a sort of second-order effect.

The critical point is determined by the relations

$$\left.\frac{\partial P}{\partial v}\right|_{v_c,T_c} = 0, \qquad \left.\frac{\partial^2 P}{\partial v^2}\right|_{v_c,T_c} = 0, \qquad P_c = P(v_c, T_c),\tag{P.9}$$

which give

$$v_c = 3b, \qquad T_c = \frac{8a}{27b}, \qquad P_c = \frac{a}{27b^2}.\tag{P.10}$$

Normalizing thermodynamical quantities with respect to their values at the critical point,

$$v^* = \frac{v}{v_c}, \qquad P^* = \frac{P}{P_c}, \qquad T^* = \frac{T}{T_c},\tag{P.11}$$

the van der Waals equation can be written in a free-parameter form,

$$P^* = \frac{8}{3}\frac{T^*}{v^* - \frac{1}{3}} - \frac{3}{(v^*)^2}.\tag{P.12}$$

The Helmholtz free energy per particle can be derived from the relation $P = -\partial f/\partial v|_T$,

$$f(v, T) = -T \ln(v - b) - \frac{a}{v} + f_0(T), \tag{P.13}$$

where $f_0(T)$ is an arbitrary function. The Gibbs free energy per particle, $g = f + Pv$, in reduced variables can be written as

$$g^* = P^* v^* - \frac{8}{3} T^* \ln\left(v^* - \frac{1}{3}\right) - \frac{3}{v^*} + f_0^*(T^*). \tag{P.14}$$

For $T^* < 1$ (i.e., $T < T_c$) the two equivalent minima of g^*, corresponding to the two coexisting phases, can be found by imposing the following conditions,

$$g^*(v_1^*) = g^*(v_2^*) \tag{P.15}$$

$$\left.\frac{\partial g^*}{\partial v^*}\right|_{v^*=v_1^*} = \left.\frac{\partial g^*}{\partial v^*}\right|_{v^*=v_2^*} = 0, \tag{P.16}$$

which are equivalent to the condition

$$P^*(v_2^* - v_1^*) = \int_{v_1^*}^{v_2^*} P^* dv^*, \tag{P.17}$$

corresponding to the so-called Maxwell construction, graphically depicted by the horizontal segment in Fig. 8.4, whose extrema determine the points on the coexistence curve.

The critical exponents can be estimated by the equation of state (P.12) close to the critical point $P^* = v^* = T^* = 1$. This can be obtained by expanding up to leading order Eq. (P.12) around the critical point, namely,

$$\delta p = \left.\frac{\partial P^*}{\partial T^*}\right|_c \delta t + \frac{1}{6}\left.\frac{\partial^3 P^*}{\partial (v^*)^3}\right|_c (\delta v)^3 = 4\delta t - \frac{3}{2}(\delta v)^3, \tag{P.18}$$

where $\delta p = P^* - 1$, $\delta t = T^* - 1$, and $\delta v = v^* - 1$. The terms of order δv and $(\delta v)^2$ in the expansion are absent, because $\left.\frac{\partial P^*}{\partial v^*}\right|_c = \left.\frac{\partial^2 P^*}{\partial (v^*)^2}\right|_c = 0$. For $T = T_c$, that is, $\delta t = 0$, Eq. (P.18) indicates that the pressure is a cubic function of the volume. Accordingly, we can conclude that close to the critical point the volumes v_1^* and v_2^* are symmetric with respect to the critical value, namely, $v_1^* = 1 + \delta v$ and $v_2^* = 1 - \delta v$. By substituting into Eq. (P.16), that is equivalent to

$$P^*(v_1^*) = P^*(v_2^*) \tag{P.19}$$

and expanding up to leading order in δv (i.e., $(\delta v)^2$), we obtain

$$\delta v \sim |\delta t|^{1/2} \sim \left(\frac{T_c - T}{T_c}\right)^{1/2}. \tag{P.20}$$

We can conclude that the scaling law of the order parameter close to the critical temperature is characterized by a critical exponent $\beta = \frac{1}{2}$, which is the classical exponent predicted by the Landau mean-field theory (see Section 5.2.3). Notice that close to the critical point, the heat capacity at constant volume $C_V = T^2 \frac{\partial^2 F}{\partial T^2}$, where $F = Nf$, cannot be determined by the Helmholtz free energy (P.13), because this is defined up to an arbitrary function $F_0(T) = N f_0(T)$. The discontinuity of C_V at the critical point predicted by the Landau mean-field theory, corresponding to a classical critical exponent $\alpha = 0$, can be recovered by assuming that the critical isothermal line ($T^* = 1$) for large values of v^* recovers the behavior of an ideal gas; that is, the van der Waals equation of state simplifies to

$$P^* = \frac{8}{3}\frac{T^*}{v^*}. \tag{P.21}$$

Derivation of the Ginzburg–Landau Free Energy

Let us consider an Ising model in general dimension d, with a ferromagnetic interaction, $J_{ij} > 0$, which depends on the distance between the lattice sites i and j,

$$\mathcal{H} = -\frac{1}{2} \sum_{i,j} J_{ij} \sigma_i \sigma_j. \tag{Q.1}$$

In a continuum spirit, we replace σ_i by $m(\mathbf{x})$, the average magnetization in a cube V of side ℓ, centered in $\mathbf{x}$,

$$m(\mathbf{x}) = \frac{M(\mathbf{x})}{\ell^d} \equiv \frac{\displaystyle\sum_{i \in V} \sigma_i}{\ell^d} \tag{Q.2}$$

and

$$\mathcal{H} = -\frac{1}{2} \int d\mathbf{x}\, d\mathbf{x}'\, J(|\mathbf{x} - \mathbf{x}'|) m(\mathbf{x}) m(\mathbf{x}') \tag{Q.3}$$

$$= \frac{1}{4} \int d\mathbf{x}\, d\mathbf{x}'\, J(|\mathbf{x} - \mathbf{x}'|) \left[(m(\mathbf{x}) - m(\mathbf{x}'))^2 - \left(m^2(\mathbf{x}) + m^2(\mathbf{x}') \right) \right] \tag{Q.4}$$

$$\simeq \frac{1}{4} \int d\mathbf{x}\, d\mathbf{r}\, J(r)\, (\nabla m \cdot \mathbf{r})^2 - \frac{1}{2} \int d\mathbf{x}\, d\mathbf{r}\, J(r) m^2(\mathbf{x}) \tag{Q.5}$$

$$= \frac{1}{4} \sum_i^d \int d\mathbf{r}\, J(r) r_i^2 \int d\mathbf{x} \left(\frac{\partial m}{\partial x_i} \right)^2 - \frac{1}{2} \int d\mathbf{r}\, J(r) \int d\mathbf{x}\, m^2(\mathbf{x}). \tag{Q.6}$$

The integral $\int d\mathbf{r}\, J(r) r_i^2$ does not depend on the direction i, so we can define the quantities

$$K = \frac{1}{2} \int d\mathbf{r}\, J(r) r_i^2 = \frac{1}{2d} \int d\mathbf{r}\, J(r) r^2 \tag{Q.7}$$

$$\tilde{J} = \int d\mathbf{r}\, J(r) \tag{Q.8}$$

and finally write

$$\mathcal{H} = \frac{K}{2} \int d\mathbf{x}\, (\nabla m)^2 - \frac{\tilde{J}}{2} \int d\mathbf{x}\, m^2(\mathbf{x}). \tag{Q.9}$$

Let us now evaluate the entropy term counting the number of microscopic configurations corresponding to the same value of $m(\mathbf{x})$. This task is easily accomplished if we observe that $M(\mathbf{x}) = N_+ - N_-$, where $N_\pm$ are the number of up and down spins in the volume V and $N = N_+ + N_- = (\ell/a)^d$ is the total number of spins in V, a being the lattice constant. In fact, the entropy of volume V is given by $S(\mathbf{x}) = \ln W$, where

$$W = \frac{N!}{N_+! N_-!}. \tag{Q.10}$$

Using the Stirling approximation $N! \simeq (N/e)^N$, we find

$$S(\mathbf{x}) = N \ln 2 - \frac{1}{2}(N + M) \ln\left(1 + \frac{M}{N}\right) - \frac{1}{2}(N - M) \ln\left(1 - \frac{M}{N}\right), \tag{Q.11}$$

where the spatial dependence of M on $\mathbf{x}$ has not been made explicit to simplify the notation. Finally, it is easier to normalize the magnetization m so that its maximum value is ± 1,

$$u(\mathbf{x}) \equiv \frac{M}{N} = a^d m(\mathbf{x}). \tag{Q.12}$$

If $S = \int d\mathbf{x} S(\mathbf{x})$ is the total entropy, the Helmholtz free energy $\mathcal{F} = \mathcal{H} - TS$ is written as

$$\begin{aligned}
\mathcal{F} &= \int \frac{d\mathbf{x}}{a^{2d}}\left[\frac{K}{2}(\nabla u)^2 - \frac{\tilde{J}}{2}u^2(\mathbf{x})\right] \\
&\quad - \frac{T}{V}\int d\mathbf{x} N\left[\ln 2 - \frac{1}{2}(1 + u)\ln(1 + u) - \frac{1}{2}(1 - u)\ln(1 - u)\right] \\
&= \int d\mathbf{x}\left[\frac{K}{2a^{2d}}(\nabla u)^2 + \mathcal{V}(u)\right],
\end{aligned} \tag{Q.13}$$

where

$$\mathcal{V}(u) = \frac{\tilde{J}}{a^{2d}}\left\{-\frac{u^2}{2} + \frac{Ta^d}{\tilde{J}}\left[-\ln 2 + \frac{1}{2}(1 + u)\ln(1 + u) + \frac{1}{2}(1 - u)\ln(1 - u)\right]\right\}. \tag{Q.14}$$

The function $\mathcal{V}(u)$ has a single minimum in $u = 0$ if $T > T_c = \tilde{J}/a^d$, and two symmetric minima if $T < T_c$. In Fig. Q.1(a), we plot $\mathcal{V}^*(u) = (a^{2d}/\tilde{J})\mathcal{V}(u)$, for different values of $T^* = T/T_c$,

$$\mathcal{V}^*(u) = -\frac{u^2}{2} + T^*\left[-\ln 2 + \frac{1}{2}(1 + u)\ln(1 + u) + \frac{1}{2}(1 - u)\ln(1 - u)\right]. \tag{Q.15}$$

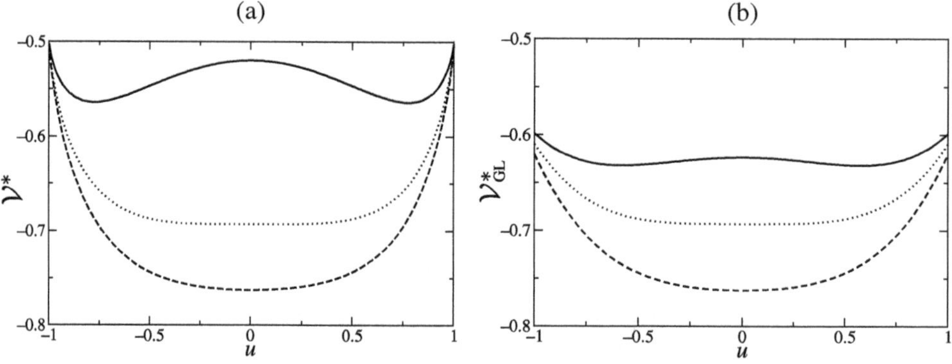

Fig. Q.1   The potentials $\mathcal{V}^*(u)$ (a) and $\mathcal{V}^*_{\mathrm{GL}}(u)$ (b) for different values of T^*: $T^* = 1.1$ (dashed line), $T^* = T_c^* = 1$ (dotted line), $T^* = 0.9$ (solid line).

It is worth noting that physically it must be $|u| \leq 1$. In the limit $u \to 1$

$$\mathcal{V}^*(u) = -\frac{1}{2} + \frac{T^*}{2}(1-u)\ln(1-u) + (1-u)\left(1 - \frac{T^*}{2}(1+\ln 2)\right) + O((1-u)^2). \quad \text{(Q.16)}$$

Therefore, $\mathcal{V}^*(1)$ is finite, but $(\mathcal{V}^*)'(1) \to +\infty$, which implements the earlier constraint $|u| \leq 1$. Furthermore, minimizing the earlier expression, we find that for small T^* the minima of the potential occur for $\bar{u} = \pm(1 - \exp(-2/T^*))$.

In the literature, the potential $\mathcal{V}^*(u)$ is rarely used. Rather, it is common to use the classical quartic potential, which comes out analytically from $\mathcal{V}^*(u)$ in the limit of small u, that is, in the vicinity of T_c. Using the Taylor expansion $\ln(1+u) = \sum_{n\geq 1}(-1)^{n+1}u^n/n$ up to fourth order, we obtain

$$\mathcal{V}^*_{\text{GL}}(u) = -T^* \ln 2 + \frac{(T^*-1)}{2}u^2 + \frac{T^*}{12}u^4, \quad \text{(Q.17)}$$

which is also plotted in Fig. Q.1(b).

If we enter this potential within Eq. (Q.13), we get

$$\mathcal{F}_{\text{GL}} = \frac{\tilde{J}}{a^{2d}} \int d\mathbf{x} \left[\frac{K}{\tilde{J}}(\nabla u)^2 + \frac{T^*-1}{2}u^2 + \frac{T^*}{12}u^4\right], \quad \text{(Q.18)}$$

where a constant in the integrand has been removed.

It is worth noting that by rescaling the space, $\mathbf{x} \to b\mathbf{x}$, and the order parameter, $u \to cu$, it is possible to rewrite $\mathcal{F}_{\text{GL}}$ so as to display one parameter only, the scale of energy:

$$\mathcal{F}_{\text{GL}} = f_0 \int d\mathbf{x} \left[\frac{1}{2}(\nabla u)^2 \pm \frac{1}{2}u^2 + \frac{1}{4}u^4\right], \quad \text{(Q.19)}$$

where the sign in front of the quadratic term is the same sign of $(T - T_c)$ and the numerical prefactors $\frac{1}{2}$ and $\frac{1}{4}$ can be set to any other positive values.

For completeness, we also rewrite the Eq. (Q.13) in rescaled variables,

$$\mathcal{F} = f_0 \int d\mathbf{x} \left\{\frac{1}{2}(\nabla u)^2 - \frac{u^2}{2} + \frac{T}{T_c}\left[-\ln 2 + \frac{1}{2}(1+u)\ln(1+u) + \frac{1}{2}(1-u)\ln(1-u)\right]\right\}. \quad \text{(Q.20)}$$

We limit ourselves to giving a few details here about the method of the dynamical renormalization group for the KPZ equation. The starting point is the linear relation between noise and interface profile in the $\mathbf{k}$-space, for the EW equation; then, using the EW propagator, that is, the ratio between profile and noise, we construct a perturbative theory for the KPZ equation. The RG is a method to get rid of the divergences appearing in the naive perturbation theory.

The linear character of the EW equation allows for a simple and direct relationship between noise and interface profile. Such a relation is better shown in the (space and time) Fourier space, by writing

$$h(\mathbf{x}, t) = \int_{-\infty}^{\infty} \frac{d\omega}{2\pi} \int_{V_{\mathbf{q}}} \frac{d\mathbf{q}}{(2\pi)^d} e^{i(\mathbf{q}\cdot\mathbf{x}+\omega t)} h(\mathbf{q}, \omega) \tag{R.1}$$

and replacing it in Eq. (7.37). We obtain

$$h(\mathbf{q}, \omega) = G_0(\mathbf{q}, \omega)\eta(\mathbf{q}, \omega), \tag{R.2}$$

where

$$G_0(\mathbf{q}, \omega) = \frac{1}{\nu q^2 + i\omega} \tag{R.3}$$

is the propagator of the EW equation. If we do the same thing for KPZ, we obtain

$$\begin{aligned}
i\omega h(\mathbf{k}, \omega) = &-\nu k^2 h(\mathbf{k}, \omega) \\
&-\frac{\lambda}{2} \int \frac{d\omega}{2\pi} \int_{V_{\mathbf{q}}} \frac{d\mathbf{q}}{(2\pi)^d} \mathbf{q}\cdot(\mathbf{k}-\mathbf{q}) h(\mathbf{q}, \Omega) h(\mathbf{k}-\mathbf{q}, \omega - \Omega) \\
&+\eta(\mathbf{k}, \omega).
\end{aligned} \tag{R.4}$$

The quantities appearing in the first and third lines are simple, because they are the Fourier transform of linear terms (time derivative, Laplacian, and noise, respectively). The term in the second line derives from the KPZ nonlinearity. Using the definition of G_0, we can rewrite Eq. (R.4) as

$$\begin{aligned}
h(\mathbf{k}, \omega) = &G_0(\mathbf{k}, \omega)\eta(\mathbf{k}, \omega) \\
&-\frac{\lambda}{2} G_0(\mathbf{k}, \omega) \int \frac{d\omega}{2\pi} \int_{V_{\mathbf{q}}} \frac{d\mathbf{q}}{(2\pi)^d} \mathbf{q}\cdot(\mathbf{k}-\mathbf{q}) h(\mathbf{q}, \Omega) h(\mathbf{k}-\mathbf{q}, \omega - \Omega).
\end{aligned} \tag{R.5}$$

This expression can be easily iterated perturbatively, getting a sequence of terms in increasing powers of λ, the perturbation parameter: It is sufficient to replace the earlier expression for $h(\mathbf{k}, \omega)$ in the λ term and repeat the same operation. The key point is that this expansion produces terms that are divergent for $d < 2$, making useless such naive approach.

The renormalization group is a method of circumventing these divergences. The idea is to change the spatial scale b of the system, integrating out small scales. As a result of this coarse-graining procedure, the system is described by new coupling parameters ν, λ and by a new strength of noise Γ, much in the same way a coarse-graining procedure of an equilibrium Ising system leads to new couplings $J_1, J_2, \ldots$ and to a new temperature. If we define, $\ell = \ln b$, the result of the dynamical RG machinery is

$$\frac{d\nu}{d\ell} = \nu \left[z - 2 + \frac{K_d(2-d)}{4d} g^2 \right] \equiv \nu f_\nu(g) \tag{R.6a}$$

$$\frac{d\lambda}{d\ell} = \lambda(\alpha + z - 2) \tag{R.6b}$$

$$\frac{d\Gamma}{d\ell} = \Gamma \left[z - d - 2\alpha + \frac{K_d}{4} g^2 \right] \equiv \Gamma f_\Gamma(g), \tag{R.6c}$$

where $K_d = S_d/(2\pi)^d$, S_d is the surface area of a d-dimensional sphere of radius equal to one, and

$$g^2 = \frac{\Gamma \lambda^2}{\nu^3}. \tag{R.7}$$

These equations define the so-called RG flow. On large scales, the system can be described by stable fixed points, which also fix the exponents α, z. Let us first consider the EW case, where $\lambda = 0 = g^2$. Eqs. (R.6) simplify to

$$\frac{d\nu}{d\ell} = \nu(z - 2) \tag{R.8a}$$

$$\frac{d\Gamma}{d\ell} = \Gamma(z - d - 2\alpha), \tag{R.8b}$$

and imposing the condition for a fixed point, $d\nu/d\ell = 0 = d\Gamma/d\ell$, we find the well-known values $z = 2$ and $\alpha = (2 - d)/2$. Furthermore, any pair ν, Γ turns out to be a fixed point, because parameters of the EW equation do *not* renormalize, because of the linear character of the equation.

When we consider the KPZ equation, we can see immediately from Eq. (R.6b) that λ does not renormalize, because the fixed-point condition, $d\lambda/d\ell = 0$, does not depend on λ and gives the general condition[1]

$$\alpha + z = 2. \tag{R.9}$$

This relation is not limited to our perturbative approach; it is an exact relation depending on a physical invariance of the KPZ equation, as discussed in Section 7.7.2.

---

[1] More precisely, the condition $d\lambda/d\ell = 0$ also has the solution $\lambda = 0$, which corresponds to the EW limit we have just discussed.

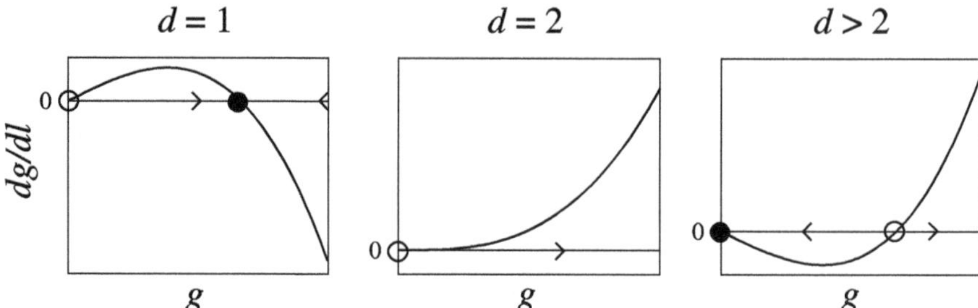

Fig. R.1   Plots of $dg/d\ell$ for different dimensions. Solid circles mean stable fixed points, open circles mean unstable fixed points.

Let us now come back to the RG flow. Eqs. (R.6a) and (R.6c) can be combined, in order to find the RG flow of the coupling g,

$$\frac{dg^2}{d\ell} = \frac{\lambda^2}{\nu^3}\frac{d\Gamma}{d\ell} - 3\frac{\lambda^2\Gamma}{\nu^4}\frac{d\nu}{d\ell}$$
$$= g^2(f_\Gamma(g) - 3f_\nu(g)). \tag{R.10}$$

Since $dg^2/d\ell = 2g(dg/d\ell)$, using the relation $(\alpha + z) = 2$, we finally obtain

$$\frac{dg}{d\ell} = \frac{2-d}{2}g + \frac{(2d-3)K_d}{4d}g^3. \tag{R.11}$$

The function $dg/d\ell$ versus g is plotted in Fig. R.1 for $d = 1$, $d = 2$, and $d > 2$. It depicts the RG flow, the resulting fixed points, and their stability. The point $g = 0$ is always a fixed point, that is, a constant solution of the RG flow, and it corresponds to the EW fixed point. For $d \leq 2$, it is unstable, as signaled by the positive coefficient of the linear term in the right-hand side of Eq. (R.11): However small is $g(0)$ (i.e., however small the coefficient λ of the KPZ nonlinearity is), a coarse graining of the system leads to increasing values of g, because the nonlinearity is relevant and dominant for $d < 2$. Instead, for $d > 2$, the coefficient of the linear term is negative, meaning that a small g is renormalized to zero under coarse graining: The KPZ nonlinearity is irrelevant and exponents α, z are the same as for the EW equation. Therefore, the RG analysis confirms the expectations based on the simple scale transformation discussed at the beginning of this section.

If we want to go beyond the small g limit, that is, make a perturbative analysis of the EW fixed point, we need to take into account the cubic term in Eq. (R.11). We focus on $d = 1$, the only relevant (integer) dimension with a stable fixed point for $g = g^* \neq 0$. Since $K_1 = \frac{1}{2\pi}$, Eq. (R.11) is written $\frac{dg}{d\ell} = \frac{1}{2}g - \frac{1}{8\pi}g^3$, which gives the stable fixed point,

$$g^* = 2\sqrt{\pi}, \qquad d = 1. \tag{R.12}$$

If we impose $f_\nu(g^*) = 0$ (see Eq. (R.6a)), because we must have $d\nu/d\ell = 0$ at the fixed point, we find

$$z_{\text{KPZ}} = \frac{3}{2}, \qquad d = 1, \tag{R.13}$$

and using the relation $(\alpha + z) = 2$,

$$\alpha_{\mathrm{KPZ}} = \frac{1}{2}, \qquad d = 1. \tag{R.14}$$

Therefore, in $d = 1$, $\alpha_{\mathrm{KPZ}} = \alpha_{\mathrm{EW}}$. This equality has a physical motivation, as explained in Section 7.7.3, but it is not true in $d > 1$. The value of $\beta = \alpha/z$ is larger, instead: $\beta_{\mathrm{KPZ}} = \frac{1}{3} > \beta_{\mathrm{EW}} = \frac{1}{4}$. For $d = 2$, we can say nothing, while for $d > 2$, the point $g = 0$ is stable, but its basin of attraction is confined to $g < g^*$. If $g(0) > g^*$, the nonlinearity is not renormalized to zero. Rather, it increases forever. A perturbative RG approach, such as the one discussed here, is unable to say more than that.

TASEP: Map Method and Simulations

S.1 Mean-fiel Solution of TASEP with the Map Method

A recursive map is a function

$$x_{n+1} = F(x_n), \tag{S.1}$$

and a fixed point x^* is a solution of the equation $x^* = F(x^*)$: If $x_0 = x^*$, $x_n = x^* \, \forall n$. If x_0 is close to x^*, we can determine its evolution via a linear analysis of Eq. (S.1): $x_n = x^* + \epsilon_n$, so $\epsilon_{n+1} = F'(x^*)\epsilon_n$, whose solution is $\epsilon_n = (F'(x^*))^n \epsilon_0$. If $|F'(x^*)| < 1$, $\epsilon_n \to 0$ exponentially and the fixed point is linearly stable. If $|F'(x^*)| > 1$, $|\epsilon_n| \to \infty$ exponentially and the fixed point is linearly unstable. In both cases, the sign of $F'(x^*)$ indicates whether ϵ_n has a constant or an oscillating sign. If $|F'(x^*)| = 1$, we must expand $F(x)$ to the next order. Let us do that explicitly for $F'(x^*) = 1$: $\epsilon_{n+1} = \epsilon_n + \frac{1}{2}F''(x^*)\epsilon_n^2$. The discrete equation cannot be iterated analytically, but we can pass to continuum and solve the resulting differential equation,

$$\frac{d\epsilon}{dn} = \frac{1}{2}F''(x^*)\epsilon^2, \tag{S.2}$$

getting

$$\epsilon(n) = \frac{1}{\dfrac{1}{\epsilon(0)} - \dfrac{1}{2}F''(x^*)n}. \tag{S.3}$$

If $\epsilon(0)$ has the same sign as $F''(x^*)$, x_n moves away from the fixed point (the unphysical change of sign of $\epsilon(n)$ at large n occurs when ϵ is so large that the above analysis is no longer valid). If $\epsilon(0)$ has the opposite sign of $F''(x^*)$, x_n approaches the fixed point, with $|\epsilon(n)|$ vanishing as $1/n$. All the qualitative features concerning the stability of fixed points can be found by an easy graphical analysis: See Fig. S.1, where we have assumed a positive slope and a negative curvature.

Let us now consider the (discrete) TASEP model with open boundary conditions, see Section 5.4.2. A steady configuration is characterized by time-independent averages. Therefore, even if $p_i = \langle n_i \rangle$ is site dependent, the current $J_{i,i+1} = \nu \langle n_i(1 - n_{i+1}) \rangle$ cannot depend on site i in the steady state, because an unbalance between $J_{i-1,i}$ and $J_{i,i+1}$ would determine a temporal variation of p_i,

$$\frac{d\langle n_i \rangle}{dt} = \nu \langle n_{i-1}(1 - n_i) \rangle - \nu \langle n_i(1 - n_{i+1}) \rangle$$
$$= \nu p_{i-1}(1 - p_i) - \nu p_i(1 - p_{i+1}), \tag{S.4}$$

where the last equality applies in a mean-field approximation. The conditions $J_{i,i+1} = J$ produce $L + 1$ coupled equations, along with an equal number of unknowns ($p_1, \ldots, p_L$ and J itself)[1]:

$$\alpha(1 - p_1) = J \tag{S.5}$$

$$\nu p_i(1 - p_{i+1}) = J, \quad i = 1, \ldots, L - 1 \tag{S.6}$$

$$\beta p_L = J. \tag{S.7}$$

It is worth noting that boundary conditions (S.5, S.7) can be rewritten in the form Eq. (S.6) if we assume fictitious sites $i = 0$ and $i = L+1$, with $p_0 = \alpha$ and $p_{L+1} = (1-\beta)$.[2] Therefore, mean-field equations correspond to a set of recursive equations. However, while in a kinetic Monte Carlo simulation, α and β are the natural input data and J is evaluated by computing the average of $n_i(1 - n_{i+1})$, when solving Eqs. (S.5) to (S.7), the line of reasoning is different. We should rather start with a test value for J and check if it is consistent: In fact, assuming some J, Eq. (S.5) provides p_1 and Eqs. (S.6) determine all p_i up to p_L. Finally, Eq. (S.7) allows us to check if the chosen value of J is correct. This way of proceeding is not practical, neither for a numerical solution of recursive equations nor for their qualitative analysis.

Recursive equation (S.6) is also called map in the language of dynamical systems, where its solution gives the discrete time i evolution of the quantity p_i. In our case, i is a site index, not a time index, and the evolution of Eq. (S.6) for increasing/decreasing i just corresponds to moving to the right/left of the system. Far from the edges, we can expect the density p_i is approximately constant: in the language of maps, that means we are in proximity of a fixed point.

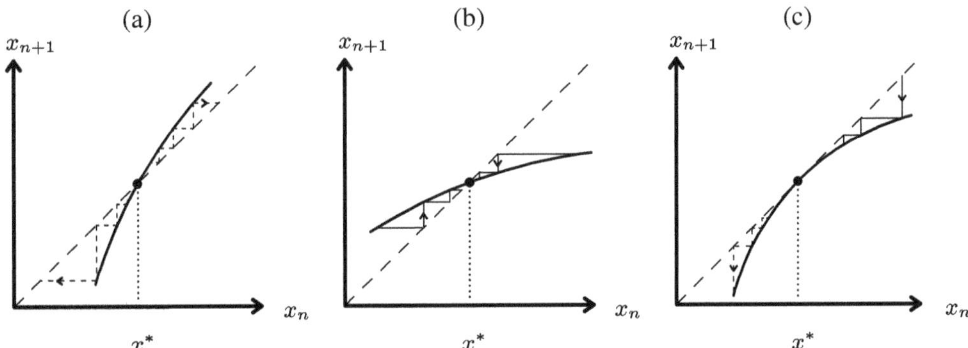

Fig. S.1　Graphical analysis of the stability of a fixed point, in the three cases $F'(x^*) > 1$ (unstable, (a), $F'(x^*) < 1$ (stable, (b), and $F'(x^*) = 1$ (marginal, (c). Solid lines are the function $F(x)$. Long dashed lines represent the diagonal $x_{n+1} = x_n$, so their intersection with the solid lines (solid circle) is the fixed point x^*. The evolution of a point is shown graphically by thin solid lines (stability) and thin dashed lines (instability).

---

[1] It is apparent from these equations that a hopping rate $\nu \neq 1$ corresponds to rescale α and β: From now on, as in the main text, we assume $\nu = 1$.

[2] Equations (S.5) and (S.7) clearly show there is no inconsistency if $\alpha > 1$ (and therefore $p_0 > 1$) or $\beta > 1$ (and therefore $p_{L+1} < 0$): A diverging α implies $p_1 \to 1^-$ and a diverging β implies a vanishing p_L. All average occupancies of *real* sites $i = 1, \ldots, L$ are between 0 and 1, as it must be.

Table S.1 Number and stability of fixed points with varying J

$J < \dfrac{1}{4}$	Two fixed points	$p^*, 1 - p^*$	$F'(p^*) > 1, F'(1 - p^*) < 1$
$J = \dfrac{1}{4}$	One fixed point	$p^* = \dfrac{1}{2}$	$F'\left(\dfrac{1}{2}\right) = 1$
$J > \dfrac{1}{4}$	No fixed points		

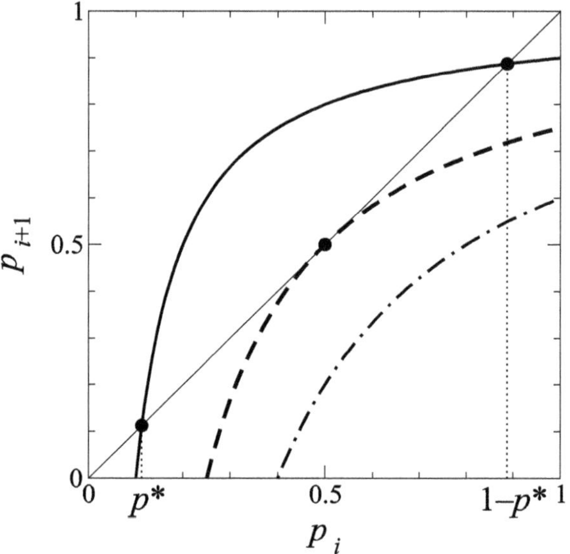

Fig. S.2 The map associated with the TASEP model, Eq. (S.8), for $J < \frac{1}{4}$ (solid line), $J = \frac{1}{4}$ (dashed line), and $J > \frac{1}{4}$ (dot-dashed line).

The map associated with the recursive equation of TASEP model, $p_i(1 - p_{i+1}) = J$, is

$$p_{i+1} = 1 - \frac{J}{p_i} \equiv F(p_i), \tag{S.8}$$

which is plotted in Fig. S.2 for different values of J. The fixed points are determined by the second-order equation $p_{\mathrm{f}}(1 - p_{\mathrm{f}}) = J$, whose solutions are $p_{\mathrm{f}} = p^* \equiv \frac{1}{2} - \sqrt{\frac{1}{4} - J}$ and $p_{\mathrm{f}} = 1 - p^*$. The full behavior of the map with varying J is reported in Table S.1 and in the flow diagram of Fig. S.3.

We are now ready to solve the recursive equations that provide the mean-field description of TASEP model, just on the basis of Fig. S.3. Note that boundary conditions, injection, and removal of particles correspond to setting $p_0 = \alpha$ and $p_{L+1} = 1 - \beta$.

For $J < \frac{1}{4}$, if we evolve Eq. (S.8) forward, we have a stable fixed point for $p = 1 - p^*$ whose basin of attraction is $p > p^*$. Therefore, any initial condition $p_0 = \alpha$ in this basin

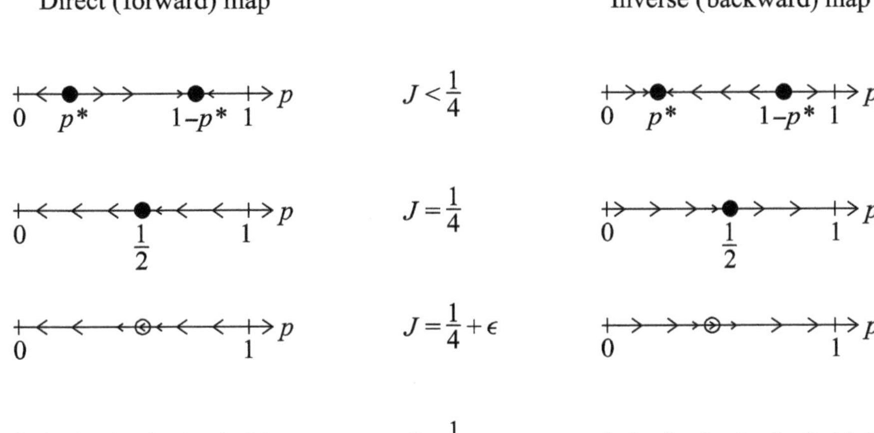

Fig. S.3 Map flows for different values of J in the TASEP model. Arrows indicate the evolution of the map: The smaller the arrow, the smaller the quantity $|p_{i+1} - p_i|$. Solid circles are fixed points. The open circle appearing for $J = \frac{1}{4} + \epsilon$ signals the memory of the fixed point for $J = \frac{1}{4}$.

will flow toward such fixed point, giving the following relations:

$$p_0 = \alpha > p^*, \qquad p_i \underset{i \nearrow}{\longrightarrow} (1 - p^*) = 1 - \beta, \qquad p^* < \frac{1}{2}. \tag{S.9}$$

Therefore, $p^* = \beta$ and the earlier solution applies in the region $\alpha > \beta$ and $\beta < \frac{1}{2}$. Except for a finite region close to the left edge, the density is equal to the stable fixed point, $p_i = 1 - \beta$, and the current is equal to $J = \beta(1 - \beta)$. Since the density is almost everywhere larger than $1/2$, this phase is called a *high-density* phase (HD, see Fig. 5.8). In simple words, this solution appears when we are injecting more particles than we are removing ($\alpha > \beta$) and removal is not fully efficient ($\beta < \frac{1}{2}$). In these conditions, particle density gets constant toward the exit. As for the profile of particle density, p_i, it may be either increasing or decreasing, as easily understood from Fig. S.3: It is increasing/decreasing if $p_0 = \alpha$ is smaller/larger than $p_{L+1} = 1 - \beta$.

The motion of particles driven to the right can be understood as the motion of holes (empty sites) driven to the left. Since the hole density is one minus the particle density, holes are injected on the right at rate β and removed on the left at rate α. This means that the right-left density profile of holes that we get for $(\alpha, \beta) = (\alpha^*, \beta^*)$ is equal to the left-right density profile of particles for $(\alpha, \beta) = (\beta^*, \alpha^*)$. In other words, the symmetric of the HD phase, $\beta > \alpha$ and $\alpha < \frac{1}{2}$, must correspond to a low-density (LD) phase, where particle density is constant close to the left edge and equal to the low density fixed point, $p = p^*$. More formally, this phase can be found iterating the inverse map (see Fig. S.3),

$$p_{L+1} = 1 - \beta < 1 - p^*, \qquad p_i \underset{i \searrow}{\longrightarrow} p^* = \alpha, \qquad p^* < \frac{1}{2}. \tag{S.10}$$

Therefore, $p^* = \alpha$ and the LD phase appears in the region $\beta > \alpha$ and $\alpha < \frac{1}{2}$, as expected.

If we follow the same lines of reasoning for $J = \frac{1}{4}$, we find that this value of J corresponds to the two lines, $\alpha > \frac{1}{2}, \beta = \frac{1}{2}$ and $\alpha = \frac{1}{2}, \beta > \frac{1}{2}$. The remaining phase $\alpha, \beta > \frac{1}{2}$ requires a few more words about the map. If J is significantly larger than $\frac{1}{4}$, the iteration of the map will always lead outside the physical domain $p = [0, 1]$ within a few steps. However, if $J - \frac{1}{4} \ll 1$, the map will stay for a long "time" close to $p = \frac{1}{2}$, a "time" that diverges when $(J - \frac{1}{4})$ vanishes. This phase is called maximal current (MC), because J reaches its greatest possible value, $J = \frac{1}{4}$, and it corresponds to $p_0 = \alpha > \frac{1}{2}$ and $p_{L+1} = 1 - \beta < \frac{1}{2}$, that is, $\beta > \frac{1}{2}$.

We can now summarize the properties of the three different phases (see also the Table 5.2 and Figs. 5.8 and 5.9): (1) the HD phase, where the density is equal to $p_f = 1 - \beta > \frac{1}{2}$ and $J = \beta(1 - \beta)$; (2) the LD phase, where the density is equal to $p_f = \alpha < \frac{1}{2}$ and $J = \alpha(1 - \alpha)$; and (3) the MC phase, where $p = \frac{1}{2}$ and $J = \frac{1}{4}$. The separation line HD/LD corresponds to a transition between the high-density fixed point and the low-density fixed point, therefore indicating a discontinuity. What happens on that line?

When $\alpha = \beta < \frac{1}{2}$, there are two separate fixed points, at $p_f = \alpha$ and $p_f = 1 - \alpha$. Since the solution of mean-field equations corresponds to the evolution of the map starting at $p_0 = \alpha$ and terminating at $p_{L+1} = 1 - \alpha$, this trajectory simply connects the unstable fixed point at $p = \alpha$ to the stable fixed point at $p = 1 - \alpha$. The resulting density profile is shown in Fig. 5.9. Instead, the two lines $(\alpha = \frac{1}{2}, \beta > \frac{1}{2})$ and $(\alpha > \frac{1}{2}, \beta = \frac{1}{2})$ represent the continuous transitions LD/MC and HD/MC, respectively. In both cases, the density goes continuously from the high-density or low-density fixed point toward $p = \frac{1}{2}$.

S.2 Simple Simulation of TASEP

TASEP model can be easily simulated as follows. If $\nu_{\max} = \max\{\alpha, \beta, \nu\}$, we can rescale transition rates and define the probabilities $p_\chi = \chi/\nu_{\max}$, where $\chi = \alpha, \beta, \nu$. A microscopic configuration is defined by the set of integers $n_i = 0, 1$, with $i = 1, \ldots, L$; $n_i = 0$ ($n_i = 1$) meaning that site i is empty (occupied).

The next step is to extract a random integer $k = 0, \ldots, L$ and attempt one of the following moves, depending on k: (1) if $k = 0$ and $n_1 = 0$, then $n_1 \leftarrow 1$ with probability p_α: (2) if $k = 1, \ldots, L - 1$, $n_k = 1$ and $n_{k+1} = 0$, then $n_k \leftarrow 0$ and $n_{k+1} \leftarrow 1$ with probability p_ν; (3) if $k = L$ and $n_L = 1$, then $n_L \leftarrow 0$ with probability p_β. The reader should have easily recognized the formalization of the three basic moves of TASEP: injection, hopping, and extraction of a particle. It is manifest that in many cases, the attempt of a move is not successful, either because the initial configuration does not allow it or because $p_\chi < 1$. In Appendix T.2, we describe a rejection-free algorithm, which however requires having and updating a list of possible moves and of their probabilities.

In this type of simulation, time is measured in terms of Monte Carlo units, where the single unit corresponds to attempt L moves.

Bridge Model: Mean-field and Simulations

T.1 The Mean-field Phase Diagram of the Bridge Model

In the main text we found that for $q = 1$, the bridge model is equivalent to two independent TASEP models (each model for a class of particles), with input/output parameters equal to (α_+, β) and (α_-, β) for positive and negative particles, respectively, where

$$\alpha_+ = \frac{J_+}{\dfrac{J_+}{\alpha} + \dfrac{J_-}{\beta}} \quad \text{and} \quad \alpha_- = \frac{J_-}{\dfrac{J_-}{\alpha} + \dfrac{J_+}{\beta}}. \tag{T.1}$$

We now assume that the bridge model is in a given phase and use the earlier equations to find if such solution exists for some values of α and β. Since symmetric solutions are easier to study, we start with them.

T.1.1 Symmetric Solutions

If $J_+ = J_- = \tilde{J}$, we also have

$$\alpha_+ = \alpha_- = \tilde{\alpha} = \frac{\alpha\beta}{\alpha + \beta}. \tag{T.2}$$

The symmetric low-density (LD) phase exists if $\tilde{\alpha} < \beta$ and $\tilde{\alpha} < \frac{1}{2}$. The former condition (see Eq. (T.2)) is always satisfied, while the latter gives

$$\boxed{\text{LD} \quad \frac{1}{\alpha} + \frac{1}{\beta} > 2} \tag{T.3}$$

We can also check formally that the symmetric high-density (HD) phase cannot exist. In fact, it would require $\tilde{\alpha} > \beta$ and $\beta < \frac{1}{2}$, but the first condition cannot be satisfied. Finally, the symmetric maximal current (MC) phase exists if $\tilde{\alpha} \geq \frac{1}{2}$ and $\beta \geq \frac{1}{2}$. The former condition gives

$$\boxed{\text{MC} \quad \frac{1}{\alpha} + \frac{1}{\beta} \leq 2} \tag{T.4}$$

so the latter is automatically satisfied.

T.1.2 The Asymmetric HD–LD Phase

We now consider the interesting possibility that one class of particles (e.g., the positive ones) is in the HD phase and the other class (the negative ones) is in the LD phase. This choice implies the following relations and conditions:

$$\oplus \text{ HD} \quad J_+ = \beta(1-\beta), \qquad \alpha_+ > \beta, \qquad \beta < \tfrac{1}{2}$$
$$\ominus \text{ LD} \quad J_- = \alpha_-(1-\alpha_-), \qquad \alpha_- < \beta, \qquad \alpha_- < \tfrac{1}{2}. \tag{T.5}$$

We immediately remark that the second condition on α_- is redundant. Then, we can replace the explicit expressions for $J_\pm$ in Eq. (T.1) and get

$$\alpha_+ = \frac{\beta(1-\beta)}{\dfrac{\beta(1-\beta)}{\alpha} + \dfrac{\alpha_-(1-\alpha_-)}{\beta}}, \qquad \alpha_- = \frac{\alpha_-(1-\alpha_-)}{\dfrac{\alpha_-(1-\alpha_-)}{\alpha} + \dfrac{\beta(1-\beta)}{\beta}}. \tag{T.6}$$

The second equation is a simple quadratic equation for α_-, whose solutions is

$$\alpha_- = \frac{(1+\alpha) \pm \sqrt{(1+\alpha)^2 - 4\alpha\beta}}{2} \tag{T.7}$$

and whose discriminant is certainly positive if $\beta < \tfrac{1}{2}$. The solution with the positive sign must be rejected, because α_- would be larger than $\tfrac{1}{2}$. Summarizing, we have

$$\alpha_- = \frac{1+\alpha - \sqrt{(1+\alpha)^2 - 4\alpha\beta}}{2}, \qquad \alpha_+ = \frac{\beta(1-\beta)}{\dfrac{\beta(1-\beta)}{\alpha} + \dfrac{\alpha_-(1-\alpha_-)}{\beta}}. \tag{T.8}$$

Let us first test the condition $\alpha_- < \beta$, that is,

$$\frac{1+\alpha - \sqrt{(1+\alpha)^2 - 4\alpha\beta}}{2} < \beta. \tag{T.9}$$

Isolating the square root, we find

$$1 + \alpha - 2\beta < \sqrt{(1+\alpha)^2 - 4\alpha\beta}. \tag{T.10}$$

Since $\beta < \tfrac{1}{2}$ and $\alpha > 0$, the left-hand side is positive and we can take the square of both terms, obtaining the condition $4\beta^2 < 4\beta$, which is manifestly satisfied for $\beta < \tfrac{1}{2}$. Therefore, we are left with the conditions $\beta < \tfrac{1}{2}$ and $\alpha_+ > \beta$. The latter inequality (see Eq. (T.8), right) is rewritten as

$$\frac{\beta(1-\beta)}{\alpha} + \frac{\alpha_-(1-\alpha_-)}{\beta} < 1 - \beta. \tag{T.11}$$

Replacing the expression of $\alpha_-(\alpha, \beta)$ given in Eq. (T.8), left, we get

$$\boxed{\text{HD–LD} \quad \beta^2(1+\alpha-\beta) + \frac{\alpha^2}{2}\left(2\beta + \sqrt{(1+\alpha)^2 - 4\alpha\beta} - \alpha - 1\right) - \alpha\beta < 0}$$

$$\tag{T.12}$$

Let us focus analytically on the case $\beta \ll 1$. In this limit, we get

$$-\left(\frac{\alpha}{1+\alpha}\right)\beta + O(\beta^2) < 0, \tag{T.13}$$

which is surely satisfied if α is not vanishing. If α also is vanishing, higher-order terms should be accounted for. The limit $\alpha \to \infty$ requires expansion of the square root in Eq. (T.12) up to α^{-2} terms and gives the condition $\beta < \frac{1}{3}$. The numerical solution of Eq. (T.12) gives the lower line of Fig. 5.11, $\beta_1(\alpha)$. Since $\beta_1(\alpha) < \frac{1}{3}$ for any α, the condition $\beta < \frac{1}{2}$ is redundant and the asymmetric LD–HD phase exists for $\beta < \beta_1(\alpha)$.

T.1.3 The Asymmetric MC–LD Phase

We assume, without loss of generality, that positive particles are in the MC phase and negative particles in the LD phase, so that

$$\begin{aligned}
\oplus \text{ MC} \quad & J_+ = \tfrac{1}{4}, & \alpha_+ \geq \tfrac{1}{2}, & \quad \beta \geq \tfrac{1}{2} \\[4pt]
\ominus \text{ LD} \quad & J_- = \alpha_-(1-\alpha_-), & \alpha_- < \beta, & \quad \alpha_- < \tfrac{1}{2}.
\end{aligned} \tag{T.14}$$

We easily remark that the first condition on α_- is redundant. We can replace the explicit expressions for $J_\pm$ in Eq. (T.1) and get

$$\alpha_+ = \frac{\frac{1}{4}}{\dfrac{1}{4\alpha} + \dfrac{\alpha_-(1-\alpha_-)}{\beta}}, \qquad \alpha_- = \frac{\alpha_-(1-\alpha_-)}{\dfrac{\alpha_-(1-\alpha_-)}{\alpha} + \dfrac{1}{4\beta}}. \tag{T.15}$$

If we define the quantity $\gamma = \alpha_-(1-\alpha_-)$, the two previous equations (T.15) can be rewritten in compact form as

$$\alpha_+^{-1} = \alpha^{-1} + 4\gamma\beta^{-1}, \qquad \alpha_-^{-1} = \alpha^{-1} + \frac{\beta^{-1}}{4\gamma}, \tag{T.16}$$

whence

$$\alpha_+^{-1} - \alpha_-^{-1} = \beta^{-1}\left(4\gamma + \frac{1}{4\gamma}\right) \geq 0, \tag{T.17}$$

where we have used that both γ and β are positive. However, from the conditions $\alpha_+ \geq \frac{1}{2}$ and $\alpha_- < \frac{1}{2}$, we deduce that $\alpha_+^{-1} - \alpha_-^{-1} < 0$, which is in contradiction with Eq. (T.17). Therefore, there are no solutions satisfying the condition for the existence of the MC–LD phase, which is forbidden.

T.1.4 The Asymmetric LD–LD Phase

As strange as it may seem, this asymmetric phase is legitimate: Both types of particles are in the LD phase, but they have different densities (which also hints at the impossibility of having an asymmetric MC–MC phase). In short, we must have

$$\begin{aligned}
\oplus \text{ LD} \quad & J_+ = \alpha_+(1-\alpha_+), & \alpha_+ < \beta, & \quad \alpha_+ < \tfrac{1}{2} \\[4pt]
\ominus \text{ LD} \quad & J_- = \alpha_-(1-\alpha_-), & \alpha_- < \beta, & \quad \alpha_- < \tfrac{1}{2}
\end{aligned} \tag{T.18}$$

with

$$\alpha_+ = \frac{\alpha_+(1 - \alpha_+)}{\dfrac{\alpha_+(1 - \alpha_+)}{\alpha} + \dfrac{\alpha_-(1 - \alpha_-)}{\beta}}, \qquad \alpha_- = \frac{\alpha_-(1 - \alpha_-)}{\dfrac{\alpha_-(1 - \alpha_-)}{\alpha} + \dfrac{\alpha_+(1 - \alpha_+)}{\beta}}, \qquad \text{(T.19)}$$

which can be simplified to the expressions

$$\alpha_+ = 1 - \frac{\alpha_+(1 - \alpha_+)}{\alpha} - \frac{\alpha_-(1 - \alpha_-)}{\beta}, \qquad \alpha_- = 1 - \frac{\alpha_-(1 - \alpha_-)}{\alpha} - \frac{\alpha_+(1 - \alpha_+)}{\beta}. \qquad \text{(T.20)}$$

Defining the quantities $S = \alpha_+ + \alpha_-$ and $D = \alpha_+ - \alpha_-$ and taking the sum and the difference of $\alpha_\pm$ as given earlier, we obtain

$$S = 1 - \left(\frac{\alpha\beta}{\alpha - \beta}\right), \qquad D^2 = (2 - S)\left(S - \frac{2\alpha\beta}{\alpha + \beta}\right). \qquad \text{(T.21)}$$

Since $S = \alpha_+ + \alpha_- < 1$, the condition $D^2 > 0$ implies $S > 2\alpha\beta/(\alpha + \beta)$, that is,

$$\boxed{\text{LD–LD} \qquad \frac{\alpha\beta}{\alpha + \beta} < \frac{1}{2}\left(1 - \frac{\alpha\beta}{\alpha - \beta}\right)} \qquad \text{(T.22)}$$

The line $\beta = \beta_2(\alpha)$, where the two sides are equal, corresponds to the vanishing of D^2, that is, $\alpha_+ = \alpha_-$. Therefore, it means we are passing toward the symmetric LD phase. As for the line $\beta = \beta_1(\alpha)$, corresponding to the vanishing of the left-hand side of Eq. (T.12), it means the transition between the HD–LD phase toward the LD–LD phase. In conclusion, the LD–LD phase is expected to exist in the (α, β) region defined by the conditions $\beta_1(\alpha) < \beta < \beta_2(\alpha)$. We avoid giving a more rigorous proof of this statement.

T.2 Kinetic Monte Carlo

We describe here the basics of a kinetic Monte Carlo simulation, using the original formulation by Bortz, Kalos, and Lebowitz and applying it to the bridge model; see Fig. T.1.

A discrete and stochastic model is first of all defined by specifying the configuration space, then by assigning the transition rates $\nu_{i \to j}$ between state i and state j. For a given microscopic state i at time t, we can think to list *all* allowed transitions, whose rates are described for simplicity by $\nu_1, \nu_2, \ldots, \nu_M$. Making explicit reference to Fig. T.1 for the initial configuration, we have $M = 6$, with $\nu_1 = \nu_2 = \nu_3 = 1$ (hops toward empty sites), $\nu_4 = q$ (exchange between a positive and a negative particle), $\nu_5 = \alpha$ (input of a negative particle), and $\nu_6 = \beta$ (removal of a negative particle). We can now define $\nu_{\text{tot}} = \sum_{k=1}^{M} \nu_k$ and represent a segment of length ν_{tot} as in Fig. T.1.

We now extract a random number, uniformly distributed in the interval $[0, \nu_{\text{tot}}]$, select the corresponding transition, and perform it. It is clear that a given transition k has the probability $p_k = \nu_k/\nu_{\text{tot}}$ to occur. This is a so-called rejection-free algorithm, because it provides an effective evolution at any time step.

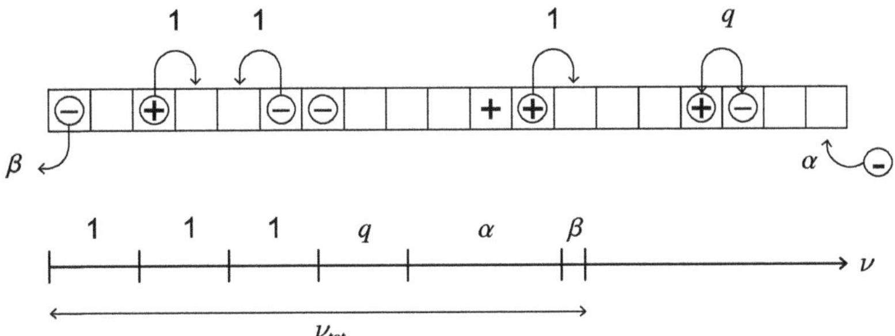

Fig. T.1 Top: Sketch of a specific microscopic configuration for the bridge model, with three positive particles and four negative particles and the indication of allowed transitions and of their rates: three hops toward empty sites (rate 1), one exchange between a positive and a negative particle (rate q), the input of a negative particle from the right (rate α), and the output of a negative particle from the left (rate β). Bottom: Each transition rate $\nu \neq 0$ is indicated by a segment of length ν. The choice of a random number in the interval $[0, \nu_{tot}]$ allows us to select (and implement) a specific transition.

The last step is to increment time by a quantity Δt that is exponentially distributed, according to

$$p(\Delta t) = \nu_{tot} \exp(-\nu_{tot} \Delta t). \tag{T.23}$$

The use of this distribution is only useful if we need to study correlations on time scales of the order or shorter than $1/\nu_{tot}$; otherwise, we can take Δt equal to its average value, $\langle \Delta t \rangle = 1/\nu_{tot}$, at any step.

The earlier dynamics is particularly effective if there are processes with transition rates that vary by orders of magnitude. This occurs, for example, in the conserved, Kawasaki dynamics of an Ising model (see the discussion in Section 8.4.2) or in the bridge model in the regime $\beta \ll \alpha$ (see Section 5.5).

Equation (8.64) is a special case of a more general equation describing the dynamics of a domain wall in the nonconserved case, whose derivation is the goal of this appendix. Let us start with Eq. (8.53),

$$\frac{\partial m}{\partial t} = \nabla^2 m - U'(m).$$

The local order parameter m varies only in proximity to a domain wall and if we move in the direction perpendicular to the wall itself. For this reason, it is useful to introduce a local unit vector $\hat{g}$, perpendicular to the wall and oriented in the direction of increasing m; see Fig. U.1. Along such direction the local coordinate is denoted s. Under the new coordinate system,

$$\nabla m = \left(\frac{\partial m}{\partial s}\right)_t \hat{g}, \tag{U.1}$$

$$\nabla^2 m = \left(\frac{\partial^2 m}{\partial s^2}\right)_t + \left(\frac{\partial m}{\partial s}\right)_t \nabla \cdot \hat{g}, \tag{U.2}$$

$$\frac{\partial m}{\partial t} = -v \left(\frac{\partial m}{\partial s}\right)_t. \tag{U.3}$$

The first equation is trivial, because m varies in space only in the $\hat{g}$ direction and the second is just its reapplication plus the fact that $\hat{g}$ itself varies along the wall. We remark that $\nabla \cdot \hat{g}$ is the curvature K of the domain wall. The time dependence of m is due to a translation

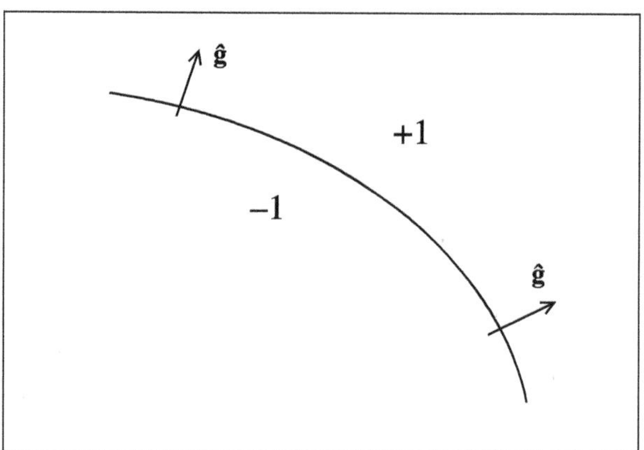

 The unit vector $\hat{g}$ is locally perpendicular to the domain wall and points toward the domain $m = +1$.

of the domain wall in the direction of $\hat{\mathbf{g}}$ or to a rotation of $\hat{\mathbf{g}}$. The distance of a point from the wall, which is what determines the value of the function m, changes linearly with an infinitesimal domain wall translation and quadratically with an infinitesimal rotation, therefore $(\partial m/\partial t)$ accounts only for the translation. The quantity v is the domain wall velocity and it is assumed positive if it moves in the $+\hat{\mathbf{g}}$ direction.

If we assemble previous formulas, we obtain the equation,

$$-v\left(\frac{\partial m}{\partial s}\right)_t = \left(\frac{\partial^2 m}{\partial s^2}\right)_t + \left(\frac{\partial m}{\partial s}\right)_t K - U'(m). \tag{U.4}$$

The steady profile $m(s)$ of a straight wall is determined by imposing the stationarity condition, $v = 0$, and the condition of straight wall, $K = 0$:

$$\left(\frac{\partial^2 m}{\partial s^2}\right)_t - U'(m) = 0. \tag{U.5}$$

We assume that dynamics is slow enough not to perturb this profile, so the earlier relation is still valid in out-of-equilibrium condition, which implies the so-called Allen–Cahn equation,

$$v = -K. \tag{U.6}$$

For a spherical domain wall of radius R, $K = (d-1)/R$ and we recover Eq. (8.64). However, Eq. (U.6) is more general and can be used, for example, to obtain valuable information on the statistics of domains; see Section 8.6.

The Gibbs–Thomson Relation

In this appendix we derive how the density of a gas in equilibrium with a droplet of the condensed phase depends on the radius R of the droplet. This result is used to write boundary conditions (8.88) in Section 8.4.1.

Let us imagine that particles are condensed in a droplet of radius R, surface S, volume V, mass M_s, and density ρ_s. The droplet is in equilibrium with a gas of mass M and density ρ. If g_s and g are the mass densities of the Gibbs free energy of the two phases and σ is the interface free energy, we have

$$G_{\text{TOT}} = g_s M_s + g M + \sigma S. \tag{V.1}$$

If we vary the mass of the droplet by a quantity δM_s, we have

$$\delta G_{\text{TOT}} = \delta M_s \left(g_s - g + \sigma \frac{1}{\rho_s} \frac{\partial S}{\partial V} \right) \tag{V.2}$$

$$= \delta M_s \left(g_s - g + \sigma \frac{d-1}{\rho_s} \frac{1}{R} \right). \tag{V.3}$$

At equilibrium we must have $\delta G_{\text{TOT}} = 0$, that is,

$$g - g_s = \frac{d-1}{\rho_s} \frac{\sigma}{R}. \tag{V.4}$$

Taking the derivative of both sides with respect to the pressure P at constant temperature T and using the thermodynamic relations

$$\frac{1}{\rho_s} = \left. \frac{\partial g_s}{\partial P} \right|_T, \qquad \frac{1}{\rho} = \left. \frac{\partial g}{\partial P} \right|_T, \tag{V.5}$$

we find

$$\frac{1}{\rho} - \frac{1}{\rho_s} = -\frac{(d-1)\sigma}{\rho_s R} \left[\frac{1}{R} \left. \frac{\partial R}{\partial P} \right|_T + \frac{1}{\rho_s} \left. \frac{\partial \rho_s}{\partial P} \right|_T \right]. \tag{V.6}$$

Assuming the condensed phase is incompressible, $\partial \rho_s / \partial P = 0$, and $\rho_s \gg \rho$, we finally get

$$\frac{1}{\rho} = -\frac{(d-1)\sigma}{\rho_s R^2} \left(\left. \frac{\partial P}{\partial R} \right|_T \right)^{-1}. \tag{V.7}$$

If we use the equation of state for the ideal gas, $\rho = mP/T$, we find

$$\frac{1}{\rho} = -\frac{(d-1)\sigma m}{\rho_s T R^2} \frac{1}{\rho'(R)}, \tag{V.8}$$

whose solution is

$$\rho(R) = \rho_\infty(T) \exp\left(\frac{(d-1)\sigma m}{\rho_s T}\frac{1}{R}\right). \tag{V.9}$$

For large R, we can approximate,

$$\rho(R) \simeq \rho_\infty(T)\left[1 + \frac{(d-1)\sigma m}{\rho_s T}\frac{1}{R}\right], \tag{V.10}$$

which is equivalent to the first boundary condition (8.88). The second boundary condition corresponds to exchanging the droplet and the gas. Within this geometry, a positive δM_s implies a negative surface term in Eq. (V.2), equivalent to changing the sign of σ in Eq. (V.10).

It is worth noting that we can interpret Eq. (V.9) as a sort of detailed balance relation: being at equilibrium the flux of detaching particles must equal the flux of attaching particles. If $e = e_G - e_c$ is the difference between the energy e_G of a gas particle and the energy e_c of a particle lying on the outer surface of the cluster, then detailed balance reads

$$\rho_s \exp(-\beta e) = \rho, \tag{V.11}$$

where e is a function of the local curvature $K = (d-1)/R$ because e_c depends on it. For large R, we can expand and find $e(K) = e(0) - e_c'(0)K$, so that

$$\begin{aligned}\rho(R) &= \rho_s \exp\left(-\beta\left(e(0) - e_c'(0)K\right)\right)\\ &= \rho_\infty(T) \exp\left(\frac{(d-1)e_c'(0)}{TR}\right). \end{aligned} \tag{V.12}$$

Equation (V.12) has the same form as Eq. (V.9) and it allows to establish the relation

$$e_c'(0) = \frac{\sigma m}{\rho_s}. \tag{V.13}$$

If the system is described by the lattice gas Hamiltonian, see Eq. (H.7), $\mathcal{H}_{LG} = -\epsilon_0 \sum_{\langle ij\rangle} n_i n_j$, $e_c(K) = -\epsilon_0 \bar{z}(K)$, where $\bar{z}(K)$ is the average number of neighboring particles for a particle lying on the outer surface of a cluster of curvature K. For a lattice gas, $\rho_s = m/a_0^3$, where a_0 is the lattice constant, and we obtain the relation

$$\bar{z}'(0) = -\frac{\sigma a_0^3}{\epsilon_0}. \tag{V.14}$$

In Fig. W.1 we sketch the geometry of the Rayleigh–Bénard instability: A fluid is confined between two horizontal plates at a distance d and kept at different temperatures, the lower plate being warmer than the upper one. Heat is transported by conduction for a small temperature difference ΔT and by convection when ΔT exceeds a critical value $(\Delta T)_c$.

The starting point for our analysis is the Navier–Stokes equations for an incompressible fluid,

$$\rho \left(\frac{\partial}{\partial t} + \mathbf{u} \cdot \nabla \right) \mathbf{u} = -\nabla p + \rho \nu \nabla^2 \mathbf{u} - \rho g \hat{\mathbf{z}}, \tag{W.1}$$

where ρ is the density, $\mathbf{u}(\mathbf{x}, t)$ is the velocity field, g is the acceleration of gravity (which is oriented downward in the vertical direction $\hat{\mathbf{z}}$), and ν is the kinematic viscosity. We do not derive the Navier–Stokes equations here and we limit ourselves to their interpretation. The left-hand side has the form of a material derivative,

$$\frac{d}{dt} \equiv \frac{\partial}{\partial t} + \mathbf{u} \cdot \nabla, \tag{W.2}$$

which corresponds to evaluating the time variation of a quantity, which is transported by the fluid. On the right-hand side, we have the forces *per unit volume* acting on a fluid parcel: the force induced by a pressure gradient, the viscous force, and the gravity force. It is worth stressing that the first two terms both derive from the stress tensor $\mathbb{P}$ whose form, for an incompressible fluid, is

$$\mathbb{P}_{ij} = -p \delta_{ij} + \rho \nu \left(\frac{\partial u_i}{\partial x_j} + \frac{\partial u_j}{\partial x_i} \right). \tag{W.3}$$

Applying the divergence to the stress tensor, we obtain the first two terms on the right-hand side.

The key physical process triggering convection is the dependence of the fluid density on temperature. Because of that, heating the system from below determines an unstable situation, where lighter slices of fluid are located below heavier slices of fluid. Therefore, we need to introduce the temperature field $T(\mathbf{x}, t)$, whose evolution is ruled by the heat equation,

$$\frac{dT}{dt} \equiv \frac{\partial T}{\partial t} + (\mathbf{u} \cdot \nabla) T = \kappa \nabla^2 T, \tag{W.4}$$

where κ is the thermal diffusivity. In conclusion, we have four scalar equations, Eqs. (W.1) and (W.4), with four unknowns, the three components of $\mathbf{u}$ and T.

Matter conservation implies

$$\frac{\partial \rho}{\partial t} + \nabla \cdot (\rho \mathbf{u}) = 0 \tag{W.5}$$

or

$$\frac{\partial \rho}{\partial t} + (\mathbf{u} \cdot \nabla)\rho + \rho \nabla \cdot \mathbf{u} \equiv \frac{d\rho}{dt} + \rho \nabla \cdot \mathbf{u} = 0. \tag{W.6}$$

Incompressibility means $d\rho/dt = 0$, so it must be that

$$\nabla \cdot \mathbf{u} = 0. \tag{W.7}$$

In the Navier–Stokes equations (W.1), the density ρ appears in the inertial term, in the gravity term, and in the viscous force. It varies very weakly with T, because the coefficient of thermal expansion, $\alpha = -\dfrac{1}{\rho}\dfrac{\partial \rho}{\partial T}$, is of order $10^{-4}\mathrm{K}^{-1}$ for a fluid. As explained earlier, the temperature dependence of ρ cannot be neglected, otherwise no convective instability would appear. However, we can assume a simple linear dependence,

$$\rho = \rho_0(1 - \alpha(T - T_0)) \tag{W.8}$$

and we can limit ourselves to considering such dependence in the gravity term, which is the primary cause of the instability. This simplification is known as Boussinesq approximation (after the French Joseph Valentin Boussinesq). In conclusion, we have to consider the following set of equations:

$$\nabla \cdot \mathbf{u} = 0 \tag{W.9}$$

$$\rho_0 \left(\frac{\partial}{\partial t} + \mathbf{u} \cdot \nabla \right)\mathbf{u} = -\nabla p + \rho_0 \nu \nabla^2 \mathbf{u} - \rho g \hat{\mathbf{z}} \tag{W.10}$$

$$\frac{\partial T}{\partial t} + (\mathbf{u} \cdot \nabla)T = \kappa \nabla^2 T \tag{W.11}$$

$$\rho = \rho_0(1 - \alpha(T - T_0)). \tag{W.12}$$

The first step is to determine the temperature and pressure profiles of the conductive state, $\mathbf{u}_c \equiv 0$. The second step will be to perturb it and analyze its linear stability spectrum. Since the conductive state is time independent and translational invariant in the (x, y)-plane,

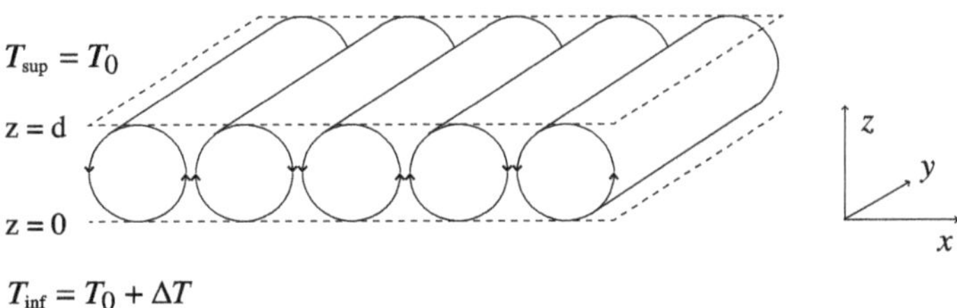

Fig. W.1   Sketch of the Rayleigh–Bénard instability.

Equation (W.11) reduces to $d^2T(z)/dz^2 = 0$, whose solution is

$$T = T_{\mathrm{c}}(z) = T_0 + \Delta T \left(1 - \frac{z}{d}\right), \tag{W.13}$$

where we have used the following boundary conditions for the temperature, $T(z = 0) = T_0 + \Delta T$ and $T(z = d) = T_0$.

The pressure profile is determined by Eq. (W.10), which reduces in the conductive state to $dp/dz = -\rho g$. Its solution is

$$p = p_c(z) = p_0 - g \int_0^z \rho(T_c(z))dz = p_0 - \rho_0 g z \left[1 - \alpha \Delta T \left(1 - \frac{z}{2d}\right)\right]. \tag{W.14}$$

Therefore, in the conductive state, the temperature profile has a linear behavior that interpolates between $T_{\mathrm{inf}} = T_0 + \Delta T$ and $T_{\mathrm{sup}} = T_0$. The pressure profile is a downward parabolic profile.

We should now perturb the conductive state, writing

$$\mathbf{u} = \mathbf{u}_c + \mathbf{u}^*(\mathbf{x}, t) \tag{W.15}$$

$$T = T_{\mathrm{c}}(z) + \theta^*(\mathbf{x}, t) \tag{W.16}$$

$$p = p_C(z) + p^*(\mathbf{x}, t) \tag{W.17}$$

and expanding the resulting equations, keeping only terms linear in starred quantities. This procedure gives

$$\rho_0 \frac{\partial \mathbf{u}^*}{\partial t} = -\nabla p^* + \alpha \rho_0 g \theta^* \hat{\mathbf{z}} + \nu \rho_0 \nabla^2 \mathbf{u}^* \tag{W.18}$$

$$\frac{\partial \theta^*}{\partial t} = \frac{\Delta T}{d} u_z^* + \kappa \nabla^2 \theta^*. \tag{W.19}$$

It is now useful to rescale all variables so as to make them dimensionless,

$$\tilde{\mathbf{x}} = \frac{\mathbf{x}}{d}, \quad \tilde{t} = \frac{\kappa t}{d^2}, \quad \tilde{\mathbf{u}} = \frac{d\mathbf{u}^*}{\kappa}, \quad \tilde{\theta} = \frac{\theta^*}{\Delta T}, \quad \tilde{p} = \frac{d^2 p^*}{\rho_0 \kappa^2}. \tag{W.20}$$

While using d for rescaling space and ΔT for rescaling temperature is straightforward, other rescaling operations deserve some comment. Thermal diffusivity κ defines a time scale $\tau_c = d^2/\kappa$ that is used to rescale t; $d/\tau_c = \kappa/d$ is used to rescale the velocity field. Finally, since p has the dimension of a density multiplied by a square velocity, it can be rescaled with $\rho_0(d/\tau_c)^2 = \rho_0 \kappa^2/d^2$.

In the following, we omit the tilde and the linear equations are written as[1]

$$\frac{\partial \mathbf{u}}{\partial t} = -\nabla p + RP\theta \hat{\mathbf{z}} + P\nabla^2 \mathbf{u} \tag{W.21}$$

$$\frac{\partial \theta}{\partial t} = u_z + \nabla^2 \theta, \tag{W.22}$$

where

$$R = \frac{\alpha \Delta T g d^3}{\nu \kappa} \quad \text{and} \quad P = \frac{\nu}{\kappa} \tag{W.23}$$

---

[1] With the new variables, the height z varies in the interval $[0, 1]$.

are, respectively, the (dimensionless) Rayleigh and the Prandtl number.

It is now possible to get rid of the pressure field by applying the curl operator to both sides of Eq. (W.21). More precisely, we apply it twice, because the exact relation

$$\nabla \times (\nabla \times \mathbf{a}) = \nabla(\nabla \cdot \mathbf{a}) - \nabla^2 \mathbf{a} \tag{W.24}$$

simplifies if $\mathbf{a}$ is a solenoidal vector field ($\nabla \cdot \mathbf{a} = 0$), which is the case for $\mathbf{u}$ and $\nabla^2 \mathbf{u}$. In conclusion, we get

$$-\frac{\partial \nabla^2 \mathbf{u}}{\partial t} = RP\left(\nabla \cdot \frac{\partial \theta}{\partial z} - \nabla^2 \theta \hat{\mathbf{z}}\right) - P\nabla^4 \mathbf{u}. \tag{W.25}$$

The $\hat{\mathbf{z}}$ component of Eq. (W.25), together with Eq. (W.22), gives a coupled system for u_z, θ,

$$\frac{\partial \nabla^2 u_z}{\partial t} = RP\nabla^2_{\|}\theta + P\nabla^4 u_z \tag{W.26}$$

$$\frac{\partial \theta}{\partial t} = u_z + \nabla^2 \theta. \tag{W.27}$$

We should now supplement earlier equations with six boundary conditions, because the highest order derivative for u_z (four) sums to the highest order derivative for θ (two), giving six. There are three boundary conditions at each plate ($z = 0$ and $z = 1$), one for θ and two for u_z. Since θ is the perturbation of a conductive solution satisfying boundary conditions for T, we must require $\theta(z = 0) = \theta(z = 1) = 0$. As for the velocity field, which vanishes in the conductive solution, the physical boundary conditions are $\mathbf{u}(x, y, 0) = \mathbf{u}(x, y, 1) = 0$. Since $\nabla \cdot \mathbf{u} = 0$, they imply the vanishing of $\partial_z u_z(z = 0, 1)$. In conclusion, θ, u_z, and $\partial_z u_z$ must vanish in $z = 0, 1$ (i.e., $z = 0, d$ in dimensional variables).

These conditions make the problem too complicated to be treated here in full detail, so we make the common choice of considering different boundary conditions for u_z: Instead of vanishing its first derivative (no-slip conditions, valid at a rigid boundary), we impose the vanishing of its second derivative (stress-free conditions, valid at a free boundary). Therefore, we are going to impose that $\theta, u_z, \partial_{zz} u_z$ all vanish for $z = 0, 1$.

The standard procedure for a linear stability analysis is to expand the perturbations in Fourier components. Translational invariance along x, y implies that any component $e^{i(k_x x + k_y y)}$ is permitted, while the z-dependence must satisfy boundary conditions, which implies a discretization of k_z because the wave must have nodes. We can finally write

$$u_z(\mathbf{x}, t) = u_n \sin(n\pi z)e^{i(k_x x + k_y y)}e^{\sigma t} + \text{c.c.} \tag{W.28}$$

$$\theta(\mathbf{x}, t) = \theta_n \sin(n\pi z)e^{i(k_x x + k_y y)}e^{\sigma t} + \text{c.c.} \tag{W.29}$$

The wavevector in the (x, y)-plane determines the (horizontal) size and orientation of emerging convection cells, while n determines the number of convection cells in the z-direction. It is reasonable to expect the first mode to destabilize the convective state, while increasing ΔT corresponds to $n = 1$, so we limit ourselves to this case. Replacing the earlier expressions for $n = 1$ in Eqs. (W.21) and (W.22), we get the linear system

$$\left(\sigma(k^2 + \pi^2) + P(k^2 + \pi^2)^2\right)u_1 - RPk^2\theta_1 = 0 \tag{W.30}$$

$$u_1 - (\sigma + k^2 + \pi^2)\theta_1 = 0. \tag{W.31}$$

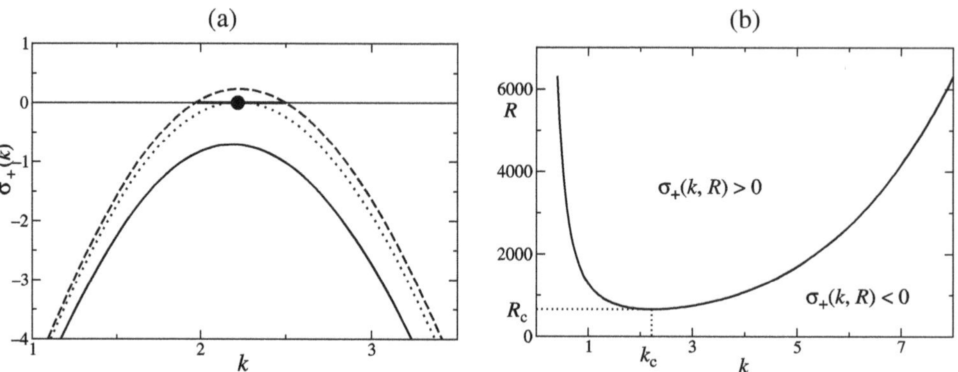

Fig. W.2 (a) The curve $\sigma_+(k)$, Eq. (W.32), for $R < R_c$ (solid line), $R = R_c$ (dotted line), and $R > R_c$ (dashed line). The thick, horizontal segment indicates the unstable interval of wavevectors, and the solid circle indicates the critical wavevector k_c. Compare this figure with Fig. (9.4). (b) The neutral curve, Eq. (W.35), separating the stable region (below the curve) from the unstable region (above the curve).

Imposing the vanishing of the determinant to assure the existence of nontrivial solutions, we can find the relation between the growth rate σ of the perturbation and its wavevector k,

$$\sigma_\pm(k) = -\frac{1}{2}(1+P)(k^2+\pi^2) \pm \sqrt{\frac{1}{4}(P-1)^2(k^2+\pi^2)^2 + \frac{RPk^2}{k^2+\pi^2}}. \tag{W.32}$$

The solution $\sigma_-(k)$ is certainly negative for all k, therefore stable. In Fig. W.2 we plot $\sigma_+(k)$ for $P = 5$ (the Prandtl number for water) and different values of R. We can observe how the growth rate starts to become positive when R is larger than a critical value of the Rayleigh number, $R_c = 27\pi^4/4$ (see later). Just above R_c, $\sigma_+(k)$ is positive in a small interval around a critical wavevector $k_c = \pi/\sqrt{2}$, therefore labeling the Rayleigh–Bénard instability as a type I instability.

Both R_c and k_c can be determined analytically in the stress-free case. However, recall that the physical picture is represented by the no-slip case, which gives

$$R_c \simeq 1708, \qquad k_c \simeq 3.117, \qquad \lambda_c = \frac{2\pi}{k_c} \simeq 2.016 \qquad \text{(no-slip case)} \tag{W.33}$$

Since λ_c is twice the size of a convection cell, the earlier result means that at the threshold, convection cells have an aspect ratio almost equal to one.

Let us now complete the stress-free case, by observing that each mode k becomes unstable at large enough R, $R > R^*$, where $\sigma_+(k, R^*) = 0$. It is interesting to check that R^* does not depend on the Prandtl number. In fact,

$$R^*(k) = \frac{(k^2+\pi^2)^3}{k^2}. \tag{W.34}$$

The sought-after values k_c, R_c correspond to the values minimizing $R^*(k)$,

$$\left.\frac{dR^*}{dk}\right|_{k_c} = 0 \qquad \text{and} \qquad R_c = R^*(k_c), \tag{W.35}$$

which give

$$R_c = \frac{27\pi^4}{4} \simeq 658, \qquad k_c = \frac{\pi}{\sqrt{2}}, \qquad \lambda_c = \frac{2\pi}{k_c} = 2\sqrt{2}, \qquad \text{(stress-free case)} \quad \text{(W.36)}$$

Let us conclude this appendix by providing a simple physical explanation of the existence of a minimal Rayleigh number, because convective cells appear. Heat conduction and convection are two distinct channels to transfer heat between the two plates at different temperature. The evidence that convection appears only above a certain threshold $(\Delta T)_c$ means that below this value, conduction is more efficient, and above this value, convection is more efficient. We can quantify their efficiency by evaluating the typical times τ_c, τ_v of the conductive and convective channel, respectively. The smaller τ is, the more efficient the channel.

The typical conductive time is easily found, because conduction is a diffusive process, so the time τ_c relevant for diffusion on a length scale d via thermal diffusivity κ is simply $\tau_c \simeq d^2/\kappa$. Evaluating the typical convection term requires instead determining the speed v_0 of a parcel of liquid. Once it is known, we can write the ballistic expression $\tau_v = d/v_0$. The velocity v_0 can be inferred by the Navier–Stokes equation, by balancing the force due to Archimedes' principle with the viscous force,

$$(\alpha \Delta T)g\rho_0 = \nu \frac{v_0}{d^2}\rho_0, \tag{W.37}$$

which gives $v_0 = (\alpha \Delta T g d^2)/\nu$. Finally, we get $\tau_v = \nu/(\alpha \Delta T dg)$. The ratio between the conductive and the convective time is nothing but the Rayleigh number,

$$\frac{\tau_c}{\tau_v} = \frac{\alpha \Delta T g d^3}{\kappa \nu} \equiv R. \tag{W.38}$$

Therefore, the condition $R > R_c$ for triggering convection cells means that τ_v is sufficiently smaller than τ_c.

The Turing instability appears when $\mathcal{D}_q$ changes sign from positive to negative (see Section 9.3.1), assuming that no instability occurs for $q = 0$. The expression of $\mathcal{D}_q(q^2)$ (see Eq. (9.39b)),

$$\mathcal{D}_q = (a_{11} - D_1 q^2)(a_{22} - D_2 q^2) - a_{12} a_{21} \tag{X.1}$$

$$= D_1 D_2 q^4 - (D_1 a_{22} + D_2 a_{11}) q^2 + (a_{11} a_{22} - a_{12} a_{21}), \tag{X.2}$$

is an upward-facing parabola, whose minimum q_{m}^2 satisfies the condition

$$\left. \frac{\partial \mathcal{D}_q}{\partial q^2} \right|_{q_{\mathrm{m}}} = 2 D_1 D_2 q_{\mathrm{m}}^2 - (D_1 a_{22} + D_2 a_{11}) = 0, \tag{X.3}$$

so that

$$q_{\mathrm{m}}^2 = \frac{D_1 a_{22} + D_2 a_{11}}{2 D_1 D_2}. \tag{X.4}$$

It is worth noting that q_{m} is an extremum of $\mathcal{D}_q$ (the other, trivial extremum being $q = 0$), and for generic values of the coefficients a_{ij}, it differs from the critical wavevector q_c, which is the maximum of the eigenvalue $\sigma_1(q)$ at the threshold, defined by the conditions $\sigma_1(q_c) = 0$ and $\sigma_1'(q_c) = 0$. However, at the threshold $q_{\mathrm{m}} = q_c$, as is clear from the expression $\mathcal{D}_q = \sigma_1(q)\sigma_2(q)$. In fact, at the threshold, $\sigma_1(q_c) = \mathcal{D}_{q_c} = 0$ and $\sigma_2(q_c) < 0$. Therefore, we also have $\mathcal{D}_{q_c}' = \sigma_1'(q_c) = 0$.

The condition $\mathcal{D}_{q_{\mathrm{m}}} = 0$ defines the criticality, and the condition

$$\mathcal{D}_{q_{\mathrm{m}}} = -\frac{(D_1 a_{22} + D_2 a_{11})^2}{4 D_1 D_2} + (a_{11} a_{22} - a_{12} a_{21}) < 0 \tag{X.5}$$

defines the parameter region, where the Turing instability may appear. We can finally summarize the conditions causing a Turing pattern to appear:

$$\begin{aligned} \mathcal{T}_0 < 0 \quad &\text{(i)} \quad a_{11} + a_{22} < 0 \\ \mathcal{D}_0 > 0 \quad &\text{(ii)} \quad a_{11} a_{22} - a_{12} a_{21} > 0 \\ \mathcal{D}_{q_{\mathrm{m}}} < 0 \quad &\text{(iii)} \quad D_1 a_{22} + D_2 a_{11} > 2\sqrt{D_1 D_2 (a_{11} a_{22} - a_{12} a_{21})}, \end{aligned} \tag{X.6}$$

with the last condition derived from Eq. (X.5), taking into account that

$$D_1 a_{22} + D_2 a_{11} = 2 D_1 D_2 q_{\mathrm{m}}^2 > 0. \tag{X.7}$$

This same equation, if rewritten as

$$(D_1 - D_2) a_{22} + D_2 (a_{22} + a_{11}) > 0, \tag{X.8}$$

provides two results: First, if $D_1 = D_2$, this condition cannot agree with (i), as already shown in the main text; second, if we assume $D_1 > D_2$, we must have $a_{22} > 0$ and consequently (i) gives $a_{11} < 0$. As a consequence of that, (ii) gives that crossing couplings a_{12}, a_{21} must have opposite signs.

Finally, the equation for eigenvalues at criticality gives Eq. (9.46),

$$(a_{11} - D_1 q_c^2)\epsilon_1^0 + a_{12}\epsilon_2^0 = 0.$$

Using the expression of q_c^2 (see Eq. (X.4)), we can evaluate the sign of the coefficient multiplying ϵ_1^0

$$a_{11} - D_1 q_c^2 = \frac{D_2 a_{11} - D_1 a_{22}}{2D_2} < 0, \tag{X.9}$$

because $a_{22} > 0$ and $a_{11} < 0$. Therefore, the fluctuations $\epsilon_1^0, \epsilon_2^0$ have the same sign if $a_{12} > 0$ and opposite signs if $a_{12} < 0$.

Steady States of the One-Dimensional TDGL Equation

The TDGL equation is written as

$$u_{xx} + ru - u^3 = 0, \tag{Y.1}$$

where the explicit dependence on the control parameter r can be removed by rescaling x and u. If

$$x = \frac{y}{\sqrt{r}}, \qquad u(x) = \sqrt{r}\,g(y) = \sqrt{r}\,g(\sqrt{r}x), \tag{Y.2}$$

we obtain

$$g_{yy} = -g - g^3, \tag{Y.3}$$

which can be integrated by multiplying both sides by g_y. We obtain

$$\frac{1}{2}g_y^2 = -\frac{g^2}{2} + \frac{g^4}{4} + E, \tag{Y.4}$$

where $E = \frac{1}{2}\bar{A}^2 - \frac{1}{4}\bar{A}^4$, $\bar{A}$ being the maximal amplitude of the oscillation (in reduced variables). For $g_y > 0$,

$$dy = \frac{dg}{\sqrt{2E - g^2 + \frac{g^4}{2}}}. \tag{Y.5}$$

The reduced wavelength of the oscillation is therefore defined as

$$\bar{\lambda} = 2 \int_{-\bar{A}}^{\bar{A}} \frac{dg}{\sqrt{2E - g^2 + \frac{g^4}{2}}} \tag{Y.6}$$

$$= \frac{4}{\sqrt{1 - \frac{\bar{A}^2}{2}}} \int_0^1 \frac{ds}{\sqrt{(1 - s^2)(1 - k^2 s^2)}} \tag{Y.7}$$

$$= 4\sqrt{\frac{2}{2 - \bar{A}^2}}\, K(k), \tag{Y.8}$$

where $k = \bar{A}/\sqrt{2 - \bar{A}^2}$ and $K(k)$ is the complete elliptic integral of the first kind.

Coming back to old variables, that is, to Eq. (Y.1), we find

$$\lambda(A) = 4\sqrt{\frac{2}{2r - A^2}}\, K\left(\frac{A}{\sqrt{2r - A^2}}\right). \tag{Y.9}$$

The equation for a one-dimensional anharmonic oscillator is

$$\ddot{x} + x = -\epsilon f(x, \dot{x}), \tag{Z.1}$$

where $f(x, \dot{x})$ is a generic function of position and velocity and ϵ is the parameter for the perturbation expansion. A naive perturbation theory simply assumes $x(t) = \sum_{n \geq 0} \epsilon^n x_n(t)$ and replaces it in Eq. (Z.1). Let us do it by limiting ourselves to zero and first-order terms,

$$\ddot{x}_0 + \epsilon \ddot{x}_1 + x_0 + \epsilon x_1 = -\epsilon f(x_0, \dot{x}_0). \tag{Z.2}$$

At zero (unperturbed) order, we have the harmonic oscillator equation, $\ddot{x}_0 + x_0 = 0$, whose solution is $x_0 = A_0 \cos(t + \phi_0)$. At first order, the nonlinear term f acts as a forcing term. To be specific, we suppose having $f(x, \dot{x}) = x^3$ (Duffing oscillator), in which case

$$\ddot{x}_1 + x_1 = -A_0^3 \cos^3(t + \phi_0) \tag{Z.3}$$

$$= -\frac{A_0^3}{4} \cos(3t + \phi_0) - \frac{3A_0^3}{4} \cos(t + \phi_0). \tag{Z.4}$$

The second term on the right-hand side is resonant with the natural frequency of the harmonic oscillator, and this determines a secular term in the solution, which grows linearly in time,

$$x_1(t) = -\frac{3A_0^3}{8} t \sin(t + \phi_0) + \text{nonsecular terms.} \tag{Z.5}$$

A linear term in the expansion means that the perturbation theory breaks down when $t \sim 1/\epsilon$, because $\epsilon x_1(1/\epsilon)$ is of the same order as $x_0(1/\epsilon)$. The multiple-scale analysis aims at a perturbation expansion without secular terms, which is obtained by removing the resonant terms. This is done by introducing multiple temporal scales, whose physical meaning is clarified by the following example.

Let us consider the damped harmonic oscillator

$$\ddot{x} + x = -\epsilon \dot{x}, \tag{Z.6}$$

whose exact solution (in complex notation) is

$$x(t) = A \exp\left(i\sqrt{1 - \frac{\epsilon^2}{4}}\, t - \frac{\epsilon}{2} t\right) = A \exp\left(i\left(t - \frac{\epsilon^2}{8} t\right) - \frac{\epsilon}{2} t + O(\epsilon^3)\right). \tag{Z.7}$$

This expression makes explicit that the solution depends on time through $t_0 = t$, $t_1 = \epsilon t$, $t_2 = \epsilon^2 t$, and so on. The "times" t_n represent different time scales over which different phenomena may occur. So, the oscillation period is of order $t_0 \approx 1$, the amplitude is dumped on a time of order $t_1 \approx 1$, the oscillation period changes on a time of order $t_2 \approx 1, \ldots$. From a formal point of view, the multiple scale analysis introduces an expansion of times in addition to the expansion of the solution, and these additional ϵ terms allow us to impose the vanishing of the resonant terms. Let us see the method at work for the Duffing oscillator.

Introducing additional times means that

$$\frac{d}{dt} = \partial_{t_0} + \epsilon \partial_{t_1} + \cdots \tag{Z.8}$$

$$\frac{d^2}{dt^2} = \partial_{t_0 t_0} + 2\epsilon \partial_{t_1 t_0} + \cdots. \tag{Z.9}$$

The zero-order perturbation does not change and the solution is the same as before, with the caveat that the integration constants A_0 and ϕ_0 now can depend on the slower time scale, $x_0(t_0, t_1) = A_0(t_1) \cos(t_0 + \phi_0(t_1))$. Instead, the first-order perturbation is written as

$$\frac{\partial^2 x_1}{\partial t_0^2} + x_1 = -2 \frac{\partial^2 x_0}{\partial t_1 \partial t_0} - A_0^3(t_1) \cos^3(t_0 + \phi_0(t_1)) \tag{Z.10}$$

$$= 2 \frac{\partial}{\partial t_1} \Big(A_0(t_1) \sin(t_0 + \phi_0(t_1)) \Big) - A_0^3(t_1) \cos^3(t_0 + \phi_0(t_1)) \tag{Z.11}$$

$$= 2\dot{A}_0 \sin(t_0 + \phi_0) + \left[2A_0\dot{\phi}_0 - \frac{3}{4}A_0^3 \right] \cos(t_0 + \phi_0) - \frac{A_0^3}{4} \cos(3(t_0 + \phi_0)),$$

where in the last expression, for ease of notation, we have not made explicit the t_1-dependence of A_0, ϕ_0, and the dot over such quantities means the derivative with respect to t_1.

We can first remark that if no slow time scales are introduced, A_0 and ϕ_0 are constant, and for $\dot{A}_0 = 0 = \dot{\phi}_0$ the above expression reduces to Eq. (Z.4), which contains resonant terms. Now we have the freedom to impose the vanishing of such terms, writing

$$\dot{A}_0 = 0, \tag{Z.12}$$

$$2A_0\dot{\phi}_0 - \frac{3}{4}A_0^3 = 0. \tag{Z.13}$$

The former condition implies that A_0 is a constant (or, better, it depends on t_2 and slower time scales, which are neglected in this first-order calculation). The latter condition implies $\phi_0(t_1) = \frac{3}{8}A_0^2 t_1 + \bar{\phi}_0(t_2)$.

We can now turn to the equation for x_1,

$$\frac{\partial^2 x_1}{\partial t_0^2} + x_1 = -\frac{A_0^3}{4} \cos(3(t_0 + \phi_0)), \tag{Z.14}$$

whose solution (if $x_1(0) = \dot{x}_1(0) = 0$) is $x_1(t_0) = \frac{A_0^3}{32} [\cos(3t_0) - \cos(t_0)]$.

Reassembling the different pieces and getting back to the variable t, we obtain, at order ϵ,

$$x(t) = A_0 \cos \left[\left(1 + \frac{3}{8}A_0^2 \epsilon \right) t \right] + \epsilon \frac{A_0^3}{32} [\cos(3t) - \cos(t)]. \tag{Z.15}$$

In this appendix we have made use of the simple anharmonic oscillator to show step by step how the method works, but in the main text we use the multiple-scale method in the more complicated context of partial differential equations. In that case, two main difficulties arise: First, we have temporal and spatial multiple scales, because the order parameter now depends on both x and t; second, the condition of absence of resonant terms might not be that easy to solve, because the unperturbed equation is itself nonlinear. This is the reason why we must conclude this section by giving a nonrigorous formulation of the Fredholm alternative theorem, which is used in the main text. In brief, let us suppose having the linear problem

$$\mathcal{L}[u] = g(x), \tag{Z.16}$$

where $\mathcal{L}$ is a linear differential operator. If the homogeneous problem, $\mathcal{L}[u_0] = 0$, has a nontrivial solution $u_0 \neq 0$, then also the adjoint problem $\mathcal{L}^\dagger[v_0] = 0$ has a nontrivial solution and Eq. (Z.16) has a solution if and only if g is orthogonal to v_0,

$$\langle g|v_0 \rangle \equiv \int dx\, g^*(x) v_0(x) = 0. \tag{Z.17}$$

Index